Carrion Ecology, Evolution, and Their Applications

Carrion Ecology, Evolution, and Their Applications

EDITED BY

M. Eric Benbow
MICHIGAN STATE UNIVERSITY, EAST LANSING, USA

Jeffery K. Tomberlin
TEXAS A&M UNIVERSITY, COLLEGE STATION, USA

Aaron M. Tarone
TEXAS A&M UNIVERSITY, COLLEGE STATION, USA

CRC Press
Taylor & Francis Group
Boca Raton London New York

CRC Press is an imprint of the
Taylor & Francis Group, an **informa** business

Front cover: A turkey vulture (*Cathartes aura*) awaits the teeth of a top predator, now absent from this ecosystem, to open the bloated carcass of a dead bovine. Albion Mountains, Minidoka Ranger District, Sawtooth National Forest (June 29, 2010). Photo copyright Robert A. Miller. Used with permission. All rights reserved.

CRC Press
Taylor & Francis Group
6000 Broken Sound Parkway NW, Suite 300
Boca Raton, FL 33487-2742

First issued in paperback 2018

© 2016 by Taylor & Francis Group, LLC
CRC Press is an imprint of Taylor & Francis Group, an Informa business

No claim to original U.S. Government works

ISBN-13: 978-1-4665-7546-2 (hbk)
ISBN-13: 978-1-138-89384-9 (pbk)

Library of Congress Cataloging-in-Publication Data

Carrion ecology, evolution, and their applications / editors, M. Eric Benbow, Jeffery K. Tomberlin, and Aaron M. Tarone.
    pages cm
  Includes bibliographical references and index.
  ISBN 978-1-4665-7546-2
  1. Animal carcasses--Biodegradation. 2. Dead animals. 3. Predation (Biology) 4. Biodegradation. 5. Biochemistry. I. Benbow, Mark Eric, 1971- editor. II. Tomberlin, Jeffery K., editor. III. Tarone, Aaron M., editor.

QP517.B5C37 2016
579.7'1--dc23   2015004566

Visit the Taylor & Francis Web site at
http://www.taylorandfrancis.com

and the CRC Press Web site at
http://www.crcpress.com

We would like to dedicate this book to Mary Fuller and Jerry Payne for their courage to

ask questions about vertebrate carrion and its related significance to systems ecology.

We would also like to dedicate this book to Dan Janzen who wrote an inspiring paper on

rotting fruit, moldy seeds, and spoiling meat in 1977, a paper that expanded our views of the process

and nature of carrion decomposition that in many ways served as the impetus for this book.

Contents

Section I   Introduction to Carrion Decomposition

Section II   Ecological Mechanisms of Carrion Decomposition

Section III Evolutionary Ecology of Carrion Decomposition

Section IV Applications of Carrion Decomposition

Acknowledgments

We are indebted to the outstanding authors of this volume and the anonymous reviewers of the chapters. They have made this one of the most comprehensive and cohesive collection of topics covering carrion ecology, evolution of associated organisms, and applications of the basic scientific understanding of it.

We would like to thank our families (Melissa, Alia, and Arielle Benbow; Laura, Celeste, and Jonah Tomberlin; Lauren Kalns and Rocco Tarone) for their support and encouragement during the many hours that we dedicated to this endeavor and for their tolerance and understanding when we have been up late working, away pursuing science, conducting collaborative research, and forming professional friendships important to accomplishing a task of this magnitude.

We would also like to acknowledge special gratitude to Jen Pechal and Jonathan Cammack for their extra efforts in editing and proofing various aspects of this book. Their efforts greatly improved the final version of this book.

Finally, we are indebted to the efforts of Rich Merritt and John Wallace for their initial support and encouragement for taking on such a time-consuming, yet important, project. Rich Merritt deserves special recognition for the educational underpinnings of this book, as he was responsible for leading Benbow into carrion ecology and forensic science and was a critical member of Tarone's PhD committee. Rich is truly admired for his lifetime achievements and mentorship in these fields, as much as for those in his true passion (and Benbow's) of aquatic entomology. Tomberlin would also like to extend his thanks to his MS advisor at Clemson University—Peter Adler. Through the years, Peter has been, and continues to be, Tomberlin's mentor, colleague, and friend.

M. Eric Benbow
Jeffery K. Tomberlin
Aaron M. Tarone

Contributors

Michael S. Allen
Department of Molecular Biology
and Genetics
University of North Texas Health Science
Center
Fort Worth, Texas

Gail S. Anderson
School of Criminology
Simon Fraser University
Burnaby, British Columbia, Canada

Philip S. Barton
Fenner School of Environment and Society
The Australian National University
Canberra, Australia

James C. Beasley
Savannah River Ecology Laboratory
University of Georgia
Aiken, South Carolina

M. Eric Benbow
Department of Entomology
and
Department of Osteopathic Medical
Specialties
Michigan State University
East Lansing, Michigan

Wolf Blanckenhorn
Institute of Evolutionary Biology
and Environmental Studies
University of Zürich
Zürich, Switzerland

Jonathan A. Cammack
Department of Entomology
Texas A&M University
College Station, Texas

David O. Carter
Division of Natural Sciences and Mathematics
Chaminade University of Honolulu
Honolulu, Hawaii

Tawni L. Crippen
Southern Plains Agricultural Research Center
United States Department of Agriculture—
Agriculture Research Service
College Station, Texas

Grant D. De Jong
GEI Consultants, Inc.
Denver, Colorado

Travis L. DeVault
National Wildlife Research Center
U.S. Department of Agriculture
Sandusky, Ohio

Shari L. Forbes
Centre for Forensic Science
University of Technology
Sydney, Australia

M.D. Hocking
Hakai Beach Institute
and
School of Environmental Studies
University of Victoria
Victoria, British Columbia, Canada

Kay E. Holekamp
Department of Zoology
Michigan State University
East Lansing, Michigan

Sarah C. Jones
Department of Zoology
Michigan State University
East Lansing, Michigan

Heather R. Jordan
Department of Microbiology
Mississippi State University
Mississippi State, Mississippi

Andreas Jürgens
School of Life Sciences
University of KwaZulu-Natal
Scottsville, South Africa

Rob Knight
BioFrontiers Institute
University of Colorado
Boulder, Colorado

Michael G. LaMontagne
Department of Biology
Missouri State University
Springfield, Missouri

Francis J. Larney
Lethbridge Research Center
Agriculture and Agri-Food Canada
Lethbridge, Alberta, Canada

Tim A. McAllister
Lethbridge Research Center
Agriculture and Agri-Food Canada
Lethbridge, Alberta, Canada

Richard W. Merritt
Department of Entomology
Michigan State University
East Lansing, Michigan

Jessica L. Metcalf
BioFrontiers Institute
University of Colorado
Boulder, Colorado

J.-P. Michaud
Serious Crimes Branch
Royal Canadian Mounted Police
Calgary, Alberta, Canada

and

Département de biologie
Université de Moncton
Moncton, New Brunswick, Canada

Rachel M. Mohr
Department of Forensic and Investigative
   Sciences
West Virginia University
Morgantown, West Virginia

Gaétan Moreau
Département de Biologie
Université de Moncton
Moncton, New Brunswick, Canada

Zach H. Olson
Department of Psychology
University of New England
Biddeford, Maine

S.M. O'Regan
Ecosystems and Management
Simon Fraser University
Burnaby, British Columbia, Canada

Jonathan J. Parrott
Department of Entomology
Texas A&M University
College Station, Texas

Jennifer L. Pechal
Department of Entomology
Michigan State University
East Lansing, Michigan

Christine J. Picard
Department of Biology
Indiana University–Purdue University
Indianapolis, Indiana

Meaghan L. Pimsler
Department of Entomology
Texas A&M University
College Station, Texas

Tim Reuter
Livestock Research Branch
Alberta Agriculture and Rural Development
Lethbridge, Alberta, Canada

Michelle R. Sanford
Harris County Institute of Forensic Sciences
Houston, Texas

Kenneth G. Schoenly
Department of Biological Sciences
California State University
Turlock, California

Adam Shuttleworth
School of Life Sciences
University of KwaZulu-Natal
Scottsville, Pietermaritzburg, South Africa

Kim Stanford
Harris County Institute of Forensic Sciences
Lethbridge, Alberta, Canada

Eli D. Strauss
Department of Zoology
Michigan State University
East Lansing, Michigan

Michael S. Strickland
Department of Biological Sciences
Virginia Polytechnic and State University
Blacksburg, Virginia

Aaron M. Tarone
Department of Entomology
Texas A&M University
College Station, Texas

Jeffery K. Tomberlin
Department of Entomology
Texas A&M University
College Station, Texas

Sherah L. VanLaerhoven
Department of Biology
University of Windsor
Windsor, Ontario, Canada

John R. Wallace
Department of Biology
Millersville University
Millersville, Pennsylvania

John W. Whale
Department of Biology
Indiana University–Purdue University
Indianapolis, Indiana

Kyle Wickings
Department of Entomology
Cornell University—New York State
    Agricultural Experimental Station
Geneva, New York

Thomas K. Wood
Department of Chemical Engineering
Penn State University
University Park, Pennsylvania

Shanwei Xu
Lethbridge Research Center
Agriculture and Agri-Food Canada
Lethbridge, Alberta, Canada

Section I

Introduction to Carrion Decomposition

1

Introduction to Carrion Ecology, Evolution, and Their Applications

M. Eric Benbow, Jeffery K. Tomberlin, and Aaron M. Tarone

CONTENTS

1.1 Introduction

1.1.1 Carrion as Decomposing Organic Matter

Detritus, defined as any source of nonliving organic matter (Swift et al. 1979), is considered the basal trophic level of many food webs (Lindeman 1942; Teal 1962; Odum 1969; Moore et al. 2004) and is an important component of recycling energy and nutrients in ecosystems (Swift et al. 1979; Hättenschwiler et al. 2005; Moore and Schindler 2008; Barton et al. 2013a, b). Given this importance, the decomposition of detritus has been intensively studied for many years with reviews and empirical studies providing ample evidence that this process is fundamental to ecosystem properties through complex ecological linkages (Swift et al. 1979; Hättenschwiler and Gasser 2005; Parmenter and MacMahon 2009; Gessner et al. 2010). By far, the most studied portion of the detrital pool has been phototrophically derived sources of nonliving organic matter (e.g. abscised leaves, grass, or decaying algae). Indeed, studies of leaf litter continue to provide fundamental understanding of the regulation of ecosystem processes such as nutrient and energy cycling, community interactions, and food web network stability and resilience (Hawlena et al. 2012; Jabiol et al. 2014; Majdi et al. 2014). Although it is vital to understand the decomposition of phototrophically derived aspects of ecosystems, there has been relatively limited research attention on the heterotrophically derived component of the detrital pool—carrion (Figure 1.1).

1.1.2 Carrion Defined

What is carrion? Some (but certainly not all) of the contemporary literature describing carrion within the context of ecosystem services can be found in Swift et al. (1979), in which carrion was considered detritus and was placed into the same compartment as any type of decaying organic matter, mostly represented by plant biomass. Others have given more ecologically relevant conceptual definitions to carrion, either as a "cadaver decomposition island" (Carter et al. 2007) or as an "ephemeral resource patch" (Doube 1987; Finn 2001). For the purposes of this book, the etymology of the word carrion and a short description are provided within the context of ecology, evolution, and their applications.

FIGURE 1.1 **(See color insert.)** Salmon carcasses provide a significant input of nutrients and energy into streams of the Pacific Northwest as heterotrophically derived organic matter. (Photo by M.E. Benbow.)

Carrion (Anglo-French *carogne*; Vulgar Latin *caronia* "carcass") has historically been defined as dead and decaying flesh or as a carcass of an animal (Merriam-Webster 2014) and has by far been most often considered as vertebrate animals. Carrion as related to its use in this text includes a broader representation of decomposing animal tissue (heterotrophically derived biomass) in ecology and evolution, including the decaying carcass of any once living, heterotrophic organism. With this definition, the carrion system also includes microscopic eukaryotes such as rotifers and cladocerans, nematodes, and macroinvertebrates (e.g. insects, crabs, and cephalopods), in addition to amphibians, reptiles, birds, and mammals. This definition is supported by previous studies on carrion that have used macroinvertebrates (Beaver 1973, 1977; Seastedt et al. 1981) and more recent studies that demonstrated the importance of grasshopper, *Melanoplus femurrubrum* (De Geer) (Orthoptera: Acrididae), carrion on ecosystem processes (Hawlena et al. 2012), and the ecological and evolutionary importance of some types of invertebrate nutrient resource pulses (Polis et al. 1997; Yang 2004; Yang et al. 2008).

Although there have been advances in understanding carrion in ecosystems, studies of vertebrate carrion in general, and ungulate carrion in particular, have provided a basis for uncovering the important but largely overlooked role of carrion in complex ecosystem structure and function (Hobbs 1996, 2006; Melis et al. 2007; Bump et al. 2009a,b). Some of this research has been developed to understand the basic ecological and evolutionary relevance of carrion (Fuller 1934; Payne 1965; Schoenly and Reid 1987; Tomberlin et al. 2011a; Barton et al. 2013a); however, a substantial portion of what is understood about these processes has come from studies with forensic (Mégnin 1894; Carter et al. 2007; Tomberlin et al. 2011b; Pechal et al. 2014) or disease significance (Miller et al. 2004; Hugh-Jones and Blackburn 2009; Villet 2011).

1.2 Ecological Mechanisms of Carrion Decomposition

With the development of new molecular sequencing and advancing computational technologies (e.g. high performance computing and geographical information systems), scientists are now better able to investigate complex ecological networks such as interactions involving nonculturable microbial species and wide-ranging vertebrate scavengers at different scales (from the microscopic to continental scales). This advancement in technological power allows for a better resolution of relationships among organisms assembling as a community around or on an ephemeral resource patch, that is, carrion. Such heterotrophically derived resource patches were once considered to be too ephemeral and isolated to play a substantial role in ecosystem processes (Swift et al. 1979). However, carrion's effect on the surrounding areas, such as soil biogeochemistry and invertebrate communities (Chapters 2, 3, and 5) or the ecology

FIGURE 1.2 **(See color insert.)** Carrion patches become resources where complex interactions occur between species, as in this photo where a praying mantis is captured eating an adult fly on carrion. (Photo with permission from C.S. Ulrich and K. Black.)

of invertebrates and vertebrates that offers significant ecosystem services (Chapters 4 and 6), amply demonstrates the substantial and wide-ranging ecological impact of carrion (Figure 1.2).

As described in Chapter 2, the process of death and decomposition is initiated by the cessation of blood (or other respiratory fluid/pigments in some invertebrates) flow in the body, serving as the catalyst to a cascade of biochemical reactions that mediate changes in microbial metabolic activity. The microbial communities associated with carrion are discussed in great detail in Chapter 3, providing a survey of the taxa found during the early process of decomposition, how they function, and how they may affect the localized soil and arthropod communities. As decomposition progresses, in most instances, blow flies (Diptera: Calliphoridae) and flesh flies (Diptera: Sarcophagidae) immediately cue onto the carrion

FIGURE 1.3 **(See color insert.)** Carrion is often quickly colonized by blow fly larvae that can take the form of huge masses that convert the carrion tissue into larval biomass, often with impressive efficiency. In this photo, an entire swine carcass was consumed by larvae that number well into the thousands: the larvae are beginning to disperse from the carcass location in order to find a place to pupate, and in this case they are even moving up a tree trunk. (Photo by M.E. Benbow.)

source and can initiate colonization within hours of its death (Figure 1.3). These arthropod communities are important consumers of the new carrion resource patch and are introduced in Chapter 4, in which the primary taxa and how the communities proceed through a successional sequence as the carrion is consumed and transformed into new arthropod biomass are described.

In Chapter 5, the authors explore the soil chemistry and biological communities that respond to the nutrient and microbial pulses that affect the soil beneath and immediately adjacent to carrion patches. During this entire process of carrion decomposition, the community of vertebrate scavenger species that are attracted to the newly available food resource plays a role in direct consumption and dispersal of associated nutrients into the landscape. Many of these scavengers are facultative and opportunistic, but this chapter explores the ecological importance of scavenging and describes how changes in carrion density may be changing the populations of some of these species. Chapter 7 provides an approach to designing strong and statistically robust field studies to better understand the ecology and evolution of carrion decomposition and its potential importance to ecosystem processes and services.

The importance of carrion to the detrital energy and nutrient foundation in ecosystems has historically been considered minimal; however, recent research has demonstrated that carrion can be significant to ecosystem processes often through indirect interactions among several trophic groups (Towne 2000; Yang 2004; Yang et al. 2008; Bump et al. 2009a; Parmenter and MacMahon 2009; Beasley et al. 2012; Hawlena et al. 2012; Barton et al. 2013a). One striking example of this is found in Chapter 21, in which a migration event of just one species of mammal results in carrion, of the order of thousands of tons, being distributed throughout a broad swath of the North American continent.

However, the impacts of carrion are not just limited to mass migration events. Hawlena et al. (2012) reported how terrestrial predators altered soil community function by indirectly affecting the quality of overlying carrion resource condition. They indicated that the quality of grasshopper (Orthoptera) carcasses was reduced when the grasshoppers were raised in the presence of predatory spiders, and although the rate of carcass decomposition remained unaltered, subsequent plant litter decomposition and nutrient mineralization in the soil associated with the carcass decomposition were threefold lower than soils with nonstressed carcasses. When scaled to the habitat based on grasshopper and spider densities and life history dynamics, the authors suggested plausible ecosystem scale effects on prairie nutrient dynamics. Thus, in at least this one documented instance, the indirect carrion legacy effects of nonconsumptive predation on ecosystem level processes were significant, even outside of large food fall events such as the effects of mass cicada emergence (Yang 2004), fish die-offs (Tiegs et al. 2009; Levi et al. 2011), or whale falls (Smith and Baco 2003). Barton et al. (2013a) provided an excellent review of carrion ecology that introduces how this discipline can be developed from a broader empirical basis from other ephemeral resources. The authors also argued effectively that carrion has significant impacts on terrestrial biodiversity and ecosystem properties through influence on soil nutrients, belowground microbial communities, plants, arthropods, and other invertebrates and vertebrates. Much of the literature on the importance of carrion to ecosystems is reviewed in Chapter 13.

Viewing the vertebrate carrion system from 10,000 m altitude would suggest that such a system is rather simple. However, like many other systems, and in comparison to an onion that can be peeled apart one layer at a time, upon closer inspection one discovers a dynamic world that serves as a conduit through which nutrients are distributed throughout the greater ecosystems that culminate globally. This second section of the book presents examples of the complexities, and value, of such systems and their use as models for understanding not only how nutrients are released into a given system, but also other questions relevant to the larger ecology and evolutionary biology fields. Animals that compete for these resources are diverse, representing most forms of life common to other systems such as plant–herbivore interaction dynamics.

Three of the chapters contain the word ecology in their title and for good reason. Chapter 8 explores the complex dynamics associated with invertebrates that compete for these resources and demonstrates how these systems can be used to explore basic questions associated with species or community-level interactions. Chapter 9 provides context to the volatiles associated with carrion systems and how they serve as mechanisms regulating species interactions, colonization, and resource utilization. Chapter 10 presents an argument for the use of the carrion model as a platform for conducting behavior research.

With a strong introduction to the community, chemical, and behavioral ecology of carrion decomposition, Chapter 11 explores how ecological theory can be used to better understand and more specifically explore mechanisms of species interactions at, on, or associated with carrion. This is followed by Chapter 12 that provides a detailed discussion of the organisms associated with carrion in aquatic ecosystems and how these organisms that range from bacteria to vertebrates interact in several aquatic habitat types. This aquatic-focussed chapter also provides a comprehensive treatment of the broader ecological importance of carrion in aquatic systems, ranging from streams to the abyssal depths of the oceans. This section of the book is concluded with an insightful review in Chapter 13 of the ecological roles of carrion to ecosystem structure and function, providing the introduction to the section of the book that covers the evolutionary ecology of carrion decomposition.

1.3 Evolutionary Ecology of Carrion

The fields of ecology and evolution have always been intimately linked. This is evident in Dobzhansky's famous essay championing evolution as the means by which all problems in biology make sense (Dobzhansky 1973). In the same document, Dobzhansky clearly considered problems in ecology, and he did so from an evolutionary perspective. This perspective becomes apparent as soon as one begins to conduct research in either discipline. To understand the evolution of an organism, one must understand its ecology. Invariably, ecological forces are the sources of selection and drift that ultimately impact the evolution of genes and species. Similarly, studying the ecology of the organism without considering the evolutionary pressures that shape its biology is not feasible.

This link between ecology and evolution is also evident in the carrion system, although historically there has been much less attention paid to the genetics and evolution of the system when compared with the ecological interactions and descriptions of carrion-associated species. With this imbalance in mind, we recently encouraged our forensically focused colleagues to consider the biology of the decomposing remains they study in the context of ecology and evolution (Tomberlin et al. 2011a).

Remains do not decompose in an ecological vacuum (Figures 1.1 and 1.2). They do so in an ecosystem, and that ecosystem exerts evolutionary pressures on all of the organisms that utilize carrion within the ecosystem. Similarly, the evolution of decomposers in their local environments will impact the important ecological processes associated with carrion. The third section of the book provides a summary of basic principles in evolutionary ecology that seem most relevant to the carrion system, though this section is not meant to be comprehensive. Rather, the intention is to help the reader develop the mindset of an evolutionary ecologist in order to assist them in applying those thought processes to the carrion system. Specifically, the aim of this section is to assist the reader in understanding the following: (1) the ecological pressures that will impact the evolution of carrion consumers (Figure 1.4) and (2) the evolutionary principles that are associated with and relevant to carrion decomposition. Chapters 14–17 will provide a basic understanding of evolutionary and genetic principles of, as well as their applications to, the evolution of carrion and dung systems. Chapters 18–20 deal with various aspects of our newly emerging understanding of how carrion is decomposed by microbes, including descriptions of some of the most recent technological advancements that have facilitated this study. The last chapter (Chapter 21) of this section provides a comprehensive assessment of the carrion consumers in Africa, allowing the reader to observe and consider these principles of evolutionary ecology at play.

1.4 Applications of Carrion Decomposition

Carrion decomposition has long been a part of societal evolution as its regulation was found to play a major role in the spread of pathogens and resulting disease. Keeping urban dwellings clean of such materials, whether it be human waste and refuse to control arthropods or rodents responsible for some of these pathogens, is now known to be an important part of hygiene and maintenance of public health standards necessary to prevent disease outbreaks. However, readers of this text will have an opportunity to learn about broader applications of decomposition ecology beyond human health.

Confined animal facilities and aquaculture represent two of the largest producers of animal protein for human consumption globally. However, as demonstrated in Chapters 22 and 23, these facilities,

FIGURE 1.4   (See color insert.) As part of the ecological process of carrion decomposition, many species of arthropods have evolved to efficiently use the resource using interesting adaptations such as dispersal away from larval conspecific competitors and predators in a dark place for pupation and the completion of their life cycle. (Photo by M.E. Benbow.)

while producing mass quantities of protein, are also responsible for the production of waste streams that must be properly managed in order to protect the surrounding populus and the environment. These chapters also provide an introduction of how something viewed negative, such as pollutants, sources of odor, and pest species (e.g. flies and rodents), can be harvested and utilized to produce materials of value (e.g. feed and fertilizer). One common application of decomposition ecology that is observed daily through media sources, mainly television, is using such science to assist with solving crimes against people and other animals. Chapter 24 discusses the use of insects associated with decomposing vertebrate carrion and how they can be used to determine crucial information about a deceased or injured individual within a forensic context (Figure 1.5). These chapters provide some evidence of the applications of decomposition ecology; however, they do not represent all aspects of applications. Hopefully, by reading these chapters, individuals will create, or discover, novel aspects of applications

(a)   (b)

FIGURE 1.5   (See color insert.) Forensic entomologists can encounter different types of insects during an investigation. These insects can be (a) commonly encountered during investigations, as seen with the North American invasive *Chrysomya rufifacies* larva (Diptera: Calliphoridae), or (b) unexpected, such as these wood ants (*Camponotus* sp.) feeding on the open wound of a swine carcass. (Photos with permission from J.L. Pechal.)

that will better society and protect the environment. For a recent global treatment of forensic entomology, see Tomberlin and Benbow (2015).

1.5 The Future of Carrion Ecology and Evolution

Recent technological advances ranging from high-throughput metagenomic sequencing to unpiloted drones to high-resolution satellite remote sensors have opened many new doors to studying the ecology and evolution of organisms from microbes to gray whales. These advances have also facilitated novel discoveries about the natural environment that were once intractable because of technological limitations. This book was developed with the growing recognition provided by recent papers using these new technologies that carrion decomposition is an important part of an ecosystem and provides significant functions that cross scales of both time and space. This volume brings together some of the leading experts from around the world who are examining questions related to carrion decomposition, whether it is a project associated with the molecular mechanisms of bacterial communication on carrion to engineering explorations on how to improve the rendering process of carcasses important to industry. Further, each chapter was reviewed by each editor in addition to undergoing an internal peer review process in which each chapter was also reviewed by another chapter author or co-author of the book, providing a rigorous examination of each topic by experts in the field.

The intent of this book is to provide a reasonable collection of studies and perspectives of how the study of carrion expands the globe and transcends systems of inquiry, hopefully broadening the awareness of this important ecosystem process to the more general scientific communities in ecology and evolution. By bringing together the various aspects of carrion biology, the other goal was to enhance the basic scientific understanding of these processes for the improvement or development of innovative applications such as carcass rendering, fisheries management, or forensics. The latest technological advancements have provided the means to explore new and historically understudied natural systems, such as carrion, in a way that will most surely uncover new genomes, species, and ways that organisms interact. These potential new discoveries will hopefully broaden the scientific interest in carrion, an area of study that is often perceived as distasteful and a reminder of our ultimate biological destiny.

REFERENCES

Barton, P.S., S. Cunningham, D. Lindenmayer, and A. Manning. 2013a. The role of carrion in maintaining biodiversity and ecological processes in terrestrial ecosystems. *Oecologia* 171: 761–772.

Barton, P.S., S.A. Cunningham, B.C.T. Macdonald et al. 2013b. Species traits predict assemblage dynamics at ephemeral resource patches created by carrion. *PLoS One* 8: e53961.

Beasley, J.C., Z.H. Olson, and T.L. Devault. 2012. Carrion cycling in food webs: Comparisons among terrestrial and marine ecosystems. *Oikos* 121: 1021–1026.

Beaver, R. 1973. The effects of larval competition on puparial size in *Sarcophaga* spp. (Diptera: Sarcophagidae) breeding in dead snails. *Journal of Entomology Series A, General Entomology* 48: 1–9.

Beaver, R. 1977. Non-equilibrium island' communities: Diptera breeding in dead snails. *The Journal of Animal Ecology* 46: 783–798.

Bump, J.K., R.O. Peterson, and J.A. Vucetich. 2009a. Wolves modulate soil nutrient heterogeneity and foliar nitrogen by configuring the distribution of ungulate carcasses. *Ecology* 90: 3159–3167.

Bump, J.K., C.R. Webster, J.A. Vucetich et al. 2009b. Ungulate carcasses perforate ecological filters and create biogeochemical hotspots in forest herbaceous layers allowing trees a competitive advantage. *Ecosystems* 12: 996–1007.

Carter, D.O., D. Yellowlees, and M. Tibbett. 2007. Cadaver decomposition in terrestrial ecosystems. *Naturwissenschaften* 94: 12–24.

Dobzhansky, T. 1973. Nothing in biology makes sense except in the light of evolution. *American Biology Teacher* 35: 125–129.

Doube, B.M. 1987. Spatial and temporal organization in communities associated with dung pats and carcasses. In *Organization of Communities: Past and Present.* J. Gee and P.S. Giller (eds.), pp. 253–280. Oxford: Blackwell Scientific Publications.

Finn, J.A. 2001. Ephemeral resource patches as model systems for diversity-function experiments. *Oikos* 92: 363–366.

Fuller, M.E. 1934. The insect inhabitants of carrion: A study in animal ecology. *Council of Scientific and Industrial Research* 82: 1–63.

Gessner, M.O., C.M. Swan, C.K. Dang et al. 2010. Diversity meets decomposition. *Trends in Ecology and Evolution* 25: 372–380.

Hättenschwiler, S. and P. Gasser. 2005. Soil animals alter plant litter diversity effects on decomposition. *Proceedings of the National Academy of Sciences* 102: 1519–1524.

Hättenschwiler, S., A.V. Tiunov, and S. Scheu. 2005. Biodiversity and litter decomposition in terrestrial ecosystems. *Annual Review of Ecology, Evolution, and Systematics* 36: 191–218.

Hawlena, D., M.S. Strickland, M.A. Bradford, and O.J. Schmitz. 2012. Fear of predation slows plant-litter decomposition. *Science* 336: 1434–1438.

Hobbs, N.T. 1996. Modification of ecosystems by ungulates. *The Journal of Wildlife Management* 60: 695–713.

Hobbs, N.T. 2006. Large herbivores as sources of disturbance in ecosystems. In *Large Herbivore Ecology, Ecosystem Dynamics and Conservation*. New York: Cambridge University Press.

Hugh-Jones, M. and J. Blackburn. 2009. The ecology of *Bacillus anthracis*. *Molecular Aspects of Medicine* 30: 356–367.

Jabiol, J., J. Cornut, M. Danger et al. 2014. Litter identity mediates predator impacts on the functioning of an aquatic detritus-based food web. *Oecologia* 176: 225–235.

Levi, P.S., J.L. Tank, S.D. Tiegs et al. 2011. Does timber harvest influence the dynamics of marine-derived nutrients in Southeast Alaska streams? *Canadian Journal of Fisheries and Aquatic Sciences* 68: 1316–1329.

Lindeman, R.L. 1942. The trophic-dynamic aspect of ecology. *Ecology* 23: 399–418.

Majdi, N., A. Boiché, W. Traunspurger, and A. Lecerf. 2014. Predator effects on a detritus-based food web are primarily mediated by non-trophic interactions. *Journal of Animal Ecology* 83: 953–962.

Mégnin, P. 1894. La fauna des cadavres: Application de l'entomologie a la médecine légale. In *Encycolopedia Scientifique des Aides-memoire*, G. Masson (ed.). pp. 4–214. Paris, France: Gauthier-Villars.

Melis, C., N. Selva, I. Teurlings et al. 2007. Soil and vegetation nutrient response to bison carcasses in Białowieża Primeval Forest, Poland. *Ecological Research* 22: 807–813.

Miller, M.W., E.S. Williams, N.T. Hobbs, and L.L. Wolfe. 2004. Environmental sources of prion transmission in mule deer. *Emerging Infectious Diseases* 10: 1003–1006.

Moore, J.C., E.L. Berlow, D.C. Coleman et al. 2004. Detritus, trophic dynamics and biodiversity. *Ecology Letters* 7: 584–600.

Moore, J.W. and D.E. Schindler. 2008. Biotic disturbance and benthic community dynamics in salmon-bearing streams. *Journal of Animal Ecology* 77: 275–284.

Odum, E.P. 1969. The strategy of ecosystem development. *Science* 164: 262–270.

Parmenter, R. and J. MacMahon. 2009. Carrion decomposition and nutrient cycling in a semiarid shrub-steppe ecosystem. *Ecological Monographs* 79: 637–661.

Payne, J.A. 1965. A summer carrion study of the baby pig *Sus scrofa* Linnaeus. *Ecology* 46: 592–602.

Pechal, J.L., T.L. Crippen, M.E. Benbow, A.M. Tarone, and J.K. Tomberlin. 2014. The potential use of bacterial community succession in forensics as described by high throughput metagenomic sequencing. *International Journal of Legal Medicine* 128: 193–205.

Polis, G.A., W.B. Anderson, and R.D. Holt. 1997. Toward an integration of landscape and food web ecology: The dynamics of spatially subsidized food webs. *Annual Review of Ecology and Systematics* 28: 289–316.

Schoenly, K.G. and W. Reid. 1987. Dynamics of heterotrophic succession in carrion arthropod assemblages: Discrete series or a continuum of change? *Oecologia* 73: 192–202.

Seastedt, T., L. Mameli, and K. Gridley. 1981. Arthropod use of invertebrate carrion. *American Midland Naturalist* 105: 124–129.

Smith, C.R. and A.R. Baco. 2003. Ecology of whale falls at the deep-sea floor. *Oceanography and Marine Biology: An Annual Review* 41: 311–354.

Swift, M.J., O.W. Heal, and J.M. Anderson. 1979. *Decomposition in Terrestrial Ecosystems*. Berkeley and Los Angeles, CA: University of California Press.

Teal, J.M. 1962. Energy flow in the salt marsh ecosystem of Georgia. *Ecology* 43: 614–624.

Tiegs, S.D., E.Y. Campbell, P.S. Levi et al. 2009. Separating physical disturbance and nutrient enrichment caused by Pacific Salmon in stream ecosystems. *Freshwater Biology* 54: 1864–1875.

Tomberlin, J.K. and Benbow, M.E. 2015. *Forensic Entomology: International Dimensions and Frontiers*. Boca Raton, FL: CRC Press.

Tomberlin, J.K., M.E. Benbow, A.M. Tarone, and R. Mohr. 2011a. Basic research in evolution and ecology enhances forensics. *Trends in Ecology and Evolution* 26: 53–55.

Tomberlin, J.K., R. Mohr, M.E. Benbow, A.M. Tarone, and S. Vanlaerhoven. 2011b. A roadmap for bridging basic and applied research in forensic entomology. *Annual Review of Entomology* 56: 401–421.

Towne, E.G. 2000. Prairie vegetation and soil nutrient responses to ungulate carcasses. *Oecologia* 122: 232–239.

Villet, M.H. 2011. African carrion ecosystems and their insect communities in relation to forensic entomology. *Pest Technology* 5: 1–15.

Yang, L.H. 2004. Periodical cicadas as resource pulses in North American forests. *Science* 306: 1565–1567.

Yang, L.H., J.L. Bastow, K.O. Spence, and A.N. Wright. 2008. What can we learn from resource pulses? *Ecology* 89: 621–634.

2

Processes and Mechanisms of Death and Decomposition of Vertebrate Carrion

Shari L. Forbes and David O. Carter

CONTENTS

2.1 Introduction

The majority of the published literature referring to carrion ecology focusses on *vertebrate decomposition* as it is these larger organism's carcasses that typically survive long enough for the processes and mechanisms to be studied. Vertebrate bodies comprise various soft and hard tissues. The soft tissues consist of organs, muscles, ligaments, tendons, and hide, whereas the hard tissues make up the skeleton and include bone, teeth, cartilage, horn, and antler (Lyman 1994). Soft tissues will typically decay faster than hard tissues, and it is the dead and decaying tissues of the animal that are referred to as carrion. The hard tissues, especially bone, can be preserved for extended periods of time. Cartilage is harder than most soft tissues, but softer than bone, and occupies internal areas of the body that are not close to the skin surface. Horn is primarily keratin, the same protein which makes up hair, hoof sheaths, and feathers while antler resembles bone (Lyman 1994). Teeth are a different material and are largely composed of enamel and dentine, two of the hardest skeletal tissues in the body. The chemical and macrostructural properties of hard tissues make them resilient against the normal decomposition processes that impact soft tissues (Chapter 23). For this reason, they are often the only materials that survive in the natural environment.

The purpose of this chapter is to define the mechanisms and processes of animal death and how they are related to the ecology of carrion decomposition. This includes a description of the chemical and biological processes of carcass decomposition and preservation, the effects of the associated environment, and the subsequent cycling of energy and nutrients in that environment. Studies on mammal decomposition are the most prevalent in the published literature and will be predominantly discussed herein.

2.2 Mode and Mechanism of Death

Death is a dynamic process, which can occur in a matter of seconds or can be prolonged over a period of time. Death results from the reduction of living matter through a process known as histolysis (i.e., the dissolution of organic tissues), or it can result from the accumulation of foreign substances within the body that are not conducive to the normal living process. The metabolic processes, which are necessary and present in the living, are absent in the deceased. Organisms have the capacity to self-regulate these mechanisms during life, as demonstrated by hibernation.

The two main causes of death for vertebrates are broadly classified as "dying" and "being killed" (Weigelt 1989; Lyman 1994). The process of dying refers to normal death, which results from sickness or old age. The process of being killed refers to animals that become victims of accidents, enemies, predators, or some other external forces. Accidental modes of death are many and varied and include death resulting from a wide range of disturbances and other processes: fire, drowning, flooding, drought, volcanic activity, hunting, freezing, falling, combat, and overpopulation (Weigelt 1989). All of these modes of death are associated with a source external to the vertebrate that dies. This is in contrast to disease, illness, and old age, all of which are considered internal to the animal. For this reason, one should distinguish between the actual cause of death and the circumstances surrounding it. For example, exhaustion is a common cause of death, particularly in migratory birds, animals hunted by predators, and weak or young animals that are left behind in a herd stampede (Weigelt 1989; Lyman 1994). The effects of exhaustion represent the cause of death, whereas the migration, hunting, and stampede represent the circumstances leading to death. The cause of death, including the circumstances surrounding death, may impact the process of carrion decomposition, particularly if the internal body temperature has increased antemortem (e.g., fever, sprinting, etc.).

2.3 Chemical Processes of Decomposition

Death is a process rather than an event. During life, cells and tissues are sustained by highly organized chemical activities that maintain organismal functions. The chemical processes of living cells are defined by biochemical pathways in the body and associated with lipid-bound membranes and organelles (Gill-King 1997). Following death, chemical disorder and failure of cellular metabolism occur as a result of the lack of cardiac pumping of oxygenated blood throughout the body (Clark et al. 1997; Gill-King 1997). As the environment within the body becomes more anoxic following circulatory stasis, the disorganization within the cellular chemistry leads to the common macroscopic signs of death and decomposition.

Immediate effects of death are evidenced by a relaxation of muscles, which causes the lower mandible to drop and the eyelids to lose their tension (Janaway 1996). The early process of decomposition is not immediately evident macroscopically, but microscopically the cellular chemistry begins to fail soon after the cessation of life. Relaxation of the muscles can lead to purging of fluids including fecal matter and gastric contents if the carcass is moved. Blood pH will decrease resulting in an acidic environment caused by the accumulation of degraded compounds. This environment initiates an intrinsic coagulation mechanism causing blood clots in the arteries and vessels (Clark et al. 1997). As the pH continues to decrease, enzymes will assist in re-liquefying the coagulated blood.

Soon after, the chemical processes of rigor mortis (rigidity of the muscles) and livor mortis (blood pooling in dependent areas) commence. Rigor mortis causes a stiffening or rigidity of the muscles as a result of the loss of adenosine triphosphate (ATP). During life, an organism's muscles require a constant supply of ATP to contract. Muscular contraction involves the shortening of muscles by the muscle

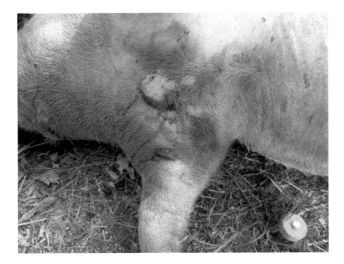

FIGURE 2.1 **(See color insert.)** Swine decomposition showing the darkened colors of livor mortis beneath the skin. (Photo by S. Forbes.)

proteins, actin and myosin (Dix and Graham 2000). After death, the production of ATP ceases, although consumption continues. Following the loss of ATP, the actin and myosin molecules become complexed, which represents the onset of rigor mortis (DiMaio and DiMaio 2001). The chemical process is reversible as the bond is only sustained until decomposition is initiated, at which point rigor mortis passes and the muscles will become flaccid once more. Rigor mortis occurs in all skeletal muscles at approximately the same time, but is typically more evident in the smaller facial muscle groups before being evidenced in the larger muscle groups in the limbs and joints. A carcass is considered to display "full rigor" when the jaw, neck, and limbs are immovable without significant force. Instances of rigor mortis immediately following death have been reported for animals killed in bullfights (Weigelt 1989). This is often referred to as "cadaveric spasm."

Livor mortis refers to the postmortem gravitational settling of blood in the dependent parts of the body. This occurs because the heart is no longer pumping oxygenated blood throughout the body, and as a result, the blood will start to pool in the lowest areas of the body (Janaway 1996; Dix and Graham 2000). It is initially evidenced by a red-to-dark-red discoloration in the capillaries beneath the skin (Figure 2.1). As the oxygen dissociates from the hemoglobin of red blood cells, deoxyhemoglobin will form and become evident as a purple discoloration (Clark et al. 1997).

When the process of livor mortis first occurs, the discoloration is not fixed. If pressure is applied to the dependent areas, the blood will be forced out of the capillaries and cause a blanching of the skin before resettling. Furthermore, if scavengers or other external forces move the carcass, the blood can resettle in the new dependent areas. Once livor mortis is fixed, however, the blood will remain where it has settled even when pressure is applied to the surface or the carcass is moved. The process is not all or nothing—the blood in some dependent areas may become fixed sooner than that in other areas (Dix and Graham 2000). If constant pressure is applied to an area, for example, through contact of one side of the carcass with a soil surface, the small blood vessels may also become compressed and the blood is prevented from settling in this area (Clark et al. 1997). This will be observed as a lack of livor mortis in those areas directly in contact with the soil. Livor mortis is generally only apparent in animals with a fine hair covering over their skin (e.g., swine) as opposed to animals with dense fur. It will persist until color changes resulting from autolysis and putrefaction become visible.

2.3.1 Autolysis

Autolysis refers to the postmortem enzymatic degradation of soft tissue by those same enzymes that assisted with metabolic function during the life of the organism. The phenomena described above are

FIGURE 2.2 **(See color insert.)** Swine decomposition showing autolytic degradation characterized by "marbling" in the torso as well as putrefactive degradation in the head. (Photo by S. Forbes.)

sometimes referred to as early autolysis as the processes can occur concomitantly. A decrease in oxygen levels and intracellular pH leads to the inhibition of aerobic metabolism and subsequent destruction of the cellular structure (Evans 1963; Janaway 1996; Clark et al. 1997; Carter et al. 2007). Many of the enzymes that contribute to the early autolytic stage of degradation are already present in the soft tissues. However, additional enzymes can be derived from microorganisms both internal and external to the remains, as well as from adult and immature insects present on the body (Carter et al. 2010). Disruption of the cell membranes releases macromolecules, particularly proteins and carbohydrates, which can be utilized by enteric microorganisms (Gill-King 1997). Lipids are also digested by hydrolytic enzymes but to a lesser degree. The first macroscopic signs of autolysis may be evidenced by discoloration and marbling of the skin, fluid blisters, and skin slippage (Dix and Graham 2000).

Reaction of the degraded hemoglobin with hydrogen sulfide in the body (discussed subsequently) forms a green pigment, sulfhemoglobin, which produces a green discoloration beneath the skin (Clark et al. 1997; Gill-King 1997). The onset of this discoloration is variable but is often evident in the abdomen and spreads across the torso. It may also be seen in those areas where livor mortis was prominent. If the pigment stains the walls of the superficial blood vessels beneath the skin, it appears as a reticulated pattern known as "marbling" (Clark et al. 1997; Dix and Graham 2000). The color may vary from green to purple or blue in the presence of deoxyhemoglobin (Figure 2.2).

The postmortem release of hydrolytic enzymes by cells also results in a loosening of the epidermis from the dermis, a phenomenon known as "skin slippage" (Gill-King 1997). Hair, fur, and other keratinaceous material (e.g., nails, hooves, horns, etc.) will also loosen and fall off with the skin or become dislodged if the carcass is moved. As decomposition proceeds, fluid may accumulate in the slipping skin producing postmortem bullae, which are more commonly referred to as "fluid blisters" (Clark et al. 1997). Larval masses may also aggregate in these areas.

2.3.2 Putrefaction

Putrefaction is the microbially driven process of fermenting proteins, carbohydrates, and lipids into simpler compounds such as acids, alcohols, amines, and gases. Putrefactive degradation is initiated as microorganisms from the intestinal tract migrate to the lymphatic system and penetrate the body tissues (Caplan and Koontz 2001). Although the organism's normal homeostatic mechanisms can prevent microbial decomposer activity during life, putrefaction supports a large microbial community while the release of decomposition fluids can also trigger vigorous growth of environmental microorganisms and plants (Tibbett and Carter 2003; Watson and Forbes 2008; Carter et al. 2010). Toward the end of autolysis, the environment within the body favors proliferation of fermentative bacteria present in the large intestine. These bacteria will eventually spread throughout the body (Janaway 1996), whereby putrefactive bacteria typically associated with the gut, such as *Proteus* spp., *Escherichia* spp., and *Lactobacillus* spp., can proliferate on the skin (Hyde et al. 2013; Metcalf et al. 2013; Pechal et al. 2014). Proliferation of these organisms is further fueled by the release of carbohydrate, protein, and lipid by-products during autolytic degradation.

Proteins are degraded through the process of proteolysis. Protein degradation in the intestinal tract, pancreas, and other organ tissues occurs early in the decomposition process, whereas the degradation of muscle protein typically follows the cessation of rigor mortis (Evans 1963). Extracellular enzymes denature the proteins into their constituent amino acids, which undergo deamination or decarboxylation by intracellular enzymes. Decarboxylation can lead to the production of a range of biogenic amines (Swann et al. 2010). Deamination of amino acids such as phenylalanine produces a carboxylic acid and ammonia. Sulfur-containing amino acids can be further degraded to yield hydrogen sulfide; these processes contribute to the visual changes of bloating and marbling (Gill-King 1997).

Lipid degradation results in the hydrolysis of triglycerides by intrinsic lipases into free fatty acids and glycerol. Water and a mildly acidic pH are typically required for this process to be initiated (Gill-King 1997). In an oxidative environment, unsaturated fatty acids will convert to aldehydes and ketones (Janaway 1996). In a reducing environment, unsaturated fatty acids will convert to saturated fatty acids, which may proceed to adipocere formation (Section 2.7). If there is sufficient moisture in the soft tissues, hydrolytic degradation of lipids will continue until completion at which point an abundance of free fatty acids remains.

Carbohydrate fermentation degrades the sugar carbon backbones to a variety of organic acids, alcohols, and gases. Under aerobic conditions, the sugars degrade to acids through a series of steps (lactic acid—pyruvic acid—acetaldehyde) and ultimately form carbon dioxide and water (Dent et al. 2004; Forbes 2008). Under anaerobic conditions, fermentation of sugars produces a range of by-products including lactic acid, propionic acid, butyric acid, ethanol, butanol, acetate, and butyrate (Dent et al. 2004; Forbes 2008; Paczkowski and Schutz 2011). The formation of acids may result in slightly lower soil pH beneath a carcass during the early stages of putrefaction, such as observed in maggot masses (Chun et al. 2014). This is often followed by a significant increase in a soil pH, usually within a week of death during warmer seasons (Spicka et al. 2011; Metcalf et al. 2013).

The microorganisms responsible for putrefaction produce an abundance of gases including simple molecules such as methane, hydrogen sulfide, and more complex volatile organic compounds (VOCs) such as dimethyl disulfide. As these gases accumulate within the intestinal and respiratory tracts, swelling and distension occur within the body, referred to as "bloating." These changes are first evident in those areas of the carcass that contain the most blood, often the regions associated with livor mortis (Clark et al. 1997). The head and neck will typically be the first areas to show signs of bloat, followed by the abdomen and other parts of the torso. Accumulation of gases and fluids within the body cavity can cause purging of fluids from the orifices (Janaway 1996; Dix and Graham 2000; Carter et al. 2007). Increased pressure in the soft tissue can also cause the skin to tear and rupture, further releasing the decomposed fluids purged from lungs, airways, and the intestinal tract and shifting the soil pH as outlined previously.

The odors most commonly associated with decomposition become evident during the bloating of the carcass and are readily detected as the soft tissue and organs begin to liquefy. Saprophytic bacteria degrade the tissues to become soft and fluid, releasing a range of gaseous products (Weigelt 1989). Initially, small molecules such as methane, hydrogen sulfide, and carbon dioxide are released as a result of carbohydrate fermentation (Gill-King 1997). Larger VOCs are also produced through the action of bacterial degradation and fermentation of amino acids, fatty acids, and polysaccharides (Dent et al. 2004; Paczkowski and Schutz 2011). During the distension of the torso, a range of short chain alcohols (e.g., C_3–C_8) and polysulfide compounds (e.g., dimethyl disulfide, dimethyl trisulfide, and dimethyl tetrasulfide) are prevalent. As the decomposition process becomes more active, short chain volatile fatty acids (e.g., C_2–C_6) and aromatics (e.g., phenol, indole, and skatole) are also released (Stadler et al. 2013). Further details relating to VOC production during carrion decomposition are provided in Chapter 9.

FIGURE 2.3 **(See color insert.)** Swine decomposition showing partial skeletonization in the head and limbs and mummification of the remaining soft tissue. (Photo by S. Forbes.)

2.3.3 Disintegration and Skeletonization

Degradation of soft tissue leads to liquefaction and disintegration of the remains, a phase commonly referred to as "wet decomposition." As soft tissues are depleted, hard tissues including bone will become exposed, leading to partial or complete skeletonization of the carcass. The process is considered complete if all of the soft tissue has been removed from the bone. Partial skeletonization results when areas of the carcass still contain soft tissue (Clark et al. 1997). Any remaining soft tissue will desiccate and collapse, assuming a dry, leathery texture that is typical of mummification (Figure 2.3). Residual organs will also desiccate and shrink (Dix and Graham 2000) and will be of little interest to scavengers, except certain carrion beetles (Anderson and Cervenka 2002; Anderson 2010). Degradation of the skin, muscle, and internal organs will also lead to disarticulation of the skeleton. The ligaments and tendons will be lost during this process (Clark et al. 1997). At this point, the carcass may be referred to as "dry" or "remains."

The continued reduction of hard tissues will depend on physical, chemical, and biological conditions (Chapter 23). The action of surrounding plants may cause physical degradation as roots penetrate and fracture the bones (Behrensmeyer 1978; Janaway 1996). Microbial activity will deplete the collagen phase of the hard tissues, subsequently altering the chemical basis for decomposition (Gill-King 1997). Once the organic phase has been lost, chemical decomposition of hydroxyapatite relies on the inorganic equilibria between the hard tissues and the surrounding medium (e.g., soil or water). In highly aerobic environments, such as on the soil surface or in shallow burials, decomposition is essentially complete (Gill-King 1997). In environments in which minimal oxygen, water, or low temperatures predominate, the complete disintegration of soft and hard tissues can take considerably longer and may lead to fossilization (Shipman 1981).

2.4 Biological Processes of Decomposition

The removal of soft tissue is not only a chemical process but also a biological process. Without biological activity, much of the chemical degradation would not occur. A great diversity of organisms is associated with carrion because it represents a valuable food source. Decomposition and decay are terms typically associated with microorganisms and invertebrates. Scavenging refers to the use of a carcass by vertebrates, although some invertebrates (e.g., crayfish or marine crustaceans) are considered to scavenge carrion (Burkepile et al. 2006, Chapters 6 and 12). All of these organisms are constantly competing for food (DeVault et al. 2003). Microbes are at their most effective in warm moist climates in which insects will likely access a body (Anderson and Cervenka 2002). Rapid bacterial and fungal growth is conducive to these climates, and arthropods further assist rapid decay by transferring decomposer organisms to the carcass as well as feeding on the carrion themselves (Carter et al. 2007). Scavengers can outcompete insects and microbes in these environments; however, they must be able to access a carcass quickly, as insects and microbes can consume 85% of a large (50 kg) carcass within 7 days postmortem (Spicka et al. 2011). Yet, vertebrates will successfully scavenge most available carcasses if they locate the remains before decomposition has proceeded to any great extent (Putman 1983, Chapters 6 and 21). A brief overview of the impact of microorganisms and macroorganisms on carrion decomposition is provided herein as further details can be found in subsequent chapters and recent publications (Barton et al. 2013a, b).

2.4.1 Microorganisms

Animals host a great diversity of microorganisms during life. These microorganisms, primarily bacteria, live in most regions of the body including the skin, mouth, and gut (Costello et al. 2009). The gut microbial biomass arguably has the greatest impact on carcass decomposition because it is responsible for the majority of putrefaction, which significantly alters the appearance, odor, and chemistry of a carcass (Gill-King 1997; Forbes 2008). It is these microorganisms that transform the body's macromolecules (carbohydrates, proteins, lipids, and nucleic acids) into simpler compounds which are subsequently released into the surrounding decomposition environment (Forbes 2008). As aerobic bacteria, predominantly from phyla Firmicutes and Bacteroidetes, consume the available oxygen within the gut, the abundance of fermentative anaerobes from genera *Bacteroides*, *Bifidobacterium*, *Clostridium*, *Enterobacter*, and *Streptococcus* increases (Kellerman et al. 1976; Melvin et al. 1984; Micozzi 1986). Recent microbiome sequencing studies observed a shift in the skin and mouth microbiology of carcasses from one that included respiring aerobes from phyla Actinobacteria and Firmicutes to one that was dominated by anaerobic bacteria that generate energy through fermentation, particularly lactic acid fermentation associated with *Lactobacillus*, *Staphylococcus*, and *Streptococcus* (Metcalf et al. 2013; Pechal et al. 2014). These studies also observed a decrease in bacterial diversity over time, probably because a decomposing carcass is a specialized habitat that selects for bacteria that can tolerate a warm, acidic, and highly reducing habitat. As putrefaction and insect activity continue, the skin will rupture, allowing oxygen to return to the internal cavity of the carcass and aerobic microbial activity will once again be initiated (Carter et al. 2007). Bacterial proliferation requires moderate temperatures and moisture; as will be discussed later in this chapter, desiccation and cooler temperatures prohibit bacterial growth (Lyman 1994).

The conditions of a decomposing carcass select for bacterial taxa while sufficient moisture and soft tissue are present. The fungi can proliferate once these resources have been depleted because hyphae can bridge gaps in pools of moisture and nutrients. Fungal growth is often observed on older carrion (Sagara 1995; Tibbett and Carter 2003; Sagara et al. 2008). As the fungi that colonize carrion are aerobic, their presence is often restricted to the skin and exposed surfaces of the carcass (Evans 1963), and it is common for fungi to distribute hyphae along the surface of carrion, including humans (Hitosugi et al. 2006). These microfungi are ubiquitous microorganisms that can actively decompose a diversity of organic matter, including hydrocarbons, skin, hair, bone, wood, and other plant materials. Decomposer organisms produce a broad array of degradative enzymes, many of which rarely occur in other fungal groups or in bacteria. A large number of microfungi produce antibiotic compounds that allow them to outcompete other microorganisms (Sagara et al. 2008). They can also more effectively tolerate the acidic pH levels typically formed by decomposition than bacteria (Janaway 1996).

In some scenarios, fungi will form fruiting structures in association with carrion (Sagara et al. 2008). These taphonomic mycota have been reviewed by Tibbett and Carter (2003) and include a diversity of fungal taxa, particularly from Ascomycota and Basidiomycota. Sometimes called the postputrefaction fungi (Sagara 1995), these organisms play an important role in terrestrial ecosystems because they can contribute to the cycling of the more recalcitrant parts of a carcass: hair, skin, and bone. Sagara (1995) viewed the role of ectomycorrhizal decomposers as a tripartite symbiotic relationship among animals, fungi, and plants.

The proliferation of the postputrefaction fungi also demonstrates that carcass decomposition has a significant effect on soil microbial communities. Soil microorganisms respond rapidly to the presence of a carcass, observed as an increase in microbial biomass (Carter et al. 2010) and carbon dioxide evolution (Putman 1978a, b). This effect on the soil microbial biomass is not surprising: carcass breakdown results in the release of significant amounts of chemicals into soils. Significantly greater concentrations of nitrogen, phosphorus, potassium, and sodium have been observed in soils associated with carrion (Benninger et al. 2008; Bump et al. 2009; Parmenter and MacMahon 2009; Van Belle et al. 2009). Parmenter and MacMahon (2009) also observed a disproportionately faster rate of carcass sulfur loss, which might also indicate an increase in soil sulfur. Additional details of microbial communities of carrion are discussed in Chapter 3.

2.4.2 Invertebrates

Bacterial and insect activities are considered two of the most influential factors for carrion decomposition (Anderson and Cervenka 2002). Arthropods are excellent agents of soft tissue removal particularly, given their prevalence in most natural environments. They hold an important ecological role as they recycle much of the biomass of carrion, making nutrients available for other organisms in the food chain. Animal remains become attractive to insects almost immediately after death; however, a succession of insects will visit the remains as the food resource modifies throughout the decomposition period (Anderson and Cervenka 2002; Anderson 2010; Kreitlow 2010). Ecological succession will vary by region based on climate, intra- and interspecies dynamics, and other random occurrences (Kreitlow 2010). The carcass, its inhabitants, and the surrounding environment are themselves an ecological community. In most ecosystems, arthropods will be present throughout the early, autolytic, putrefactive, and disintegrative processes, with some carrion beetles sustaining into the skeletonization stage.

Arthropods are attracted to carrion within minutes of death, and the specific odors that attract carrion insects are detailed in Chapter 9. It is predominantly the blow flies (Diptera: Calliphoridae) and flesh flies (Diptera: Sarcophagidae) that arrive during autolysis as key volatile compounds are released from the orifices of the carcass. Fly oviposition is important at this stage of decomposition before the tissue desiccates and becomes inhospitable to larval development. Without invertebrate or vertebrate scavenging, bacterial decay will be the prominent degradative action and decay will occur at a substantially slower rate (Payne 1965; Carter et al. 2007; Anderson 2010). The early-arriving insects will deposit eggs or larvae in the moist regions of the carcass, so that the remains can be used as a food source throughout larval development (Kreitlow 2010). Larval feeding assists to rupture the skin and liquefy soft tissue during the putrefactive stage (Carter et al. 2007). Other Diptera are attracted to the remains as the internal soft tissue becomes exposed and bacterial decay proceeds. Predacious Coleoptera, such as the rove beetles (Coleoptera: Staphylinidae), will also arrive during putrefaction if not already present. Rapid biomass loss is often evidenced during putrefaction as larval masses feed voraciously (Figure 2.4). As putrefaction and disintegration of soft tissue proceed, successions of insect larvae will migrate away from the carcass to pupate and emerge as adults (Greenberg 1990; Lewis and Benbow 2011). A range of Coleoptera species may persist on the remains as they become skeletonized (Kreitlow 2010). Most species do not feed on bone but rather consume the dry, mummified tissue that adheres to the bones. Dermestid beetles and their larvae are able to perforate mummified and desiccated tissue, assisting with disarticulation of the skeleton (Galloway 1997). Larvae of the moth *Tinea deperdella* have also been reported feeding on the organic components of horn (Behrensmeyer 1978). Many of these organisms will also contribute to the nutrient pool of a habitat through the deposition of puparia, exuvia, and insect carcasses. These activities play an important role because they contribute to the cycling of recalcitrant resources.

2.4.3 Vertebrates

Vertebrates that feed on carrion are referred to as scavengers, distinct from predators that kill their prey in order to consume it. As predators normally guard their prey or consume it quickly, scavengers rely on animals dying from malnutrition, disease, exposure, parasites, and accidents to provide their food source (DeVault et al. 2003). The location of animal death is important for scavenger accessibility, and the

FIGURE 2.4 **(See color insert.)** Larval masses contributing to rapid biomass loss after 6 days during swine decomposition. (Photo by S. Forbes.)

amount of carrion available can vary significantly among ecosystems (Carter et al. 2007). Larger carrion are usually more accessible to scavengers when compared with smaller carrion that may die in cavities, burrows, and other concealed locations (Chapter 6). This section focusses on terrestrial scavenging, with aquatic scavenging addressed elsewhere (Chapter 12).

The value of carrion to a vertebrate scavenger decreases over time. Once microorganisms begin to digest the soft tissue, potentially toxic chemical compounds are produced and those compounds useful to vertebrates may have already been metabolized. Vertebrates will generally only scavenge animal remains when they can outcompete microbial decay by locating and feeding on the remains prior to decomposition (Janzen 1977). However, scavenger activity has been reported on carcasses displaying advanced decomposition, mummification, and skeletonization in certain geographical climates, for example, southwest USA (Galloway 1997). Obligate scavengers will be less abundant in an ecosystem, but several animals such as hyenas, the red fox, and vultures do utilize carrion frequently as their main food source (DeVault et al. 2003) (Chapter 21).

Removal of soft tissue can be rapid, depending on the species of scavenger. Turkey vultures have been shown to scavenge 97% of brown-feathered chickens within 3 days of their placement in forests in Venezuela and Panama (Houston 1986, 1988). Grizzly bears, wolves, and ravens each scavenged more than 50% of large mammal carcasses placed in the Alaskan tundra (Magoun 1976). A number of studies investigating small bird scavenging have demonstrated a 25%–99% efficiency based on habitat type, location, and scavenging guilds (see Table 1 in DeVault et al. 2003), with the majority of scavenging occurring in 5 days or less. Rodent scavenging appears to be dependent on season as is the case for most scavenging behaviors (Simonetti et al. 1984; DeVault and Rhodes 2002).

Hill (1979), Haynes (1980, 1982), Haglund et al. (1989), and Haglund (1997a, b) have described soft tissue consumption and disarticulation for canid scavengers, including the domestic dog (*Canis familiaris*), coyote (*C. latrans*), and wolf (*C. lupus*). In large mammals, skin and muscle are typically the first soft tissues to be removed from the head, face, and neck. Feeding will then proceed to the thoracic cage where evisceration and separation of the head from the cervical and lumbar vertebrae may occur. The scapula and forelegs may also be detached early in the scavenging process. The pelvis may remain attached to the sacrum and to some of the lumbar vertebrae, or the lower limbs may be removed in association with the pelvic girdle and vertebrae. Long bones often become disarticulated and the ends damaged by gnawing. If feeding progresses to the final stage, all bones will become disarticulated, many will be scattered, and will show extensive damage resulting from gnawing.

Rodents are excellent small scavengers and predominantly gnaw on skeletal material. They are capable of accelerating the decay of bone following skeletonization (Galloway 1997) and are well-known "hoarders" of skeletal material in their dens or nests. Rodents cited as bone gnawers include the African porcupine (*Hystrix africaeustralis*), gerbil (*Desmodillus* sp.), mouse (*Peromyscus manaculatus*), as well as squirrels and rats (Haglund 1992, 1997a, b; Klippel and Synstelien 2007). Evidence of gnawing results from the rodent's tooth marks on the bone surface, the artifact often described as channels, striae, or grooves (Sorg 1985; Johnson 1985). The patterns produced by rodent gnawing will vary based on their chewing behavior and the type of bone being modified (Haglund 1997a, b).

2.5 Effect of the Environment on Carrion Decomposition

Biological factors including microorganisms, invertebrate and vertebrate scavengers, are by far the greatest contributors to the decomposition of soft tissue. Other factors that can influence the process of decomposition include environmental effects such as temperature, precipitation, humidity, and solar radiation, among others.

Ambient temperature is the most influential of the environmental factors as warm conditions encourage the proliferation of microorganisms both within the carcass and in the soil environment. Enhanced enzymatic and microbial degradation will lead to the rapid accumulation and subsequent release of gases from the carcass, which will attract invertebrate and vertebrate scavengers (Carter et al. 2007). Fly activity on a carcass is more prominent in warmer temperatures, and fly eggs will better survive allowing development through their larval stages. Increased larval insect and scavenging activity will assist in the rapid digestion of soft tissue, thus accelerating the process of decomposition (Mann et al. 1990; Bass 1997; Forbes and Dadour 2010).

In contrast, cold temperatures are less conducive to decomposition due to limited microbial, invertebrate, and vertebrate activity. Temperatures below freezing slow bacterial proliferation and eggs and larvae are unable to survive (Mann et al. 1990). The decomposition process is inhibited, resulting in a lack of putrefactive gases to attract vertebrate scavengers. The result is the preservation of the soft tissue indefinitely or until the environment once more becomes conducive to decomposition (e.g., seasonal changes in temperature) (Rosendahl 2010).

Moisture is equally important to the survival and proliferation of bacteria and insect larvae (Bass 1997). Humidity and precipitation are therefore important environmental factors that can mediate these biological activities. Although insects can still be present in arid environments, a lack of moisture will result in the desiccation of fly eggs and inhibition of larval growth. The rapid loss of moisture is associated with desiccation of the soft tissue, leading to preservation by mummification (Section 2.7). Mummified tissue is typically of little interest to vertebrate scavengers.

The degree of solar radiation to which the remains are exposed will impact the location of larval masses on the carcass. When the carcass is in a shaded environment, larval masses will be visible on the surface of the carcass. However, when the carcass is situated in direct sunlight, larvae will typically source a darker environment within the body cavity (Bass 1997). Larvae will feed on the internal tissues, leaving the skin as a protective barrier from the sunlight. The resultant effect will be an outer mummified layer of skin resembling the carcass body shape, with few, if any, tissues or organs remaining in the cavity.

In addition to climatic conditions, the decomposition environment will also impact the process of soft tissue degradation. Many animals will die and decompose in a terrestrial (surface) environment. However, some scavengers will bury carrion to ensure that the food source cannot be utilized by other scavengers and is therefore available for a longer period of time. The impact of burial is a reduction in the biological and chemical processes of decomposition. Depending on the depth of burial, temperatures can be lower than ambient temperature as the soil acts as a barrier to solar radiation (Rodriguez 1997). Access to the remains by invertebrates may be limited or completely prevented (Rodriguez and Bass 1985; Janaway 1996). The soil environment is integral to the process as it surrounds the remains. The microbial community, soil pH, oxygen content, and water-holding and ion-exchange capacities will become influential factors, which can impact the degree of decomposition or preservation of the carcass. Typically, animals that are buried will decompose at a slower rate than those that die on the surface (Rodriguez 1997).

Terrestrial animals that drown or whose remains end up in a water body will be subjected to a different decomposition environment. A similar process will also impact aquatic animals following death. Similar to soil burials, decomposition in aquatic environments is considerably slower than that on the soil surface (Rodriguez 1997; Madea et al. 2010). This reduced rate of decomposition results from cooler temperatures and differences in dissolved oxygen which impacts microbial degradation (Chapter 12). Additionally, invertebrate scavenging is minimal in these types of environments, although scavenging by aquatic animals can still occur. As air escapes from the lungs, the carcass will sink toward the bottom of the aquatic environment in which early postmortem changes will commence. Once gases accumulate within the body cavity, the carcass will rise to the surface and float (Allison and Briggs 1991; Lyman 1994; Sorg et al. 1997). Decomposition will proceed with the exposure to air, and insects or avian scavengers may be attracted to the remains at this time. Their feeding is predominantly limited to the exposed portions of the body, and their impact on the decomposition rate is noticeably reduced when compared with carcasses on a soil surface. Once the gases are released and the soft tissue is weakened by decay, the carcass will again sink (Haglund and Sorg 2002). At this time, the soft tissue will continue to undergo degradation and disarticulation or may be preserved by adipocere formation (Section 2.7).

Caves represent another environment in which animals may die, particularly sick or injured animals that are utilizing the location for respite or recovery. Caves typically exhibit uniform temperature and humidity year round (Rosendahl 2010). Air movement results from the network of passages that eventually open to the natural environment. Minimal light will penetrate deep within a cave, which can limit bacterial species survival to the extreme microorganisms. Although decomposition may commence, the dehydration of soft tissue is more common, leading to preservation of the remains. This process takes time but is possible because the remains are not affected by normal degradative influences such as temperature fluctuations and scavenging (Rosendahl 2010).

Differential decomposition of carrion can occur if parts of the carcass are exposed to two different environments. For example, if an animal drowns and is washed or pulled ashore, part of the body will be exposed to the natural soil or sand environment, while the other part may still be partially submerged. The part exposed to air may undergo normal decomposition involving microbial, invertebrate, and vertebrate scavenging, whereas the submerged part of the body may be preserved by adipocere formation. It is not unusual for differential decomposition to be observed in any of the environments discussed earlier.

2.6 Effect of Carrion Decomposition on the Environment

The presence of a carcass can have a significant effect on the environment. The magnitude of this effect is related to carcass size, but even a mouse carcass can result in a significant localization of insect activity for several hours (Putman 1978a,b). As discussed earlier, the flush of compounds that a carcass releases into the soil and the atmosphere has effects on the chemistry of the area around a carcass; however, it can also influence the biology of an area, including vegetation (Yang 2004). Ultimately, carcasses can serve as disturbances that trigger secondary succession within an ecosystem (Hawlena et al. 2012).

Every carcass releases nutrients into the environment. Carcasses, even the size of guinea pigs (Bornemissza 1957) and neonatal swine (Spicka et al. 2011), will have a significant effect on associated soil. The release of nitrogen, phosphorus, and potassium into soil stimulates a pulse of microbial activity and a significant change in the chemistry of an area for a significant period (Anderson et al. 2013). Larger carcasses can result in a disturbance that persists for a significant period of time. Anderson et al. (2013) observed significantly elevated nitrogen associated with swine carrion (*Sus scrofa domesticus*) for up to 1 year postmortem. Towne (2000) observed elevated concentrations of soil inorganic nitrogen and phosphorus associated with bison carrion (*Bos bison*) for up to 3 years postmortem. This level of disturbance typically results in the death of all associated plant life (Towne 2000; Anderson et al. 2013), a process that has been observed in several studies (Chapter 13).

The mechanism of plant death during carcass decomposition has not been studied extensively. Plants are often exposed to significant shifts in pH and concentrations of nutrients not typically experienced, one of the major reasons why carcass decomposition is viewed as a disturbance. Larger carcasses will be associated with chlorotic and dead vegetation by the time of maggot migration. All of these will eventually die, and the patch will eventually support the vigorous growth of new vegetation. Oftentimes, pioneer species are able to outcompete their autochthonous counterparts. This reestablishment of vegetation can occur after a period of 1 year for large carcasses and within 1 year for smaller carcasses (Towne 2000; Anderson et al. 2013). The smallest carcasses, such as mice, might not have any effect on the surrounding vegetation. They might simply be too small or nutrients might be taken up by ants and flies rather than released into the soil. Whatever the exact mechanisms, carcasses can cause disturbances within a plant community that facilitates the establishment of pioneer species that would otherwise not be present.

Volatile compounds are released into the atmosphere and the soil. It has long been established that even small carcasses release a significant amount of carbon dioxide into the atmosphere (Putman 1978a; Carter et al. 2010). More recently observed is the vast array of VOCs released into the atmosphere (Dent et al. 2004; Paczkowski and Schutz 2011). Short chain alcohols (e.g., C_3–C_8), polysulfide compounds (e.g., dimethyl disulfide, dimethyl trisulfide, and dimethyl tetrasulfide), volatile fatty acids (e.g., C_2–C_6), and aromatics (e.g., phenol, indole, and skatole) are also released (Stadler et al. 2013); Chapter 9 presents further details.

Although many details of carrion decomposition have been resolved in recent years, the fundamental effect of a carcass has long been understood: the decomposition of dead animals contributes to the cycling of nutrients within and between habitats (Yang 2004; Bump et al. 2009). Many roles of the biological component of carcass decomposers have been previously discussed, but the overarching concept is that carrion acts as an island of fertility within an environment and, therefore, contributes to biodiversity by acting as a habitat for species that use carcass material for food and/or shelter. Succession associated with a carcass is driven by changes in the environmental and edaphic conditions as resources are consumed. Facilitation is a developmental sequence in which each stage, through its activity, facilitates the development of subsequent stages. For example, the initial oviposition of blow fly eggs leads to the development of a population of first instar blow fly larvae. The presence of these larvae supports the development of a population of larval feeding insects such as the rove beetles. In contrast, inhibition occurs when the colonization of a resource by one species inhibits the entrance of other organisms into a sere. This might occur through the predation of one species by another, decreasing resources to a level that does not support subsequent species, or competition through the use of toxins. Lastly, tolerance is a developmental sequence in which some species colonize an area and become established, independent of the presence or absence of other species. The subsequent sere is the product of competitive exclusion. An example of this can occur on a carcass in geographical regions where a carcass is colonized by the predatory blow fly *Chrysomya rufifacies* (Diptera: Calliphoridae). Larvae of *C. rufifacies* can prevent the colonization and development of other blow fly species through predation (Wells and Greenberg 1992).

2.7 Inhibitory Effects on Decomposition

The chemical processes described previously (autolysis, putrefaction, and disintegration) lead to the dissolution of soft tissue to the point that only hard tissue and other recalcitrant tissues (cartilage, keratin, etc.) remain. Under certain environmental conditions, the carcass can, however, be preserved. Preservation processes reported for animal remains include mummification, adipocere formation, and the tanning processes of peat bogs, as outlined subsequently.

2.7.1 Mummification

Under certain conditions, the chemical processes of decomposition can be halted, even after they have been initiated. In the later stages of decomposition, when the body has almost completely disintegrated and the skeleton becomes exposed, it was previously noted that any remaining tissue can dry out and shrink, forming a leathery texture. This process of moisture loss, referred to as desiccation, is not simply restricted to the later postmortem period but can also commence soon after death. In a dry or drying environment, desiccation results from the rapid loss of moisture by evaporation (Clark et al. 1997). The product of desiccation is referred to as *mummification* and can occur in both hot and cold environments in which conditions are arid (Micozzi 1986, 1991, 1997; Lyman 1994; Sledzik and Micozzi 1997). The term is often used to refer to the preservation of hard tissue only; however, in the context of carrion decomposition, it is better described as the preservation of some soft tissues as well as bones (Figure 2.2) (Lyman 1994). Natural mummification occurs when the carcass loses most of its moisture before the normal processes of decomposition can proceed (Weigelt 1989).

In hot, arid regions, the early decomposition stage may occur concomitantly with the dehydration of the outer layer of the carcass, particularly the skin (Galloway 1997). Although the skin dries out and forms a shell over the remains, the underlying soft tissue may still contain sufficient moisture for decomposition to proceed. The shell provides a protective barrier to the loss of moisture from the underlying tissue, thereby allowing biological activity (i.e., microorganisms and insects) to progress. Eventually, dehydration of the remaining soft tissue will also occur, and the effects of bacteria and larval activity will wane, thus slowing down the process of decomposition. Dermestid beetles may continue to feed on the mummified remains, but few macroscopic changes will be visible (Galloway 1997). At this point, the mummified remains can be preserved for years, decades, or millennia.

Desiccation can occur rapidly in extremely hot, dry climatological conditions such as deserts, where the rapid dehydration of soft tissue prevents any degradative action by microorganisms, soil bacteria, or other decomposer organisms (Sledzik and Micozzi 1997). Desiccation can affect the entire body or only those parts of the body exposed to the drying conditions. The activity of larval masses can also cause mummification in discrete regions of the carcass as they utilize the moisture during their feeding process. Drying and shrinkage of skin usually occur in those areas that contain minimal fluids such as the extremities and face. The internal organs will typically be the last tissues to desiccate (Sledzik and Micozzi 1997).

Animals that die in frozen climates may become "freeze-dried"—a process whereby the body freezes and the water sublimes directly to water vapor within the remains (Micozzi 1986, 1991, 1997; Janaway 1996). Sublimation should not be confused with the natural desiccation that occurs in cold, dry environments of low temperatures and low humidity. Under these conditions, it is the effect of both the temperature and lack of moisture that inhibits biological activity. Low temperatures inhibit bacterial growth and therefore the chemical process of putrefaction, by affecting the cell division time. Temperatures below 12°C severely retard most bacterial reproduction, whereas temperatures between 0°C and 5°C halt bacterial multiplication completely, and the time for cell division to occur approaches infinity (Micozzi 1986, 1991, 1997). The lack of moisture in both the soft tissue and the surrounding environment further restricts the action of microorganisms. Very little entomological activity will be observed in cold environments, although scavengers may still be present. Depending on latitude, the animal remains may be subjected to freeze–thaw cycles through the transition of seasons. It has been suggested that animals that have been frozen and then commence thawing undergo external decomposition initially as the soil microorganisms invade the carcass. Internal carcass temperature must increase sufficiently before enteric microorganisms become active (Micozzi 1986, 1991, 1997). This is in contrast to a carcass decomposing in a temperate environment whereby internal decomposition commences almost immediately after death (Vass et al. 2002).

2.7.2 Adipocere Formation

Adipocere formation results from the hydrolysis and hydrogenation of lipids in the body to saturated fatty acids. Degradation of lipids by enzymes and microorganisms is required to release the fatty acids from the glycerol backbone in the constituent triglycerides (Janaway 1996). Typically, the environment must be relatively anaerobic to prevent oxidation of the fatty acids into ketones and aldehydes. In anaerobic conditions, the unsaturated fatty acids (predominantly oleic, linoleic, and palmitoleic acids) undergo β-oxidation and hydrogenation to form saturated fatty acids (predominantly myristic, palmitic, and stearic acids) (Forbes et al. 2005a,b,c; Frund and Schoenen 2009). This action results from bacterial enzymes present in both the tissue and the surrounding soil (Fiedler and Graw 2003) and can lead to the formation of hydroxy and oxo-fatty acids in adipocere (Takatori 2001). The fatty acids can also conjugate with bivalent ions present in the body or soil to form insoluble salts, including calcium and sodium salts of fatty acids (Notter et al. 2009).

Adipocere formation is typically seen in moist environments, for example, when an animal has died or been moved to a water body such as a creek, river, or ocean. However, the moisture content of fat cells can offset the lack of moisture in a dry environment (Gill-King 1997), and adipocere may be observed on a carcass decomposing in an arid region, typically under the dehydrated outer layer of skin (Evans 1963). In such instances, both mummification and adipocere can be observed concurrently. Similar to mummification, adipocere can form over the entire body or may only be present in discrete regions of the carcass. The fatty regions of the body are most conducive to adipocere formation, although the transloca-tion of liquefied fat can cause adipocere to form in almost any part of the body.

Adipocere formation can occur over a broad range of temperatures, but conditions that are too warm will result in the liquefaction of the tissue before saturated fatty acids can form into the characteristic solid, white substance of adipocere (O'Brien 1997). Cold environments slow decomposition, which also impacts the likely formation of adipocere. However, it has been shown that adipocere can form in tem-peratures below those considered optimal for its formation (Forbes et al. 2011). Adipocere formation does not inhibit bacterial activity but appears to shift the bacterial community away from those organ-isms conducive to putrefaction (Pfieffer et al. 1998; Ueland et al. 2014). The altered microbial environ-ment coupled with a reduced pH has been suggested as the major reason for tissue preservation where adipocere forms. Once formed, adipocere can preserve a carcass for years, decades, or even centuries (Fiedler and Graw 2003; Dent et al. 2004; Rosendahl 2010).

2.7.3 Preservation by Peat Bogs

The preservation of animals in peat bogs is variable, although a large number of remains have been recovered from the peat bogs of northern Europe (Janaway 1996). The action of the humic acids in the peat is believed to aid preservation of soft tissue. Heath humus, moor, and swamp soils are often rich in humic acid (Weigelt 1989). Humic acid causes the tanning of soft tissue, leads to bone mineraliza-tion, and results in the reddening of the hair if preserved (Madea et al. 2010). Along with the presence of humic acid and tannins in the bog, reduced oxygen in the peat suppresses putrefaction, resulting in a degree of preservation of the remains.

Different tissue types will demonstrate varying levels of preservation. Those tissues containing high levels of collagen and keratin are typically better preserved than the internal muscles and organs, while the skeleton will become extensively decalcified causing the dissolution and loss of bone structure (Janaway 1996; Micozzi 1997). It is thought that the collagen fibers of the skin and other connective tissues are preserved through the process of tanning by polysaccharide sphagnan present in the peat. Keratinaceous materials, in contrast, are likely preserved due to the exclusion of the microorganisms required to degrade keratin.

2.8 Conclusions

The mechanism of death of vertebrate animals can vary greatly, depending on the circusmtances sur-rounding their death and whether the cause of death is natural or due to external forces. The postmortem process of decomposition is heavily dependent on the location of death and the impact of the surrounding environment on the chemical and biological processes of decay. The environment will dictate the degree of decomposition and, in some cases, the degree of preservation, particularly when biological activity is suppressed or eliminated. Although the chemical processes of autolyis, putrefaction, and disintegration will typically occur in a predicatable order, the rate of these chemical processes can vary considerably and is an inherent limitation of understanding decomposition processes. Continued research in different ecozones across the world is required to assist in better understanding the complex chemical and biologi-cal processes of carrion decomposition.

REFERENCES

Allison, P.A. and D.E.G. Briggs. 1991. *Taphonomy: Releasing the Data Locked in the Fossil Record*, vol. 9. New York: Plenum Press.

Anderson, G.S. 2010. Factors that influence insect succession on carrion, in: *Forensic Entomology: The Utility of Arthropods in Legal Investigations*, 2nd edition, J.H. Byrd and J.L. Castner (eds.), pp. 201–250. Boca Raton: CRC Press.

Anderson, G.S. and V.J. Cervenka. 2002. Insects associated with the body: Their use and analyses, in: *Advances in Forensic Taphonomy: Method, Theory, and Archaeological Perspectives*, W.D. Haglund and M.H. Sorg (eds.), pp. 173–200. Boca Raton: CRC Press.

Anderson, B., J. Meyer, and D.O. Carter. 2013. Dynamics of ninhydrin-reactive nitrogen and pH in gravesoil during the extended postmortem interval. *Journal of Forensic Sciences* 58: 1348–1352.

Barton, P.S., S.A. Cunningham, D.B. Lindenmayer, and A.D. Manning. 2013a. The role of carrion in maintaining biodiversity and ecological processes in terrestrial ecosystems. *Oecologia* 171: 761–772.

Barton, P.S., S.A. Cunningham, B.C.T. Macdonald, S. McIntyre, D.B. Lindenmayer, and A.D. Manning. 2013b. Species traits predict assemblage dynamics at ephemeral resource patches created by carrion. *PLoS One* 8: 1–8.

Bass, W.M. 1997. Outdoor decomposition rates in Tennessee, in: *Forensic Taphonomy: The Postmortem Fate of Human Remains*, W.D. Haglund and M.H. Sorg (eds.), pp. 181–186. Boca Raton: CRC Press.

Behrensmeyer, A.K. 1978. Taphonomic and ecologic information from bone weathering. *Paleobiology* 4: 150–162.

Benninger, L.A., D.O. Carter, and S.L. Forbes. 2008. The biochemical alteration of soil beneath a decomposing carcass. *Forensic Science International* 180: 70–75.

Bornemissza, G.F. 1957. An analysis of arthropod succession in carrion and the effect of its decomposition on the soil fauna. *Australian Journal of Zoology* 5: 1–12.

Bump, J.K., C.R. Webster, J.A. Vucetich, R.O. Peterson, J.M. Shields, and M.D. Powers. 2009. Ungulate carcasses perforate ecological filters and create biogeochemical hotspots in forest herbaceous layers allowing trees a competitive advantage. *Ecosystems* 12: 996–1007.

Burkepile, D.E., J.D. Parker, C.B. Woodson, H.J. Mills, J. Kubanek, P.A. Sobecky, and M.E. Hay. 2006. Chemically mediated competition between microbes and animals: Microbes as consumers in food webs. *Ecology* 87: 2821–2831.

Caplan, M.J. and F.P. Koontz. 2001. *Postmortem Microbiology*. Washington, DC: American Society for Microbiology Press.

Carter, D.O., D. Yellowlees, and M. Tibbett. 2007. Cadaver decomposition in terrestrial ecosystems. *Naturwissenschaften* 94: 12–24.

Carter, D.O., D. Yellowlees, and M. Tibbett. 2010. Moisture can be the dominant environmental parameter governing cadaver decomposition in soil. *Forensic Science International* 200: 60–66.

Chun, L., M. Miguel, S.L. Forbes, and D.O. Carter. 2014. Postmortem microbiology: Culturable bacteria of the maggot mass. *Proceedings of the 66th Annual Meeting of the American Academy of Forensic Sciences* 20: 358–359.

Clark, M.A., M.B. Worrell, and J.E. Pless. 1997. Postmortem changes in soft tissues, in: *Forensic Taphonomy: The Postmortem Fate of Human Remains*, W.D. Haglund and M.H. Sorg (eds.), pp. 151–164. Boca Raton: CRC Press.

Costello, E.K., C.L. Lauber, M. Hamaday, N. Fierer, J.L. Gordon, and R. Knight. 2009. Bacterial community variation in the human body habitats across space and time. *Science* 326: 1694–1697.

Dent, B.B., S.L. Forbes, and B.H. Stuart. 2004. Review of human decomposition processes in soil. *Environmental Geology* 45: 576–585.

DeVault, T.L. and O.E. Rhodes Jr. 2002. Identification of vertebrate scavengers of small mammal carcasses in a forested landscape. *Acta Theriologica* 47: 185–192.

DeVault, T.L., O.E. Rhodes, and J.A. Shivik. 2003. Scavenging by vertebrates: Behavioral, ecological, and evolutionary perspectives on an important energy transfer pathway in terrestrial ecosystems. *Oikos* 102: 225–234.

DiMaio, D. and V.J.M. DiMaio. 2001. *Forensic Pathology*, 2nd edition. Boca Raton: CRC Press.

Dix, J. and M. Graham. 2000. *Time of Death, Decomposition and Identification: An Atlas*. Boca Raton: CRC Press.

Evans, W.E.D. 1963. *The Chemistry of Death*. Illinois: Charles C. Thomas.

Fiedler, S. and M. Graw. 2003. Decomposition of buried corpses, with special reference to the formation of adipocere. *Naturwissenschaften* 90: 291–300.

Forbes, S.L. 2008. Decomposition chemistry in a burial environment, in: *Soil Analysis in Forensic Taphonomy: Chemical and Biological Effects of Buried Human Remains*, M. Tibbett and D.O. Carter (eds.), pp. 203–224. Boca Raton: CRC Press.

Forbes, S.L. and I.R. Dadour. 2010. The soil environment and forensic entomology, in: *Forensic Entomology: The Utility of Arthropods in Legal Investigations*, 2nd edition, J.H. Byrd and J.L. Castner (eds.), pp. 407–426. Boca Raton: CRC Press.

Forbes, S.L., B.B. Dent, and B.H. Stuart. 2005a. The effect of soil type on adipocere formation. *Forensic Science International* 154: 35–43.

Forbes, S.L., B.H. Stuart, and B.B. Dent. 2005b. The effect of the burial environment on adipocere formation. *Forensic Science International* 154: 24–34.

Forbes, S.L., B.H. Stuart, and B.B. Dent. 2005c. The effect of the method of burial on adipocere formation. *Forensic Science International* 154: 44–52.

Forbes, S.L., M.E.A. Wilson, and B.H. Stuart. 2011. Examination of adipocere formation in a cold water environment. *International Journal of Legal Medicine* 125: 643–650.

Frund, H.C. and D. Schoenen. 2009. Quantification of adipocere degradation with and without access to oxygen and to the living soil. *Forensic Science International* 188: 18–22.

Galloway, A. 1997. The process of decomposition: A model from the Arizona-Sonoran desert, in: *Forensic Taphonomy: The Postmortem Fate of Human Remains*, W.D. Haglund and M.H. Sorg (eds.), pp. 139–150. Boca Raton: CRC Press.

Gill-King, H. 1997. Chemical and ultrastructural aspects of decomposition, in: *Forensic Taphonomy: The Postmortem Fate of Human Remains*, W.D. Haglund and M.H. Sorg (eds.), pp. 93–108. Boca Raton: CRC Press.

Greenberg, B. 1990. Behavior of postfeeding larvae of some Calliphoridae and a Muscid (Diptera). *Annals of the Entomological Society of America* 83: 1210–1214.

Haglund, W.D. 1992. Contributions of rodents to postmortem artifacts of bone and soft tissue. *Journal of Forensic Sciences* 37: 1459–1465.

Haglund, W.D. 1997a. Dogs and coyotes: Postmortem involvement with human remains, in: *Forensic Taphonomy: The Postmortem Fate of Human Remains*, W.D. Haglund and M.H. Sorg (eds.), pp. 367–382. Boca Raton: CRC Press.

Haglund, W.D. 1997b. Rodents and human remains, in: *Forensic Taphonomy: The Postmortem Fate of Human Remains*, W.D. Haglund and M.H. Sorg (eds.), pp. 405–414. Boca Raton: CRC Press.

Haglund, W.D., D.T. Reay, and D.R. Swindler. 1989. Canid scavenging/disarticulation sequence of human remains in the Pacific Northwest. *Journal of Forensic Sciences* 34: 587–606.

Haglund, W.D. and M.H. Sorg. 2002. Human remains in water environments, in: *Advances in Forensic Taphonomy: Method, Theory, and Archaeological Perspectives*, W.D. Haglund and M.H. Sorg (eds.), pp. 201–218. Boca Raton: CRC Press.

Hawlena, D., M.S. Strickland, M.A. Bradford, and O.J. Schmitz. 2012. Fear of predation slows plant-litter decomposition. *Science* 336: 1434–1438.

Haynes, G. 1980. Prey, bones and predators: Potential ecological information from analysis of bone sites. *Ossa* 7: 75–97.

Haynes, G. 1982. Utilization and skeletal disturbances of North American prey carcasses. *Arctic* 35: 266–281.

Hill, A.P. 1979. Disarticulation and scattering of mammal skeletons. *Paleobiology* 5: 261–274.

Hitosugi, M., I. Kiyoshi, T. Yaguchi, Y. Chigusa, A. Kurosa, M. Kido, T. Nagai, and S. Tokudome. 2006. Fungi can be a useful forensic tool. *Legal Medicine (Tokyo)* 8: 240–242.

Houston, D.C. 1986. Scavenging efficiency of Turkey vultures in tropical forest. *Condor* 88: 318–323.

Houston, D.C. 1988. Competition for food between neotropical vultures in forest. *Ibis* 130: 402–417.

Hyde, E.R., D.P. Haarmann, A.M. Lynne, S.R. Bucheli, and J.F. Petrosino. 2013. The living dead: Bacterial community structure of a cadaver at the onset and end of the bloat stage of decomposition. *PLoS One* 8: e77733.

Janaway, R.C. 1996. The decay of buried human remains and their associated materials, in: *Studies in Crime: An Introduction to Forensic Archaeology*, J. Hunter, C. Roberts, and A. Martin (eds.), pp. 58–85. London: B.T. Batsford Ltd.

Janzen, D.H. 1977. Why fruits rot, seeds mold, and meat spoils. *The American Naturalist* 111: 691–713.

Johnson, E. 1985. Current developments in bone technology, in: *Advances in Archaeological Method and Theory*, vol. 8, M.B. Schiffer (ed.), pp. 157–235. New York: Academic Press.

Kellerman, G.D., N.G. Waterman, and L.F. Scharfenberger. 1976. Demonstration *in vitro* of postmortem bacterial transmigration. *American Journal of Clinical Pathology* 66: 911–915.

Klippel, W.E. and J.A. Synstelien. 2007. Rodents as taphonomic agents: Bone gnawing by brown rats and gray squirrels. *Journal of Forensic Sciences* 52: 765–773.

Kreitlow, K.L.T. 2010. Insect succession in a natural environment, in: *Forensic Entomology: The Utility of Arthropods in Legal Investigations*, 2nd edition, J.H. Byrd and J.L. Castner (eds.), pp. 251–270. Boca Raton: CRC Press.

Lewis, A.J. and M.E. Benbow. 2011. When entomological evidence crawls away: *Phormia Regina En Masse* larval dispersal. *Journal of Medical Entomology* 48: 1112–1119.

Lyman, R.L. 1994. *Vertebrate Taphonomy*. Cambridge: Cambridge University Press.

Madea, B., J. Preuss, and F. Musshoff. 2010. From flourishing life to dust—The natural cycle of growth and decay, in: *Mummies of the World*, A. Wieczorek and W. Rosendahl (eds.), pp. 14–29. Munich: Prestel Verlag.

Magoun, A.J. 1976. *Summer Scavenging Activity in Northeastern Alaska*. MS thesis, University of Alaska, Fairbanks, AK, USA.

Mann, R.W., W.M. Bass, and L. Meadows. 1990. Time since death and decomposition of the human body: Variables and observations in case and experimental field studies. *Journal of Forensic Sciences* 35: 103–111.

Melvin, J.R., L.S. Cronholm, L.R. Simson, and A.M. Isaacs. 1984. Bacterial transmigration as an indicator of time of death. *Journal of Forensic Sciences* 29: 412–417.

Metcalf, J.L., L. Wegener-Parfrey, A. Gonzales, C. Lauber, D. Knights, G. Ackermann, G.C. Humphrey et al. 2013. A microbial clock provides an accurate estimate of the postmortem interval in a mouse model system. *eLife* DOI: 10.7554/eLife.01104.

Micozzi, M.S. 1986. Experimental study of postmortem change under field conditions: Effects of freezing, thawing, and mechanical injury. *Journal of Forensic Sciences* 31: 953–961.

Micozzi, M.S. 1991. *Postmortem Change in Human and Animal Remains: A Systematic Approach*. Illinois: Charles C Springfield.

Micozzi, M.S. 1997. Frozen environments and soft tissue preservation, in: *Forensic Taphonomy: The Postmortem Fate of Human Remains*, W.D. Haglund and M.H. Sorg (eds.), pp. 171–180. Boca Raton: CRC Press.

Notter, S.J., B.H. Stuart, R. Rowe, and N. Langlois. 2009. The initial changes of fat deposits during the decomposition of human and pig remains. *Journal of Forensic Sciences* 54: 195–201.

O'Brien, T.G. 1997. Movement of bodies in lake Ontario, in: *Forensic Taphonomy: The Postmortem Fate of Human Remains*, W.D. Haglund and M.H. Sorg (eds.), pp. 559–566. Boca Raton: CRC Press.

Paczkowski, S. and S. Schutz. 2011. Post-mortem volatiles of vertebrate tissue. *Applied Microbiology and Biotechnology* 91: 917–935.

Parmenter, R.R. and J.A. MacMahon, 2009. Carrion decomposition and nutrient cycling in a semiarid shrub-steppe ecosystem. *Ecological Monographs* 79: 637–661.

Payne, J.A. 1965. A summer carrion study of the baby pig *Sus scrofa* Linnaeus. *Ecology* 46: 592–602.

Pechal, J.L., T.L. Crippen, M.E. Benbow, A.M. Tarone, S. Dowd, and J.K. Tomberlin. 2014. The potential use of bacterial community succession in forensics as described by high throughput metagenomic sequencing. *International Journal of Legal Medicine* 128: 193–205.

Pfeiffer, S., S. Milne, and R.M. Stevenson. 1998. The natural decomposition of adipocere. *Journal of Forensic Sciences* 43: 368–370.

Putman, R.J. 1978a. Patterns of carbon dioxide evolution from decaying carrion: Decomposition of small mammal carrion in temperate systems, Part 1. *Oikos* 31: 47–57.

Putman, R.J. 1978b. Flow of energy and organic matter from a carcass during decomposition: Decomposition of small mammal carrion in temperate systems, Part 2. *Oikos* 31: 58–68.

Putman, R.J. 1983. *Carrion and Dung: The Decomposition of Animal Wastes*. London: Edward Arnold.

Rodriguez, W.C. 1997. Decomposition of buried and submerged bodies, in: *Forensic Taphonomy: The Postmortem Fate of Human Remains*, W.D. Haglund and M.H. Sorg (eds.), pp. 459–468. Boca Raton: CRC Press.

Rodriguez, W.C. and W.M. Bass. 1985. Decomposition of buried bodies and methods that may aid in their location. *Journal of Forensic Sciences* 30: 836–852.

Rosendahl, W. 2010. Natural mummification—Rare, but varied, in: *Mummies of the World*, A. Wieczorek and W. Rosendahl (eds.), pp. 30–41. Munich: Prestel Verlag.

Sagara, N. 1995. Association of ectomycorrhizal fungi with decomposed animal wastes in forest habitats: A cleaning symbiosis? *Canadian Journal of Botany* 73: S1423–S1433.

Sagara, N., T. Yamanaka, and M. Tibbett. 2008. Soil fungi associated with graves and latrines: Toward a forensic mycology, in: *Soil Analysis in Forensic Taphonomy: Chemical and Biological Effects of Buried Human Remains*, M. Tibbett and D.O. Carter (eds.), pp. 67–107. Boca Raton: CRC Press.

Shipman, P. 1981. *Life History of a Fossil: An Introduction to Taphonomy and Paleoecology.* Cambridge: Harvard University Press.

Simonetti, J.A., J.L. Yanez, and E.R. Fuentes. 1984. Efficiency of rodent scavengers in central Chile. *Mammalia* 48: 608–609.

Sledzik, P.S. and M.S. Micozzi. 1997. Autopsied, embalmed, and preserved human remains: Distinguishing features in forensic and historic contexts, in: *Forensic Taphonomy: The Postmortem Fate of Human Remains*, W.D. Haglund and M.H. Sorg (eds.), pp. 483–496. Boca Raton: CRC Press.

Sorg, M.H. 1985. Scavengers modifications of human remains. *Current Research in the Pleistocene* 2: 37–38.

Sorg, M.H., J.H. Dearborn, E.I. Monahan, H.F. Ryan, K.G. Sweeney, and E. David. 1997. Forensic taphonomy in marine contexts, in: *Forensic Taphonomy: The Postmortem Fate of Human Remains*, W.D. Haglund and M.H. Sorg (eds.), pp. 567–604. Boca Raton: CRC Press.

Spicka, A., R. Johnson, J. Bushing, L.G. Higley, and D.O. Carter. 2011. Carcass mass can influence rate of decomposition and release of ninhydrin-reactive nitrogen into gravesoil. *Forensic Science International* 209: 80–85.

Stadler, S., P.H. Stefanuto, M. Brokl, S.L. Forbes, and J.F. Focant. 2013. Characterization of volatile organic compounds from human analogue decomposition using thermal desorption followed by comprehensive two-dimensional gas chromatography—Time-of-flight mass spectrometry. *Analytical Chemistry* 85: 998–1005.

Swann, L.M., S.L. Forbes, and S.W. Lewis. 2010. A capillary electrophoresis method for the determination of selected biogenic amines and amino acids in mammalian decomposition fluid. *Talanta* 81: 1697–1702.

Takatori, T. 2001. The mechanism of human adipocere formation. *Legal Medicine (Tokyo)* 3: 193–204.

Tibbett, M. and D.O. Carter. 2003. Mushrooms and taphonomy: The fungi that mark woodland graves. *Mycologist* 17: 20–24.

Towne, E.G. 2000. Prairie vegetation and soil nutrient responses to ungulate carcasses. *Oecologia* 122: 232–239.

Ueland, M., H. Breton, and S.L. Forbes. 2014. Bacterial populations associated with early stage adipocere formation in lacustrine waters. *International Journal of Legal Medicine* 28: 379–387.

Van Belle, L.E., D.O. Carter, and S.L. Forbes. 2009. Measurement of ninhydrin-reactive nitrogen influx into gravesoil during aboveground and belowground carcass (*Sus scrofa*) decomposition. *Forensic Science International* 193: 37–41.

Vass, A.A., S.A. Barshick, G. Sega, J. Caton, J.T. Skeen, J.C. Love, and J.A. Synstelien. 2002. Decomposition chemistry of human remains: A new methodology for determining the postmortem interval. *Journal of Forensic Sciences* 47: 542–552.

Watson, C.J. and S.L. Forbes. 2008. An investigation of the vegetation associated with grave sites in southern Ontario. *Canadian Society of Forensic Science Journal* 41: 199–207.

Weigelt, J. 1989. *Recent Vertebrate Carcasses and Their Paleobiological Implications,* [Translation of *Rezente Wirbeltierleichen und ihre Palaobiologische Bedeutung*, 1927] Translated by J. Schaefer. Chicago: University of Chicago Press.

Wells, J.D. and B. Greenberg. 1992. Interaction between *Chrysomya rufifacies* and *Cochliomyia macellaria* (Diptera: Calliphoridae): The possible consequences of an invasion. *Bulletin of Entomological Research* 82: 133–137.

Yang, L.H. 2004. Periodical cicadas as resource pulses in North American forests. *Science* 306: 1565–1567.

3

Microbial Interactions during Carrion Decomposition

Tawni L. Crippen, M. Eric Benbow, and Jennifer L. Pechal

CONTENTS

3.1 Introduction

Ecological studies on the decomposition of nonliving organic matter have historically focused on plant systems (Swift et al. 1979; Moore et al. 2004; Hättenschwiler et al. 2005; Gessner et al. 2010). A fundamental component of any type of detrital decomposition is the microbial community, an assemblage of organisms representing all three domains of life (Woese et al. 1990), including prokaryotes of the bacteria and archaea domains and several representative taxonomic groups of the eukarya domain (e.g., fungi, protozoans, heterotrophic algae, microarthropods, and some worms). Research describing and quantifying the microbial importance to decomposition, nutrient cycling, and energy elaboration has been predominately from studies on photosynthetically derived decaying organic matter (e.g., leaf litter) and has largely neglected the functional contribution of heterotrophically derived decomposition (e.g., carrion) to ecosystem processes (Chapter 1). Because of this difference in historical study between photosynthetically derived and heterotrophically derived biomass decomposition, this chapter will briefly review the microbial communities important to plant litter decomposition (detritus) and then concentrate on providing a conceptual foundation for those neglected communities, processes, and functions associated with carrion decomposition.

For years, the microbial community structure was considered a "black box" in ecological studies of detritus. This paradigm was predominately a function of technological limitations and the inadequate ability to characterize the true diversity within microbial communities by using culturing techniques (Barton and Northrup 2011). Because of this limitation, the history of detritus decomposition studies was essentially limited to measuring rates of organic matter breakdown in different ecosystems, habitats, and under a variety of abiotic and biotic conditions (Swift et al. 1979; Hättenschwiler et al. 2005; Gessner et al. 2010; Woodward et al. 2012). The inability to identify the spectra of microbial decomposers on plant litter resources from different ecosystems has proven to be challenging. It has proven difficult to describe or predict the general patterns of decomposition correlated with specific resource type breakdown and its associated effects on trophic cascades and biodiversity–ecosystem function relationships without the use of metagenomic techniques (Moore et al. 2004; Hättenschwiler et al. 2005; Hladyz et al. 2011; Woodward et al. 2012). However, recent advances in high-throughput next generation sequencing (NGS) technologies have made it easier to comprehensively describe microbial communities with greater confidence in taxonomic identification within any ecosystem (Lozupone and Knight 2007; Barton and Northrup 2011; Caporaso et al. 2011). For instance, this technology has led to a rapid increase in our understanding of the human microbiome (Costello et al. 2009; The Human Microbiome Consortium 2010, 2012), specifically identifying what species are present and how they function within the vertebrate body. This improvement in technological advances allows for novel studies to better understand microbial community variation within the black box of ecology, including those communities associated with the decomposition of both photosynthetically and heterotrophically derived organic matter; the communities of the latter are described as a component of the necrobiome (Benbow et al. 2013).

In this chapter, a brief summary of microbial decomposition ecology based on culture-only or genomic diversity (discussed subsequently) community descriptions will precede more recent studies that have employed high-throughput NGS approaches to more comprehensively study microbial diversity variation in plant detritus decomposition (see also Chapter 19). This summary of plant litter microbial ecology provides the conceptual and methodological introduction for a comprehensive discussion of the microbial community structure and function important to carrion decomposition. This chapter will describe the taxonomy and phylogeny of carrion microbial communities; how they assemble and function to decompose heterotrophically derived biomass; how they interact with individual species and communities of necrophagous invertebrate communities; and how these cross-domain interactions mediate the carrion decomposition process with effects on soil communities. One must also consider the challenges posed to bacteria that rely on the utilization of an ephemeral resource such as carrion. These bacteria must somehow get to the resource and then must compete with other microbes and fauna that also desire to exploit this rich nutrient supply and finally must produce viable progeny. Other chapters provide more thorough discussions on how carrion affects large-scale ecological processes (see Chapters 6, 13, 21, and 22). But, this chapter will discuss other research applications of carrion microbial ecology and how the carrion necrobiome can be used as a model system to explore novel complex questions related to the core importance of microbes within community and ecosystem dynamics. The black box that has limited some aspects of microbial ecology research for so long is finally opening in the genomic age.

3.1.1 Techniques of Studying Microbial Communities: Lessons from Plant Decomposing Organic Matter

The microbial communities associated with plant detritus (e.g., leaves or woody debris) are dependent on several factors (Hättenschwiler et al. 2005; Kominoski and Pringle 2009; Chapman and Newman 2010; Kominoski et al. 2011; Chapman et al. 2013). These factors include, but are not limited to, the following: the plant species of the leaves individually or mixed with other species; timing of leaf abscission; leaf composition and quality; disease or infestation by other organisms; soil type, habitat, and microhabitat physicochemical conditions during decomposition; interactions with macroorganism conditioning the breakdown of the leaf; and the season (i.e., abiotic). Similar factors likely drive the microbial ecology of carrion decomposition. Although these factors are known to influence the microbial diversity of plant detritus, most studies to date have not employed NGS techniques to describe the associated communities. Rather, they have used nontaxon-specific techniques such as phospholipid fatty acid (PLFA) profiles, terminal restriction fragment length polymorphisms (tRFLPs), automated ribosomal intergenic spacer analysis (ARISA), and denaturing gradient gel electrophoresis (DGGE) (Frostegård and Bååth

1996; Tiedje et al. 1999; Chapman et al. 2013; Fernandes et al. 2013; Frossard et al. 2013). Other studies have used traditional DNA-sequencing methods, such as Sanger sequencing (Sanger et al. 1977), but this approach is time-consuming, expensive, and requires creating clone libraries that can provide some degree of taxonomic resolution in limited microbial communities (unlike complex, multispecies environmental communities). Because of these limitations, Sanger sequencing has not been used in large-scale ecological studies (Shokralla et al. 2012).

The approaches described earlier have greatly contributed to understanding microbial decomposition of plant litter and other forms of organic matter. However, much remains unresolved in this global ecosystem process, including how to acquire more precise microbial diversity estimates and better understanding the associated functional importance; how biodiversity affects both single and mixed plant litter species decomposition rates; and how cascading interactions affect the ecosystem, beginning with the leaf–soil interface and amplifying to the ecosystem detrital food web (Moore et al. 2004; Chapman and Newman 2010; Kominoski et al. 2011; Chapman et al. 2013).

Fundamental work investigating plant litter diversity effects on decomposition (Hättenschwiler et al. 2005; Gessner et al. 2010) has been enhanced by recent studies exploring how mixed species litter assemblages affects decomposition through additive and nonadditive effects (Hättenschwiler and Gasser 2005; Kominoski et al. 2007; Chapman et al. 2013). For instance, the leaf litter decomposition rate of mixed conifer species increased up to 50%, compared with the decomposition of a single species after 10 months (Chapman et al. 2013). Furthermore, the microbial diversity present throughout the decomposition process positively correlated with the increase in leaf litter diversity (Chapman et al. 2013). Other research has focused on how the diversity of species affects decomposition rates and balances with top-down controls of detritivore consumers (Srivastava et al. 2009). However, there is a dearth of studies investigating fluctuations in microbial diversity during terrestrial or aquatic litter decomposition. Few studies that have been recently conducted could only report general patterns of diversity change over time or among plant litter types. This limitation is due to their use of microbial characterization techniques (e.g., DGGE and ARISA) that cannot provide fine resolution community taxonomic descriptions (e.g., family, genus, and species) and only reported diversity as operational taxonomic units (OTUs) (Fernandes et al. 2013). Although useful, OTUs are limited because bacterial sequence diversity occurs more frequently with increasing 16S rRNA gene copy numbers (Vetrovsky and Baldrian 2013). Hence, some taxa, such as Firmicutes and Gammaproteobacteria, have higher copy numbers and thus more variation (Vetrovsky and Baldrian 2013). Additionally, 16S rRNA sequences from the same species can differ, and conversely, multiple bacterial species may contain common sequences. Therefore, estimates of bacterial diversity, abundance, and community structure by measurement of 16S rRNA can be skewed and should be recognized as only approximations of diversity (see Chapter 18 for further discussion of such limitations). This limitation is also true for the internal transcribed spacer region of rRNA used to identify fungal organisms (Amend et al. 2010). These OTU-based patterns of microbial communities often change over time or among litter species and can be inconsistent and dependent on numerous factors such as habitat type, soil, and the pre-existing plant communities in which the litter is being decomposed.

The environmental influence on litter microbial communities makes developing generalizations difficult. Early studies of plant litter microbial communities, as reviewed in Swift et al. (1979), used culture-based techniques to provide broad taxonomic descriptions of microbial communities, which included taxonomic groups in bacteria, such as the myxobacteriales, cytophagales, spirochaetales, and actinomycetales, and fungi, including chytridiomycetes, trichomycetes, and hyphomycetes (see Table 3.2 in Swift et al. [1979]). However, in the area of understanding fruit and seed decomposition, often in relation to agricultural and economic importance, there were more specific taxonomic descriptions given to both bacteria and fungi communities: these are eloquently reviewed in Janzen (1977), who proposed that microbes caused food to spoil as a means of competing with larger animals for such resources.

Based on the current understanding of environmental microbial diversity, these taxa lists are woefully incomplete considering the current state of technology, which estimates that less than 1% of the bacteria can be effectively cultured (Amann et al. 1995; Pace 1997). In the past, emphasis has been put on developing culture-based techniques for microbes of health, agricultural, and industrial importance, leaving insufficient knowledge of the true diversity and function of environmental microbes, even those found in the most extreme environments (Barns et al. 1994).

The ability to describe microbial communities from complex environmental samples, such as carrion, was revolutionized in 2005 by the introduction of NGS technologies. With these techniques, it is now possible to analyze entire communities of microbes from complex environmental samples (Shokralla et al. 2012), overcoming some of the limitations of more traditional culture and biochemical-based approaches. However, shortfalls still exist, in that sequence comparisons and annotation are restricted to those genes and species presently available in databases such as NBCI Sequence Read Archive (http://www.ncbi.nlm.nih.gov/Traces/sra/sra.cgi), DDBJ Sequence Read Archive (http://trace.ddbj.nig.ac.jp/dra/index_e.html), and EMBL-EBI European Nucleotide Archive (http://www.ebi.ac.uk/ena/). Public input of sequences from a variety of environmental sources continually augments the expansion of these databases; therefore, this shortcoming will diminish with time. In conjunction with public input, the sequence quality must be stringently controlled to insure the accuracy of added sequences for all who benefit from these important resources (see Chapter 18 for more information).

As the quantity of data increases as a result of the use of NGS, so too must computational power and memory of computing hardware and software. Bioinformatic challenges to synthesizing and managing these huge quantities of sequence data are a constant dilemma (Caporaso et al. 2011). Nevertheless, NGS approaches to describing environmental microbial communities are opening exciting new research directions in decomposition, biodiversity–function relationships, and general mechanisms of ecosystem function processes, including those important for studying the ecology and evolution of carrion decomposition. To date, there are few studies using NGS to better understand microbial biodiversity in plant litter decomposition or other forms of organic matter, including carrion. This limitation is changing as ecologists become more familiar with molecular biology, especially for fungi (Lim et al. 2010; McGuire et al. 2012; Štursová et al. 2012; Kuramae et al. 2013; Voříšková and Baldrian 2013), but those evaluating bacterial communities are still rare (DeAngelis et al. 2013; Pechal et al. 2014b). Soil scientists have been some of the first ecologists to embrace NGS approaches, such as in the recent review by Prescott and Grayston (2013), discussing the influence tree species, such as broadleaf versus coniferous, have on the composition of the microbial communities within the decomposing litter, soils, and forest floor. These studies are beginning to tease out the extent of the relationship between microbes and macrobes, setting an optimistic stage for future microbial decomposition research using NGS technologies.

3.1.2 Microbial Communities of Carrion and Soil

Descriptions of microbial communities associated with carrion, or the soil beneath, have in the past suffered the same methodological limitations as those studies on plant detritus. Some of the earlier works on microbial communities of the necrobiome were conducted by researchers initially interested in biochemical changes that occur following death, usually within a forensic context focused on human cadaver decomposition (Vass 2001; Vass et al. 2002). In a more popular article, Vass (2001) described his attempts to understand microbial succession on human cadavers in an effort to make estimations of time since death. Presumably using culture-based techniques, the author reported several bacterial genera that occurred in sequence, including *Staphylococcus*, *Candida*, *Malasseria*, *Bacillus*, and *Streptococcus* spp. during early decomposition. The community then transitioned into anaerobic bacterial representatives including micrococci, coliforms, diptheriods, *Clostridium* spp., *Serratia* spp., *Klebsiella* spp., *Proteus* spp., *Salmonella* sp., *Cytophaga*, additional pseudomonads, and flavobacteria. As decomposition progressed, Vass noted that these necrotic communities began mixing with bacterial taxa found in the surrounding environment and those exogenous taxa brought in by colonizing insects such as blow flies (Diptera: Calliphoridae) (see below for fly–microbe interactions). Given the complexity of these communities, the author was unable to continue the study. After a thorough literature search, no peer-reviewed publications by Vass describing these results in additional detail were found, and so his 2001 publication remains one of the only sources of information regarding culture-based bacterial community succession on decomposing human remains. Recently, a description of the internal gut bacterial community from human cadavers during the process of bloating in decomposition has been obtained using pyrosequencing (Hyde et al. 2013); this study provided an internal snapshot and identification of the bacteria phyla within a cadaver. Other researchers have gone forward using fungi taxa succession to make estimates of the decomposition time using both nonhuman and human carrion models (Carter and Tibbett 2003;

Hitosugi et al. 2006; Ishii et al. 2006), whereas still others have used algae and diatoms to make determinations about the length of time carrion has been submerged in water (Zimmerman and Wallace 2008).

Since the early descriptions by Vass (2001), there has been a paucity of work examining bacterial succession on carrion, but there have been studies on how soils beneath carcasses change throughout and after carrion decomposition in a localized area (Hopkins et al. 2000; Parkinson et al. 2009; Howard et al. 2010; Lenz and Foran 2010). The importance of microbial decomposition of carrion can have profound impacts on soil microbial communities, nutrient transformation, and plant succession as part of complex aboveground and belowground biotic interactions (Wardle et al. 2004; Hawlena et al. 2012). Wardle et al. (2004) suggested that greater than 50% of the net primary productivity (NPP) can be returned to the soil via labile fecal material in fertile terrestrial habitats that support high herbivore densities. Considering the potential for large numbers of herbivore mammal deaths (up to 1% of the total organic matter input in some terrestrial ecosystems) estimated for some regions of the world (Chapters 6, 13, and 23) (Houston 1985; Carter et al. 2007), there is a plausible scenario for a substantial return of nutrients and NPP to soil communities via carrion deposition, as documented by Parmenter and MacMahon (2009) and reviewed by others (Beasley et al. 2012; Barton et al. 2013a). However, the rate and dynamics of this return depend on microbial processing and decomposition of these complex and ephemeral resources.

The significance of carrion or human cadaver decomposition on soil microbial communities has been the subject of study by several researchers, with some studies using changes in soil communities for forensic purposes as previously mentioned (Hopkins et al. 2000; Parkinson et al. 2009), whereas others more generally evaluated the ecological consequences of carrion decomposition on soils (Parmenter and MacMahon 2009; Howard et al. 2010). Despite differences in focus (e.g., basic theory or forensic application), they all provide data for developing a mechanistic understanding of soil microbial responses to carrion inputs.

There have been forensically related studies that measured microbial biomass and diversity to evaluate the effects of heterotrophic tissue decomposition on soils. These studies determined that carrion decomposition can have significant and lasting effects on soil microbial communities, including increases in microbial biomass and function (e.g., net nitrogen mineralization) (Hopkins et al. 2000; Carter and Tibbett 2006; Carter et al. 2007) and notable successional shifts in bacterial and fungal communities during the human decomposition process (Parkinson et al. 2009). In the latter study, the authors used t-RFLP and PLFA approaches to explore soil bacterial and fungal diversity, respectively, beneath three human cadavers. Using multidimensional statistical analyses, they found that both communities changed gradually over decomposition, rather than in punctuated shifts; however, they also reported substantial differences in soil community responses between the three sets of remains primarily based on the size of the corpse (Parkinson et al. 2009). Additional work is necessary to better understand the role of carrion type (e.g., different species or sexes and sizes within a single species) for use in forensic investigations. Some of this information is coming from basic research on the ecology of carrion decomposition, a first step in bridging of basic and applied science as advocated by Tomberlin et al. (2011) and Tomberlin and Benbow (2015).

In addition to forensically oriented studies of soil microbial community changes associated with carrion decomposition, there have been several basic science studies investigating the importance of carrion in soil ecology. Although not directly assessing microbial communities, Parmenter and MacMahon (2009) investigated soil nutrient changes underlying 11 species of vertebrate carrion, providing one of the first comprehensive studies to measure the importance of carrion species (type) variation in soil nutrient responses. They found profound decreases in soil nitrogen, phosphorus, and sodium after carcass deposition that depended on carcass species, generally following a positive relationship with carcass size. Given the importance of nutrient ratios to microbes, localized nutrient pulses that vary by carrion species (and quality) may have significant effects on the structure and function of soil microbial communities, with additional bottom-up effects on local plant communities such as those reviewed in Chapters 6 and 13 and documented in other studies (Parmenter and MacMahon 2009; Barton et al. 2013a).

In a fascinating study published in *Science*, Hawlena et al. (2012) reported that when the quality of carrion was mediated by the fear of predation, there were significant cascade effects on leaf litter decomposition by changing the soil microbial structure and function. They found that carcasses of grasshoppers exposed to the threat of predation prior to death changed in nutrient quality, and that this shift in quality corresponded to changes in soil microbial communities and function. The soil microbial shifts were associated with accelerated plant litter decomposition by three-fold, thus documenting that carcass quality can potentially alter plant detritus decomposition both at localized habitat and ecosystem scales. Soil microbial communities act as ecological conduits for nutrients and energy that begin with primary producers, that are assimilated and transformed during herbivory and predation, and released by other carrion-associated microbes upon death of consumers, feeding back to the primary producers and completing nutrient cycling loops. Additional review and discussion on carrion effects on soil microbiology and associated biochemistry can be found in Chapter 5.

In recent years, research examining microbial changes to soil resulting from carrion decomposition has seen increased attention. However, few studies have evaluated the microbes actually associated with the decomposing carcass. Having an understanding of the composition of carrion microbial communities, whether endogenous (associated with the organism upon death) or exogenous (brought in by other organisms such as vertebrate scavengers), is vital to forensics, agriculture, and ecology, as well as veterinary and human health. The lack of knowledge in this field is presumably attributed to expense, a deficiency in high-resolution identification technologies and possibly the necessity during experimentation to directly interface with the unappealing ecological stages of death.

Current technological advances have led to renewed interests in using molecular techniques, such as those described earlier for plant litter detritus, to evaluate bacterial communities on carrion. Dickson et al. (2010) used 16S rRNA gene cloning to make estimates of the postmortem submersion interval of swine heads from bacterial community succession in a marine habitat. The authors reported different communities of bacteria that represented 15 orders, 21 families, and 39 genera on different parts of the head (i.e., cheek, snout, and wound), which also changed during the decomposition progression and during different seasons. They suggested that the bacterial communities observed early in decomposition were different from those present later in decomposition (see Figure 3 in Dickson et al. 2010) and thus might be useful for time since immersion calculations. For example, the genus *Fusobacterium* was collected only on the first sampling date (sampling every 2–4 days over an 11–21-day period) from the snout and not found in any later samples. Some genera of the order Bacteroidales appeared on the carcass snout only after a particular period of submersion (after 5 days) and then persisted on the carcass; whereas some, such as *Moritella*, *Psychromonas*, *Marinomonas*, and *Cowellia*, were present on the snout only later (day 10–11) in the decomposition process. Bacteria such as *Pseudoalteromonas*, *Acrobacter*, *Psychrobacter*, and *Vibrio* were considered reoccurring bacteria that were found at various times throughout the decomposition period and thus were unlikely to be good indicator species. However, this study did not employ any statistical analysis to provide a quantitative measure of potential indicator taxa for different postmortem submersion intervals. So, the question of whether bacterial species quantity differences could function as indicators remained open.

For the first time in carrion decomposition studies, Pechal et al. (2014b) used 454-pyrosequencing to more comprehensively and statistically describe the epinecrotic bacterial succession on decomposing swine carcasses in a terrestrial habitat. They followed the bacterial communities on three replicate carcasses and recovered 137,181 sequences that represented four phlya and 20 families (Tables 3.1 and 3.2). Using multivariate statistics, the epinecrotic communities were found to be significantly different over days of decomposition, but were not different among replicate carcasses. They also reported that Bacteroidaceae and Moraxellaceae were significant familial level indicators of the communities found at initial death, compared with Bacillaceae and Clostridiales Incertae Sedis XI that represented the bacterial communities 5 days after death. Lastly, they provided models to describe the relationship between the physiological time (accumulated degree hours) and the epinecrotic bacterial community composition (Pechal et al. 2014b).

In an associated experiment of the Pechal et al. (2014b) study, the authors investigated the epinecrotic bacterial succession on swine carcasses where initial necrophagous insect access and colonization were prevented for the first 5 days of decomposition. They reported that at the phylum level of taxonomic resolution, there were distinct successional changes in carrion bacterial communities over decomposition (Figure 3.1) (Pechal et al. 2014a). They also determined that necrophagous insect colonization shifted this bacterial succession trajectory. There was variability in the initial (immediately after death) bacterial communities among the three replicate carcasses in the treatment (excluded from insect access) and the

TABLE 3.1

Relative Abundance of Bacterial Phyla and Rare Taxa (Phyla with <3% Relative Abundance) Used to Describe the Epinecrotic Communities over Days of Decomposition of Swine Carcasses Where Insects Were Excluded for 5 Days (EXC) Compared with Carcasses with Initial Insect Access (ACC) ($n = 3$ Carcasses per Treatment) Using Pyrosequencing Analyses (Mean [±SD])

ADH Treatment	Day 0 (0) EXC		ACC		Day 1 (541) EXC		ACC		Day 3 (1535) EXC		ACC		Day 5 (2692) EXC		ACC	
Proteobacteria	59.1	±32.9	70.5	±1.7	37.4	±29.4	41.9	±10.3	62.6	±13.0	31.2	±44.8	81.3	±8.1	2.7	±2.3
Firmicutes	36.2	±32.9	20.1	±6.6	37.9	±30.9	36.1	±6.9	28.0	±5.9	57.9	±35.6	16.6	±7.3	96.6	±1.3
Actinobacteria	2.7	±4.6	3.2	±5.6	12.2	±12.7	10.4	±4.6	7.4	±4.4	10.4	±9.3	0.0	±0.0	0.0	±0.0
Bacteroidetes	0.0	±0.0	3.8	±3.6	11.9	±20.6	10.3	±17.9	1.7	±2.9	0.0	±0.0	0.0	±0.0	0.0	±0.0
Rare taxa (<3%)	4.6	±3.0	3.5	±0.3	0.6	±0.3	1.3	±1.8	0.4	±0.4	0.7	±0.5	2.0	±2.1	0.7	±1.1

TABLE 3.2

Relative Abundance of Bacterial Families and Rare Taxa (Families with <3% Relative Abundance) Used to Describe the Epinecrotic Communities over Days of Decomposition of Swine Carcasses Where Insects Were Excluded for 5 Days (EXC) Compared with Carcasses with Initial Insect Access (ACC) (*n* = 3 Carcasses per Treatment) Using 454-Pyrosequencing Analyses (Mean [±SD])

	0 ADH (Day 0)	541 ADH (Day 1)	1535 ADH (Day 3)	2692 ADH (Day 5)
Aerococcaceae	6.8 ± 3.0	15.3 ± 6.9	30.5 ± 26.8	0.0 ± 0.0
Alcaligenaceae	0.0 ± 0.0	1.1 ± 1.9	0.0 ± 0.0	0.0 ± 0.0
Clostridiaceae 1	0.0 ± 0.0	0.0 ± 0.0	0.0 ± 0.0	6.5 ± 8.2
Clostridiales Incertae Sedis XI	0.0 ± 0.0	0.0 ± 0.0	0.7 ± 1.1	26.3 ± 17.2
Comamonadaceae	0.0 ± 0.0	1.4 ± 2.5	0.0 ± 0.0	0.0 ± 0.0
Corynebacteriaceae	0.3 ± 0.6	0.0 ± 0.0	4.8 ± 5.9	0.0 ± 0.0
Enterobacteriaceae	7.9 ± 13.8	23.5 ± 20.5	21.3 ± 36.7	0.0 ± 0.0
Enterococcaceae	0.0 ± 0.0	0.0 ± 0.0	0.4 ± 0.6	0.0 ± 0.0
Flavobacteriaceae	0.4 ± 0.6	12.2 ± 18.7	0.0 ± 0.0	0.0 ± 0.0
Hydrogenophilaceae	0.0 ± 0.0	0.0 ± 0.0	0.4 ± 0.7	0.0 ± 0.0
Lactobacillaceae	0.0 ± 0.0	1.6 ± 1.8	3.2 ± 5.5	0.0 ± 0.0
Micrococcaceae	0.3 ± 0.4	7.2 ± 2.2	2.3 ± 3.9	0.0 ± 0.0
Moraxellacea	51.3 ± 23.3	11.2 ± 5.5	0.0 ± 0.0	0.0 ± 0.0
Pasteurellaceae	8.9 ± 4.9	0.0 ± 0.0	0.0 ± 0.0	0.0 ± 0.0
Peptostreptococcaceae	0.0 ± 0.0	0.0 ± 0.0	0.0 ± 0.0	1.4 ± 2.4
Planococcaceae	2.3 ± 2.4	2.8 ± 2.5	2.5 ± 2.4	49.2 ± 12.3
Porphyromonadaceae	0.9 ± 1.6	0.0 ± 0.0	0.0 ± 0.0	0.0 ± 0.0
Staphylococcaceae	0.9 ± 0.8	6.5 ± 0.6	11.3 ± 15.0	0.0 ± 0.0
Streptococcaceae	4.1 ± 2.2	0.0 ± 0.0	0.0 ± 0.0	0.0 ± 0.0
Xanthomonadaceae	1.3 ± 2.3	2.1 ± 3.6	14.4 ± 14.0	8.6 ± 3.5
Rare taxa	12.7 ± 3.5	15.2 ± 0.8	8.5 ± 4.4	8.0 ± 3.3

Source: Modified from Table S2 in Pechal, J.L. et al. 2014a. *Ecosphere* 5: art45; Pechal, J.L. et al. 2014b. *International Journal of Legal Medicine* 128: 193–205.

control (immediate insect access to the remains). After 5 days of decomposition, the carcasses open to insect colonization were dominated by the phylum Firmicutes, whereas carcasses excluded from insects were dominated by Proteobacteria. These results suggest that there are important interactions between necrophagous insects (e.g., blow flies) and epinecrotic communities that require future investigation. In another study (Metcalf et al. 2013), using pyrosequencing to document the bacterial community succession patterns on small vertebrate decomposing remains (i.e., mouse carcasses), Firmicutes and Proteobacteria also dominated the bacterial communities during decomposition. However, Firmicutes was the predominate taxon of the community in the abdominal cavity during the early decomposition process, whereas Proteobacteria dominated the abdominal cavity community in later phases of decomposition and was the predominate taxon on skin communities throughout the decomposition process (Metcalf et al. 2013).

Bacterial community assemblages on specific carrion sites diverge over the decomposition process (Metcalf et al. 2013; Pechal et al. 2013, 2014b). Studies, such as those previously discussed, suggest possible key interactions between specific bacteria taxa, for instance, between the members of the phyla Firmicutes and Proteobacteria. What drives these community shifts among taxa on carrion has yet to be elucidated. Differences in the bacterial community composition between resource types (e.g., swine vs.

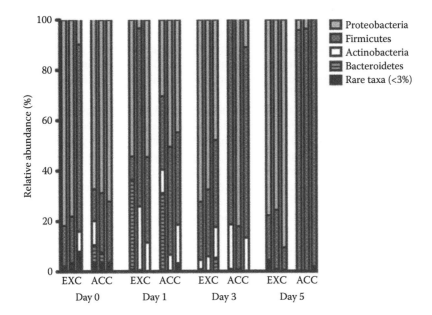

FIGURE 3.1 **(See color insert.)** Successional changes in carrion bacterial communities over decomposition. Phylum-level taxonomic relative abundance of the microbial communities based on 454-pyrosequencing between carcasses where insect access was allowed (ACC) or excluded (EXC). Initial field placement (day 0) was followed by subsequent sample collections on days 1, 3, and 5. Rare taxa include any phyla with <3% of the total relative abundance. (Reprinted from Pechal, J.L. et al. 2013. *PLoS ONE* 8: e79035. With permission.)

mouse) could contribute. Once the body ruptures and spills its contents into the surrounding environment, soil composition, pH, and nutrient changes must certainly affect which microbes might prosper and which will not (Meyer et al. 2013). The carcass size and endogenous bacteria, influenced by factors such as diet, would also impact community composition, as would the location of the remains: in the case of these studies a field versus a laboratory setting. Certainly confounding seasonal factors for field studies, the location ecology, soil type and inhabitants, and ambient conditions surrounding the remains are just some of the considerations that will complicate a comparison of any two studies. With the ever expanding in-depth analysis allowed by NGS techniques, future experimentation may discover trends in community structure that specifically correlate with decomposition processes. In the short time that has elapsed between the attempt by Vass (2001) to describe microbial succession on human cadavers and the studies reviewed here, metagenomic and computational advances have been the driving forces that have allowed researchers to peer more deeply into the once impenetrable world, that is, carrion microbial ecology.

3.2 Evolution in Concert with Eukaryotes

Darwin observed phylogenetic species changes (variation) and natural selection (adaptation) (Darwin 1859). These changes were later attributed to random genetic mutation creating new variations and to the ecological pressures of survival facilitating retention of useful mutations, while suppressing preservation of unsuccessful mutations by inherently limiting the quantity of descendants (Grant and Grant 2002, 2006; Abzhanov et al. 2004, 2006). Thus, beneficial physiological, anatomical, and behavioral adaptations are retained over the course of time, whereas harmful characteristics are selected against and purged from the population. Additionally, a portion of variation is due to nonadaptive evolutionary processes, such as random drift of mutant alleles, leading to variation within and between species at the molecular level (Kimura 1968; Duret 2008). Such genetic drift can impact nondetrimental (neutral) alleles, as well as advantageous and weakly deleterious alleles, resulting in the fixation of some, while others may eventually disappear. Notably, in the microbial world, evolutionary processes occur at an accelerated rate in comparison to the macroorganism world due to short reproductive cycles, high mutation rates, and the ease of gene exchange that occurs between microbes, such as bacteria (Lujan et al. 2011).

Microbes existed on earth during the Precambrian era, nearly 3.5 billion years ago (Waggoner 1996). Thus far, the oldest animal fossil found is of a primitive sponge-grade metazoan that existed during the Neoproterozoic era at the end of the Cryogenian period, about 635 million years ago (Maloof et al. 2010). Therefore, the multicellular animals of the Earth have always shared their space with microbes, as their appearance on Earth occurred later. Microbes have often been viewed as villains to be controlled with disinfectants, antibiotics, and vaccines. However, a more in-depth understanding has revealed that many microbes can be allies that live in, on, and around every animal and are actually required for many essential processes of life, such as the metabolic production of essential compounds, protection from disease, and the ability to fully digest food. Therefore, animals and microbes actually constitute a consortium that not only cohabitates, but also have coevolved (Pennisi 2013). Subsequently, microbes both exert influence on animals when they are alive (Ezenwa et al. 2012) and contribute to their decay after an animal's demise. This influence, posed by microbes on their surrounding environment and fellow fauna, may be more extensive than once thought. In fact, bacteria facilitate the evolution of a diverse macroorganism community (Beasley et al. 2012). The pressures of survival have led to evolutionary alterations facilitating a well-established exchange of information between microbes and other organisms often to the benefit of both. However, this exchange can also be used to one or another's advantage. Burkepile et al. (2006) determined that bacteria on a carrion resource produced noxious chemicals that deterred animal consumers: a strategy that allowed bacteria to compete with larger scavengers for this valuable resource. This publication serves as an example of Janzen's theory of microbe–macrobe competition, recently modeled by Ruxton et al. (2014).

3.3 Development through Environmental Interactions

3.3.1 The Biofilm

The majority of the bacteria on Earth live as interlinked communities of cells attached to a substrate, known as a biofilm (Costerton et al. 1978, 1999). Early investigations neglected the biofilm state and focused on bacteria that existed in the planktonic state (those existing as single cells surrounded by an aqueous solution). The biofilm is now recognized as the predominant state found ubiquitously in nature and in which bacteria can endure the stresses of the environment (Stoodley et al. 2001; Hall-Stoodley et al. 2004; Hoffman et al. 2005; An and Parsek 2007; Karatan and Watnick 2009). The scientific community has recognized the importance of bacterial biofilms in many areas of human, animal, and environmental health. Biofilms are found in the gut, on skin and chronic wounds, as plaque on teeth, colonizing indwelling medical catheters, and biofouling industrial surfaces such as boat hulls and water pipes (Jones et al. 1969; Donlan 2002; Davis et al. 2008; James et al. 2008; Stickler 2008; Percival et al. 2012).

Biofilms comprised a single species monolayer or a complex multiorganism community that may involve higher-level organisms such as nematodes and protozoans (Cloete et al. 2009). In general, biofilm formation begins with planktonic microorganisms (e.g., bacteria, fungi, and algae) that gather on a substrate and then adhere to the surface, where they exude extracellular polymeric substances forming a matrix for enhanced attachment (Lemon et al. 2008). The community proceeds through five major developmental phases: attachment, formation, microcolony development, maturation, and detachment/dispersal. Initially, attachment to a surface is reversible, but as time progresses in a supportive environment, the community will increase in both diversity and quantity of taxa present and affix more securely (Stoodley et al. 2002). The community secretes extrapolymeric substances, consisting of polysaccharides, known as glycocalyx (hydrated polymeric slimy matrices) and lipopolysaccharides, that enable the bacterium to encapsulate itself on the surface. Gram-negative bacteria are generally more proficient at forming biofilms because of their superior ability to produce glycocalyx and lipopolysaccharides (Costerton et al. 1981; Dunne 2002). With time, this extracellular matrix becomes a more complex and stronger structure comprised of proteins, polysaccharides, extracellular DNA, and the various organisms. Microcolonies form as the biofilm matures, the community expands, and the biofilm becomes a three-dimensional structure containing cells packed in clusters with connecting channels and voids. These channels serve as portals for exchange of sustenance, waste, oxygen, metabolites, and enzymes between bacteria deeper within the strata of the biofilm. Eventually, the bacteria adhere very strongly to the surface and, in this mature state, use energy predominantly to produce exopolysaccharides, which the cells use as nutrients. In this state, cell division is uncommon (Watnick and Kolter 2000). Such a community allows a variety of microbes to persist under otherwise uninhabitable environmental conditions.

Existing in the sessile, multicellular community of the biofilm provides numerous advantages for a bacterium increasing survival chances when confronted with unpredictable alterations in nutrient levels, pH, oxygen levels, shear forces, ultraviolet (UV) exposure, metal toxicity, acid exposure, dehydration, and salinity (Sheffield et al. 2009; Sheffield and Crippen 2012). Moreover, the biofilm state brings about changes in bacterial behavior and responses to changing environmental conditions and their interactions with other microbes (Sbordone and Bortolaia 2003). The bacteria within the biofilm can produce a variety of compounds such as antibiotics or metabolites that change pH (Sturgen and Casida 1962; Raaijmakers and Mazzola 2012), which can be beneficial not only to the bacterium that produced the compound, but also to neighboring microorganisms within the biofilm. Members within the biofilm are thus provided protection without suffering any of the fitness consequences typically associated with producing such compounds. If circumstances become too unfavorable for the biofilm, then to sustain the community, sections can detach to relocate and initiate new biofilms in more hospitable surroundings.

A small fraction of essentially invulnerable cells within the biofilm, called persister cells, is also important to the survival and continuance of biofilm in an ecosystem (Lewis 2010). They are not impacted by bactericidal agents and can even exhibit antibiotic tolerance (Lewis 2005). Biofilms, in general, are slow growing, but persisters foregoing propagation making them less vulnerable to lethal factors, such as antibiotics, that target the physiological mechanisms of rapidly dividing cells. These nondividing bacteria represent only a small fraction of the community population and are akin to spores, a dormant and highly resistant bacterial form that preserves an individual cell's genetic material in times of extreme stress. Similar to spores, persister cells also endure under adverse environmental conditions to act as traveling forms of the biofilm population that safeguard survival.

Recently, it has been shown that biofilms are present on carrion during decomposition. These microbial communities assist in the fossilization of vertebrates and plants by mineralizing soft tissues and directly enhancing their preservation, thereby mediating the rate and processes of nutrient release and translocation within the environment (Peterson et al. 2010; Dunn et al. 2013). Certain bacteria within biofilms enhance metal binding and mineral formation, making tissues more likely to be converted to fossils (Dunn et al. 2013). Thus, the lack of biofilms may result in tissue that does not mineralize.

Even more intriguing is the possible role of bacterial quorum sensing occurring during the decomposition process. Quorum sensing occurs when a bacterial population is dense enough to initiate changes in the group genetic expression through a chemical mediator. It was first described in the marine bacterium *Vibrio fischeri*, which will fluoresce once a sufficient bacterial density is sensed (Nealson and Hastings 1979; Bassler 1999). In order for bacteria to live in a communal environment, such as a biofilm, they must be able to communicate with one another; this communication occurs using chemical signals. Some of these signals are exported through the outer membrane and are interpreted not only by members of the same bacterial species, but also by other bacterial species present in the biofilm. The ability to communicate is enhanced by the inherent communal situation within a biofilm. In planktonic populations, a chemical signal may be less effective when the bacterial cells are not in adequate concentration, and the chemical is diluted by the surrounding aqueous environment. In biofilms, the matrix material holds the cells in close proximity, allowing for the increased density of bacteria and signaling molecules. The finding that bacterial cells are capable of coordinated behavior gives new insight into their complexity and has led to the speculation that other species may be eavesdropping on their communiqué for their own exploitation of the carrion resource (Ma et al. 2012; Tomberlin et al. 2012). In other words, microbes and necrophagous fauna may interact on a much more extensive and intimate level than previously postulated.

3.3.2 The Biofilm and Arthropods

The movement of bacteria can be facilitated by larger organisms such as arthropods. Many insects such as flies utilize and develop in bacteria-laden decomposing substrates, and in fact, some require associated bacteria for development (Chang and Wang 1958; Perotti et al. 2001). In the case of flies, they can easily acquire bacteria from the environment or a biofilm community that they have encountered. The bacteria can then be dispersed as the flies move from a bacteria-laden site to subsequent locations by mechanical dislodgement from the exoskeleton or fecal deposition or regurgitation of internalized bacteria. However, the scope to which insects contribute to bacterial dispersal and investigation into the occurrence of gene exchange and evolution from bacterial interactions with arthropods has been mostly overlooked.

Bacteria in biofilms exhibit adaptive strategies in the form of phenotypic diversification and promote evolution through group-beneficial behavior such as the economy of resource use (Kreft 2004). An increased incidence of gene transfer occurs in the biofilm, which increases the genetic diversity leading to increased evolution via the acquisition of beneficial mutations, such as those conferring protection (Mena et al. 2008). Consequently, bacteria are different when they exist within a biofilm and undergo dramatic genetic shifts between the planktonic and biofilm states. As much as 40% of their genes may undergo up- or downregulation in the transition from the planktonic to the biofilm state (Hall-Stoodley et al. 2004; Karatan and Watnick 2009). This also allows differential genetic expression that can lead to the formation of physiologically distinct subpopulations to carry out different activities within the biofilm, in effect, a division of labor (Mai-Prochnow et al. 2006).

So, these simple, single-cell microorganisms lead a more complicated life when existing within a biofilm colony: a life that is phenotypically distinct from their planktonic form. Such adaptive strategies, although currently not well understood, are no doubt important for the survival and persistence of microbes associated with carrion and may be useful for the biotic challenges associated with exogenous microbes that are disseminated to a carcass via transmission by necrophagous insects.

The close association between insects and bacteria is thought to facilitate gene exchange (Ahmad et al. 2007; Crippen and Poole 2009; Poole and Crippen 2009). Arthropods, such as crickets, grasshoppers, cockroaches, and beetles, harbor large bacterial concentrations ($10^8–10^{11}$ cfu/mL) in their gut (Cazemier et al. 1997). An alarming finding is the high frequency of horizontally transferred genes between different bacterial species inhabiting the insect gut (Hoffmann et al. 1998; Watanabe and Sato 1998; Akhtar et al. 2009; Crippen and Poole 2009; Poole and Crippen 2009). The ability of bacteria to evolve by acquisition of genes from other bacteria undoubtedly increases their survival capabilities in the environment, but the transfer of detrimental genes (i.e., antibiotic-resistant genes) could have serious impact on animal and public health.

3.3.3 Carrion Microbial Assemblages and the Ecosystem

Microbial assemblages are important for many ecosystem processes (Hättenschwiler et al. 2005; Parmenter and MacMahon 2009; Nemergut et al. 2011). Approximately 90% of all decomposition in terrestrial systems is a result of microbial activity (Swift et al. 1979). Bacterial community composition may mediate biotic interactions throughout the decomposition of carrion; yet, there remains a paucity of data related to bacterial communities associated with carrion (Metcalf et al. 2013; Pechal et al. 2014b).

Bacterial communities have previously been assumed to be selected by "filters" or environmental conditions, which in turn regulate bacterial composition (Baas Becking 1934). However, no two naturally occurring communities appear to be the same (Curtis et al. 2002). Therefore, other biotic interactions, such as competition with necrophagous insects and vertebrate scavengers, could be important to community assembly and succession (Hooper et al. 2005). Many studies have explored the importance of microbial assemblages in terrestrial, for example, soil and leaf litter (Bell et al. 2009; Redford and Fierer 2009), and aquatic habitats (Jones and McMahon 2009; Burke et al. 2011; Kominoski et al. 2011), but fewer have attempted to understand these communities during carrion decomposition.

Empirical data describing the community dynamics of microbes on carrion are currently sparse (Vass 2001; Zak et al. 2003; Chung et al. 2007; Rohlfs 2008; Strickland et al. 2009). Microbial species richness may predict soil decomposition; therefore, most (~70%) of the soil models include microbial communities as a component (Zak et al. 2003; Manzoni and Porporato 2009; McGuire et al. 2012). Aggregation models have been proposed to describe how competitors (e.g., blow flies and bacteria) coexist on ephemeral resources (Woodcock et al. 2002; Fiene et al. 2014). These models of coexistence state that several species can utilize the same resource simultaneously, and studies performed using fruit-breeding Diptera (i.e., *Drosophila* [Diptera: Drosophilidae]) on decomposing organic matter indicate that a single species may never completely exclude another species (Shorrocks et al. 1979; Atkinson and Shorrocks 1981). This demonstration of coexistence may result from grouping of individual species consuming one resource patch and no other patches (Hartley and Shorrocks 2002; Abos et al. 2006), thus affecting the interactions between microbial communities and invertebrates.

Environmental conditions are vital in the structuring of communities for both arthropod and microbial organisms. For example, arthropods central to the decomposition of heterotrophic organic material into the ecosystem, such as blow flies, have temperature-dependent development. Cooler temperatures slow the growth rate of a larva from instar to instar, whereas warmer temperatures accelerate the development. Thus temperature thresholds regulate this process. Temperatures that either exceed or do not meet the developmental thresholds can force an insect to diapause or be lethal. There are also temperature thresholds for microorganisms; these thresholds may be known for some individual organisms, but they are not well defined for multiple species assemblages. Some authors have suggested using a minimum growth temperature threshold of 0°C for the microbial communities (Pechal et al. 2014b), but in reality, environmental effects on communities of microorganisms have not been extensively investigated. These effects are undoubtedly important for microbial function during carrion decomposition.

There have been only limited studies performed to understand how variation in the local environment and in types of vertebrate resources affects microbial colonization of carrion. Successional data vary for a given region and season, and both spatial and temporal ecological processes governing insect colonization on carrion must be considered (Barton et al. 2013b; Fiene et al. 2014). Variation in ecological metrics (i.e., Shannon–Weaver diversity, Simpson's diversity, richness, and evenness) between years was reported when monitoring insect succession patterns and bacterial community composition during carrion decomposition in the same habitat (Pechal et al. 2014a,b). The yearly variation in insect community structure attracted to and utilizing replicate decomposing swine carcasses in a Midwestern temperate forest, as described in Pechal et al. (2014a), could be attributed to: changes in annual environmental conditions such as weather; differences in resource size (Braack 1987); or priority effects of initial colonizers altering subsequent community structure (Chase 2010; Fukami and Nakajima 2011; Brundage et al. 2014). Alternatively, variation in necrophagous insect community structure (asynchronous blow fly succession) on carrion has been documented on alligator, *Alligator mississippiensis*, carcasses, with *Lucilia coeruleiviridis* arriving early during decomposition and *Chrysomya* spp., *Cochliomyia macellaria*, and *Phormia regina* arriving later in decomposition (Nelder et al. 2009). These insect arrival patterns have been hypothesized to be affected by the initial microbial communities. This suggests that a niche theory is mediating colonization patterns of necrophagous flies. These dynamics are suspected to influence how microbial communities develop and fluctuate on a carcass.

Many species of archaea thrive in extreme and harsh environments with high salt, high and low temperatures, and even extreme pH, such as hot springs, salt lakes, marshlands, oceans, and the gut of vertebrates (extremophiles) (Pikuta et al. 2007). They have unique physiological characteristics distinct from bacteria, such as a thermophilic lipid bilayer membrane that enables some species to function at high temperatures (Koga 2012). It is not known whether these microorganisms or others might be present during decomposition in extreme environments. Many methods of mummification of remains have occurred, some purposeful and some accidental: quick drying in the sun, with smoke or with chemicals such as salt; permanent freezing; or placing a body in an oxygen-free environment such as a peat bog (NOVA 2000, Sivrev et al. 2005). In general, mummification is assumed to occur by hindering or removing microorganisms that cause decay by limiting the air or moisture that aids bacterial function (Lynnerup 2007). What role, if any, microbes might play in the mummification of remains has not been studied.

3.4 Functions and Eukaryotic Interactions

3.4.1 Basic Bacterial Physiology during Carrion Decomposition

The decomposition of organic matter by microbes is a highly variable process dependent on many environmental factors. In particular, this process is dependent on moisture, oxygen, and the diversity of bacterial taxa, contributed both from the decomposing organism and from the surrounding environment. Although specific areas of an individual's microbiome are unique, there is also overlap of bacterial composition or at least function (Fierer et al. 2008, 2010; Hyde et al. 2013). A study comparing the decomposition of skeletal muscle from several vertebrate species, including humans, demonstrated similar decomposition dynamics for microbial activity and mass of tissue loss (Stokes et al. 2013). Also, repetitive use of the same deposition study site with different individuals resulted in similar soil nutrient changes and an increased rate of decomposition likely due to enhancement of soil microbes able to utilize carrion (Carter and Tibbett 2008; Damann et al. 2012). Thus, even if the particular microbial species is unique to the individual, if it participates in the decompositional process, it likely serves a requisite function that several species can undertake due to redundancy in bacterial metabolic abilities in different bacteria. Thus, functional measurements could be useful as an evidentiary element in establishing time since death during forensic taphonomic analyses (Metcalf et al. 2013; Pechal et al. 2014b).

Prior to death, bacteria make up approximately 1%–3% of the human body mass (MacDougall 2012). After death, many of those bacteria are instrumental in the breakdown of tissues; however, without suitable moisture levels and proper oxygen concentrations or pH conditions, microbial metabolic activity is slowed and bacterial proliferation decreased. Aerobic bacteria, which expedite the rapid degradation of carrion, require oxygen for respiration and energy production. In wet and aquatic environments, anaerobic conditions may exist that considerably retard decomposition (Figure 3.2). In arid regions, bacteria may dry out or be exposed to lethal UV levels from the sun, thus limiting their participation in the

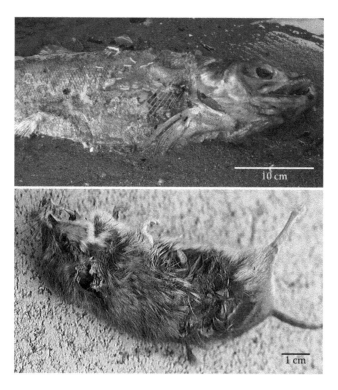

FIGURE 3.2   Salmon decomposing in an aquatic environment (top) (Photo courtesy of M. Eric Benbow.). Field mouse decomposing in an arid environment (bottom) (Photo courtesy of Tawni L. Crippen. All rights reserved.).

exploitation of the available carrion resources (Figure 3.2). Therefore, abiotic effects on bacterial well-being have direct consequences on the rates and extent of carrion decomposition.

Putrefaction is the process of proliferation of bacteria found within the body. Using humans as an example, the gastrointestinal tract holds an estimated 10^{11} bacteria/g, the species diversity of which is adapted to an individual's diet and environment (Moore and Holdeman 1974). Even those individuals typically assumed to be genetically similar (i.e., identical twins) can have different gut microbiota. For example, the gut bacterial community of a set of identical adult twins living <5 km apart was assessed using pyrosequencing, and there was only a 17% similarity of genes (16s rRNA) shared between the individuals (Turnbaugh et al. 2010). Under normal conditions, the breakdown of tissues is initiated by the release of hydrolytic enzymes from cell death. This results in the autolytic degradation of tissues generating carbohydrates, proteins, fats, and minerals for utilization by microorganisms. Proteolysis of the proteins of epithelial and neuronal tissues occurs early in this process, prompting the linings of the body's system to degrade, thus freeing bacteria from their normal biological containment within the digestive and respiratory systems and allowing their relocation and colonization outside of these barriers (Dent et al. 2004).

Decomposition continues in the form of growth and reproduction of dispersed bacteria, which are augmented by the utilization of tissue breakdown products, thus facilitating the further decomposition of organic tissues (Dent et al. 2004; Paczkowski and Schutz 2011). Aerobic and facultative bacteria abound in the beginning of this autolytic process, but as oxygen is rapidly depleted and not replenished due to the nonfunctional circulatory system, anaerobic species prevail (Vass 2001; Parkinson et al. 2009).

There is an impressive array of bacterial species in the gastrointestinal system enabling a multitude of metabolic capacities, allowing the utilization of a vast assortment of different nutrient sources (Suchodolski et al. 2008; Handl et al. 2011). The proliferation of these bacteria is responsible for the majority of the putrefactive processes that occur in carrion and results in the release of many compounds as products of their metabolism (Schultz and Dickschat 2007). Lipolytic bacteria include species from the genera *Micrococcus, Staphylococcus, Pseudomonas, Alteromonas,* and *Flavobacterium* (Ray and Bhunia 2007). These bacteria hydrolyze triglycerides by producing extracellular lipases. Saccharolytic bacteria include species from the genera *Bacillus, Clostridium, Aeromonas, Pseudomonas,* and *Enterobacter* (Ray and Bhunia 2007); these bacteria hydrolyze complex carbohydrates. Proteolytic bacteria include species from the genera *Micrococcus, Staphylococcus, Bacillus, Clostridium, Pseudomonas, Alteromonas, Flavobacterium,* and *Alcaligenes,* along with a few from the families Enterobacteriaceae and Brevibacteriaceae (Ray and Bhunia 2007); these bacteria hydrolyze proteins by producing extracellular proteinases.

Putrefaction generates the second process of decomposition, bloat (Carter et al. 2007) (Figure 3.3). As the aerobic organisms flourish, oxygen is utilized and quickly depleted. The accumulation of gases from bacterial metabolism creates a favorable environment for the proliferation of anaerobic organisms already present in the gastrointestinal tract and respiratory system (Dent et al. 2004; Goff 2009). Hyde et al. (2013) used 16S rRNA sequencing to demonstrate a definitive shift from aerobic to anaerobic bacteria in human cadaver samples taken at the onset and at the end stage of bloat. Results from a 48-day study using a mouse model also demonstrated that endogenous and facultative anaerobes common to the gut, such as those from the families Lactobacillaceae and Bacteroidaceae, predominated during the bloat stage (Metcalf et al. 2013). The bacteria metabolize the available carbohydrates, lipids, and proteins into products such as organic acids (i.e., propionic, lactic, sulfurous, butyric, acetic, pyruvic, stearic, oleic, and palmitic), phenolic compounds (i.e., indole and skatole), diamines (i.e., putrescine and cadaverine), alcohols (i.e., butyl, ethyl, and acetone), and other gases (i.e., ammonia, carbon dioxide, hydrogen sulfide, and methane). The gases trapped within the body accumulate, causing an expansion of the body cavities that manifest as bloat (Figure 3.3) (Janaway 1996; Montllor et al. 2002). Continued breakdown of the cells in combination with the pressure from accumulated gases can eventually cause skin ruptures. This exposes the internal milieu to oxygen, as well as fungi and bacteria from a multitude of sources, such as the skin, soil, vertebrate scavengers, arthropods, and the surrounding environment. These external microbes infiltrate the carcass and become involved in the decomposition process. The exposure to oxygen causes the anaerobic bacterial taxa to decrease in abundance, and aerobes and facultative anaerobes increase, such as from the orders Rhizobiales and Enterbacteriaceae, respectively (Metcalf et al. 2013).

FIGURE 3.3   Bloat stage (top) of swine carrion decomposing in the Post Oak Savannah region of Texas (Photo courtesy of Tawni L. Crippen.). Bacteria-laden adipocere (bottom) released after rupture of swine carrion decomposing in the Post Oak Savannah region of Texas (Photo courtesy of C. Chin Heo. All rights reserved.).

Adipocere is produced by anaerobic bacterial hydrolysis of fat (saponification) in tissue that converts triglycerides in fats into a wax-like saturated fatty acid substance (Figure 3.3). It is typically facilitated by degradative anaerobes such as *Clostridium perfringens* and *C. frigidicanes* (Fiedler et al. 2009). The production of adipocere occurs best in the absence of oxygen and in high moisture levels such as wet ground or a body of water (Forbes et al. 2005, 2011). In the absence of air, adipocere can persist for extended periods and can actually help to preserve tissue for decades (Mellen et al. 1993; Fiedler et al. 2009) and even up to 120 years in some instances (Pfeiffer et al. 1998). A body exposed to the elements in a warm, dry, aerobic environment is not as likely to form adipocere. In a study of the formation of adipocere during the early postmortem stage of submerged swine tissue, it was determined that an abundance of Gram-positive aquatic bacteria was required to assist in the lipolyzing of lipids into free fatty acids and the formation of adipocere (Ueland et al. 2014). Degradative anaerobes secrete toxins containing proteinases and phospholipases that can destroy the cell membrane of the adipocyte (fat cell). This facilitates in the hydrolysis that absorbs water into the cell and the hydrogenation required for the formation of adipocere, eventually rupturing the cell and releasing its cellular components into the outside environment (O'Brien and Kuehner 2007). The chemical reactions can also create an alkaline environment that inhibits the growth of some bacteria active in purification or stimulates sporulation or the production of a tough, nonreproductive structure that lays dormant to endure harsh environments (Mohan Kumar et al. 2009). Additional details of biochemical changes during decomposition are discussed in Chapter 2.

In addition to the microbial communities directly associated with the remains (e.g., gut bacteria), those in the surrounding habitat can also affect the decomposition process. In terrestrial environments, the local soil type influences decomposition rates and soil microbial communities undergo changes, given that they have access to a new, ephemeral nutrient pulse (Tumer et al. 2013). *Clostridium* is a ubiquitous soil inhabitant, as well as being found in the gastrointestinal tract of many organisms (Haagsma 1991). They produce a diverse battery of strong hydrolytic enzymes, such as lecithinase, that can degrade extracellular components (Matsushita and Okabe 2001; Montllor et al. 2002). These enzymes likely evolved to expand the nutritional sources available for exploitation by *Clostridium*. These bacteria augment the carrion decomposition processes resulting in the leaching of fluids into the surrounding environment (Dent et al. 2004). Increased fungal and bacterial growth often occurs on a carcass, and the metabolites produced, such as ammonia and nitrate, become available to the indigenous soil inhabitants. These compounds may promote microbial and plant growth in the area of a carcass. However, the extent to which this occurs is highly influenced by ambient environmental conditions such as soil pH, temperature, and moisture content. The impact of an ephemeral decomposing carrion resource on surrounding microbial communities remains understudied.

Utilization of an ephemeral resource, such as carrion, by microbes offers some unique challenges and rewards. The first challenge is in locating an exploitable resource. Many bacteria are ubiquitous and, in addition, can utilize varied metabolic pathways in order to obtain sustenance from diverse environmental sources. So, the sudden appearance of a rich, if ephemeral, carrion resource offers a fortuitous prize to be exploited. Many Gram-negative bacteria have developed a strategy to persist in the environment as a tough, dormant, spore structure that can survive years in unfavorable conditions between availability of adequate resources to reproduce. This has allowed for many such species to take advantage of ephemeral carrion resources.

Microbes are not self-mobile on the same scale as vertebrate animals. Most are nonmotile, whereas others can move locally via the rotation of an external apparatus known as a flagellum. In contrast, helical bacteria have an internal axial filament that helps the organism move via a spiral rotation, and gliding bacteria are able to swarm by the secretion of and movement through a slime "highway." Still, others are adapted to rely on eukaryotic organisms to assist them in dispersal over any large distance. Despite the risks of being dispersed to an inadequate environment, the rewards for encountering an ephemeral carrion resource by a microbe are high in terms of quantity of bacterial generations that can be produced from such a rich nutritional supply.

The second challenge is the fierce competition for a transient resource. Unlike animal users of carrion, microbes can rapidly produce multiple generations from one set of remains. Microbial decomposers utilize a resource to produce a generational scale massively superior from competing eukaryotes (Burkepile et al. 2006; Barton et al. 2013a). Consequently, a resource pulse can benefit many more prokaryotes, but they are also most challenged with encountering it. Additionally, microbes are constantly adapting and invading new environments for resources: the rapid growth rate and flexibility of their genome allow them to evolve to inhabit a wide spectrum of environments from ordinary to extreme (Pikuta et al. 2007). Wiser et al. (2013) measured the fitness trajectories (growth rate) for 12 *Escherichia coli* populations over 21 years (50,000 generations) in a supported environment of constant conditions. They determined that contrary to the expected, adaptive evolution never plateaued, but rather mean fitness continued to increase. In fact, several populations in the study became hypermutators, developing accelerated point mutation rates: an alteration allowing an augmented adaptation rate over ancestral mutations. Therefore, even under constant conditions, bacterial organisms do not cease to evolve, but instead continually improve their capabilities even if only by small increments. This constant evolution permits acclimatization to an array of environmental conditions resulting in the multitude of bacterial species known today and still being discovered. The ability to constantly evolve may allow microbes to more efficaciously utilize rare pulse resources such as carrion, and this would be of immense benefit to microbes that may spend the majority of their existence in a very sparse environment (such as the soil) and then must rapidly adjust to compete for a rich resource whose occurrence is sporadic and unpredictable.

3.4.2 Microbes—Arthropod Codependence

Arthropods constantly interact with microorganisms either in the environment or within their internal microbiome. Insects have adapted to mate, reproduce, and develop in association with these microorganisms. Bacterial communities utilizing vertebrate carrion undergo population shifts through-out the decomposition process with the communities being most similar to that of the host in the early stages of decomposition, going through succession into bacterial communities that resemble those of the surrounding habitat by the end of decomposition (Pechal et al. 2013; Pechal et al. 2014b). Concurrently, these bacterial communities interact with arthropods that are attracted to and use carrion for habitat, reproduction, or food, leading one to ask how do insects develop in such a microbe-laden environment?

A classic example is that of the leaf-cutter ants that cultivate decomposing fungal gardens for a direct food source (Quinlan and Cherrett 1979). Microbial communities can also influence food availability for arthropods by increasing the quantity and nutritional value of the food or being used directly as a food source (Bärlocher 1985). However, this beneficial effect on arthropods is not universal across all microbial taxa. Differential effects on dipteran growth have been correlated to specific bacterial taxa being present in the diet, as demonstrated in stable flies (*Stomoxys calcitrans* [L.] [Diptera: Muscidae]) where *Citrobacter freundii* and *Serratia fanticola* resulted in faster larval development rates (Romero et al. 2006), whereas house flies, *Musca domestica* L. (Diptera: Muscidae), feeding on colonies of *E. coli* O157:H7 and *Providencia* sp. had slower developmental rates and smaller adult size (Ahmad et al. 2006). Other insects, such as termites (Isoptera), have evolved a symbiotic microbial gut commu-nity that allows previously nondigestible or unusable resources to be consumed by performing cellulose and xylan hydrolysis (Warnecke et al. 2007). Insects can also manipulate microbial communities in the environment by facilitating competition between bacterial and fungal communities; for example, the bacterium *Klebsiella oxytoca* is an oviposition cue. It can suppress fungal growth detrimental to *M. domestica* fly eggs and subsequent larval development on an ephemeral carrion resource (Lam et al. 2009, 2010). Fungal communities on carrion can also hinder carrion beetle (Coleoptera: Silphidae) growth (Suzuki 2001), and consequently, these insects have developed antimicrobial strategies to inhibit fungal growth on the resources used for their offspring's development (Hoback et al. 2004; Rozen et al. 2008). Although not extensively studied in carrion insects, there is a wealth of examples demonstrating the importance of interkingdom communication and widespread relationships between microbes and insects, providing precedence that similar relationships may govern the community ecology of carrion (see also Chapter 20).

Microorganisms have long been intertwined with immune responses and can influence an insect's immune system. The immune system regulates many life-history traits, and microbial infections can alter host behaviors and reproduction. Activation of the innate immune response can inhibit develop-mental processes, which demonstrates a pleiotrophic response to the immunity genes (Ye et al. 2009). Developmental genes encoding for dorsal–ventral patterning in the embryo (e.g., imaginal disk growth factor 3) are upregulated during bacterial infections (Irving et al. 2001). Male damselflies, *Calopteryx virgo* L. (Odonata: Calopterygidae), treated with a juvenile hormone (JH) analog have reduced phe-noloxidase (PPO) activity, but have increased aggression behaviors (Contreras-Garduño et al. 2009). This tradeoff between reproduction and survival has also been demonstrated in the beetle *Tenebrio molitor* L. (Coleoptera: Tenebrionidae), which had decreased PPO levels when a JH inhibitor was pres-ent (Rolff and Siva-Jothy 2002). *Wolbachia* endosymbionts in *Drosophila* facilitate increased fecundity (Fry et al. 2004) and *Rickettsia* sp. nr. *bellii* infected whiteflies, *Bemisia tabaci* (Gennadius) (Hemiptera: Sternorrhyncha) produced almost double the number of adult progeny over their lifetime in comparison to uninfected whiteflies (Himler et al. 2011). A combination of reduced bacterial growth and increased immune response has been demonstrated to increase *Drosophila* longevity (Linder et al. 2008). However, mosquitoes that resisted malaria infection resulted in less fecund females (Yan et al. 1997). Immune defenses that evolved in response to opportunistic pathogens resulted in reduced longevity and larval viability, thus demonstrating a fitness cost for immune responses (Ye et al. 2009).

One of the most interesting behavior–immunity interactions of microbes and insects involves altera-tions in host behavior due to parasitoids (Libersat et al. 2009). Wood ants, *Fomica lugubris* Zetterstedt (Hymenoptera: Formicidae), incorporate solidified conifer resin into their nests to inhibit microbial growth (Christe et al. 2003). Alternative host-feeding preferences to combat pathogens and parasites are well known in humans and other vertebrates (e.g., chimpanzees and birds) (Clayton and Wolfe 1993; Karban and English-Loeb 1997; Huffman and Caton 2001). For example, in invertebrates, worker honey bees (*Apis mellifera*, L. [Hymenoptera: Apidae]) will synthesize higher levels of antibiotic secretions and remove waste, bee corpses, and larvae infected with *Varroa jacobsoni* Oudemans (Mesostigmata: Varroidae) mites from the brood to ward off infection for the colony (Boecking and Drescher 1992; Rosengaus et al. 2000; Hart and Ratnieks 2001). This response demonstrates the power of behavioral changes not only in individuals, but also in communities.

Similar microbe-mediated behavioral effects in necrophagous flies have recently been investigated (Ma et al. 2012; Tomberlin et al. 2012), suggesting that these interactions are important to the ecology and evolution of carrion decomposition. Self-medication can occur in an infected insect and result in a shifting of its behavior or feeding preferences (Karban and English-Loeb 1997; Smilanich et al. 2011). Diet shifts have been studied in caterpillars, *Platyprepia virginalis* (Boisduval) (Lepidoptera: Arctiidae), infected with *Thelaira americana* Brooks (Diptera: Tachinidae), a parasitoid fly, in which the caterpillar survives parasitism at higher rates if feeding on poison hemlock versus lupin (Karban and English-Loeb 1997). The shift in food choice also resulted in an indirect fitness effect for the parasitoid fly, as pupae that emerged from caterpillars that had been reared on hemlock were heavier than those emerging from lupin-fed caterpillars (Karban and English-Loeb 1997). Insects can control the temporal onset self-medication. The caterpillar, *Grammia incorrupta* (H. Edwards) (Lepidoptera: Erebidae), infected with tachinids increased consumption of plant toxins (i.e., pyrrolizidine alkaloids) 96 h after tachinid oviposition (Smilanich et al. 2011). However, during the early stages of infection, the caterpillars reduced their level of plant toxin consumption, thus demonstrating delayed self-medication (Smilanich et al. 2011).

The cost of immune responses to microbes can have positive or negative fitness effects on an organism. Microbial infection can affect food intake in insects (Shirasu-Hiza and Schneider 2007); for example, *Manduca sexta* (L.) (Lepidoptera: Sphingidae) becomes anorexic after *Serratia marcescens* infections (Adamo et al. 2007). Effects on the morphological appearances (e.g., pigmentation) will alter behavioral patterns; melanism correlates strongly to lepidopteran pathogen resistance, and darker male damselflies have greater immunity and parasite resistance than lighter-colored males (Rantala et al. 2000; Wilson et al. 2001; True 2003). Resource allocation from larval diets affects adult immunity (Fellous and Lazzoro 2010); diets with higher protein levels govern immune gene transcriptional responses in adults, leading to increased levels of defensive antimicrobial peptides (Fellous and Lazzoro 2010). *Drosophila* selected for parasitoid resistance was less competitive for feeding (Kraaijeveld and Godfray 1997). Also, flies were less capable of resisting desiccation and starvation when the immune system was activated (Hoang 2001).

Mating behaviors have also been affected by immune responses to microbes. Male crickets with more chirps in their calling song had a higher concentration of hemocytes (Ryder 2000). However, parasitized male crickets had their calling signal eliminated (Fedorka et al. 2004). Female crickets responded to those males with strong immune responses (e.g., encapsulation ability) (Fedorka et al. 2004). Territorial behaviors are also linked to the immune function. Male dragonflies that lose their territory will have decreased immune functions in comparison to nonterritorial males (Contreras-Garduño et al. 2006), thus demonstrating selection behaviors of mating with individuals with higher fitness (e.g., stronger immune system). These studies provide convincing precedent that microbial communities are important to insect fitness and ecology. Considering that necrophagous insects are primary consumers of carrion, these cross-domain interactions should be considered fundamental processes in carrion ecology (Chapter 20). Future research that focuses on understanding the mechanisms of these interactions is vital to elucidating the codependence of microbes and arthropods, the metabolic processes of both, and their influence on carrion decomposition.

3.5 Mechanisms of Interactions during Carrion Decomposition

3.5.1 Carrion Decomposition, Microbial Metabolism, and Volatile Organic Compounds

Prokaryote–eukaryote interactions are common in nature. Microbes have many types of relationships with higher organisms ranging from mutualism to commensalism to pathogenesis. Since before Janzen (1977), it has been readily recognized that bacteria and other microbial species play a major role in the diet and life history of many insects and other arthropods (e.g., crustaceans). Consequently, insects have evolved ways not only to detect and evaluate resources colonized by microbial communities, but also to differentiate between specific taxa within these communities, or perhaps their associated metabolic byproducts or signaling molecules (see Chapters 9 and 20). For terrestrial species that use carrion as a food or mating resource, it is expected that individuals can quickly detect minute concentrations of distinctive volatile organic compounds from the air. There is precedent that insects detect many types of chemicals and mixtures (Vickers 2000).

Microbial species inhabiting carrion use the resource to fuel their own metabolism requirements. During the breakdown of this organic matter by bacteria, the metabolic reactions result in the production of microbial volatile organic compounds (MVOCs). These MVOCs (discussed in more detail in Chapters 9 and 10) often function as semiochemicals for communication between bacteria, but are also utilized by other species, such as arthropods, for their own purposes (Davis et al. 2013). This type of interkingdom communication presumably results in a wide variety of effects on the receiving species. These MVOCs can facilitate success during interspecies competition, and the lack of these compounds may lower survivability and fecundity by decreasing important outcomes such as resource location success. Therefore, recipients that have evolved to perceive this exchange of information from a distance may have survival advantages when engaged in competition for limited resources such as carrion. Unfortunately, these insights remain minimally studied.

If MVOCs influence other species' physiology and behavior, then a particular suite of MVOCs could potentially shape the community structure of local species through downstream interkingdom communication (Chapter 8). Ultimately, microbes influence carrion decomposition directly by utilizing the tissues, as well as indirectly by influencing the occurrence of other microbes and macroorganisms consuming the carrion. This has important implications for understanding the ecology of carrion decomposition.

3.5.2 Microbial Communities Driving Insect Community Assembly

Changes in the temporal arthropod community structure on carrion are partly explained as a process of competition between species for a resource (Norris 1965; Hanski 1977). Changes in the decomposition rate are more apparent under the influence of insects, specifically necrophagous insects utilizing carrion as a food source (Denno and Cothran 1975). MVOCs have been proposed as governing the mechanisms of necrophagous insect colonization on carrion with species-specific cues from the resource dictating successional patterns (Dekeirsschieter et al. 2009). Microbial communities associated with the carcasses could be altering the quality of the resource and thus mediating insect community assembly as demonstrated with fungi governing phytophagous insect composition in a plant-based system (Tack et al. 2012). In turn, as insects consume the resource, many substrates are made available for utilization by microbes, which then produce more and varied MVOCs (see Chapter 9 for more information) (Boumba et al. 2008, 2012).

Blow flies may also directly impact microbial species through chemical secretions while consuming carrion tissue (Sherman et al. 2000; Mumcuoglu et al. 2001). Subsequent larval development on a resource may disrupt established microbial communities through direct or indirect competitive interactions on the carcass. Blow fly oviposition can be induced by bacteria and their associated semiochemicals; specifically, *Cochliomyia hominivorax* (Coquerel) (Diptera: Calliphoridae) has a preference for five out of eight Enterobacteriaceae bacteria, such as *Proteus* spp., in oviposition assays (Chaudhury et al. 2010), and 10–12-day-old gravid females are attracted to *Providencia* sp. (Hammack et al. 1987). Therefore, the blow fly species that secondarily colonize remains may be responding to cues released by the bacteria that subsequently stimulate oviposition on the resource (Ashworth and Wall 1994).

Speculations of competition for ephemeral resources between micro- and macroorganisms are documented, and there is substantial variation in decomposition rates after taking into account the influence of substrate type and microbial and insect communities (Burkepile et al. 2006; Davis et al. 2013). *Drosophila melanogaster* (Meigen) (Diptera: Drosophilidae) and fungi (*Aspergillus* spp.) clash over fruit resources, with larval mortality correlating with the age, species, and possibly species density, of fungi present on the resource (Trienens et al. 2010). Microbial and insect communities facilitate nutrient cycling and decomposition rates in some leaf litter systems, whereas in other systems, microbial and insect communities have little or no influence on decomposition rates (Hieber and Gessner 2002; Srivastava et al. 2009; Kominoski et al. 2011). Insects have many strategies for combating detrimental microbial communities, including initiating immune responses after pathogen detection (Gottar et al. 2002; Ferrandon et al. 2007; Gerardo et al. 2010). Pathogen avoidance is transpiring in honey bee colonies rejecting nest mates whom have been infected or parasitized (Wilson-Rich et al. 2009) and insects producing secretions containing antibiotic properties (Kerridge et al. 2005; Nigam et al. 2006; Rozen et al. 2008). Alternatively, microbes also have strategies to fulfill their needs. The bacterial pathogen, *Candidatus Liberibacter asiaticus*, mediated the manipulation of insect vector behavior to facilitate its spread by the psyllid, *Diaphorina citri* (Kuwayama) (Hemiptera: Psyllidae), using volatiles to differentially signal attraction between infected plants and noninfected psyllids, as well as between noninfected plants and infected psyllids (Mann et al. 2012). An interkingdom (or interdomain) struggle is occurring between insects and bacteria, in which both are exploiting aspects of one another's occurrence and physiology for their own ecological benefit. So, do microbial communities drive insect community assembly or do insect communities drive microbial community assembly? Perhaps it is a little of both.

3.5.3 Microbial Community Interactions with Vertebrate Scavengers

Vertebrate scavengers are also consumers of dead organic matter, such as carrion in an ecosystem, and play an important functional role in the removal of carrion from the environment (Wilson and Wolkovich 2011) (see also Chapters 6 and 21). The mention of vertebrate scavengers typically evokes images of a wake of vultures or a murder of crows ripping at a dead carcass, but scavengers can include animals such as foxes, coyotes, raccoons, feral swine, hyenas, crocodiles, sharks, and so on (Polis et al. 1989; DeVault and Rhodes 2002; DeVault et al. 2004). These scavengers generally have olfactory and/or visual searching abilities to enhance their efficiency of finding a carrion resource (DeVault et al. 2003).

When these scavengers locate remains, microbial communities already utilizing the remains are not defenseless against them and can produce toxins, such as botulism, to deter competition by many vertebrate scavengers (Janzen 1977). If microbes are allowed to proliferate uninterrupted, there is an upper threshold in the usability of a microbe-laden carcass by other scavengers. Burkepile et al. (2006) placed fresh and microbe-laden fish carrion in a coastal marine ecosystem and then monitored for utilization by large scavengers. The microbe-laden carrion was four times less likely to be utilized by some large consumers. A similar threshold has been documented when fungal growth deterred benthic macroinvertebrates from scavenging the remains of post-spawning anadromous clupeid fish (*Alosa* spp.) (Garman 1992).

On the flip side, although Burkepile et al. (2006) demonstrated that some large scavengers are sensitive to and deterred by microbe-laden remains, the study also showed that other scavengers were not. Many vertebrate scavengers have well-developed and fascinating mechanisms to compete against the microbial communities present on the remains (DeVault et al. 2003). But because these scavengers regularly feed on deceased animals that may have succumbed to microbial infections, they are suspected of facilitating the dispersal of diseases to other wildlife, humans, and domestic livestock by mechanisms such as regurgitation or complete passage and voiding from the alimentary tract in urine and fecal waste. Mechanical transfer of pathogens can also occur passively by carriage on the feet, fur, or feathers; for example, the griffon vulture (*Gyps fulvus*) baths and preens to clean itself after a meal, which could result in mechanical contamination of a water source (Houston and Cooper 1975).

Yet, these same scavengers are often resistant to the pathogens they carry and the damage caused by the toxins they may produce. Some scavengers, such as the American black vulture (*Coragyps atratus*), are theorized to carry an indigenous gut microbiota that excludes pathogens by forming a barrier to colonization or through competitive exclusion (Carvalho et al. 2003). Turkey vultures (*Cathartes aura*), coyotes (*Canis latrans*), and crows (*Corvus brachyrhynchos*) have been found to carry naturally occurring antibodies against pathogens, such as *Clostridium botulinum* toxins (Ohishi et al. 1979). The stomach of the whiteback griffon vulture (*Gyps africanus*) is a very acidic environment (pH 1–2), in which few bacteria can survive, even the vegetative form of *Bacillus anthracis* perishes at such a low pH. Although some bacterial spores could survive such extremes, the short time interval that the carrion resource exists prior to complete decomposition may preclude their occurrence (Houston and Cooper 1975). The blood of the American alligator (*A. mississippiensis*), the saltwater crocodile (*Crocodylus porosus*), the freshwater crocodile (*Crocodylus johnstoni*), the Siamese crocodile (*Crocodylus siamensis*), and the Komodo dragon (*Varanus komodoensis*) show antibacterial activity against pathogenic bacteria (Merchant and Britton 2006; Merchant et al. 2006, 2012; Kommanee et al. 2012). So viewed from a different perspective, such antimicrobial defenses by scavengers may actually help to limit the spread of pathogens by destroying much of bacteria before it can be dispersed to other fauna. Such an ecological role for their antimicrobial defenses has not been well studied. Although these limited studies suggest important interactions between microbes and vertebrate scavengers, future research that focuses on understanding the mechanisms of these interactions is needed to better describe their complex relationships and their influence on carrion decomposition and pathogen epidemiology.

3.6 Conclusions and Future of Microbial Communities in Carrion Ecology

Based on the current understanding of microbial communities related to carrion decomposition (Metcalf et al. 2013; Pechal et al. 2013, 2014b), microbial interactions with necrophagous fauna, such as insects, are important for carrion nutrients to be released into the underlying soil and the surrounding ecosystem (Carter et al. 2007). Considering that there can be an estimated 5000 kg of mammal carrion biomass per km^2 in some ecosystems (Houston 1985; Carter et al. 2007), the release of the associated nutrients is influential over large spatial scales. These localized nutrients are known to have broader effects in the surrounding ecosystem (see Chapters 6, 13, and 21) (Yang 2004; Yang et al. 2008; Parmenter and MacMahon 2009; Barton et al. 2013a). Therefore, the role of microbial community ecology and its interkingdom interactions during carrion decomposition go beyond influencing the community structure of local species and have cascading importance within entire ecosystems. Additionally, the quality of carrion can be influenced by indirect biotic interactions such as the fear of predation, and those predatory-stressed animal carcasses have significant legacy effects on subsequent soil function and leaf litter decomposition (Hawlena et al. 2012). Thus, if carrion-associated microbial communities have interactions with vertebrates and invertebrates that alter its nutrient quality, it is quite plausible that these microscopic organisms drive complex ecological dynamics over multiple spatial and temporal scales. As the microbial "black box" continues to be opened and defined, it is apparent that the complexities of microbial life on Earth are much more intricate and vastly influential on the ecology of life. There are generations of scientific questions yet to be explored about microbial carrion decomposition.

REFERENCES

Abos, C.P., F. Lepori, B.G. McKie, and B. Malmqvist. 2006. Aggregation among resource patches can promote coexistence in stream-living shredders. *Freshwater Biology* 51: 545–553.

Abzhanov, A., W.P. Kuo, C. Hartmann, B.R. Grant, P.R. Grant, and C.J. Tabin. 2006. The calmodulin pathway and evolution of elongated beak morphology in Darwin's finches. *Nature* 442: 563–567.

Abzhanov, A., M. Protas, B.R. Grant, P.R. Grant, and C.J. Tabin. 2004. Bmp4 and morphological variation of beaks in Darwin's finches. *Science* 305: 1462–1465.

Adamo, S.A., T.L. Fidler, and C.A. Forestell. 2007. Illness-induced anorexia and its possible function in the caterpillar, *Manduca sexta. Brain, Behavior, and Immunity* 21: 292–300.

Ahmad, A., A. Broce, and L. Zurek. 2006. Evaluation of significance of bacteria in larval development of *Cochliomyia macellaria* (Diptera: Calliphoridae). *Journal of Medical Entomology* 43: 1129–1133.

Ahmad, A., T.G. Nagaraja, and L. Zurek. 2007. Transmission of *Escherichia coli* O157:H7 to cattle by house flies. *Preventive Veterinary Medicine* 80: 74–81.

Akhtar, M., H. Hirt, and L. Zurek. 2009. Horizontal transfer of the tetracycline resistance gene tetM mediated by pCF10 among *Enterococcus faecalis* in the house fly (*Musca domestica* L.) alimentary canal. *Microbial Ecology* 58: 509–518.

Amann, R.I., W. Ludwig, and K.H. Schleifer. 1995. Phylogenetic identification and *in situ* detection of individual microbial cells without cultivation. *Microbiological Reviews* 59: 143–169.

Amend, A.S., K.A. Seifert, and T.D. Bruns. 2010. Quantifying microbial communities with 454 pyrosequencing: Does read abundance count? *Molecular Ecology* 19: 5555–5565.

An, D. and M.R. Parsek. 2007. The promise and peril of transcriptional profiling in biofilm communities. *Current Opinions in Microbiology* 10: 292–296.

Ashworth, J.R. and R. Wall. 1994. Response of the sheep blowflies *Lucilia sericata* and *L. cuprina* to odour and the development of semiochemical baits. *Medical and Veterinary Entomology* 8: 303–309.

Atkinson, W.D. and B. Shorrocks. 1981. Competition on a divided and ephemeral resource: A simulation model. *Journal of Animal Ecology* 50: 461–471.

Baas Becking, L. 1934. *Geobiologie of inleidung tot de milieukunde.* Den Haag: WP van Stockum & Zoon.

Bärlocher, F. 1985. The role of fungi in the nutrition of stream invertebrates. *Botanical Journal of the Linnean Society* 91: 83–94.

Barns, S.M., R.E. Fundyga, M.W. Jeffries, and N.R. Pace. 1994. Remarkable archaeal diversity detected in a Yellowstone National Park hot spring environment. *Proceedings of the National Academy of Sciences* 91: 1609–1613.

Barton, P.S, S.A. Cunningham, D. Lindenmayer, and A. Manning. 2013a. The role of carrion in maintaining biodiversity and ecological processes in terrestrial ecosystems. *Oecologia* 171: 761–772.

Barton, P.S., S.A. Cunningham, B.C.T. Macdonald, S. McIntyre, D.B. Lindenmayer, and A.D. Manning. 2013b. Species traits predict assemblage dynamics at ephemeral resource patches created by carrion. *PLoS ONE* 8: e53961.

Barton, L.L. and D.E. Northrup. 2011. *Microbial Ecology.* Hoboken, New Jersey: John Wiley & Sons, Inc.

Bassler, B.L. 1999. How bacteria talk to each other: Regulation of gene expression by quorum sensing. *Current Opinions in Microbiology* 2: 582–587.

Beasley, J.C., Z.H. Olson, and T.L. DeVault. 2012. Carrion cycling in food webs: Comparisons among terrestrial and marine ecosystems. *Oikos* 121: 1021–1026.

Bell, T.G., M.O. Gessner, R.I. Griffiths, J.R. McLaren, P.J. Morin, M. Heijden, and W.H.V.D. Putten. 2009. Microbial biodiversity and ecosystem functioning under controlled conditions and in the wild. In *Biodiversity, Ecosystem Functioning, and Human Wellbeing: An Ecological and Economic Perspective*, S. Naeem, D.E. Bunker, A. Hector, M. Loreau, and C. Perrings (eds), pp. 121–133. New York: Oxford University Press.

Benbow, M.E., A.J. Lewis, J.K. Tomberlin, and J.L. Pechal. 2013. Seasonal necrophagous insect community assembly during vertebrate carrion decomposition. *Journal of Medical Entomology* 50: 440–450.

Boecking, O. and W. Drescher. 1992. The removal response *Apis mellifera* L. colonies to brood in wax and plastic cells after artificial and natural infestation with *Varroa jacobsoni* Oud. and to freeze-killed brood. *Experimental and Applied Acarology* 16: 321–329.

Boumba, V.A., V. Economou, N. Kourkoumelis, P. Gousia, C. Papadopoulou, and T. Vougiouklakis. 2012. Microbial ethanol production: Experimental study and multivariate evaluation. *Forensic Science International* 215: 189–198.

Boumba, V.A., K.S. Ziavrou, and T. Vougiouklakis. 2008. Biochemical pathways generating post-mortem volatile compounds co-detected during forensic ethanol analyses. *Forensic Science International* 174: 133–151.

Braack, L.E.O. 1987. Community dynamics of carrion-attendant arthropods in tropical African woodlands. *Oecologia* 72: 402–409.

Brundage, A., M. Benbow, and J. Tomberlin. 2014. Priority effects on the life-history traits of two carrion blow fly (Diptera, Calliphoridae) species. *Ecological Entomology* 39: 539–547.

Burke, C., P. Steinberg, D. Rusch, S. Kjelleberg, and T. Thomas. 2011. Bacterial community assembly based on functional genes rather than species. *Proceedings of the National Academy of Sciences* 108: 14288–14293.

Burkepile, D., J. Parker, C. Woodson, H. Mills, J. Kubanek, P. Sobecky, and M. Hay. 2006. Chemically mediated competition between microbes and animals: Microbes as consumers in food webs. *Ecology* 87: 2821–2831.

Caporaso, J.G., C.L. Lauber, W.A. Walters, D. Berg-Lyons, C.A. Lozupone, P.J. Turnbaugh, N. Fierer, and R. Knight. 2011. Global patterns of 16S rRNA diversity at a depth of millions of sequences per sample. *Proceedings of the National Academy of Sciences* 108: 4516–4522.

Carter, D.O. and M. Tibbett. 2003. Taphonomic mycota: Fungi with forensic potential. *Journal of Forensic Sciences* 48: 168–171.

Carter, D.O. and M. Tibbett. 2006. Microbial decomposition of skeletal muscle tissue (*Ovis aries*) in a sandy loam soil at different temperatures. *Soil Biology and Biochemistry* 38: 1139–1145.

Carter, D.O. and M. Tibbett. 2008. Does repeated burial of skeletal muscle tissue (*Ovis aries*) in soil affect subsequent decomposition? *Applied Soil Ecology* 40: 529–535.

Carter, D.O., D. Yellowlees, and M. Tibbett. 2007. Cadaver decomposition in terrestrial ecosystems. *Naturwissenschaften* 94: 12–24.

Carvalho, L.R.D., L.M. Farias, J.R. Nicoli, M.C.F. Silva, A.T.S.M. Corsino, L.A.D. Lima, R.A.F. Redondo, P.C.P. Ferreira, and M.E.B.M. Pinto. 2003. Dominant culturable bacterial microbiota in the digestive tract of the American black vulture (*Coragyps atratus* Bechstein 1793) and search for antagonistic substances. *Brazilian Journal of Microbiology* 34: 218–224.

Cazemier, A.E., J.H.P. Hackstein, H.J.M. Op Den Camp, J. Rosenberg, and C. Van Der Drift. 1997. Bacteria in the intestinal tract of different species of arthropods. *Microbial Ecology* 33: 189–197.

Chang, J.T. and M.Y. Wang. 1958. Nutritional requirements of the common housefly, *Musca domestica vicina* Macq. *Nature* 181: 566.

Chapman, S.K. and G.S. Newman. 2010. Biodiversity at the plant–soil interface: Microbial abundance and community structure respond to litter mixing. *Oecologia* 162: 763–769.

Chapman, S.K., G.S. Newman, S.C. Hart, J.A. Schweitzer, and G.W. Koch. 2013. Leaf litter mixtures alter microbial community development: Mechanisms for non-additive effects in litter decomposition. *PLoS ONE* 8: e62671.

Chase, J.M. 2010. Stochastic community assembly causes higher biodiversity in more productive environments. *Science* 328: 1388–1391.

Chaudhury, M., S. Skoda, A. Sagel, and J. Welch. 2010. Volatiles emitted from eight would-isolated bacteria differentially attracted gravid screwworms (Diptera: Calliphoridae) to oviposit. *Journal of Medical Entomology* 47: 349–354.

Christe, P., A. Oppliger, F. Bancalà, G. Castella, and M. Chapuisat. 2003. Evidence for collective medication in ants. *Ecology Letters* 6: 19–22.

Chung, H., D.R. Zak, P.B. Reich, and D.S. Ellsworth. 2007. Plant species richness, elevated CO_2, and atmospheric nitrogen deposition alter soil microbial community composition and function. *Global Change Biology* 13: 980–989.

Clayton, D.H. and N.D. Wolfe. 1993. The adaptive significance of self-medication. *Trends in Ecology and Evolution* 8: 60–63.

Cloete, T.E., M.S. Thantsha, M.R. Maluleke, and R. Kirkpatrick. 2009. The antimicrobial mechanism of electrochemically activated water against *Pseudomonas aeruginosa* and *Escherichia coli* as determined by SDS–PAGE analysis. *Journal of Applied Microbiology* 107: 379–384.

Contreras-Garduño, J., J. Canales-Lazcano, and A. Córdoba-Aguilar. 2006. Wing pigmentation, immune ability, fat reserves and territorial status in males of the rubyspot damselfly, *Hetaerina americana*. *Journal of Ethology* 24: 165–173.

Contreras-Garduño, J., A. Córdoba-Aguilar, H. Lanz-Mendoza, and A. Cordero Rivera. 2009. Territorial behaviour and immunity are mediated by juvenile hormone: The physiological basis of honest signalling? *Functional Ecology* 23: 157–163.

Costello, E.K., C.L. Lauber, M. Hamady, N. Fierer, J.I. Gordon, and R. Knight. 2009. Bacterial community variation in human body habitats across space and time. *Science* 326: 1694–1697.

Costerton, J., G. Geesey, and K.-J. Cheng. 1978. How bacteria stick. *Scientific American* 238: 86–95.

Costerton, J.W., R.T. Irvin, and K.J. Cheng. 1981. The bacterial glycocalyx in nature and disease. *Annual Review of Microbiology* 35: 299–324.

Costerton, J.W., P.S. Stewart, and E.P. Greenberg. 1999. Bacterial biofilms: A common cause of persistent infections. *Science* 284: 1318–1322.

Crippen, T.L. and T.L. Poole. 2009. Conjugative transfer of plasmid-located antibiotic resistance genes within the gastrointestinal tract of lesser mealworm larvae, *Alphitobius diaperinus* (Coleoptera: Tenebrionidae). *Foodborne Pathogens and Disease* 6: 907–915.

Curtis, T.P., W.T. Sloan, and J.W. Scannell. 2002. Estimating prokaryotic diversity and its limits. *Proceedings of the National Academy of Sciences* 99: 10494–10499.

Damann, F.E., A. Tanittaisong, and D.O. Carter. 2012. Potential carcass enrichment of the University of Tennessee Anthropology Research Facility: A baseline survey of edaphic features. *Forensic Science International* 222: 4–10.

Darwin, C. 1859. *On the Origin of Species by Means of Natural Selection, or the Preservation of Favoured Races in the Struggle for Life*. London, UK: John Murry Publisher.

Davis, T., T. Crippen, R. Hofstetter, and J. Tomberlin. 2013. Microbial volatiles as arthropod semiochemicals. *Journal of Chemical Ecology* 39: 840–859.

Davis, S.C., C. Ricotti, A. Cazzaniga, E. Welsh, W.H. Eaglstein, and P.M. Mertz. 2008. Microscopic and physiologic evidence for biofilm-associated wound colonization *in vivo*. *Wound Repair and Regeneration* 16: 23–29.

Deangelis, K.M., D. Chivian, J.L. Fortney, A.P. Arkin, B. Simmons, T.C. Hazen, and W.L. Silver. 2013. Changes in microbial dynamics during long-term decomposition in tropical forests. *Soil Biology and Biochemistry* 66: 60–68.

Dekeirsschieter, J., F.J. Verheggen, M. Gohy, F. Hubrecht, L. Bourguignon, G. Lognay, and E. Haubruge. 2009. Cadaveric volatile organic compounds released by decaying pig carcasses (*Sus domesticus* L.) in different biotopes. *Forensic Science International* 189: 46–53.

Denno, R.F. and W.R. Cothran. 1975. Niche relationships of a guild of necrophagous flies. *Annals of the Entomological Society of America* 68: 741–754.

Dent, B., S. Forbes, and B. Stuart. 2004. Review of human decomposition processes in soil. *Environmental Geology* 45: 576–585.

DeVault, T.L., I.L. Brisbin, and O.E. Rhodes. 2004. Factors influencing the acquisition of rodent carrion by vertebrate scavengers and decomposers. *Canadian Journal of Zoology—Revue Canadienne De Zoologie* 82: 502–509.

DeVault, T. and O.E. Rhodes. 2002. Identification of vertebrate scavengers of small mammal carcasses in a forested landscape. *Acta Theriologica* 47: 185–192.

DeVault, T., O. Rhodes, and J. Shivik. 2003. Scavenging by vertebrates: Behavioral, ecological, and evolutionary perspectives on an important energy transfer pathway in terrestrial ecosystems. *Oikos* 102: 225–234.

Dickson, G.C., R.T. Poulter, E.W. Maas, P.K. Probert, and J.A. Kieser. 2010. Marine bacterial succession as a potential indicator of postmortem submersion interval. *Forensic Science International* 209: 1–10.

Donlan, R.M. 2002. Biofilms: Microbial life on surfaces. *Emerging Infectious Diseases* 8: 881–890.

Dunn, K.A., R.J.C. McLean, J.G.R. Upchurch, and R.L. Folk. 2013. Enhancement of leaf fossilization potential by bacterial biofilms. *Geology* 25: 1119–1122.

Dunne, W. 2002. Bacterial adhesion: Seen any good biofilms lately? *Clinical Microbiology Reviews* 15: 155–166.

Duret, L. 2008. Neutral theory: The null hypothesis of molecular evolution. *Nature Education* 1: 218.

Ezenwa, V.O., N.M. Gerardo, D.W. Inouye, M. Medina, and J.B. Xavier. 2012. Animal behavior and the microbiome. *Science and Culture* 338: 198–199.

Fedorka, K.M., M. Zuk, and T.A. Mousseau. 2004. Immune suppression and the cost of reproduction in the ground cricket, *Allonemobius socius*. *Evolution* 58: 2478–2485.

Fellous, S. and B.P. Lazzaro. 2010. Larval food quality affects adult (but not larval) immune gene expression independent of effects on general condition. *Molecular Ecology* 19: 1462–1468.

Fernandes, I., S. Duarte, F. Cássio, and C. Pascoal. 2013. Effects of riparian plant diversity loss on aquatic microbial decomposers become more pronounced with increasing time. *Microbial Ecology* 66: 763–772.

Ferrandon, D., J.-L. Imler, C. Hetru, and J.A. Hoffmann. 2007. The *Drosophila* systemic immune response: Sensing and signalling during bacterial and fungal infections. *Nature Reviews Immunology* 7: 862–874.

Fiedler, S., F. Buegger, B. Klaubert, K. Zipp, R. Dohrmann, M. Witteyer, M. Zarei, and M. Graw. 2009. Adipocere withstands 1600 years of fluctuating groundwater levels in soil. *Journal of Archaeological Science* 36: 1328–1333.

Fiene, J.G., G.A. Sword, S.L. Van Laerhoven, and A.M. Tarone. 2014. The role of spatial aggregation in forensic entomology. *Journal of Medical Entomology* 51: 1–9.

Fierer, N., M. Hamady, C.L. Lauber, and R. Knight. 2008. The influence of sex, handedness, and washing on the diversity of hand surface bacteria. *Proceedings of the National Academy of Sciences* 105: 17994–17999.

Fierer, N., C.L. Lauber, N. Zhou, D. McDonald, E.K. Costello, and R. Knight. 2010. Forensic identification using skin bacterial communities. *Proceedings of the National Academy of Sciences* 107: 6477–6481.

Forbes, S.L., B.H. Stuart, and B.B. Dent. 2005. The effect of the burial environment on adipocere formation. *Forensic Science International* 154: 24–34.

Forbes, S.L., M.E. Wilson, and B.H. Stuart. 2011. Examination of adipocere formation in a cold water environment. *International Journal of Legal Medicine* 125: 643–650.

Frossard, A., L. Gerull, M. Mutz, and M.O. Gessner. 2013. Litter supply as a driver of microbial activity and community structure on decomposing leaves: A test in experimental streams. *Applied and Environmental Microbiology* 79: 4965–4973.

Frostegård, A. and E. Bååth. 1996. The use of phospholipid fatty acid analysis to estimate bacterial and fungal biomass in soil. *Biology and Fertility of Soils* 22: 59–65.

Fry, A.J., M.R. Palmer, and D.M. Rand. 2004. Variable fitness effects of *Wolbachia* infection in *Drosophila melanogaster*. *Heredity* 93: 379–389.

Fukami, T. and M. Nakajima. 2011. Community assembly: Alternative stable states or alternative transient states? *Ecology Letters* 14: 973–984.

Garman, G.C. 1992. Fate and potential significance of postspawning anadromous fish carcasses in an Atlantic coastal river. *Transactions of the American Fisheries Society* 121: 390–394.

Gerardo, N.M., B. Altincicek, C. Anselme, H. Atamian, S.M. Barribeau, M.D. Vos, E.J. Duncan et al. 2010. Immunity and other defenses in pea aphids, *Acyrthosiphon pisum*. *Genome Biology* 11: R21.

Gessner, M.O., C.M. Swan, C.K. Dang, B.G. McKie, R.D. Bardgett, D.H. Wall, and S. Hättenschwiler. 2010. Diversity meets decomposition. *Trends in Ecology and Evolution* 25: 372–380.

Goff, M.L. 2009. Early post-mortem changes and stages of decomposition in exposed cadavers. *Experimental and Applied Acarology* 49: 21–36.

Gottar, M., V. Gobert, T. Michel, M. Belvin, G. Duyk, J.A. Hoffmann, D. Ferrandon, and J. Royet. 2002. The *Drosophila* immune response against Gram-negative bacteria is mediated by a peptidoglycan recognition protein. *Nature* 416: 640–644.

Grant, P.R. and B.R. Grant. 2002. Unpredictable evolution in a 30-year study of Darwin's finches. *Science* 296: 707–711.

Grant, P.R. and B.R. Grant. 2006. Evolution of character displacement in Darwin's finches. *Science* 313: 224–226.

Haagsma, J. 1991. Pathogenic anaerobic bacteria and the environment. *Revue Scientifique et Technique (International Office of Epizootics)* 10: 749–764.

Hall-Stoodley, L., J.W. Costerton, and P. Stoodley. 2004. Bacterial biofilms: From the natural environment to infectious diseases. *Nature Reviews Microbiology* 2: 95–108.

Hammack, L., M. Bromel, F.M. Duh, and G. Gassner. 1987. Reproductive factors affecting responses of the screwworm fly, *Cochliomyia hominivorax* (Diptera: Calliphoridae), to an attractant of bacterial origin. *Annals of the Entomological Society of America* 80: 775–780.

Handl, S., S. Dowd, J. Garcia-Mazcorro, J. Steiner, and J. Suchodolski. 2011. Massive parallel 16S rRNA gene pyrosequencing reveals highly diverse fecal bacterial and fungal communities in healthy dogs and cats. *FEMS Microbiology Ecology* 76: 301–310.

Hanski, I. 1977. An interpolation model of assimilation by larvae of the blowfly *Lucilia illustris* (Calliphoridae) in changing temperatures. *Oikos* 28: 187–195.

Hart, A.G. and F.L.W. Ratnieks. 2001. Task partitioning, division of labour and nest compartmentalisation collectively isolate hazardous waste in the leafcutting *Atta cephalotes*. *Behavioral Ecology and Sociobiology* 49: 387–392.

Hartley, S. and B. Shorrocks. 2002. A general framework for the aggregation model of coexistence. *Journal of Animal Ecology* 71: 651–662.

Hättenschwiler, S. and P. Gasser. 2005. Soil animals alter plant litter diversity effects on decomposition. *Proceedings of the National Academy of Sciences* 102: 1519–1524.

Hättenschwiler, S., A.V. Tiunov, and S. Scheu. 2005. Biodiversity and litter decomposition in terrestrial ecosystems. *Annual Review of Ecology, Evolution, and Systematics* 36: 191–218.

Hawlena, D., M.S. Strickland, M.A. Bradford, and O.J. Schmitz. 2012. Fear of predation slows plant-litter decomposition. *Science* 336: 1434–1438.

Hieber, M. and M. Gessner. 2002. Contribution of stream detrivores, fungi and bacteria to leaf breakdown based on biomass estimates. *Ecology and Conservation Studies* 83: 1026–1038.

Himler, A.G., T. Adachi-Hagimori, J.E. Bergen, A. Kozuch, S.E. Kelly, B.E. Tabashnik, E. Chiel et al. 2011. Rapid spread of a bacterial symbiont in an invasive whitefly is driven by fitness benefits and female bias. *Science* 332: 1–4.

Hitosugi, M., K. Ishii, T. Yaguchi, Y. Chigusa, A. Kurosu, M. Kido, T. Nagai, and S. Tokudome. 2006. Fungi can be a useful forensic tool. *Legal Medicine* 8: 240–242.

Hladyz, S., K. Åbjörnsson, E. Chauvet, M. Dobson, A. Elosegi, V. Ferreira, T. Fleituch et al. 2011. Stream ecosystem functioning in an agricultural landscape: The importance of terrestrial–aquatic linkages. In *Advances in Ecological Research*, W. Guy (ed.), pp. 211–276. Waltham, Massachusetts: Academic Press.

Hoang, A. 2001. Immune response to parasitism reduces resistance of *Drosophila melanogaster* to desiccation and starvation. *Evolution* 55: 2353–2358.

Hoback, W.W., A.A. Bishop, J. Kroemer, J. Scalzitti, and J.J. Shaffer. 2004. Differences among antimicrobial properties of carrion beetle secretions reflect phylogeny and ecology. *Journal of Chemical Ecology* 30: 719–729.

Hoffman, L.R., D.A. D'Argenio, M.J. Maccoss, Z. Zhang, R.A. Jones, and S.I. Miller. 2005. Aminoglycoside antibiotics induce bacterial biofilm formation. *Nature* 436: 1171–1175.

Hoffmann, A., T. Thimm, M. Droge, E.R. Moore, J.C. Munch, and C.C. Tebbe. 1998. Intergeneric transfer of conjugative and mobilizable plasmids harbored by *Escherichia coli* in the gut of the soil microarthropod *Folsomia candida* (Collembola). *Applied and Environmental Microbiology* 64: 2652–2659.

Hooper, D.U., F.S. Chapin, J.J. Ewel, A. Hector, P. Inchausti, S. Lavorel, J.H. Lawton et al. 2005. Effects of biodiversity on ecosystem functioning: A consensus of current knowledge. *Ecological Monographs* 75: 3–35.

Hopkins, D.W., P.E.J. Wiltshire, and B.D. Turner. 2000. Microbial characteristics of soils from graves: An investigation at the interface of soil microbiology and forensic science. *Applied Soil Ecology* 14: 283–288.

Houston, D.C. 1985. Evolutionary ecology of Afrotropical and Neotropical vultures in forests. In *Neotropical Ornithology, American Ornithologists' Union Monograph 36*, M. Foster (ed.), pp. 856–864. Washington, D.C.: University of California Press for the American Ornithologists' Union.

Houston, D.C. and J. Cooper. 1975. The digestive tract of the whiteback griffon vulture and its role in disease transmission among wild ungulates. *Journal of Wildlife Diseases* 11: 306–313.

Howard, G.T., B. Duos, and E.J. Watson-Horzelski. 2010. Characterization of the soil microbial community associated with the decomposition of a swine carcass. *International Biodeterioration and Biodegradation* 64: 300–304.

Huffman, M.A. and J.M. Caton. 2001. Self-induced increase of gut motility and the control of parasitic infections in wild chimpanzees. *International Journal of Primatology* 22: 329–346.

Hyde, E.R., D.P. Haarmann, A.M. Lynne, S.R. Bucheli, and J.F. Petrosino. 2013. The living dead: Bacterial community structure of a cadaver at the onset and end of the bloat stage of decomposition. *PLoS ONE* 8: e77733.

Irving, P., L. Troxler, T.S. Heuer, M. Belvin, C. Kopczynski, J.-M. Reichhart, J.A. Hoffmann, and C. Hetru. 2001. A genome-wide analysis of immune responses in *Drosophila*. *Proceedings of the National Academy of Sciences* 98: 15119–15124.

Ishii, K., M. Hitosugi, M. Kido, T. Yaguchi, K. Nishimura, T. Hosoya, and S. Tokudome. 2006. Analysis of fungi detected in human cadavers. *Legal Medicine* 8: 188–190.

James, G.A., E. Swogger, R. Wolcott, E. Pulcini, P. Secor, J. Sestrich, J.W. Costerton, and P.S. Stewart. 2008. Biofilms in chronic wounds. *Wound Repair and Regeneration* 16: 37–44.

Janaway, R.C. 1996. The decay of human buried remains and their associated materials. In *Studies in Crime: An Introduction to Forensic Archaeology*, J. Hunter, C.A. Roberts, and A. Martin (eds), pp. 58–85. Batsford: Psychology Press.

Janzen, D.H. 1977. Why fruits rot, seeds mold, and meat spoils. *The American Naturalist* 111: 691–713.

Jones, S.E. and K.D. McMahon. 2009. Species-sorting may explain an apparent minimal effect of immigration on freshwater bacterial community dynamics. *Environmental Microbiology* 11: 905–913.

Jones, H.C., I.L. Roth, and W.M. Sanders III. 1969. Electron microscopic study of a slime layer. *Journal of Bacteriology* 99: 316–325.

Karatan, E. and P. Watnick. 2009. Signals, regulatory networks, and materials that build and break bacterial biofilms. *Microbiology and Molecular Biology Reviews* 73: 310–347.

Karban, R. and G. English-Loeb. 1997. Tachinid parasitoids affect host plant choice by caterpillars to increase caterpillar survival. *Ecology* 78: 603–611.

Kerridge, A., H. Lappin-Scott, and J.R. Stevens. 2005. Antibacterial properties of larval secretions of the blowfly, *Lucilia sericata*. *Medical and Veterinary Entomology* 19: 333–337.

Kimura, M. 1968. Evolutionary rate at the molecular level. *Nature and Insects* 217: 624–626.

Koga, Y. 2012. Thermal adaptation of the archaeal and bacterial lipid membranes. *Archaea* 2012: 1–6, Article ID 789652.

Kominoski, J.S., L.B. Marczak, and J.S. Richardson. 2011. Riparian forest composition affects stream litter decomposition despite similar microbial and invertebrate communities. *Ecology* 92: 151–159.

Kominoski, J.S. and C.M. Pringle. 2009. Resource and consumer diversity: Testing the effects of leaf litter species diversity on stream macroinvertebrate communities. *Freshwater Biology* 54: 1461–1473.

Kominoski, J.S., C.M. Pringle, B.A. Ball, M.A. Bradford, D.C. Coleman, D.B. Hall, and M.D. Hunter. 2007. Nonadditive effects of leaf litter species diversity on breakdown dynamics in a detritus-based stream. *Ecology* 88: 1167–1176.

Kommanee, J., S. Preecharram, S. Daduang, Y. Temsiripong, A. Dhiravisit, Y. Yamada, and S. Thammasirirak. 2012. Antibacterial activity of plasma from crocodile (*Crocodylus siamensis*) against pathogenic bacteria. *Annals of Clinical Microbiology and Antimicrobials* 11: 22.

Kraaijeveld, A.R. and H.C.J. Godfray. 1997. Trade-off between parasitoid resistance and larval competitive ability in *Drosophila melanogaster*. *Nature* 389: 278–280.

Kreft, J.-U. 2004. Biofilms promote altruism. *Microbiology* 150: 2751–2760.

Kuramae, E.E., R.H.E. Hillekens, M. De Hollander, M.G.A. Van Der Heijden, M. Van Den Berg, N.M. Van Straalen, and G.A. Kowalchuk. 2013. Structural and functional variation in soil fungal communities associated with litter bags containing maize leaf. *FEMS Microbiology Ecology* 84: 519–531.

Lam, K., K. Thu, M. Tsang, M. Moore, and G. Gries. 2009. Bacteria on housefly eggs, *Musca domestica*, suppress fungal growth in chicken manure through nutrient depletion or antifungal metabolites. *Naturwissenschaften* 96: 1127–1132.

Lam, K., M. Tsang, A. Labrie, R. Gries, and G. Gries. 2010. Semiochemical-mediated oviposition avoidance by female house flies, *Musca domestica*, on animal feces colonized with harmful fungi. *Journal of Chemical Ecology* 36: 141–147.

Lemon, K.P., A.M. Earl, H.C. Vlamakis, C. Aguilar, and R. Kolter. 2008. Biofilm development with an emphasis on *Bacillus subtilis*. *Current Topics in Microbiology and Immunology* 322: 1–16.

Lenz, E.J. and D.R. Foran. 2010. Bacterial profiling of soil using genus-specific markers and multidimensional scaling. *Journal of Forensic Sciences* 55: 1437–1442.

Lewis, K. 2005. Persister cells and the riddle of biofilm survival. *Biochemistry (Mosc)* 70: 267–274.

Lewis, K. 2010. Persister cells. *Annual Review of Microbiology* 64: 357–372.

Libersat, F., A. Delago, and R. Gal. 2009. Manipulation of host behavior by parasitic insects and insect parasites. *Annual Review of Entomology* 54: 189–207.

Lim, Y., B. Kim, C. Kim, H. Jung, B.-S. Kim, J.-H. Lee, and J. Chun. 2010. Assessment of soil fungal communities using pyrosequencing. *The Journal of Microbiology* 48: 284–289.

Linder, J.E., K.A. Owers, and D.E.L. Promislow. 2008. The effects of temperature on host–pathogen interactions in *D. melanogaster*: Who benefits? *Journal of Insect Physiology* 54: 297–308.

Lozupone, C.A. and R. Knight. 2007. Global patterns in bacterial diversity. *Proceedings of the National Academy of Sciences* 104: 11436–11440.

Lujan, A.M., M.D. Macia, L. Yang, S. Molin, A. Oliver, and A.M. Smania. 2011. Evolution and adaptation in *Pseudomonas aeruginosa* biofilms driven by mismatch repair system-deficient mutators. *PLoS ONE* 6: e27842.

Lynnerup, N. 2007. Mummies. *Yearbook of Physical Anthropology* 50: 162–190.

Ma, Q., A. Fonseca, W. Liu, A.T. Fields, M.L. Pimsler, A.F. Spindola, A.M. Tarone, T.L. Crippen, J.K. Tomberlin, and T.K. Wood. 2012. *Proteus mirabilis* interkingdom swarming signals attract blow flies. *International Society of Microbial Ecology Journal* 6: 1356–1366.

MacDougall, R. 2012. NIH Human Microbiome Project defines normal bacterial makeup of the body. *Journal* (June 13, 2012): http://www.nih.gov/news/health/jun2012/nhgri-13.htm.

Mai-Prochnow, A., J.S. Webb, B.C. Ferrari, and S. Kjelleberg. 2006. Ecological advantages of autolysis during the development and dispersal of *Pseudoalteromonas tunicata* biofilms. *Applied and Environmental Microbiology* 72: 5414–5420.

Maloof, A.C., C.V. Rose, R. Beach, B.M. Samuels, C.C. Calmet, D.H. Erwin, G.R. Poirier, N. Yao, and F.J. Simons. 2010. Possible animal-body fossils in pre-Marinoan limestones from South Australia. *Nature Geoscience* 3: 653–659.

Mann, R.S., J.G. Ali, S.L. Hermann, S. Tiwari, K.S. Pelz-Stelinski, H.T. Alborn, and L.L. Stelinski. 2012. Induced release of a plant-defense volatile "deceptively" attracts insect vectors to plants infected with a bacterial pathogen. *PLoS Pathogens* 8: e1002610.

Manzoni, S. and A. Porporato. 2009. Soil carbon and nitrogen mineralization: Theory and models across scales. *Soil Biology and Biochemistry* 41: 1355–1379.

Matsushita, O. and A. Okabe. 2001. Clostridial hydrolytic enzymes degrading extracellular components. *Toxicon* 39: 1769–1780.

McGuire, K., N. Fierer, C. Bateman, K. Treseder, and B. Turner. 2012. Fungal community composition in neotropical rain forests: The influence of tree diversity and precipitation. *Microbial Ecology* 63: 804–812.

Mellen, P.F., M.A. Lowry, and M.S. Micozzi. 1993. Experimental observations on adipocere formation. *Journal of Forensic Sciences* 38: 91–93.

Mena, A., E.E. Smith, J.L. Burns, D.P. Speert, S.M. Moskowitz, J.L. Perez, and A. Oliver. 2008. Genetic adaptation of *Pseudomonas aeruginosa* to the airways of cystic fibrosis patients is catalyzed by hypermutation *Journal of Bacteriology* 190: 7910–7917.

Merchant, M. and A. Britton. 2006. Characterization of serum complement activity of saltwater (*Crocodylus porosus*) and freshwater (*Crocodylus johnstoni*) crocodiles. *Comparative Biochemistry and Physiology. Part A, Molecular and Integrative Physiology* 143: 488–493.

Merchant, M., D. Henry, R. Falconi, B. Muscher, and J. Bryja. 2012. Characterization of serum complement activity in serum of the Komodo dragon (*Varanus komodoensis*). *Advances in Biological Chemistry* 2: 353–359.

Merchant, M.E., N. Leger, E. Jerkins, K. Mills, M.B. Pallansch, R.L. Paulman, and R.G. Ptak. 2006. Broad spectrum antimicrobial activity of leukocyte extracts from the American alligator (*Alligator mississippiensis*). *Veterinary Immunology and Immunopathology* 110: 221–228.

Metcalf, J.L., L. Wegener Parfrey, A. Gonzalez, C.L. Lauber, D. Knights, G. Ackermann, G.C. Humphrey, M.J. Gebert, W. Van Treuren, and D. Berg-Lyons. 2013. A microbial clock provides an accurate estimate of the postmortem interval in a mouse model system. *eLife* 2: e01104.

Meyer, J., B. Anderson, and D.O. Carter. 2013. Seasonal variation of carcass decomposition and gravesoil chemistry in a cold (Dfa) climate. *Journal of Forensic Sciences* 58: 1175–1182.

Mohan Kumar, T.S., F.N. Monteiro, P. Bhagavath, and S.M. Bakkannavar. 2009. Early adipocere formation: A case report and review of literature. *Journal of Forensic and Legal Medicine* 16: 475–477.

Montllor, C.B., A. Maxmen, and A.H. Purcell. 2002. Facultative bacterial endosymbionts benefit pea aphids *Acyrthosiphon pisum* under heat stress. *Ecological Entomology* 27: 189–195.

Moore, J.C., E.L. Berlow, D.C. Coleman, P.C. De Ruiter, Q. Dong, A. Hastings, N.C. Johnson et al. 2004. Detritus, trophic dynamics and biodiversity. *Ecology Letters* 7: 584–600.

Moore, W. and L. Holdeman. 1974. Human fecal flora–normal flora of 20 Japanese-Hawaiians. *Applied Microbiology and Biotechnology* 27: 961–979.

Mumcuoglu, K.Y., J. Miller, M. Mumcuoglu, M. Friger, and M. Tarshis. 2001. Destruction of bacteria in the digestive tract of the maggot of *Lucilia sericata* (Diptera: Calliphoridae). *Journal of Medical Entomology* 38: 161–166.

Nealson, K.H. and J.W. Hastings. 1979. Bacterial bioluminescence: Its control and ecological significance. *Microbiological Reviews* 43: 496–518.

Nelder, M.P., J.W. McCreadie, and C.S. Major. 2009. Blow flies visiting decaying alligators: Is succession synchronous or asynchronous? *Psyche* 2009: 1–7.

Nemergut, D.R., E.K. Costello, M. Hamady, C. Lozupone, L. Jiang, S.K. Schmidt, N. Fierer et al. 2011. Global patterns in the biogeography of bacterial taxa. *Environmental Microbiology* 13: 135–144.

Nigam, Y., A. Bexfield, S. Thomas, and N.A. Ratcliffe. 2006. Maggot therapy: The science and implication for CAM—Part II—Maggots combat infection. *Evidence-Based Complementary and Alternative Medicine* 3: 303–308.

Norris, K.R. 1965. The bionomics of blow flies. *Annual Review of Entomology* 10: 47–68.

O'Brien, T. and A. Kuehner. 2007. Waxing grave about adipocere: Soft tissue change in an aquatic context. *Journal of Forensic Sciences* 52: 294–301.

Ohishi, I., G. Sakaguchi, H. Riemann, D. Behymer, and B. Hurvell. 1979. Antibodies to clostridium botulinum toxins in free-living birds and mammals. *Journal of Wildlife Diseases* 15: 3–9.

Pace, N.R. 1997. A molecular view of microbial diversity and the biosphere. *Science* 276: 734–740.

Paczkowski, S. and S. Schutz. 2011. Post-mortem volatiles of vertebrate tissue. *Applied Microbiology and Biotechnology* 91: 917–935.

Parkinson, R.A., K.-R. Dias, J. Horswell, P. Greenwood, N. Banning, M. Tibbett, and A.A. Vass. 2009. Microbial community analysis of human decomposition on soil. *Criminal and Environmental Soil Forensics*, K. Ritz, L. Dawson, and D. Miller (eds), pp. 379–394. The Netherlands: Springer.

Parmenter, R. and J. MacMahon. 2009. Carrion decomposition and nutrient cycling in a semiarid shrub-steppe ecosystem. *Ecological Monographs* 79: 637–661.

Pechal, J.L., M.E. Benbow, T.L. Crippen, A.M. Tarone, and J.K. Tomberlin. 2014a. Delayed insect access alters carrion decomposition and necrophagous insect community assembly. *Ecosphere* 5: art45.

Pechal, J.L., T.L. Crippen, M.E. Benbow, A.M. Tarone, S. Dowd, and J.K. Tomberlin. 2014b. The potential use of bacterial community succession in forensics as described by high throughput metagenomic sequencing. *International Journal of Legal Medicine* 128: 193–205.

Pechal, J.L., T.L. Crippen, A.M. Tarone, A.J. Lewis, J.K. Tomberlin, and M.E. Benbow. 2013. Microbial community functional change during vertebrate carrion decomposition. *PLoS ONE* 8: e79035.

Pennisi, E. 2013. How do microbes shape animal development. *Science* 340: 1159–1160.

Percival, S.L., K.E. Hill, D.W. Williams, S.J. Hooper, D.W. Thomas, and J.W. Costerton. 2012. A review of the scientific evidence for biofilms in wounds. *Wound Repair and Regeneration* 20: 647–657.

Perotti, M., T. Lysyk, L. Kalischuk-Tymensen, L. Yanke, and L. Selinger. 2001. Growth and survival of immature *Haematobia irritans* (Diptera: Muscidae) is influenced by bacteria isolated from cattle manure and conspecific larvae. *Journal of Medical Entomology* 38: 180–187.

Peterson, J.E., M.E. Lenczewski, and R.P. Scherer. 2010. Influence of microbial biofilms on the preservation of primary soft tissue in fossil and extant archosaurs. *PLoS ONE* 5: e13334.

Pfeiffer, S., S. Milne, and R. Stevenson. 1998. The natural decomposition of adipocere. *Journal of Forensic Sciences* 43: 368–370.

Pikuta, E., R. Hoover, and J. Tang. 2007. Microbial extremophiles at the limits of life. *Critical Reviews in Microbiology* 33: 183–209.

Polis, G.A., C.A. Myers, and R.D. Holt. 1989. The ecology and evolution of intraguild predation: Potential competitors that eat each other. *Annual Review of Ecology and Systematics* 20: 297–330.

Poole, T.L. and T.L. Crippen. 2009. Conjugative plasmid transfer between *Salmonella enterica* Newport and *Escherichia coli* within the gastrointestinal tract of the lesser mealworm beetle, *Alphitobius diaperinus* (Coleoptera: Tenebrionidae). *Poultry Science* 88: 1553–1558.

Prescott, C.E. and S.J. Grayston. 2013. Tree species influence on microbial communities in litter and soil: Current knowledge and research needs. *Forest Ecology and Management* 309: 19–27.

Quinlan, R. and J. Cherrett. 1979. The role of fungus in the diet of the leaf-cutting ant *Atta cephalotes* (L.). *Ecological Entomology* 4: 151–160.

Raaijmakers, J.M. and M. Mazzola. 2012. Diversity and natural functions of antibiotics produced by beneficial and plant pathogenic bacteria. *Annual Review of Phytopathology* 50: 403–424.

Rantala, M.J., J. Koskimaki, J. Taskinen, K. Tynkkynen, and J. Suhonen. 2000. Immunocompetence, developmental stability and wingspot size in the damselfly *Calopteryx splendens* L. *Proceedings of the Royal Society of London Series B—Biological Sciences* 267: 2453–2457.

Ray, B. and A. Bhunia. 2007. *Fundamental Food Microbiology*. Boca Raton: CRC Press.

Redford, A. and N. Fierer. 2009. Bacterial succession on the leaf surface: A novel system for studying successional dynamics. *Microbial Ecology* 58: 189–198.

Rohlfs, M. 2008. Host–parasitoid interaction as affected by interkingdom competition. *Oecologia* 155: 161–168.

Rolff, J. and M.T. Siva-Jothy. 2002. Copulation corrupts immunity: A mechanism for a cost of mating in insects. *Proceedings of the National Academy of Sciences* 99: 9916–9918.

Romero, A., A. Broce, and L. Zurek. 2006. Role of bacteria in the oviposition behaviour and larval development of stable flies. *Medical and Veterinary Entomology* 20: 115–121.

Rosengaus, R.B., M.L. Lefebvre, and J.F.A. Traniello. 2000. Inhibition of fungal spore germination by *Nasutitermes*: Evidence for a possible antiseptic role of soldier defensive secretions. *Journal of Chemical Ecology* 26: 21–39.

Rozen, D.E., D.J.P. Engelmoer, and P.T. Smiseth. 2008. Antimicrobial strategies in burying beetles breeding on carrion. *Proceedings of the National Academy of Sciences* 105: 17890–17895.

Ruxton, G.D., D.M. Wilkinson, H.M. Schaefer, and T.N. Sherratt. 2014. Why fruit rots: Theoretical support for Janzen's theory of microbe–macrobe competition. *Proceedings of Biological Sciences/The Royal Society* 281: 20133320.

Ryder, J.J. 2000. Male calling song provides a reliable signal of immune function in a cricket. *Proceedings of the Royal Society of London Series B—Biological Sciences* 267: 1171–1175.

Sanger, F., S. Nicklen, and A.R. Coulson. 1977. DNA sequencing with chain-terminating inhibitors. *Proceedings of the National Academy of Sciences* 74: 5463–5467.

Sbordone, L. and C. Bortolaia. 2003. Oral microbial biofilms and plaque-related diseases: Microbial communities and their role in the shift from oral health to disease. *Clinical Oral Investigations* 7: 181–188.

Schultz, S. and J. Dickschat. 2007. Bacterial volatiles: The smell of small organisms. *Natural Product Reports* 24: 814–842.

Sheffield, C. and T. Crippen. 2012. Invasion and survival of Salmonella in the environment: The role of biofilms. *Salmonella—A Diversified Superbug,* Y. Kumar (ed.), pp. 1–28. Rijeka, Croatia: Intech Open Access Publishers.

Sheffield, C.L., T.L. Crippen, K. Andrews, R.J. Bongaerts, and D.J. Nisbet. 2009. Planktonic and biofilm communities from 7-day-old chicken cecal microflora cultures: Characterization and resistance to *Salmonella* colonization. *Journal of Food Protection* 72: 1812–1820.

Sherman, R.A., M.J.R. Hall, and S. Thomas. 2000. Medicinal maggots: An ancient remedy for some contemporary afflictions. *Annual Review of Entomology* 45: 55–81.

Shirasu-Hiza, M.M. and D.S. Schneider. 2007. Confronting physiology: How do infected flies die? *Cellular Microbiology* 9: 2775–2783.

Shokralla, S., J.L. Spall, J.F. Gibson, and M. Hajibabaei. 2012. Next-generation sequencing technologies for environmental DNA research. *Molecular Ecology* 21: 1794–1805.

Shorrocks, B., W. Atkinson, and P. Charlesworth. 1979. Competition on a divided and ephemeral resource. *Journal of Animal Ecology* 48: 899–908.

Sivrev, D., M. Miklosova, A. Georgieva, and N. Dimitrov. 2005. Modern day plastination techniques—Successor of ancient embalmment methods. *Trakia Journal of Sciences* 3: 48–51.

Smilanich, A., P. Mason, L. Sprung, T. Chase, and M. Singer. 2011. Complex effects of parasitoids on pharmacophagy and diet choice of a polyphagous caterpillar. *Oecologia* 165: 995–1005.

Srivastava, D.S., B.J. Cardinale, A.L. Downing, J.E. Duffy, C. Jouseau, M. Sankaran, and J.P. Wright. 2009. Diversity has stronger top-down than bottom-up effects on decomposition. *Ecology* 90: 1073–1083.

Stickler, D.J. 2008. Bacterial biofilms in patients with indwelling urinary catheters. *Nature Clinical Practice—Urology* 5: 598–608.

Stokes, K.L., S.L. Forbes, and M. Tibbett. 2013. Human versus animal: Contrasting decomposition dynamics of mammalian analogues in experimental taphonomy. *Journal of Forensic Sciences* 58: 583–591.

Stoodley, P., K. Sauer, D.G. Davies, and J.W. Costerton. 2002. Biofilms as complex differentiated communities. *Annual Reviews in Microbiology* 56: 187–209.

Stoodley, P., S. Wilson, L. Hall-Stoodley, J.D. Boyle, H.M. Lappin-Scott, and J.W. Costerton. 2001. Growth and detachment of cell clusters from mature mixed-species biofilms. *Applied and Environmental Microbiology* 67: 5608–5613.

Strickland, M.S., C. Lauber, N. Fierer, and M.A. Bradford. 2009. Testing the functional significance of microbial community composition. *Ecology* 90: 441–451.

Sturgen, N.O. and L.E. Casida, Jr. 1962. Antibiotic production by anaerobic bacteria. *Applied Microbiology* 10: 55–59.

Štursová, M., L. Žifčáková, M.B. Leigh, R. Burgess, and P. Baldrian. 2012. Cellulose utilization in forest litter and soil: Identification of bacterial and fungal decomposers. *FEMS Microbiology Ecology* 80: 735–746.

Suchodolski, J.S., J. Camacho, and J.R.M. Steiner. 2008. Analysis of bacterial diversity in the canine duodenum, jejunum, ileum, and colon by comparative16S rRNA gene analysis. *FEMS Microbiology Ecology* 66: 567–578.

Suzuki, S. 2001. Suppression of fungal development on carcasses by the burying beetle *Nicrophorus quadripunctatus* (Coleoptera: Silphidae). *Entomological Science* 4: 403–406.

Swift, M.J., O.W. Heal, and J.M. Anderson. 1979. *Decomposition in Terrestrial Ecosystems,* 509pp. Berkeley and Los Angeles, CA: University of California Press.

Tack, A.J.M., S. Gripenberg, and T. Roslin. 2012. Cross-kingdom interactions matter: Fungal-mediated interactions structure an insect community on oak. *Ecology Letters* 15: 177–185.

The Human Microbiome Consortium. 2010. A catalog of reference genomes from the human microbiome. *Science* 328: 994–999.

The Human Microbiome Consortium. 2012. A framework for human microbiome research. *Nature* 486: 215–221.

Tiedje, J.M., S. Asuming-Brempong, K. Nüsslein, T.L. Marsh, and S.J. Flynn. 1999. Opening the black box of soil microbial diversity. *Applied Soil Ecology* 13: 109–122.

Tomberlin, J. and M. Benbow. 2015. *Forensic Entomology: International Dimensions and Frontiers.* Boca Raton, FL: CRC Press.

Tomberlin, J.K., M.E. Benbow, A.M. Tarone, and R. Mohr. 2011. Basic research in evolution and ecology enhances forensics. *Trends in Ecology and Evolution* 26: 53–55.

Tomberlin, J.K., T.L. Crippen, A.M. Tarone, B. Singh, K. Adams, Y.H. Rezenom, M.E. Benbow et al. 2012. Interkingdom responses of flies to bacteria mediated by fly physiology and bacterial quorum sensing. *Animal Behaviour* 84: 1449–1456.

Trienens, M., N.P. Keller, and M. Rohlfs. 2010. Fruit, flies and filamentous fungi—Experimental analysis of animal–microbe competition using *Drosophila melanogaster* and *Aspergillus* mould as a model system. *Oikos* 119: 1765–1775.

True, J.R. 2003. Insect melanism: The molecules matter. *Trends in Ecology and Evolution* 18: 640–647.

Tumer, A.R., E. Karacaoglu, A. Namli, A. Keten, S. Farasat, R. Akcan, O. Sert, and A.B. Odabasi. 2013. Effects of different types of soil on decomposition: An experimental study. *Legal Medicine* 15: 149–156.

Turnbaugh, P.J., C. Quince, J.J. Faith, A.C. McHardy, T. Yatsunenko, F. Niazi, J. Affourtit et al. 2010. Organismal, genetic, and transcriptional variation in the deeply sequenced gut microbiomes of identical twins. *Proceedings of the National Academy of Sciences* 107: 7503–7508.

Ueland, M., H.A. Breton, and S.L. Forbes. 2014. Bacterial populations associated with early-stage adipocere formation in lacustrine waters. *International Journal of Legal Medicine* 128: 379–387.

Vass, A.A. 2001. Beyond the grave: Understanding human decomposition. *Microbiology Today* 28: 190–192.

Vass, A.A., S.A. Barshick, G. Sega, J. Caton, J.T. Skeen, J.C. Love, and J.A. Synstelien. 2002. Decomposition chemistry of human remains: A new methodology for determining the postmortem interval. *Journal of Forensic Sciences* 47: 542–553.

Vetrovsky, T. and P. Baldrian. 2013. The variability of the 16S rRNA gene in bacterial genomes and its consequences for bacterial community analyses. *PLoS ONE* 8: e57923.

Vickers, N. 2000. Mechanisms of animal navigation in odor plumes. *The Biological Bulletin* 198: 203–212.

Voříšková, J. and P. Baldrian. 2013. Fungal community on decomposing leaf litter undergoes rapid successional changes. *The ISME Journal* 7: 477–486.

Waggoner, B. 1996. Bacteria and protists from Middle Cretaceous amber of Ellsworth County, Kansas. *PaleoBios* 17: 20–26.

Wardle, D.A., R.D. Bardgett, J.N. Klironomos, H. Setälä, W.H. Van Der Putten, and D.H. Wall. 2004. Ecological linkages between aboveground and belowground biota. *Science* 304: 1629–1633.

Warnecke, F., P. Luginbühl, N. Ivanova, M. Ghassemian, T.H. Richardson, J.T. Stege, M. Cayouette, A.C. McHardy, G. Djordjevic, and N. Aboushadi. 2007. Metagenomic and functional analysis of hindgut microbiota of a wood-feeding higher termite. *Nature* 450: 560–565.

Watanabe, K. and M. Sato. 1998. Plasmid-mediated gene transfer between insect-resident bacteria, *Enterobacter cloacae*, and plant-epiphytic bacteria, *Erwinia herbicola*, in guts of silkworm larvae. *Current Microbiology* 37: 352–355.

Watnick, P. and R. Kolter. 2000. Biofilm, city of microbes. *Journal of Bacteriology Proceedings of the National Academy of Sciences* 182: 2675–2679.

Wilson, K., S.C. Cotter, A.F. Reeson, and J.K. Pell. 2001. Melanism and disease resistance in insects. *Ecology Letters* 4: 637–649.

Wilson, E.E. and E.M. Wolkovich. 2011. Scavenging: How carnivores and carrion structure communities. *Trends in Ecology and Evolution* 26: 129–135.

Wilson-Rich, N., M. Spivak, N.H. Fefferman, and P.T. Starks. 2009. Genetic, individual, and group facilitation of disease resistance in insect societies. *Annual Review of Entomology* 54: 405–423.

Wiser, M., N. Ribeck, R. Lenski, and N. Science. 2013. Long-term dynamics of adaptation in asexual populations. *Science* 10.1126: 1–5. Doi: 10.1126/Science.

Woese, C.R., O. Kandler, and M.L. Wheelis. 1990. Towards a natural system of organisms: Proposal for the domains Archaea, Bacteria, and Eucarya. *Proceedings of the National Academy of Sciences* 87: 4576–4579.

Woodcock, B.A., A.D. Watt, and S.R. Leather. 2002. Aggregation, habitat quality and coexistence: A case study on carrion fly communities in slug cadavers. *Journal of Animal Ecology* 71: 131–140.

Woodward, G., M.O. Gessner, P.S. Giller, V. Gulis, S. Hladyz, A. Lecerf, B. Malmqvist et al. 2012. Continental-scale effects of nutrient pollution on stream ecosystem functioning. *Science* 336: 1438–1440.

Yan, G., D.W. Severson, and B.M. Christensen. 1997. Costs and benefits of mosquito refractoriness to malaria parasites: Implications for genetic variability of mosquitoes and genetic control of malaria. *Evolution* 51: 441–450.

Yang, L.H. 2004. Periodical cicadas as resource pulses in North American forests. *Science* 306: 1565–1567.

Yang, L., J. Bastow, K. Spence, and A. Wright. 2008. What can we learn from resource pulses? *Ecology* 89: 621–634.

Ye, Y.H., S.F. Chenoweth, and E.A. McGraw. 2009. Effective but costly, evolved mechanisms of defense against a virulent opportunistic pathogen in *Drosophila melanogaster*. *PLoS Pathogens* 5: e1000385.

Zak, D.R., W.E. Holmes, D.C. White, A.D. Peacock, and D. Tilman. 2003. Plant diversity, soil microbial communities, and ecosystem function: Are there any links? *Ecology* 84: 2042–2050.

Zimmerman, K.A. and J.R. Wallace. 2008. The potential to determine a postmortem submersion interval based on algal/diatom diversity on decomposing mammalian carcasses in brackish ponds in Delaware. *Journal of Forensic Sciences* 53: 935–941.

4

Arthropod Communities in Terrestrial Environments

Richard W. Merritt and Grant D. De Jong

CONTENTS

4.1 Introduction

Carrion in terrestrial environments usually attracts a rather predictable assemblage of arthropods, and several studies have shown that a very large number of taxa can be acquired simply by visiting animal carcasses. As examples, Payne (1965) reported 522 arthropod species collected from pig (*Sus scrofa* L.) carcasses in South Carolina, USA; Reed (1958) reported 240 taxa from dog (*Canis familiaris* L.) carcasses in Tennessee, USA; Goff et al. (1986) reported 149 taxa from several kinds of carrion in the Hawaiian Islands; Braack (1986) reported 227 arthropod taxa at impala (*Aepyceros melampus* (Lichtenstein)) carcasses in South Africa; and Kočárek (2003) recorded 145 species of Coleoptera at rat (*Rattus rattus* L.) carcasses in the Czech Republic. Dozens of studies have reported smaller numbers of taxa (25–100), usually because fewer carcasses were used, the fauna was not intensively sampled, or the climate may have restricted regional taxa richness.

This arthropod community associated with carrion, similar to almost every other ecological community, is primarily shaped by spatial and temporal constraints. Some arthropod taxa (if considered at the family level) seem to be found nearly everywhere, whereas other carrion arthropod taxa are found only in certain situations; yet, all are in fact guided by these constraints. Temporal constraints include seasonal, successional, and circadian components, whereas spatial constraints include geographical and ecological components. Each of these constraints will be discussed first to provide an ecological background for these communities.

4.1.1 Temporal Constraints

4.1.1.1 Seasonal Constraints

Seasonal constraints involve the changing seasons, whether that is winter–spring–summer–fall in temperate regions of both hemispheres or wet–dry in tropical regions. Being heterothermic organisms, most carrion arthropods are present and actively colonize carrion during the warmer seasons such as spring, summer, and fall (Johnson 1975; Nuorteva 1977; Tabor et al. 2004; Shi et al. 2009; Anton et al. 2011; Horenstein and Linhares 2011). Because of this, most carrion research studies in the temperate regions take place in these warmer months, perhaps emplacing a foundational bias in how we view carrion arthropod communities (see review by Anderson 2010).

Few arthropods have been found to be actively colonizing carrion in the colder seasons. The dipteran families Heleomyzidae, Trichoceridae, and Sepsidae have been reported actively colonizing carrion in the winter (Erzinçlioğlu 1980; Anton et al. 2011), whereas in Colorado, USA, *Piophila casei* (L.) (Diptera: Piophilidae) and *Dermestes* spp. (Coleoptera: Dermestidae) are the primary species on carrion during winter (Adair and Kondratieff 1996). Sometimes, other arthropods may be present on carrion because they colonized during the fall and had not completed their life cycle before winter arrived.

Some locations, such as the Front Range of Colorado, can have alternating warm and cold temperature windows during the winter, such that opportunities may exist for some carrion arthropods to colonize during this time. For example, a raccoon (*Procyon lotor* L.) carcass exposed in February in Denver, Colorado experienced colonization by *Phormia regina* (Meigen) (Diptera: Calliphoridae) during a warm window in early March, but a cat (*Felis catus* L.) carcass exposed in December of a different year was not colonized by any arthropods until late April despite numerous warm windows of opportunity (GDD, personal observation). As noted by Anderson (2010), the time of colonization for some species of carrion arthropods may relate less to time since death and more to season.

4.1.1.2 Succession and Circadian Constraints

Successional constraints involve the community assembly over the time that carrion is exposed, with different communities present during the life cycle of carrion decomposition, whether considered in stages or as a continuum of change (Schoenly and Reid 1987, 1989), even remaining in the soil for some time after the carrion is removed (Saloña et al. 2010). Circadian constraints fall into an even smaller time frame and are defined by the time of day in which arthropods are present (Schoenly 1983; Baldridge et al. 2006; Zurawski et al. 2009). Some species are diurnal, nocturnal, or crepuscular, and a few species may arrive or be present at any time of day.

This topic is of considerable interest to forensic entomologists as it can help estimate postmortem interval (PMI; e.g., Nuorteva 1977; Smith 1986; Catts and Goff 1992; VanLaerhoven and Anderson 1999; Archer 2003; Merritt and Benbow 2009; Wells and LaMotte 2010; Villet and Amendt 2011) and is further discussed theoretically and practically in Chapters 8 and 24 of this book.

4.1.2 Spatial Constraints

4.1.2.1 Geographical Constraints

Geographical constraints involve the biogeographical distribution of various arthropod taxa. Some species are considered to be tropicopolitan, Holarctic, Nearctic, Palaearctic, Oriental, Sub-Saharan, or otherwise widespread. Other species are limited in their geographical distribution, perhaps being limited to the Arctic or Antarctic latitudes, arid or mesic regions, or islands, including oceanic islands and mountaintop islands. For example, *Calliphora quadrimaculata* (Swederus) (Diptera: Calliphoridae) is restricted to New Zealand and is unlikely to be found anywhere else; reports of this species even in nearby Australia have been demonstrated to be incorrect (Norris 1997). The Silphidae (Coleoptera) are absent from the Hawaiian Islands, USA, although they are often conspicuous members of the carrion arthropod assemblage on all of the continental land masses except Antarctica. However, one species of silphid beetle, *Thanatophilus coloradensis* (Wickham), is restricted to sites at or near the tree line in the Rocky Mountains from Alaska to Colorado, USA, in the Nearctic (Peck and Anderson 1982). Nonetheless, even when two carrion succession studies are conducted in similar climates and in close proximity to each other, significant differences in the carrion fauna can be evident, indicating an aggregation effect (Shi et al. 2009; Anderson 2010).

Carrion researchers must recognize that some carrion arthropod species are well known for invasiveness and greatly expanding their ranges. For example, the sap beetle *Nitidula flavomaculata* Rossi (Coleoptera: Nitidulidae) is a European native, but has been found in North Africa, the Near East, and, recently, the Western USA (Adair and Kondratieff 1996). Likewise, the blow fly genus *Chrysomya* (Diptera: Calliphoridae) is native to the Old World, but three species (the African *Chrysomya albiceps* (Wiedemann) and *C. putoria* (Wiedemann) and the Oriental *C. megacephala* (F.)) were introduced to Brazil ca. 1975, and a fourth species (the Oriental *C. rufifacies* (Macquart)) was introduced to Costa Rica ca. 1978 (Baumgartner and Greenberg 1984; Peris 1987). Subsequently, all four species have since spread throughout the Americas, with *C. megacephala* and *C. rufifacies* appearing to invade the Nearctic region seasonally as far north as New Mexico, USA and Ontario, Canada, respectively (De Jong 1995; Shahid et al. 2000; Whitworth 2006; Rosati and VanLaerhoven 2007), and *C. megacephala*, *C. albiceps*, and *C. putoria* found as far south as Argentina (Olea et al. 2011). *C. megacephala* has also spread throughout southern Asia (Verves 2002) and various regions of Africa (Kurahashi 1978; Prins 1979; Verves 2002) and Europe (Martínez-Sánchez et al. 2001). These invasions may be responsible for declines in native species within those ranges (Wells and Kurahashi 1997; Faria et al. 1999). Conversely, the New World muscid *Hydrotaea aenescens* (Wiedemann) (Diptera: Muscidae), the black dump fly, has invaded Europe as far east as Russia and Turkey (Vikhrev 2008).

4.1.2.2 Ecological Constraints

In most modern carrion succession research programs, a baseline study using exposed carrion is usually conducted first and subsequent research involves testing ecological constraints. Investigators have studied shaded versus sun-exposed locations, terrestrial versus aquatic habitats, freshwater versus marine habitats, lotic (flowing water) versus lentic (standing water) habitats, exposed on the ground versus buried versus suspended, urban versus rural, and indoor versus outdoor (e.g., Goff 1991; Shean et al. 1993; Avila and Goff 1998; Davis and Goff 2000; Anderson 2010; Merritt and Wallace 2010; Voss et al. 2011; Pastula and Merritt 2013). Carrion of one sort or another has been exposed in these ecological habitats, with the goal of determining differences in communities and succession patterns among these habitats.

As a result of these kinds of studies, it has been determined, for example, that there are some fly species that are more synanthropic than others and are thus more likely to be found near towns than in rural fields or woodlands (Greenberg 1971). Dekeirsschieter et al. (2011) did not find silphid beetles at urban sites in Belgium, but found six and seven species in agricultural and forest biotypes, respectively. Nuorteva et al. (1967) described four forensic case studies from bodies discovered indoors where ecological differences proved to be an important observation: *Calliphora vicina* (Robineau-Desvoidy) (as *Calliphora erythrocephala* [Meigen] [Diptera: Calliphoridae]) preferred shaded locations and was commonly found as expected, but specimens of the generally heliophilic *Lucilia sericata* (Meigen) (Diptera: Calliphoridae) were found on a corpse that had been laid near a sunny window.

In many cases, differences in communities among ecological habitats are discovered, and some of these differences are discussed elsewhere in this book; however, given that experimental research into these ecological constraints continues to this day, experimental exposures should be well planned in preparation for proper statistical analysis (Michaud et al. 2012; Tomberlin et al. 2012, Chapter 7 of this book).

A final type of ecological constraint is the type and size of the carrion (Kuusela and Hanski 1982; Kneidel 1984a,b; Hewadikaram and Goff 1991). Although many members of the carrion arthropod community appear to be rather catholic in their carrion preferences, others appear to be more specialized, particularly in the choice between vertebrate and invertebrate carrion. Buenaventura and Pape (2013) summarized the known carrion preferences of several species of the sarcophagid (Diptera: Sarcophagidae) flies of the genus *Peckia*, in which some actively preferred invertebrate carrion, whereas others preferred vertebrate carrion. Likewise, Groenewald and Fonds (2000) showed that some oceanic crustacean scavengers in the North Sea, such as lysianoid amphipods, preferred invertebrate carrion, whereas others showed no distinct preference for fish or invertebrate carrion. Conversely, *Orchomene* spp. (Lysianassidae) can skeletonize a pig (*S. scrofa* L.) carcass in a few days in coastal British Columbia, Canada (Gail Anderson, personal communication).

Smaller-sized vertebrate carrion often, but not always, attracts primarily sarcophagid flies, whereas calliphorid flies seem to ignore those baits. Larger carcasses attract calliphorid flies much more than sarcophagid flies. In a study using dead rats (*R. rattus*), De Jong and Hoback (2006) found equal numbers of adults of calliphorid and sarcophagid species and individuals attracted to the carrion, but, except in a single carcass where *Cochliomyia macellaria* (F.) (Diptera: Calliphoridae) was successful, only *Sarcophaga* larvae were successful in completing their life cycle.

4.2 "Normal" Carrion Arthropod Communities

The most conspicuous arthropod taxa of almost any carrion community on land include flies and beetles. Of these, there are some family-level taxa that would be expected in nearly all situations, (even though, as discussed earlier, species-level differences may exist). This chapter serves as an introduction to the carrion arthropod fauna and cannot be comprehensive or exhaustive in its review. Other families and genera than those discussed here can—and will—be found on carrion; however, these are the taxa that comprise what most carrion researchers would describe as a "normal" carrion arthropod community.

4.2.1 Diptera

Whenever movie or television producers want to show that a body is dead, they frequently have a few flies buzzing around. Flies are the stereotypical arthropods associated with carrion, both in the adult and in the larval life stages. In fact, in the absence of vertebrate scavengers, flies generally represent the largest amount of arthropod biomass supported in a carcass and can consume over half of the original carcass mass (Putman 1978; Greenberg 1991; De Jong and Hoback 2006; De Jong et al. 2011); however, the total number of families regularly associated with carrion is relatively small.

Calliphoridae, Sarcophagidae, and Muscidae are the usual fly families that would be considered normal carrion fauna, although there is a lot of intrafamilial ecological variation, such that these families also include members who are coprophagous or even carnivorous rather than strictly saprophagous as larvae, with adults even more varied in their habits (Ferrar 1987). In addition to those three families, there are a host of other families that are represented at carrion (e.g., Heleomyzidae, Ulidiidae, Dryomyzidae, Platystomatidae, Dolichopodidae, and Ptychopteridae), although these families are usually limited in the extent of how many species, how many individuals, and in which studies they are actually involved (Smith 1986; Anton et al. 2011; Martín-Vega and Baz 2013).

4.2.1.1 Calliphoridae

The Calliphoridae is a moderately diverse family, with approximately 1100 species, the majority of which are likely to be found on carrion (Figure 4.1). This family, as currently defined, is paraphyletic (Rognes 1997), but the sarcosaprophagous taxa within the Calliphoridae fall pretty well in most of

FIGURE 4.1 **(See color insert.)** Calliphoridae. Clockwise, from top right: adult *Calliphora vicina* Robineau-Desvoidy; adult *Cochliomyia macellaria* Fabricius; adult *Chrysomya rufifacies* (Macquart); adult *Lucilia coeruliviridis* Macquart; adult *Pollenia rudis* (Fabricius); and larva of Calliphoridae. (All photos by Steve Marshall.)

the subfamilial groupings that would remain in the Calliphoridae were it to be fully revised. These are the most typical members of the carrion fly community throughout the world, with species in the Calliphorinae, Phormiinae, and Chrysomyiinae associated with carrion as adults and larvae (e.g., Greenberg 1991; Villet 2011; Whitworth 2006, 2010; Szpila et al. 2013).

Calliphorids typically arrive very early in succession, often within hours (Greenberg 1991; De Jong 1994; Anton et al. 2011), and some species lay their eggs in characteristic patterns. By far, calliphorids are the best studied group of carrion fauna, such that much is known and currently being studied about adult carrion preferences, oviposition behaviors, egg and larval morphology, development, behaviors, and pupal development sequences—not to say that there is not a universe of biological knowledge yet to be discovered about these creatures (Greenberg and Kunich 2002). Most of this research has been carried out in the context of forensic entomology (Haskell and Williams 2008; Byrd and Castner 2010a,b), but two species, *C. vicina* and *P. regina*, are frequently used as model laboratory organisms, so much is known about the physiology of these flies, as well as their ecological relations with carrion.

Surprisingly, it is not uncommon to find members of the Polleniinae and Rhiniinae (=Rhiniidae) on carrion as well. *Pollenia* spp. (Figure 4.1), in particular, are commonly found on carrion, even though they are parasites of free-living annelids (Thomson and Davies 1973a,b; Jewiss-Gaines et al. 2012). Their function in the decomposition of carrion is unclear, because no notable associations with the carrion (e.g., oviposition) other than presence have been described (Baz et al. 2007; Anton et al. 2011). Other calliphorids include the Tricycleinae, which is a small subfamily of obligate vertebrate parasites, and the Bengaliinae, which is a small subfamily of termitophilous species.

4.2.1.2 Sarcophagidae

The Sarcophagidae contains about 2500 species (Pape 1987), but generally only the subfamily Sarcophaginae is represented in the carrion community (Figure 4.2). Although a few species of the other two subfamilies (Miltogramminae and Paramacronychiinae) have been reported on carrion, they

FIGURE 4.2   (See color insert.) Sarcophagidae and Muscidae. Clockwise, from top right: adult *Sarcophaga rohdendorfi* Salem; adult and larvae of *Ravinia* spp.; larvae of *Musca domestica* L.; adult female *M. domestica* L.; and adult *Hydrotaea aenescens* (Wiedemann). (All photos by Steve Marshall.)

are generally parasitic on Hymenoptera or generally feed on the provisions that wasps had paralyzed for their own young (Tabor et al. 2005; Szpila et al. 2010). The genus *Eumacronychia* has several species that have been reared from reptilian eggs, and the genus *Wohlfahrtia* causes cuticular myiasis in mammals (James 1947).

Sometimes, sarcophagids have specific host preferences, such as within the genus *Peckia* in the New World tropics and subtropics; Buenaventura and Pape (2013) summarized which species had been collected at different kinds of carrion. Many, if not all, sarcophagids that are part of the carrion fauna are larviparous, meaning that females deposit early larval instars instead of eggs on the carrion, thus theoretically giving them a "jump start" on some of the other carrion flies in succession on the carrion. Identification of sarcophagids has historically been incredibly difficult (Sanjean 1957; de Carvalho and Mello-Patiu 2008; Stamper et al. 2012): in many genera, only males can be identified and then only after dissection of genitalia (Vairo et al. 2011). Recent advances in DNA barcoding have made identification of some sarcophagids less difficult (e.g., Sperling et al. 1994; Zehner et al. 2004; Draber-Mońko et al. 2009; Meiklejohn et al. 2013).

4.2.1.3 Muscidae and Fanniidae

The Muscidae contains about 4500 species and the Fanniidae contains about 300 species (de Carvalho 1989; de Carvalho et al. 2005), and between them only a few genera, including *Hydrotaea, Musca, Muscina, Synthesiomyia,* and *Fannia,* are found on carrion (Skidmore 1985). Muscids (Figure 4.2) and fanniids (Figure 4.3) can be among the dominant flies in the carrion fauna (Shi et al. 2009), often arriving a bit later than the calliphorids and sarcophagids, but still before the carrion completely dries out. Because a few members of the group tend to be highly synanthropic and the fact that most of the muscids that are found on carrion are also common flies on dung (Merritt 1976), they can constitute a major health hazard. For example, it is not uncommon for a house fly, *Musca domestica* L. (Figure 4.2),

FIGURE 4.3 **(See color insert.)** Fanniidae, Stratiomyidae, Sepsidae, Scathophagidae. Clockwise, from top right: larvae of *Fannia canicularis* L.; adult *F. canicularis* L.; adult *Hermetia illucens* L.; adult *Themira annulipes* (Meigen); and adult male and female *Scathophaga stercoraria* (L.). (Photo of *S. stercoraria* by Robert Armstrong; all other photos by Steve Marshall.)

to feed on carrion or fecal material, pick up a few million bacteria, and then mechanically transfer them to our food or skin (e.g., Cohen et al. 1991; Graczyk et al. 2001). The lesser house fly, *Fannia canicularis* L. (Fanniidae) (Figure 4.3), is somewhat smaller than the house fly, and the larvae are easily identified by their characteristic shape; the larvae are often common inhabitants of decaying organic matter, including carrion.

4.2.1.4 Sepsidae

The Sepsidae, or black scavenger flies, is a small family with about 300 species worldwide (Pont and Meier 2002) (Figure 4.3). These small flies nearly always breed in decaying organic matter, particularly dung (Merritt 1976), but also vertebrate and invertebrate carrion, decaying vegetation, slime molds, and algae. The adults likewise can usually be collected near such materials, usually arriving while the material is not yet dry. *Sepsis* spp. and *Themira putris* (L.) can frequently be collected on carrion in the United States (De Jong and Chadwick 1999).

4.2.1.5 Stratiomyidae

Of the approximately 3000 species of Stratiomyidae in the world (Woodley 2001), the most commonly found species represented in the carrion community is *Hermetia illucens* (L.) (Figure 4.3). This species of soldier fly can breed in many types of organic materials, particularly dung (Tomberlin and Sheppard 2001; Myers et al. 2008), but it has been reported from carrion in warm temperate and tropical regions in the New World (Tomberlin et al. 2005). The larvae have been used extensively in poultry breeding facilities for composting manure, as an efficient competitor for muscids and calliphorids, and as food for domestic animals (Sheppard 1983, 1992; Sheppard et al. 1994).

4.2.1.6 Scathophagidae

About 500 species of Scathophagidae are known, almost exclusively from the Holarctic. Only a few species, particularly the geographically widespread *Scatophaga stercoraria* (L.) (Figure 4.3), the yellow dung fly, are collected on decaying materials, mainly dung (Merritt and Anderson 1977), but also carrion (Byrd and Castner 2010a,b). Females spend most of their time foraging for other small insects, often calliphorids, in the surrounding vegetation and only visit dung pats to mate and oviposit on the dung surface. Šifner (2008) reported that some arctic scathophagids are commonly collected on carrion.

4.2.1.7 Phoridae

The Phoridae is a large family (approximately 4000 species) of small, hump-backed flies (Figure 4.4). Some of the most charismatic species are those in the genus *Pseudacteon*, known for their ant-decapitating skills and the resultant potential for biocontrol of pestiferous *Solenopsis* spp. (Porter 1998). Many Phoridae, however, are found in decaying organic matter (especially plants), dung, fungi, and, commonly, around poorly draining sewer pipes. On carrion, the genus *Conicera* and the hyperdiverse genus *Megaselia* (Figure 4.4) are particularly common (Disney 1994; Greenberg and Wells 1998). As will be discussed later, *Conicera tibialis* Schmitz is known as the "coffin fly," sometimes being found in human corpses buried up to 2 m below ground (Merritt et al. 2007).

4.2.1.8 Piophilidae

A small family with about 75 species, Piophilidae is represented by several species in the carrion community; however, the most frequently noted species are *P. casei* (L.) (Figure 4.4) and *Stearibia nigriceps* (Meigen), usually found on carrion in later stages of decomposition (Martín-Vega 2011). *P. casei* is commonly known as the "cheese skipper," due to its frequent presence in foods with high fatty acid and caseic product concentrations and the ability of the mature larvae to jump or "skip." Other piophilid

FIGURE 4.4 **(See color insert.)** Phoridae, Piophilidae, Heleomyzidae, Sphaeroceridae. Clockwise, from top right: larvae of Phoridae; adult *Megaselia* spp.; adult *Piophila casei* (L.); adult male and female *Neoleria* spp.; and adult *Lotophila atra* (Meigen). (Photos of phorid larvae by Richard Merritt; all other photos by Steve Marshall.)

genera collected at carrion, often in later decay stages but sometimes earlier, include *Arctopiophila*, *Centrophlebomyia*, *Liopiophila*, *Parapiophila*, *Prochyliza*, and *Thyreophora* (De Jong and Chadwick 1999; Anton et al. 2011; Martín-Vega 2011).

4.2.1.9 Heleomyzidae

This small family has approximately 650 species, and the larvae generally occur in decaying plant materials and mushrooms. The genus *Neoleria* (Figure 4.4) is considered to be necrophagous, and Martín-Vega and Baz (2013) showed that males of *Neoleria* congregate around carrion to mate with arriving females, which then oviposit on the carrion. In contrast, the genus *Suillia* does not utilize carrion for oviposition, but females may use it as a protein source, and males are uncommonly collected (Martín-Vega and Baz 2013). Other heleomyzid genera encountered at carrion, particularly during winter, include *Borboroides*, *Heleomicra*, *Heleomyza*, and *Scoliocentra* (Gill and Peterson 1987; Anton et al. 2011).

4.2.1.10 Anthomyiidae

This is a moderately sized family of small flies that includes significant agricultural pests, such as the genus *Delia*, and the "seaweed flies" or "kelp flies" of the genus *Fucellia*. Other species feed on decaying plant materials or are scavengers in bird nests and other organically enriched environments such as carrion. Mollusk carrion has hosted *Subhylemyia longula* (Fallén), *Lasiomma octoguttatum* (Zetterstedt), and *Craspedochoeta pullula* (Zettersted) in the United Kingdom and *Anthomyia illocata* Walker in Southeast Asia (Beaver 1969, 1986). In the mountains of Colorado, USA, anthomyiid flies are occasionally the most common adult dipterans collected at vertebrate carrion (GDD, personal observation; De Jong and Chadwick 1999); however, their role within the vertebrate carrion community remains unknown, because there was no particular association observed with the carrion (i.e., feeding, oviposition, etc.).

4.2.1.11 Sphaeroceridae

There are more than 1300 species of Sphaeroceridae and most are known to be microbial grazers as larvae, living in moist, decomposing organic materials such as decaying plants, fungi, dung, and carrion (Merritt 1976, Roháček 2001) (Figure 4.4). Sphaerocerids have been reared from both vertebrate and invertebrate carrion, and some species appear to specialize in carrion that is inaccessible (e.g., buried) or unfavorable (e.g., invertebrate carrion) to the larger carrion flies such as Calliphoridae and Sarcophagidae (Buck 1997). Buck (1997) listed 34 species known to have been bred from carrion, finding that some were primarily necrophagous whereas others used carrion facultatively.

4.2.2 Coleoptera

One out of every four species on Earth is a beetle, so it stands to reason that the beetles should be at least fairly well represented in the carrion arthropod community. Again, though, as with the Diptera, there are relatively few families that are regularly encountered on carrion. The most conspicuous beetles on carrion include Silphidae, Dermestidae, Trogidae, Histeridae, and Staphylinidae, but several other smaller families are regularly represented by at least a few species (Payne and King 1970).

4.2.2.1 Silphidae

The Silphidae contains about 200 species in two distinct subfamilies: the Nicrophorinae and the Silphinae (Sikes et al. 2002) (Figure 4.5). Although some species are rather drab in black or brown, many species in both subfamilies bear striking red or orange coloration patterns. In some geographical regions, particularly in North America, the fauna is well known, and easy-to-use regional identification guides are readily available.

Most Nicrophorinae are in the genus *Nicrophorus* (Figure 4.5) and are known as sexton beetles or burying beetles (Sikes et al. 2002). *Nicrophorus* species generally prefer smaller vertebrate carrion such as small rodents, birds, lizards, and snakes, which they bury and in which they rear their larvae (Pukowski 1933; Milne and Milne 1976; Kočárek 2003), but they can also be found on larger carrion (Peck 1986). Silphinae do not appear to have a preference for carcass size, or they prefer larger carcasses, they tend to arrive later in succession than Nicrophorinae, and they can be abundant (De Jong and

FIGURE 4.5 **(See color insert.)** Silphidae. Clockwise, from top right: larva of *Nicrophorus* spp. on pink salmon carcass; larva of *Nicrophorus* spp. on pink salmon carcass; adult *Nicrophorus tomentosus* Weber; and adult *N. investigator* Zetterstedt. (Photos by Robert Armstrong; except for the *Nicrophorus tomentosus* which was by Steve Marshall.)

Chadwick 1999; Anton et al. 2011). The ethology of parental care of larvae has been well studied in many species of *Nicrophorus* (Pukowski 1933; Milne and Milne 1976).

The American burying beetle, *Nicrophorus americanus* Olivier, is listed as a critically endangered species in the United States because its range has been reduced from the eastern two-thirds of the country to a handful of sites in Nebraska, South Dakota, Oklahoma, Arkansas, and Rhode Island. A good review of its natural history is presented in Kozol et al. (1988). Reasons given for its severe demise include habitat destruction (for both beetles and small mammal hosts), widespread pesticide use, urban illumination (distracting the adult beetles), and even the demise of the passenger pigeon (*Ectopistes migratorius* (L.), a previously common host).

4.2.2.2 Histeridae

Histeridae is a large family of about 4300 species (Figure 4.6). They have been introduced to poultry houses to control fly populations, and they are efficient predators of calliphorid and other cyclorrhaphan Diptera larvae (Nuorteva 1970). Numerous genera have been reported from carrion, particularly in the African and Neotropical fauna (Villet 2011; Aballay et al. 2013). In the Northern Hemisphere, histerids (Figure 4.6) are frequently reported in carrion succession studies, albeit generally in small numbers and often from only a couple of genera, particularly *Saprinus* (Anderson and VanLaerhoven 1996; Anton et al. 2011).

FIGURE 4.6 **(See color insert.)** Histeridae, Staphylinidae, Dermestidae, Scarabaeidae, Carabidae; clockwise, from top left: adult of Histeridae (Photo by Richard W. Merritt.); adult *Creophilus maxillosus* (L.) feeding on Diptera larvae (Photo by Richard W. Merritt.); adult Staphylinidae feeding on Diptera eggs (Photo by Robert Armstrong.); adult *Dermestes lardarius* L. (Photo by Steve Marshall.); adult *Aphodius* spp. (Photo by Richard W. Merritt.); and adult *Pterostichus* spp. with mites (Photo by Robert Armstrong.).

4.2.2.3 Staphylinidae

Staphylinidae is purportedly the largest family of beetles with approximately 50,000 species (and uncounted thousands yet to be identified), but only a few genera of staphylinids have been reported from carrion (Figure 4.6). *Creophilus maxillosus* (L.) and *Ontholestes cinqulatus* (Gravenhorst) are common, large, easily identifiable staphylinids on carrion in the Holarctic, whereas species of *Philonthus*, *Oxytelus*, and other difficult-to-identify genera are regularly found on carrion and dung all over the world (Merritt and Anderson 1977). Much work is currently being done worldwide to alleviate the poor taxonomic understanding of Staphylinidae (Brunke et al. 2012).

4.2.2.4 Dermestidae

Dermestidae is a moderately large family with about 600 species, of which likely half can be encountered on carrion (Figure 4.6). Numerous species in *Dermestes*, *Attagenus*, and *Anthrenus* are common members of the carrion community. These beetles often arrive later in succession and are commonly characteristic fauna of those later stages (Anderson and VanLaerhoven 1996), but can arrive even in early stages (De Jong and Hoback 2006). In outdoor markets, particularly in the tropics, dermestids can often infest dried fish and meats, and the rest of the Dermestidae that do not feed on carrion instead feed on dried plant material, in which they are considered to be stored products pests (e.g., Kingsolver 1987; Metcalf and Metcalf 1993).

4.2.2.5 Scarabaeidae

Most Scarabaeidae are phytophagous, but many scarab beetles (Figure 4.6), particularly in the Scarabaeinae, are frequently found on vertebrate and invertebrate carrion throughout Africa (Braack 1986; Villet 2011; F.-T. Krell, personal communication). Because scarab beetles seem to be found more commonly on the carcasses of large, herbivorous mammals, some researchers have hypothesized that their presence is not necessarily due to the carrion itself, but rather to materials in the gastrointestinal tract as it becomes exposed during carcass decomposition (Midgley et al. 2012).

4.2.2.6 Trogidae

Trogidae is a small family of about 300 species worldwide, all of which are associated with dried carcasses, arriving very late in succession, or in bird nests and small mammal burrows, where sloughed feathers, hairs, or feces may abound (Vaurie 1955, 1962). Many species can be found on owl pellets, and similar to dermestids and clothes moths, they have occasionally been reported from objects made from animal products, such as wool sweaters, carpet, horsehair automobile seats, and felt hats. Some species appear to be somewhat host-specific, such as *Trox laticollis* LeConte, which has only been found in fox (*Vulpes vulpes* (L.)) dens (Vaurie 1955).

4.2.2.7 Cleridae

Two cosmopolitan species of *Necrobia* are commonly found on carrion and are distinct members of the carrion community. Unlike most clerids, which are generally regarded as predators in both the larval and adult forms (Payne and King 1970), the *Necrobia* spp. appear to be both necrophagous and predaceous (Anderson and VanLaerhoven 1996; Horenstein and Linhares 2011). Other clerids found at carrion include *Phloeocopus* in southern Africa (Braack 1986).

4.2.2.8 Agyrtidae and Leiodidae

The Agyrtidae is a small family that has some species found on vertebrate and invertebrate carrion, primarily in the Pacific Northwest of North America (Peck 1990). In the Leiodidae, most of the carrion-feeding forms are in the Cholevinae, which are often called the "small carrion beetles." *Catops* is a genus commonly found on carrion in the Holarctic, being reported from all types of vertebrate carrion (Jeannel 1936; Chapman and Sankey 1955; Kentner and Streit 1990; Dillon 1997; De Jong and Chadwick 1999).

4.2.2.9 Nitidulidae

This moderately sized family of small beetles occurs worldwide, with many genera that are considered to be cosmopolitan (Parsons 1943). Species in the subfamily Cateretinae are exclusively phytophagous, but most species in the other subfamilies live in decaying fruits and plants or on fungi, and some have more specialized habits; however, *Carpophilus* spp., *Glischrochilus* spp., *Nitidula* spp., and *Omosita* spp. are often found on carrion throughout the world (Braack 1986; Adair and Kondratieff 1996; Anderson and VanLaerhoven 1996; De Jong and Chadwick 1999; Saloña et al. 2010). Nitidulidae are often found in later stages of carrion decomposition (Payne and King 1970; Anderson and VanLaerhoven 1996).

4.2.2.10 Carabidae

Carabidae (Figure 4.6) have regularly been reported from carrion in South Africa, where they may be considered a normal part of the carrion fauna (Prins 1984). In other parts of the world, however, they are not commonly encountered on carrion, and their presence is either not discussed or explained as incidental, even if they are abundant (De Jong and Chadwick 1999; Ma et al. 2000; Honda et al. 2008; Anton et al. 2011).

4.2.3 Hymenoptera

4.2.3.1 Formicidae

Fuller (1934) believed that ants were not a regular part of the carrion community in Australia, because they seemed to be found only on carrion that was near ant mounds; however, more recent research around the world suggests otherwise. Payne et al. (1968), Payne and Mason (1971), and Anderson and VanLaerhoven (1996) found ants on most animal carcasses and in all stages of decomposition, indicating that they are truly part of the normal carrion fauna. Likewise, Chin et al. (2009) collected six species of ants from three subfamilies (Formicinae, Myrmicinae, and Ponerinae) that were associated with pig (*S. scrofa*) carcasses placed on the ground in Malaysia. These ants were observed feeding on fly eggs, larvae, pupae, and adults and were found at all stages of decomposition starting from fresh until dry remains. The red imported fire ant, *Solenopsis invicta* Buren, is now considered a normal part of the carrion community throughout its naturalized range in the southeastern United States (Stoker et al. 1995; Lindgren et al. 2010). GDD (personal observation) has commonly found *Pheidole* spp. on invertebrate carrion in the Gulf Coast states of the United States, for example, on stranded *Clibanarius vittatus* (Bosc) hermit crabs in Biloxi, Mississippi and on dead *Periplaneta americana* (L.) cockroaches in Pensacola, Florida. In addition to feeding on the carrion itself, ants can have a considerable effect on the rest of the carrion arthropod community through predation as well (Wells and Greenberg 1994; Stoker et al. 1995). De Jong and Hoback (2006) found that rat (*R. rattus*) carcasses near *Myrmica* mounds were generally devoid of dipteran maggots, although active predation was not observed.

4.2.3.2 Vespidae

Some yellowjackets, hornets, and other vespine wasps rely on proteinaceous foods such as other insects, discarded food from humans, and carrion to sustain their colonies (Akre et al. 1981). They are particularly aggressive toward organisms they perceive as intruders in the fall when they are actively searching for protein sources, including carrion. In addition to cutting away small pieces from the carrion itself, vespids also capture adult flies and maggots in the vicinity. In British Columbia, Kennedy (2002) reports having watched yellowjackets completely dismantle a mouse (*Mus musculus* L.) carcass and haul it off to their nest. In southeastern Brazil, Moretti et al. (2011) observed several species of vespid wasps directly feeding on carrion baits and preying on adult insects collected in carrion traps.

4.2.3.3 Apidae

The three species of bees in the Neotropical *Trigona hypogea* Silvestri group contravene all conventional wisdom about bees, as they are obligate necrophages and use carrion, instead of pollen, as their source of protein (Camargo and Roubik 1991). The adults congregate at carrion and masticate the tissues, after which they return to their underground nest to regurgitate and feed their developing larvae (Roubik 1982).

4.2.3.4 Parasitic Hymenoptera

Reports of Hymenoptera parasitic on Diptera are widespread in the agricultural literature, and several species are of interest in carrion ecology because the parasites or parasitoids attack Diptera on carrion. For example, Rodrigues-Guimarães et al. (2006) reported *Aphaereta laviuscula* (Spinola), a braconid wasp, and *Nasonia vitripennis* (Walker), a pteromalid wasp, on the screwworm, *Cochliomyia hominivorax* (Coquerel), in Brazil. Frederickx et al. (2013) reported and/or reared six species of parasitic Hymenoptera from pig carrion and associated Diptera pupae, including *Aspilota fuscicornis* Haliday (Braconidae), *Alysia manducator* Panzer (Braconidae), *N. vitripennis*, *Tachinaephagus zealandicus* Ashmead (Encyrtidae), *Trichopria* spp. (Diapriidae), and *Figites* spp. (Figitidae). *N. vitripennis* and *A. manducator* are the most common parasitoids that have been reported from dipteran pupae on carrion (Disney and Munk 2004; Grassberger and Frank 2004) and have even been considered as biological control agents for flies around stockyards and slaughterhouses.

4.2.4 Acari

Mites are often present on carrion, including soil mites and phoretic mites, which are brought in on the bodies of other carrion arthropods, and aquatic mites (Braig and Perotti 2009; Proctor 2009; Perotti and Braig 2010; Perotti et al. 2010). Even though the presence of mites on carrion has been known, and used forensically, as far back as the late 1800s (Mégnin 1894; Perotti 2009), serious investigations into the mite fauna on carrion are fairly recent, compared with the relatively longer record of studies using insects.

4.2.4.1 Phoretic Mites

Phoresy is an interspecific relationship based on the transportation of one or more organisms (phoronts) by another (host). Many mites associated with carrion use this type of displacement, being especially frequent in the Parasitidae and Macrochelidae (Figure 4.7). Most Parasitidae are geophilic (Krantz and Ainscough 1990) and become excluded from the soil fauna when carrion is present (Anderson and VanLaerhoven 1996), but a well-known phoretic relationship has developed between one parasitid genus, *Poecilochirus* (Figure 4.7), and burying beetles of the genus *Nicrophorus*

FIGURE 4.7   (See color insert.) Acari. Left to right: deutonymph of Parasitidae on pink salmon carcass (Photo by Robert Armstrong.) and deutonymph of *Poecilochirus austrooasiaticus* Vitzthum on human corpse. (Photo by Alejandro Medino.)

(Springett 1968; Wilson 1983; González et al. 2013). The exact role that *Poecilochirus* has in the decomposition of carrion has not been fully elucidated as studies seem to contradict; however, it generally appears that the mites feed on dipteran eggs and first instar larvae, aiding the beetles by removing that potential source of competition. Compared with the remarkable host specificity of *Poecilochirus* species, *Macrocheles* mites (Acari, Macrochelidae) show a very broad range of phoresy hosts, both Diptera and Coleoptera (Axtell 1964; Costa 1969; Niogret et al. 2006). Phoretic relationships are known between other mite species and saprophagous invertebrates and constitute a very wide field of study (Perotti et al. 2010).

4.2.4.2 Other Acari

Most of the normal soil-dwelling mites disappear when carrion is present (Bornemissza 1957; Anderson and VanLaerhoven 1996), whereas other mites are attracted. However, many other families of mites have been found associated with carrion, although, in many cases, their specific association is unknown. Over 180 species of mites are known to be associated with coprophagous Scarabaeoidea (Costa 1969), many of which, as mentioned earlier, are attracted to carrion, particularly in Africa. The dynamics of these soil invertebrates in carrion decomposition is discussed in more depth in Chapter 5.

Domestic mites (including house dust mites, stored product mites, and others, especially in the suborder Astigmata) are often associated with carrion indoors (Solarz 2009). Whatever attracts the mites to cause their colonization on carrion is unclear; however, some species seem to be related to cadaveric mycota (González Medina, A., pers. comm.), which creates an interesting link between forensic entomology and forensic mycology. The short period of time needed to complete their life cycles and their ubiquity in synanthropic environments make these arthropods a potentially useful forensic tool when calculating the period of arthropod activity and, by conceptual relationship, the PMI.

4.3 Ecological Guilds

The previous section examined the normal carrion community from a taxonomic viewpoint, but an alternate way of understanding the carrion community is from the perspective of ecological guilds. The ecological guilds comprised by the common carrion arthropods are primarily necrophages, predators, parasites/parasitoids, and omnivores (VanLaerhoven 2010; Villet 2011). Some species are restricted to one ecological guild, whereas many others are more plastic; however, rather than relegating all of the more plastic taxa to the generalist "omnivore" guild, we instead divide the carrion arthropod fauna according to the most prevalent guild. Additionally, ecological circumstances may switch some taxa from one guild to another (e.g., density resulting in crowding or starvation in *Chrysomya*; Goodbrod and Goff 1990).

4.3.1 Necrophages

Obligate necrophages, which feed directly on the carrion itself, include primarily the larvae of Calliphoridae and Sarcophagidae and adults and larvae of Silphidae, Dermestidae, and Trogidae. The necrophagous flies are much more common early in succession, when soft tissues are easily liquefied to allow access to the abundant nutrients in the carrion. Dermestids, trogids, and piophilids tend to be more abundant late in succession, specializing on the drier portions of carcasses, such as tendons and mummified skin. Tineidae (clothes moths) are among the few representatives of the Lepidoptera that could be considered part of the carrion fauna, with the larvae feeding on keratin-based carrion portions such as hair, wool, claws, or horn.

4.3.2 Predators

Obligate predators, feeding on other members of the carrion arthropod fauna, include adults and larvae of some beetles in the families Staphylinidae and Histeridae and adults of some Silphidae. There appear to be few obligate predators within the carrion fauna; most predators adopt this feeding mode faculta-tively (and so are relegated to their more prevalent guild, usually necrophages) or utilize both necrophagy and predation equally and are termed omnivores.

4.3.3 Parasites and Parasitoids

Specialized parasites and parasitoids of the other members of the carrion fauna generally come from the Hymenoptera, particularly families such as the Braconidae and Pteromalidae (Frederickx et al. 2013), but the staphylinid beetle genus *Aleochara* is also a parasitoid of dipteran larvae (Prins 1984). Host ranges (some wide and others narrow) and temperature-dependent development of these parasites and parasitoids also make them ideal for forensic interpretation (Floate et al. 1999; Ferreira de Almeida et al. 2002; Voss et al. 2009, 2010).

4.3.4 Grazers

This term is being introduced here to tentatively reflect those organisms that appear to feed on the microbiological biofilms associated with carrion rather than the carrion itself or other macroorgan-isms, a phenomenon which has only recently been documented to occur (Pechal et al. 2013, 2014). For example, larvae of most Sphaeroceridae are microbial grazers when they are found in association with decomposing organic materials such as decaying plants, fungi, dung, and carrion (Roháček 2001), and domestic Acari appear to be attracted to cadaveric mycota (González Medina, A., pers. comm.). The microbial communities on carrion and the interkingdom relations are discussed in much more detail in Chapters 3 and 19 of this book.

4.3.5 Omnivores

Omnivores feed on both the carrion and the other carrion-attendant fauna opportunistically; this group includes ants, termites, cockroaches, wasps, and many kinds of beetles in the terrestrial realm and cray-fish and other crustaceans in the aquatic realm. These organisms, although part of the carrion arthropod fauna, are more generalist than the other guilds. In most cases, these are necrophages that are faculta-tively predaceous (VanLaerhoven 2010).

In contrast, most calliphorids are primarily necrophages, but some species of *Chrysomya* (e.g., *C. rufi-facies* and *C. albiceps*) are well adapted to cross over to predation on other maggots, including conspecif-ics, inhabiting carrion (Goodbrod and Goff 1990; Faria et al. 1999). Among the beetles, most silphids are necrophages, but they also do not seem to hesitate to move into facultative predation. Most Staphylinidae that are found in the carrion arthropod community are acting as predators, but may feed on the decom-posing tissues, whereas some, as mentioned earlier, are parasitic on dipteran pupae.

4.4 Specialized Habitats and Specialized Communities

4.4.1 Buried

Volatile cadaveric compounds emanating through soil may attract flies and beetles to the vicinity of bur-ied carrion (von Hoermann et al. 2011), but they might not be able to reach it. For example, in Colorado, USA, statistically larger numbers of calliphorid adults were collected in Malaise traps erected over pig (*S. scrofa*) carcasses buried 0.5 m deep compared with control traps, but it is unlikely that the flies were able to access the carrion (France et al. 1992). A few centimeters of soil appear to deter most arthropods from gaining access to carrion. In shallowly buried carrion, a thin layer of soil appears to effectively prevent beetles from accessing carrion. Shubeck and Blank (1982) and Shubeck (1985) demonstrated that the silphid beetles *Necrophila americana* (L.) and *Oiceoptoma noveboracense* (Förster) can still find carrion buried up to about 4 cm. Other species of silphid beetles encountered were only found on carrion buried up to 1 cm.

Flies appear to be less deterred than beetles when locating shallowly buried carrion. Several fly species can colonize carrion buried up to 60 cm (Rodriguez and Bass 1985; Gunn and Bird 2011). Studies by VanLaerhoven and Anderson (1999) showed the presence of both Calliphoridae and Sarcophagidae in pig (*S. scrofa*) burials at a depth of approximately 1 ft (30.48 cm); however, maggot masses did not form, and carcass temperatures were found to be comparable to soil temperatures (VanLaerhoven and Anderson 1999; Forbes and Dadour 2010). The muscid flies *Muscina stabulans* (Fallèn) and *Muscina prolapsa* (Harris) colonized remains buried in loose soil at a depth of 40 cm (Gunn and Bird 2011). In Michigan, USA, *Sarcophaga bullata* (Parker) and *Hydrotaea* spp. (Diptera: Muscidae) were the first Diptera found colonizing buried pig (*S. scrofa*) carrion 5 days after burial at 30 cm, followed by calliphorids. At 60 cm depth, *Hydrotaea* spp. and the phorid fly, *Megaselia scalaris* (Loew), were the only flies collected 7 days after burial (Pastula and Merritt 2013). Notably, if as little as 5 h of access is given prior to burial of carrion, at least some of the "normal" surface succession community (e.g., calliphorid and muscid flies) can survive burial up to 571 accumulated degree days (Bachmann and Simmons 2010). Buck (1979) noted that some sphaerocerid flies appear to specialize on carrion that is buried and generally inaccessible to calliphorids.

The "coffin fly" *C. tibialis* is able to colonize cadavers buried 2 m below the ground surface and is one of the few species recognized as a forensic indicator in deeply buried carrion (Colyer 1954; Leclercq 1999; Merritt et al. 2007; Martín-Vega et al. 2011). Collembolans and mites are also occasionally encountered in deeply buried carrion (Merritt et al. 2007). Szpila et al. (2010) found that miltogrammine flesh flies, particularly the Nearctic *Eumacronychia persolla* (Reinhard) and the Palaearctic *Phylloteles pictipennis* (Loew), represented the primary, and sometimes exclusive, arthropod colonizer of carrion buried 40–66 cm deep in dry soils. Bachmann and Simmons (2010) also reported four species of Carabidae in the exhumed soil around buried rabbit (*Oryctolagus cuniculus* (L.)) carcasses.

A special arthropod community exists when the arthropods themselves bury the carrion. Several species in the silphid beetle genus *Nicrophorus* are well known as "sexton beetles" because the males and females bury the carcasses of small vertebrates to reduce competition with flies, other beetles, and mold (Suzuki 2000, 2001). Although often cooperating in the burial, males and females also may bury carrion separately (Kozol et al. 1988) or communally (Scott et al. 2007). The resulting, simple, arthropod community consists only of *Nicrophorus* larvae and adults and *Poecilochirus* mites that are phoretic on the adult beetles. Springett (1968) showed that this association is beneficial to both *Nicrophorus* and *Poecilochirus*, because the mites feed on fly eggs and reproduce in the carcass, and the beetles are often unable to reproduce in competition with fly larvae. However, Satou et al. (2000) in Japan demonstrated that *Nicrophorus quadripunctatus* Kraatz is capable of successful reproduction in carrion, despite the loss of phoretic mites and the presence of *Chrysomya pinguis* (Walker) larvae. Lindgren et al. (2010) reported a case study wherein fire ants buried an exposed portion of a human cadaver, thereby temporarily excluding the fly community from colonizing.

4.4.2 Antemortem Communities

Right after death of a host, parasites may still be present, but they are usually leaving in search of a new host. These parasites generally include ticks, fleas, bed bugs, and lice and are often host-specific. The time frame that these organisms are present on carrion is usually very short but, as such, can provide useful information on time of death. Conversely, *Dermacentor variabilis* (Say), a common North American tick, has been observed at carrion even up to 2 months post-mortem, presumably attracted by gases emanating from the carcass (Kneidel 1984a,b; Carroll and Grasela 1986; McNemee et al. 2003). Furthermore, antemortem blood meals taken by parasites such as human lice (*Phthiris humanus* L.) and bed bugs (*Cimex lectularis* L.) can be retained in their digestive tracts for up to 2 months after death of the host (Mumcuoglu et al. 2004; Szalanski et al. 2006).

Of considerable importance is myiasis prior to death. Several species of Calliphoridae and other forensically important Diptera are known to invade wounds and natural orifices even before death of their host (James 1947). This phenomenon has been well studied in sheep-herding countries such as England, South Africa, Australia, and New Zealand. Myiasis can be important when considering neglect of animals, children, or elderly people and can confound the estimation of PMI when using larval growth patterns as evidence (Benecke 2010; De Jong 2014).

4.4.3 Invertebrate Carrion

Little research has gone into the arthropod communities feeding on invertebrate carrion (Seastedt et al. 1981). Especially rare are studies in which the invertebrate carrion is a single organism, as there often is such small mass that a carrion community is not easily attracted, or wind or a single scavenging individual can actually displace the carcass. Nevertheless, very small scavengers can be found on larger invertebrate carrion. The literature on this topic is often relegated to notes in scattered journals, but what follows is a sample of the types of studies that have been performed on arthropod use of invertebrate carrion.

Beaver (1969, 1972, 1986) reported rearing of sarcophagid, phorid, psychodid, anthomyiid, and muscid flies from dead snails in the United Kingdom and Hong Kong. Buck (1979) reported on numerous species of sphaerocerids on invertebrate carrion in southern Germany, with a review of many more records from around the world. Seastedt et al. (1981) reported numerous taxa of ants, scarab beetles, diptera larvae, psocopterans, collembolans, and mites on carcasses of the house cricket, *Acheta domestica* (L.), tethered to the ground or placed in litter bags in Georgia, USA. Norrbom (1983) reared an ephydrid, a chloropid, and two sphaerocerid species from dead horseshoe crabs (*Limulus polyphemus* (L.)) that had washed up on the shore in New Jersey, USA. Sarcophagids were more common than calliphorids on dead snails in Egypt (Hegazi et al. 1991). Woodcock et al. (2002) described the arthropod communities on *Arion* slugs. Krell (2004) reported that several dung beetle species of the genus *Onthophagus* from throughout Africa and in Borneo are attracted to the quinone defensive secretions of large millipedes, which they bury after the millipedes are dead. De Jong and Krell (2011) reared the calliphorid *Hemipyrellia fernandica* (Macquart) from a carcass of the scarabaeid *Heliocopris antenor* (Olivier) in Uganda. The muscid *Lispe binotata* Becker scavenges dead arthropods in forest streams in China (Vikhrev 2011). Tan and Corlett (2012) reported on the arthropods associated with scavenging of the carrion of five invertebrate species in Singapore, noting that 42 species of ants represented the majority of scavengers on this kind of carrion, with flies, cockroaches, and other arthropod taxa less commonly represented. A labidurid earwig, rather than ants, however, was the dominant scavenger of invertebrate carrion at night in grasslands. Buenaventura and Pape (2013) summarized the range of invertebrate carrion that can be used by the common sarcophagid genus *Peckia* in the New World tropics and subtropics, including snails, semiterrestrial crabs, and squid. Méndez and Pape (2002) showed that *Peckia gulo* (Fabricius) can successfully breed in dead crabs in Panamá, but apparently not in vertebrate carrion.

Special mention must be made of the arthropods that attack museum insect collections, particularly dermestid beetles. Although *Dermestes* colonies have been used beneficially by curators of vertebrate specimens, these little beetles (particularly *Anthrenus*, *Attagenus*, and *Trogoderma*) are the bane of many museum curators, as they can easily destroy insect specimens on pins. They also attack vertebrate taxidermy specimens, while a whole suite of stored product pests, including dermestids, can infest museum collections of any organic material.

4.4.4 Incidental Taxa

There are essentially three groups of incidental taxa: (1) those that actively live on the carcass, but do not utilize it except as shelter, in which case the carrion might as well have been a rock or a log; (2) a staging location for predaceous forays on suddenly abundant prey; and (3) those taxa that stochastically occur (i.e., happen to fly by the carrion).

These taxa can be just about any arthropod in the vicinity of a carcass. Several of the broad successional studies published in the literature have listed hundreds of taxa that were collected from carrion, most of which were considered to be incidental taxa (Reed 1958; Braack 1986; Goff et al. 1986; Pechal et al. 2011). These organisms were not attracted to the carcass *per se*, but were present on the surrounding vegetation, or the trap they were collected in was set up in a natural flyway. These taxa usually do not have much direct importance either within the carrion community or in applied forensic use of the community, but they have been used in forensic contexts—such as the grasshopper leg in a pants cuff that demonstrated that the body had been moved, or spider webs that can indicate a carcass has been present long enough for the spider to take up residence.

It is possible that some of these taxa that are considered to be incidental are, in fact, using the high-quality resources afforded by decomposing carrion. Baz et al. (2010) listed 114 species of normally phytophagous insects in the Hymenoptera, Lepidoptera, and Heteroptera that were collected in carrion-baited traps in Spain. They also provided an excellent review of the literature of phytophagous insects and their relations with carrion, citing examples from both tropical and temperate latitudes.

4.5 Conclusions

When analyzing the arthropod community on a carcass, it is essential to know what arthropod taxa to expect. And, although some members of the carrion fauna are rather predictable in their presence, others may be present or absent for no apparent reason. When certain taxa are present or absent, those facts can provide ecological clues to help answer many of the questions asked of the carrion arthropod community. To accurately predict the arthropod fauna on carrion, consideration must be given to the temporal and spatial constraints on the community, as well as the life histories of the arthropods themselves.

Acknowledgments

We would like to acknowledge Greg Dahlem (Northern Kentucky University) and Alejandro Ganzález Medina (Institute of Legal Medicine, Granada, Spain) for reviewing the section of this chapter on sarcophagid flies and mites, respectively. The book editors and Gail Anderson (Simon Fraser University) reviewed the chapter in its entirety. Also, we would like to thank AGM for the use of his mite photograph. We would especially like to thank Stephen A. Marshall (University of Guelph) for providing photographs of many carrion arthropods for use in this chapter. Most of these photos were taken from his excellent book, titled *Flies: The Natural History and Diversity of Diptera* (Firefly Books Ltd, Ontario, Canada, 2012).

REFERENCES

Adair, T.W. and B.C. Kondratieff. 1996. The occurrence of *Nitidula flavomaculata* (Coleoptera: Nitidulidae) on a human corpse. *ENT News* 107: 233–236.

Akre, R.D., A. Greene, J.F. MacDonald, P.J. Landolt, and H.G. Davis. 1981. *The Yellowjackets of America North of Mexico*. Agriculture Handbook No. 552, U.S. Department of Agriculture, Washington, D.C.

Aballay, F.H., G. Arriagada, G.E. Flores, and N.D. Centeno. 2013. An illustrated key to and diagnoses of the species of Histeridae (Coleoptera) associated with decaying carcasses in Argentina. *Zookeys* 261: 61–84.

Anderson, G.S. 2010. Factors that influence insect succession on carrion. In: Byrd, J.H. and J.L. Castner (eds.), *Forensic Entomology: The Utility of Arthropods in Legal Investigations*, 2nd edition. CRC Press, Boca Raton, FL, pp. 201–250.

Anderson, G.S. and S.L. VanLaerhoven. 1996. Initial studies on insect succession on carrion in southwestern British Columbia. *Journal of Forensic Sciences* 41: 617–625.

Anton, E., S. Niederegger, and R.G. Beutel. 2011. Beetles and flies collected on pig carrion in an experimental setting in Thuringia and their forensic implications. *Med Vet Entomol* 25: 353–364.

Archer, M.S. 2003. Annual variation in arrival and departure times of carrion insects at carcasses: Implications for succession studies in forensic entomology. *Aust J Zool* 51: 569–576.

Avila, F. and M.L. Goff. 1998. Arthropod succession patterns onto burnt carrion in two contrasting habitats in the Hawaiian Islands. *J Forensic Sci* 43: 581–586.

Axtell, R.C. 1964. Phoretic relationship of some common manure-inhabiting Macrochelidae (Acarina: Mesostigmata) to the house fly. *Ann Entomol Soc Am* 57: 584–587.

Bachmann, J. and T. Simmons. 2010. The influence of preburial insect access on the decomposition rate. *J Forensic Sci* 55: 893–900.

Baldridge, R.S., S.G. Wallace, and R. Kirkpatrick. 2006. Investigation of nocturnal oviposition by necrophilous flies in central Texas. *J Forensic Sci* 51: 125–126.

Baumgartner, D.L. and B. Greenberg. 1984. The genus *Chrysomya* (Diptera: Calliphoridae) in the New World. *J Med Entomol* 21: 105–113.

Baz, A., B. Cifrián, L.M. Díaz-Aranda, and D. Martín-Vega. 2007. The distribution of the adult blow-flies (Diptera: Calliphoridae) along an altitudinal gradient in central Spain. *Ann Soc Entomol France* 43: 289–296.

Baz, A., B. Cifrián, D. Martín-Vega, and M. Baena. 2010. Phytophagous insects captured in carrion-baited traps in central Spain. *Bull Insectol* 63: 21–30.

Beaver, R.A. 1969. Anthomyiid and muscid flies bred from snails. *Entomol Mon Mag* 105: 25–26.

Beaver, R.A. 1972. Ecological studies on Diptera breeding in dead snails. 1. Biology of the species found in *Cepaea nemoralis* (L.). *Entomologist* 105: 41–52.

Beaver, R.A. 1986. Biological studies of muscoid flies (Diptera) breeding in mollusk carrion in Southeast Asia. *Jpn J Sanit Zool* 37: 205–211.

Benecke, M. 2010. Cases of neglect involving entomological evidence. In: Byrd, J.H. and J.L. Castner (eds.), *Forensic Entomology: The Utility of Arthropods in Legal Investigations*, 2nd edition. CRC Press, Boca Raton, FL, pp. 627–635.

Bornemissza, G.F. 1957. An analysis of arthropod succession in carrion and the effects of its decomposition on the soil fauna. *Austral J Zool* 5: 1–12.

Braack, L.E.O. 1986. Arthropods associated with carcasses in the northern Kruger National Park. *S Afr J Wildl Res* 16: 91–98.

Braig, H.R. and M.A. Perotti. 2009. Carcasses and mites. *Exp Appl Acarol* 49: 45–84.

Brunke, A., J. Klimaszewski, and R.S. Anderson. 2012. Present taxonomic work on Staphylinidae (Coleoptera) in Canada: Progress against all odds. *Zookeys* 186: 1–5.

Buck, M. 1997. Sphaeroceridae (Diptera) reared from various types of carrion and other decaying substrates in southern Germany, including new faunistic data on some rarely collected species. *Eur J Entomol* 94: 137–151.

Buenaventura, E. and T. Pape. 2013. Revision of the New World genus *Peckia* Robineau-Desvoidy (Diptera: Sarcophagidae). *Zootaxa* 3622: 1–87.

Byrd, J.H. and J.L. Castner (eds.) 2010a. *Forensic Entomology: The Utility of Arthropods in Legal Investigations*. CRC Press, Inc., Boca Raton, FL.

Byrd, J.H. and J.L. Castner. 2010b. Insects of forensic importance. In: Byrd, J.H. and J.L. Castner (eds.), *Forensic Entomology: The Utility of Arthropods in Legal Investigations*, 2nd edition. CRC Press, Boca Raton, FL, pp. 39–126.

Camargo, J.M.F. and D.W. Roubik. 1991. Systematics and bionomics of the apoid obligate necrophages: The *Trigona hypogea* group (Hymenoptera: Apidae, Meliponinae). *Biol J Linn Soc* 44: 13–39.

Carroll, J.F. and J.J. Grasela. 1986. Occurrence of adult American dog tick, *Dermacentur variabilis* (Say), around small mammal traps and vertebrate carcasses. *Proc Entomol Soc Wash* 88: 77–82.

Catts, E.P. and M.L. Goff. 1992. Forensic entomology in criminal investigations. *Ann Rev Entomol* 37: 253–272.

Chapman, R.F. and J.H.P. Sankey. 1955. The larger invertebrate fauna of three rabbit carcasses. *J Anim Ecol* 24: 395–402.

Chin, H.C., M.A. Marwi, R. Hashim, N.A. Abdullah, C.C. Dhang, J. Jeffery, H. Kurahashi et al. 2009. Ants (Hymenoptera: Formicidae) associated with pig carcasses in Malaysia. *Asian Pac J Trop Biomed* 26: 106–109.

Cohen, D., M. Green, C. Block, R. Slepon, R. Ambar, S.S. Wasserman, and M.M. Levine. 1991. Reduction of transmission of shigellosis by control of houseflies (*Musca domestica*). *Lancet* 337: 993–997.

Colyer, C.N. 1954. The "coffin" fly, *Conicera tibialis* Schmitz (Dipt., Phoridae). *J Soc Br Entomol* 4: 203–206.

Costa, M. 1969. The association between mesostigmatic mites and coprid beetles. *Acaralogia* 11: 411–428.

Davis, J. and M.L. Goff. 2000. Decomposition patterns in terrestrial and intertidal habitats on Oahu Island and Coconut Island, Hawaii. *J Forensic Sci* 45: 836–842.

de Carvalho, C.J.B. 1989. Classificação de Muscidae (Diptera): uma proposta através da análise cladística. *Rev Bras Zool* 6: 627–648.

de Carvalho, C.J.B., M.S. Couri, A.C. Pont, D.M. Pamplona, and S.M. Lopes. 2005. A catalogue of the Muscidae (Diptera) of the neotropical region. *Zootaxa* 860: 1–282.

de Carvalho, C.J.B. and C.A. Mello-Patiu. 2008. Key to the adults of the most common forensic species of Diptera in South America. *Rev Bras Entomol* 52: 390–406.

De Jong, G.D. 1994. An annotated checklist of the Calliphoridae (Diptera) of Colorado, with notes on carrion associations and forensic importance. *J Kans Entomol Soc* 67: 378–385.

De Jong, G.D. 1995. Report of *Chrysomya megacephala* (Diptera: Calliphoridae) in northern New Mexico. *Entomol News* 106: 192.

De Jong, G.D. 2014. Field study on the attraction and development of insects on human meconium and breast-fed infant feces. *J Forensic Sci* 59: 1394–1396.

De Jong, G.D. and J.W. Chadwick. 1999. Decomposition and arthropod succession on exposed rabbit carrion during summer at high altitudes in Colorado, USA. *J Med Entomol* 36: 833–845.

De Jong, G.D. and W.W. Hoback. 2006. Effect of investigator disturbance in experimental forensic entomology: Succession and community composition. *Med Vet Entomol* 20: 248–258.

De Jong, G.D., W.W. Hoback, and L.G. Higley. 2011. Effect of investigator disturbance in experimental forensic entomology: Carcass biomass loss and temperature. *J Forensic Sci* 56: 143–149.

De Jong, G.D. and F.-T. Krell. 2011. Development of the blowfly *Hemipyrellia fernandica* (Macquart) (Diptera: Calliphoridae) on a carcass of the giant dung beetle *Heliocopris antenor* (Olivier) (Coleoptera: Scarabaeidae) in Uganda. *Proc Entomol Soc Wash* 113: 77–78.

Dekeirsschieter, J., F.J. Verheggen, E. Haubruge, and Y. Brostaux. 2011. Carrion beetles visiting pig carcasses during early spring in urban, forest and agricultural biotopes on western Europe. *J Insect Sci* 11: 73.

Dillon, L.C. 1997. Insect succession on carrion in three biogeoclimatic zones of British Columbia. MS thesis, Simon Fraser University, Burnaby, BC.

Disney, R.H.L. 1994. *Scuttle Flies: The Phoridae*. Chapman & Hall, London, UK.

Disney, R. and T. Munk. 2004. Potential use of Braconidae (Hymenoptera) in forensic cases. *Med Vet Entomol* 18: 442–444.

Draber-Mońko, A., T. Malewski, J. Pomorski, M. Łoś, and P. Ślipiński. 2009. On the morphology and mitochondrial DNA barcoding of the flesh fly *Sarcophaga* (*Liopyga*) *argyrostoma* (Robineau-Desvoidy, 1830) (Diptera: Sarcophagidae)—An important species in forensic entomology. *Ann Zool* 59: 465–493.

Erzinçlioğlu, Y.Z. 1980. On the role of *Trichocera* larvae (Diptera, Trichoceridae) in the decomposition of carrion in winter. *Naturalist* 105: 133–134.

Faria, L.D.B., L. Orsi, L.A. Trinca, and W.A.C. Godoy. 1999. Larval predation by *Chrysomya albiceps* on *Cochliomyia macellaria*, *Chrysomya megacephala* and *Chrysomya putoria*. *Entomol Exp Appl* 90: 149–155.

Ferrar, P. 1987. *A Guide to the Breeding Habits and Immature Stages of Diptera Cyclorrhapha*. Parts 1 and 2, Entomonograph 8, Brill and Scandinavian Science Press, Leiden, Belgium.

Ferreira de Almeida, M.A., A. Pires do Prado, and C.J. Geden. 2002. Influence of temperature on development time and longevity of *Tachinaephagus zealandicus* (Hymenoptera: Encyrtidae), and effects of nutrition and emergence order on longevity. *Biol Control* 31: 375–380.

Floate, K., B. Khan, and G. Gibson. 1999. Hymenopterous parasitoids of filth fly (Diptera: Muscidae) pupae in cattle feedlots. *Can Entomol* 131: 347–362.

Forbes, S.L. and I. Dadour. 2010. The soil environment and forensic entomology. In: Byrd, J.H. and J.L. Castner (eds.), *Forensic Entomology: The Utility of Arthropods in Legal Investigations*, 2nd edition. CRC Press, Boca Raton, FL, pp. 407–426.

France, D.L., T.J. Griffin, J.G. Swanburg, J.W. Lindemann, G.C. Davenport, V. Trammell, C.T. Armbrust et al. 1992. A multidisciplinary approach to the detection of clandestine graves. *J Forensic Sci* 37: 1445–1458.

Frederickx, C., J. Dekeirsschieter, F.J. Verheggen, and E. Haubruge. 2013. The community of Hymenoptera parasitizing necrophagous Diptera in an urban biotope. *J Insect Sci* 13: 32.

Fuller, M.E. 1934. The insect inhabitants of carrion: A study in animal ecology. *CSIRO Bull* 82: 5–62.

Gill, G.D. and B.V. Peterson. 1987. Heleomyzidae. In: McAlpine, J.F. (ed.). *Manual of Nearctic Diptera*, Volume 2, Agriculture Canada, Ottawa, Canada, pp. 973–980.

Goff, M.L. 1991. Comparison of insect species associated with decomposing human remains recovered inside dwellings and outdoors on the island of Oahu, Hawaii. *J Forensic Sci* 36: 748–753.

Goff, M.L., M. Early, C.B. Odom, and K. Tullis. 1986. A preliminary checklist of arthropods associated with exposed carrion in the Hawaiian Islands. *Proc Hawaii Entomol Soc* 26: 53–57.

González Medina, A., L.G. Herrera, M.A. Perotti, and G.J. Ríos. 2013b. Occurrence of Poecilochirus austroasiaticus (Acari: Parasitidae) in forensic autopsies and its application on postmortem interval estimation. *Exp Appl Acarol* 59: 297–305.

Goodbrod, J.R. and M.L. Goff. 1990. Effects of larval population density on rates of development and interactions between two species of *Chrysomya* (Diptera: Calliphoridae) in laboratory culture. *J Med Entomol* 27: 338–343.

Graczyk, T., R. Knight, R.H. Gilman, and M.R. Cranfield. 2001. The role of non-biting flies in the epidemiology of human infectious diseases. *Microbes Infect* 3: 231–235.

Grassberger, M. and C. Frank. 2004. Initial study of arthropod succession on pig carrion in a central European urban habitat. *J Med Entomol* 41: 511–523.

Greenberg, B. 1971. *Flies and Disease*, vol. 2. Princeton University Press, Princeton, NJ.

Greenberg, B. 1991. Flies as forensic indicators. *J Med Entomol* 28: 565–577.

Greenberg, B. and J.C. Kunich. 2002. *Entomology and the Law*. Cambridge University Press, Cambridge, UK.

Greenberg, B. and J.D. Wells. 1998. Forensic use of *Megaselia abdita* and *M. scalaris* (Phoridae: Diptera): Case studies, development rates, and egg structure. *J Med Entomol* 35: 205–209.

Groenewold, S. and M. Fonds. 2000. Effects on benthic scavengers of discards and damaged benthos produced by the beam-trawl fishery in the southern North Sea. *ICES J Marine Sci* 57: 1395–1406.

Gunn, A. and J. Bird. 2011. The ability of the blowflies *Calliphora vomitoria* (Linnaeus), *Calliphora vicina* (Rob-Desvoidy) and *Lucilia sericata* (Meigen) (Diptera: Calliphoridae) and the muscid flies *Muscina stabulans* (Fallén) and *Muscina prolapsa* (Harris) (Diptera: Muscidae) to colonise buried remains. *Forensic Sci Int* 207: 198–204.

Haskell, N.H. and R.E. Williams (eds.). 2008. *Entomology and Death: A Procedural Guide*, 2nd edition. Forensic Entomology Partners, Clemson, SC.

Hegazi, E.M., M.A. Shaaban, and E. Sabry. 1991. Carrion insects of the Egyptian western desert. *J Med Entomol* 28: 734–739.

Hewadikaram, K.A. and M.L. Goff. 1991. Effect of carcass size on rate of decomposition and arthropod succession patterns. *Am J Forensic Med Pathol* 12: 235–240.

Honda, J.Y., A. Brundage, C. Happy, S.C. Kelly, and J. Melinek. 2008. New records of carrion feeding insects collected on human remains. *Pan-Pac Entomol* 84: 29–32.

Horenstein, M.B. and A.X. Linhares. 2011. Seasonal composition and temporal succession of necrophagous and predator beetles on pig carrion in central Argentina. *Med Vet Entomol* 25: 395–401.

James, M.T. 1947. *The Flies that Cause Myiasis in Man*. Miscellaneous Publication 631, U.S. Department of Agriculture, Washington, D.C.

Jeannel, R. 1936. Monographie des Catopidae. *Mem Mus Natl Hist Nat* 1: 1–433.

Jewiss-Gaines, A., S.A. Marshall, and T.L. Whitworth. 2012. Cluster flies (Calliphoridae: Polleniinae: Pollenia) of North America. *Can J Arthropod Ident* 19: 22.

Johnson, M.D. 1975. Seasonal and microseral variations in the insect populations on carrion. *Am Midl Nat* 93: 79–90.

Kennedy, D. 2002. *Living Things We Love to Hate*. 10th Anniversary Edition, Whitecap Books, Vancouver, Canada.

Kentner, E. and B. Streit. 1990. Temporal distribution and habitat preference of congeneric insect species found at rat carrion. *Pedobiologia* 34: 347–359.

Kingsolver, J.M. 1987. Dermestid beetles (Dermestidae, Coleoptera). In: Gorham, J.R. (ed.), *Insect and Mite Pests in Food: An Illustrated Key*. Agriculture Handbook No. 655, U.S. Department of Agriculture, Washington, D.C., pp. 115–136.

Kneidel, K.A. 1984a. Influence of carcass taxon and size on species composition of carrion-breeding Diptera. *Am Midl Nat* 111: 57–63.

Kneidel, K.A. 1984b. Carrion as an attractant to the American dog tick, *Dermacentur variabilis* (Say). *J N Y Entomol Soc* 92: 405–406.

Kočárek, P. 2003. Decomposition and Coleoptera succession on exposed carrion of small mammal in Opava, the Czech Republic. *Eur J Soil Biol* 39: 31–45.

Kozol, A.J., M.P. Scott, and J.F.A. Traniello. 1988. The American burying beetle, *Nicrophorus americanus*: Studies on the natural history of a declining species. *Psyche* 95: 167–176.

Krantz, G.W. and B.D. Ainscough. 1990. Acarina: Mesostigmata (Gamasida). In: Dillon, D.L. (ed.), *Soil Biology Guide*. John Wiley & Sons, New York, NY, pp. 583–665.

Krell, F.-T. 2004. East African dung beetles (Scarabaeidae) attracted by defensive secretions of millipedes. *J E Afr Nat Hist* 93: 69–73.

Kurahashi, H. 1978. The Oriental latrine fly: *Chrysomya megacephala* (Fabricius) newly recorded from Ghana and Senegal, West Africa. *Kontyû* 46: 432.

Kuusela, S. and I. Hanski. 1982. The structure of carrion fly communities: The size and the type of carrion. *Hol Ecol* 5: 337–348.

Leclercq, M. 1999. Entomologie et medicine légale: Importance des Phorides (Diptères) sur cadavres humains. *Ann Soc Entomol France* 35 (Suppl.): 566–568.

Lindgren, N.K., S.R. Bucheli, A.D. Archambeault, and J.A. Bytheway. 2010. Exclusion of forensically important flies due to burying behavior by the red imported fire ant (*Solenopsis invicta*) in southeast Texas. *Forensic Sci Int* 204: e1–e3.

Ma, Y.-K., C. Hui, and J.-X. Min. 2000. A preliminary study on the constitution and succession of insect community on pig carcasses in Hangzhou District. *Acta Entomol Sin* 43: 388–393.

Martín-Vega, D. 2011. Skipping clues: Forensic importance of the family Piophilidae (Diptera). *Forensic Sci Int* 212: 1–5.

Martín-Vega, D. and A. Baz. 2013. Sex-biased captures of sarcosaprophagous Diptera in carrion-baited traps. *J Insect Sci* 13: 14.

Martín-Vega, D., A. Gómez-Gómez, and A. Baz. 2011. The "coffin fly" *Conicera tibialis* (Diptera: Phoridae) breeding on buried human remains after a postmortem interval of 18 years. *J Forensic Sci* 56: 1654–1656.

Martínez-Sánchez, A., M.A. Marcos-García, and S. Rojo. 2001. First collection of *Chrysomya megacephala* (Fabr.) in Europe (Diptera: Calliphoridae). *Pan-Pac Entomol* 77: 240–243.

McNemee, R.B., Jr, W.J. Sames, IV, and F.A. Maloney, Jr. 2003. Occurrence of *Dermacentor variabilis* (Acari: Ixodidae) around a porcupine (Rodentia: Erthethizontidae) carcass at Camp Ridley, Minnesota. *J Med Entomol* 40: 108–111.

Mégnin, P. 1894. *La faune des cadavres. Application de l'entomologie à la médecine légale.* G. Masson and Gauthier-Villars et Fils, Paris.

Meiklejohn, K.A., J.F. Wallman, and M. Dowton. 2013. DNA barcoding identifies all immature life stages of a forensically important flesh fly (Diptera: Sarcophagidae). *J Forensic Sci* 58: 184–187.

Méndez, J. and T. Pape. 2002. Biology and immature states of *Peckia gulo* (Fabricius, 1805) (Diptera: Sarcophagidae). *Stud Dipterol* 9: 371–374.

Merritt, R.W. 1976. A review of the food habits of the insect fauna inhabiting cattle droppings in north central California. *Pan-Pac Entomol* 52: 13–22.

Merritt, R.W. and J.R. Anderson. 1977. The effects of different pasture and rangeland ecosystems on the annual dynamics of insects in cattle droppings. *Hilgardia* 45: 31–71.

Merritt, R.W. and M.E. Benbow. 2009. Entomology. In: Jamieson, A. and A. Moenssens (eds.), *Wiley Encyclopedia of Forensic Science.* John Wiley & Sons, Ltd, New Jersey, pp. 1–12.

Merritt, R.W., R. Snider, J.L. De Jong, M.E. Benbow, R.K. Kimbirauskas, and R.E. Kolar. 2007. Collembola of the grave: A cold case history involving arthropods 28 years after death. *J Forensic Sci* 52: 1359–1361.

Merritt, R.W. and J.R. Wallace. 2010. The role of aquatic insects in forensic investigations. In: Byrd, J.H. and J.L. Castner (eds.), *Forensic Entomology: The Utility of Arthropods in Legal Investigations*, 2nd edition. CRC Press, Boca Raton, FL, pp. 271–319.

Metcalf, R.L. and R.A. Metcalf. 1993. *Destructive and Useful Insects: Their Habits and Control*, 5th edition. McGraw Hill, Inc., New York, NY.

Michaud, J.-P., K.G. Schoenly, and G. Moreau. 2012. Sampling flies or sampling flaws? Experimental design and inference strength in forensic entomology. *J Med Entomol* 49: 1–10.

Midgley, J.M., I.J. Collett, and M.H. Villet. 2012. The distribution, habitat, diet and forensic significance of the scarab *Frankenbergerius forcipatus* (Harold, 1881) (Coleoptera: Scarabaeidae). *Afr Invert* 53: 745–749.

Milne, L.J. and M.J. Milne. 1976. The social behavior of burying beetles. *Sci Am* 235(2): 84–89.

Moretti, T.C., E. Giannotti, P.J. Thyssen, D.R. Solis, and W.A.C. Godoy. 2011. Bait and habitat preferences, and temporal variability of social wasps (Hymenoptera: Vespidae) attracted to vertebrate carrion. *J Med Entomol* 48: 1069–1075.

Mumcuoglu, K.Y., N. Gallili, A. Reshef, P. Brauner, and H. Grant. 2004. Use of human lice in forensic entomology. *J Med Entomol* 41: 803–906.

Myers, H., J.K. Tomberlin, B. Lambert, and D. Kattes. 2008. Development of the black soldier fly on dairy manure. *Environ Entomol* 37: 11–15.

Niogret, J., J.P. Lumaret, and M. Bertrand. 2006. Review of the phoretic association between coprophilous insects and macrochelid mites (Acari: Mesostigmata) in France. *Elytron* 20: 99–121.

Norrbom, A.L. 1983. Four acalyptrate Diptera reared from dead horseshoe crabs. *Entomol News* 94: 117–121.

Norris, K.R. 1997. Supposed Australian record of the New Zealand blowfly *Calliphora quadrimaculata* (Swederus) (Diptera: Calliphoridae). *Entomologist* 116: 37–39.

Nuorteva, P., 1970. Histerid beetles as predator of blowflies (Diptera, Calliphoridae) in Finland. *Ann Zool Fennici* 7: 195–198.

Nuorteva, P. 1977. Sarcosaprophagous insects as forensic indicators. In: Tedeschi, C.G., W.G. Eckert, and L.G. Tedeschi (eds.), *Forensic Medicine, a Study in Trauma and Environmental Hazards*. Vol. II. Physical Trauma. Saunders, Philadelphia, PA, pp. 1072–1095.

Nuorteva, P., M. Isokoski, and K. Laiho. 1967. Studies on the possibilities of using blowflies (Dipt.) as medico-legal indicators in Finland. I. Report of four indoor cases from the city of Helsinki. *Ann Entomol Fennici* 33: 217–225.

Olea, M.S., M.J. Dantur Juri, and N. Centeno. 2011. First report of *Chrysomya megacephala* (Diptera: Calliphoridae) in northwestern Argentina. *Fla Entomol* 94: 345–346.

Parsons, C.T. 1943. A revision of Nearctic Nitidulidae (Coleoptera). *Bull Mus Comp Zool* 92: 1–278.

Pape, T. 1987. The Sarcophagidae (Diptera) of Fennoscandia and Denmark. *Fauna Entomol Scand* 19: 1–203.

Pastula, E.C. and R.W. Merritt. 2013. Insect arrival pattern and succession on buried carrion in Michigan. *J Med Entomol* 50: 432–439.

Payne, J.A. 1965. A summer carrion study of the baby pig *Sus scrofa* Linneaus. *Ecology* 46: 592–602.

Payne, J.A. and E.W. King. 1970. Coleoptera associated with pig carrion. *Entomol Mon Mag* 105: 224–232.

Payne, J.A. E.W. King, and G. Reinhart. 1968. Arthropod succession and decomposition of buried pigs. *Nature* 219: 1180–1181.

Payne, J.A. and W.R.M. Mason. 1971. Hymenoptera associated with pig carrion. *Proc Entomol Soc Wash* 73: 132–141.

Pechal, J.L., M.E. Benbow, and J.K. Tomberlin. 2011. *Merope tuber* (Mecoptera: Meropidae) collected in association with carrion in Greene County, Ohio, USA: An infrequent collection of a rare species. *The American Midland Naturalist* 166(2): 453–457.

Pechal, J.L., T.L. Crippen, M.E. Benbow, A.M. Tarone, and J.K. Tomberlin. 2014. The potential use of bacterial community succession in forensics as described by high throughput metagenomic sequencing. *Int J Legal Med* 128: 193–205.

Pechal, J.L., T.L. Crippen, A.M. Tarone, A.J. Lewis, J.K. Tomberlin, and M.E. Benbow. 2013. Microbial community functional change during vertebrate carrion decomposition. *PLoS ONE* 8: e79035.

Peck, S.B. 1986. *Nicrophorus* (Silphidae) can use large carcasses for reproduction. *Coleopt Bull* 40: 44.

Peck, S.B. 1990. Insecta: Coleoptera: Silphidae and the associated families Agyrtidae and Leiodidae. In: Dindal, D.L. (ed.). *Soil Biology Guide*, Wiley, New York, NY, pp. 1113–1136.

Peck, S.B. and R.S. Anderson. 1982. The distribution and ecology of the alpine-tundra carrion beetle *Thanatophilus coloradensis* (Wickham) in North America (Coleoptera: Silphidae). *Coleopt Bull* 36: 112–115.

Peris, S.V. 1987. La invasión de las especies de *Chrysomya* en América (Dipt. Calliphoridae). *Graellsia* 43: 205–210.

Perotti, M.A. 2009. Mégnin re-analyzed: The case of the newborn baby girl, Paris, 1878. *Exp Appl Acarol* 49: 37–44.

Perotti, M.A. and H.R. Braig. 2010. Acarology and criminolegal investigations: The human acarofauna during life and death, chapter 21. In: Byrd, J.H. and J.L. Castner (eds.), *Forensic Entomology: The Utility of Arthropods in Legal Investigations*, 2nd edition. CRC Press, Boca Raton, FL, pp. 637–649.

Perotti, M.A., H.R. Braig, and M.L. Goff. 2010. Phoretic mites and carcasses: Acari transported by organisms associated with animal and human decomposition. In: Amendt, J., C.P. Campobasso, M. Grassberger, and M.L. Goff (eds.), *Current Concepts in Forensic Entomology*. Springer, New York, NY, pp. 69–92.

Pont, A.C. and R. Meier. 2002. The Sepsidae (Diptera) of Europe. *Fauna Entomol Scand* 37: 1–231.

Porter, S.D. 1998. Biology and behavior of *Pseudacteon* decapitating flies (Diptera: Phoridae) that parasitize *Solenopsis* fire ants (Hymenoptera: Formicidae). *Fla Entomol* 81: 292–309.

Prins, A.J. 1979. The discovery of the Oriental latrine fly, *Chrysomyia megacephala* (Fab.) along the south-western coast of South Africa. *Ann S Afr Mus* 78: 39–47.

Prins, A.J. 1984. Morphological and biological notes on some South African arthropods associated with decaying organic matter. II: The predatory families Carabidae, Hydrophilidae, Histeridae, Staphylinidae, and Silphidae (Coleoptera). *Ann S Afr Mus* 94: 203–304.

Proctor, H.C. 2009. Can freshwater mites act as forensic tools? *Exp Appl Acarol* 48: 161–169.

Pukowski, E. 1933. Ökologische untersuchungen an *Necrophorus* F. *Z Morphol Okol Tiere* 27: 518–586.

Putman, R.J. 1978. Decomposition of small mammal carrion in temperate systems 2. Flow of energy and organic matter from a carcase during decomposition. *Oikos* 31: 58–68.

Reed, H.B., Jr. 1958. A study of dog carcass communities in Tennessee, with special reference to the insects. *Am Midl Nat* 59: 213–245.

Rodrigues-Guimarães, R., R.R. Guimarães, R.W. de Carvalho, A.J. Mayhé-Nunes, and G.E. Moya-Borja. 2006. Registo de *Aphaereta laeviuscula* (Spinola) (Hymenoptera: Braconidae) e *Nasonia vitripennis* (Walker) (Hymenoptera: Pteromalidae) como parastóide de *Cochliomyia hominivorax* (Coquerel) (Diptera: Calliphoridae), no Estado de Rio de Janeiro. *Neotrop Entomol* 35: 402–407.

Rodriguez, W.C. and W.M. Bass. 1985. Decomposition of buried bodies and methods that may aid in their location. *J Forensic Sci* 30: 836–852.

Rognes, K. 1997. The Calliphoridae (Blowflies) (Diptera: Oestroidea) are not a monophyletic group. *Cladistics* 13: 27–66.

Roháček, J. 2001. *World Catalog of Sphaeroceridae (Diptera)*. Slezské Zemské Muzeum, Opava, Czech Republic.

Rosati, J.Y. and S.L. VanLaerhoven. 2007. New record of *Chrysomya rufifacies* (Diptera: Calliphoridae) in Canada: Predicted range expansion and potential effects on native species. *Can Entomol* 139: 670–677.

Roubik, D.W. 1982. Obligate necrophagy in a social bee. *Science* 217: 1059–1060.

Saloña, M.I., M.L. Moraza, M. Carles-Tolrá, V. Iraola, P. Bahillo, T. Yélamos, R. Onterelo, and R. Alcaraz. 2010. Searching the soil: Forensic importance of edaphic fauna after the removal of a corpse. *J Forensic Sci* 55: 1652–1655.

Sanjean, J. 1957. Taxonomic studies of *Sarcophaga* larvae of New York, with notes on the adults. *Mem Cornell Univ Agric Exp Stat* 349: 1–115.

Satou, A., T. Nisimura, and H. Numata. 2000. Reproductive competition between the burying beetle *Nicrophorus quadripunctatus* without phoretic mites and the blow fly *Chrysomya pinguis*. *Entomol Sci* 3: 265–268.

Schoenly, K. 1983. Microclimate observations and diel activities of certain carrion arthropods in the Chihuahuan Desert. *J N Y Entomol Soc* 91: 342–347.

Schoenly, K. and W. Reid. 1987. Dynamics of heterotrophic succession in carrion arthropod assemblages: Discrete seres or a continuum of change? *Oecologia* 73: 192–202.

Schoenly, K. G. and W. Reid. 1989. Dynamics of heterotrophic succession in carrion revisited. *Oecologia* 79: 140–142.

Scott, M.P., W.-J. Lee, and E.D. van der Reijden. 2007. The frequency and fitness consequences of communal breeding in a natural population of burying beetles: A test of reproductive skew. *Ecol Entomol* 32: 651–661.

Seastedt, T.R., L. Mameli, and K. Gridley. 1981. Arthropod use of invertebrate carrion. *Am Midl Nat* 105: 124–129.

Shahid, S.A., R.D. Hall, N.H. Haskell, and R.W. Merritt. 2000. *Chrysomya rufifacies* (Macquart) (Diptera: Calliphoridae) established in the vicinity of Knoxville, Tennessee, USA. *J Forensic Sci* 45: 896–897.

Shean, B., L. Messinger, and M. Papworth. 1993. Observations of differential decomposition on sun exposed versus shaded pig carrion in coastal Washington State. *J Forensic Sci* 38: 938–949.

Sheppard, D.C. 1983. House fly and lesser house fly control utilizing the black soldier fly in manure management systems for caged laying hens. *Environ Entomol* 12: 1439–1442.

Sheppard, D.C. 1992. Large-scale feed production from animal manure with a non-pest native fly. *Food Insects Newsl* 5(2): 3.

Sheppard, D.C., G.L. Newton, S.A. Thompson, and S. Savage. 1994. A value added manure management system using the black soldier fly. *Biores Technol* 50: 275–279.

Shi, Y.-W., X.-S. Liu, H.-Y. Wang, and R.-J. Zhang. 2009. Seasonality of insect succession on exposed rabbit carrion in Guangzhou, China. *Insect Sci* 16: 425–439.

Shubeck, P.P. 1985. Orientation of carrion beetles to carrion buried under shallow layers of sand (Coleoptera: Silphidae). *Entomol News* 96: 163–166.

Shubeck, P.P. and D.L. Blank. 1982. Carrion beetle attraction to buried fetal pig carrion (Coleoptera: Silphidae). *Coleopt Bull* 36: 240–245.

Šifner, F. 2008. A catalogue of the Scathophagidae (Diptera) of the Palaearctic Region, with notes on their taxonomy and faunistics. *Acta Entomol Mus Nat Prague* 48: 111–196.

Sikes, D.S., R.B. Madge, and A.F. Newton. 2002. A catalog of the Nicrophorinae (Coleoptera: Silphidae) of the world. *Zootaxa* 65: 1–304.

Skidmore, P. 1985. *The Biology of the Muscidae of the World*. Junk, Dordrecht, The Netherlands.

Smith, K.V.G. 1986. *A Manual of Forensic Entomology*. British Museum (Natural History), Cornstock Publishing Associates, London, UK.

Solarz, K. 2009. Indoor mites and forensic acarology. *Exp Appl Acarol* 49: 135–142.

Sperling, F.A.H., G.S. Anderson, and D.A. Hickey. 1994. A DNA-based approach to the identification of insect species used for postmortem interval estimation. *J Forensic Sci* 39: 418–427.

Springett, B.P. 1968. Aspects of the relationships between burying beetles, *Necrophorus* spp., and the mite *Poecilochirus necrophori*. *J Anim Ecol* 37: 417–424.

Stamper, T., G.A. Dahlem, C. Cookman, and R.W. Debry. 2012. Phylogenetic relationships of flesh flies in the sub-family Sarcophaginae based on three mtDNA fragments (Diptera: Sarcophagidae). *Syst Entomol* 38: 35–44.

Stoker, R.L., W.E. Grant, and S.B. Vinson. 1995. *Solenopsis invicta* (Hymenoptera: Formicidae) effect on invertebrate decomposers of carrion in central Texas. *Environ Entomol* 24: 817–822.

Suzuki, S. 2000. Carrion burial by *Nicrophorus vespilloides* (Coleoptera: Silphidae) prevents fly infestation. *Entomol Sci* 3: 269–272.

Suzuki, S. 2001. Suppression of fungal development on carcasses by the burying beetle *Nicrophorus quadripunctatus* (Coleoptera: Silphidae). *Entomol Sci* 4: 403–405.

Szalanski, A.L., J.W. Austin, J.A. McKern, T. McCoy, C.D. Steelman, and D.M. Miller. 2006. Time course analysis of bed bug, *Cimex lectularius* L., (Hemiptera: Cimicidae) blood meals with the use of polymerase chain reaction. *J Agric Urban Entomol* 23: 237–241.

Szpila, K., M.J.R. Hall, K.L. Sukontason, and T.I. Tantawi. 2013. Morphology and identification of first instars of the European and Mediterranean blowflies of forensic importance. Part I. Chrysomyinae. *Med Vet Entomol* 27: 181–193.

Szpila, K., J.G. Voss, and T. Pape. 2010. A new dipteran forensic indicator in buried bodies. *Med Vet Entomol* 24: 278–283.

Tabor, K.L., C.C. Brewster, and R.D. Fell. 2004. Analysis of the successional patterns of insects on carrion in southwest Virginia. *J Med Entomol* 41: 785–795.

Tabor, K.L., R.D. Fell, and C.C. Brewster. 2005. Insect fauna visiting carrion in Southwest Virginia. *Forensic Sci Int* 150: 73–80.

Tan, C.K.W. and R.T. Corlett. 2012. Scavenging of dead invertebrates along an urbanization gradient in Singapore. *Insect Conserv Divers* 5: 138–145.

Thomson, A.J. and D.M. Davies. 1973a. The biology of *Pollenia rudis*, the cluster fly (Diptera: Calliphoridae). Host location by first-instar larvae. *Can Entomol* 105: 335–341.

Thomson, A.J. and D.M. Davies. 1973b. The biology of *Pollenia rudis*, the cluster fly (Diptera: Calliphoridae). Larval feeding behaviour and host specificity. *Can Entomol* 105: 985–990.

Tomberlin, J.K., J.H. Byrd, J.R. Wallace, and M.E. Benbow. 2012. Assessment of decomposition studies indicates need for standardized and repeatable research methods in forensic entomology. *Forensic Res* 3: 147.

Tomberlin, J.K. and D.C. Sheppard. 2001. Lekking behavior of the black soldier fly (Diptera: Stratiomyidae). *Fla Entomol* 84: 729–730.

Tomberlin, J.K., D.C. Sheppard, and J.A. Joyce. 2005. Colonization of pig carrion by the black soldier fly (Diptera: Stratiomyidae). *J Forensic Sci* 50: 152–153.

Vairo, K.P.E., C.A. de Mello-Patiu, and C.J.B. de Carvalho. 2011. Pictorial identification key for species of Sarcophagidae (Diptera) of potential forensic importance in southern Brazil. *Rev Bras Entomol* 55: 333–347.

VanLaerhoven, S.L. 2010. Ecological theory and its application in forensic entomology. In: Byrd, J.H. and J.L. Castner (eds.), *Forensic Entomology: The Utility of Arthropods in Legal Investigations*, 2nd edition. CRC Press, Boca Raton, FL, pp. 493–517.

VanLaerhoven, S.L. and G. Anderson. 1999. Insect succession on buried carrion in two biogeoclimatic zones of British Columbia. *J Forensic Sci* 44: 32–41.

Vaurie, P. 1955. A revision of the genus *Trox* in North America (Coleoptera, Scarabaeidae). *Bull Am Mus Nat Hist* 106: 1–89.

Vaurie, P. 1962. A revision of the genus *Trox* in South America (Coleoptera, Scarabaeidae). *Bull Am Mus Nat Hist* 124: 1–167.

Verves, Y.G. 2002. An annotated check-list of Calliphoridae (Diptera) of the Russian Far East. *Far East Entomol* 116: 1–14.

Vikhrev, N. 2008. New data on the distribution and biology of the invasive species *Hydrotaea aenescens* (Wiedemann, 1830) (Diptera, Muscidae). *Zookeys* 4: 47–53.

Vikhrev, N. 2011. Review of the Palaearctic members of the *Lispe tentaculata* species-group (Diptera, Muscidae): Revised key, synonymy and notes on ecology. *Zookeys* 84: 59–70.

Villet, M.H. 2011. African carrion ecosystems and their insect communities in relation to forensic entomology. *Pest Technol* 5: 1–15.

Villet, M.H. and J. Amendt. 2011. Advances in entomological methods for death time estimation, chapter 11. In: Turk, E.E. (ed.), *Forensic Pathology Reviews* 6. Springer Science + Business Media LLC, Philadelphia, PA, pp. 213–237.

von Hoermann, C., J. Ruther, S. Reibe, B. Madea, and M. Ayasse. 2011. The importance of carcass volatiles as attractants for the hide beetle *Dermestes maculatus* (De Geer). *Forensic Sci Int* 212: 173–179.

Voss, S.C., D.F. Cook, and I.R. Dadour. 2011. Decomposition and insect succession of clothed and unclothed carcasses in Western Australia. *Forensic Sci Int* 211: 67–75.

Voss, S.C., H. Spafford, and I.R. Dadour. 2009. Hymenopteran parasitoids of forensic importance: Host associations, seasonality, and prevalence of parasitoids of carrion flies in Western Australia. *J Med Entomol* 46: 1210–1219.

Voss, S.C., H. Spafford, and I.R. Dadour. 2010. Temperature-dependent development of the parasitoid *Tachinaephagus zealandicus* on five forensically important carrion fly species. *Med Vet Entomol* 24: 189–198.

Wells, J.D. and B. Greenberg. 1994. Effect of the red imported fire ant (Hymenoptera: Formicidae) and carcass type on the daily occurrence of postfeeding carrion-fly larvae (Diptera: Calliphoridae, Sarcophagidae). *J Med Entomol* 31: 171–174.

Wells, J.D. and H. Kurahashi. 1997. *Chrysomya megacephala* (Fabr.) is more resistant to attack by *Chrysomya rufifacies* (Macquart) in laboratory arena than is *Cochliomyia macellaria* (Fabr.) (Diptera: Calliphoridae). *Pan-Pac Entomol* 73: 16–20.

Wells, J.D. and L.R. LaMotte. 2001. Estimating the postmortem interval. In: Byrd, J.H. and J.L. Castner (eds.)., *Forensic Entomology: The Utility of Arthropods in Legal Investigations*, 2nd edition. CRC Press, Boca Raton, FL, pp. 263–285.

Whitworth, T. 2006. Keys to genera and species of blow flies of America north of Mexico. *Proc Entomol Soc Wash* 108: 689–725.

Whitworth, T. 2010. Keys to the blow fly species (Diptera: Calliphoridae) of America, north of Mexico. In: Byrd, J.H. and J.L. Castner (eds.), *Forensic Entomology: The Utility of Arthropods in Legal Investigations*, 2nd edition. CRC Press, Boca Raton, FL, pp. 582–625.

Wilson, D.S. 1983. The effect of population structure on the evolution of mutualism: A field test involving burying beetles and their phoretic mites. *Am Nat* 121: 851–870.

Woodcock, B.A., A.D. Watt, and S.R. Leather. 2002. Aggregation, habitat quality, and coexistence: A case study on carrion fly communities in slug cadavers. *J Anim Ecol* 71: 131–140.

Woodley, N.E. 2001. A world catalog of Stratiomyidae (Insecta: Diptera). *Myia* 11: 1–475.

Zehner, R., J. Amendt, S. Schütt, J. Sauer, R. Krettek, and D. Povolný. 2004. Genetic identification of forensically important flesh flies (Diptera: Sarcophagidae). *Int J Legal Med* 118: 245–247.

Zurawski, K.N., M.E. Benbow, J.R. Miller, and R.W. Merritt. 2009. An examination of nocturnal blow fly (Diptera: Calliphoridae) oviposition on pig carcasses in mid-Michigan. *J Med Entomol* 46: 671–679.

5

Carrion Effects on Belowground Communities and Consequences for Soil Processes

Michael S. Strickland and Kyle Wickings[*]

CONTENTS

5.1 Introduction: The Soil Environment

Soil is both a chemically and biologically complex environment of great ecological importance. It is the primary store of terrestrial carbon (C) and is one of the most actively cycling pools of carbon (Schlesinger 1997; Manzoni and Porporato 2009). Additionally, it is a reservoir for plant nutrients, which fuel primary production (Epstein and Bloom 2005). Yet, not all soils are equal regarding their chemical and physical properties. For example, the underlying parent material of a soil can greatly impact its properties (e.g., pH, clay type, and soil texture). Additionally, soil mineralogy influences cation exchange capacity, which is an important determinant of traits such as soil fertility (Evangelou and Phillips 2005).

Humans can also have a marked impact on soil properties, and many soil characteristics are sensitive to land-use and management intensity. For example, conventional cultivation leads to a depletion of organic matter and soil fertility, requiring inorganic inputs of fertilizers to maintain plant biomass (Lal 2008). Intensive cultivation can also lead to long-term shifts in plant litter decay rates and patterns of litter chemistry over the course of decomposition (Grandy and Robertson 2007; Wickings et al. 2011).

Differences in the physical and chemical attributes of soil represent the prime determinants of the size and structure of belowground food webs. Soil microbial and invertebrate communities are heavily shaped by a legacy of resource quality, nutrient availability, soil pH, texture, and porosity (Petersen and Luxton 1982; Behan-Pelletier 1999; Callaham et al. 2006; Fierer and Jackson 2006; Salamon et al. 2006; Lauber et al. 2008, 2009; Strickland et al. 2009b; Wickings and Grandy 2011). Yet, the relationship between soil organisms and soil properties is bidirectional and involves numerous feedbacks. Although existing soil traits determine microbial and invertebrate community structure, differences in community structure can have significant functional consequences for soil attributes and processes. For example, change in soil microbial communities and the presence or absence of microarthropods can influence both rates of carbon cycling and decomposition (Strickland et al. 2009a, 2012; Wickings and Grandy 2011). Structural changes in soil food webs can also impact the chemistry of organic matter inputs, which may have long-term implications for soil organic matter (SOM) formation and stability (Wickings et al.

---

[*] The authors contributed equally to this work.

2012). Soil organisms also shape soil structure. For example, macroinvertebrates, including earthworms, ants, and termites, alter soil porosity, organic matter distribution, and soil aggregates (Lavelle et al. 1997).

Organic matter inputs serve as the basal resource for the soil food web, and it is well recognized that the quality of these inputs can affect the size and structure of soil biological communities (Haynes 1999). Recent studies have also shown that the quality of organic matter inputs is an important constraint on the function of soil microbes and invertebrates (Smith and Bradford 2003; Strickland et al. 2009b; Yang and Chen 2009). Carrion represents a very high-quality input to soil and may have substantial effects on soil processes that are mediated by soil biota. Historically, the potential importance of these inputs has been overlooked as a very ephemeral resource pulse, with no long-term consequences for soils or ecosystems (but see Putnam 1978). Yet, there is a growing awareness that high-quality inputs such as carrion may play an important role in long-term soil processes such as the formation and stability of SOM (Barton et al. 2013; Cotrufo et al. 2013).

5.2 Carrion as a Resource Input

The bulk of inputs entering soil is plant-derived, consisting of both leaf litter and root-derived inputs (Swift et al. 1979; Lavelle et al. 1993; Aerts 1997; van Hees et al. 2005). These inputs contribute the most to SOM formation and fuel the bulk of biological activity in soil (Aerts 1997; van Hees et al. 2005; Pollierer et al. 2007). The chemical quality of plant-derived resources is, under similar climate, a major determinant of their decomposition rates. Often, this is graded using lignin: nitrogen or carbon: nitrogen ratios, with a greater ratio either indicative of a more recalcitrant input or slower breakdown rate.

Animal-derived inputs, however, have often been overlooked. The rationale is that such inputs decompose very rapidly, represent a relatively small input to any system, and likely have little-to-no long-term effect on soil processes (Parmenter and MacMahon 2009). Although this may be true, there is little-to-no empirical evidence supporting this view and recent studies are beginning to challenge this idea (Barton et al. 2013; Macdonald et al. 2014). Additionally, although such inputs are rapidly decomposed, this does not indicate that they are unimportant or that their impacts are ephemeral. In fact, recent research indicates that the more labile an input is the greater its long-term influence on soil properties such as soil organic carbon (Yang and Janssen 2002; Strickland et al. 2010; Bradford et al. 2013; Cotrufo et al. 2013).

This idea is termed the microbial efficiency-matrix stabilization (MEMS) framework (Cotrufo et al. 2013) and is based on the microbial assimilation efficiency of a given compound. The more efficient the assimilation, the more likely the resource will be converted into microbial biomass and/or metabolites (Yang and Janssen 2002; Cotrufo et al. 2013). These microbial products are then more apt to lead to the formation of stable SOM (Yang and Janssen 2002). Alternatively, such labile compounds may actually stimulate microbial activity and subsequently the loss of SOM (Dalenberg and Jager 1981; Fontaine and Barot 2005; Bradford et al. 2008). This is commonly referred to as priming (Dalenberg and Jager 1981). The possibility that priming occurs is often dependent on the amount of labile compounds entering a system. If labile compounds represent a small input relative to more recalcitrant inputs, then priming is likely to occur (Fontaine and Barot 2005; Bradford et al. 2008). Alternatively, if labile compounds represent a relatively high input, then preferential substrate utilization (akin to MEMS) will occur (Wu et al. 1993; Fontaine and Barot 2005). Often, the balance between these two processes is dependent on both the quality and availability of resources at a given site.

These recent hypotheses related to SOM formation might aid us in better understanding the influence carrion inputs potentially have on soil ecosystems. For instance, the quality of most carrion inputs is high when compared to most plant-derived inputs (Figure 5.1). These inputs are stoichiometrically similar to microbial biomass (Figure 5.2) and should lead to high biological assimilation efficiency (Guggenberger et al. 1999; Sterner and Elser 2002). The amount of input and the context of the surrounding resource quality are then likely to strike the balance between priming and preferential substrate utilization

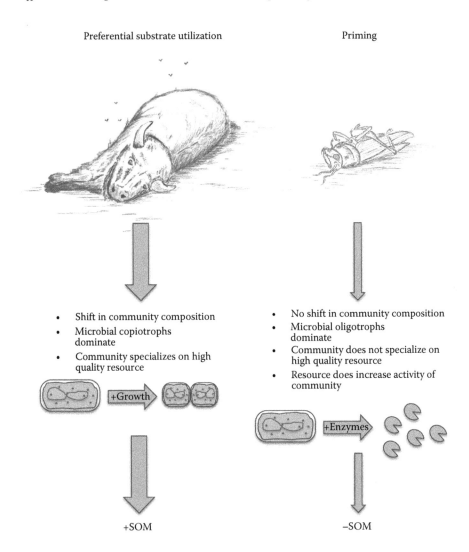

FIGURE 5.1   Effects of carrion inputs on soil communities and ecosystem processes. Large quantities of carrion inputs may lead to a shift in soil microbial community composition toward copiotrophic organisms. These organisms specialize on labile substrates and decompose little SOM. As these organisms grow and die, their biomass forms the precursors for stable SOM, leading to an increase in SOM. In contrast, small quantities of carrion inputs are not likely to lead to a change in the microbial community but may alleviate the energy demands on the current, oligotrophic community. This may then lead to increased extracellular enzyme production by that community, which induces a priming effect and subsequent loss of SOM.

(Bradford et al. 2008). For instance, a bison carcass is likely to represent a large labile input, relative to the dominant recalcitrant plant litter, and leads to a marked shift in the soil community (discussed subsequently) toward organisms that primarily consume that carcass. This occurrence could result in preferential substrate utilization and potentially an increase in SOM. In contrast, small carrion inputs, such as arthropod carcasses, may induce a priming effect in the community. Dispersal of the carrion by scavengers (see Chapter 6 on Scavengers) or mass die-offs (see Chapter 13 on Ecosystems) may alter these outcomes. That is, spreading a large carcass across a landscape may lead to a shift from preferential substrate utilization to priming.

The impacts that carrion will have on soil processes are likely to vary over time (Putnam 1978; Parmenter and MacMahon 2009; Metcalf et al. 2013; Macdonald et al. 2014). Initially, changes to the soil environment should be relatively minor, with low carrion inputs entering the soil during the fresh

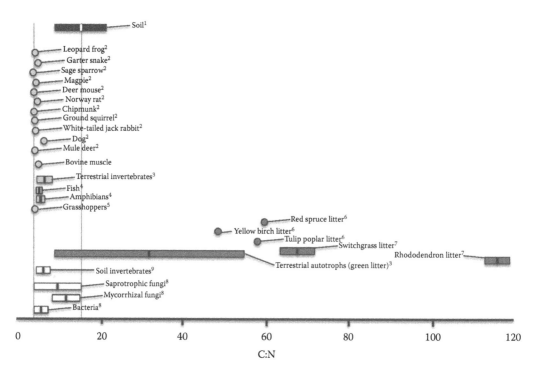

FIGURE 5.2 The carbon:nitrogen ratios of decomposer organisms (white symbols), plants (dark gray symbols), animals (light gray symbols), and soil (black symbol). Bars represent the mean (solid line) and standard deviation. Circles represent single point estimates for a given organism. Vertical dotted lines indicate the range of C:N ratios associated with decomposer organisms. The data were sourced from the following publications with corresponding numbers: [1](Fierer, N. et al. 2009. *Ecology Letters* 12: 1238–1249), [2](Parmenter, R.R. and J.A. Macmahon. 2009. *Ecological Monographs* 79: 637–661; *Note*: %C was assumed to be 45%), [3](Elser, J.J. et al. 2000. *Nature* 408: 578–580), [4](Vanni, M.J. et al. 2002. *Ecology Letters* 5: 285–293), [5](Hawlena, D. and O.J. Schmitz. 2010. *American Naturalist* 176: 537–556), [6](Keiser, A.D. et al. 2011. *Biogeosciences* 8: 1477–1486), [7](Strickland, M.S. et al. 2009a. *Ecology* 90: 441–451, Strickland, M.S. et al. 2009b. *Functional Ecology* 23: 627–636), [8](Strickland, M.S. and J. Rousk. 2010. *Soil Biology and Biochemistry* 42: 1385–1395), [9](Ferris, H., R.C. Venette, and S.S. Lau. 1997. *Soil Biology and Biochemistry* 29: 1183–1194).

and bloat stages of decomposition, although colonization of the carrion by soil organisms at the soil–carrion interface is likely to accelerate input rates.

Once active decay (and the process of putrefaction) begins, a flush of resources enters the soil, leading to the greatest alterations in soil processes and the soil community. It is likely, during this stage, when the balance between priming and preferential substrate utilization is struck. It is also during this time when changes in the soil physiochemical environment are the most dramatic, with rapid increases in soil moisture and pH (due to carrion fluids) (Metcalf et al. 2013). These changes probably continue through to the dry stage of decomposition, with an overall decrease in soil community activity. Although there is little empirical evidence for such temporal effects, Metcalf et al. (2013) demonstrated that soil microbial community composition and soil pH tended to follow this pattern for mouse carcasses and Vass et al. (1992) found that soil solution characteristics also followed this pattern for human cadavers.

The effects of carrion inputs on the soil, though, are likely to be long lasting, with either gains or losses of SOM and most likely increased soil nutrients (Towne 2000; Macdonald et al. 2014). The long-term consequences of these inputs are relatively unknown (Barton et al. 2013), but research has demonstrated that carrion inputs can alter future processes, such as leaf litter decomposition, in soil ecosystems (Hawlena et al. 2012).

5.3 Carrion Inputs: Soil Microbial Community Dynamics

Microbes (see Chapter 3 on Microbial Communities) constitute the dominant heterotrophic biomass in soil, and they are responsible for the bulk of elemental cycling in terrestrial ecosystems (Zak et al. 2006; Fierer et al. 2009). Microbial communities vary with regard to the potential influence that they have on these processes as well as how they respond to alterations in resource inputs (Strickland et al. 2009a, b). Although carrion inputs represent a short-term pulse of a high-quality resource, this input potentially leads to shifts in the function of soil microbial communities.

Typically, the expectation is that soil microbial communities in bulk soil are dominated by oligotrophic organisms (Fierer et al. 2007). Oligotrophs are organisms that tend to have slow growth rates and high substrate affinities, meaning that they are adapted to low-resource environments (Koch 2001). But, these communities are punctuated by copiotrophic organisms around high-resource sites (e.g., roots). Copiotrophs are organisms that grow rapidly when resources are plentiful and have low substrate affinity (Koch 2001). This dichotomy is akin to the well-known ecological definition of life-history traits (i.e., r- vs. K-strategists) (Fierer et al. 2007). Although this differentiation only highlights two extremes in life history, it does provide a useful metric for testing hypotheses regarding the functional implications of carrion inputs.

For instance, carrion inputs of significant quantity are likely to lead to a more copiotrophic community. Metcalf et al. (2013) found evidence that this occurs with a decrease in the relative abundance of Acidobacteria, a primarily oligotrophic phylum in soil, and an increase in Proteobacteria, a primarily copiotrophic phylum in soil, under decomposing mouse carcasses. Under this scenario, microbial activity is likely to increase during the course of carrion decomposition and then decrease as the carrion substrate is depleted (Carter et al. 2007, 2008; Pechal et al. 2013). That is, microbial activity with time potentially elicits a unimodal response (Figure 5.3). Additionally, the shift in functional groups within the microbial community may follow a pattern in which copiotrophs increase in dominance up to the peak of activity and subsequently decline, being replaced by more oligotrophic organisms (Figure 5.3). Such changes in the functional attributes of the microbial community have been proposed during litter decomposition but not for carrion decomposition (Moorhead and Sinsabaugh 2006). However, the expectations should remain the same but at a much shorter time scale.

Shifts between copiotrophic and oligotrophic life histories have implications for soil ecosystem functioning (Fierer et al. 2007). Oligotrophs tend to degrade more recalcitrant SOM but are often energy (i.e., C) limited in this endeavor. Thus, carrion inputs may alleviate this energy demand and lead to increased SOM decomposition. This occurrence is likely to happen during the initial or later stages of carrion decomposition if carrion input is of relatively large quantity and high quality and when copiotrophs are less abundant. If the carrion input is initially of low quality, then copiotrophs may never increase in abundance and oligotrophs will utilize the carrion input to subsidize SOM decomposition, potentially leading to declines in SOM. Of course, gradients in the carrion effect on SOM may occur. For example, Hawlena et al. (2012) added grasshopper carrion, which varied in quality (C:N of 4.00 and 3.85), to soils and then subsequently determined how this input altered leaf litter decomposition. In both instances, an increase in litter decomposition was found, but this effect was much greater for the higher quality grasshopper carrion. This suggests that fine-scale gradation in the quality of carrion inputs, brought about by factors such as stress, has marked impacts on ecosystem function.

Another metric used to assess the function of soil microbial communities is the ratio of fungi to bacteria. In many ways, this metric resembles the copiotroph–oligotroph dichotomy, with bacteria largely described as the former and fungi the latter (Guggenberger et al. 1999; Strickland and Rousk 2010). However, one key distinction is the potential difference in stoichiometry and growth form between fungi and bacteria (Strickland and Rousk 2010). Based solely on stoichiometry, bacteria are likely to dominate the decomposition of most carrion but fungi may be important decomposers of many invertebrate species, due to greater stoichiometric similarity and may also dominate during the latter stages of decomposition (Figure 5.2). Additionally, most large vertebrate carrion inputs are associated with an increase in soil pH, which is known to favor bacteria over fungi (Rousk et al. 2010). However, although carrion inputs are apt to favor the bacterial community, fungi may play an important role as well. Because of

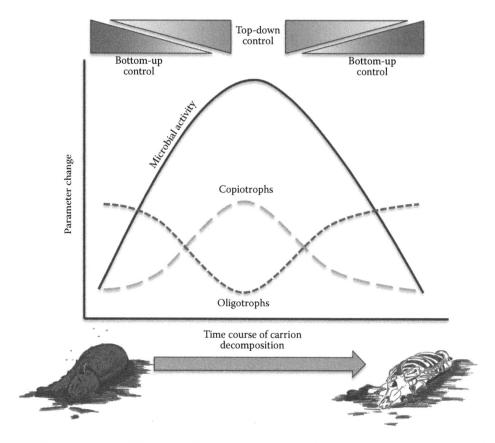

FIGURE 5.3 (**See color insert.**) Change in soil community parameters during the time course of carrion decomposition. Microbial biomass and activity are likely to increase until active decay at which point it will begin to decrease. During this time, the typically oligotrophic soil community will give way to a more copitrophic one until carrion decomposition begins to slow. Finally, regulation of the microbial community will also change during carrion decomposition, with bottom-up regulation being more important during the early and later stages and top-down regulation being more important during active carrion decay.

the mycelial growth form, fungi may relocate nutrients derived from carrion to locations throughout the mycelial network potentially stimulating decomposition across a much larger area (Strickland and Rousk 2010). In the case of mycorrhizal fungi, translocation is likely to stimulate plant growth (Sagara 1995). The overall zone of influence that carrion inputs have on soil microbial communities and processes is unknown; however, soil fungi may play an important role in extending the carrion footprint beyond its direct contact with soil.

Another important consideration is the relationship between the extant microbial community and the carrion-derived community. Soil is a very different environment from carrion itself and, although there are broadly similar groups of microbes living in both environments, it seems unlikely that the carrion community could survive in soil for an extended period of time. However, this does not mean that the carrion community does not influence the soil community. In fact, during active decay, a huge influx of carrion-derived microbes enters the soil. These organisms will possibly survive for a period of time, but will ultimately be replaced by extant soil microbes. It is probable that the carrion resource, potentially due to the introduction of novel organisms or genes, may have a marked influence on the function of the soil community in the future. In fact, Metcalf et al. (2013) found that some taxa of bacteria (e.g., Rhizobiales) increase in abundance in response to carcass inputs.

Resource history of soil microbial communities has been shown to have a marked influence on community function (Ayres et al. 2009; Strickland et al. 2009a, b). One well-recognized example is the home-field advantage hypothesis, which states that soil microbes are better able to degrade a resource that they have had historical exposure to (Gholz et al. 2000; Ayres et al. 2009). This has been well documented in the leaf litter decomposition literature (Ayres et al. 2009). In fact, Carter and Tibbett (2008) observed a similar occurrence during the decomposition of skeletal muscle tissue. In the context of carrion decomposition, this may mean that systems that are frequently exposed to carrion inputs may be geared toward their decomposition. Examples of such systems may be grasslands, in which annual die-offs of grasshoppers are common. Such systems likely maintain a community capable of rapidly utilizing carrion pulses. If this is the case, then there may be implications to the overall function of the system. For example, Strickland et al. (2009a, b) found that systems that tended to be more fertile also tended to decompose labile compared with recalcitrant leaf litter more rapidly. If carrion improves site fertility, then this ephemeral resource may have long-term consequences for microbial function.

Soil microbial communities also interact with other soil organisms and form the base of the decomposer food web. Typically, these organisms are expected to be bottom-up regulated with the overall low quality of plant-derived inputs expected to control total microbial biomass (Hairston et al. 1960; Allison 2006). However, carrion inputs may remove this bottom-up control, making top-down control (i.e., microbivory) a more important mechanism. If this is the case, then microbivores, such as microarthropods, nematodes, and protists, should flourish (Bonkowski 2004). Increased predation will stimulate microbial turnover and mobilization of soil nutrients, making these nutrients available for plant uptake and/or loss from the system (Bonkowski 2004). Top-down control would be strongest during active decay of the carrion and bottom-up control would be strongest during the fresh and dry stages of decomposition.

Although little is known regarding the effect carrion has on the soil microbial community, current theory related to resource acquisition strategies of the microbial community may provide some of the foundational hypotheses to begin understanding these effects. Once carrion was thought as an insignificant, ephemeral input to soil (Swift et al. 1979), but now it is being realized that the legacy of vertebrate carrion inputs can markedly impact microbial community function (Hawlena et al. 2012). Future research in this realm should not just employ short-term measures of a changing microbial community but also assess the long-term functional implications of carrion inputs.

5.4 Carrion Inputs: Soil Invertebrate Community Dynamics

Invertebrates (see Chapter 4 on Arthropod Communities) play diverse and important roles in the decomposition of organic matter, including carrion. Invertebrate contributions to carrion decay have been recognized for centuries, and, today, invertebrates are a critical component in modern forensic diagnostics. When it comes to soil invertebrates, however, knowledge of the invertebrate–carrion relationship is scarce. Past research emphasizes the direct colonization of the carrion resource, and only a handful of studies have focussed on how carrion inputs affect invertebrates in bulk soil (Braig and Perotti 2009).

Although it is difficult to make generalizations about them as a whole, soil invertebrate communities may exhibit some predictable responses to carrion inputs in the landscape. Soil invertebrates, such as microarthropods, are generally thought to be nutrient-limited due to the low nutritional quality of dead plant matter serving as their basal resource (Hairston et al. 1960; Allison 2006). Carrion inputs have drastic effects on soil nutrient availability and may serve as resource islands (Carter et al. 2007), ultimately dampening the negative consequences of nutrient limitation on soil invertebrate communities. Although nutrient effects on soil invertebrates have not been studied extensively, some fairly consistent responses can be found within the existing literature. In particular, detritivorous and microbivorous soil invertebrates have been shown to decrease in density in response to nutrient additions, whereas predator and herbivore densities often increase (Throop and Lerdau 2004; Gan et al. 2013; Grandy et al. 2013). The positive effect of nutrient addition on herbivores is well recognized and is typically attributed to improved plant nutrition (Mattson 1980). Similarly, positive predator responses to nutrient amendment may be driven by shifts in prey abundance. The opposing response of decomposers and microbivores to nutrient addition, however, may be related to the importance of microbes in the interaction between arthropods and detritus. Soil invertebrates commonly rely on microbes for deriving nutrition from low-quality plant tissues. They accomplish this via many routes, the most common of which is microbivory, or the direct feeding on microbial cells. Soil invertebrates also benefit from soil microbes indirectly, by feeding on plant detritus that has undergone extensive microbial decay. Whether direct or indirect, invertebrates benefit from soil microbes via the same mechanism: harnessing the ability of soil microbes to access and degrade recalcitrant organic matter. Thus, changes in decomposer and microbivore density in response to nutrient addition reflect changes in microbial biomass and/or enzyme production. Past studies have shown that microbial biomass and activity in soil are sensitive to carrion inputs (Benninger et al. 2008), suggesting that carrion may have strong indirect effects on microbivorous invertebrate populations.

Thus, taken solely as a nutrient pulse, carrion may have substantial effects on soil arthropod communities. However, carrion differs from other nutrient inputs in many ways, and its impacts on soil invertebrates may not be as predictable as those derived from inorganic nutrient additions, which have been the focus of past work. For example, in addition to the mineralized nutrients released from carrion during decay, many labile organic molecules are also released into soil (Macdonald et al. 2014). Additionally, compared with plant detritus, particulate organic matter from carrion may be directly assimilated by soil invertebrates. Past studies support this assertion, showing that some soil invertebrates will feed on cadaver tissue (Rusek 1998; Von Berg et al. 2012). This behavior has been attributed to the overall nutrient quality of carrion combined with the high degree of opportunistic feeding that occurs in soil food webs and suggests that carrion can potentially serve as a significant food source for soil invertebrates.

Carrion may also have many effects on soil invertebrate communities related to shifts in climatic and physical factors. For example, fresh carrion has high water content and can serve as a significant moisture source for soil food webs. Large cadavers may also provide a shading effect on soil, decreasing water loss, and attracting cryptic invertebrate taxa that prefer shaded, moist environments. In addition to moisture, carrion inputs can cause a significant increase in the temperature of nearby soil during the initial stages of decomposition (Wilson et al. 2007). Soil tunneling by carrion-inhabiting fly larvae seeking pupation sites also changes the physical pore structure of soil, which can have marked effects on soil invertebrate communities (Nielsen et al. 2008). These changes in physical and climatic factors due to carrion inputs may lead to increased soil invertebrate abundance and activity.

Carrion can have many other effects on soil chemistry. Past studies have shown changes in soil pH below decomposing carrion corresponding with the activity of carrion-feeding larvae and a flush of nutrients from cadaver to soil, and more recent work suggests that the impact of carrion on soil pH varies over the course of decay (Carter et al. 2008). This shift in soil pH is likely driven by changes in the form of available nitrogen over the course of decay; however, its magnitude may vary with soil type. Soil invertebrates typically exhibit a negative response to increasing soil pH; however, the response varies based on the type of substrate causing the change (e.g., fertilizer, wood ash, and insect waste) and can be site-dependent (Liiri et al. 2002). Carrion also introduces many other compounds to soil during its decay, including cadaverine, putrescine, various acids, and phenolic compounds. Although few studies have explored relationships between cadaver-derived compounds and soil invertebrates in the field, it is recognized that these, and other related compounds, can suppress the density and activity of invertebrates in soil (Neuhauser and Hartenstein 1978; Katase et al. 2009; Asplund and Wardle 2013; Dekeirsschieter et al. 2013).

Shifts in pH and the introduction of repellant or toxic compounds in soil beneath decomposing carrion are assumed to have a negative impact on soil invertebrate communities, and past studies have observed that during active decay the soil fauna are substantially reduced (Bornemissza 1957). This finding, however, is not always consistent, and, in a recent review, Braig and Perotti (2009) highlight studies showing an enrichment of soil invertebrate communities under decomposing cadavers. Additional studies are needed to better understand these conflicting reports.

The functional significance of shifts in soil invertebrate communities is still poorly understood among soil ecologists. Hence, predicting how carrion-induced changes in soil invertebrate communities will influence belowground processes is a difficult task. Still, past studies assessing the influence of substrate quality on soil invertebrate function may prove insightful. Osler and Sommerkorn (2007) provide a framework for predicting the impact of soil invertebrates on carbon and nitrogen cycling based on stoichiometric differences between microbivorous invertebrates and their prey. Invertebrates with high assimilation efficiency (e.g., bacteria feeders) are predicted to contribute greatest to nitrogen availability, whereas those with low assimilation efficiency (e.g., detritivores and fungivores) should contribute greatest to the availability of dissolved organic matter. Although this framework remains to be tested in the field, past studies confirm that resource quality constrains the influence that soil invertebrates exert over soil processes such as decomposition (Smith and Bradford 2003; Yang and Chen 2009). Ultimately, changes in microbial communities (e.g., between fungal and bacterial dominance) in response to carrion inputs may modify the role of invertebrates in soil carbon and nutrient cycling. Invertebrate contributions to soil processes under decomposing carrion may also be highly sensitive to successional changes in compounds introduced to soil during decay.

Finally, it is worth considering the potential direct contribution of soil invertebrates as a carrion input. Although invertebrate biomass is dwarfed by microbial biomass (Fierer et al. 2009), soil invertebrate carrion constitutes a substantial organic matter input to soil. In their comprehensive paper, Petersen and Luxton (1982) report total soil invertebrate biomasses from different habitats ranging from roughly 2 to 8 g dry wt. m^{-2}. In extreme cases, total invertebrate biomass can reach over 20 g dry wt. m^{-2} (Paoletti and Bressan 1996; Lavelle et al. 1997). Within this range, soil macrofauna (large earthworms, millipedes, centipedes, predatory beetles, fly larvae, spiders, snails, and ants) comprised anywhere from 5% to 80% of all biomasses, whereas soil mesofauna (mites, collembolans, and enchytraeid worms) comprised anywhere from 2% to 62%. Given their low C:N ratio (Figure 5.2), and the fact that many taxa have high turnover rates, soil invertebrates may represent a significant carrion input (Barton et al. 2013).

It is clear that carrion inputs to soil can influence the biomass and composition of soil invertebrates. However, to date, the effects of carrion on soil invertebrate communities and their role in soil processes have received limited attention. Although the responses of soil invertebrates to carrion in previous studies have been mixed, recent work suggests that carrion-induced changes in soil invertebrate communities may persist well beyond the physical input of carrion itself (Salona et al. 2010). Improving our understanding of soil invertebrate responses to carrion inputs will require interdisciplinary studies of soil microbe and invertebrate interactions during, and well after, carrion decay. These studies must also identify the underlying chemical and physical drivers of biological responses across different decay stages.

5.5 Conclusions: Do Carrion Inputs Affect Soil Processes?

Do carrion inputs influence soil processes? The answer to that question is likely yes. In the short term, carrion effects can be quite dramatic: from a rapid increase in soil pH to increased available soil nitrogen (Benninger et al. 2008). However, whether carrion inputs can have marked influence on long-term soil processes and/or at the ecosystem scale is a topic that has only recently begun to be discussed in depth (Barton et al. 2013).

Depending on what is considered long term, there is evidence that carrion effects can have lasting impacts on soil properties for months (Benninger et al. 2008; Macdonald et al. 2014). Additionally, changes in the availability of carbon due to carrion inputs may also have long-term impacts on SOM. For instance, priming may elicit a rapid loss in soil carbon that may take years to decades to regain. The possibility of this and the balance between SOM formation and loss due to carrion inputs have yet to be tested.

Additionally, the ecosystem level implications of carrion inputs as they relate to soil processes are not well known (see Chapter 13 on Ecosystems). The marked increase in nutrients associated with carrion can ultimately stimulate primary production and a shift in plant community composition (Yang 2004). Additionally, carrion inputs, depending on quality and quantity, are likely to accelerate litter decomposition (Hawlena et al. 2012). Such impacts on processes may simply equate to point disturbances (i.e., carrion decomposition island) in the case of organisms for which death is a spatially and temporally random event (Carter et al. 2007). However, the death of other organisms may equate to a pulse of more evenly distributed resources and may have more widespread ecosystem consequences (Hawlena et al. 2012). Yet, even the former of these two types of carrion inputs may contribute to the heterogeneity in soil properties and process.

In the future, research needs to address the potential ecosystem scale consequences of carrion inputs on soil processes. It must also determine how the quantity, quality, and distribution of carrion inputs influence soil properties and communities. Although carrion inputs have often been thought to be of little significance, there is a growing realization that this may not be the case. Furthermore, contemporary theory in soil ecology, while not focussed explicitly on carrion, also suggests that carrion inputs may play a significant role in structuring soil communities and the processes that they drive, especially legacy effects. All that is left to do is explore this possibility.

REFERENCES

Aerts, R. 1997. Climate, leaf litter chemistry and leaf litter decomposition in terrestrial ecosystems: A triangular relationship. *Oikos* 79: 439–449.

Allison, S.D. 2006. Brown ground: A soil carbon analogue for the green world hypothesis? *The American Naturalist* 167: 619–627.

Asplund, J. and D.A. Wardle. 2013. The impact of secondary compounds and functional charateristics on lichen palatability and decomposition. *Journal of Ecology* 101: 689–700.

Ayres, E., H. Steltzer, B.L. Simmons, R.T. Simpson, J.M. Steinweg, M.D. Wallenstein, N. Mellor, W.J. Parton, J.C. Moore, and D.H. Wall. 2009. Home-field advantage accelerates leaf litter decomposition in forests. *Soil Biology and Biochemistry* 41: 606–610.

Barton, P.S., S.A. Cunningham, D.B. Lindenmayer, and A.D. Manning. 2013. The role of carrion in maintaining biodiversity and ecological processes in terrestrial ecosystems. *Oecologia* 171: 761–772.

Behan-Pelletier, V.M. 1999. Oribatid mite biodiversity in agroecosystems: Role for bioindication. *Agriculture Ecosystems and Environment* 74: 411–423.

Benninger, L.A., D.O. Carter, and S.L. Forbes. 2008. The biochemical alteration of soil beneath a decomposing carcass. *Forensic Science International* 180: 70–75.

Bonkowski, M. 2004. Protozoa and plant growth: The microbial loop in soil revisited. *New Phytologist* 162: 617–631.

Bornemissza, G.F. 1957. An analysis of arthropod succession in carrion and the effect of its decomposition on the soil fauna. *Australian Journal of Zoology* 5: 1–13.

Bradford, M.A., N. Fierer, and J.F. Reynolds. 2008. Soil carbon stocks in experimental mesocosms are dependent on the rate of labile carbon, nitrogen and phosphorus inputs to soils. *Functional Ecology* 22: 964–974.

Bradford, M.A., A.D. Keiser, C.A. Davies, C.A. Mersmann, and M.S. Strickland. 2013. Empirical evidence that soil carbon formation from plant inputs is positively related to microbial growth. *Biogeochemistry* 113: 271–281.

Braig, H.R. and M.A. Perotti. 2009. Carcases and mites. *Experimental and Applied Acarology* 49: 45–84.

Callaham, M.A., D.D. Richter, D.C. Coleman, and M. Hofmockel. 2006. Long-term land-use effects on soil invertebrate communities in southern piedmont soils, USA. *European Journal of Soil Biology* 42: S150–S156.

Carter, D.O. and M. Tibbett. 2008. Does repeated burial of skeletal muscle tissue (*Ovis aries*) in soil affect subsequent decomposition? *Applied Soil Ecology* 40: 529–535.

Carter, D.O., D. Yellowlees, and M. Tibbett. 2007. Cadaver decomposition in terrestrial ecosystems. *Naturwissenschaften* 94: 12–24.

Carter, D.O., D. Yellowlees, and M. Tibbett. 2008. Temperature affects microbial decomposition of cadavers (*Rattus rattus*) in contrasting soils. *Applied Soil Ecology* 40: 129–137.

Cotrufo, M.F., M.D. Wallenstein, C.M. Boot, K. Denef, and E. Paul. 2013. The microbial efficiency-matrix stabilization (mems) framework integrates plant litter decomposition with soil organic matter stabilization: Do labile plant inputs form stable soil organic matter? *Global Change Biology* 19: 988–995.

Dalenberg, J.W. and G. Jager. 1981. Priming effect of small glucose additions to c-14-labeled soil. *Soil Biology and Biochemistry* 13: 219–223.

Dekeirsschieter, J., C. Frederickx, G. Lognay, Y. Brostaux, F.J. Verheggen, and E. Haubruge. 2013. Electrophysiological and behavioral responses of *Thanatophilus* sinuatus fabricius (Coleoptera: Silphidae) to selected cadaveric volatile organic compounds. *Journal of Forensic Sciences* 58: 917–923.

Elser, J.J., W.F. Fagan, R.F. Denno, D.R. Dobberfuhl, A. Folarin, A. Huberty, S. Interlandi et al. 2000. Nutritional constraints in terrestrial and freshwater food webs. *Nature* 408: 578–580.

Epstein, E. and A.J. Bloom. 2005. *Mineral Nutrition of Plants: Principles and Perspectives*. Sunderland, MA, USA: Sinauer Associates, Inc.

Evangelou, V. and R. Phillips. 2005. Cation exchange in soils. In: *Chemical Processes in Soils*, A. Tabatabai, and D. Sparks (eds.). Madison, WI, USA: Soil Science Society of America, pp. 343–410.

Ferris, H., R.C. Venette, and S.S. Lau. 1997. Population energetics of bacterial-feeding nematodes: Carbon and nitrogen budgets. *Soil Biology and Biochemistry* 29: 1183–1194.

Fierer, N., M.A. Bradford, and R.B. Jackson. 2007. Toward an ecological classification of soil bacteria. *Ecology* 88: 1354–1364.

Fierer, N. and R.B. Jackson. 2006. The diversity and biogeography of soil bacterial communities. *Proceedings of the National Academy of Sciences* 103: 626–631.

Fierer, N., M.S. Strickland, D. Liptzin, M.A. Bradford, and C.C. Cleveland. 2009. Global patterns in belowground communities. *Ecology Letters* 12: 1238–1249.

Fontaine, S. and S. Barot. 2005. Size and functional diversity of microbe populations control plant persistence and long-term soil carbon accumulation. *Ecology Letters* 8: 1075–1087.

Gan, H., D.R. Zak, and M.D. Hunter. 2013. Chronic nitrogen deposition alters the structure and function of detrital food webs in a northern hardwood ecosystem. *Ecological Applications* 23: 1311–1321.

Gholz, H.L., D.A. Wedin, S.M. Smitherman, M.E. Harmon, and W.J. Parton. 2000. Long-term dynamics of pine and hardwood litter in contrasting environments: Toward a global model of decomposition. *Global Change Biology* 6: 751–765.

Grandy, A.S. and G.P. Robertson. 2007. Land-use intensity effects on soil organic carbon accumulation rates and mechanisms. *Ecosystems* 10: 58–73.

Grandy, A.S., D.S. Salam, K. Wickings, M.D. McDaniel, S.W. Culman, and S.S. Snapp. 2013. Soil respiration and litter decomposition responses to nitrogen fertilization rate in no-till corn systems. *Agriculture Ecosystems and Environment* 179: 35–40.

Guggenberger, G., S.D. Frey, J. Six, K. Paustian, and E.T. Elliott. 1999. Bacterial and fungal cell-wall residues in conventional and no-tillage agroecosystems. *Soil Science Society of America Journal* 63: 1188–1198.

Hairston, N., F. Smith, and L. Slobodkin. 1960. Community structure, population control, and competition. *The American Naturalist* 94: 421–425.

Hawlena, D. and O.J. Schmitz. 2010. Physiological stress as a fundamental mechanism linking predation to ecosystem functioning. *The American Naturalist* 176: 537–556.

Hawlena, D., M.S. Strickland, M.A. Bradford, and O.J. Schmitz. 2012. Fear of predation slows plant-litter decomposition. *Science* 336: 1434–1438.

Haynes, R.J. 1999. Size and activity of the soil microbial biomass under grass and arable management. *Biology and Fertility of Soils* 30: 210–216.

Katase, M., C. Kubo, S. Ushio, E. Ootsuka, T. Takeuchi, and T. Mizukubo. 2009. Nematicidal activity of volatile fatty acids generated from wheat bran in reductive soil disinfestation. *Nematological Research* 39: 53–62.

Keiser, A.D., M.S. Strickland, N. Fierer, and M.A. Bradford. 2011. The effect of resource history on the functioning of soil microbial communities is maintained across time. *Biogeosciences* 8: 1477–1486.

Koch, A.L. 2001. Oligotrophs versus copiotrophs. *Bioessays* 23: 657–661.

Lal, R. 2008. Carbon sequestration. *Philosophical Transactions of the Royal Society B—Biological Sciences* 363: 815–830.

Lauber, C.L., M. Hamady, R. Knight, and N. Fierer. 2009. Pyrosequencing-based assessment of soil pH as a predictor of soil bacterial community structure at the continental scale. *Applied and Environmental Microbiology* 75: 5111–5120.

Lauber, C.L., M.S. Strickland, M.A. Bradford, and N. Fierer. 2008. The influence of soil properties on the structure of bacterial and fungal communities across land-use types. *Soil Biology and Biochemistry* 40: 2407–2415.

Lavelle, P., D. Bignell, M. Lepage, V. Wolters, P. Roger, P. Ineson, O.W. Heal, and S. Dhillion. 1997. Soil function in a changing world: The role of invertebrate ecosystem engineers. *European Journal of Soil Biology* 33: 159–193.

Lavelle, P., E. Blanchart, A. Martin, S. Martin, and A. Spain. 1993. A hierarchical model for decomposition in terrestrial ecosystems–Application to soils of the humid tropics. *Biotropica* 25: 130–150.

Liiri, M., J. Haimi, and H. Setala. 2002. Community composition of soil microarthropods of acid forest soils as affected by wood ash application. *Pedobiologia* 46: 108–124.

Macdonald, B.C.T., M. Farrell, S. Tuomi, P.S. Barton, S.A. Cunningham, and A.D. Manning. 2014. Carrion decomposition causes large and lasting effects on soil amino acid and peptide flux. *Soil Biology and Biochemistry* 69: 132–140.

Manzoni, S. and A. Porporato. 2009. Soil carbon and nitrogen mineralization: Theory and models across scales. *Soil Biology and Biochemistry* 41: 1355–1379.

Mattson, W.J.J. 1980. Herbivory in relation to plant nitrogen content. *Annual Review of Ecology and Systematics* 11: 119–161.

Metcalf, J., L.W. Parfrey, A. Gonzalez, C.L. Lauber, D. Knights, G. Ackermann, G.C. Humphrey et al. 2013. A microbial clock provides an accurate estimate of the postmortem interval in a mouse model system. *eLIFE* 2: 1–19.

Moorhead, D.L. and R.L. Sinsabaugh. 2006. A theoretical model of litter decay and microbial interaction. *Ecological Monographs* 76: 151–174.

Neuhauser, E.F. and R. Hartenstein. 1978. Phenolic content and palatability of leaves and wood to soil isopods and diplopods. *Pedobiologia* 18: 99–109.

Nielsen, U.N., G.H.R. Osler, R. Van Der Wal, C.D. Campbell, and D.F.R.P. Burslem. 2008. Soil pore volume and the abundance of soil mites in two contrasting habitats. *Soil Biology and Biochemistry* 40: 1538–1541.

Osler, G.H.R. and M. Sommerkorn. 2007. Toward a complete soil C and N cycle: Incorporating the soil fauna. *Ecology* 88: 1611–1621.

Paoletti, M.G. and M. Bressan. 1996. Soil invertebrates as bioindicators of human disturbance. *Critical Reviews in Plant Sciences* 15: 21–62.

Parmenter, R.R. and J.A. MacMahon. 2009. Carrion decomposition and nutrient cycling in a semiarid shrub-steppe ecosystem. *Ecological Monographs* 79: 637–661.

Pechal, J.L., T.L. Crippen, A.M. Tarone, A.J. Lewis, J.K. Tomberlin, and M.E. Benbow. 2013. Microbial community functional change during vertebrate carrion decomposition. *PLoS ONE* 8: e79035.

Petersen, H. and M. Luxton. 1982. A comparative analysis of soil fauna populations and their role in decomposition. *Oikos* 39: 288–388.

Pollierer, M.M., R. Langel, C. Korner, M. Maraun, and S. Scheu. 2007. The underestimated importance of belowground carbon input for forest soil animal food webs. *Ecology Letters* 10: 729–736.

Putnam, R.J. 1978. Flow of energy and organic matter from a carcass during decomposition: Decomposition of small mammal carrion in temperate systems 2. *Oikos* 31: 58–68.

Rousk, J., E. Bååth, P.C. Brookes, C.L. Lauber, C. Lozupone, J.G. Caporaso, R. Knight, and N. Fierer. 2010. Soil bacterial and fungal communities across a pH gradient in an arable soil. *ISME Journal* 4: 1340–1351.

Rusek, J. 1998. Biodiversity of Collembola and their functional role in the ecosystem. *Biodiversity and Conservation* 7: 1207–1219.

Sagara, N. 1995. Association of ectomycorrhizal fungi with decomposed animal wastes in forest habitats: A cleaning symbiosis? *Canadian Journal of Botany* 73: 1423–1433.

Salamon, J.-A., J. Alphei, A. Ruf, M. Schaefer, S. Scheu, K. Schneider, A. Sührig, and M. Maraun. 2006. Transitory dynamic effects in the soil invertebrate community in a temperate deciduous forest: Effects of resource quality. *Soil Biology and Biochemistry* 38: 209–221.

Salona, M.I., M.L. Moraza, M. Carles-Tolrá, V. Iraola, P. Bahillo, T. Yélamos, R. Outerelo, and R. Alcaraz. 2010. Searching the soil: Forensic importance of edaphic fauna after the removal of a corpse. *Journal of Forensic Sciences* 55: 1652–1655.

Schlesinger, W. 1997. *Biogeochemistry: An Analysis of Global Change*. San Diego, CA, USA: Academic Press.

Smith, V.C. and M.A. Bradford. 2003. Litter quality impacts on grassland litter decomposition are differently dependent on soil fauna across time. *Applied Soil Ecology* 24: 197–203.

Sterner, R. and J. Elser. 2002. *Ecological Stoichiometry: The Biology of Elements from Molecules to the Biosphere*. Princeton, NJ, USA: Princeton University Press.

Strickland, M.S., M.A. Callaham, C.A. Davies, C.L. Lauber, K. Ramirez, D.D. Richter, N. Fierer, and M.A. Bradford. 2010. Rates of *in situ* carbon mineralization in relation to land-use, microbial community and edaphic characteristics. *Soil Biology and Biochemistry* 42: 260–269.

Strickland, M.S., C. Lauber, N. Fierer, and M.A. Bradford. 2009a. Testing the functional significance of microbial community composition. *Ecology* 90: 441–451.

Strickland, M.S., E. Osburn, C. Lauber, N. Fierer, and M.A. Bradford. 2009b. Litter quality is in the eye of the beholder: Initial decomposition rates as a function of inoculum characteristics. *Functional Ecology* 23: 627–636.

Strickland, M.S. and J. Rousk. 2010. Considering fungal: Bacterial dominance in soils—Methods, controls, and ecosystem implications. *Soil Biology and Biochemistry* 42: 1385–1395.

Strickland, M.S., K. Wickings, and M.A. Bradford. 2012. The fate of glucose, a low molecular weight compound of root exudates, in the belowground foodweb of forests and pastures. *Soil Biology and Biochemistry* 49: 23–29.

Swift, M.J., O.W. Heal, and J.M. Anderson. 1979. *Decomposition in Terrestrial Ecosystems*. Berkeley, CA, USA: University of California Press.

Throop, H.L. and M.T. Lerdau. 2004. Effects of nitrogen deposition on insect herbivory: Implications for community and ecosystem processes. *Ecosystems* 7: 109–133.

Towne, E.G. 2000. Prairie vegetation and soil nutrient reposes to ungulate carcasses. *Oecologia* 122: 232–239.

van Hees, P.A.W., D.L. Jones, R. Finlay, D.L. Godbold, and U.S. Lundstomd. 2005. The carbon we do not see—The impact of low molecular weight compounds on carbon dynamics and respiration in forest soils: A review. *Soil Biology and Biochemistry* 37: 1–13.

Vanni, M.J., A.S. Flecker, J.M. Hood, and J.L. Headworth. 2002. Stoichiometry of nutrient recycling by vertebrates in a tropical stream: Linking species identity and ecosystem processes. *Ecology Letters* 5: 285–293.

Vass, A., W. Bass, J. Wolt, J. Foss, and J. Ammons. 1992. Time since death determinations of human cadavers using soil solution. *Journal of Forensic Science* 37: 1236–1253.

Von Berg, K., M. Traugott, and S. Scheu. 2012. Scavenging and active predation in generalist predators: A mesocosm study employing DNA-based gut content analysis. *Pedobiologia* 55: 1–5.

Wickings, K. and A.S. Grandy. 2011. The oribatid mite *Scheloribates moestus* (Acari: Oribatida) alters litter chemistry and nutrient cycling during decomposition. *Soil Biology and Biochemistry* 43: 351–358.

Wickings, K., A.S. Grandy, S. Reed, and C. Cleveland. 2011. Management intensity alters decomposition via biological pathways. *Biogeochemistry* 104: 365–379.

Wickings, K., A.S. Grandy, S.C. Reed, and C.C. Cleveland. 2012. The origin of litter chemical complexity during decomposition. *Ecology Letters* 15: 1180–1188.

Wilson, A.S., R.C. Janaway, A.D. Holland, H.I. Dodson, E. Baran, A.M. Pollard, D.J. Tobin. 2007. Modelling the buried human body environment in upland climes using three contrasting field sites. *Forensic Science International* 169: 6–18.

Wu, J., P.C. Brookes, and D.S. Jenkinson. 1993. Formation and destruction of microbial biomass during the decomposition of glucose and ryegrass in soil. *Soil Biology and Biochemistry* 25: 1435–1441.

Yang, L.H. 2004. Periodical cicadas as resource pulses in North American forests. *Science* 306: 1565–1567.

Yang, X. and J. Chen. 2009. Plant litter quality influences the contribution of soil fauna to litter decomposition in humid tropical forests, southwestern China. *Soil Biology and Biochemistry* 41: 910–918.

Yang, H.S. and B.H. Janssen. 2002. Relationship between substrate initial reactivity and residues ageing speed in carbon mineralization. *Plant and Soil* 239: 215–224.

Zak, D.R., C.B. Blackwood, and M.P. Waldrop. 2006. A molecular dawn for biogeochemistry. *Trends in Ecology and Evolution* 21: 288–295.

6

Ecological Role of Vertebrate Scavengers

James C. Beasley, Zach H. Olson, and Travis L. DeVault

CONTENTS

6.1 Introduction

Scavenging, or the consumption of dead animal matter, has been documented for a wide array of vertebrate taxa by naturalists and researchers for centuries. Yet, until recently, the importance of scavenging-derived nutrients to many vertebrate species has been largely unknown (DeVault et al. 2003; Beasley et al. 2012b; Barton et al. 2013). As a result, the role vertebrate scavenging plays in food-web dynamics relative to that of microbes and invertebrates has been greatly underestimated (Wilson and Wolkovich 2011) and has thus become an important area of research. Indeed, in some ecosystems, vertebrates have been documented to assimilate as much as 90% of the available carrion (Houston 1986; DeVault et al. 2011). Such substantive acquisition of carrion resources by vertebrates challenges the traditional paradigm of microbial and invertebrate dominance of nutrient recycling in food-web theory and suggests that intensive interkingdom competition exists for access to carrion nutrients.

The lack of quantitative measures of vertebrate scavenging behavior, particularly among facultative scavengers, largely stems from human aversion to decomposing matter and difficulties in assessing foraging behavior for many species (DeVault et al. 2003). However, advances in technology and an increased awareness of the ecological (Bump et al. 2009; Ogada et al. 2012; Barton et al. 2013) and economic (Markandya et al. 2008) importance of scavenging by vertebrates have sparked an abundance of scavenging research during the past decade. These studies have greatly advanced our understanding of the importance of scavenging to both individual species and as an ecosystem service, but we are only beginning to appreciate the complexity of energy recycling via scavenging pathways and the cascading impacts anthropogenic activities can have on the disruption of these processes.

This chapter highlights the growing body of evidence supporting the importance of scavenging-derived nutrients to a multitude of vertebrate scavengers in both terrestrial and aquatic ecosystems, as well as the complex interactions among microbes, invertebrates, and vertebrates for access to carrion resources. The role of carrion in structuring vertebrate scavenging communities and the subsequent ecosystem services provided by this pervasive feeding strategy will also be discussed, drawing examples from the literature to highlight advancements along this front and where additional research is particularly needed. The chapter will conclude with a discussion of the documented and potential effects of various anthropogenic activities (e.g., climate change, habitat loss and fragmentation, loss of apex predators, and pollution) on vertebrate scavengers and the important ecosystem services scavengers provide.

6.2 Evolution of Vertebrate Scavengers

Species that scavenge can be separated into two unequal groups: those that rely on carrion for survival and reproduction ("obligate scavengers") and those species that will scavenge, but do not depend solely on carrion for their survival or reproduction ("facultative scavengers"). Vultures (families: Accipitridae and Cathartidae) feed extensively and in some cases exclusively on carrion, and these birds are believed to be the only obligate vertebrate scavengers (DeVault et al. 2003), although some benthic scavengers (e.g., hagfish: family Myxinidae) rely on necrophagy for a large portion of their diet and may indeed be obligate scavengers (Smith and Baco 2003; Beasley et al. 2012b). Although interesting for their adaptations to carrion-feeding and monopolization, obligate scavenging species are greatly outnumbered by a diverse assemblage of facultative scavengers. For example, most carnivorous species fall among the ranks of facultative scavengers; even species we recognize primarily as predators are regularly documented consuming carrion as part of their food habits, including the bobcat (*Lynx rufus*; Platt et al. 2010), barred owl (*Strix varia*; Kapfer et al. 2011), and many snake species (DeVault and Krochmal 2002). However, facultative scavengers are a more diverse group than just carnivores, as scavenging activity appears to be pervasive across the animal kingdom (DeVault et al. 2003) (Figure 6.1). In fact, a surprising array of animals will forage on carrion, including the hippopotamus (*Hippopotamus amphibious*; Dudley 1996), white-tailed deer (*Odocoileus virginianus*; Rooney and Waller 2003; Olson et al. 2012), pileated woodpecker (*Dryocopus pileatus*; Servín et al. 2001), and various lizards (Huijbers et al. 2013).

Acquisition of carrion by scavengers has been described as a function of their ability to detect carcasses, and thus obligate scavengers are primarily limited by the efficiency with which they can locate meals (Ruxton and Houston 2004; Shivik 2006). Consequently, obligate vertebrate scavengers are all large, soaring birds in terrestrial ecosystems (Houston 1986; Shivik 2006). Because of this specialization, Ruxton and Houston (2004) proposed that obligate scavenging differs from other trophic relationships such as predation and herbivory. They conclude that the evolutionary costs of being a predator are in large part related to the energetic and physical demands associated with handling prey (Ruxton and Houston 2004). Herbivores, in contrast, expend much of their energy processing and overcoming the chemical defenses of the plants that they eat (Freeland and Janzen 1974).

In addition to efficient locomotion, obligate scavenging birds exhibit spectacular adaptations to a lifestyle dependent on carrion. Diminished feather coverage on the head presumably helps protect against fouling (Houston 1979). Vultures also exhibit highly acidic guts, which maintain the dual benefits of speeding digestion during foraging bouts and also presumably of protecting against virulent pathogens encountered at carrion (Houston and Cooper 1975). Excellent visual and olfactory perception and extreme efficiency of travel via soaring all increase scavenging efficiency, as these animals have adapted to scale-up their search area to overcome the fine-scale spatial and temporal unpredictability of most carrion (Wilmers et al. 2003b; Ruxton and Houston 2004). The temporal unpredictability of carrion has also seemed to select for larger body sizes that can sustain some periods of time without food (Ruxton and Houston 2004). Similarly, hagfish are highly mobile marine scavengers that are able to survive a year or more between meals and have evolved sensitive chemoreceptive abilities, which allow them to

FIGURE 6.1 **(See color insert.)** Although few vertebrates are considered obligate scavengers, most species appear to utilize carrion resources facultatively. Results of experimental scavenging trials showing coyote, *Canis latrans* (top left) and black vulture, *Coragyps atratus* (top right) scavenging of a feral pig—*Sus scrofa*—carcass, scavenging of a cane toad—*Rhinella marina*—carcass in Hawaii by an invasive small Asian mongoose—*Herpestes javanicus* (bottom left), and scavenging of a rat carcass by a gray fox—*Urocyon cinereoargenteus* (bottom right).

detect distant carrion resources in benthic ecosystems (Smith 1985). In combination, these traits are the hallmark of the group of organisms best adapted to scavenging as a way of life.

Outside of this specialized group, the highly diverse assemblage of facultative scavengers varies in how frequently each species engages in scavenging activity. These animals do not require carrion for survival or reproduction in general, and so differences in scavenging frequency are likely driven by a number of factors specific to particular phylogenies, populations, and individuals. For example, variance in the frequency of scavenging by facultative scavengers can be related to the relative ability of each species to tolerate microbes and the microbial by-products of decomposition associated with carrion (Janzen 1977; Shivik 2006). Moreover, facultative scavengers that utilize carrion frequently must also possess the ability to detect and acquire carrion resources (Ruxton and Houston 2003) either by superior perceptive ability or travel efficiency. Indeed, there has been a discussion in the literature surrounding the idea that efficient distance running in early humans may have evolved to allow our ancestors to capitalize on nonpredator killed carcasses to supply their diets with protein (e.g., Bramble and Lieberman 2004; Ruxton and Wilkinson 2011). Alternatively, a facultative scavenging species may be able to consume a disproportionate amount of carrion resources simply by occurring in very high abundances relative to other species on the landscape (DeVault et al. 2011; Ruxton and Wilkinson 2012). All of these factors interact to form scavenger guilds at carcasses and across landscapes, but disentangling how each factor affects guild formation represents a challenge in scavenging ecology. What is clear is that there are substantial benefits to vertebrates for participating in scavenging behavior.

The benefits associated with feeding on carrion might be apparent for the obligate scavengers as they require carrion for survival and reproduction. However, the benefits of scavenging may also be substantial for the facultative scavengers. The food resources provided by carrion are a critical subsidy for many vertebrates, particularly in temperate climates during colder seasons when carcasses are less susceptible to decomposition (Gese et al. 1996; Fuglei et al. 2003; Selva et al. 2005; Blázquez et al. 2009;

Killengreen et al. 2012). In fact, grizzly bears (*Ursus arctos*) eliminate much of their annual energy debt by utilizing carrion after they emerge from hibernation (Green et al. 1997; Mattson 1997). Carrion use may also enhance the survival of certain individuals of a species at times of the year when other food resources are generally less abundant. Juveniles or low-dominance individuals may be outcompeted for preferred resources in the presence of larger, more experienced, or more dominant individuals, but may subsist in part by using carrion resources (Gese et al. 1996; Bennetts and McClelland 1997; Shivik and Clark 1999). However, the degree to which facultative scavengers actually require carrion resources is difficult to address in natural systems. This certainly represents a growth edge for discovery in scavenging ecology.

6.3 Interkingdom Competition among Vertebrates, Invertebrates, and Microbes

Because carcasses represent a rich source of nutrients to any organism that can utilize the resource, there is a race for organisms to find and consume carrion before other organisms can monopolize it (Janzen 1977). This race can lead to sometimes intense competition among microbes, invertebrates, and vertebrate scavengers for the resources sequestered in a carcass (DeVault et al. 2003). Competition among these groups is shaped by factors associated with the carcass itself, but also by factors associated with the environment surrounding the carcass (Payne 1965; DeVault et al. 2004b).

During the successive stages of decomposition, the competitive landscape at a carcass shifts as the process of degradation occurs (DeVault et al. 2003; Carter et al. 2007). The first organisms to compete at a carcass are the endogenous microbes that existed in and on the animal before it died (Putman 1978a; Carter et al. 2007). As the microbial community begins to break down the carcass, they compete with one another for space and nutrients, but also begin producing odiferous byproducts of their metabolism (Janzen 1977; Brown et al. 2009). The production of these volatile compounds signals the availability of a carcass to species of invertebrates, which are able to detect even minute quantities of volatiles from a volume of air (Paczkowski et al. 2012). Upon the arrival of the first invertebrate scavengers, competition for the resources in a carcass shifts from interactions among microbes to include interactions between microbes and invertebrates (Scott 1998; Burkepile et al. 2006; Rozen et al. 2008) and interactions among invertebrates (Denno and Cothran 1976; Hanski and Kuusela 1980). Degradation of a carcass during invertebrate feeding, primarily due to the action of their larvae, dramatically increases the rate of carcass attenuation (Payne 1965; Putman 1978b; Carter et al. 2007). However, during the early phases of carcass decay, the volatile compounds produced during microbial metabolism may also alert vertebrate scavengers to the potential for a meal (Smith and Paselk 1986; DeVault et al. 2003).

Competition among vertebrates for carcasses is often spectacular and has received much attention, particularly among scavenging birds (Kruuk 1967; Wallace and Temple 1987) and African mammals (Kruuk 1972; Houston 1979). However, competition for carcasses also occurs among microbes, invertebrates, and vertebrates at this stage in carcass decomposition, and interkingdom competition for carrion has had profound effects on the evolution and ecology of scavengers. For example, microbes involved in this competition use chemicals to ward off crustaceans in marine ecosystems (Burkepile et al. 2006), and some species of burying beetle (e.g., *Nicrophorus vespilloides* [Coleoptera: Silphidae]) produce an antimicrobial compound to protect carrion used in reproduction from decomposition by microbes (Rozen et al. 2008). Vertebrate scavenging may ultimately supersede some of these other mechanisms of retaining or monopolizing a carcass, because carcasses consumed by vertebrates may act as ecological sinks for some invertebrate species when the entire local population is consumed. Janzen (1977) was the first to discuss the idea of interkingdom competition for carcasses in detail, and much interesting work has added to our knowledge base since then.

Factors outside the carcass itself also affect competition for carrion resources. For example, vertebrate scavengers appear to be disadvantaged when humidity and temperature favor microbial and invertebrate reproduction (between 21°C and 38°C, Vass 2001; Zhou and Byard 2011). In a study by DeVault et al. (2004b), it was found that at temperatures above 20°C vertebrates were able to detect and consume only

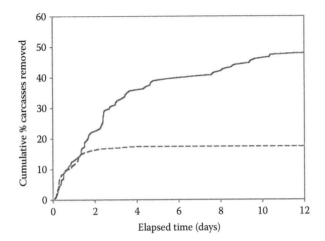

FIGURE 6.2 The competition for carrion between vertebrate scavengers and decomposers is heavily influenced by ambient air temperature. Here, the cumulative percentage of carcasses removed by vertebrates as a function of time during cool (5.8–17.0°C; solid line) and warm weather (22.6–27.8°C; dashed line) is shown for experimental rodent carcasses from South Carolina, USA. (Modified from DeVault, T.L., I.L. Brisbin Jr, and O.E. Rhodes Jr. 2004b. *Canadian Journal of Zoology* 82:502–509.)

19% of small-mammal carcasses, whereas at temperatures below 18°C, vertebrates consumed 49% of carcasses (Figure 6.2). Interestingly, early work by Houston (1986) indicated that the presence of a highly efficient vertebrate scavenger could alter the outcome of interkingdom competition for carrion nutrients. Namely, 71 of 74 carcasses were consumed by vultures at that equatorial study site, despite relatively high temperature and humidity (Houston 1986). Along similar lines, DeVault et al. (2011) found that temperature effects could be mediated by scavenging community composition in a fragmented, agricultural landscape. Specifically, facultative scavenging mammalian mesopredators (i.e., midtrophic level predators) dominated carcass acquisition over invertebrate and microbial competitors, likely due to the elevated abundances of mesopredators in their study landscape (DeVault et al. 2011). Similarly, highly abundant scavengers from one habitat may recruit into adjacent habitats when carcasses are available (Killengreen et al. 2012). Despite the focus on the outcome of interkingdom competition in the context of temperature and community structure, there is evidence to suggest that in some ecosystems the timing of carcass disappearance is similar, regardless of whether carcass assimilation is dominated by vertebrate or invertebrate communities (Sugiura et al. 2013).

Information regarding the influence of habitats on the competition for carrion is more equivocal and is often conflated with climate differences (i.e., temperature and relative humidity). For example, the available literature suggests that carcass decomposition occurs more rapidly in warm, moist habitats when compared with more xeric habitats (see Carter et al. 2007; Parmenter and MacMahon 2009). Competition for carrion can also be affected by fine-scale or microsite differences in habitat. DeVault et al. (2004b) found that carcass acquisition by vertebrate scavengers was higher in lowland hardwood habitats compared with upland conifer habitats within the same study site. Parmenter and MacMahon (2009) extensively investigated microsite effects on decomposition rates of carcasses and found that carcasses located underground in burrows decomposed faster in their semiarid study site during spring and summer months. However, more frequently, carcass burial is negatively correlated with decomposition rates in the forensic literature (Carter et al. 2007). The location of a carcass in relation to habitat attributes on the landscape may allow particular species to colonize more easily. For example, forest openings and other open habitats may facilitate carcass location by vertebrate scavengers that rely largely on vision to find carrion (Houston 1988; Selva et al. 2003). The habitat occupied by a carcass has also been shown to affect the time of appearance for carrion-associated insects on pig carcasses in Europe, although these differences did not translate to an overall shift in the successional pattern of colonization by insects (Matuszewski et al. 2011).

In addition to the effects of habitat type on the competition for carrion, habitat fragmentation appears to affect the balance in interkingdom competition for carcasses. Gibbs and Stanton (2001) demonstrated that the abundance of carrion beetles decreased, taxon richness decreased, and community structure shifted in areas of fragmented versus more contiguous forests in New York, USA. However, they were unable to link these changes in an important group of carrion specialists to carrion abundance or other factors (Gibbs and Stanton 2001). The work of DeVault et al. (2011) may provide such a link. DeVault et al. (2011) discovered that in an agriculturally fragmented habitat, most (88%) small-mammal carcasses were consumed by vertebrates, which was a substantial increase in carrion acquisition by vertebrates over earlier work conducted in more contiguous habitats (i.e., DeVault et al. 2004b). They argued that an increase in mesopredator abundance, primarily raccoons (*Procyon lotor*) and Virginia opossums (*Didelphis virginiana*), accounted for this disparity (DeVault et al. 2011). Moreover, they proposed that the occurrence of a highly efficient suite of vertebrate competitors for the same carcasses used by carrion beetles might explain the declines of carrion beetles that Gibbs and Stanton (2001) reported in similarly fragmented habitats. An interesting area of future research associated with habitat fragmentation might include an assessment of competition for carcasses along the front of northward-expanding vulture populations in North America, an expansion that may be due, at least in part, to greater carcass availability from vehicular-related road-kills (Houston et al. 2011).

6.4 Ecosystem Effects of Carrion Use by Vertebrates

Carrion use by vertebrates has interesting implications for ecosystems. From a physical perspective, the molecules that make up an animal's body become progressively less aggregated after that animal dies (see Chapter 2 for details). The speed and the extent to which these molecules are dispersed are explicitly related to the scavengers that feed on the carcass. Microbial decomposition creates well-defined and well-documented cadaver decomposition islands (CDIs; Johnson 1975) as nutrients released through microbial action are incorporated into the detrital pathway immediately below and near where the carcass decomposes (Moore et al. 2004; Melis et al. 2007). Subsequently, those nutrients are mobilized by the adjacent flora creating localized effects (Towne 2000; Danell et al. 2002; Bump et al. 2009). For large carcasses, elevated nutrient levels coincident with a CDI have been detected in soil and foliar samples for years after the carcass decomposed (Danell et al. 2002; Melis et al. 2007).

The addition of invertebrate scavenging to the effects of microbial action increases the final nutrient dispersion from a carcass farther into the surrounding environment as the arthropods pupate and disperse (Carter et al. 2007). However, in general, vertebrate scavenging represents the widest dispersal of nutrients and energy from carcasses as vertebrate movement scales away to the broader landscape (Payne and Moore 2006; Barton et al. 2013). In this process, vertebrate scavengers diminish the formation of CDIs by scattering and incorporating portions of the carcass into the surrounding ecosystem (Kjorlien et al. 2009; Reeves 2009). Thus, the spatial heterogeneity of resources that carcasses contribute to the landscape depends in part on the identity of the scavengers that feed on them. Carcasses, then, in some instances, can contribute nutrients and energy to initiate resource hot-spots that add to landscape complexity (Payne and Moore 2006; Bump et al. 2009; Parmenter and MacMahon 2009), but in other instances, carrion is effectively recycled among higher trophic levels by scavengers with limited direct inputs to the detrital pathway. We are just beginning to understand the ecological effects that carcasses have on the landscape (Bump et al. 2009), and further work is needed to determine how vertebrate scavengers influence those effects.

In addition to impacting landscape heterogeneity by the disruption of decomposition patterns, scavengers can also shape the structure of vertebrate communities through interactions with predators. In systems in which the predator is smaller than its prey, some portion of each fresh kill often cannot be immediately consumed by the predator. For example, the gleanings from partially consumed, or usurped, predator kills on the African savannah sustain an entire guild of vertebrate scavengers (Hunter et al. 2007), although predator kills are probably not the primary source of carrion in this ecosystem (Houston 1979). Scavenging activity such as this is often grouped with kleptoparasitism in the literature, and it places a limit on the amount of resources the original predator can gain from each kill with potentially cascading effects (Vucetich et al. 2004). In Isle Royale National Park, USA, raven scavenging of wolf (*Canis lupus*)-killed moose (*Alces alces*) carcasses is apparently extensive enough to make increased pack sizes beneficial, despite the smaller share of each moose that an individual wolf receives when hunting in these larger packs (Vucetich et al. 2004). Eurasian Lynx (*Lynx lynx*) in areas of Europe have been shown to increase their kill rates when the carcasses they cache are scavenged by grizzly bears (*U. arctos*; Krofel et al. 2012). Further, their review of the literature revealed lower but still significant scavenging of felid-killed carcasses from Europe, North America, and Africa (Krofel et al. 2012), indicating that the phenomenon is not restricted to any particular ecosystem. Presumably, scavenging by grizzly bears also increases ungulate kill rates of wolves in Yellowstone National Park, USA (Hebblewhite and Smith 2010). Increased kill rates by top predators may represent a little acknowledged marginal cost to the vertebrate community, directly attributable to scavenging activity. Given the importance of top-down effects in many ecosystems, even a minor alteration to predation rates as driven by vertebrate scavengers may cause a significant flux in community composition.

Another possible effect of vertebrate scavengers on vertebrate communities is related to the fact that many facultative scavengers are also predators (DeVault et al. 2003). Cortés-Avizanda et al. (2009) found that the abundance of prey species (i.e., hares—*Lepus* spp. and squirrels—*Sciurus* spp. in this case) decreased in sectors containing a carcass based on evidence from tracks in snow. An interesting hypothesis emerged, in which the scavengers that are recruited to a carcass may have temporarily played the dual role of increasing predator abundance near each carcass (Cortés-Avizanda et al. 2009). It is unclear from this study whether tracks of the prey species might have declined because incidental predation by facultative scavengers reduced the abundance of prey species near carcasses or because of nonconsumptive effects such as altered behavior by prey species in the vicinity of carcasses due to the presence of predatory scavengers (Cortés-Avizanda et al. 2009). However, the second possibility adds an interesting dimension to the "landscape of fear" hypothesis, in which movements and foraging decisions by prey species are influenced by a perceived risk of predation (Lima and Dill 1990; Blumstein 2006).

Some of the most visible and well-studied effects of vertebrate scavengers on communities can be seen in the interactions of scavengers at carrion. Dominance hierarchies and structural differences within and among species of vultures in the Old World (Kruuk 1967; Houston 1975; Alvarez et al. 1976) and the New World (Wallace and Temple 1987) have established competitive frameworks that delineate resource acquisition at carcasses. Superficially, larger species tend to dominate smaller species for access to carcasses (Petrides 1959; Alvarez et al. 1976), but larger species also facilitate access to carcasses for a variety of less-specialized and smaller species by breaking through the thick hides of larger carrion (Kruuk 1967; Selva et al. 2005; Blázquez et al. 2009).

6.5 Ecosystem Services Provided by Vertebrate Scavengers

Vertebrate scavengers provide various ecosystem services, including cultural (e.g., spiritual value), supporting (nutrient cycling), and regulating services (carcass removal from the landscape) (Millennium Ecosystem Assessment 2003; Wenny et al. 2011; DeVault et al. 2016). In this section, four of the most prominent are highlighted: critical linkages in food webs, nutrient distribution within and among ecosystems, economic benefits related to sanitary measures, and altered disease dynamics. This section documents how vertebrate scavengers, both obligate (e.g., vultures) and facultative (e.g., crows—*Corvus* spp., raccoons), play underappreciated but pivotal roles in maintaining healthy ecosystem function.

6.5.1 Vertebrate Scavengers Provide Critical Linkages in Food Webs

Historically, the prevalence of scavenging activities has been greatly underestimated. However, upon recognition that (1) in most ecosystems, a large number of animals die from causes other than predation and thus become available to scavengers; (2) most carcasses are scavenged by vertebrates before they are completely decomposed by arthropods and bacteria; and (3) almost all carnivorous animals are facultative scavengers, the importance of scavenging in food webs seems unsurprising (DeVault et al. 2003). In fact, Wilson and Wolkovich (2011) estimated that scavenging links are underrepresented in food-web research by 16-fold. The omission of these connections in ecological models is striking, especially considering the role that the number (Dunne et al. 2002) and strength of trophic connections (weak links, McCann et al. 1998; McCann 2000) are known play in promoting the persistence and stability of ecological communities and the ecosystem services they deliver. The omission of scavenging activities from food-web models has largely been the result of oversimplification; that is, the treatment of all types of detritus (from low-quality dead plant material to high-quality animal carrion) as a single resource pool (Swift et al. 1979; Wilson and Wolkovich 2011). However, food webs are increasingly recognized as complex and highly interconnected (e.g., Polis 1991; Polis and Strong 1996), and the importance of detritus, especially animal carrion as a distinct resource, is becoming widely accepted (DeVault et al. 2003; Moore et al. 2004; Wilson and Wolkovich 2011; Barton et al. 2013).

Scavenging activities may be especially important for the resiliency of food webs. The stabilizing nature primarily results from the high number of interspecies links from scavenging (scavengers often feed on carrion from many species, making webs more interconnected; Wilson and Wolkovich 2011). Also, the use of carrion as a supplemental food resource during prey shortages for species that are primarily predators might add to the stabilizing nature of scavenging (McCann et al. 1998; DeVault et al. 2003). As ecosystems are increasingly subject to multiple stressors from human activities, it is important to gain a better understanding of the intrinsic properties of food webs that promote stability, and carrion use by vertebrates appears to be one such factor.

6.5.2 Distribution of Nutrients within and among Ecosystems

In addition to acute visual and/or olfactory abilities for detecting carcasses, one of the key attributes of the most successful scavengers is the ability to quickly and efficiently travel great distances in search of carrion, which in many cases is unpredictable and ephemeral (Houston 1979; Ruxton and Houston 2004; Shivik 2006). As a result, obligate scavengers such as vultures generally have very large home ranges (e.g., Houston 1974; DeVault et al. 2004a). These scavengers, as well as some facultative scavengers such as certain bottom-dwelling marine species (Smith 1985; Beasley et al. 2012b), often disperse assimilated carrion across large areas. This dispersion of carrion biomass by vertebrates is especially evident when carrion is initially concentrated spatially. For example, carcasses produced from fishing by-catch (Furness 2003), salmon (Salmonidae) die-offs (Hewson 1995), forest fires (Blanchard and Knight 1990), and single large carcasses (e.g., whales—Cetacea; Smith and Baco 2003) are often visited by multiple scavengers that range widely and therefore transport the nutrients from those carcasses over large distances. In particular, salmon represent a significant annual pulse of marine-derived nutrients that can be disseminated from aquatic to terrestrial ecosystems through vertebrate scavenging. Movement of salmon carcasses to terrestrial habitats, usually by bears (*Ursus* spp.) or other large mammals, also links terrestrial invertebrate communities to marine-derived nutrients by providing substantial carrion subsidies to ovipositing flies and terrestrial invertebrate scavengers unable to access carrion in aquatic habitats (Meehan et al. 2005). Cross-habitat nutrient transport can produce a variety of important outcomes in recipient systems (e.g., Polis et al. 2004), and scavengers can play a significant role in moving "ecological subsidies" between habitats. For example, the use of ocean-derived carrion by terrestrial mammals (Rose and Polis 1998) and birds (Schlacher et al. 2013) is extensive and may strongly influence dynamics of coastal food webs.

6.5.3 Economic Benefits Related to Sanitary Measures

Traditionally, in Europe, carcasses of free-ranging livestock have been left at the site of death to be consumed by avian scavengers (Donázar et al. 1997; Margalida et al. 2010). The arrangement, which persisted for centuries, was mutually beneficial, as this ecosystem service saved farmers in the European Union €0.97–1.60 million annually (Margalida and Colomer 2012) and provided vultures with a vital food resource (Donázar et al. 1997; Margalida et al. 2011). However, more restrictive sanitary policies enacted in Europe after the emergence of bovine spongiform encephalopathy ("mad cow disease") in cattle between 1996 and 2000 required that carcasses be disposed of in authorized facilities (Margalida et al. 2010). Because of the resultant food shortages for avian scavengers, several species in the region experienced decreased breeding success and other demographic problems (Margalida et al. 2010). Recently, however, new guidelines have been enacted that again allow farmers to abandon dead livestock in the field (Margalida et al. 2012). Even so, it is unclear how these new regulations will be applied by various governments and subsequently how vulture populations will respond to any new management scenarios that may emerge (DeVault et al. 2016).

6.6 The Role of Scavengers in Disease Ecology

Large carcasses can serve as incubators for many types of infectious materials. Because many mammalian and avian species often visit single large carcasses (Selva et al. 2003; Jennelle et al. 2009), and scavengers often have large home ranges (discussed earlier), some have suggested that vultures and other wide-ranging vertebrate scavengers might facilitate the spread of pathogens across large areas. For example, vultures might harbor infectious materials on their feet and feathers and introduce them across the landscape as they forage at widely spaced carcasses (Houston and Cooper 1975). Further, VerCauteren et al. (2012) showed that infectious scrapie prions survived passage through the digestive system of American crows (*Corvus brachyrhynchos*) and suggested that crows might spread prion diseases. Also, it has been suggested that scavengers might exacerbate production of anthrax spores by opening carcasses and thus suppressing the ability of anaerobic bacteria residing inside carcasses to antagonize vegetative anthrax cells and thus impede sporulation (Bellan et al. 2013 and references therein).

However, vultures are generally very resistant to diseases, a trait common among many successful scavengers (Shivik 2006). Houston and Cooper (1975) concluded that the digestive tract of griffon vultures is likely to kill most pathogenic bacteria, given that the pH of the stomach ranges from 1 to 2. Thus, as vultures forage, they probably reduce the proliferation of diseases, at least at the local scale, by removing infected carcasses from the landscape (see also Ogada et al. 2012). Also, in experimental work using electrified cage exclosures, Bellan et al. (2013) determined that vertebrate scavenging was not critical for the production of anthrax spores at carcass sites.

It is thus unclear the extent to which scavengers remove infectious materials from the landscape or, alternatively, spread pathogens across large areas. Even so, it seems likely that the identity of the vertebrate species scavenging at carcasses determines, at least in part, whether or not diseases are proliferated or impeded. For example, the near-extirpation of several vulture species in south Asia due to the use of toxic livestock chemicals (Green et al. 2004; Oaks et al. 2004) allowed cattle carcasses to remain in the landscape for longer time periods and thus were made available for consumption by feral dogs and rats, which apparently resulted in population increases in those species (Pain et al. 2003; Prakash et al. 2003). Markandya et al. (2008) estimated that the total costs to human health (including rabies cases from dog bites) that resulted from severe vulture declines totaled over $34 billion from 1993 to 2006. Also, Ogada et al. (2012) determined that the exclusion of vultures from large animal carcasses in Kenya resulted in a tripling of carcass decomposition times. In addition, the number of mammals scavenging carcasses, the average time spent at carcasses by mammals, and the number of contacts between mammals increased substantially in the absence of vultures. Such increases in inter- and intraspecific interactions due to increased persistence times of carcasses could increase the probability of disease spread within and among species (especially mammals), particularly for some important zoonotic diseases including rabies. Clearly, the role of vertebrate scavengers in disease proliferation is complex, and more research is needed to elucidate factors that influence disease dynamics with regard to scavenging ecology (Jennelle et al. 2009).

6.7 Challenges to Vertebrate Scavengers in Modern Society

There is a growing body of evidence that facultative scavenging is widespread among vertebrates, and resources provided through scavenging are likely critical to many species (DeVault et al. 2003; Fuglei et al. 2003). Consequently, any shift in the availability or distribution of carrion could have a profound impact on the composition of scavenging communities and distribution of carrion resources throughout

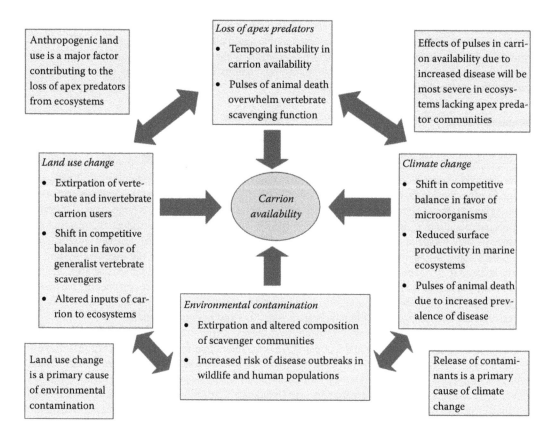

FIGURE 6.3 Generalized diagram illustrating the effects of anthropogenic disturbances on the fate of carrion and the organisms that use these resources. In many instances, ecosystems may face a multitude of human disturbances simultaneously that may interact to exacerbate impacts to scavenger communities. For example, the effects of temporal shifts in carrion availability due to climate change likely will be most severe in ecosystems lacking apex predators as these species play a critical role in the temporal stabilization of carrion resources to facultative scavengers.

food webs. Although numerous factors are involved in modulating carrion availability, climate (DeVault et al. 2004b; Selva et al. 2005; Parmenter and MacMahon 2009), trophic integrity (Wilmers et al. 2003a, b; Wilmers and Post 2006), disease (Wilson and Wolkovich 2011), and habitat availability (DeVault et al. 2011) play integral roles in regulating the fate of carrion, as well as the spatial and temporal distributions of these resources in ecosystems (Figure 6.3). Thus, in light of the numerous anthropogenic threats currently impacting ecosystems across the globe (e.g., climate change, pollution, trophic downgrading, habitat loss and fragmentation, and invasive species), there is a growing need to quantify the potential impacts of such changes to the structure and function of ecosystems.

Scavenging communities, in particular, are highly sensitive to a variety of anthropogenic disturbances to ecosystems due to their reliance on the availability and distribution of carrion resources. Although scavengers face a multitude of challenges due to human activities, below we highlight several that are globally relevant and for which sufficient literature exists to suggest that these activities are having measureable impacts to scavenging dynamics, although this list certainly is not inclusive.

6.7.1 Potential Climate Change Effects on Vertebrates

Competition among vertebrates, invertebrates, and microbes can be influenced by a variety of factors, but is often modulated by abiotic conditions. In particular, acquisition of carrion by vertebrates is highly influenced by temperature, with reduced scavenging efficiency as temperature increases due to increased microbial and invertebrate activity (DeVault et al. 2004b, 2011). As a result, altered temperature and precipitation patterns due to climate change likely will alter competitive interactions among vertebrates, invertebrates, and microbes, disrupting the flow of energy within and among ecosystems. Slight changes in temperature could have profound impacts on energy flow within ecosystems as carrion decomposition rates roughly double for every 10°C increase in ambient temperature (Vass et al. 1992; Parmenter and MacMahon 2009). Consequently, the availability of carrion to vertebrate scavengers could be reduced by as much as 20%–40% over the next century based on projections from current climate change models (Beasley et al. 2012b). Reduced access to carrion by vertebrates is not trivial as facultative scavengers serve important roles in stabilizing food webs by maintaining energy resources high within food webs of many ecosystems (Rooney et al. 2006). Moreover, in some ecosystems, vertebrates consume as much as 90% of the available carrion biomass and thus a reduction in carrion availability could have a profound effect on the distribution and abundance of vertebrate scavengers, as well as a multitude of other organisms (Houston 1986; DeVault et al. 2011).

In addition to reductions in carrion availability due to increased microbial and invertebrate activity, altered temperature and precipitation patterns resulting from climate change are likely to shift the spatial and temporal distribution of carrion resources through increased incidence and geographic range of many diseases (Patz et al. 1996; Harvell et al. 2002; Wilson and Wolkovich 2011). In particular, vector-borne pathogens are likely to increase in both abundance and geographic range in response to rising temperatures. Although data are not available for many pathogens, such shifts in their distribution have recently been observed for several vector-borne human and livestock diseases including malaria, Lyme disease, tick-borne encephalitis, plague, blue tongue viruses, and African horse sickness (Harvell et al. 2002).

Rather than produce a steady increase in carrion availability, such increases in disease, particularly epidemics, will likely produce pulses of animal death, disrupting the spatial and temporal availability of carrion within ecosystems. As a result, the temporal aggregation of carrion resources may reduce the diversity and evenness of carrion consumption among scavengers (Wilmers et al. 2003b; Cortés-Avizanda et al. 2012). Although the long-term impacts of truncated carrion availability to obligate scavengers are unknown, given that most obligate vertebrate scavengers currently are threatened with extinction (Ogada et al. 2012), any reductions in carrion availability or distribution could contribute to further population declines of these species. Similarly, facultative scavengers that rely on carrion resources for overwinter survival could be negatively impacted by a shift in carrion availability due to increased incidence and aggregation of disease (Fuglei et al. 2003).

Changes in global climate are also expected to substantially alter the availability of carrion resources in marine ecosystems as surface production of organic material could decline by 50% or more (Smith et al. 2008). Such a drastic reduction in productivity undoubtedly will impact a multitude of ecosystem processes, including the availability and distribution of carrion within marine ecosystems. However, the effects of reduced carrion subsidies may be most acute in benthic ecosystems as benthic scavengers are inextricably linked to carrion subsidies provisioned from the euphotic zone, and such reductions in carrion resources may potentially reduce the biodiversity of scavengers in deep-sea environments. Indeed, populations of facultative benthic scavengers already have declined in some regions in response to a decline in food resources from the surface and concurrent increases in water temperature due to increases in global temperatures (Bergmann et al. 2011). In addition to altered marine scavenging communities, changes in ocean surface productivity may also impact terrestrial scavenging communities in coastal areas as many species rely on carrion subsidies washed on shore from marine environments (Schlacher et al. 2013).

6.7.2 Effects of Habitat Loss, Fragmentation, and Urbanization

Habitat loss and fragmentation due to agriculture, urbanization, and other anthropogenic land uses are growing and pervasive issues that can have substantive impacts on the distribution, ecology, and population dynamics of numerous wildlife species across the globe (Foley et al. 2005; Beasley et al. 2011). Although the direct effects of anthropogenic land use to scavenging dynamics are not well defined, the composition and efficiency of vertebrate scavengers appear to be highly altered in landscapes modified by humans. For example, in a highly fragmented agricultural ecosystem in northern Indiana, USA,

(a)   Indiana, USA

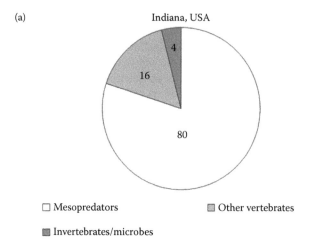

☐ Mesopredators   ▨ Other vertebrates

▨ Invertebrates/microbes

(b)   South Carolina, USA

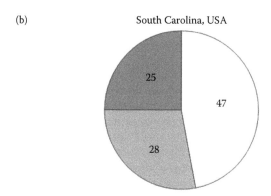

FIGURE 6.4 Results of small-mammal scavenging trials conducted in (a) highly fragmented landscape in northern Indiana, USA and (b) contiguously forested site in South Carolina, USA showing the dominance of carrion acquisition by mesopredators and reduced species diversity of scavengers in ecosystems substantially impacted by anthropogenic activities. (Data from DeVault, T.L., I.L. Brisbin Jr, and O.E. Rhodes Jr. 2004b. *Canadian Journal of Zoology* 82:502–509 and DeVault, T.L. et al. 2011. *Basic and Applied Ecology* 12:268–274.)

DeVault et al. (2011) observed substantially lower species richness of vertebrate scavengers compared to similar studies in more contiguous landscapes (Figure 6.4). Given that human-modified landscapes generally support truncated vertebrate communities comprised of generalist species (Swihart et al. 2003), it is not surprising that fragmented landscapes support impoverished scavenging communities dominated by invasive or generalist species. However, despite the reduced diversity of scavengers, the efficiency of scavenging by these species actually was greatly elevated in the highly fragmented landscape, with nearly 90% of mouse carcasses scavenged compared with 30% in a more contiguously forested site (DeVault et al. 2011). This high frequency of scavenging represents some of the highest efficiencies reported for vertebrate scavengers (Houston 1986; DeVault et al. 2003).

Interestingly, facultative mesopredators comprised 93% of the scavenging events observed by DeVault et al. (2011), and turkey vultures, the only obligate vertebrate scavenger in their study area, failed to acquire a single carcass out of 266 trials. Turkey vultures were observed regularly within the study area and were recorded scavenging on 20%–43% of rabbit and raccoon carcasses, respectively, within the same study sites (Olson et al., unpublished data). Thus, it appears that for small carcasses that can easily be consumed by a single individual, dense populations of mesopredators are able to competitively exclude most other species, including obligate avian scavengers, from these resources. Similarly, Huijbers et al. (2013) also observed highly disparate patterns of carrion removal by vertebrates between urbanized and rural beaches on the east coast of Australia. In particular, although a similar number of species were detected at both study sites, scavenging communities within rural beaches were dominated by native raptors, whereas urban beach scavengers primarily comprised generalist invasive or feral mammals (Huijbers et al. 2013). Such striking differences in the structure of the scavenging guild in disturbed ecosystems suggest that changes in land use can alter fundamental aspects of scavenging dynamics.

In addition to the direct effects of altered habitat composition and fragmentation on carrion availability, such landscapes often support high densities of generalist species, many of which are efficient scavengers that may monopolize carrion resources in these ecosystems (DeVault et al. 2011). An elevated rate of carrion consumption by mesopredators implies that fewer carrion resources likely are available to invertebrate scavengers and microbes in highly modified landscapes. Although the competitive interaction between mesopredators and invertebrate or microbial scavengers has not been rigorously studied in fragmented ecosystems, the effects of the rapid and nearly complete attenuation of rodent carcasses by mesopredators observed by DeVault et al. (2011) suggest that they are likely not trivial.

6.7.3 Effects of Trophic Downgrading (Loss of Apex Predators)

The loss of apex predators from ecosystems, referred to as trophic downgrading, can have cascading impacts on the composition and function of ecosystems. Over the last few centuries, trophic cascades due to the loss of apex consumers have become pervasive, having now been documented in all of the world's major biomes and in freshwater, marine, and terrestrial ecosystems (Estes et al. 2005). In terms of scavenging, apex predators serve a critical role in modulating the availability of carrion to other vertebrate, invertebrate, and microbial consumers of this resource by reducing or eliminating pulses of death due to starvation and disease epidemics and thus stabilizing the availability of carrion throughout the year (Wilmers et al. 2003a). This stabilization of reliable high-energy food resources may promote biodiversity and undoubtedly facilitates increased survival and reproductive output in some scavenger species (Fuglei et al. 2003; Barton et al. 2013). Predator-killed carrion resources also appear to be used by a more diverse community of vertebrate scavengers than natural or human-provided carrion subsidies (Wilmers et al. 2003a; Selva et al. 2005), although such preferences do not appear to hold for small carrion items (e.g., mice, rats, and rabbits) or in landscapes where apex predators have been extirpated (DeVault et al. 2004b, 2011).

The presence of apex predators within an ecosystem also may indirectly influence carrion availability by regulating the abundance and behavior of mesopredators. In the absence of apex predators, mesopredator populations can reach exorbitant densities (Prugh et al. 2009; Beasley et al. 2011), altering a multitude of ecological processes, including energy flow, by monopolizing carrion resources and greatly reducing the availability of these resources to other scavengers in these ecosystems, including obligate carrion consumers (Olson et al. 2012). Mesopredator release and trophic cascades due to the removal of apex predators are not unique to terrestrial environments, and there are numerous examples from both freshwater and marine ecosystems. For example, overharvesting of apex predatory sharks has led to a significant increase in elasmobranch mesopredators in many regions, particularly the cownose ray (*Rhinoptera bonasus*). In the Chesapeake Bay region alone, there may now be an estimated 40 million cownose rays, more than an order of magnitude increase since the mid-1970s (Myers et al. 2007). Although the effects of increased ray populations to marine scavengers are unknown, increased ray populations have led to the collapse of bay scallop (*Argopecten irradians*) populations and other bivalves within this region, and thus their effects to scavengers are likely profound.

The reintroduction of gray wolves to Yellowstone National Park, USA has served as a unique natural experiment to characterize the role apex predators play in modulating carrion resources to other scavengers in terrestrial ecosystems. Prior to the reintroduction of wolves, the majority of ungulate mortality, and thus carrion availability, occurred in a winter pulse coinciding with peak snow depth, ranging from 0 to >500 kg of carrion per week (Gese et al. 1996). Had wolves been present, Wilmers et al. (2003a) estimated that carrion subsidies would have been relatively stable throughout this period, providing a multitude of scavengers access to high-energy food resources on a consistent basis throughout the winter. Indeed, during November–May, established wolf packs in the northern range of Yellowstone provide an estimated ~13,000 kg of elk carrion to other scavengers in this ecosystem, accounting for wolf consumption of carcasses (Wilmers et al. 2003b). The temporal distribution of this carrion is consistent throughout the winter and utilized by a much more diverse community of vertebrate scavengers than human hunter-provided carrion that is much more truncated in distribution, both spatially and temporally (Wilmers et al. 2003b).

Although climate change, habitat loss, and trophic downgrading all may play a role in altering the distribution and flow of carrion resources through food webs, each of these effects is integrally linked and when combined may exacerbate or moderate impacts to scavenging dynamics. For example, reduced inter- and intra-annual variability in carrion availability and provisioning of carrion subsidies throughout the winter in terrestrial ecosystems in which apex predators are present likely serve as an important buffer to climate change and other anthropogenic effects for scavengers (Wilmers et al. 2003a; Wilmers and Post 2006). In contrast, habitat loss and fragmentation are usually followed by the extirpation of apex predators and mesopredator release (Estes et al. 2011), which can alter the efficiency and composition of scavenging communities (Olson et al. 2012).

6.7.4 Pollution, Heavy Metals, Veterinary Pharmaceuticals, and Other Anthropogenic Toxicants

Although the aforementioned factors undoubtedly have had a significant impact on vertebrate scavengers, poisoning, whether deliberate or accidental, has probably played the most substantial role in the allocation of protection for this important group of species. Vertebrate scavengers are particularly vulnerable to toxic substances as many species rely on carrion as a significant source of food, numerous individuals and/or species often can feed on a single carcass, and scavenging facilitates feeding above an organism's trophic rank, allowing for increased exposure to toxins that may have bioaccumulated within the tissues of an individual prior to death. For centuries, humans have recognized and exploited these characteristics and the susceptibility of scavengers to toxic substances. For example, the widespread use of carrion laced with strychnine, 1080, and other toxins to control predator populations throughout the last century resulted in the near extirpation of many carnivores where such practices were widely employed (Langley and Yalden 1977; Whitfield et al. 2004). However, such eradication campaigns were often not species-specific, and in addition to mammalian carnivores, numerous other scavengers were killed as collateral damage of these eradication efforts, particularly during the use of 1080, including vultures, eagles, and California condors (*Gymnogyps californianus*). Indeed, in many places, targeted predators often are affected less by poison baiting than nontarget species (Eason and Spurr 1995; Berny 2007; Berny and Gaillet 2008; Márquez et al. 2013).

Although use of toxins to control carnivore populations has been banned or highly regulated throughout much of the world today, such practices continue to be used illegally in some regions to protect livestock and manage game species (Hernández and Margalida 2008, 2009a; Ogada et al. 2012; Márquez et al. 2013) and likely remain the most widespread cause of vulture poisoning worldwide (Donázar 1993; Margalida 2012; Ogada et al. 2012). Today, managers continue to exploit the scavenging tendencies of many vertebrates as a means of controlling invasive species and managing the spread of infectious disease (Slate et al. 2005; Page et al. 2011; Beasley et al. 2012a).

In addition to deliberate poisoning, scavengers can inadvertently be exposed to a multitude of other toxins through consumption of tissues containing pollutants, environmental contaminants, veterinary drugs, or other anthropogenic compounds. In a profound example that highlights the vulnerability of vertebrate scavengers to accidental exposure to toxins, over the last two decades, populations of *Gyps* vultures in Asia have precipitously declined to 3%–5% of their original population size due to accidental poisoning through consumption of livestock treated with diclofenac, a nonsteroidal anti-inflammatory drug (Green et al. 2004; Oaks et al. 2004; Shultz et al. 2004). Although use of diclofenac has been banned in many countries and numerous recovery efforts are underway, some species continue to decline at alarming rates, and full recovery of these species may take centuries (Cuthbert et al. 2011).

Accidental exposure to lead from pellets or bullet fragments also remains a threat to some vertebrate scavengers (Hunt et al. 2006; Kelly et al. 2011; Lambertucci et al. 2011). Indeed, unintentional lead poisoning is the leading cause of death for the California condor and remains one of the factors limiting the recovery of this species (Cade 2007). Similarly, elevated lead exposure has been documented for a wide range of obligate (e.g., turkey vultures—*Cathartes aura*, Egyptian vultures—*Neophron percnopterus*, and bearded vultures—*Gypaetus barbatus*) and facultative (great horned owls—*Bubo virginianus*, red-tailed hawks—*Buteo jamaicensis*, golden eagles—*Aquila chrysaetos*, and bald eagles—*Haliaeetus leucocephalus*) scavengers across the globe (Clark and Scheuhammer 2003; Hernández and Margalida 2009b; Kelly and Johnson 2011; Kelly et al. 2011). Although exposure to many toxicants can cause direct mortality, particularly in high doses, scavengers often may be exposed to low doses of toxicants that have sublethal effects such as reduced bone mineralization (Gangoso et al. 2009), reduced muscle and fat concentrations (Carpenter et al. 2003), organ damage and internal lesions (Pattee et al. 1981), and reduced hatching success (Steidl et al. 1991). Such effects often go unnoticed and thus are likely underreported in many species (Senthil Kumar et al. 2003).

6.8 Conclusions and Perspectives

Despite the fact that there are few obligate vertebrate scavengers, carrion use by vertebrates has evolved as a pervasive foraging strategy among most vertebrate taxa (DeVault et al. 2003). Yet, we have only begun to recognize and appreciate the important role vertebrate scavengers play in maintaining the stability and structure of food webs (Wilson and Wolkovich 2011). Indeed, in some ecosystems, vertebrates may assimilate as much or more carrion biomass than invertebrates or microbes (Houston 1986; DeVault et al. 2011), maintaining carrion-derived energy resources higher within food webs. Scavenging itself is also an activity that spans trophic levels and can link species through feeding relationships that otherwise would not be associated in food webs. Thus, effects of vertebrate scavengers should not be overlooked in nutrient cycling models or by researchers studying invertebrate and microbial decomposers.

As highlighted, carrion removal by vertebrates is not only important in maintaining biodiversity, but scavenging guilds also play an important role in provisioning numerous ecosystem services, including the regulation of some diseases (Markandya et al. 2008). Current anthropogenic activities threaten the stability and persistence of many of these communities, which in turn may diminish the important ecosystem services provided by carrion consumers. Such activities may not only affect vertebrate scavengers, but also directly impact invertebrate and microbial decomposers by altering carrion availability and may indirectly impact these species by disrupting competitive interactions among vertebrates, microbes, and invertebrates.

Despite the numerous advancements achieved in food-web research over the last few decades, the importance of carrion to the survival and reproduction of facultative scavengers remains unknown for many species. Moreover, the impact of changes in carrion availability on ecosystem-level dynamics and interkingdom competitive interactions remains an important area of research. In particular, a comprehensive evaluation of the fate of carrion, and the nutrients sequestered therein, in response to various biotic and abiotic alterations to ecosystems to due anthropogenic activities, remains an area of critical need in scavenging research.

REFERENCES

Alvarez, F., L. de Reyna, and F. Hiraldo. 1976. Interactions among avian scavengers in southern Spain. *Ornis Scandinavica* 7:215–226.

Barton, P.S., S.A. Cunningham, D.B. Lindenmayer, and A.D. Manning. 2013. The role of carrion in maintaining biodiversity and ecological processes in terrestrial ecosystems. *Oecologia* 171:761–772.

Beasley, J.C., W.S. Beatty, T.C. Atwood, S.R. Johnson, and O.E. Rhodes, Jr. 2012a. A comparison of methods for estimating raccoon abundance: Implications for disease vaccination programs. *Journal of Wildlife Management* 76:1290–1297.

Beasley, J.C., Z.H. Olson, and T.L. DeVault. 2012b. Carrion cycling in food webs: Comparisons among terrestrial and marine ecosystems. *Oikos* 121:1021–1026.

Beasley, J.C., Z.H. Olson, G. Dharmarajan, T.S. Eagan II, and O.E. Rhodes, Jr. 2011. Spatio-temporal variation in the demographic attributes of a generalist mesopredator. *Landscape Ecology* 26:937–950.

Bellan, S.E., P.C.B. Turnbull, W. Beyer, and W.M. Getz. 2013. Effects of experimental exclusion of scavengers from carcasses of anthrax-infected herbivores on *Bacillus anthracis* sporulation, survival, and distribution. *Applied Environmental Microbiology* 79:3756–3761.

Bennetts, R.E. and B.R. McClelland. 1997. Influence of age and prey availability on bald eagle foraging behavior at Glacier National Park, Montana. *Wilson Bulletin* 109:393–409.

Bergmann, M., T. Soltwedel, and M. Klages. 2011. The interannual variability of megafaunal assemblages in the Arctic deep sea: Preliminary results from the HAUSGARTEN observatory (79°N). *Deep-Sea Research Part I—Oceanographic Research Papers* 58:711–723.

Berny, P. 2007. Pesticides and the intoxication of wild animals. *Journal of Veterinary Pharmacology and Therapeutics* 30:93–100.

Berny, P. and J.R. Gaillet. 2008. Acute poisoning of red kites (*Milvus milvus*) in France: Data from the SAGIR network. *Journal of Wildlife Diseases* 44:417–426.

Blanchard, B.M. and R.R. Knight. 1990. Reactions of grizzly bears, *Ursus arctos horribilis*, to wildfire in Yellowstone National Park, Wyoming. *Canadian Field-Naturalist* 104:592–594.

Blázquez, M., J.A. Sánchez-Zapata, F. Botella, M. Carrete, and S. Eguía. 2009. Spatio-temporal segregation of facultative avian scavengers at ungulate carcasses. *Acta Oecologica* 35:645–650.

Blumstein, D.T. 2006. Developing an evolutionary ecology of fear: How life history and natural history traits affect disturbance tolerance in birds. *Animal Behaviour* 71:389–399.

Bramble, D.M. and D.E. Lieberman. 2004. Endurance running and the evolution of *Homo*. *Nature* 432:345–352.

Brown, S.P., R. Fredrik Inglis, and F. Taddei. 2009. Synthesis: Evolutionary ecology of microbial wars: Within-host competition and (incidental) virulence. *Evolutionary Applications* 2:32–39.

Bump, J.K., R.O. Peterson, and J.A. Vucetich. 2009. Wolves modulate soil nutrient heterogeneity and foliar nitrogen by configuring the distribution of ungulate carcasses. *Ecology* 90:3159–3167.

Burkepile, D., J. Parker, and C. Woodson. 2006. Chemically mediated competition between microbes and animals: Microbes as consumers in food webs. *Ecology* 87:2821–2831.

Cade, T.J. 2007. Exposure of California condors to lead from spent ammunition. *Journal of Wildlife Management* 71:2125–2133.

Carpenter, J.W., O.H. Pattee, S.H. Fritts, B.A. Rattner, S.N. Wiemeyer, J.A. Royle, and M.R. Smith. 2003. Experimental lead poisoning in turkey vultures (*Cathartes aura*). *Journal of Wildlife Diseases* 39:96–104.

Carter, D.O., D. Yellowlees, and M. Tibbett. 2007. Cadaver decomposition in terrestrial ecosystems. *Naturwissenschaften* 94:12–24.

Clark, A.J. and A.M. Scheuhammer. 2003. Lead poisoning in upland-foraging birds of prey in Canada. *Ecotoxicology* 12:23–30.

Cortés-Avizanda, A., R. Jovani, M. Carrete, and J.A. Donázar. 2012. Resource unpredictability promotes species diversity and coexistence in an avian scavenger guild: A field experiment. *Ecology* 93:2570–2579.

Cortés-Avizanda, A., N. Selva, M. Carrete, and J.A. Donázar. 2009. Effects of carrion resources on herbivore spatial distribution are mediated by facultative scavengers. *Basic and Applied Ecology* 10:265–272.

Cuthbert, R., M.A. Taggart, V. Prakash, M. Saini, D. Swarup, S. Upreti, R. Mateo et al. 2011. Effectiveness of action in India to reduce exposure of *Gyps* vultures to the toxic veterinary drug diclofenac. *PLoS ONE* 6:e19069.

Danell, K., D. Berteaux, and K.A. Bråthen. 2002. Effect of muskox carcasses on nitrogen concentration in tundra vegetation. *Arctic* 55:389–392.

Denno, R.F. and W.R. Cothran. 1976. Competitive interactions and ecological strategies of Sarcophagid and Calliphorid flies inhabiting rabbit carrion. *Annals of the Entomological Society of America* 69:109–113.

DeVault, T.L., J.C. Beasley, Z.H. Olson, M. Moleón, M. Carrete, A. Margalida, and J.A. Sánchez-Zapata. 2016. Ecosystem services provided by avian scavengers. In: C.H. Şekercioğlu, D.G. Wenny, and C.J. Whelan (eds), *Why Do Birds Matter? Avian Ecological Function and Ecosystem Services*. University of Chicago Press (Forthcoming).

DeVault, T.L., I.L. Brisbin Jr, and O.E. Rhodes Jr. 2004b. Factors influencing the acquisition of rodent carrion by vertebrate scavengers and decomposers. *Canadian Journal of Zoology* 82:502–509.

DeVault, T.L. and A.R. Krochmal. 2002. Scavenging by snakes: An examination of the literature. *Herpetologica* 58:429–436.

DeVault, T.L., Z.H. Olson, J.C. Beasley, and O.E. Rhodes Jr. 2011. Mesopredators dominate competition for carrion in an agricultural landscape. *Basic and Applied Ecology* 12:268–274.

DeVault, T.L., B.D. Reinhart, I.L. Brisbin, Jr, and O.E. Rhodes, Jr. 2004a. Home ranges of sympatric black and turkey vultures in South Carolina. *Condor* 106:706–711.

DeVault, T.L., O.E. Rhodes, Jr, and J.A. Shivik. 2003. Scavenging by vertebrates: Behavioral, ecological, and evolutionary perspectives on an important energy transfer pathway in terrestrial ecosystems. *Oikos* 102:225–234.

Donázar, J.A. 1993. *Los buitresibéricos*. Ed. Quercus, Madrid.

Donázar, J.A., M.A. Naveso, J.L. Tella, and D. Campión. 1997. Extensive grazing and raptors in Spain. In: D. Pain and M.W. Pienkowsky (eds), *Farming and Birds in Europe*. Academic Press, London, pp. 117–149.

Dudley, J.P. 1996. Record of carnivory, scavenging and predation for *Hippopotamus amphibius* in Hwange National Park, Zimbabwe. *Mammalia* 60:486–488.

Dunne, J.A., R.J. Williams, and N.D. Martinez. 2002. Network structure and biodiversity loss in food webs: Robustness increases with connectance. *Ecology Letters* 5:558–567.

Eason, C.T. and E.B. Spurr. 1995. Review of the toxicity and impacts of brodifacoum on non-target wildlife in New Zealand. *New Zealand Journal of Zoology* 22:371–379.

Estes, J.A., J. Terborgh, J.S. Brahsares, M.E. Power, J. Berger, W.J. Bond, S.R. Carpenter et al. 2011. Trophic downgrading of planet earth. *Science* 333:301–306.

Foley, J.A., R. DeFries, G.P. Asner, C. Barford, G. Bonan, S.R. Carpenter, F.S. Chapin et al. 2005. Global consequences of land use. *Science* 309:570–574.

Freeland, W.J. and D.H. Janzen. 1974. Strategies in herbivory by mammals: The role of plant secondary compounds. *The American Naturalist* 108:269–289.

Fuglei, E., N.A. Øritsland, and P. Prestrud. 2003. Local variation in arctic fox abundance on Svalbard, Norway. *Polar Biology* 26:93–98.

Furness, R.W. 2003. Impacts of fisheries on seabird communities. *Scientia Marina* 67:33–45.

Gangoso, L., P. Álvarez-Lloret, A.A.B. Rodríguez-Navarro, R. Mateo, F. Hiraldo, and J.A. Donázar. 2009. Long-term effects of lead poisoning on bone mineralization in vultures exposed to ammunition sources. *Environmental Pollution* 157:569–574.

Gese, E.M., R.L. Ruff, and R.L. Crabtree. 1996. Foraging ecology of coyotes (*Canis latrans*): The influence of extrinsic factors and a dominance hierarchy. *Canadian Journal of Zoology* 74:769–783.

Gibbs, J.P. and E.J. Stanton. 2001. Habitat fragmentation and arthropod community change: Carrion beetles, phoretic mites, and flies. *Ecological Applications* 11:79–85.

Green, G.I., D.J. Mattson, and J.M. Peek. 1997. Spring feeding on ungulate carcasses by grizzly bears in Yellowstone National Park. *Journal of Wildlife Management* 61:1040–1055.

Green, R.E., M.A. Taggart, D. Das, D.J. Pain, C.S. Kumar, A.A. Cunningham, and R. Cuthbert. 2004. Diclofenac poisoning as a cause of vulture population declines across the Indian subcontinent. *Journal of Applied Ecology* 41:793–800.

Hanski, I. and S. Kuusela. 1980. The structure of carrion fly communities: Differences in breeding seasons. *Annales Zoologici Fennici* 17:185–190.

Harvell, C.D., C.E. Mitchell, J.R. Ward, S. Altizer, A.P. Dobson, R.S. Ostfeld, and M.D. Samuel. 2002. Climate warming and disease risks for terrestrial and marine biota. *Science* 296:2158–2162.

Hebblewhite, M. and D. Smith. 2010. Wolf community ecology: Ecosystem effects of recovering wolves in Banff and Yellowstone National Park. In: M.P.C. Musiani and P.C. Paquet (eds), *The Wolves of the World: New Perspectives on Ecology, Behavior, and Policy*. University of Calgary Press, Calgary, Alberta, pp. 69–120.

Hernández, M. and A. Margalida. 2008. Pesticide abuse in Europe: Effects on the Cinereous vulture (*Aegypius monachus*) population in Spain. *Ecotoxicology* 17:264–272.

Hernández, M. and A. Margalida. 2009a. Poison-related mortality effects in the endangered Egyptian vulture (*Neophron percnopterus*) population in Spain. *European Journal of Wildlife Research* 55:415–423.

Hernández, M. and A. Margalida. 2009b. Assessing the risk of lead exposure for the conservation of the endangered Pyrenean bearded vulture (*Gypaetus barbatus*) population. *Environmental Research* 109:837–842.

Hewson, R. 1995. Use of salmonid carcasses by vertebrate scavengers. *Journal of Zoology* 235:53–65.

Houston, D.C. 1974. Food searching in griffon vultures. *East African Wildlife Journal* 12:63–77.

Houston, D.C. 1979. The adaptation of scavengers. In: A.R.E. Sinclair and N. Griffiths (eds), *Serengeti, Dynamics of an Ecosystem*. University of Chicago Press, Chicago, pp. 263–286.

Houston, D.C. 1986. Scavenging efficiency of turkey vultures in tropical forest. *The Condor* 88:318–323.

Houston, D.C. 1988. Competition for food between Neotropical vultures in forest. *Ibis* 130:402–417.

Houston, D. and J. Cooper. 1975. The digestive tract of the whiteback griffon vulture and its role in disease transmission among wild ungulates. *Journal of Wildlife Diseases* 11:306–313.

Houston, C.S., P.D. McLoughlin, J.T. Mandel, M.J. Bechard, M.J. Stoffel, D.R. Barber, and K.L. Bildstein. 2011. Breeding home ranges of migratory turkey vultures near their northern limit. *Wilson Journal of Ornithology* 123:472–478.

Huijbers, C.M., T.A. Schlacher, D.S. Schoeman, M.A. Weston, and R.M. Connolly. 2013. Urbanisation alters processing of marine carrion on sandy beaches. *Landscape and Urban Planning* 119:1–8.

Hunt, W.G., W. Burnham, C.N. Parish, K.K. Burnham, B. Mutch, and J.L. Oaks. 2006. Bullet fragments in deer remains: Implications for lead exposure in avian scavengers. *Wildlife Society Bulletin* 34:167–170.

Hunter, J.S., S.M. Durant, and T.M. Caro. 2007. Patterns of scavenger arrival at cheetah kills in Serengeti National Park Tanzania. *African Journal of Ecology* 45:275–281.

Janzen, D. 1977. Why fruits rot, seeds mold, and meat spoils. *The American Naturalist* 111:691–713.

Jennelle, C., M.D. Samuel, C.A. Nolden, and E.A. Berkley. 2009. Deer carcass decomposition and potential scavenger exposure to chronic wasting disease. *Journal of Wildlife Management* 73:655–662.

Johnson, M.D. 1975. Seasonal and microseral variations in the insect populations on carrion. *American Midland Naturalist* 93:79–90.

Kapfer, J., D. Gammon, and J. Groves. 2011. Carrion-feeding by barred owls (*Strix varia*). *The Wilson Journal of Ornithology* 123:646–649.

Kelly, T.R., P.H. Bloom, S.G. Torres, Y.Z. Hernandez, R.H. Poppenga, W.M. Boyce, and C.K. Johnson. 2011. Impact of the California lead ammunition ban on reducing lead exposure in golden eagles and turkey vultures. *PLoS ONE* 6:e17656.

Kelly, T.R. and C.K. Johnson. 2011. Lead exposure in free-flying turkey vultures is associated with big game hunting in California. *PLoS ONE* 6:e15350.

Killengreen, S.T., E. Strømseng, N.G. Yoccoz, and R.A. Ims. 2012. How ecological neighbourhoods influence the structure of the scavenger guild in low arctic tundra. *Diversity and Distributions* 18:563–574.

Kjorlien, Y.P., O.B. Beattie, and A.E. Peterson. 2009. Scavenging activity can produce predictable patterns in surface skeletal remains scattering: Observations and comments from two experiments. *Forensic Science International* 188:103–106.

Krofel, M., I. Kos, and K. Jerina. 2012. The noble cats and the big bad scavengers: Effects of dominant scavengers on solitary predators. *Behavioral Ecology and Sociobiology* 66:1297–1304.

Kruuk, H. 1967. Competition for food between vultures in East Africa. *Ardea* 55:171–193.

Kruuk, H. 1972. *The Spotted Hyena: A Study of Predation and Social Behavior*. University of Chicago Press, Chicago.

Lambertucci, S.A., J.A. Donázar, A.D. Huertas, B. Jiménez, M. Sáez, J.A. Sanchez-Zapata, and F. Hiraldo. 2011. Widening the problem of lead poisoning to a South-American top scavenger: Lead concentrations in feathers of wild Andean condors. *Biological Conservation* 144:1464–1471.

Langley, P.J.W. and D.W. Yalden. 1977. The decline of rarer carnivores in Great Britain during the nineteenth century. *Mammal Review* 7:95–162.

Lima, S.L. and L.M. Dill. 1990. Behavioral decisions made under risk of predation: A review and prospectus. *Canadian Journal of Zoology* 68:619–640.

Margalida, A., M. Carrete, J.A. Sánchez-Zapata, and J.A. Donázar. 2012. Good news for European vultures. *Science* 335:284.

Margalida, A. and M.A. Colomer. 2012. Modelling the effects of sanitary policies on European vulture conservation. *Scientific Reports* 2:753.

Margalida, A., M.A. Colomer, and D. Sanuy. 2011. Can wild ungulate carcasses provide enough biomass to maintain avian scavenger populations? An empirical assessment using a bio-inspired computational model. *PLoS ONE* 6:e20248.

Margalida, A., J.A. Donázar, M. Carrete, and J.A. Sánchez-Zapata. 2010. Sanitary versus environmental policies: Fitting together two pieces of the puzzle of European vulture conservation. *Journal of Applied Ecology* 47:931–935.

Markandya, A., T. Taylor, A. Longo, M.N. Murty, S. Murty, and K. Dhavala. 2008. Counting the cost of vulture decline—An appraisal of the human health and other benefits of vultures in India. *Ecological Economics* 67:194–204.

Márquez, C., J.M. Vargas, R. Villafuerte, and J.E. Fa. 2013. Understanding the propensity of wild predators to illegal poison baiting. *Animal Conservation* 16:118–129.

Mattson, D.J. 1997. Use of ungulates by Yellowstone grizzly bears *Ursus arctos*. *Biological Conservation* 81:161–177.

Matuszewski, S., D. Bajerlein, S. Konwerski, and K. Szpila. 2011. Insect succession and carrion decomposition in selected forests of Central Europe. Part 3: Succession of carrion fauna. *Forensic Science International* 207:150–163.

McCann, K. 2000. The diversity-stability debate. *Nature* 405:228–233.

McCann, K., A. Hastings, and G.R. Huxel. 1998. Weak trophic interactions and the balance of nature. *Nature* 395:794–798.

Meehan, E.P., E.E. Seminet-Reneau, and T.P. Quinn. 2005. Bear predation on Pacific salmon facilitates colonization of carcasses by fly maggots. *American Midland Naturalist* 153:142–151.

Melis, C., N. Selva, I. Teurlings, C. Skarpe, J.D.C. Linnell, and R. Andersen. 2007. Soil and vegetation nutrient response to bison carcasses in Bialowieza Primeval Forest, Poland. *Ecological Research* 22:807–813.

Millennium Ecosystem Assessment. 2003. *Ecosystems and Human Well-Being: A Framework for Assessment*. Millennium Ecosystem Assessment, Washington, D.C.

Moore, J.C., E.L. Berlow, D.C. Coleman, P.C. deRuiter, Q. Dong, A. Hastings, N.C. Johnson et al. 2004. Detritus, trophic dynamics and biodiversity. *Ecology Letters* 7:584–600.

Myers, R.A., J.K. Baum, T.D. Sheperd, S.P. Powers, and C.H. Peterson. 2007. Cascading effects of the loss of apex predatory sharks from a coastal ocean. *Science* 315:1846–1850.

Oaks, J.L., M. Gilbert, M.Z. Virani, R.T. Watson, C.U. Meteyer, B.A. Rideout, H.L. Shivaprasad et al. 2004. Diclofenac residues as the cause of vulture population decline in Pakistan. *Nature* 427:630–633.

Ogada, D.L., M.E. Torchin, M.F. Kinnaird, and V.O. Ezenwa. 2012. Effects of vulture declines on facultative scavengers and potential implications for mammalian disease transmission. *Conservation Biology* 26:453–460.

Olson, Z.H., J.C. Beasley, T.L. DeVault, and O.E. Rhodes, Jr. 2012. Scavenger community response to the removal of a dominant scavenger. *Oikos* 121:77–84.

Paczkowski, S., F. Maibaum, M. Paczkowska, and S. Schütz. 2012. Decaying mouse volatiles perceived by *Calliphora vicina* Rob.-Desv. *Journal of Forensic Sciences* 57:1497–1506.

Page, K., J.C. Beasley, Z.H. Olson, T.J. Smyser, M. Downey, K.F. Kellner, S.E. McCord et al. 2011. Reducing *Baylisascaris procyonis* roundworm larvae in raccoon latrines. *Emerging Infectious Diseases* 17:90–93.

Pain, D.J., A.A. Cunningham, P.F. Donald, J.W. Duckworth, D.C. Houston, T. Katzner, J. Parry-Jones et al. 2003. Causes and effects of temporospatial declines of *Gyps* vultures in Asia. *Conservation Biology* 17:661–671.

Parmenter, R.R. and J.A. MacMahon. 2009. Carrion decomposition and nutrient cycling in a semiarid shrub-steppe ecosystem. *Ecological Monographs* 79:637–661.

Pattee, O.H., S.N. Wiemeyer, B.M. Mulhern, L. Sileo, and J.W. Carpenter. 1981. Experimental lead-shot poisoning in Bald Eagles. *Journal of Wildlife Management* 45:806–810.

Patz, J.A., P.R. Epstein, T.A. Burke, and J.M. Balbus. 1996. Global climate change and emerging infectious diseases. *Journal of the American Medical Association* 275:217–223.

Payne, J.A. 1965. A summer carrion study of the baby pig *Sus scrofa* Linnaeus. *Ecology* 592–602.

Payne, L.X. and J.W. Moore. 2006. Mobile scavengers create hotspots of freshwater productivity. *Oikos* 115:69–80.

Petrides, G.A. 1959. Competition for food between five species of East African vultures. *The Auk* 76:104–106.

Platt, S.G., G.T. Salmon, S.M. Miller, and T.R. Rainwater. 2010. Scavenging by a bobcat, *Lynx rufus*. *Canadian Field Naturalist* 124:265–267.

Polis, G.A. 1991. Complex trophic interactions in deserts: An empirical critique of food-web theory. *The American Naturalist* 138:123–155.

Polis, G.A., M.E. Power, and G.R. Huxel (eds). 2004. *Food Webs at the Landscape Level*. University of Chicago Press, Chicago, IL, USA, 548pp.

Polis, G.A. and D.R. Strong. 1996. Food web complexity and community dynamics. *The American Naturalist* 147:813–846.

Prakash, V., D.J. Pain, A.A. Cunningham, P.F. Donald, N. Prakash, A. Verma, R. Gargi et al. 2003. Catastrophic collapse of Indian white-backed *Gyps bengalensis* and long-billed *Gyps indicus* vulture population. *Biological Conservation* 109:381–390.

Prugh, L.R., C.J. Stoner, C.W. Epps, W.T. Bean, W.J. Ripple, A.S. Laliberte, J.S. Brashares. 2009. The rise of the mesopredator. *Bioscience* 59:779–791.

Putman, R.J. 1978a. Flow of energy and organic matter from a carcase during decomposition: Decomposition of small mammal carrion in temperate systems 2. *Oikos* 31:58–68.

Putman, R.J. 1978b. Patterns of carbon dioxide evolution from decaying carrion decomposition of small mammal carrion in temperate systems 1. *Oikos* 31:47–57.

Reeves, N.M. 2009. Taphonomic effects of vulture scavenging. *Journal of Forensic Sciences* 54:523–528.

Rooney, N., K. McCann, G. Gellner, and J.C. Moore. 2006. Structural asymmetry and the stability of diverse food webs. *Nature* 442:265–269.

Rooney, T.P. and D.M. Waller. 2003. Direct and indirect effects of white-tailed deer in forest ecosystems. *Forest Ecology and Management* 181:165–176.

Rose, M.D. and G.A. Polis. 1998. The distribution and abundance of coyotes: The effects of allochthonous food subsidies from the sea. *Ecology* 79:998–1007.

Rozen, D.E., D.J.P. Engelmoer, and P.T. Smiseth. 2008. Antimicrobial strategies in burying beetles breeding on carrion. *Proceedings of the National Academy of Sciences* 105:17890–17895.

Ruxton, G.D. and D.C. Houston. 2003. Could *Tyrannosaurus rex* have been a scavenger rather than a predator? An energetics approach. *Proceedings of the Royal Society B—Biological Sciences* 270:731–733.

Ruxton, G.D. and D.C. Houston. 2004. Energetic feasibility of an obligate marine scavenger. *Marine Ecology Progress Series* 266:59–63.

Ruxton, G.D. and D.M. Wilkinson. 2011. Thermoregulation and endurance running in extinct hominins: Wheeler's models revisited. *Journal of Human Evolution* 61:169–175.

Ruxton, G.D. and D.M. Wilkinson. 2012. Endurance running and its relevance to scavenging by early hominins. *Evolution* 67:861–867.

Schlacher, T.A., S. Strydon, R.M. Connolly, and D. Schoeman. 2013. Donor-control of scavenging food webs at the land–ocean interface. *PLoS ONE* 8:e68221.

Scott, M.P. 1998. The ecology and behavior of burying beetles. *Annual Review of Entomology* 43:595–618.

Selva, N., B. Jedrzejewska, W. Jedrzejewski, and A. Wajrak. 2003. Scavenging on European bison carcasses in Bialowieza Primeval Forest (eastern Poland). *Ecoscience* 10:303–311.

Selva, N., B. Jedrzejewska, W. Jedrzejewski, and A. Wajrak. 2005. Factors affecting carcass use by a guild of scavengers in European temperate woodland. *Canadian Journal of Zoology* 83:1590–1601.

Senthil Kumar, K.S., W.W. Bowerman, T.L. DeVault, T. Takasuga, O.E. Rhodes, Jr, I.L. Brisbin, Jr, and S. Masunaga. 2003. Chlorinated hydrocarbon contaminants in blood of black and turkey vultures from Savannah River Site, South Carolina, USA. *Chemosphere* 53:173–182.

Servín, J., S.L. Lindsey, and B.A. Loiselle. 2001. Pileated woodpecker scavenges on a carcass in Missouri. *Wilson Bulletin* 113:249–250.

Shivik, J.A. 2006. Are vultures birds, and do snakes have venom, because of macro- and microscavenger conflict? *Bioscience* 56:819–823.

Shivik, J.A. and L. Clark. 1999. Ontogenetic shifts in carrion attractiveness to brown tree snakes (*Boiga irregularis*). *Journal of Herpetology* 33:334–336.

Shultz, S., H.S. Baral, S. Charman, A.A. Cunningham, D. Das, G.R. Ghalsasi, M.S. Goudar et al. 2004. Diclofenac poisoning is widespread in declining vulture populations across the Indian subcontinent. *Proceedings of the Royal Society B—Biological Sciences* 271:S458–S460.

Slate, D., C.E. Rupprecht, J.A. Rooney, D. Donovan, D.H. Lein, and R.B. Chipman. 2005. Status of oral rabies vaccination in wild carnivores in the United States. *Virus Research* 111:68–76.

Smith, C.R. 1985. Food for the deep sea: Utilization, dispersal, and flux of nekton falls at the Santa Catalina Basin floor. *Deep-Sea Research I* 32:417–442.

Smith, C.R. and A.R. Baco. 2003. Ecology of whale falls at the deep-sea floor. *Oceanography and Marine Biology* 41:311–354.

Smith, C.R., F.C. De Leo, A.F. Bernardino, A.K. Sweetman, and P.M. Arbizu. 2008. Abyssal food limitation, ecosystem structure and climate change. *Trends in Ecology and Evolution* 23:518–528.

Smith, S. and R. Paselk. 1986. Olfactory sensitivity of the turkey vulture (*Cathartes aura*) to three carrion-associated odorants. *The Auk* 103:586–592.

Steidl, R.J., C.R. Griffin, and L.J. Niles. 1991. Contaminant levels of osprey eggs and prey reflect regional differences in reproductive success. *Journal of Wildlife Management* 55:601–608.

Sugiura, S., R. Tanaka, H. Taki, and N. Kanzaki. 2013. Differential responses of scavenging arthropods and vertebrates to forest loss maintain ecosystem function in a heterogeneous landscape. *Biological Conservation* 159:206–213.

Swift, M.J., O.W. Heal, and J.M. Anderson (eds). 1979. *Decomposition in Terrestrial Ecosystems*. University of California Press, Los Angeles, CA.

Swihart, R.K., T.M. Gehring, M.B. Kolozvary, and T.E. Nupp. 2003. Responses of "resistant" vertebrates to habitat loss and fragmentation: The importance of niche breadth and range boundaries. *Diversity and Distributions* 9:1–18.

Towne, E.G. 2000. Prairie vegetation and soil nutrient responses to ungulate carcasses. *Oecologia* 122:232–239.

Vass, A.A. 2001. Beyond the grave—Understanding human decomposition. *Microbiology Today* 28:190–193.

Vass, A.A., W.M. Bass, J.D. Wolt, J.E. Foss, and J.T. Ammons. 1992. Time since death determinations of human cadavers using soil solution. *Journal of Forensic Science* 37: 1236–1253.

VerCauteren, K.C., J.L. Pilon, P.B. Nash, G.E. Phillips, and J.W. Fischer. 2012. Prion remains infectious after passage through digestive system of American crows (*Corvus brachyrhynchos*). *PLoS ONE* 7:e45774.

Vucetich, J.A., R.O. Peterson, and T.A. Waite. 2004. Raven scavenging favours group foraging in wolves. *Animal Behaviour* 67:1117–1126.

Wallace, M.P. and S.A. Temple. 1987. Competitive interactions within and between species in a guild of avian scavengers. *The Auk* 104:290–295.

Wenny, D.G., T.L. DeVault, M.D. Johnson, D. Kelly, C.H. Sekercioglu, D.F. Tomback, and C.J. Whelan. 2011. The need to quantify ecosystem services provided by birds. *The Auk* 128:1–14.

Whitfield, D.P., A.H. Fielding, D.R.A. McLeod, and P.F. Haworth. 2004. Modelling the effects of persecution on the population dynamics of golden eagles in Scotland. *Biological Conservation* 119:319–333.

Wilmers, C.C., R.L. Crabtree, D.W. Smith, K.M. Murphy, and W.M. Getz. 2003a. Trophic facilitation by introduced top predators: Grey wolf subsidies to scavengers in Yellowstone National Park. *Journal of Animal Ecology* 72:909–916.

Wilmers, C.C. and E. Post. 2006. Predicting the influence of wolf-provided carrion on scavenger community dynamics under climate change scenarios. *Global Change Biology* 12:403–409.

Wilmers, C.C., D.R. Stahler, R.L. Crabtree, D.W. Smith, and W.M. Getz. 2003b. Resource dispersion and consumer dominance: Scavenging at wolf-and hunter-killed carcasses in greater Yellowstone, USA. *Ecology Letters* 6:996–1003.

Wilson, E.E. and E.M. Wolkovich. 2011. Scavenging: How carnivores and carrion structure communities. *Trends in Ecology and Evolution* 26:129–135.

Zhou, C. and R.W. Byard. 2011. Factors and processes causing accelerated decomposition in human cadavers—An overview. *Journal of Forensic and Legal Medicine* 18:6–9.

7

Design and Analysis of Field
Studies in Carrion Ecology

Kenneth G. Schoenly, J.-P. Michaud, and Gaétan Moreau

CONTENTS

> ...field experiments in ecology [usually] either have no replication, or have so few replicates as to have very little sensitivity...
>
> **L.L. Eberhardt (1978, cited in Hurlbert 1984, p. 187)**

> ...you must be thinking that pseudoreplication is a minefield that will cause you no end of trouble. Not so; forewarned is forearmed...
>
> **G.D. Ruxton and N. Colegrave (2011, p. 51)**

7.1 Introduction

As experimental units (EUs), carcasses make promising subjects for testing ecological concepts and theories. Ecological theory and experiments have shown that subdividing populations into spatially isolated patches (i.e., carcasses), linked by dispersing populations, can favor the coexistence of competitors (Atkinson and Shorrocks 1981; Kneidel 1985; Hanski 1987a; Barton et al. 2013) and form metacommunities (Leibold et al. 2004) and resource pulses (Rose and Polis 1998; Yang 2006; Yang et al. 2008).

Because carrion communities assemble from a regional species pool, they offer more realistic diversity gradients than hand-picked communities for investigating diversity–function relationships (Finn 2001). Indeed, given the potential of carcasses to attract up to 150 invertebrate families (Payne and Crossley 1966) and several vertebrate species (Wilson and Wolkovich 2011), a nearly inexhaustible number of community combinations are possible. Carrion has been the model for investigating nonequilibrium island dynamics (Beaver 1977), which embodies rapid successional changes before carcass tissues become exhausted. Rates of decomposition follow a Q_{10} (temperature coefficient) of around 2, representing a doubling rate of decay for every 10°C, at least for unconcealed carcasses in semi-arid ecosystems (Parmenter and MacMahon 2009). Such rapid decay permits both within- and between-season comparisons (Schoenly and Reid 1987), while avoiding space-for-time substitution (Pickett 1989), a practical but controversial method (Johnson and Miyanishi 2008; Walker et al. 2010) for analyzing communities with slow successional dynamics (e.g., oak-pine forest). Enclosing carcasses in screened cages (i.e., to prevent vertebrate scavenging) keeps them intact, simplifying sampling and observation. Carcasses are also easy to acquire and are affordable, thus permitting high replication of uniform size or along a size series (of the same or different species). In sum, as patchy, ephemeral, permutable, and replicable units, carcasses are model systems for investigating population and community interactions (Barton et al. 2013), aggregation and coexistence (Atkinson and Shorrocks 1981; Kneidel 1985; Hanski 1987a), successional dynamics (Schoenly and Reid 1987; Boulton and Lake 1988; Moura et al. 2005), diversity–function relationships (Finn 2001), metacommunity concepts (Leibold et al. 2004), resource pulses (Yang 2006; Yang et al. 2008), and forensic applications (Byrd and Castner 2010).

Long before Elton and Miller (1954) recognized carrion as an ecological habitat, nineteenth century medical examiners conducted field studies on human corpses and advanced the first formal definition and testable mechanism of ecological succession (Mégnin 1883, 1887, 1894; Johnston and Villeneuve 1897; Motter 1898). Despite this auspicious beginning, however, carrion ecology became a descriptive science and, since the 1970s, has mostly emphasized forensic applications. Taken together, this body of literature includes many valuable insights on biological agents, soft tissue decomposition, dismemberment, and weathering of vertebrate (including human) remains, but progress toward understanding ecological processes, such as community dynamics, languished. Unlike other biological disciplines (e.g., conservation biology and paleobiology), carrion ecology is not limited to observational studies and modeling; however, the discipline does depend on the availability of vertebrate animals, including donated human remains. Consequently, it may be difficult logistically and ethically to always conduct manipulative experiments with high replication, at several study sites, and in multiple years.

The aim of this chapter is to review field experiments, underscrutinized topics, and quantitative methods used by carrion ecologists (i.e., anthropologists, entomologists, and taphonomists). We limit our attention to variables, methods, and models that inform arthropod succession and decomposition of vertebrate (including human) remains. As we did for forensic entomology (Michaud et al. 2012; Moreau et al. 2015), we identify design flaws and knowledge gaps and suggest remedies for increasing inference strength and statistical power. We end the chapter with some forensic applications of carrion ecology. Thus, this chapter is both retrospective and proactive, in that we review the enormous gains researchers have made, while urging a stronger mechanistic and statistical framework for the future.

7.2 Study Designs and Field Practices

The publication of Sir Ronald Fisher's "The Arrangement of Field Experiments" in 1926 and his 1935 sequel, "The Design of Experiments," ushered in a new era of scientific analysis that led to widespread acceptance of four cornerstones of experimental design: randomization, replication, independence, and controls (Fisher 1926, 1935). This mix of design elements requires that an experiment be configured to minimize bias and to limit the effects of sampling error (Whitlock and Schluter 2009). Bias is minimized by using controls (contemporaneous or historical), randomly assigning EUs to treatments, and through blinding, whereas sampling error is reduced by replication, blocking, and a balanced design (Whitlock and Schluter 2009).

In manipulative experiments, treatment assignment is directly under investigator control; thus, treatment levels (e.g., effects of different fixed temperatures on insect development) can be randomized across EUs and formed into blocks (i.e., physical groupings that contain every combination of treatments). In contrast, observational studies lack investigator control of treatments (e.g., comparative study of postmortem decay rates at different sites), but can incorporate all other elements of manipulative experiments (i.e., replication, simultaneous controls, balance, blocking, and blinding). Advances in technology (e.g., remote sensing, microscopy, and computing) and the need to study ecological phenomena over wide spatiotemporal scales (afforded by museum records and old photographs) have brought greater integration of observational and experimental approaches (Sagarin and Pauchard 2010).

Ecological experiments impose inevitable tradeoffs between realism, precision, and generality (Morin 1998). For example, although field experiments have the advantage of realism, they suffer from low precision due to high between-replicate variation that can result from financial and logistical limitations. Similarly, laboratory experiments often have high precision and internal validity, but come at the expense of realism. In either situation, generality becomes elusive because high replication is required, conducted over many species, seasons, and sites, to disentangle common patterns (Morin 1998).

To achieve adequate inference strength (i.e., generalizability of conclusions to another time, place, and context), carrion ecologists must comply with three basic design rules: adequate replication of EUs, spatial and temporal independence of EUs, and efficient deployment of EUs to capture a representative range of natural variability.

7.2.1 Replication versus Pseudoreplication

An EU is the smallest unit to which a treatment is applied (Hurlbert 1984). Thus, when a treatment is applied to at least two independent EUs, it is "replicated." However, whenever a treatment is unreplicated (though samples may be), the study falls victim to *simple pseudoreplication* (Figure 7.1a); if the study is replicated, but samples are pooled without considering group structure in the data, the study falls victim to *sacrificial pseudoreplication* (Figure 7.1b), and if the replicates are statistically interdependent in space or time, the study falls victim to *spatial* or *temporal pseudoreplication* (Figure 7.1c), respectively (Hurlbert 1984).

7.2.2 Independence of EUs

In general, the more independent replicates in a study are, the greater their ability to detect treatment effects (Karban and Huntzinger 2006) and to reduce type 1 error rates. For succession and decomposition studies, independence of EUs implies an adequate spacing distance between simultaneously deployed carcasses and accounting for repetitive sampling of the same carcass over time (Michaud et al. 2012; Moreau et al. 2015). Moreover, carcasses that are closely spaced can become linked by dispersing organisms and favor the development of metacommunities (Leibold et al. 2004) that can synchronize regional dynamics.

7.2.3 Carcass Deployment in Heterogeneous Landscapes

To capture a representative range of natural variability in postmortem decomposition or succession for an entire ecosystem or geographical area, carcasses need to be deployed in multiple sites over multiple seasons. Deploying carcasses within long narrow rectangles (i.e., transects) will likely capture a wider range of biological variability and habitat heterogeneity than square or circular deployments of equal area (Goodall 1952; Krebs 2000; Pearson and Ruggiero 2003). Identifying ecotones in heterogeneous landscapes can help inform the meaning of "site" in carrion ecology. What constitutes an ecotone to humans, such as a transitional zone between two plant communities (e.g., prairie-forest ecotone), may be perceived differently by other organisms. Probability-based boundary detection methods (Cornelius and Reynolds 1991), combined with gradient-directed transects (Gillison and Brewer 1985), use community dissimilarity (or distance) measures to detect taxon-identified ecotones that are independent of

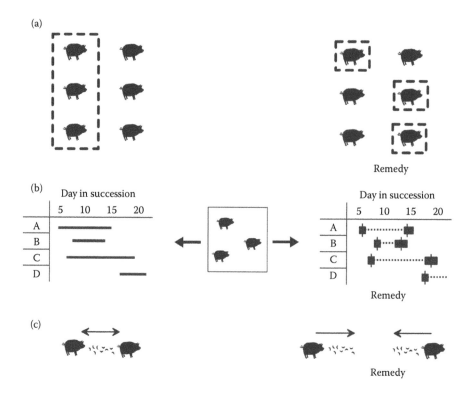

FIGURE 7.1 Schematic representation of three pseudoreplication errors discovered in the carrion literature (anthropology, entomology, and taphonomy). Panels on the right are suggested remedies: (a) simple pseudoreplication, (b) sacrificial pseudoreplications, and (c) spatial pseudoreplication.

human perceptions (Figure 7.2a through d). For example, if a fence restricts movements of walking and crawling animals, species identities and abundances on both sides of the fence will be different enough to constitute an ecotone, whereas flying animals may perceive the same region as a single habitat. Consequently, these methods could be used to help researchers pinpoint or concentrate carcass deployments in large heterogeneous landscapes. Using the same (carcass-by-species) data, bias-corrected collection curves (Colwell and Coddington 1994) could show how steep (or shallow) species richness rises as more carcasses are successively added to the collector's curve. Furthermore, the quantitative power of collectors' curves can be extended using extrapolation-based richness models (Colwell 2013) to estimate how many uncollected species have gone undetected by the researcher's sampling methods. Using the likelihood-ratio statistic and Fisher's exact test, LaMotte and Wells (2000) determined that future field studies would require a minimum of 25 carcasses to achieve a 10% level of significance to carry out succession-based estimates of time since death. This approach was recently field-tested in north-central Indiana, USA and found to be forensically promising when larvae of two blow fly species and adults of one rove beetle species were used singly and in pairs to render time-since-death estimates (Perez et al. 2014). These approaches differ from standard statistical methods that estimate minimum sample sizes (i.e., carcasses) needed for different sites or treatments for sufficient statistical power (Section 7.2.5).

7.2.4 Critique of Published Field Studies

The experimental design of field studies in the anthropology and taphonomy literature was reviewed, as was performed for entomology (Michaud et al. 2012; Moreau et al. 2015), post-Hurlbert (Hurlbert 1984) through 2013, using keyword and citation searches in Google Scholar, JSTOR, Thomson Scientific (ISI),

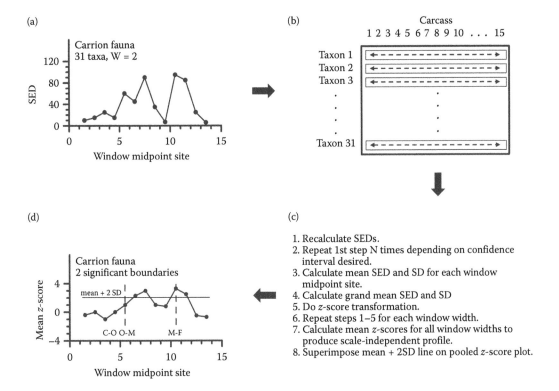

FIGURE 7.2 Bootstrap procedure for statistically locating taxon-defined boundaries, based on a hypothetical 15-carcass transect, using the algorithm of Cornelius and Reynolds (1991). In (a) standardized Euclidean distances are plotted for each window width (*W*); (b) species abundances are reshuffled; (c) statistical significance is calculated; and (d) the observed and expected *z*-scores are plotted. In this hypothetical example, carrion arthropods partition the rural landscape into three zones (C-O, crop field-orchard; M, meadow; F, forest), justifying future carcass deployments in each of those sites. (Redrawn from Zhang, W.J. and K.G. Schoenly. 1999. *Biodiversity Software Series III. BOUNDARY: A Program for Detecting Boundaries in Ecological Landscapes.* IRRI Technical Bulletin No. 3, Manila, Philippines: International Rice Research Institute.)

the Digital Commons, and other sources. Although the search included both ecological and forensic contributions, the latter group constituted the bulk of the carrion ecology literature published since 1984. The search sought to identify observational or manipulative field studies on postmortem decomposition in vertebrate (including human) remains from one or more habitats, seasons, or ecological circumstances. After evaluation, each study was placed into one of six categories (Michaud et al. 2012; Moreau et al. 2015): (1) adequate analysis and design (i.e., inferential study with a valid design and analysis); (2) descriptive study (i.e., study that did not draw inferences from the results); (3) pseudoreplication suspected (i.e., a study with ambiguously defined methodology in terms of identifying EUs); (4) simple pseudoreplication (see definition mentioned earlier); (5) sacrificial pseudoreplication (see definition mentioned earlier); and (6) inadequate analysis or design (i.e., study with a design or analysis that is inappropriate to answer the question).

In total, 160 studies, 19 for anthropology, 103 for entomology, and 38 for taphonomy, were reviewed. Over the 30-year period, studies in all three disciplines fell victim to pseudoreplication in a majority of cases (Figure 7.3), including two of the authors' own studies (Schoenly et al. 2007; Michaud and Moreau 2009). Although 11% of the anthropology field studies had adequate analysis and design, 68% had design flaws affecting inferential strength, whereas 21% were descriptive studies (Figure 7.3, left). Similarly, 19% of the entomology field studies had adequate analysis and design, but 77% had design flaws, whereas 4% were descriptive studies (Figure 7.3, middle). Finally, 13% of the taphonomy studies

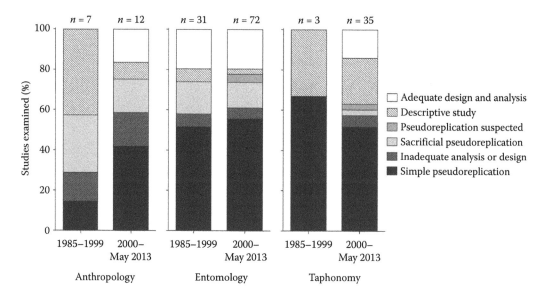

FIGURE 7.3 Frequency, in the anthropology (left), entomology (middle), and taphonomy (right) literatures, of studies on vertebrate (including human) remains displaying adequate analysis and design, descriptive results only, suspected cases of simple pseudoreplication, sacrificial pseudoreplication, inadequate analysis or design, and simple pseudoreplication.

had adequate analysis and design, but another 71% had flaws affecting inferential strength, whereas 16% were descriptive studies (Figure 7.3, right). In sum, the three disciplines had similar pseudoreplication frequencies (range: 68%–77%) but differed in the frequency of description-only content (range: 4%–21%). The descriptive studies that were reviewed did not allow for inferences, and their methodology prevented statistical assessments of effects.

For all three disciplines, simple pseudoreplication was due to the absence of treatment replication (Figure 7.1a); thus, it would be impossible to salvage those studies using linear mixed-effects models (Chaves 2010; Behm et al. 2013). Consequently, the only remedy for simple pseudoreplication is to repeat the study. In contrast, sacrificial pseudoreplication can be reversed by analyzing unpooled replicates (Figure 7.1b).

7.2.5 Replication, Effect Size, and Statistical Power

Even well-executed studies may lack sufficient power to detect significant differences between treatments. Statistical power is proportional to the number of replicates, effect size, and the significance level (i.e., alpha) and is inversely proportional to the variance of the population (Thomas and Juanes 1996). Including more replicates is often the easiest way to increase statistical power. High replication can also rescue imprecise (though not biased) measurements due to the properties of the central limit theorem (Zschokke and Ludin 2001). Given the variability of the study materials being estimated (e.g., carcass weight, arthropod abundance and presence/absence, and odor composition), a researcher should aim for a replicate size that will achieve adequate statistical power, usually 80% (or 0.8) or larger (Cohen 1988). Using published data from Kashyap and Pillay (1989), it was previously reported (Michaud et al. 2012; Moreau et al. 2015) that 16 cadavers were inadequate to detect a significant difference between three methods for estimating time since death (power = 0.38) using a two-tailed, one-sample t-test (H_o: $\mu = 0$, $\alpha = 0.05$, effect size = 0.44). At this effect size, 43 cadavers would be required to reach the desired power of 0.8 (Figure 7.4a). However, if the authors' results had yielded a larger effect size, say 0.8, only 15 cadavers (one less than was taken) would be required to achieve the desired power (Figure 7.4b). Also, analyzing the same data with different statistical models can yield different power results. For example, a power analysis of the authors' two-tailed, parametric correlation test (Pearson's r, H_o: $r = 0$, $\alpha = 0.05$,

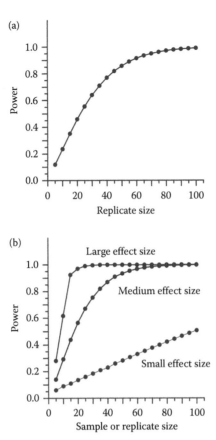

FIGURE 7.4 Statistical power as a function of (a) replicate size and (b) for three effect sizes (difference between PMI estimation methods; Kashyap and Pillay [1989]) using a two-tailed, one-sample *t*-test. In (a), the effect size was 0.44 (observed in the study) and the significance level was $\alpha = 0.05$. In (b), small, medium, and large effect sizes were set at 0.2, 0.5, and 0.8, respectively. In both figures, statistical power was determined using G*Power 3 software. (With kind permission from Springer Science+Business Media *Behavior Research Methods*, G*Power 3: A flexible statistical power analysis program for the social, behavioral, and biomedical sciences, 39, 2007, 175–191, Faul, F., E. Erdfelder, A.-G. Lang, and A. Buchner.)

observed $r = 0.813$) showed that 16 cadavers were sufficient to reject the null hypothesis (power = 0.987). Analyses such as these, when performed *a priori* (i.e., prospective power analysis), are instructive when planning future experiments and will increase the study's inferential strength. Freeware programs that perform power calculations are available on the internet, including Piface (Lenth 2006–2009) and G*Power (Faul et al. 2007).

7.3 Underscrutinized Topics

7.3.1 Decay Stages

Originally developed by medical examiners in the late 1800s, decay stages reflected the view that distinct changes in postmortem decay and arthropod succession occurred in human corpses in a deterministic sequence (i.e., the eight "squads" of Mégnin 1894). More recently, decay stages have been used as convenient "descriptors" for summarizing physicochemical changes in decomposing remains (e.g., Mann et al. 1990; Anderson and Van Laerhoven 1996; Smith and Baco 2003; Carter and Tibbett 2008) and as "reference points" to educate juries in the courtroom (Goff 1993). In another context, Michaud and Moreau (2011) showed that decay stages could be predicted from accumulated degree-days (ADDs). Today, the decay-stage view remains a dominant paradigm in the carrion literature, including entomology (e.g., Goff 1993), anthropology (e.g., Galloway et al. 1989), taphonomy (e.g., Vass et al. 1992; Tibbett and Carter 2009), and marine biology (e.g., whale falls [Smith and Baco 2003]).

However, decay stages represent typological thinking (Birks 1987) that encourages researchers to see boundaries where none exists or to cluster events around stages that defy easy classification (Peters 1991). Indeed, frustration with such typological thinking shifted ecological succession theory in the early 1900s (Simberloff 1980; McIntosh 1985; Peters 1991) away from the "superorganism" or stage-based view (Clements 1916, 1936) to the "individualistic" or continuum view (Gleason 1917, 1926). In addition, description and agreement on the number and duration of named decay stages vary between researchers, even among studies that share similar ecological and sampling features (Schoenly and Reid 1987). For example, Fuller (1934) described three decay stages, Reed (1958) described four stages, Bornemissza (1957) described five, and Payne (1965) six. Using univariate and multivariate statistics, carrion ecologists rebuked the stage-based view of arthropod succession (Schoenly and Reid 1987; Boulton and Lake 1988; Moura et al. 2005) when they found few spikes of community dissimilarity at purported stage boundaries. In actuality, carcass decomposition is a dynamic and continuous process that starts at death and continues until all soft tissues have been cycled—first aboveground and then belowground—into the larger landscape (e.g., Tibbett and Carter 2009). As such, decay stages evoke a stepwise process that may misinform more continuous succession and decomposition processes that occur on the carcass.

In our view, far too much journal space has been devoted to stage descriptions and agreement/disagreement with other authors, a situation that has slowed progress in testing ecological mechanisms and models. If decay stages are to have scientific utility, they should be verified statistically (Ordóñez et al. 2008; Lefebvre and Gaudry 2009; Michaud and Moreau 2011; Moreau et al. 2015), not just inferred or carried over from related studies or disciplines.

7.3.2 Cage and Trap Effects

Fuller (1934) was among the first carrion researchers to enclose carcasses inside screened cages to prevent vertebrate scavenging. More recently, Payne (1965) used cages with different mesh sizes to create insect-proof and insect-open cages to study insect succession and decomposition in dead pigs. Payne showed that carrion decomposition, based on weight loss measurements, accelerated in the presence of insects. Meanwhile, agricultural and medical entomologists were developing carrion-baited traps to monitor pest species (Bishopp 1916; Vogt and Havenstein 1974; Broce et al. 1977; Suenaga and Kurahashi 1994), which were modified for use by carrion ecologists and forensic entomologists (Erzinçlioğlu 1980; Schoenly 1981; Jirón 1984; Schoenly et al. 1991; Prado e Castro 2009).

Despite the widespread appeal of exclosure cages and bait traps in entomology and ecology, results from these studies should be interpreted with caution because they have the potential to affect the outcome of experiments (Luck et al. 1988; Hairston 1989). Any commercial product used in cage or trap construction will have a slight, moderate, or significant effect on one or more microclimate variables (Peterson 1953). In general, screening around cages and traps will shade the carcass inside and alter interior microclimates. For example, in an empty cube-shaped cage (183 cm on each side) wrapped by 20-mesh screen, Hand and Keaster (1967) found that solar radiation, rainfall, wind speed, and evaporation rates were reduced by 19%, 16%, 52%, and 20%, respectively, of outside conditions. Only daily air temperature and relative humidity differed insignificantly between the two environments (Hand and Keaster 1967). However, Schoenly et al. (1991) found that air and soil temperatures inside an empty coffin-shaped trap were, on average, 8°C and 9°C cooler, respectively, than outside conditions. Relative humidity and evaporation rates were also lower inside the coffin trap compared with outside conditions. Light penetration tests showed that the trap's screen mesh reduced light intensity to an average of 48% of outside readings (Schoenly et al. 1991). Similarly, Ordóñez et al. (2008) found significant differences in air temperature and absolute humidity when microclimates of carrion-baited traps were compared with outside conditions. The use of newer screening materials, such as those made from polyester resins (e.g., Lumite®), may offer higher light penetration while maintaining a tight and uniform mesh. Nonetheless, if cage or trap microclimates are underestimated or ignored, the result could lead to "undesirable properties of otherwise well-designed experiments" (Hairston 1989).

Cages and traps also have the potential for altering carcass decay rates and the behavior of necrophagous species and their natural enemies. For example, cooler temperatures inside cages and traps could reduce development times and activity levels of arthropod species. Although subdued light conditions inside cages are not known to prevent successful oviposition or larviposition by carrion-breeding flies, some calliphorid species are positively heliotropic (e.g., *Lucilia* spp.), preferring open sunlight to shade (Smith 1986). Moreover, exclosure cages designed as insect-free are not free of insects (Luck et al. 1988), as several carrion researchers have found (Payne 1965; Pechal et al. 2014; Comstock et al. 2015), even if cages are treated first with insecticides (van den Bosch et al. 1969; Irwin et al. 1974). This is because some insects will find (or make their own) openings in the screen (e.g., ants) or have immature stages small enough to penetrate the mesh. Whether exclosure cages affect predator–prey, parasitoid–host, or microbe–host interactions in carrion systems remains an open question.

Unlike a cage-enclosed carcass that is periodically hand-sampled for insects, carrion-baited traps collect insects automatically and continuously as the carcass slowly dries out or is consumed. The hand sample represents a "snapshot" (i.e., record of species present at that time) that depends on the collector's experience level and close contact with and manipulation of the carcass. Such "handling effects" may produce a biased and incomplete inventory that results in an inaccurate picture of carrion–insect colonization and succession (Ordóñez et al. 2008). In addition, few carrion studies have used unsampled control carcasses to test for qualitative differences in decomposition rates (Schoenly et al. 2007) or investigated the effect of variable sampling effort on rates of carcass decomposition and arthropod succession (Michaud and Moreau 2013).

In comparing carrion-fly catches from hand samples and bait traps, Ordóñez et al. (2008) found that trap collections were more repeatable (i.e., similar to one another) and caught more individuals and species than collections derived from hand samples. Moreover, the hand-sampled community was a completely nested subset of the trap-caught community and had three times more reoccurring taxa (i.e., gaps in occurrence times; Schoenly 1992) than trap collections. However, most carrion-baited traps obscure from view important events that occur on the carcass that are more visible in cage-enclosed carcasses (e.g., oviposition, maggot masses, predator–prey interactions, carcass changes, and skeletal disarticulation). Moreover, unless emerging insects can emigrate from traps or cages (Schoenly 1981), their confinement could increase predation and parasitism rates (Luck et al. 1988), which could slow carcass decay rates. Additional comparative studies are needed that test carcasses (and their constituent communities) in unenclosed, cage-enclosed, and trap-enclosed settings.

7.4 Analyzing Carrion-Arthropod Succession

7.4.1 Patterns of Succession

In his classic text, Mégnin (1894) described eight squads of arthropods (insects and mites) on human corpses that succeeded one another at different times during decomposition. More recently, Schoenly and Reid (1987) analyzed trends in succession in 11 published reports and found repeatable patterns in day-to-day changes in species richness, new arrivals, and local extinctions. In a second meta-analysis, Schoenly (1992) analyzed 23 carrion-arthropod studies and provided a set of testing protocols that incorporated null models and Monte Carlo methods to test hypotheses. The analysis revealed that some taxa (e.g., histerids, staphylinids, sarcophagids, muscids, phorids, and calliphorids) "reoccur" in the succession, with adults reoccurring more often than larvae (Schoenly 1992; Matuszewski et al. 2010). For ecological modeling and postmortem interval (PMI) estimation, residence time and high repeatability of occurrence on carcasses are important traits for choosing indicator taxa (Matuszewski et al. 2010). Reoccurring taxa may be the result of sampling inefficiency (Schoenly 1992; Moura et al. 2005; Anderson 2007), seasonality (Moura et al. 2005), weather disturbances (Matuszewski et al. 2010), gaps between generations (Matuszewski et al. 2010), or predation effects (Wells and Greenberg 1994). Whatever the cause, this finding underscores the need to disaggregate carcass records to reveal hidden variation (Michaud and Moreau 2009; Matuszewski et al. 2010). In comparing species accumulation curves for observed and simulated data, Schoenly (1992) showed that a majority of 23 studies followed a "clumped" model of carrion-arthropod succession (rather than "uniform" or "random" models), and within the "clumped" group, a majority followed a "clumped, early" model (rather than "clumped, mid-term" or "clumped, late"). The prevalence of the "clumped, early" model reflects the rapid accumulation of arthropod taxa that colonize carrion and is consistent with natural selection favoring specialized species with rapid dispersal abilities and short life cycles (Beaver 1984; Hanski 1987b).

Anderson (2007) analyzed temporal patterns in 62 plant and animal succession sequences, of which 17 came from carrion-arthropod reports. Across these systems, the author found that colonization of carrion-arthropod species increased rapidly during early succession, reached an asymptote, and decreased when species gains dropped below losses, as carrion researchers have found (Schoenly and Reid 1987; Schoenly 1992). The author also concluded that patterns of community assembly were consistent with both Gleasonian succession (Gleason 1917) and community stability (Odum 1969) and proposed that dispersal limitation governs succession and decomposition rates in carrion communities. However, dispersal studies on carrion-frequenting flies show that adults can disperse several kilometers between habitat patches (Section 7.2.2). For carrion-arthropod systems, a more likely driver of succession rates is trophic interaction (also proposed by Anderson 2007). For example, in two carrion studies that compared presence and absence of fire ants, ant-exposed carcasses had significantly more gaps in their successional timetables (Wells and Greenberg 1994) and had smaller population sizes of fly and beetle larvae as well as slower rates of carcass decomposition (Stoker et al. 1995), compared with unexposed (control) carcasses.

7.4.2 Mechanisms and Models

In the early 1900s, ecologists viewed a community undergoing succession as a "superorganism" that progressed stepwise and deterministically toward a stable or "mature" complex (Clements 1916, 1936). The Clementsian school emphasized the closed, sequential, directional, climactic, and predictable nature of the successional process. Vigorous debate followed, and since the 1930s, ecologists accepted a more "individualistic" view (Gleason 1917, 1926) that emphasized environmental stochasticity and life histories and interactions of individual species (Tansley 1939; Whittaker 1953; Connell and Slatyer 1977; McIntosh 1985). This paradigm shift, led by plant ecologists, has not resonated with carrion ecologists (Michaud et al. 2015). Indeed, the Clementsian (stage-based) view remains a dominant paradigm among entomologists, anthropologists, and taphonomists, especially in regard to the directional and predictable nature of succession and the emphasis put on decomposition stages (often referred to as microseres or seres, terms borrowed from plant ecologists).

For degradative and heterotrophic systems such as carrion, facilitation has been viewed as a likely succession mechanism (Connell and Slatyer 1977; Schoenly and Reid 1987; Smith and Baco 2003), in which the establishment of late arrivals is conditional upon the modifying actions of pioneer species. However, this is not supported by experimental evidence. Other mechanisms developed by Connell and Slatyer (1977) for plant and marine communities, such as tolerance and inhibition, may also be compatible with degradative, heterotrophic systems, but their "transferability" to carcasses requires experimental verification. Thus, all the carrion disciplines could greatly benefit from experimental tests of these basic mechanisms of succession (Michaud et al. 2015).

Markov chains (e.g., Norris 1998) are another class of succession models that have found appeal among plant and marine ecologists (reviewed by Hill et al. 2004), but not carrion ecologists. A Markov chain (or process) is a sequence of states (e.g., days in succession) of a system (e.g., coexisting populations), such that each future state (e.g., day 2) depends on the previous state (e.g., day 1) and the transitional probabilities between all possible states (see worked examples in Donovan and Welden 2002). The result is a transition matrix that captures future successional events (i.e., colonization, replacement, displacement, and persistence; Hill et al. 2004). Such models may also disentangle competing succession mechanisms (Usher 1987; Hill et al. 2004). However, because carcasses degrade over time and are studied retrospectively to obtain time-since-death estimates, time-reversed Markov models (Solow and Smith 2006) may find greater appeal among carrion ecologists.

Computer-intensive, Monte Carlo methods (Manly 1991; Gotelli and Graves 1996) can also elucidate succession and decomposition patterns in carrion. These methods recombine the original data (i.e., succession or decomposition timetables), either through reshuffling or resampling, to produce computer-generated pseudovalues. Each method uses the resulting pseudovalues to calculate statistics of central tendency or dispersion about the parameter of interest. Using bootstrapping and an algorithm for detecting patterns in flowering phenologies (Poole and Rathcke 1979), Schoenly (1992) quantified the degree to which arthropod communities in carrion overlap each other in successional time. Simulation results revealed a tendency of carrion-arthropod communities to form a "clumped, early" arrival pattern. Using the bootstrap and an algorithm for detecting differences in species co-occurrences on different carcass pairs, Nelder et al. (2009) found that blow fly succession was partly synchronous (i.e., predictable) on three alligator carcasses, depending on which species and carcass combinations were compared. Randomization tests can also determine the extent to which arthropod sequences differ between replicates or different subjects (e.g., human vs. pig). Comparing many (e.g., 10,000) random draws of the same taxa from different subjects could provide uncertainty and repeatability information (Schoenly et al. 2003).

Ecological null models, combined with randomization tests (Harvey et al. 1983; Colwell and Winkler 1984), have also been underexplored by carrion researchers. When properly constructed, an ecological null model is a pattern-generating model based on reshuffling observed data designed around an ecological mechanism (e.g., interspecific competition, facilitation, nutrient cycling). Although their utility and flaws were hotly debated in the 1970s, null models have regained popularity among ecologists (e.g., Gotelli and Graves 1996; Gotelli et al. 2011), and software is available (Entsminger 2012) to test ecological mechanisms.

7.4.3 ADD Models

ADD models have long been utilized in insect pest management (e.g., Bryant et al. 1998) and as a tool to assess plant development (e.g., Idso et al. 1978). ADD models calculate the total amount of heat required, between the lower and upper development thresholds, for an organism to develop from one life stage to the next. Recently, ADD models have also carried over into decomposition ecology (e.g., Parmenter and MacMahon 2009) and the forensic disciplines (e.g., Megyesi et al. 2005; Michaud and Moreau 2009, 2011; Simmons et al. 2010a, b; Matuszewski 2011, 2012) for predicting decomposition-related phenomena such as the PMI. ADD models incorporate time and heat, the two variables that affect decomposition most, and are arguably better than those using days alone in succession matrices (Megyesi et al. 2005; Carter et al. 2007). We are aware of no study that has applied ADD models to predict whole carrion–arthropod communities, although the output of single-species models for producing probabilities of decompositional progress can be combined to indirectly obtain such a model (Michaud and Moreau 2009).

In an ADD method based on decay stages, Megyesi et al. (2005) devised a point system for calculating a total body score (TBS) using photographs from a sample of 68 human death cases. The TBS depicts the total amount of accumulated decomposition identified from three body regions (head and neck, trunk, and limbs). Because soft tissue decay is positively correlated with ambient temperature (Mann et al. 1990; Carter and Tibbett 2006; Carter et al. 2007; Parmenter and MacMahon 2009), the authors converted the TBS to ADD. Their results showed that the ADD predicted the actual PMI (estimated from insect evidence) better than if TBS was used alone. However, because the TBS is based on the accumulation of categorical scores from decay stages, it is expressed without measurement error. Since this study, at least two other researchers have used this method and obtained mixed results (Dautartas 2009; Parks 2011). Until repeatability studies reveal the magnitude of inter-scorer variation, as noted by the authors, this method implies precision and quantification that may not exist.

Composite ADD indices were developed by Michaud and Moreau (2009, 2011) as continuous variables in logistic regression to predict insect visitation patterns (Michaud and Moreau 2009) and decay stages (Michaud and Moreau 2011). The chief advantages of this approach are that it can generate statistical probabilities and confidence intervals for PMI estimates with just a few carcasses (i.e., as few as four), and it can account for repeated measures and within-season, between-season, and inter-annual variability. The model is also flexible enough to handle other decomposition-related phenomena such as microbial succession, species abundance data, carcass temperature, and so on. However, this model has yet to be verified in other geographical areas or validated in casework.

7.5 Forensic Applications of Carrion Ecology

7.5.1 Can Human Corpses Be Replicated?

Citing the obvious challenges, Tibbett and Carter (2009) argued that it would be impossible to replicate human corpses in forensic research. Presently, several U.S. universities (through their Anthropology departments) have active outdoor facilities that conduct decomposition research on donated human bodies. The oldest of these is the Forensic Anthropology Center (FAC) at the University of Tennessee (UT), Knoxville, founded in 1980. Over its 30-plus year history, FAC has received more than 1200 donations (Shirley et al. 2011). To answer the replication question, we requested and received selected fields of the FAC database for 1991–2012 (through July), kindly provided by UT anthropologists.

Our analysis of the FAC database for years 1991–2012 shows that donations were received on 773 (10%) of the 7843 days for a total of 962 donations. Multiple donations (not more than six on the same day) were received on 18% (142) of the 773 days (Figure 7.5a). A month-by-month analysis reveals that these data marginally fit a uniform distribution, indicating that FAC received single (G-fit = 20.0, df = 11, $P < 0.045$) and multiple donations (G-fit = 12.6, df = 11, $P < 0.317$) in similar numbers throughout the year (Figure 7.5b). As annual donations increased, the waiting period between multiple donations shortened from a median of 56 days in 2001 to 20 days in 2011 (Figure 7.5c). Variability in postmortem condition, however, made some donations unsuitable as potential "replicates." Although FAC rarely takes embalmed bodies (D. Steadman, personal communication, October 8, 2012), donations varied in wholeness (i.e., intact, autopsied, or dismembered) and postmortem age (i.e., 46% died within 48 h of acquisition). To slow decay, most donations were kept in coolers between death and arrival (L. Jantz, personal communication, October 17, 2012) and remained in refrigerated storage at the FAC until field placement (e.g., Rodriguez and Bass 1983, 1985; Srnka 2003; Dautartas 2009). Although freeze–thawing animal tissues accelerates their external decay and disarticulation rates (Micozzi 1986), forensic entomologists have failed to find significant differences in blow fly oviposition, initial succession, or maggot dispersal behavior between freshly euthanized, refrigerated, and freeze–thawed pig carcasses (Grisbaum et al. 1995; Bugajski et al. 2011).

If design criteria set in early FAC studies (Rodriguez and Bass 1983, 1985; Schoenly et al. 2007) are accepted as thresholds for "replication" (i.e., death within 48 h of acquisition, intact, unautopsied, unembalmed, and refrigerated), then replicate corpses have been available to FAC researchers for over two decades. Over the years 1991–2012, our analysis shows that replicate cadavers were available on 25 of the 142 multiple-donation days (not more than three on the same day). Over that same period, the median waiting time shortened from 56 to 30 days (Figure 7.5c, black bars). Relaxing the age threshold to 4 days postmortem, as Rodriguez and Bass (1983) and Vass et al. (1992) did, added another 32 multiple-donations days (for a total of 57) and shortened the median waiting period further (Figure 7.5c, gray bars). With 2000+ individuals currently registered as future donors at the FAC (Shirley et al. 2011), "replicate" corpses for decomposition research will likely be available in the future. Whether such replication results in adequate statistical power to detect significant treatment differences, however, is a separate issue (Section 7.2.5).

7.5.2 Enrichment Effects at Human Decomposition Facilities

Experiments and correlational evidence show that long-term input of nutrients by natural or anthropogenic agents impoverishes biological communities with fewer (species) but larger population sizes (e.g., Kempton 1979; Cole 1994; Molles 1999). For carrion research, especially forensic studies, it is important to know whether cadaver enrichment at human decomposition facilities, such as the 30-plus year-old FAC in Tennessee, has altered biotic or abiotic variables at these sites, compared with surrounding (nonenriched) sites.

In a test of the "arthropod saturation hypothesis," Shahid et al. (2003) asked whether the FAC has altered carrion decay rates or arthropod community structure, compared with those at three nonenriched sites various distances away. Over the 12 days of the study, the authors found that pig carcasses at the FAC did not decay faster, nor did sarcosaprophagous taxa occur there in abnormally higher numbers,

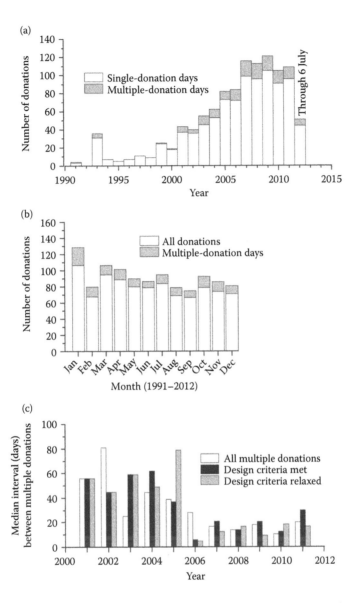

FIGURE 7.5 History of cadaver donations for human decomposition research at the FAC, University of Tennessee, Knoxville, over the period 1991–2012. Donations are presented as (a) frequencies per year, (b) frequencies per month, and (c) in terms of intervals between donations. In (c), the category "design criteria met" refers to thresholds required for the replication definition to be met (i.e., death within 48 h of acquisition, intact, unautopsied, unembalmed, and refrigerated); the category "design criteria relaxed" allows death within 96 h of acquisition, but all other criteria must be met (e.g., intact, unautopsied, unembalmed, and refrigerated).

or colonize carcasses faster than those in surrounding sites. In a second study that tested the same hypothesis with the predatory and parasitic fauna (e.g., rove, clown and carrion beetles, mites, ants, and mosquitoes), Schoenly et al. (2005) found that the four sites were comparable with respect to species composition, colonization rates, and evenness of pitfall-trap catches on a per carcass basis, but not taxonomic evenness and ranked abundances of sweep-net taxa. Taking both studies into account, it appears that the FAC is more representative of surrounding sites with respect to the sarcosaprophagous fauna than to the predatory/parasitic fauna.

Carter and Tibbett (2008) found that repeated burial of skeletal muscle enhanced belowground decomposition rates by "conditioning" the soil microbial community. Using eight soil variables, Damann et al. (2012) tested whether nutrient enrichment at the FAC altered the soil environment, relative to an unenriched site 1 km away. The authors found that samples at the FAC were significantly different from those at the unenriched site for all variables except two (i.e., total carbon content and extractable DNA), suggesting that soil variables vary in their resistance and resilience to this three-decades long "press" experiment (Bender et al. 1984).

Given the demonstrated potential for arthropod and soil variables to respond to cadaver enrichment, we join Damann et al. (2012) in urging that personnel at decomposition facilities incorporate periodic tests of enrichment effects into their long-term goals. We also recommend that researchers at newly established decomposition facilities conduct pre-impact surveys of soils, plants, arthropods, and vertebrates on-and-off site before such facilities become fully operational. Effects of cadaver enrichment can be analyzed using before–after, control–impact (BACI) designs (e.g., Levin and Tolimieri 2001; Steinbeck et al. 2005), which combine before–after and control–impact comparisons to detect impacts beyond natural (i.e., spatial and temporal) variability. BACI can also test for reversal of enrichment effects in the event that site retirement (Michener 1997) is planned in the future.

7.6 Concluding Remarks

In the life sciences, two overarching goals of experimental design are to minimize random variation and to account for confounding factors (Ruxton and Colegrave 2011). However, over the last 30 years (and possibly longer), many carrion researchers have confused observational and experimental units (EUs), which has led to inflated type I error rates and pseudoreplicated results (Hurlbert 1984). Although replication in human decomposition research is not always possible, every experiment involving nonprimate models (e.g., pig carcasses) should be replicated. Moreover, independence of EUs in time and space, as required by most statistical tests, implies that time (i.e., repeated measurements) is a within-carcass factor and that space (i.e., distance between carcasses) is sufficient to prevent cross-contamination. However, it is rare to find repeated-measures designs or minimum distance tests in the carrion literature (Michaud et al. 2012; Moreau et al. 2015). In addition, carcass decay stages—which misinform the more continuous processes of succession and decomposition—have been uncritically accepted by many carrion researchers. Consequently, throughout the twentieth century, many carrion researchers have missed opportunities to improve their field practices, use and update statistical models, and test succession mechanisms (Michaud et al. 2015). Although much more work lies ahead, as momentum grows to return the carrion disciplines to their former ecological roots (Tomberlin et al. 2011a, b; Barton et al. 2013), we hope that carrion—as an experimental subject—will become a model system for stimulating future crossover and mainstream ecological research.

Acknowledgments

We would like to thank Professors Dawnie Steadman, Lee Meadows Jantz, and Richard Jantz of the University of Tennessee's Anthropology Department for kindly sharing their FAC database with us and for their prompt and detailed clarifications to our queries. We also thank the editors for choosing to include a chapter on basic experimental design in this textbook and for inviting us to write it. We appreciate the suggestions and diligence of the editors and one anonymous reviewer.

REFERENCES

Anderson, K.J. 2007. Temporal patterns in rates of community change during succession. *The American Naturalist* 169: 780–793.

Anderson, G.S. and S.L. Van Laerhoven. 1996. Initial studies on insect succession on carrion in southwestern British Columbia. *Journal of Forensic Sciences* 41: 617–625.

Atkinson, W.D. and B. Shorrocks. 1981. Competition on a divided and ephemeral resource: A simulation model. *Journal of Animal Ecology* 50: 461–471.

Barton, P.S., S.A. Cunningham, D.B. Lindermayer, and A.D. Manning. 2013. The role of carrion in maintaining biodiversity and ecological processes in terrestrial ecosystems. *Oecologia* 171: 761–772.

Beaver, R.A. 1977. Non-equilibrium "island" communities: Diptera breeding in dead snails. *Journal of Animal Ecology* 46: 783–798.

Beaver, R.A. 1984. Insect exploitation of ephemeral habitats. *South Pacific Journal of Natural Science* 6: 3–47.

Behm, J.E., D.E. Edmonds, J.P. Harmon, and A.R. Ives. 2013. Multilevel statistical models and the analysis of experimental data. *Ecology* 94: 1479–1486.

Bender, E.A., T.J. Case, and M.E. Gilpin. 1984. Perturbation experiments in community ecology: Theory and practice. *Ecology* 65: 1–13.

Birks, H.J.B. 1987. Recent methodological developments in quantitative descriptive biogeography. *Annals Zoologici Fennici* 24: 165–178.

Bishopp, F.C. 1916. Fly traps and their operation. *USDA Farmer's Bulletin* 734: 1–13.

Bornemissza, G.F. 1957. An analysis of arthropod succession in carrion and the effect of its decomposition on the soil fauna. *Australian Journal of Zoology* 5: 1–12.

Boulton, A.J. and P.S. Lake. 1988. Dynamics of heterotrophic succession in carrion arthropod assemblages: A comment on Schoenly and Reid (1987). *Oecologia* 76: 477–480.

Broce, A.B., J.L. Goodenough, and J.R. Coppedge. 1977. A wind oriented trap for screwworm flies. *Journal of Economic Entomology* 70: 413–419.

Bryant, S.R., J.S. Bale, and C.D. Thomas. 1998. Modification of the triangle method of degree-day accumulation to allow for behavioral thermoregulation in insects. *Journal of Applied Ecology* 35: 921–927.

Bugajski, K.N., C.C. Seddon, and R.E. Williams. 2011. A comparison of blow fly (Diptera: Calliphoridae) and beetle (Coleoptera) activity on refrigerated only versus frozen-thawed pig carcasses in Indiana. *Journal of Medical Entomology* 48: 1231–1235.

Byrd, J.H. and J.L. Castner. 2010. *Forensic Entomology: The Utility of Arthropods in Legal Investigations*, 2nd edition. Boca Raton: CRC Press.

Carter, D.O. and M. Tibbett. 2006. Microbial decomposition of skeletal muscle tissue (*Ovis aries*) in a sandy loam soil at different temperatures. *Soil Biology and Biochemistry* 38: 1139–1145.

Carter, D.O. and M. Tibbett. 2008. Cadaver decomposition and soil: Processes. In: *Soil Analysis in Forensic Taphonomy: Chemical and Biological Effects of Buried Human Remains*, M. Tibbett and D.O. Carter (eds), pp. 29–51. Boca Raton: CRC Press.

Carter, D.O., D. Yellowlees, and M. Tibbett. 2007. Cadaver decomposition in terrestrial ecosystems. *Naturwissenschaften* 94: 12–24.

Chaves, L.F. 2010. An entomologist guide to demystify pseudoreplication: Data analysis of field studies with design constraints. *Journal of Medical Entomology* 47: 291–298.

Clements, F.E. 1916. *Plant Succession: Analysis of the Development of Vegetation*. Carnegie Institution of Washington, Publication No. 242, Washington, D.C.

Clements, F.E. 1936. Nature and structure of the climax. *Journal of Ecology* 24: 252–284.

Cohen, J. 1988. *Statistical Power Analysis for the Behavioral Sciences*, 2nd edition. New York: Taylor and Francis.

Cole, G.A. 1994. *Textbook of Limnology*, 4th edition. Prospect Heights: Waveland Press.

Colwell, R.K. 2013. *Estimates: Statistical Estimation of Species Richness and Shared Species from Samples*. Version 9. Persistent URL: <http://purl.oclc.org/estimates>.

Colwell, R.K. and J.A. Coddington. 1994. Estimating terrestrial biodiversity through extrapolation. *Philosophical Transactions of the Royal Society of London B* 345: 101–118.

Colwell, R.K. and D.W. Winkler. 1984. A null model for null models in biogeography. In: *Ecological Communities: Conceptual Issues and the Evidence*, D.R. Strong, D. Simberloff, and L.G. Abele (eds), pp. 344–359. Princeton: Princeton University Press.

Comstock, J.L., J.-P. Desaulniers, H.N. LeBlanc, and S.L. Forbes. 2015. New decomposition stages to describe scenarios involving the partial and complete removal of insects. *Canadian Society of Forensic Science Journal* 48: 1–19.

Connell, J.H. and R.O. Slatyer. 1977. Mechanisms of succession in natural communities and their role in community stability and organization. *The American Naturalist* 111: 1119–1144.

Cornelius, J.M. and J.F. Reynolds. 1991. On determining the statistical significance of discontinuities within ordered ecological data. *Ecology* 72: 2057–2070.

Damann, F.E., A. Tanittaisong, and D.O. Carter. 2012. Potential carcass enrichment of the University of Tennessee Anthropology Research Facility: A baseline survey of edaphic features. *Forensic Science International* 222: 4–10.

Dautartas, A.M. 2009. The effect of various coverings on the rate of human decomposition. MA Thesis, University of Tennessee, Knoxville.

Donovan, T.M. and C.W. Welden. 2002. *Spreadsheet Exercises in Ecology and Evolution*. Sunderland: Sinauer Press.

Eberhardt, L.L. 1978. Appraising variability in population studies. *Journal of Wildlife Management* 42: 207–238.

Elton, C.S. and R.S. Miller 1954. The ecological survey of animal communities: With a practical system of classifying habitats by structural characters. *Journal of Ecology* 42: 460–496.

Entsminger, G.L. 2012. EcoSim Professional: Null Modeling Software for Ecologists, Version 1. Acquired Intelligence Inc., Kesey-Bear & Pinyon Publishing. http://www.garyentsminger.com/ecosim/index.htm.

Erzinçlioğlu, Y.Z. 1980. A new trap for carrion flies. *Entomologist's Record and Journal of Variation* 92: 219.

Faul, F., E. Erdfelder, A.-G. Lang, and A. Buchner. 2007. G*Power 3: A flexible statistical power analysis program for the social, behavioral, and biomedical sciences. *Behavior Research Methods* 39: 175–191.

Finn, J.A. 2001. Ephemeral resource patches as model systems for diversity–function experiments. *Oikos* 92: 363–366.

Fisher, R.A. 1926. The arrangement of field experiments. *Journal of the Ministry of Agriculture* 33: 503–513.

Fisher, R.A. 1935. *The Design of Experiments*. Edinburgh and London: Oliver and Boyd.

Fuller, M.E. 1934. The insect inhabitants of carrion, a study of animal ecology. *Bulletin of the Council for Scientific and Industrial Research (Australia)* 82: 5–62.

Galloway, A., W.H. Birkby, A.M. Jones, T.E. Henry, and B.O. Parks. 1989. Decay rates of human remains in an arid environment. *Journal of Forensic Sciences* 34: 607–616.

Gillison, A.N. and K.R.W. Brewer. 1985. The use of gradient direct transects or gradsects in natural resource surveys. *Journal of Environmental Management* 20: 103–127.

Gleason, H.A. 1917. The structure and development of the plant association. *Bulletin of the Torrey Botanical Club* 43: 463–481.

Gleason, H.A. 1926. The individualistic concept of the plant association. *Bulletin of the Torrey Botanical Club* 53: 7–26.

Goff, M.L. 1993. Estimation of post mortem interval using arthropod development and successional patterns. *Forensic Science Review* 5: 81–94.

Goodall, D.W. 1952. Quantitative aspects of plant distributions. *Biological Reviews* 27: 194–245.

Gotelli, N.J. and G.R. Graves. 1996. *Null Models in Ecology*. Washington, D.C.: Smithsonian Institution. Press.

Gotelli, N.J., W. Ulrich, and F.T. Maestre. 2011. Randomization tests for quantifying species importance to ecosystem function. *Methods in Ecology and Evolution* 2: 634–642.

Grisbaum, G.A., C.L. Meek, and J.W. Tessmer. 1995. Effects of initial post-mortem refrigeration of animal carcasses on necrophilous adult fly activity. *Southwestern Entomologist* 20: 165–169.

Hairston, N.G., Sr. 1989. *Ecological Experiments: Purpose, Design, and Execution*. Cambridge: Cambridge University Press.

Hand, L.F. and A.J. Keaster. 1967. The environment of an insect field cage. *Journal of Economic Entomology* 60: 910–915.

Hanski, I. 1987a. Carrion fly community dynamics: Patchiness, seasonality and coexistence. *Ecological Entomology* 12: 257–266.

Hanski, I. 1987b. Colonization of ephemeral habitats. In: *Colonization, Succession, and Stability*, A.J. Gray, M.J. Crawley, and P.J. Edwards (eds), pp. 155–185. Oxford: Blackwell Scientific Publications.

Harvey, P.H., R.K. Colwell, J.W. Silvertown, and R.M. May. 1983. Null models in ecology. *Annual Review of Ecology and Systematics* 14: 189–211.

Hill, M.H., J.D. Witman, and H. Caswell. 2004. Markov chain analysis of succession in a rocky subtidal community. *The American Naturalist* 164: E46–E61.

Hurlbert, S.H. 1984. Pseudoreplication and the design of ecological field experiments. *Ecological Monographs* 54: 187–211.

Idso, S.B., R.D. Jackson, and R.J. Reginato. 1978. Extending the degree-day concept of plant phenological development to include water stress effects. *Ecology* 59: 431–433.

Irwin, M.E., R.W. Gill, and D. Gonzalez. 1974. Field cage studies of native egg predators of the pink bollworm in southern California cotton. *Journal of Economic Entomology* 67: 193–196.

Jirón, L.F. 1984. A blowfly trap for standardized field sampling. *Entomological News* 95: 202–206.

Johnson, E.A. and K. Miyanishi. 2008. Testing the assumptions of chronosequences in succession. *Ecology Letters* 11: 419–431.

Johnston, W. and G. Villeneuve. 1897. On the medico-legal application of entomology. *Montreal Medical Journal* 26: 81–90.

Karban, R. and M. Huntzinger. 2006. *How to do Ecology: A Concise Handbook*. Princeton: Princeton University Press.

Kashyap, V.K. and V.V. Pillay. 1989. Efficacy of entomological method in estimation of postmortem interval: A comparative analysis. *Forensic Science International* 40: 245–250.

Kempton, R.A. 1979. The structure of species abundance and measurement of diversity. *Biometrics* 35: 307–321.

Kneidel, K.A. 1985. Patchiness, aggregation, and the coexistence of competitors for ephemeral resources. *Ecological Entomology* 10: 441–448.

Krebs, C.J. 2000. *Ecological Methodology*, 2nd edition. Menlo Park: Addison-Wesley Longman.

LaMotte, L.R. and J.D. Wells. 2000. *P*-values for post mortem intervals from arthropod succession data. *Journal of Agricultural Biological Environmental Statistics* 5: 58–68.

Lefebvre, F. and E. Gaudry. 2009. Forensic entomology: A new hypothesis for the chronological succession pattern of necrophagous insect on human corpses. *Annales de la Société Entomologique de France* 45: 377–392.

Leibold, M.A., M. Holyoak, N. Mouquet, P. Amarasekare, J.M. Chase, M.F. Hoopes, R.D. Holt et al. 2004. The metacommunity concept: A framework for multi-scale community ecology. *Ecology Letters* 7: 601–613.

Lenth, R.V. 2006–2009. *Java Applets for Power and Sample Size*. http://www.stat.uiowa.edu/ ~ rlenth/Power.

Levin, P.S. and N. Tolimieri. 2001. Differences in the impacts of dams on the dynamics of salmon populations. *Animal Conservation* 4: 291–299.

Luck, R.F., B.M. Shepard, and P.E. Kenmore. 1988. Experimental methods for evaluating arthropod natural enemies. *Annual Review of Entomology* 33: 367–391.

Manly, B.F.J. 1991. *Randomization and Monte Carlo Methods in Biology*. New York: Chapman and Hall.

Mann, R.W., W.M. Bass, and L. Meadows. 1990. Time since death and decomposition of the human body: Variables and observations in case and experimental field studies. *Journal of Forensic Sciences* 35: 103–111.

Matuszewski, S. 2011. Estimating the pre-appearance interval from temperature in *Necrodes littoralis* L. (Coleoptera: Silphidae). *Forensic Science International* 212: 180–188.

Matuszewski, S. 2012. Estimating the preappearance interval from temperature in *Creophilus maxillosus* L. (Coleoptera: Staphylinidae). *Journal of Forensic Sciences* 57: 136–145.

Matuszewski, S., D. Bajerlein, S. Konwerski, and K. Szpila. 2010. Insect succession and carrion decomposition in selected forests of Central Europe. Part 2: Composition and residency patterns of carrion fauna. *Forensic Science International* 195: 42–51.

McIntosh, R.P. 1985. *The Background of Ecology: Concept and Theory*. Cambridge: Cambridge University Press.

Mégnin, P. 1883. De l'application de l'entomologie à la médicine légale. *Gazette hebdomadaire de médicine et de chirurgie* 29: 480–482.

Mégnin, P. 1887. La faune des tombeaux. *Compte Rendu Hebdomadaire des Séances de l'Académie des Sciences* 105: 948–951.

Mégnin, P. 1894. *La Faune des cadavres. Encyclopédie Scientifique des Aides-Mémoires. G. Masson*. Paris: Gauthier-Villars et Fils.

Megyesi, M.S., S.P. Nawrocki, and N.H. Haskell. 2005. Using accumulated degree-days to estimate the postmortem interval from decomposed human remains. *Journal of Forensic Sciences* 50: 1–9.

Michaud, J.-P. and G. Moreau. 2009. Predicting the visitation of carcasses by carrion-related insects under different rates of degree-day accumulation. *Forensic Science International* 185: 78–83.

Michaud, J.-P. and G. Moreau. 2011. A statistical approach based on accumulated degree-days to predict decomposition-related processes in forensic studies. *Journal of Forensic Sciences* 56: 229–232.

Michaud, J.-P. and G. Moreau. 2013. Effect of variable rates of daily sampling of fly larvae on decomposition and carrion insect community assembly: Implications for forensic entomology field study protocols. *Journal of Medical Entomology* 50: 890–897.

Michaud, J.-P., K.G. Schoenly, and G. Moreau. 2012. Sampling flies or sampling flaws? Experimental design and inference strength in forensic entomology. *Journal of Medical Entomology* 49: 1–10.

Michaud, J.-P., K.G. Schoenly, and G. Moreau. 2015. Rewriting ecological succession history: Did carrion ecologists get there first? *The Quarterly Review of Biology* 90: 45–66.

Michener, W.K. 1997. Quantitatively evaluating restoration experiments: Research design, statistical analysis, and data management considerations. *Restoration Ecology* 5: 324–337.

Micozzi, M.S. 1986. Experimental study of postmortem change under field conditions: Effects of freezing, thawing, and mechanical injury. *Journal of Forensic Sciences* 31: 953–961.

Molles, M.C., Jr. 1999. *Ecology: Concepts and Applications*. Boston: WCB McGraw-Hill.

Moreau, A., J.-P. Michaud, and K.G. Schoenly. 2015. Experimental design, inferential statistics, and computer modeling. In: *International Dimensions and Frontiers of Forensic Entomology*, M.E. Benbow and J.K. Tomberlin (eds), pp. 205–229. Boca Raton: CRC Press.

Morin, P.J. 1998. Realism, precision, and generality in experimental ecology. In: *Experimental Ecology: Issues and Perspectives*, W.J. Resetarits and J. Bernardo (eds), pp. 50–70. Oxford: Oxford University Press.

Motter, M.G. 1898. A contribution to the study of the fauna of the grave. A study of one hundred and fifty disinterments, with some additional experimental observations. *Journal of the New York Entomological Society* 6: 201–231.

Moura, M.O., E.L.A. Monteiro-Filho, and C.J.B. Carvalho. 2005. Heterotrophic succession in carrion arthropod assemblages. *Brazilian Archives of Biology and Technology* 48: 477–486.

Nelder, M.P., J.W. McCreadie, and C.S. Major. 2009. Blow flies visiting decaying alligators: Is succession synchronous or asynchronous? *Psyche* 2009: 575362.

Norris, J.R. 1998. *Markov Chains*. Cambridge: Cambridge University Press.

Odum, E. 1969. The strategy of ecosystem development. *Science* 164: 262–270.

Ordóñez, A., M.D. Garcia, and G. Fagua. 2008. Evaluation of efficiency of Schoenly trap for collecting adult sarcosaprophagous dipterans. *Journal of Medical Entomology* 45: 522–532.

Parks, C.L. 2011. A study of the human decomposition sequence in central Texas. *Journal of Forensic Sciences* 56: 19–22.

Parmenter, R.R. and J.A. MacMahon. 2009. Carrion decomposition and nutrient cycling in a semiarid shrub-steppe ecosystem. *Ecological Monographs* 79: 637–661.

Payne, J.A. 1965. A summer carrion study of the baby pig (*Sus scrofa*). *Ecology* 46: 592–602.

Payne, J.A. and D.A. Crossley. 1966. *Animal Species Associated with Pig Carrion. ORNL-TM 1432*. Oak Ridge: Oak Ridge National Laboratory.

Pearson, D.E. and L.F. Ruggiero. 2003. Transect versus grid trapping arrangements for sampling small-mammal communities. *Wildlife Society Bulletin* 31: 454–459.

Pechal, J.L., M.E. Benbow, T.L. Crippen, A.M. Tarone, and J.K. Tomberlin. 2014. Delayed insect access alters carrion decomposition and necrophagous insect community assembly. *Ecosphere* 5(4): article 45.

Perez, A., N.H. Haskell, and J.D. Wells. 2014. Evaluating the utility of hexapod species for calculating a confidence interval about a succession based postmortem interval estimate. *Forensic Science International* 241: 91–95.

Peters, R.H. 1991. *A Critique for Ecology*. Cambridge: Cambridge University Press.

Peterson, A. 1953. *Entomological Techniques: How to Work with Insects*, 10th edition. Los Angeles: Entomological Reprint Specialists.

Pickett, S.T.A. 1989. Space-for-time substitution as an alternative to long-term studies. In: *Long-Term Studies in Ecology: Approaches and Alternatives*, G.E. Likens (ed.), pp. 110–135. New York: Springer-Verlag.

Poole, R.W. and B.J. Rathcke. 1979. Regularity, randomness and aggregation in flowering phenologies. *Science* 203: 470–471.

Prado e Castro, C. 2009. A modified version of Schoenly trap for collecting sarcosaprophagous arthropods. Detailed plans and construction. *Anales de Biología* 31: 1–6.

Reed, H.B. 1958. Study of dog carcass communities in Tennessee, with special reference to the insects. *American Midland Naturalist* 59: 213–245.

Rodriguez, W.C. and W.M. Bass. 1983. Insect activity and its relationship to decay rates of human cadavers in east Tennessee. *Journal of Forensic Sciences* 28: 423–432.

Rodriguez, W.C. and W.M. Bass. 1985. Decomposition of buried bodies and methods that may aid in their location. *Journal of Forensic Sciences* 30: 836–852.

Rose, M.D. and G.A. Polis. 1998. The distribution and abundance of coyotes: The effects of allochthonous food subsidies from the sea. *Ecology* 79: 998–1007.

Ruxton, G.D. and N. Colegrave. 2011. *Experimental Design for the Life Sciences*, 3rd edition. Oxford: Oxford University Press.

Sagarin, R. and A. Pauchard. 2010. Observational approaches in ecology open new ground in a changing world. *Frontiers in Ecology* 8: 379–386.

Schoenly, K. 1981. Demographic bait trap. *Environmental Entomology* 10: 615–617.

Schoenly, K. 1992. A statistical analysis of successional patterns in carrion-arthropod assemblages: Implications for forensic entomology and determination of the postmortem interval. *Journal of Forensic Sciences* 37: 1489–1513.

Schoenly, K.G., K. Griest, and S. Rhine. 1991. An experimental field protocol for investigating the postmortem interval using multidisciplinary indicators. *Journal of Forensic Sciences* 36: 1395–1415.

Schoenly, K.G., N.H. Haskell, and R.D. Hall. 2003. *Within- and Between-Subject Repeatability of Succession-Based PMI Statistics in Human and Pig Remains.* Paper presented at the first annual meeting of the North American Forensic Entomology Conference, Las Vegas.

Schoenly, K.G., N.H. Haskell, R.D. Hall, and J.R. Gbur. 2007. Comparative performance and complementarity of four sampling methods and arthropod preference tests from human and porcine remains at the Forensic Anthropology Center in Knoxville, Tennessee. *Journal of Medical Entomology* 44: 881–894.

Schoenly, K.G. and W.H. Reid. 1987. Dynamics of heterotrophic succession in carrion arthropod assemblages: Discrete seres or a continuuum of change? *Oecologia* 73: 192–202.

Schoenly, K.G., S.A. Shahid, N.H. Haskell, and R.D. Hall. 2005. Does carcass enrichment alter community structure of predaceous and parasitic arthropods? A second test of the arthropod saturation hypothesis at the Anthropology Research Facility in Knoxville, Tennessee. *Journal of Forensic Sciences* 50: 134–142.

Shahid, S.A., K.G. Schoenly, N.H. Haskell, R.D. Hall, and W. Zhang. 2003. Carcass enrichment does not alter rates of arthropod community structure: A test of the arthropod saturation hypothesis at the Anthropology Research Facility in Knoxville, Tennessee. *Journal of Medical Entomology* 40: 559–569.

Shirley, N.R., R.J. Wilson, and L.M. Jantz. 2011. Cadaver use at the University of Tennessee's Anthropology Research Facility. *Clinical Anatomy* 24: 372–380.

Simberloff, D. 1980. A succession of paradigms in ecology: Essentialism to materialism and probalism. In: *Conceptual Issues in Ecology*, E. Saarinen (ed.), pp. 139–153. Dordrecht: D. Reidel.

Simmons, T., R.E. Adlam, and C. Moffatt. 2010a. Debugging decomposition data—Comparative taphonomic studies and the influence of insects and carcass size on decomposition rate. *Journal of Forensic Sciences* 55: 8–13.

Simmons, T., P.A. Cross, R.E. Adlam, and C. Moffatt. 2010b. The influence of insects on decomposition rate in buried and surface remains. *Journal of Forensic Sciences* 55: 889–892.

Smith, K.G.V. 1986. *A Manual of Forensic Entomology.* London: British Museum and Cornell University Press.

Smith, C.R. and A.R. Baco. 2003. Ecology of whale falls at the deep-sea floor. *Oceanography and Marine Biology: An Annual Review* 41: 311–354.

Solow, A.R. and W.K. Smith. 2006. Using Markov chain successional models backwards. *Journal of Applied Ecology* 43: 185–188.

Srnka, C.F. 2003. *The Effects of Sun and Shade on the Early Stages of Human Decomposition.* MA Thesis, University of Tennessee, Knoxville.

Steinbeck, J.R., D.R. Schiel, and M.S. Foster. 2005. Detecting long-term change in complex communities: A case study from the rocky intertidal zone. *Ecological Applications* 15: 1813–1832.

Stoker, R.L., W.E. Grant, and S.B. Vinson. 1995. *Solenopsis invita* (Hymenoptera: Formicidae) effect on invertebrate decomposers of carrion in central Texas. *Environmental Entomology* 24: 817–822.

Suenaga, O. and H. Kurahashi. 1994. Improved types of the horse meat baited fly trap and the fly emergence trap. *Tropical Medicine* 36: 65–70.

Tansley, A.G. 1939. British ecology during the past quarter-century: The plant community and the ecosystem. *Journal of Ecology* 27: 513–530.

Thomas, L. and F. Juanes. 1996. The importance of statistical power analysis: An example from *Animal Behavior. Animal Behaviour* 52: 856–859.

Tibbett, M. and D.O. Carter. 2009. Research in forensic taphonomy: A soil-based perspective. In: *Criminal and Environmental Soil Forensics*, K. Ritz, L. Dawson, and D. Miller (eds), pp. 317–331. New York: Springer.

Tomberlin, J.K., M.E. Benbow, A.M. Tarone, and R.M. Mohr. 2011a. Basic research in evolution and ecology enhances forensics. *Trends in Ecology and Evolution* 26: 53–55.

Tomberlin, J.K., R. Mohr, M.E. Benbow, A.M. Tarone, and S. VanLaerhoven. 2011b. A roadmap for bridging basic and applied research in forensic entomology. *Annual Review of Entomology* 56: 401–421.

Usher, M.B. 1987. Modeling successional processes in ecosystems. In: *Colonization, Succession, and Stability*, A.J. Gray, M.J. Crawley, and P.J. Edwards (eds), pp. 31–55. Oxford: Blackwell Scientific Publications.

van den Bosch, R., T.F. Leigh, D. Gonzalez, and R.E. Stinner. 1969. Cage studies on predators of the bollworm in cotton. *Journal of Economic Entomology* 62: 1486–1489.

Vass, A.A., W.C. Bass, J.D. Wolt, J.E. Foss, and J.T. Ammons. 1992. Time since death determinations of human cadavers using soil solution. *Journal of Forensic Sciences* 37: 1236–1253.

Vogt, W.G. and D.E. Havenstein. 1974. A standardized bait trap for blowfly studies. *Journal of the Australian Entomological Society* 13: 249–253.

Walker, L.R., D.A. Wardle, R.D. Bardgett, and B.D. Clarkson. 2010. The use of chronosequences in studies of ecological succession and soil development. *Journal of Ecology* 98: 725–736.

Wells, J.D. and B. Greenberg. 1994. Effect of the red imported fire ant (Hymenoptera: Formicidae) and carcass type on the daily occurrence of postfeeding carrion-fly larvae (Diptera: Calliphoridae, Sarcophagidae). *Journal of Medical Entomology* 31: 171–174.

Whitlock, M.C. and D. Schluter. 2009. *The Analysis of Biological Data*. Greenwood Village, CO: Roberts and Company.

Whittaker, R.H. 1953. A consideration of climax theory: The climax as a population and pattern. *Ecological Monographs* 23: 41–78.

Wilson, E.E. and E.M. Wolkovich. 2011. Scavenging: How carnivores and carrion structure communities. *Trends in Ecology and Evolution* 26: 129–135.

Yang, L.H. 2006. Interactions between a detrital resource pulse and a detritivore community. *Oecologia* 147: 522–532.

Yang, L.H., J.L. Bastow, K.O. Spence, and A.N. Wright. 2008. What can we learn from resource pulses? *Ecology* 89: 621–634.

Zhang, W.J. and K.G. Schoenly. 1999. *Biodiversity Software Series III. BOUNDARY: A Program for Detecting Boundaries in Ecological Landscapes*. IRRI Technical Bulletin No. 3, Manila, Philippines: International Rice Research Institute.

Zschokke, S. and E. Ludin 2001. Measurement accuracy: How much is necessary? *Bulletin of the Ecological Society of America* 82: 237–243.

Section II

Ecological Mechanisms of Carrion Decomposition

8

Community and Landscape Ecology of Carrion

M. Eric Benbow, Jennifer L. Pechal, Rachel M. Mohr

CONTENTS

8.1 Introduction

Decomposing carrion are resource patches and nutrient pulses to interacting organisms, including microbial, invertebrate, plant, and vertebrate communities. From the end of life arises the beginning of energy and nutrient transformation into new life, an important ecosystem process driven by interacting communities of organisms that span all three domains of life: Archaea, Bacteria, and Eukarya. Recognized as far back as late nineteenth century by Yovanovitch (1888), the community ecology of carrion decomposition is dynamic:

> il existe deux choses après la mort: la putrefaction, dont l'agent est le microbe; la disparition des cadavres, dont l'agent est l'insecte
> there are two things after death: putrefaction, the work of microbes, and disappearance, the work of insects.

These important community dynamics and interactions can be studied and described using principles of community ecology to better understand and predict: (1) how, when, and by what means organisms (at all scales, from microbes to hyenas) respond to resource pulses, such as heterotrophically derived biomass in the form of animal carcasses or carrion (Yang 2004; Holt 2008) and (2) how organisms convert the once-living biomass into new sources of energy and nutrients. In the carrion system, energy and nutrient transformation processes are initiated by bacterial (and other microbial) species associated with the carrion resource. These groups of organisms have been described as the epinecrotic microbial community of carrion (Pechal et al. 2014b). The species of these groups begin a series of cascading community interactions spanning from intra- and interspecific competition and facilitation to complex interdomain (or interkingdom) relationships (Chapter 20) within a network of organisms associated with carrion that has been defined as the necrobiome (*sensu* Benbow et al. 2013). This interacting, multidomain network of species associated with carrion does not operate in isolation, but rather influences and is influenced by the surrounding habitat and ecosystem in ways that scientists are only beginning to understand and mechanistically explore. The interacting networks of species are also understood to drive ecosystem services in ways that are elaborated upon in this chapter.

Ecosystem processes affect the transfer and cycling of energy and nutrients within a given ecosystem. A critical link in these processes is nutrient recycling via carrion decomposition. The spatial and temporal distribution of species, nutrients, and energy define the mosaic structure of ecosystems; the extent and rate of cycling within defined geographic boundaries is an important aspect of landscape ecology (Urban et al. 1987; Forman 1995a,b) within which carrion metacommunity processes operate.

Landscape ecology can be broadly defined as the study of energy, nutrient, or organism (biomass) transfer among habitats (Polis and Hurd 1996). The community ecology of carrion decomposition is inherently a part of the landscape, driven by patch dynamics and processes of metapopulation and metacommunity ecology (Hanski 1991, 1998; Leibold et al. 2004) and island biogeography (MacArthur and Wilson 1967). Carrion, as resource pulses, can occur as an isolated patch within the environment (e.g., deer carcass in the forest) or as a mass die-off of hundreds or thousands of organisms (e.g., salmon runs or cicada [Orthoptera: Cicadidae] emergences); the way carrion becomes available within an ecosystem influences the necrophagous organisms that colonize and use the resource, thus influencing metacommunity assembly from the bottom-up. In addition, the frequency of carrion availability and position (both spatially and temporally) in the landscape can dictate top-down effects on the resource community assembly through interactions with the density and species composition of surrounding scavenging vertebrates (e.g., vultures or hyenas). Thus, depending on the scale of study (McGill 2010), the community ecology of carrion decomposition is likely dictated by synergistic bottom-up and top-down pathways. However, many more studies are needed to empirically test this and other hypotheses within the broader context of metapopulation and metacommunity theory (Hanski et al. 1995; Levin 1995; Hanski 1998; Leibold et al. 2004).

The goal of this chapter is to provide a general introduction to the overall community ecology of carrion systems while emphasizing small-scale interactions of species directly associated with the resource (the necrobiome). The chapter will first describe the major communities of organisms that use carrion in some way: the microbial, invertebrate, and vertebrate communities. What is known about the succession of these communities associated with carrion will also be described for relating to metacommunity ecology theory. While there are some recent carrion microbial community successional studies and a rich history of necrophagous invertebrate succession, there is less understood about the succession of vertebrate scavenger communities associated with carrion. For more information on the ecological dynamics of vertebrates scavenging carrion, the reader is directed to Chapters 6 and 21. In addition, how these carrion resources are ecologically integrated into the larger ecosystem will also be reviewed (Figure 8.1). The former treatment of succession is explored in the context of principles that define community assembly and succession, whereas the latter is discussed in relation to landscape ecology, patch mosaics, and metacommunity dynamics. Further, a detailed assessment of a framework derived from basic ecological principles to evaluate organism interactions on carrion is provided, leading to a call for future research directions in carrion decomposition community ecology that concludes this chapter.

8.2 Microbial, Invertebrate, and Vertebrate Communities

8.2.1 Microbial Communities

The initial biochemical processes of death are quite obvious to anyone that has smelled a dead animal along a roadway. The odors associated with death are the result of biochemical processes that are described in detail within Chapter 2, while the microbial organisms that are a part of these processes

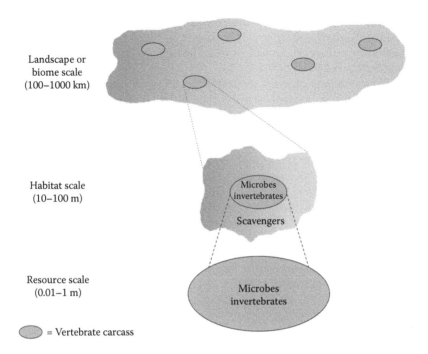

FIGURE 8.1 (See color insert.) The scale at which an ecologist studies carrion decomposition is key to elucidating ecological interactions among the necrobiome or organisms using carrion such as microbes, invertebrates, and vertebrate scavengers. At the landscape or biome scale, there can be many resources (or carrion patches) available for consumption by members of the necrobiome. At the habitat scale, the focus is on a single resource within the landscape and its imme-diate (<100 m) surrounding habitat; there are fewer vertebrate scavengers present at this scale and thus there is a shift in focus towards the microbes and invertebrates members of the necrobiome. Finally, at the resource scale or a single vertebrate carcass, the emphasis is only on the interactions occurring between the microbes and invertebrates throughout decomposition.

are discussed in Chapter 3. In this chapter, a brief introduction to the microbial organisms that take bio-chemical control of the once-living biomass is given to provide the context within which different species of the necrobiome interact in ways that define carrion community ecology.

Epinecrotic microbial communities (i.e., bacteria, fungi, protists, and algae) associated with carrion that play a role in the initial attraction cues and subsequent colonization of carrion by necrophagous insects (Lam et al. 2007; Simmons et al. 2010a; Ma et al. 2012; Zheng et al. 2013), thus influencing species sort-ing and succession of these communities. The ability of necrophagous insects to use a variety of patchy, unpredictable, and ephemeral resources contributes to their ability to persist within the environment and "appear out of nowhere" when a resource becomes available, often within seconds to hours after death (Payne 1965; Grunbaum 2002; Watson and Carlton 2003; DeJong and Hoback 2006; Gruner et al. 2007; Bugajski et al. 2011; Mohr and Tomberlin 2014). However, exactly how these insects initially detect and respond to a newly available dead animal is not well understood, despite the importance of these insect communities in applications such as forensic entomology (Tomberlin et al. 2011a). Microbial communi-ties associated with carrion, their importance to decomposition, and how necrophagous insects respond to them can be found in detail within Chapters 9, 10, 19, and 20. However, relevant to this chapter are recent studies that have described microbial succession of carrion using next-generation high-throughput metagenomic sequencing, which have provided some initial insights into the community ecology of these epinecrotic communities (Hyde et al. 2013; Metcalf et al. 2013; Pechal et al. 2013, 2014b).

One example is a study by Pechal et al. (2014b) that described the epinecrotic bacterial succession on decomposing swine carcasses in a terrestrial habitat; four bacterial phyla and 20 families were deter-mined to be associated with replicated swine remains ($n = 3$) throughout decomposition. Significantly different epinecrotic bacterial communities were documented during the decomposition process; there was a shift in the dominant phyla from Proteobacteria (early decomposition) to Firmicutes (late decomposition), with the families Bacteroidaceae and Moraxellaceae representing significant indicators of the bacterial community at recent death compared to Bacillaceae and Clostridiales incertae sedis XI at 5 days postmortem (Pechal et al. 2014b).

In another study using mouse carcasses, Metcalf et al. (2013) reported slightly different bacterial communities compared to those documented by Pechal et al. (2014b) during the decomposition process; however, the predominate phyla was (alpha-, beta-, and gamma-) Proteobacteria. For abdominal cavity communities, Firmicutes was a predominant phyla during the early decomposition process, switching to Proteobacteria being the dominant in later times of decomposition, which is inverse to the community present on the external surfaces as reported by Pechal et al. (2014b). This study also reported significant microbial eukaryotic community changes in the soils beneath the carcasses (Metcalf et al. 2013), as discussed in Chapters 3 and 19. One general observation from these and earlier studies is quite evident: microbial communities both associated in or on carrion and in the soil beneath carrion follow patterns that appear to resemble more general successional and community assembly processes initially described in sand dunes, forests, and prairies (Cowles 1899; Gleason 1927; Clements 1936), albeit at a smaller more localized scale. However, much more research is needed to explicitly test ecological theory with these communities associated with carrion. Additional studies would allow for more detailed investigations into how microbial communities influence the next trophic level: the necrophagous eukaryotic species, as discussed in Sections 8.7 and 8.8. It is also suggested that epinecrotic microbial communities play an important role in shaping attraction to and consumption of a carrion patch by organisms such as insects and vertebrate scavengers.

8.2.2 Invertebrate and Vertebrate Scavenger Communities

The majority of ecological investigation related to carrion has taken the form of necrophagy-focused studies on insect community succession (Reed 1958; Payne 1965; Early and Goff 1986; Anderson and VanLaerhoven 1996; Watson and Carlton 2003, 2005; Pechal et al. 2014a). The most well-studied group of invertebrate colonizers of carrion are the calyptrate flies in the families Calliphoridae, Sarcophagidae, and Muscidae (Catts and Goff 1992) (see Chapter 4). These flies thrive in decomposing vertebrate remains, organic materials (e.g., garbage), and vertebrate wastes (e.g., manure). Some species survive in living hosts as in the case of myiasis (Hall and Wall 1995; Sherman et al. 2000). Certain families of beetles (Coleoptera: Dermestidae, Silphidae, and Trogidae, among others) have also evolved to exploit these ephemeral decomposing organic resources. From the aforementioned studies, a general trend has emerged for the macroinvertebrate communities in terrestrial habitats: necrophagous Diptera, such as blow flies (Calliphoridae), are early colonizing taxa on fresh carcasses (Figure 8.2), followed by a mix of

(a)   (b)

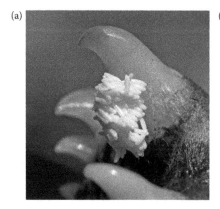

FIGURE 8.2   **(See color insert.)** Blow flies (Calliphoridae) are often the first insects to colonize carrion: (a) Blow fly egg mass on the tooth of a salmon carcass and (b) An adult Callihporidae (*Calliphora* sp.). (Photo with permission from Bob Armstrong.)

(a)  (b)

FIGURE 8.3 **(See color insert.)** Coleoptera photographed with salmon carrion in southeast Alaska, carrying phoretic mites: (a) Carabidae and (b) Silphidae (*Nicrophorus* sp.). (Photo with permission from Bob Armstrong.)

other Diptera, such as cheese skippers (Diptera: Piophillidae), flesh flies (Diptera: Sarcophagidae), and Coleoptera as decomposition progresses; finally leaving only certain species of Coleoptera and mites (Acari) inhabiting a dry set of remains (Reed 1958; Payne 1965; Michaud and Moreau 2009; Byrd and Castner 2010). The relationships among invertebrates occupying a carcass can be quite complex, as exemplified by phoretic mites that occupy adult beetles, which are attracted to carrion in order to feed on dipteran eggs and larvae (Figure 8.3).

In a recent study, entomological community assembly patterns were evaluated using replicate swine carcasses during four seasons in Ohio, USA (Benbow et al. 2013). The relative abundance and assembly of the five major insect taxa differed significantly among four seasons (see Figure 2 in Benbow et al. 2013), following a pattern of differences in taxon richness and diversity. This study was one of the first to use multivariate statistics to simultaneously evaluate all taxa collected, allowing the report of statistically significant differences in community assembly on carrion among different seasons. During spring months in Ohio, *Cochliomyia macellaria* (Fabricius) (Diptera: Calliphoridae) was the predominant early colonizer followed by *Phormia regina* Meigen (Diptera: Calliphoridae), both of which were replaced by Staphylinidae and Silphidae at approximately 14,000–17,500 accumulated degree hours. For both summer and autumn seasons, *Lucilia coeruleiviridis* (Macquart) (Diptera: Calliphoridae) was the predominant dipteran species early in decomposition; however, Muscidae dipterans dominated in the winter months, where early decomposition only lasted approximately 10,000 accumulated degree hours (ADH). A more detailed discussion of necrophagous arthropod communities can be found in Chapter 4.

Invertebrates are not the only consumers of carrion. Vertebrate species, such as raccoons, vultures, hyenas, and coyotes, are thought to compete with invertebrates for carrion. Vertebrates have been found to scavenge rodent carcasses approximately one-third of the time in forested habitats (DeVault et al. 2003). An example of vertebrate scavenging is shown in Figure 8.4; however, a more detailed discussion of vertebrate scavenger communities can be found in Chapter 6. Microbes, invertebrate, and vertebrates all interact in complex ways that define the ecology of carrion decomposition. However, the mechanisms and relative strengths of these interactions require much more additional study that may be guided by community assembly theory, which considers how particular groupings of species come together at a particular resource.

8.3 Ecological Metacommunity Theory

Community assembly theorists run the gamut, from explicit rules limiting competition and niche overlap (Diamond 1975) to treating all trophically similar species as effectively interchangeable, with species diversity arising randomly (Hubbell 2001). One means of reconciling these wildly different hypotheses is through the lens of the metacommunity, as discussed in the excellent review by Leibold et al. (2004). Local populations of species are potentially connected to each other by interactions such as dispersal to form larger communities and thus form a collective metacommunity. This metacommunity serves as the

FIGURE 8.4 **(See color insert.)** There is a diverse community of vertebrate scavengers that can include vultures and hyenas, among many other species. (Photo with permission from Richard Merritt.)

source of the individuals that will immigrate onto each new resource patch (i.e., carrion). Immediately after the death of an individual, a succession of microorganisms, localized invertebrate, and vertebrate scavengers assemble in association with that new carrion patch as a nutrient pulse (Horn 1974; Yang et al. 2008). The mechanism and variation in this assembly process can be explained and studied using several emerging paradigms.

According to Leibold et al. (2004), metacommunity theory integrates previous thoughts on community assembly to provide four basic paradigms for describing and understanding this assembly process: patch dynamics, species-sorting, mass effect, and neutral. These four paradigms explain how and why communities assemble, which is vital to understanding both landscape and spatial ecology. In carrion ecology, there are few studies to date that have explicitly explored metacommunity dynamics of either invertebrate (Hanski and Kuusela 1977; Hanski 1981, 1987) or vertebrate (Wilson and Wolkovich 2011) necrophagous communities. New studies that test how necrobiome communities are structured and changed in both space and time could fill this gap in community ecology. For instance, metacommunity ecology theory could be used to test the following competing hypotheses related to each paradigm: (1) if identical patches (Poza-Carrion et al. 2013) are capable of maintaining communities with diversity limited by dispersal ability—patch dynamics; (2) if varying patch types (niche) along a resource gradient have a greater influence on community assembly than dispersal ability—species sorting; (3) if the strength of immigration and emigration drives assembly on a single patch and surrounding patches—mass effect; and (4) if community assembly patterns on any patch are a result of random events—neutral.

The often ephemeral nature of carrion provides an ideal circumstance for facilitating interactions among species that are rapid, brief, and carry significant evolutionary consequences. These features make the carrion ecosystem a powerful natural laboratory to explore community assembly theory. As an example, adding an introduced species to the carrion decomposer community can drive changes in both diversity and abundance (Davis et al. 2005). The blow fly *Chrysomya rufifacies* (Macquart) (Diptera: Calliphoridae) was introduced to North America in the early 1980s (Gagné 1981; Richard and Gerrish 1983), and by 2007 had spread as far north as Canada (Rosati and VanLaerhoven 2007). While adult *Ch. rufifacies* superficially behave much like native fly species, the larvae are facultative predators of other dipteran larvae (Baumgartner 1993). When a carcass is colonized by both *Ch. rufifacies* and other dipteran larvae, *Ch. rufifacies* has a competitive advantage, leading to the possibility that native biodiversity might be reduced (Wells and Greenberg 1992a). Since the adults appear to be ecologically equivalent to other blow fly species, neutral theory would seem to apply. Neutral theory would predict that while native blow fly species abundance might change, widespread extinction from *Ch. rufifacies* is unlikely

(Davis 2003). However, since *Ch. rufifacies* can be predatory as larvae (Baumgartner 1993), the assumption of ecological equivalence in neutral theory might be violated. Under the patch dynamics concept, the dispersal ability of *Ch. rufifacies* would give it a distinct advantage at both capturing and colonizing potential resource patches that otherwise would be used by native species, thus preventing coexistence (Baumgartner and Greenberg 1984). Similarly, the inability of larval blow flies to disperse between patches under contest would make coexistence unlikely under the mass-effect paradigm. If the native species respond to the *Ch. rufifacies* invasion by changing their own resource use, species-sorting would predict continued native persistence (Leibold et al. 2004). Additionally, these community dynamics are possibly driven by aggregative effects as recently reviewed (Fiene et al. 2014). Further, when abiotic (e.g., temperatures below the larval activity threshold) or biotic factors (e.g., wrappings, burial) delay insect access to remains, the patterns of initial community assembly and subsequent succession may be altered, as described in Figure 8.5 (Pechal et al. 2014a).

The appearance of new carrion in the environment requires ongoing invasion and exploitation of each new patch. Being dead, carrion is not considered regenerative, responsive, nor homeostatic. All of the changes occurring on (or associated with) carrion during the decay process are driven by mixed assemblies of microbiota and macrobiota, their interactions, and the environment (e.g., habitat). The changes may either facilitate or hinder exploitation of the patch by new individuals or species. These interactions, as species across many trophic levels assemble on the patch, alter it, and depart from it, form the backbone of succession and community assembly.

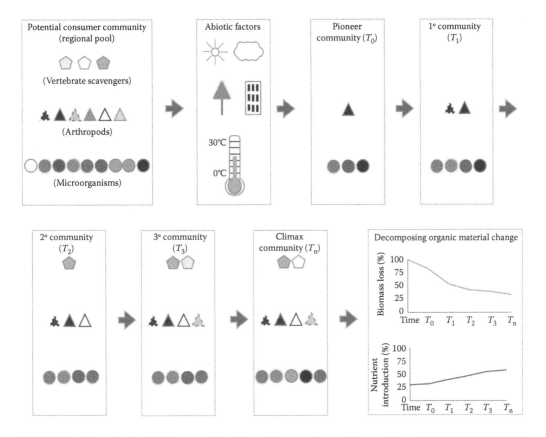

FIGURE 8.5 (See color insert.) The succession of necrobiome members on decomposing organic material such as carrion is determined by abiotic (e.g., habitat type and temperature) and biotic interactions (e.g., competition, species sorting, and landscape patch dynamics). Specifically, the potential community of consumers including microorganisms, arthropods, and vertebrate scavengers will go through many community assemblages from the time a resource is made available (T_0) until climax community (T_n) is reached. The resulting rate of carrion resource biomass loss and nutrient introduce into the environment will depend on the consumer community assembly changes throughout time.

8.4 Community Assembly of Carrion

8.4.1 Microbial Community Assembly

Microbially associated mechanisms of carrion decomposition begin immediately after death and are a driving force for the conversion of a once-living organism to a resource of energy and nutrients (see Chapters 2 and 3). Commensal and transient microflora co-evolve and co-exist on and within their host and cover skin and mucosal surfaces (Turnbaugh et al. 2007; The Human Microbiome Consortium 2012). The living host supplies resident microbes with adequate surface area, nutrients, oxygen, temperature, and pH, and these microflora grow to a maximum growth rate (Cotter and Hill 2003). Fluctuations of these parameters affect growth and activate microbial stress responses (Schimel et al. 2007). Bacteria respond to stress and dynamic environmental conditions by modulating gene expression that controls the cell's structure and function, and an extreme stress will force some microbes into dormancy or death (Schimel et al. 2007). These stress responses can result from both abiotic and biotic factors that differentially affect each microbial species and leading to competition, facilitation, or both, depending on the community composition. These intra- and interspecific interactions lead to successional changes in the communities that represent the postmortem microbiome of the dead host.

There are several factors known to affect microbial community succession during vertebrate carrion decomposition in terrestrial habitats including temperature, moisture and humidity, tissue type, surrounding vegetation, access by necrophagous insects, and soil pH (see Chapters 2, 3, and 20). However, there are a variety of challenges for researchers studying microbial assemblages of these systems. Literature exists but does not explicitly examine the amount of variation in the number and transition of microbial species present or the relative abundance of specific species (Kong 2011), which can affect microbial community assembly during decomposition.

One of the first studies to investigate postmortem microbial changes during vertebrate carrion decomposition using next-generation high-throughput metagenomic sequencing technology to describe successional changes in the postmortem epinecrotic bacterial communities in a terrestrial habitat was by Pechal et al. 2014b. The goal was to understand relationships with bacterial succession and decomposition physiological time (h°C) in an effort to more accurately estimate the PMI (Figure 1 in Pechal et al. 2014b). In that study, replicate swine carcasses had samples taken from the buccal cavity and skin at 0, 1, 3, and 5 days after placement in forested conditions. The authors reported a significant difference of bacterial communities over the course of decomposition.

Two additional studies were published the same year: Metcalf et al. (2013) used Illumina MiSeq to describe and model changes in epinecrotic and soil microbial communities associated with mouse carcasses, demonstrating significant shifts in microbial communities on and in the soil beneath carcasses. Further, Hyde et al. (2013) used pyrosequencing to measure microbial community richness and diversity from oral, rectal, and internal organs of two human cadavers during the decomposition processes occurring pre- and postbloat. Results of this study indicated that richness increased from the upper gastrointestinal tract (mouth, stomach, and small intestine) to the lower (colon and rectal/fecal). Additional details of these studies can be found in Section 8.2.1 and a discussion of how microbial succession may affect the colonization and subsequent succession of arthropod communities is given in Chapter 20.

8.4.2 Arthropod Community Assembly

Succession patterns are governed by various environmental factors within the landscape, with biodiversity resulting from spatial and temporal heterogeneity of temperature, geology, humidity, and other variables (Grinnell 1917; Elton 1927; Hutchinson 1961; Schoenly 1992; Leibold 1995). Alternatively or concurrently, succession can be governed by immigration and extinction rates of species within a habitat (Hubbell 2001). Studies of species interactions occurring on carrion have primarily focused on describing relationships amongst blow fly species. As the resource decomposes, the number of organisms and species utilizing the resource initially increases, as does the complexity of their interactions (Jiron and Cartin 1981). However, based on species diversity data for insect succession, species interactions on carrion probably follow a normal (bell-shaped) distribution pattern across the process of decomposition. Fewer species utilize carrion during fresh decomposition, increasing to maximum diversity during active decomposition, and decreasing again to a reduced number of specialists present during advanced decomposition (Payne 1965). Insect access to the remains also can affect the insect taxon richness; Pechal et al. (2014a) demonstrated that richness increased in a linear pattern over decomposition time (e.g., ADH) for carcasses immediately exposed to insects, while carcasses excluded from insect access for 5 days had a unimodal pattern or linear decrease in taxon richness over the decomposition progression. Some researchers have suggested colonization and succession patterns on a carcass are random (Braack 1987). Indeed, substantial variation occurs during colonization in similar habitats.

Further, asynchronous blow fly succession has been documented on alligator, *Alligator mississippiensis*, carcasses with *L. coeruleiviridis* arriving during early stages of decomposition and *Chrysomya* spp., *C. macellaria*, and *P. regina* arriving at later stages of decomposition, which suggests drivers of the niche paradigm may be driving colonization patterns (Nelder et al. 2009). Clearly, there may be an overlap in ecological paradigms (i.e., niche vs. neutral) explaining insect assembly on carrion. This overlap and variation is likely a function of the position and timing of carrion within the larger landscape, with the responses being driven by landscape and patch dynamic properties that could influence species aggregation patterns (Fiene et al. 2014).

These community processes and patterns have been studied for necrophagous insects with a range of outcomes. Previous studies have described blow fly species composition (Wells and Greenberg 1992b; Faria et al. 1999), densities (Goodbrod and Goff 1990), priority effects (Hanski and Kuusela 1977; Brundage et al. 2014), and competition (Burkepile et al. 2006) on population dynamics in controlled laboratory and field experiments. Disturbances of natural communities have also been proposed as a mechanism to promote species coexistence and influence community assembly (Hutchinson 1961; Horn 1974; Kuusela and Hanski 1982) or priority effects of initial colonizers altering subsequent community structure (Chase 2010; Fukami and Nakajima 2011). The aggregation model of coexistence states that species competing for ephemeral patches can coexist if they exhibit irregular aggregation in the environment. When large numbers of a given population are present on any given patch, other patches in the environment remain available for use by other species (Hartley and Shorrocks 2002; Abos et al. 2006). Despite the number of populations competing for a patch, one species rarely establishes superiority and will completely exclude another (Shorrocks et al. 1979; Atkinson and Shorrocks 1984). Kneidel (1985) found that increased patchiness supports coexistence because aggregations of a single species will compete amongst themselves for a resource rather than face interspecific competition, when working with small carrion (<1 kg). It is apparent that coexistence on carrion resources is common, but the introduction and exclusion of species impact on energy flow among the trophic levels is still unclear, especially with regard to the importance that carrion size may play in community assembly.

Differences in resource size can dramatically impact the diversity and community composition of arthropods associated with decomposing remains within a landscape and metacommunity context. For example, in an African habitat, medium and large carcasses, such as impala (*Aepyceros melampus*), were dominated by blow fly larvae, while flesh fly larvae were absent; however, flesh fly larvae dominated the larval dipteran communities on smaller carcasses, such as birds (Braack 1987). In the United States, very small carrion, such as slugs, had a substantially higher Shannon–Wiener diversity index than rodent carrion (Kneidel 1984). However, this tendency is not universal, as Kuusela and Hanski (1982) found no greater differences in insect community structure among carcasses of different sizes than there were between replicates of each carcass size.

8.4.3 Potential Applications of Community Assembly Theory

Because of its relatively small scale (e.g., within 1 m²) and ephemeral nature, the carrion system allows for replicated studies of microbial, invertebrate, and vertebrate community ecology theory that can be performed in a variety of habitats and environmental conditions. Further, this basic science research approach can have profound applications in a variety of disciplines like pest management (Walter 2005) or food security (Andreescu and Marty 2006). Forensic entomology, in particular, would benefit from basic theory studies of microbial and invertebrate community assembly (Byrd and Castner 2010;

Tomberlin and Benbow 2015). For decades (if not centuries, see Benecke 2001), forensic entomologists have used life-history traits (e.g., larval development) and community assemblage patterns (i.e., succession) of necrophagous arthropods for estimating the minimum postmortem interval (PMI_{min}) ranges (Byrd and Castner 2010), a metric based on the duration of insect activity on a carcass. This duration is also known as the period of insect activity (PIA), albeit with some variation in the exact definition and its relevance to the application of forensic entomology (see Campobasso and Introna 2001, 2014; Amendt et al. 2007; Tomberlin et al. 2011b; Villet 2011; Michaud et al. 2014; Tarone et al. 2014; Wells 2014).

As a construct based on necrophagous insect biology and ecology, the PIA could be further studied in detail to investigate the ecological processes occurring on carrion, which may influence the precision of PMI_{min} range estimates. Entomological evidence can impact the lives of suspects, perpetrators of criminal acts, and the family or friends of a victim. Therefore, better understanding the ecological dynamics that influence insect arrival and colonization patterns has the potential to increase confidence in the courts (e.g., judges and juries) when interpreting such entomologically based evidence in criminal proceedings. Beyond the macroinvertebrate level, recent studies have shown the potential use of microbial community succession to estimate PMI_{min} using next-generation metagenomic sequencing technology (Hyde et al. 2013; Metcalf et al. 2013; Pechal et al. 2014b), as previously discussed. Thus, understanding the ecological processes and mechanisms of carrion community assembly has both basic and applied importance. In many cases, as with the blow fly examples given above, the operating ecological paradigms may overlap explaining insect assembly on carrion (e.g., niche versus neutral). This overlap and variation most probably is a function of the position and timing of carrion within the larger landscape, being driven by landscape and patch dynamic properties.

8.5 Landscape and Community Ecology

Landscape ecology is defined as how and in what ways organisms function and disperse within their habitat (or habitats of multiple communities) (Forman 1995a). Community composition, including biodiversity, dominant species, keystone species, and interactions amongst species in a landscape, influences community function (Hooper et al. 2005). For example, the introduction of ungulate carrion into a terrestrial system facilitates localized succession of insect colonizers like blow flies (Beasley et al. 2012). However, overall community structure of the ecosystem is considered stable because localized decomposition events do not significantly change the total community species composition (Horn 1974). Instead, carrion resources typically represent discrete patches of energy and nutrient pulses in the landscape separated by distance and time (Figure 8.6) (Yang 2006; Yang et al. 2008; Parmenter and MacMahon 2009). While the effect of large, regular carrion pulses, such as mass emergence of cicadas or salmon runs, have been well studied (Yang 2004; Hocking et al. 2009; Moore and Schindler 2010), relatively less is understood about the landscape ecology of relatively unpredictable vertebrate carrion, such as terrestrial ungulates (Margalida et al. 2011; Villet 2011) or oceanic cetaceans, also known as whale falls (Smith and Baco 2003; Lundsten et al. 2010; Amon et al. 2013) (see also Chapter 12). Carrion input to the habitat at small (e.g., rabbit carcasses) or large (e.g., salmon runs) magnitudes is important to local- (Carter et al. 2007), regional-, and ecosystem-scale processes (Parmenter and MacMahon 2009). A recent review by Barton et al. (2013b) describes the effects of discrete carrion patches that are considered nutrient pulses on the landscape within an ecosystem, specifically how ephemeral resource patches influence the biodiversity of organisms and ecological processes occurring within the habitat. For additional information on this concept, please see Chapter 13.

As much as the introduction of carrion into an ecosystem represents a resource pulse for scavengers, it also represents one for microbes, with the soil microbial community immediately underneath the carcass responding throughout decomposition (Hopkins 2008; Stokes et al. 2009). Introducing carrion into an ecosystem can impact the associated soil microbial community (Hopkins 2008; Stokes et al. 2009) by influencing soil chemistry and composition, which may lead to diverging soil communities (Post and Kwon 2000). In turn, vegetation growing directly at a carcass site has been demonstrated to have significantly different species richness and density when compared to zones radiating outward from the carcass site (Towne 2000). Vertebrate animals that die in their burrows may therefore have a significant

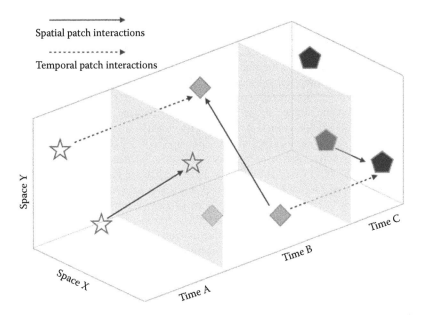

FIGURE 8.6 Carrion resources can represent small, discrete patches within a landscape separated by distance and through time. Each patch can have a distinct necrobiome community assemblage (e.g., bacteria and blow flies), which either interacts with other communities within a given spatial area (solid arrows) or across time (dotted arrows). These potentially interacting communities (each shape represents a community on a resource patch) form the metacommunity within a given habitat and can influence the assemblage patterns of consumers utilizing carrion within an environment.

impact on the vegetation of a habitat, as they often go undiscovered during ecological investigation; these typically concealed remains may also serve as distributed sources of carrion arthropods in the larger landscape. However, it is not well understood how the larger-scale distribution of carrion and the associated localized soil and diversity effects may collectively scale-up to the broader landscape. Recent studies indicate that these localized effects may significantly impact ecosystem processes such as leaf litter decomposition (Hawlena et al. 2012), providing additional evidence that even though vertebrate carrion is considered to be at low density in the landscape, the localized impact may have cascading effects at the ecosystem scale (Barton et al. 2013a), especially in the case of large resource pulses such as salmon or cicadas (Yang 2004; Yang et al. 2008).

Carrion is a high-quality form of detritus in comparison to low-quality material, such as plant detritus, which can also be used by a variety of microbes, invertebrates, and vertebrates (Moore et al. 2004; Wilson and Wolkovich 2011). Selectivity and aggregation of colonizers could be based on resource quality, as has been demonstrated in costal marine habitats where fresh carcasses attract approximately twice as many consumers as microbe-laden carcasses (Burkepile et al. 2006). Microbial communities associated with the carcasses could be altering the quality of the resource, thus mediating invertebrate community assembly, similar to how fungi govern phytophagous insect composition in a plant-based system (Tack et al. 2012).

Microbes have multiple roles within an ecosystem, such as acting as pathogens, mutualists, food sources, and most commonly as decomposers within the community (Cochran-Starifa and Ende 1998; Burkepile et al. 2006). Carrion is part of the decaying organic matter of most ecosystems, a primary level of energy flow (DeVault et al. 2003). And more recently bacteria have been established as competitors against animals for ephemeral resources (Janzen 1977; DeVault et al. 2004; Carter et al. 2007). Polis and Strong (1996) stated that there is limited energy transfer between resources and microbes despite the changes in population densities and inter- and intraspecific interactions. It is interesting to note the lack of consistency and homogeneity in energy transfer and microbes among tropic-level interactions, as there are only a few carrion-focused food-web studies (Cornaby 1974; Barton et al. 2013b; Schlacher et al. 2013).

Arthropod community structure on carrion over time has often been described as a process of competition between species for a resource (Norris 1965; Hanski and Kuusela 1977). Studies examining this process primarily focus on the "observable" data (e.g., arthropods) that can be collected, with little regard for what might be occurring at the microbial scale. Primary colonizers utilize the resource as nutrition, mating, or an oviposition site. Subsequent larval development may disrupt established microbial communities through direct or indirect competitive interactions on the carcass. Insects facilitate the decomposition of carrion, which may directly impact microbial species via chemical secretions from feeding fly larvae (Payne 1965; Sherman et al. 2000; Mumcuoglu et al. 2001; Simmons et al. 2010a,b). Arthropods arriving from disparate habitats of the landscape to colonize carrion could mechanically transport a variety of exogenous bacteria, or introduce their own internal microbial community (Asgari et al. 1998; Nayduch et al. 2002; Conn et al. 2007), thereby affecting the community assembly of the necrobiome in the landscape (Tomberlin et al. 2011a). Flies are well-known vectors of pathenogenic bacteria: *Musca domestica* L. (Diptera: Muscidae) alone carries over 100 known pathogens including *Escherichia coli* 0157:H7 (Greenberg and Klowden 1972; Alam and Zurek 2004). The introduction of insect-associated microbial communities may influence carrion microbial community function via competition, which in turn alters the metacommunity dynamics and biogeography of microbial communities in the landscape (Jones and McMahon 2009; Langenheder and Szekely 2011).

The spatial and temporal availability of carrion in the landscape can also alter vertebrate scavenger biodiversity. There is a universality of facultative scavengers when high-quality resources abound (Wilson and Wolkovich 2011). Individual carcasses placed in an open field on random days, or "unpredictable" carcasses, had great diversity of members from the avian scavenger guild (e.g., Griffon vulture, Egyptian vulture, black kites, and common ravens); however, more than 50% of "predictable" carcasses, or those consistently placed in the same location at regular time intervals, were consumed solely by the Griffon vulture (Cortes-Avizanda et al. 2009). As vulture "restaurants," these supplemental feeding locations can be used as conservation strategy to maintain the population numbers of endangered avian scavengers (García-Ripollés et al. 2004; Gilbert et al. 2007), but the timing of food presentation should be carefully considered prior to implementation. Utilization of remains by scavengers can vary across ecosystems. In beach habitats, Ghost crabs (*Ocypode* spp.) have adapted to find and consume carrion resources in less than a day, demonstrating faster and more aggressive scavenging behaviors than found in other terrestrial environments (Schlacher et al. 2013).

Additional work is needed to better understand the landscape ecology effects of variable carrion densities and community assembly processes important to ecosystem services. Pechal and Benbow (2015) provide a table (Table 27.1) that lists several questions that could be addressed in future studies of carrion landscape ecology. Next, a discussion of a framework is provided to guide future community ecology studies of carrion decomposition. This framework dissects the biological and ecological mechanisms inherent to the attraction and ultimate colonization and use of carrion by consumers, the importance of these events to succession and ultimately to the rate and variation in decomposition and its effects on the local habitat and surrounding ecosystem. The conceptual background for understanding the community ecology of carrion at the local scale processes that ultimately affect the metacommunity dynamics of carrion at the landscape and ecosystem scale are also provided.

8.6 A Framework for Community Ecology Studies of Carrion Decomposition

Tomberlin et al. (2011b) proposed a model framework for dividing the decomposition process into five phases, punctuated by four major decomposer–carrion interactions. While initially described for explaining arthropod use of vertebrate carrion, this same framework could, in part, apply to other organisms (e.g., microbes and vertebrates) as well. These phases can be useful in defining how ecological community interactions affect the overall decomposition process, and so we briefly introduce the phases and discuss how principles of community ecology define the carrion decomposition process. The model is used to provide a discussion of how the initial physiological responses of an insect play a role in the timing and magnitude of exploitation of that resource, the growth rates and competitive abilities of the insects, and ultimate successional changes and interactions with other scavengers and consumers.

However, these are not the only carrion–decomposer interactions. This framework is initially divided into conceptual sections of the precolonization interval (pre-CI) and postcolonization interval (post-CI). Within each of these intervals, the process of decomposition along with the biological and ecological interactions is described for each specific phase.

The initial detection, location, and colonization of remains by specific necrophagous insect species is thought to mediate species sorting and defines the initial species that colonize a resource from a regional pool, perhaps following one of the four metacommunity assembly processes previously discussed (Leibold et al. 2004). Volatile organic compounds (VOCs), derived from tissue autolysis and changes in microbial communities of the remains, may play a role in defining which invertebrate species will initially detect, locate, and colonize a resource. This hypothesis was recently supported in laboratory studies, suggesting that interkingdom interactions between eukaryotes and microbes during the pre-CI may be important for defining the downstream community assembly patterns of the post-CI (Ma et al. 2012; Tomberlin et al. 2012). Furthermore, there is recent evidence that symbiotic bacteria may affect the social structure of hyenas, important scavengers of large mammal carrion in African ecosystems (Theis et al. 2012, 2013). However, this line of research and enquiry will need to be expanded to different microbial communities, invertebrates and vertebrates, especially using field-based studies and experiments.

8.7 Precolonization Interval

The initiation of the pre-CI is based largely on the host seeking and selection model of parasitoid insects (Vinson 1998), which involves many small steps to find a suitable resource for offspring development, as illustrated in Figure 8.7. In each step, the insect must process information about both its own physiological state and the state or quality of the carrion. Depending on the outcome of the processing, the organism either continues or alters its behavior and moves to the next step in resource exploitation. The behavioral outcome is thus shaped by the biology and ecology of the organism. By focusing on these intersections of biology and ecology, the model actively integrates basic sciences in explaining both proximate and ultimate reasons for insect behavior (see Chapter 10 for more details).

This model is quite flexible for describing the potential biological and ecological mechanisms for how decomposers find and exploit a carrion patch. This concept can apply to a vertebrate scavenger looking directly to consume the carrion, a blow fly adult searching for an oviposition site, or an entire guild of specialist beetles searching for a food resource. The model is also temporally flexible, in that the stages last from interaction to interaction rather than for a specific amount of time.

Of particular importance to the model is insects' complex information processing capability. While a particular behavior, such as probing with the ovipositor, might be a stereotypic action, it is induced by a complex set of releasing and inhibitory stimuli. These stimuli can be either external (allothetic) or internal (idiothetic) or a combination of both. Allothetic stimuli are primarily environmental, either abiotic cues such as light intensity, day length, direction of gravity, temperature, or biotic cues such as mate-calling songs, or the odor of food or predators. Idiothetic stimuli are primarily biological, reflecting the current physiological state of the insect (e.g., egg load or nutritional need) and any learned information. As these states change over time, so do behavioral responses (Visser 1988).

An organism may have absolute limitations for perceptions of either allothetic or idiothetic stimuli based on the number or tuning of their sensory apparatuses. In the case of odor perception in insects, sensitivity is dictated by the number of olfactory receptor neurons (ORNs) and odorant-binding proteins (ORBs) expressed in the insect sensillae. ORBs allow hydrophobic odor molecules to move through the hemolymph of the sensillae, where the compound can then bind to the dendrites of the ORN. In turn, the ORN converts a chemical reaction between VOCs and the neuron into an electrical signal delivered to the central nervous system (CNS; Hansson 2002). The combination of ORBs and ORNs allow extreme sensitivity to a wide range of environmental conditions. ORBs may complex with only a particular odor molecule, such as a pheromone, or to a more general family of environmental chemicals, such as monoterpenes (Vogt et al. 1991). In *Drosophila* (Diptera: Drosophilidae), over 50 ORBs are known that are associated with 16 different classes of ORNs (Vogt et al. 1991; de Bruyne et al. 2001; Hansson and

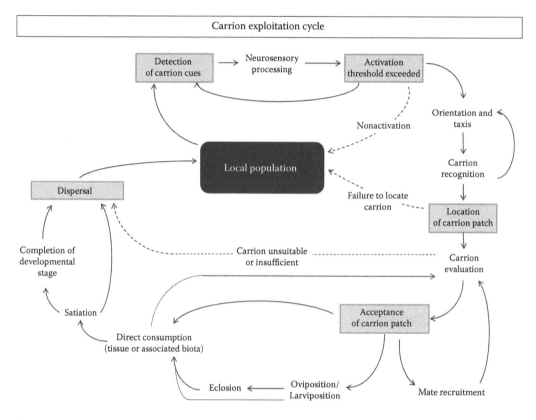

FIGURE 8.7 The generalized carrion exploitation cycle. All arthropods in the local population begin the cycle unattenuated to any carrion. Once carrion is exposed in the environment and begins to emit detectable cues, arthropods enter the cycle at the upper left. Each state must be completed in order to proceed; otherwise, individuals return to the unattenuated population (dotted lines). Grey boxes represent the control points for the conceptual framework. Death of the arthropod may also occur at any point within the cycle, and these pathways were omitted for clarity. (From Tomberlin, J.K. et al. 2011a. *Annual Review of Entomology* 56: 401–421.)

Stensmyr 2011). Each class responds to a particular set of odorants, and if exposed to a molecule from the wrong set, may not fire the CNS. Even if it does fire, the CNS may not respond with a behavioral action (Matthews and Matthews 2010). Because an insect exists in an airborne soup of chemical noise, only the correct blend of chemical signals at the correct signal intensities will cause the CNS to induce a behavioral response. This specific, filtered response allows the insect to avoid wasting energy responding to spurious cues (Bruce et al. 2005).

On the idiothetic side, the response to a particular odor blend may vary substantially with physiology. For example, *P. regina* stimulate their abdominal stretch receptors with a large meal of sugar water and will cease responding to further sugar cues at a later time (Dethier 1955). In *Lucilia sericata* (Meigen) (Diptera: Calliphoridae), gravid females are much more likely to be caught in carrion-baited traps or respond to liver odor plumes than either males or nongravid females (Wall and Warnes 1994). A similar phenomenon is seen in parasitoids, where experienced females and ones with high egg load are more responsive to chemical cues than naïve, nulliparous ones (Vinson 1998). Some of these changes may be due to learning. There is some evidence that female blow flies use multiple carrion patches throughout their lifetime: their natal patch, a protein meal patch, and an oviposition patch (Mohr 2012). Experience with multiple types of carrion might alter their idiothetic control of acceptance, as it does with parasitoids (Visser 1988). Although there is some evidence that tissue type affects larval development (Day and Wallman 2006; Tarone and Foran 2006; Boatright and Tomberlin 2010), the effect on either naïve or experienced female choice has yet to be tested. For parasitoids, experienced females have much shorter seeking and handling times (Wajnberg 2006). If the same holds true with necrophilous species, cues learned from a natal host could substantially shorten the pre-CI.

At a deeper ecological level, foraging theory helps explain some of this variation in behavioral response to stimuli (see also Chapter 10). The fitness benefits of an optimal response (i.e., optimal foraging theory) to a resource mix encourages decomposers to make the same range of choices under the same conditions—to search for carrion, to accept it as a resource, to avoid competition and predators, and to leave when it no longer serves their needs (Schmitz et al. 1997). These complex mechanisms and interactions within the insect behavioral response to stimuli affect every interaction between the carrion source and insect.

Each individual insect, species, guild, or sere must also interact with one another and with the abiotic environment. As in the example in Section 8.4 exploring community assembly through the invasion of *Ch. rufifacies*, the rapid convergence, consumption, and departure from carrion patches make it an excellent model system to explore several important ecological concepts. With the simultaneous presence of many insect larvae consuming a carrion patch, inter- and intraspecific competitive strategies have a major impact on fitness (Chapter 10). From a systemic perspective, studying succession patterns on a swine carcass is much easier than fumigating an entire mangrove island à la MacArthur and Wilson (1967)! Many authors have explored this concept, from early studies such as Reed (1958) and Payne (1965) to the more geographically focused work in British Columbia (Anderson and VanLaerhoven 1996), Southwest Virginia (Tabor et al. 2004), and Louisiana (Watson and Carlton 2003, 2005), and contemporary work integrating the entire necrobiome (Section 8.2). When these kinds of succession investigations are paired with the insect physiology, important interaction points between insects and a carrion patch arise: *detection* of a patch, *activation* to search for that patch, *location* of the patch in space, evaluation and *acceptance* of the carrion, *colonization*, and finally *dispersal* from the patch. At each of these points, the interaction between carrion and insect changes fundamentally, making them a convenient way to divide decomposition into discrete phases.

8.7.1 Exposure Phase

The exposure phase is defined as the duration between death of an organism and its detection by arthropods (Tomberlin et al. 2011a). Before scavengers can interact with carrion, it must both exist and signal its presence within their environment. The first condition is brought into being at the biological death of an organism—permanent cardiac and pulmonary arrest (DiMaio and DiMaio 2001). In the special case of myiasis, "death" can be read not as death of a complete organism, but rather as the cellular death and necrosis of some portion of that organism (e.g., an ulcer). To meet the second condition, the carrion must be accessible to at least one arthropod sensory domain—visual, chemosensory, auditory, or tactile. For most animal deaths in the natural environment, the bodies are accessible to insects immediately.

If invertebrates are excluded from a body, microbial and cellular activity begins the process of decomposition. After death, deoxygenation of cells causes a chain of events leading to autolysis. Aerobic adenosine triphosphate production fails, and anaerobic fermentation of lactic acid reduces cellular pH. Membrane permeability increases, releasing hydrolytic enzymes from lysosomes, which break down cellular components. Depending on the pH and availability of enzymes, different body tissues may undergo nearly complete liquifactive necrosis (e.g., brain, liver, and stomach) or incomplete coagulative necrosis (e.g., skeletal muscle and skin) (Gill-King 1997; Tuomisto et al. 2013).

Autolysis is followed by putrefaction, a primarily anaerobic process. During life, the gastrointestinal tract controls the population of enteric flora (DiMaio and DiMaio 2001). In the absence of this control, not only does the overall microbial population explode, but highly competitive taxa can proliferate exponentially (Sekirov et al. 2010). Some species can even migrate through the body (Tuomisto et al. 2013). These bacteria exploit the nutrients liberated by autolytic processes, as well as producing enzymes, which further break down body biomolecules, particularly proteins. The by-products of this metabolism predominantly take the form of gasses like hydrogen sulfide, methane, ammonia, and carbon dioxide. When trapped inside the body, they cause bloating and distention; when released outside the body, they produce the characteristic odors of decomposition (Powers 2005). However, these compounds and their composition is likely a result of the microbial community within the gut. Vertebrates, such as humans, have a highly variable gut flora, and the mechanisms dictating the prevalence of a given species is not yet understood. However, it is a topic of much contemporary research, as variability in enteric flora has major effects on overall health (Sekirov et al. 2010).

There is some evidence, however, that the exposure phase might be estimable based on the bacterial activity and soil changes (Hyde et al. 2013; Metcalf et al. 2013; Pechal et al. 2014b). Looking strictly at the community composition of bacteria found on the external skin and internal abdominal cavity, Metcalf et al. (2013) was able to predict true PMI within about 3 days. That study, however, excluded insect access, which significantly changes microbial community composition and decreases the decomposition rate (Benbow et al. 2013; Pechal et al. 2013, 2014a). Several studies have investigated the soil changes that accompany decomposition, including moisture pH, nitrogen, and phosphorus levels (Petersen and Luxton 1982; Dent et al. 2004; Carter et al. 2010). These studies have shown that decomposing bodies produce a predictable spike in available nutrients, and a drop in pH. These soil changes can also be used to estimate a PMI up to 100 days long (Benninger et al. 2008), although, as with insects, soil conditions do not begin to react to a cadaver until it is exposed to the ground (Forbes 2009).

8.7.2 Detection Phase

Once a set of remains is exposed and available to decomposers, they must detect it. However, there are several challenges to locating a carrion source. Necrophilous arthropods are unable to use many of the major live host-seeking cues used by host-seeking Diptera (e.g., mosquitoes [Diptera: Culicidae]) such as motion, shape, or the long distance chemical cue of exhaled carbon dioxide (Gibson and Torr 1999). Furthermore, animals do not necessarily die in any particular habitat, so necrophilous organisms cannot automatically use habitat cues as a means of resource location. Buried remains may be intentionally cryptic. The onset of the detection phase, therefore, is limited by both the production of carrion cues and the physiological limitations of the insect sensory system.

Presumably, the main sensory domain for necrophilous insects is olfactory/chemosensory, much as it is for parasitoids and herbivores. For example, a decaying human body produces upwards of 428 volatile chemical odors (Vass et al. 2008), with substantial odor profile changes over the course of decomposition (Archer and Elgar 2003) (see also Chapters 2 and 9). The components of the odor profile also vary by species, age of individual at death, soil texture, soil type, moisture level, and temperature (Vass et al. 2004, 2008; Vass 2012). Apneumone production by cell lysis and bacterial proliferation begins almost immediately after death (LeBlanc et al. 2009), which may explain the very short detection phases reported in the literature—as little as 30 s (Gruner et al. 2007).

Many of these decomposition odors, such as carbon dioxide, methane, ammonia, hydrogen sulfide, and ethanol (Dent et al. 2004; Vass 2012) are very common chemicals present in the environment in varying concentrations. Others, such as 1,4-dimethyl benzene, heptane, methenamine, sulfur dioxide, hexadecanoic acid, decanal, trichloromonofluoromethane, and 1-methyl naphthalene, are more specific to carrion decomposition (Vass et al. 2004). Based on this profusion of carrion chemicals, necrophilous insects appear to follow the herbivorous model and only react behaviorally to a one strong carrion signal supplied with an appropriate mixture of secondary odor cues. Some authors working with *L. sericata* have shown that neurons respond powerfully to 1-octen-3-ol, dimethyldisulfide, and 2-phenylethanol (Park and Cork 1999). However, this species does not exhibit a preference for pure chemicals such as cadaverine, putrescine, indole, or skatole over decaying meat emitting those same compounds in lower concentrations (Easton and Feir 1991). This dramatic difference in response suggests that a complex of carrion odors is responsible for initial attraction. Similarly, the parasitoid wasp *Nasonia vitripennis* (Walker) (Hymenoptera: Pteromalidae) demonstrated a neurosensory response to seven human remains-associated odors, but only responds to one, methyldisulfanylmethane, thus indicating the wasp is using a highly specific VOC from the corpse to find its prey—cyclorraphan fly pupae (Frederickx et al. 2014). Bacterial odors probably play an important role in affecting the blend of carrion odors, such as those of *Pseudomonas* and *Bacillus* that have been found to be strongly attractive to *L. sericata* (Emmens and Murray 1982; Wall and Warnes 1994). For additional information on the chemical ecology of carrion, see Chapter 9.

The changes in carrion odor over the decomposition process may be the controlling factor in organizing succession and resource partitioning. As the carrion patch is altered by the biota, its odor profile also changes, both in terms of the actual makeup of the plume and the relative abundance of each component (Vass 2011). For species with ORNs that are attuned to later-evolving compound blends, their detection phase would be greatly extended, as is the case of *Nicrophorus vespilloides* Herbst (Coleoptera: Silphidae). In this beetle, females are only responsive to odor plumes of piglets in advanced decay, indicating that odor composition and not searching capacity leads these beetles to arrive during the later portions of decomposition (Rozen et al. 2008). Similarly, while most blow flies are generalists with regard to carcass type, at least a few species, such as *Calliphora vicina* (R.-D.) (Diptera: Calliphoridae), *Ca. vomitoria* (L.) (Diptera: Calliphoridae), and *Lucilia richardsi* Collin (Diptera: Calliphoridae) do show distinct carcass preferences (Kneidel 1984; Anderson 2010). *Lucilia silvarum* Meigen (Diptera: Calliphoridae) is also known to have a preference for amphibians in both North America and Europe (Fremdt et al. 2012). Ergo, evolution of ORN sensitivity may be a means of niche partitioning with specialists adapting compound-specific ORNs much more quickly than generalists (Vieira et al. 2007).

Once the chemical signature of an object of interest has been perceived, the next step to searching/ location behavior is orientation (Mittelstaedt 1962). Because most insects have bilateral ORNs on the antennae, they can integrate signals. They orient themselves into an odor plume when both antennae are equally stimulated, before initiating locomotion (Murlis et al. 1992). Once in motion, many insects display a looping path back and forth across an odor plume. The combination of odor gradient and unequal antennal stimulation helps the insect hone in on the object of interest (Visser 1988). A constant, straight-line motion is not possible, because the odor molecules are not distributed in a steady stream on the air current. Instead, they are delivered in "packets" of odor separated by background air or by competing odor plumes (Matthews and Matthews 2010). This fragmentation has major implications for host finding, as odor plumes that are highly fragmented can cause searching insects to fail to locate their source or to substantially delay source location. Fragmentation is increased by low odor intensity, increased wind speed, and by the density of three-dimensional structures in the environment, for example, forest versus grassland (Murlis et al. 1992).

Once a necrophilous organism close range of the target, different stimuli may become important in arresting seeking behavior, whether it is an insect or vertebrate. These stimuli might take the form of lower volatility chemical compounds, as seen in *L. sericata*. Although the highly volatile ammonium carbonate has been shown to serve as an activation cue, lower volatility sulfide and indole-based cues seem to serve as close-range location and acceptance cues (Ashworth and Wall 1994). For species that oviposit collectively, such as *M. domestica* or *C. macellaria*, one source of attractive volatiles is bacteria proliferating on the surface of intra- and interspecfic eggs (Lam et al. 2007; Brundage 2012). On the other hand, conspecific presence can also be a deterrent, as demonstrated by reduced adult *L. coeruleiviridis* activity once larvae are present on the carrion (Ives 1991). For many host-seeking Diptera, the visual cues of their host are particularly important. Tsetse flies (*Glossina* spp. [Diptera: Muscidae]) have reduced ORP production relative to *Drosophila*, instead of relying on visual receptor strongly attracted to the colors blue and black (International Glossina Genome Initiative 2014). For carrion-associated species, the presence of bloat, marbling, lividity, or a belly-up position might be a visual indicator of death. In some rare cases, the environment itself might serve as an indicator of carrion presence, much as parasitoids use the environment to help them locate hosts (Visser 1988). For cyclical events like salmon runs or cicada pulses, carrion consumers could use habitat cues to facilitate locating an infrequent, bountiful resource.

Not all chemical cues are attractive, many serve as repellents. Dethier (1956) proposed that behavioral activation only occurs when repellent or aversive stimuli are overcome by appetitive or positive ones. Natural repellent cues might come from carcasses in a suboptimal state of decomposition or semiochemicals from competitor or predator species. For forensic entomologists, artificial chemical signals can also be important. Citronella, patchouli hydrochloric acid, and paradichlorobenzene all repel *Ca. vicina* from mouse carcasses, and application of gasoline, patchouli, or citronella can delay colonization by several days (Charabidze et al. 2009).

Not all insects that search will find their odor source, particularly if it is hidden or removed (Dukas 2008). In the carrion system, some carcasses are lost due to vertebrate scavenging (Chapter 6) with odor cues that persist afterwards. For clandestine human body disposal, bodies are often buried, wrapped, or enclosed in an automobile. These physical barriers may reduce the intensity of odor plumes, making them more difficult to locate. They may also serve as a physical barrier to reaching the carrion. For earthen burials, the coffin fly, *Conicera tibialis* Schmitz (Diptera: Phoridae) can crawl through up to 2 m of soil to reach a body (Martín-Vega et al. 2011). In vehicles or homes, necrophilous insects must find holes large enough to permit entrance, or as in the case of *Dermestes maculatus* De Greer (Coleoptera: Dermestidae), chew their way through a 2 mm plastic bag (Mohr personal observation).

If the carrion has been consumed, or if it cannot be located, insects will give up searching rather than continue to waste energy on fruitless foraging (Houston et al. 2011). This cessation is dependent on physiological status, as a hungry or gravid insect will persist longer than a satiated or nongravid one (Singer et al. 1992; Vinson 1998). Insects that have given up go back to their normal noncued foraging strategy within the environment. This normal foraging strategy could have great effect on colonization order or success. Given their need for one or more carrion patches in their lifespan, it seems likely that they engage in some degree of nondirected/noncued foraging. This is likely somewhere along a continuum from "sit and wait" to extended nondirected flight. Depending on the distribution of carrion within the environment, either strategy would be more effective. When carrion are plentiful, or in an area with few competitors, it would be better for the adults to wait for a guaranteed patch. When carrion is scarce, or there are many competitors, it would be better to maximize the chance of encountering a carrion cue by occupying a large spatial territory and reaching the carrion patch early. This differential foraging success could play a role in both succession order and species composition under any of the four metacommunity assembly paradigms, but the exact nature and magnitude of the effect does not yet seem to be known (Pyke 1984; Ward 1987; Singer et al. 1992; Wajnberg 2006; Houston et al. 2011).

8.7.3 Acceptance Phase

Once an arthropod has located a potential resource, it must evaluate whether or not the resource can be exploited. The acceptance phase of the model is largely based on the amount of processing time it requires an insect to commit to a discrete relationship with the carrion patch or to reject it and disperse back into the environment. During this phase, the insect perceives a large number of extremely short-range cues. Unlike the previous two phases, the acceptance phase triggers more than just the antennal ORN. By making contact with the carrion patch, the tarsal gustatory receptors and chemosensors on the mouthparts and ovipositor can all be used to collect and evaluate information about the carrion patch (Browne 1960). For *P. regina*, sensory receptors on the ovipositor play a critical role in mediating oviposition (Wallis 1962), and therefore the beginning of arthropod community assembly of carrion. If a gravid female cannot bring her ovipositor into contact with the corpse or corpse products, oviposition is highly unlikely. In general, if the insect cannot make immediate contact with the carrion, it may take longer for it to accept the resource. For example, wrapping of a body is known to delay colonization by 2–5 days (Goff 1992). While this delay might be due to suppression of odor cues, it seems more likely that the physical barrier prevents carrion evaluation. Larvipositing sarcophagids, however, will release their offspring even if they cannot physically contact the carrion patch, which can provide a competitive advantage when physical barriers are present (Denno and Cothran 1975, 1976). It has been posited that aggregation pheromone may play a role in inducing group oviposition in *L. sericata* (Barton-Browne et al. 1969); however, it seems more likely that the eggs and their associated microbial communities are the more likely source of attraction (Lam et al. 2007; Brundage 2012). On the other hand, *Chrysomya albiceps* (Wiedemann) (Diptera: Calliphoridae), *Ch. megacephala* and *Lucilia eximia* Wiedemann (Diptera: Calliphoridae) prefer carrion with no other larvae present, thus avoiding other fly species (Giao and Godoy 2007). Among abiotic factors influencing acceptance, moisture is probably the most significant. Viability of *Lucilia cuprina* Wiedemann (Diptera: Calliphoridae) eggs declines dramatically as moisture level decreases from saturation, and not even a dense egg mass can prevent desiccation (Vogt and Woodburn 1980). This is similar behavior to filth flies from family Muscidae, which do not tend to oviposit if overall moisture level is below approximately 40% (Stafford and Bay 1987; Farkas et al. 1998; Mullens et al. 2002).

A good case example is that of the beetle *D. maculatus*, which has a pronounced preference for muscle tissue over that of fat or bone marrow (Woodcock et al. 2013). Females were given the choice of the three materials that had been aged 0, 14, or 21 days for an oviposition site. The choices were presented for six straight days, and with the exception of the freshest meats on the first day of presentation, muscle tissue was always preferred. Although there was no difference in overall viability, larvae reared from the muscle tissue pupated more quickly and produced larger adults. The adult females apparently had a mean of assessing and accepting the resource type best suited for their offspring. This type of assessment ability has also been found in *Ch. rufifacies*, which will preferentially choose liver containing eggs or egg-associated microbes of the potential prey species *C. macellaria* (Brundage et al. 2014).

During the acceptance phase, selection pressure on behavior is high, because once a female accepts a patch of carrion for oviposition, her choice has major fitness ramifications. The carrion ecosystem is a classic example of time-limited disperser: mobile adults, sedentary larvae, and a short window of availability (Mayhew 1997). Under the optimum oviposition theory, a female will apportion her eggs among potential patches based on quality and frequency of encounters (Scheirs and De Bruyn 2002). If she chooses a poor carrion patch, one that is overcrowded, too small, not nutritionally appropriate, or predator-attractive, she will pay a large fitness penalty (Scheirs and De Bruyn 2002). As seen in Gripenberg et al. (2010), a meta-analysis of phytophagous insects demonstrated the female preference is particularly important for insects with aggregative offspring. Of course, the presumption of female choice does assume that insects have the ability to rank hosts and to apportion their egg load. The former assumption is not always borne out, both by research into parasitoids (Mayhew 1997) and by blow flies' tendency toward communal oviposition, which can lead to deposition of more eggs than an individual patch can support (Greenberg 1991). The question is whether these egg masses result from stereotypical actions in the presence of a certain set of environmental conditions (Visser 1988); if the actions of the first fly to oviposit produces additional stimulatory cues for later ones (Lam et al. 2007) or if carrion overloading produces an increase in temperature that accelerates development (Catts 1992). Ultimately, it depends on how well blow flies can assess patch suitability. Some evidence comes from an unlikely source: when insects are intensively sampled and removed from the resource, the volume of eggs deposited increases, thus demonstrating a change in reproductive behavior in response to local conditions (Michaud and Moreau 2013). As to the ability to apportion their egg load: if laboratory colony flies are denied an oviposition source, they will eventually begin laying small egg batches on the inner surfaces of their cage (Mohr, personal observation). These batches may have a very small likelihood of reaching adulthood, but it is nonzero.

At the population level, how females allocate their eggs among patches helps shape the stability of species coexistence. When the patch landscape consists of many small patches, not every patch will be colonized by multiple species, providing a refuge for less-competitive species (Atkinson and Shorrocks 1984; Hanski 1987). At a glance, this seems like a straightforward confirmation of neutral theory, but with a twist of species-sorting theory as well. If persistence in a finely divided resource pool is an effect of random encounter, then neutral theory is correct; but if it is simply the effect of optimal oviposition strategy for a small patch, then species-sorting has provided the correct prediction. Since female flies develop oocytes during adulthood, their major oviposition constraint is their own oocyte development rate. They should accept any patch they find to the maximum degree because they can develop more oocytes for a potential future patch (Rosenheim et al. 2000). If any patch is exploited, then it is simply the random nature of the encounter shaping assembly on the patch.

If the insect is looking to consume the carrion directly, rather than use it as an oviposition site, foraging theory becomes important to the ecological outcomes important to the community assembly of that carrion patch. The insect must strike a trade-off between the value of the resource at hand versus the likelihood of finding a better resource in the future, balanced by the availability of the current resource (in the case of competition) or the likelihood of predation/parasitism during extended feeding (Pyke 1984; Ward 1987; Singer et al. 1992; Wajnberg 2006; Houston et al. 2011). Moisture level and tissue toughness might also affect the likelihood of carrion acceptance by insects, much as they do for herbivores (Ibanez et al. 2013). Microbial metabolism of the carrion substrate probably plays an important role in communicating resource value. Since no Diptera have chewing mouthparts as adults, preliminary breakdown, metabolism, and liberation of biomolecules by the microbes may be of great value (Gill-King 1997). The exodigestive action of microbes could also benefit first instars, with their small mouth hooks. Other factors besides nutrient availability may also be sensed. Carcass size might play a role, particularly for larger species of fly which would not develop well on a small carcass such as a slug (Woodcock et al. 2002), while a larger carcass size allows avoidance and resource partitioning much more easily than a small one, which can provide a refuge from predatory species like *Ch. rufifacies* (Wells and Greenberg 1992a).

8.8 Postcolonization Interval

The act of acceptance and colonization marks a dramatic shift in the interaction between insects and carrion and ultimately the community assembly patterns. Immigration and dispersal abilities inform species assembly patterns. After initial colonization, however, the interaction of insects with the carrion source is dominated by the more localized interactions such as competition and predation, and resource partitioning dictate the remaining dynamics of the entomological succession, and therefore broader metacommunity dynamics with the landscape and ecosystem.

Historically, the post-CI has played an important role in forensic entomology, as the age of an immature insect can be estimated given the temperature and size or state of development, a finding recognized in the nineteenth century with authors such as Bergeret, Broudel, Megnin, and Yovanovich (Keh 1985). With the development of the accumulated degree days (ADD)/ADH models of thermally dependent insect development in the mid-twentieth century, estimations became much more precise, though their calculation only required a thermal history and state of development (Higley et al. 1986; Higley and Haskell 2010). With additional research, many environmental, biological, and ecological factors have been found to affect insect development, reviewed extensively by Keh (1985), Catts and Goff (1992), and Wells and LaMotte (2010). Among other factors, environmental conditions such as season (Moretti et al. 2011), habitat (Tomberlin and Adler 1998), and presence of exogenous chemicals (Gosselin et al. 2011) have all been found to affect development rate of insects. These same conditions also influence carrion arthropod succession (Benbow et al. 2013; Pechal et al. 2014a). More recent findings underscore the importance of population and quantitative genetics, with significant differences in development time among larval environments and geographically separated populations of *L. sericata* (Tarone and Foran 2006, 2008; Gallagher et al. 2010; Tarone et al. 2011). Integrating these findings has led to at least one proposed replacement for the accumulated degree system which would include the metrics of length, weight, developmental stage or instar, and population membership in a nonlinear general model (Tarone and Foran 2008). An additional modifying metric, depending on species, would be the type of tissue from which it was collected. *Calliphora vicina* development is sensitive to rearing on different beef tissues (El-Moaty and Kheirallah 2013) or to decomposed versus fresh tissue (Richards et al. 2013). This finding is not universal; *C. macellaria* shows no significant differences when reared on equine versus swine muscle (Boatright and Tomberlin 2010). Tissue type/quality sensitivity may be species-specific, much like host sensitivity or resource partitioning strategies.

Although the genetics and physiological side are important drivers of the post-CI, interactions within the necrobiome can also have significant effect on its duration, such as species composition (LaMotte and Wells 2000; Shiao and Yeh 2008), succession pattern (Schoenly 1992), predation (Early and Goff 1986; Wells and Greenberg 1994), and nonconsumptive effects (Mohr 2012). Just as the pre-CI makes an excellent ecological laboratory for investigating community assembly, the post-CI makes for an excellent ecological laboratory for investigating competition and cooperation, ecosystem service, and the conflicting evolutionary pressures of development time versus body size.

8.8.1 Consumption Phase

During the consumption phase, the tissue of the corpse is chemically and mechanically broken down into "carrion soup" (Greenberg and Kunich 2002). Chemically, a variety of enzymes are produced by the insects including trypsin/chymotrypsin-like proteases, amylase, collagenase, and lipases (Price 1974, 1975; Pendola and Greenberg 1975; Bowles 1988; Young et al. 1996; Chambers et al. 2003). Mechanically, the mouth hooks are scraped across the surface of the carrion to release small particulates into the salivary exudate (Schoofs and Spiess 2007; Schoofs et al. 2009). Several species, most notably *L. sericata*, are able to finely differentiate between decaying and live tissue, allowing them to be used medicinally for wound debridement (Sherman et al. 2000). How the larvae make this distinction remains unclear, but it is likely via the chemosensory sensillae on the cephalic lobes (Schoofs et al. 2009). Chemical signals from autolytic and/or putrefying tissue plus the associated microbial products are the most likely controllers for necrotic-only tissue feeding. Experiments with species that do not make such a distinction, such as *Cochliomyia hominvorax* (Coquerel) (Diptera: Calliphoridae), could resolve this question.

Interactions between fly larvae are both mutalistic and competitive. The collective activity of a simultaneously feeding larval mass (i.e., hundreds to thousands of individuals) can raise the temperature of the larval mass by 10–30°C above ambient either through motion friction, food metabolism, or bacterial fermentation (reviewed extensively by Rivers et al. 2011). Since increased heat improves nutrient assimilation and development speed (Hanski 1976, 1977), the offspring of early-arriving blow flies may be at a particular advantage in the larval mass (Catts 1992). Furthermore, the presence of many larvae may dilute the likelihood of individual death from predation, parasitism, or disease (Rohlfs and Hoffmeister 2004). Additionally, there are significant downsides to collective feeding.

While collective feeding can dilute predator effects on an individual, the mass production of odor cues may attract a larger number of predators/parasitoids than individual feeding events (Rivers et al. 2011). Collective feeding also induces higher levels of inter- and intraspecific competition. In many species, increasing larval density on a resource results in some loss of adult fitness, for example, body size, life span, and/or fecundity (Prinkkila and Hanski 1995; Smith and Wall 1997; Wall and Smith 1997; dos Reis et al. 1999; Green et al. 2003; Shiao and Yeh 2008; Rivers et al. 2011). Secondary colonizers, faced with resource depletion, may face even higher levels of competition (Ullyett 1950). As a result, secondary colonizers may engage in alternate resource-use strategies. For example: when early third instar *C. macellaria* larvae are presented with a meat resource that has been exposed to the excretions and secretions of late third instar *Ch. rufifacies* larvae, they are more likely to reject it and die of starvation than if presented meat exposed only to water (Mohr 2012). This is an example of a nonconsumptive predatory effect, which may be more widespread among successional waves on a carrion patch than currently appreciated. The complex chemical mix of autolytic and putrefactive cellular process, the other insects, quorum-sensing molecules from microbes, and fermentative products all can provide important information to a colonizer, informing their resource exploitation strategies. Secondary colonizers are not always at a disadvantage, though. For example, *Ch. rufifacies* is typically a secondary colonizer known to drive other species of larvae off a carcass (Fuller 1934).

One often underappreciated participant in the consumption phase is the vertebrate scavenger, recently and extensively reviewed by Barton et al. (2013b) and DeVault et al. (2003) (see Chapter 6 for additional details). Vertebrate scavengers are capable of arriving rapidly and consuming the entirety of small carcasses. In a review of 22 studies, DeVault et al. (2003) found that an average of 75% of carrion is consumed by vertebrates, often very quickly after death. A later study found that 88% of rodent carcasses were scavenged within a week of presentation (DeVault et al. 2004). Similarly, 97% of carrion is scavenged within 36 h on roadsides in Florida, USA (Antworth et al. 2005). Large carrion, such as swine, tend to be initially fed upon by larger scavengers, such as vultures and big cats, which break up and scatter the carcass and allow smaller mesoscavengers to finish consuming the edible tissue (Barton et al. 2013a; Demo et al. 2013; Moreno-Opo and Margalida 2013; Pereira et al. 2014). Environment seems to play a significant role in vertebrate scavenger community makeup and success rates: areas with higher species diversity see faster scavenging (Sebastian-Gonzalez et al. 2013), while areas with excluded scavengers see a significant drop in scavenging efficacy (Olson et al. 2012). Urban areas see much more scavenging by feral dog and cat than rural areas with more diverse animal populations (Huijbers et al. 2013). DeVault et al. (2004) also speculated that the amount of canopy and overgrowth for a given habitat will limit access of vertebrate scavengers away from visually oriented bird species in favor of odor-oriented mammal and reptile scavengers.

This great rate of carrion consumption would seem to put vertebrates, arthropods, and bacteria in direct competition for carrion patches. At first evaluation, the microbes would seem to be at a severe disadvantage. Several authors cite the Janzen (1977) hypothesis that microbial populations secrete toxic or revulsive compounds as a specific defense mechanism against consumption. Sherratt et al. (2006), however, offered a persuasive rebuttal based on the idea that such a defense can only arise if such production is cost-free and there is very low population dispersal. More likely, these noxious compounds are regularly produced secondary metabolites which have the secondary benefit of deterring macrofaunal consumption (Burkepile et al. 2006). Indeed, in the deep ocean, where bacteria populations are low, there is a greater diversity and number of scavenging macrofauna (Beasley et al. 2012).

Insects, particularly fly larvae, contend with microbes and their associated compounds through the excretion and secretion of antimicrobial compounds such as lucifensin (Čeřovský et al. 2010; El Shazely et al. 2013; Valachova et al. 2013). Fly larvae can disrupt biofilm production in *Staphylococcus aureus* and *Pseudomonas aeruginosa* (van der Plas et al. 2008), or larvae can simply digest certain bacteria, such as *Esherichia coli* (Mumcuoglu et al. 2001). This raises an interesting ecological question: do the fly larvae and microbes have a pathogen/host, predator/prey, commensal, competitive, or even mutualistic relationship with one another? At least for the bacteria *S. aureus* and the larval *L. sericata*, the relationship seems to be commensal: increasing the density of the bacteria had no deleterious effects on the larvae and the larvae did not increase antimicrobial secretion production (Barnes and Gennard 2011). Among the burying beetles (Silphidae: Nicrophorinae), antimicrobial investment is particularly high. Burying beetles, such as the endangered *Nicrophorus americanus* (Oliver) (Coleoptera: Silphidae), locate and bury a small animal carcass to use as a larval food source (Kozol et al. 1988). As part of the preparation process, *Nicrophorus* spp. treat the carrion with oral and anal secretions containing antimicrobial peptides (Hall et al. 2011; Arce et al. 2013), which slow microbial breakdown. These antimicrobial effects are not perfect and adults will avoid heavily putrefied carcasses in favor of fresh (Rozen et al. 2008). The cost of these antimicrobial compounds is reflected phylogenetically: members of the subfamily Silphinae, which are predators of fly larvae, do not produce these compounds (Hoback et al. 2004).

During the consumption phase, the carrion patch itself is the focus of a variety of intense multispecies, multiguild, even multikingdom interactions as each member of the necrobiome attempts to exploit the patch's physical resources. However, these interactions last only as long as the carrion patch itself, and so these organisms are required to disperse in order to find the best suited available carrion patch.

8.8.2 Dispersal Phase

For an insect or vertebrate, dispersal from a carrion patch occurs under only a few basic conditions, as illustrated in Figure 8.4: they have completed a carrion-dependent stage of development or because the carrion patch has become degraded or insufficient for their needs. Not all instances of carrion degradation are the result of consumption. As Rivers et al. (2013) point out, "overcrowding and heat production … can lead to physiological stress, including food shortage and heat stress. Such stressors commonly have the profound effect of producing smaller-sized pupae." Stressors can also come in the form of non-consumptive predator or pseudopredator effects, as discussed above. Some species, such as *L. sericata*, seem capable of coping with such stressor better than others (e.g., *L. illustris* [Meigen] [Diptera: Calliphoridae]), producing average-sized pupae even under severe larval overcrowding (Rivers et al. 2013). For direct carrion feeders or predators, the "choice" to disperse or not disperse is a fairly simple feedback loop: as long as the food is available and acceptable, they will continue to eat until satiation (Dugatkin 2009). For less mobile species such as microbes, the organisms must "hitch a ride" in order to leave the carrion patch.

For the insects using a carrion patch as a larval development site, they are confronted with an essential trade-off between development time and body size. A larger body size does confer fitness benefits, so insects should generally maximize growth as possible (Saunders et al. 1999; Grassberger and Reiter 2001; Wells and King 2001; Chown and Gaston 2010). Larger body size usually requires a longer development time, which increases risk of predation, parasitism, disease, or mid-development resource depletion (Nijhout 2003). Therefore, the selection pressures on body size and development time are opposing (Davidowitz et al. 2005), except in rare cases where the fastest to grow are also the largest in body size (Tarone et al. 2011). Unlike other interactions within the necrobiome, maximum duration of feeding does not seem to be so much a matter of neurosensory as one of genetics and physiology. Among dipterans, such as *Drosophila melanogaster* Meigen (Diptera: Drosophilidae), growth and development are linked by nutrient sensing, hormone release, and developmental thresholds minimum viable weight (MVW) and critical weight (CW) (Mirth and Riddiford 2007). The first of these thresholds, MVW is defined as the smallest size at which an immature insect has a 50/50 chance of developing to the next developmental stage (Nijhout 1975; Mirth et al. 2005). Insects reach MVW at variable rates, depending on factors such as nutrient quality, feeding rate, and food conversion efficiency (Mirth and Riddiford 2007). If forced to depart a food resource early by predatory behavior or lack of resource after reaching MVW, the normal developmental rate may be significantly extended (Davidowitz et al. 2003). CW differs from MVW in that the threshold of CW is intimately connected to the release of prothoracicotropic hormone (PTTH) and the degradation of juvenile hormone (JH) during the final larval stadium (Davidowitz et al. 2003). Prior to reaching CW, the insect larvae have high titers of JH in their hemolymph, which prevents molting to pupa, and then adulthood. When CW is attained, JH is cleared from the hemolymph and PTTH is released (Riddiford 2008). Once PTTH is released, the larva is committed to pupariation, regardless of current body size or continued availability of food (Nijhout and Williams 1974; Davidowitz et al. 2005; Mirth and Riddiford 2007). However, if the larvae does have access to food, it will eat and grow during a "terminal growth period" as the prepupariation physiological changes finish (Shingleton et al. 2007). In this way, CW divides final larval stage into nutrition-dependent and nutritionally independent periods, making it a possible source of the observed variability in insect size, mass, and development time (Shingleton et al. 2007). CW also connects environmentally variable traits (e.g., feeding rate) to genetically programmed traits (e.g., nutrient sensing), which allows the development of genotype × environment (G × E) interaction (Davidowitz et al. 2004). Identifying these G × E interactions has a major impact for applied forensic entomology, as inconstant phenotype or behavior must be compensated in any kind of interval estimation (Tomberlin et al. 2011b).

Regardless of developmental plasticity, part of the normal developmental trajectory of cyclorraphan flies is the cessation of feeding, and the formation of the puparium prior to pupation (Smith 1986). Usually, they migrate away from the carrion patch before pupating, up to 100′ away in the case of *Lucilia* spp. and *Calliphora* spp. (Green 1951) (as cited in Greenberg 1990). This tendency to migrate is not universal: when allowed to pupariate in loose wood shavings, 99% of *L. sericata* larvae and 84.5% of *Ca. vicina* dispersed at least 3 m, while 98% of *P. regina*, 90% of *Muscina stabulans* (Fallén) (Diptera: Muscidae), and 84% of *Ch. rufifacies* remained at the carrion site (Greenberg 1990). Clumping, however, can be affected by exposure to parasitoids during the larval (nonsusceptible) stage (Cammack et al. 2010), possibly another antipredator dilution effect strategy.

Once the macrobiota have departed from the carrion patch, the dry parts of the carrion remain for an extended time. Microbes and fungi are responsible for most of the biological (as opposed to soil-based) decomposition of human and animal bone (Jans et al. 2004). In human bones, bacteria typically attack from the inside-out, suggesting that the causative species are the same species of enteric bacteria which cause putrefaction (Child 1995). The microbial component of the necrobiome is present from the moment of death until the last bone is completely decomposed and may indeed remain as a legacy within the soil microbial community.

8.9 Conclusions and Future of Carrion Community Ecology

Much of our understanding of carrion community ecology has come from descriptive studies of both basic questions and those from the applied field of forensic entomology. There is a need to couple laboratory and field experiments to more explicitly test questions related to the mechanisms and importance of carrion decomposition. The carrion system is an excellent model to test community ecology theory at various spatial scales, including intra- and interspecies interactions and those that are derived from communications between kingdoms or domains of life (see Chapter 20). The system can be used to test community assembly theory, consumptive versus nonconsumptive effects on fitness and to ask questions related to coevolutionary relationships between microbes and insects and scavenging vertebrates. The framework described in detail within this chapter may be a way to guide specific questions or test mechanistic hypotheses of ecological and evolutionary importance to decomposition. It is also important for better understanding of the ecological mechanisms and interactions that drive larger scale patterns of nutrient availability and transfer into the surrounding ecosystems, either as discrete and small-scale pulses or larger-scale resource pulses in such systems as salmon spawning and periodical cicada emergence followed by mass die-off events. The future of carrion community ecology is at a place where there is a need to become more quantitative with the use of more powerful and widely available computational technology (e.g., network analyses) to ask hypothesis-driven questions. Embracing these new techniques will allow researchers to ask innovative and potentially transformative questions in ecology. Carrion community and landscape ecology offer a tremendous opportunity to test novel hypotheses and general theory that can more mechanistically connect genes to ecosystems and the biosphere.

REFERENCES

Abos, C.P., F. Lepori, B.G. Mckie, and B. Mamlqvist. 2006. Aggregation among resource patches can promote coexistence in stream-living shredders. *Freshwater Biology* 51: 545–553.

Alam, M.J. and L. Zurek. 2004. Association of *Escherichia coli* O157:H7 with houseflies on a cattle farm. *Applied and Environmental Microbiology* 70: 7578–7580.

Amendt, J., C.P. Campobasso, E. Gaudry, C. Reiter, H.N. Leblanc, and M.J.R. Hall. 2007. Best practice in forensic entomology—Standards and guidelines. *International Journal of Legal Medicine* 121: 90–104.

Amon, D.J., A.G. Glover, H. Wiklund, L. Marsh, K. Linse, A.D. Rogers, and J.T. Copley. 2013. The discovery of a natural whale fall in the Antarctic deep sea. *Deep Sea Research Part II: Topical Studies in Oceanography* 92: 87–96.

Anderson, G.S. 2010. Factors that influence insect succession on carrion. In: Byrd, J.H. and J.L. Castner (eds), *Forensic Entomology: The Utility of Arthropods in Legal Investigations*. Boca Raton, FL: CRC Press.

Anderson, G.S. and S.L. Vanlaerhoven. 1996. Initial studies on insect succession on carrion in southwestern British Columbia. *Journal of Forensic Sciences* 41: 617–625.

Andreescu, S. and J.-L. Marty. 2006. Twenty years research in cholinesterase biosensors: From basic research to practical applications. *Biomolecular Engineering* 23: 1–15.

Antworth, R.L., D.A. Pike, and E.E. Stevens. 2005. Hit and run: Effects of scavenging on estimates of road-killed vertebrates. *Southeastern Naturalist* 4: 647–656.

Arce, A.N., P.T. Smiseth, and D.E. Rozen. 2013. Antimicrobial secretions and social immunity in larval burying beetles, *Nicrophorus vespilloides*. *Animal Behaviour* 86: 741–745.

Archer, M.S. and A.A. Elgar. 2003. Effects of decomposition on carcass attendance in a guild of carrion-breeding flies. *Medical and Veterinary Entomology* 17: 263–271.

Asgari, S., J.R.E. Hardy, R.G. Sinclair, and B.D. Cooke. 1998. Field evidence for mechanical transmission of rabbit haemorrhagic disease virus (RHDV) by flies (Diptera: Calliphoridae) among wild rabbits in Australia. *Virus Research* 54: 123–132.

Ashworth, J.R. and R. Wall. 1994. Responses of the sheep blowflies *Lucilia sericata* and *L. cuprina* to odor and the development of semiochemical baits. *Medical and Veterinary Entomology* 8: 303–309.

Atkinson, W.D. and B. Shorrocks. 1984. Aggregation of larval Diptera over discrete and ephemeral breeding sites: The implications for coexistence. *American Naturalist* 124: 336–351.

Barnes, K. and D. Gennard. 2011. The effect of bacterially-dense environments on the development and immune defences of the blowfly *Lucilia sericata*. *Physiological Entomology* 36: 96–100.

Barton, P.S., S.A. Cunningham, D.B. Lindenmayer, and A.D. Manning. 2013a. The role of carrion in maintaining biodiversity and ecological processes in terrestrial ecosystems. *Oecologia* 171: 761–772.

Barton, P.S., S.A. Cunningham, B.C.T. Macdonald, S. Mcintyre, D.B. Lindenmayer, and A.D. Manning. 2013b. Species traits predict assemblage dynamics at ephemeral resource patches created by carrion. *PLoS ONE* 8: e53961.

Barton-Browne, L., R.J. Bartell, and H.H. Shorey. 1969. Pheromone-mediated behaviour leading to group oviposition in blowfly *Lucilia*. *Journal of Insect Physiology* 15: 1003–1014.

Baumgartner, D.L. 1993. Review of *Chrysomya rufifacies* (Diptera, Calliphoridae). *Journal of Medical Entomology* 30: 338–352.

Baumgartner, D.L. and B. Greenberg. 1984. The genus *Chrysomya* (Diptera: Calliphoridae) in the New World. *Journal of Medical Entomology* 21: 105–113.

Beasley, J.C., Z.H. Olson, and T.L. Devault. 2012. Carrion cycling in food webs: Comparisons among terrestrial and marine ecosystems. *Oikos* 121: 1021–1026.

Benbow, M.E., A.J. Lewis, J.K. Tomberlin, and J.L. Pechal. 2013. Seasonal necrophagous insect community assembly during vertebrate carrion decomposition. *Journal of Medical Entomology* 50: 440–450.

Benecke, M. 2001. A brief history of forensic entomology. *Forensic Science International* 120: 2–14.

Benninger, L.A., D.O. Carter, and S.L. Forbes. 2008. The biochemical alteration of soil beneath a decomposing carcass. *Forensic Science International* 180: 70–75.

Boatright, S.A. and J.K. Tomberlin. 2010. Effects of temperature and tissue type on the development of *Cochliomyia macellaria* (Diptera: Calliphoridae). *Journal of Medical Entomology* 47: 917–923.

Bowles, V.M. 1988. Characterization of proteolytic and collagenolytic enzymes from the larvae of *Lucilia cuprina*, the sheep blowfly. *Australian Journal of Biological Sciences* 41: 269–278.

Braack, L.E.O. 1987. Community dynamics of carrion-attendant arthropods in tropical African woodland. *Oecologia* 72: 402–409.

Browne, B.L. 1960. The role of olfaction in the stimulation of oviposition in the blowfly, *Phormia regina*. *Journal of Insect Physiology* 5: 16–22.

Bruce, T.J.A., L.J. Wadhams, and C.M. Woodcock. 2005. Insect host location: A volatile situation. *Trends in Plant Science* 10: 269–274.

Brundage, A.L. 2012. Fitness effects of colonization time of *Chrysomya rufifacies and Cochliomyia macellaria*, and their response to intra- and inter-specific eggs and egg-associated microbes. (PhD dissertation). College Station, TX: Texas A&M University.

Brundage, A., M. Benbow, and J. Tomberlin. 2014. Priority effects on the life-history traits of two carrion blow fly (Diptera, Calliphoridae) species. *Ecological Entomology* 39: 539–547.

Bugajski, K.N., C.C. Seddon, and R.E. Williams. 2011. A comparison of blow fly (Diptera: Calliphoridae) and beetle (Coleoptera) activity on refrigerated only versus frozen-thawed pig carcasses in Indiana. *Journal of Medical Entomology* 48: 1231–1235.

Burkepile, D.E., J.D. Parker, C.B. Woodson, H.J. Mills, J. Kubanek, P.A. Sobecky, and M.E. Hay. 2006. Chemically mediated competition between microbes and animals: Microbes as consumers in food webs. *Ecology* 87: 2821–2831.

Byrd, J.H. and J.L. Castner. 2001. *Forensic Entomology: The Utility of Arthropods in Legal Investigations*. Boca Raton, FL: CRC Press.

Cammack, J.A., P.H. Adler, J.K. Tomberlin, Y. Arai, and W.C. Bridges. 2010. Influence of parasitism and soil compaction on pupation of the green bottle fly, *Lucilia sericata. Entomologia Experimentalis Et Applicata* 136: 134–141.

Campobasso, C.P. and F. Introna. 2001. The forensic entomologist in the context of the forensic pathologist's role. *Forensic Science International* 120: 132–139.

Campobasso, C.P. and F. Introna. 2014. Reply: Commentary on letter to the editor from Jeffrey Wells. *Journal of Medical Entomology* 51: 492–494.

Carter, D., D. Yellowlees, and M. Tibbett. 2007. Cadaver decomposition in terrestrial ecosystems. *Naturwissenschaften* 94: 12–23.

Carter, D.O., D. Yellowlees, and M. Tibbett. 2010. Moisture can be the dominant environmental parameter governing cadaver decomposition in soil. *Forensic Science International* 200: 60–66.

Catts, E. 1992. Problems in estimating the postmortem interval in death investigations. *Journal of Agricultural Entomology* 9: 245–255.

Catts, E.P. and M.L. Goff. 1992. Forensic entomology in criminal investigations. *Annual Review of Entomology* 37: 253–272.

Čeřovský, V., J. Žďárek, V. Fučík, L. Monincová, Z. Voburka, and R. Bém. 2010. Lucifensin, the long-sought antimicrobial factor of medicinal maggots of the blowfly *Lucilia sericata. Cellular and Molecular Life Sciences* 67: 455–466.

Chambers, L., S. Woodrow, A. Brown, P. Harris, D. Phillips, M. Hall, J. Church, and D. Pritchard. 2003. Degradation of extracellular matrix components by defined proteinases from the greenbottle larva *Lucilia sericata* used for the clinical debridement of non-healing wounds. *British Journal of Dermatology* 148: 14–23.

Charabidze, D., B. Bourel, V. Hedouin, and D. Gosset. 2009. Repellent effect of some household products on fly attraction to cadavers. *Forensic Science International* 189: 28–33.

Chase, J.M. 2010. Stochastic community assembly causes higher biodiversity in more productive environments. *Science* 328: 1388–1391.

Child, A.M. 1995. Towards an understanding of the microbial decomposition of archaeological bone in the burial environment. *Journal of Archaeological Science* 22: 165–174.

Chown, S.L. and K.J. Gaston. 2010. Body size variation in insects: A macroecological perspective. *Biological Reviews* 85: 139–169.

Clements, F.E. 1936. Nature and structure of the climax. *Journal of Ecology* 24: 252–284.

Cochran-Starifa, D.L. and C.N.V. Ende. 1998. Integrating bacteria into food webs: Studies with *Sarracenia purprea* inquilines. *Ecology* 79: 880–898.

Conn, D.B., J. Weaver, L. Tamang, and T.K. Graczyk. 2007. Synanthropic flies as vectors of *Cryptosporidium* and *Giardia* among livestock and wildlife in a multispecies agricultural complex. *Vector-Borne and Zoonotic Diseases* 7: 643–652.

Cornaby, B.W. 1974. Carrion reduction by animals in contrasting tropical habitats. *Biotropica* 6: 51–63.

Cortes-Avizanda, A., N. Selva, M. Carrete, and J.A. Donazar. 2009. Effects of carrion resources on herbivore spatial distribution are mediated by facultative scavengers. *Basic and Applied Ecology* 10: 265–272.

Cotter, P.D. and C. Hill. 2003. Surviving the acid test: Responses of Gram-positive bacteria to low pH. *Microbiology and Molecular Biology Reviews* 67: 429–453.

Cowles, H.C. 1899. The ecological relations of the vegetation on the sand dunes of Lake Michigan. (PhD dissertation). Chicago, IL: University of Chicago.

Davidowitz, G., L.J. D'Amico, and H.F. Nijhout. 2003. Critical weight in the development of insect body size. *Evolution & Development* 5: 188–197.

Davidowitz, G., L.J. D'Amico, and H.F. Nijhout. 2004. The effects of environmental variation on a mechanism that controls insect body size. *Evolutionary Ecology Research* 6: 49–62.

Davidowitz, G., D.A. Roff, and H.F. Nijhout. 2005. A physiological perspective on the response of body size and development time to simultaneous directional selection. *Integrative and Comparative Biology* 45: 525–531.

Davis, M.A. 2003. Biotic globalization: Does competition from introduced species threaten biodiversity? *Bioscience* 53: 481–489.

Davis, M.A., K. Thompson, and J. Philip Grime. 2005. Invasibility: The local mechanism driving community assembly and species diversity. *Ecography* 28: 696–704.

Day, D.M. and J.F. Wallman. 2006. A comparison of frozen/thawed and fresh food substrates in development of *Calliphora augur* (Diptera: Calliphoridae) larvae. *International Journal of Legal Medicine* 120: 391–394.

De Bruyne, M., K. Foster, and J.R. Carlson. 2001. Odor coding in the *Drosophila* antenna. *Neuron* 30: 537–552.

Dejong, G.D. and W.W. Hoback. 2006. Effect of investigator disturbance in experimental forensic entomology: Succession and community composition. *Medical and Veterinary Entomology* 20: 248–258.

Demo, C., E.R. Cansi, C. Kosmann, and J.R. Pujol-Luz. 2013. Vultures and others scavenger vertebrates associated with man-sized pig carcasses: A perspective in forensic taphonomy. *Zoologia* 30: 574–576.

Denno, R.F. and W.R. Cothran. 1975. Niche relationships of a guild of necrophagous flies. *Annals of the Entomological Society of America* 68: 741–754.

Denno, R.F. and W.R. Cothran. 1976. Competitive interactions and ecological strategies of sarcophagid and calliphorid flies inhabiting rabbit carrion. *Annals of the Entomological Society of America* 69: 109–113.

Dent, B.B., S. Forbes, and B. Stuart. 2004. Review of human decomposition processes in soil. *Environmental Geology* 45: 576–585.

Dethier, V. 1955. Mode of action of sugar-baited fly traps. *Journal of Economic Entomology* 48: 235–239.

Dethier, V.G. 1956. Repellents. *Annual Review of Entomology* 1: 181–202.

Devault, T.L., I.L. Brisbin, and O.E. Rhodes. 2004. Factors influencing the acquisition of rodent carrion by vertebrate scavengers and decomposers. *Canadian Journal of Zoology-Revue Canadienne De Zoologie* 82: 502–509.

Devault, T.L., O.E. Rhodes, and J.A. Shivik. 2003. Scavenging by vertebrates: Behavioral, ecological, and evolutionary perspectives on an important energy transfer pathway in terrestrial ecosystems. *Oikos* 102: 225–234.

Diamond, J.M. 1975. Assembly of species communities. In: Cody, M.L. and J.M. Diamond (eds), *Ecology and Evolution of Communities*. Cambridge, Belknap Press, pp. 342–344.

Dimaio, D. and V.J. Dimaio. 2001. *Forensic Pathology*. Boca Raton, FL: CRC Press.

dos Reis, S.F., C.J. Von Zuben, and W.C. Godoy. 1999. Larval aggregation and competition for food in experimental populations of *Chrysomya putoria* (Wied.) and *Cochliomyia macellaria* (F.) (Dipt., Calliphoridae). *Journal of Applied Entomology* 123: 485–489.

Dugatkin, L.A. 2009. *Principles of Animal Behavior*. New York: W.W. Norton.

Dukas, R. 2008. Evolutionary biology of insect learning. *Annual Review of Entomology* 53: 145–160.

Early, M. and M.L. Goff. 1986. Arthropod succession patterns in exposed carrion on the island of O'Ahu, Hawaiian Islands, USA. *Journal of Medical Entomology* 23: 520–531.

Easton, C. and D. Feir. 1991. Factors affecting the oviposition of *Phaenicia sericata* (Meigen) (Diptera: Calliphoridae). *Journal of the Kansas Entomological Society* 64: 287–294.

El-Moaty, Z.A. and A.M. Kheirallah. 2013. Developmental variation of the blow fly *Lucilia sericata* (Meigen, 1826) (Diptera: Calliphoridae) by different substrate tissue types. *Journal of Asia-Pacific Entomology* 16: 297–300.

El Shazely, B., V. Veverka, V. Fucik, Z. Voburka, J. Zdarek, and V. Cerovsky. 2013. Lucifensin II, a defensin of medicinal maggots of the blowfly *Lucilia cuprina* (Diptera: Calliphoridae). *Journal of Medical Entomology* 50: 571–578.

Elton, C. 1927. *Animal Ecology*. London: Sidgwick and Jackson.

Emmens, R.L. and M.D. Murray. 1982. The role of bacterial odors in oviposition by *Lucilia cuprina* (Wiedemann) (Diptera, Calliphoridae), the Australian sheep blowfly. *Bulletin of Entomological Research* 72: 367–375.

Faria, L.D.B., L. Orsi, L.A. Trinca, and W.a.C. Godoy. 1999. Larval predation by *Chrysomya albiceps* on *Cochliomyia macellaria*, *Chrysomya megacephala* and *Chrysomya putoria*. *Entomologia Experimentalis Et Applicata* 90: 149–155.

Farkas, R., J.A. Hogsette, and L. Börzsönyi. 1998. Development of *Hydrotaea aenescens* and *Musca domestica* (Diptera: Muscidae) in poultry and pig manures of different moisture content. *Environmental Entomology* 27: 695–699.

Fiene, J., G. Sword, S. Vanlaerhoven, and A. Tarone. 2014. The role of spatial aggregation in forensic entomology. *Journal of Medical Entomology* 51: 1–9.

Forbes, S.L. 2009. Potential determinants of postmortem and postburial interval of buried remains. *Soil Analysis in Forensic Taphonomy: Chemical and Biological Effects of Buried Human Remains*. Boca Raton, FL: CRC Press, pp. 225–246.

Forman, R.T. 1995a. Some general principles of landscape and regional ecology. *Landscape Ecology* 10: 133–142.

Forman, R.T.T. 1995b. *Land Mosaics: The Ecology of Landscapes and Regions*. Cambridge, UK: Cambridge University Press.

Frederickx, C., J. Dekeirsschieter, F.J. Verheggen, and E. Haubruge. 2014. Host–habitat location by the parasitoid, *Nasonia vitripennis* Walker (Hymenoptera: Pteromalidae). *Journal of Forensic Sciences* 59: 242–249.

Fremdt, H., K. Szpila, J. Huijbregts, A. Lindstrom, R. Zehner, and J. Amendt. 2012. *Lucilia silvarum* Meigen, 1826 (Diptera: Calliphoridae)—A new species of interest for forensic entomology in Europe. *Forensic Science International* 222: 335–339.

Fukami, T. and M. Nakajima. 2011. Community assembly: Alternative stable states or alternative transient states? *Ecology Letters* 14: 973–984.

Fuller, M. 1934. The insect inhabitants of carrion, a study in animal ecology. *Australian Council for Scientific and Industrial Research*, Bulletin 82: 1–62.

Gagné, R.J. 1981. *Chrysomya* spp., Old World blow flies (Diptera: Calliphoridae), recently established in the Americas. *Bulletin of the ESA* 27: 21–22.

Gallagher, M.B., S. Sandhu, and R. Kimsey. 2010. Variation in developmental time for geographically distinct populations of the common green bottle fly, *Lucilia sericata* (Meigen). *Journal of Forensic Sciences* 55: 438–442.

García-Ripollés, C., P. López-López, and F. García-López. 2004. Management and monitoring of a vulture restaurant in Castellón Province, Spain. *Vulture News* 50: 5–14.

Giao, J.Z. and W.a.C. Godoy. 2007. Ovipositional behavior in predator and prey blowflies. *Journal of Insect Behavior* 20: 77–86.

Gibson, G. and S. Torr. 1999. Visual and olfactory responses of haematophagous Diptera to host stimuli. *Medical and Veterinary Entomology* 13: 2–23.

Gilbert, M., R.T. Watson, S. Ahmed, M. Asim, and J.A. Johnson. 2007. Vulture restaurants and their role in reducing diclofenac exposure in Asian vultures. *Bird Conservation International* 17: 63–77.

Gill-King, H. 1997. Chemical and ultrastructural aspects of decomposition. *Forensic Taphonomy: The Postmortem Fate of Human Remains*. Boca Raton, FL: CRC Press, pp. 93–108.

Gleason, H.A. 1927. Further views on the succession-concept. *Ecology* 8: 299–326.

Goff, M.L. 1992. Problems in estimation of postmortem interval resulting from wrapping of the corpse—A case study from Hawaii. *Journal of Agricultural Entomology* 9: 237–243.

Goodbrod, J.R. and M.L. Goff. 1990. Effects of larval population density on rates of development and interactions between two species of *Chrysomya* (Diptera: Calliphoridae) in laboratory culture. *Journal of Medical Entomology* 27: 338–343.

Gosselin, M., S.M.R. Wille, M.D.R. Fernandez, V. Di Fazio, N. Samyn, G. De Boeck, and B. Bourel. 2011. Entomotoxicology, experimental set-up and interpretation for forensic toxicologists. *Forensic Science International* 208: 1–9.

Grassberger, M. and C. Reiter. 2001. Effect of temperature on *Lucilia sericata* (Diptera: Calliphoridae) development with special reference to the isomegalen- and isomorphen-diagram. *Forensic Science International* 120: 32–36.

Green, A. 1951. The control of blowflies infesting slaughter-houses I. Field observations of the habits of blowflies. *Annals of Applied Biology* 38: 475–494.

Green, P.W.C., M.S.J. Simmonds, and W.M. Blaney. 2003. Diet nutriment and rearing density affect the growth of black blowfly larvae, *Phormia regina* (Diptera : Calliphoridae). *European Journal of Entomology* 100: 39–42.

Greenberg, B. 1990. Behavior of postfeeding larvae of some Calliphoridae and a muscid (Diptera). *Annals of the Entomological Society of America* 83: 1210–1214.

Greenberg, B. 1991. Flies as forensic indicators. *Journal of Medical Entomology* 28: 565–577.

Greenberg, B. and M. Klowden. 1972. Enteric bacterial interactions in insects. *American Journal of Clinical Nutrition* 25: 1459–1466.

Greenberg, B. and J.C. Kunich. 2002. *Entomology and the Law: Flies as Forensic Indicators.* Cambridge, New York: Cambridge University Press.

Grinnell, J. 1917. The niche-relations of the California thrasher. 34: *Auk* 427–433.

Gripenberg, S., P.J. Mayhew, M. Parnell, and T. Roslin. 2010. A meta-analysis of preference–performance relationships in phytophagous insects. *Ecology Letters* 13: 383–393.

Grunbaum, D. 2002. Predicting availability to consumers of spatially and temporally variable resources. *Hydrobiologia* 480: 175–191.

Gruner, S.V., D.H. Slone, and J.L. Capinera. 2007. Forensically important Calliphoridae (Diptera) associated with pig carrion in rural north-central Florida. *Journal of Medical Entomology* 44: 509–515.

Hall, C.L., N.K. Wadsworth, D.R. Howard, E.M. Jennings, L.D. Farrell, T.S. Magnuson, and R.J. Smith. 2011. Inhibition of microorganisms on a carrion breeding resource: The antimicrobial peptide activity of burying beetle (Coleoptera: Silphidae) oral and anal secretions. *Environmental Entomology* 40: 669–678.

Hall, M. and R. Wall. 1995. Myiasis of humans and domestic animals. *Advances in Parasitology* 35: 257–334.

Hanski, I. 1976. Assimilation by *Lucilia illustris* (Diptera) larvae in constant and changing temperatures. *Oikos* 27: 288–299.

Hanski, I. 1977. Interpolation model of assimilation by larvae of blowfly *Lucilia illustris* (Calliphoridae) in changing temperatures. *Oikos* 28: 187–195.

Hanski, I. 1981. Coexistence of competitors in patchy environment with and without predation. *Oikos* 37: 306–312.

Hanski, I. 1987. Carrion fly community dynamics: Patchiness, seasonality and coexistence. *Ecological Entomology* 12: 257–266.

Hanski, I. 1991. Single-species metapopulation dynamics: Concepts, models and observations. *Biological Journal of the Linnean Society* 42: 17–38.

Hanski, I. 1998. *Metapopulation Ecology.* Oxford, UK: Oxford University Press.

Hanski, I. and S. Kuusela. 1977. An experiment on competition and diversity in the carrion fly community. *Annales Entomologici Fennici* 43: 108–115.

Hanski, I., J. Poyry, T. Pakkala, and M. Kuussaari. 1995. Multiple equilibria in metapopulation dynamics. *Nature* 377: 618–621.

Hansson, B.S. 2002. A bug's smell—Research into insect olfaction. *Trends in Neurosciences* 25: 270–274.

Hansson, B.S. and M.C. Stensmyr. 2011. Evolution of insect olfaction. *Neuron* 72: 698–711.

Hartley, S. and B. Shorrocks. 2002. A general framework for the aggregation model of coexistence. *Journal of Animal Ecology* 71: 651–662.

Hawlena, D., M.S. Strickland, M.A. Bradford, and O.J. Schmitz. 2012. Fear of predation slows plant–litter decomposition. *Science* 336: 1434–1438.

Higley, L.G. and N.H. Haskell. 2010. Insect dvelopment and forensic entomology. In: Byrd, J.H. and J.L. Castner (eds), *Forensic Entomology: The Utility of Arthropods in Legal Investigations*. Boca Raton, FL: CRC Press.

Higley, L.G., L.P. Pedigo, and K.R. Ostlie. 1986. DEGDAY: A program for calculating degree-days, and assumptions behind the degree-day approach. *Environmental Entomology* 15: 999–1016.

Hoback, W.W., A.A. Bishop, J. Kroemer, J. Scalzitti, and J.J. Shaffer. 2004. Differences among antimicrobial properties of carrion beetle secretions reflect phylogeny and ecology. *Journal of Chemical Ecology* 30: 719–729.

Hocking, M., R. Ring, and T. Reimchen. 2009. The ecology of terrestrial invertebrates on Pacific salmon carcasses. *Ecological Research* 24: 1091–1100.

Holt, R.D. 2008. Theoretical perspectives on resource pulses. *Ecology* 89: 671–681.

Hooper, D.U., F.S. Chapin, J.J. Ewel, A. Hector, P. Inchausti, S. Lavorel, J.H. Lawton et al. 2005. Effects of biodiversity on ecosystem functioning: A consensus of current knowledge. *Ecological Monographs* 75: 3–35.

Hopkins, D.W. 2008. The role of soil organisms in terrestrial decomposition. In: Tibbett, M. and D.O. Carter (eds), *Soil Analysis in Forensic Taphonomy: Chemical and Biological Effects of Buried Human Remains*. Boca Raton, FL: CRC Press, pp. 53–66.

Horn, H.S. 1974. The ecology of secondary succession. *Annual Review of Ecology and Systematics* 5: 25–37.

Houston, A.I., A.D. Higginson, and J.M. Mcnamara. 2011. Optimal foraging for multiple nutrients in an unpredictable environment. *Ecology Letters* 14: 1101–1107.

Hubbell, S. 2001. *The Unifed Neutral Theory of Biodiversity and Biogeography*. Princeton, NJ: Princeton University Press.

Huijbers, C.M., T.A. Schlacher, D.S. Schoeman, M.A. Weston, and R.M. Connolly. 2013. Urbanisation alters processing of marine carrion on sandy beaches. *Landscape and Urban Planning* 119: 1–8.

Hutchinson, G.E. 1961. The paradox of the plankton. *The American Naturalist* 95: 137–145.

Hyde, E.R., D.P. Haarmann, A.M. Lynne, S.R. Bucheli, and J.F. Petrosino. 2013. The living dead: Bacterial community structure of a cadaver at the onset and end of the bloat Stage of decomposition. *PLoS ONE* 8: e77733.

Ibanez, S., S. Lavorel, S. Puijalon, and M. Moretti. 2013. Herbivory mediated by coupling between biomechanical traits of plants and grasshoppers. *Functional Ecology* 27: 479–489.

International Glossina Genome Initiative. 2014. Genome sequence of the tsetse fly (*Glossina morsitans*): Vector of African trypanosomiasis. *Science* 344: 380–386.

Ives, A.R. 1991. Aggregation and coexistence in a carrion fly community. *Ecological Monographs* 61: 75–94.

Jans, M.M.E., C.M. Nielsen-Marsh, C.I. Smith, M.J. Collins, and H. Kars. 2004. Characterisation of microbial attack on archaeological bone. *Journal of Archaeological Science* 31: 87–95.

Janzen, D.H. 1977. Why fruits rot, seeds mold, and meat spoils. *American Naturalist* 111: 691–713.

Jiron, L.F. and V.M. Cartin. 1981. Insect succession in the decomposition of a mammal in Costa Rica. *Journal of the New York Entomological Society* 89: 158–165.

Jones, S.E. and K.D. Mcmahon. 2009. Species-sorting may explain an apparent minimal effect of immigration on freshwater bacterial community dynamics. *Environmental Microbiology* 11: 905–913.

Keh, B. 1985. Scope and applications of forensic entomology. *Annual Review Entomology* 30: 137–154.

Kneidel, K.A. 1984. Influence of carcass taxon and size on species composition of carrion-breeding Diptera. *American Midland Naturalist* 111: 57–63.

Kneidel, K.A. 1985. Patchiness, aggregation, and the coexistence of competitors for ephemeral resources. *Ecological Entomology* 10: 441–448.

Kong, H.H. 2011. Skin microbiome: Genomics-based insights into the diversity and role of skin microbes. *Trends in Molecular Medicine* 17: 320–328.

Kozol, A.J., M.P. Scott, and J.F. Traniello. 1988. The American burying beetle, *Nicrophorus americanus*: Studies on the natural history of a declining species. *Psyche* 95: 167–176.

Kuusela, S. and I. Hanski. 1982. The structure of carrion fly communities: The size and the type of carrion. *Holarctic Ecology* 5: 337–348.

Lam, K., D. Babor, B. Duthie, E.M. Babor, M. Moore, and G. Gries. 2007. Proliferating bacterial symbionts on house fly eggs affect oviposition behaviour of adult flies. *Animal Behaviour* 74: 81–92.

Lamotte, L.R. and J.D. Wells. 2000. *p*-Values for postmortem intervals from arthropod succession data. *Journal of Agricultural Biological and Environmental Statistics* 5: 58–68.

Langenheder, S. and A.J. Szekely. 2011. Species sorting and neutral processes are both important during the initial assembly of bacterial communities. *The ISME Journal* 5: 1086–1094.

Leblanc, H.N., J.G. Logan, J. Amendt, M.L. Goff, C.P. Campobasso, and M. Grassberger. 2009. Exploiting insect olfaction in forensic entomology. In: Amendt, J., M.L. Goff, C.P. Campobasso, and M. Grassberger (eds), *Current Concepts in Forensic Entomology*. The Netherlands: Springer, pp. 205–221.

Leibold, M.A. 1995. The niche concept revisited: Mechanistic models and community context. *Ecology* 76: 1371–1382.

Leibold, M.A., M. Holyoak, N. Mouquet, P. Amarasekare, J.M. Chase, M.F. Hoopes, R.D. Holt et al.2004. The metacommunity concept: A framework for multi-scale community ecology. *Ecology Letters* 7: 601–613.

Levin, D.A. 1995. Metapopulations: An arena for local speciation. *Journal of Evolutionary Biology* 8: 635–644.

Lundsten, L., K.L. Schlining, K. Frasier, S.B. Johnson, L.A. Kuhnz, J.B.J. Harvey, G. Clague, and R.C. Vrijenhoek. 2010. Time-series analysis of six whale-fall communities in Monterey Canyon, California, USA. *Deep Sea Research Part I: Oceanographic Research Papers* 57: 1573–1584.

Ma, Q., A. Fonseca, W. Liu, A.T. Fields, M.L. Pimsler, A.F. Spindola, A.M. Tarone, T.L. Crippen, J.K. Tomberlin, and T.K. Wood. 2012. *Proteus mirabilis* interkingdom swarming signals attract blow flies. *ISME Journal* 6: 1356–1366.

Macarthur, R.H. and E.O. Wilson. 1967. *The Theory of Island Biogeography*. Princeton, NJ: Princeton University Press.

Margalida, A., M.À. Colomer, and D. Sanuy. 2011. Can wild ungulate carcasses provide enough biomass to maintain avian scavenger populations? An empirical assessment using a bio-inspired compuational model. *PLoS ONE* 6: e20248.

Martín-Vega, D., A. Gómez-Gómez, and A. Baz. 2011. The "coffin fly" *Conicera tibialis* (Diptera: Phoridae) breeding on buried human remains after a postmortem interval of 18 years. *Journal of Forensic Sciences* 56: 1654–1656.

Matthews, R.W. and J.R. Matthews. 2010. *Insect behavior*. New York, NY: Springer.

Mayhew, P.J. 1997. Adaptive patterns of host–plant selection by phytophagous insects. *Oikos* 79: 417–428.

McGill, B.J. 2010. Matters of scale. *Science* 328: 575–576.

Metcalf, J.L., L.W. Parfrey, A. Gonzalez, C.L. Lauber, D. Knights, G. Ackermann, G.C. Humphrey et al. 2013. A microbial clock provides an accurate estimate of the postmortem interval in a mouse model system. *eLife* 2: e01104.

Michaud, J.-P. and G. Moreau. 2009. Predicting the visitation of carcasses by carrion-related insects under different rates of degree-day accumulation. *Forensic Science International* 185: 78–83.

Michaud, J.-P. and G. Moreau. 2013. Effect of variable rates of daily sampling of fly larvae on decomposition and carrion insect community assembly: Implications for forensic entomology field study protocols. *Journal of Medical Entomology* 50: 890–897.

Michaud, J.-P., G. Moreau, and K.G. Schoenly. 2014. Reply: On throwing out the baby with the bathwater: A reply to Wells. *Journal of Medical Entomology* 51: 494–495.

Mirth, C.K. and L.M. Riddiford. 2007. Size assessment and growth control: How adult size is determined in insects. *Bioessays* 29: 344–355.

Mirth, C., J.W. Truman, and L.M. Riddiford. 2005. The role of the prothoracic gland in determining critical weight to metamorphosis in *Drosophila melanogaster*. *Current Biology* 15: 1796–1807.

Mittelstaedt, H. 1962. Control systems of orientation in insects. *Annual Review of Entomology* 7: 177–198.

Mohr, R.M. 2012. Female blow fly (Diptera: Calliphoridae) arrival patterns and consequences for larval development on ephemeral resources. (PhD dissertation). College Station, TX: Texas A&M University.

Mohr, R.M. and J.K. Tomberlin. 2014. Environmental factors affecting early carcass attendance by four species of blow flies (Diptera: Calliphoridae) in Texas, USA. *Journal of Medical Entomology* 551: 702–708.

Moore, J.C., E.L. Berlow, D.C. Coleman, P.C. De Ruiter, Q. Dong, A. Hastings, N.C. Johnson et al. 2004. Detritus, trophic dynamics and biodiversity. *Ecology Letters* 7: 584–600.

Moore, J.W. and D.E. Schindler. 2010. Spawning salmon and the phenology of emergence in stream insects. *Proceedings of the Royal Society B* 277: 1695–1703.

Moreno-Opo, R. and A. Margalida. 2013. Carcasses provide resources not exclusively to scavengers: Patterns of carrion exploitation by passerine birds. *Ecosphere* 4: art105.

Moretti, T.D., V. Bonato, and W.a.C. Godoy. 2011. Determining the season of death from the family composition of insects infesting carrion. *European Journal of Entomology* 108: 211–218.

Mullens, B.A., C.E. Szijj, and N.C. Hinkle. 2002. Oviposition and development of *Fannia* spp. (Diptera: Muscidae) on poultry manure of low moisture levels. *Environmental Entomology* 31: 588–593.

Mumcuoglu, K.Y., J. Miller, M. Mumcuoglu, M. Friger, and M. Tarshis. 2001. Destruction of bacteria in the digestive tract of the maggot of *Lucilia sericata* (Diptera: Calliphoridae). *Journal of Medical Entomology* 38: 161–166.

Murlis, J., J.S. Elkinton, and R.T. Carde. 1992. Odor plumes and how insects use them. *Annual Review of Entomology* 37: 505–532.

Nayduch, D., G.P. Noblet, and F.J. Stutzenberger. 2002. Vector potential of houseflies for the bacterium *Aeromonas caviae*. *Medical and Veterinary Entomology* 16: 193–198.

Nelder, M.P., J.W. Mccreadie, and C.S. Major. 2009. Blow flies visiting decaying alligators: Is succession synchronous or asynchronous? *Psyche* 2009: 1–7.

Nijhout, H.F. 1975. Threshold size for metamorphosis in tobacco hornworm, *Manduca sexta* (L). *Biological Bulletin* 149: 214–225.

Nijhout, H.F. 2003. The control of body size in insects. *Developmental Biology* 261: 1–9.

Nijhout, H.F. and C.M. Williams. 1974. Control of molting and metamorphosis in tobacco hornworm, *Manduca sexta* (L)—Cessation of juvenile-hormone secretion as a trigger for pupation. *Journal of Experimental Biology* 61: 493–501.

Norris, K.R. 1965. The bionomics of blow flies. *Annual Review of Entomology* 10: 47–68.

Olson, Z.H., J.C. Beasley, T.L. Devault, and O.E. Rhodes. 2012. Scavenger community response to the removal of a dominant scavenger. *Oikos* 121: 77–84.

Park, K.C. and A. Cork. 1999. Electrophysiological responses of antennal receptor neurons in female Australian sheep blowflies, *Lucilia cuprina*, to host odours. *Journal of Insect Physiology* 45: 85–91.

Parmenter, R. and J. Macmahon. 2009. Carrion decomposition and nutrient cycling in a semiarid shrub-steppe ecosystem. *Ecological Monographs* 79: 637–661.

Payne, J.A. 1965. A summer carrion study of the baby pig *Sus scrofa* Linnaeus. *Ecology* 46: 592–602.

Pechal, J.L. and M.E. Benbow. 2015. Community ecology. In: Tomberlin, J. and M. Benbow (eds), *Forensic Entomology: International Dimensions and Frontiers*. Boca Raton, FL: CRC Press, pp. 347–360.

Pechal, J.L., M. Benbow, T. Crippen, A. Tarone, and J. Tomberlin. 2014a. Delayed insect access alters carrion decomposition and necrophagous insect community assembly. *Ecosphere* 5: art45.

Pechal, J.L., T.L. Crippen, M.E. Benbow, A.M. Tarone, S. Dowd, and J.K. Tomberlin. 2014b. The potential use of bacterial community succession in forensics as described by high throughput metagenomic sequencing. *International Journal of Legal Medicine* 128: 193–205.

Pechal, J.L., T.L. Crippen, A.M. Tarone, A.J. Lewis, J.K. Tomberlin, and M.E. Benbow. 2013. Microbial community functional change during vertebrate carrion decomposition. *PLoS ONE* 8: e79035.

Pendola, S. and B. Greenberg. 1975. Substrate-specific analysis of proteolytic enzymes in the larval midgut of *Calliphora vicina*. *Annals of the Entomological Society of America* 68: 341–345.

Pereira, L.M., N. Owen-Smith, and M. Moleon. 2014. Facultative predation and scavenging by mammalian carnivores: Seasonal, regional and intra-guild comparisons. *Mammal Review* 44: 44–55.

Petersen, H. and M. Luxton. 1982. A comparative analysis of soil fauna populations and their role in decomposition processes. *Oikos* 39: 288–388.

Polis, G.A. and S.D. Hurd. 1996. Linking marine and terrestrial food webs: Allochthonous input from the ocean supports high secondary productivity on small islands and coastal land communities. *American Naturalist* 147: 396–423.

Polis, G.A. and D.R. Strong. 1996. Food web complexity and community dynamics. *The American Naturalist* 147: 813–846.

Post, W.M. and K.C. Kwon. 2000. Soil carbon sequestration and land-use change: Processes and potential. *Global Change Biology* 6: 317–327.

Powers, R.H. 2005. The decomposition of human remains. In: Rich, J., D.E. Dean, and R.H. Powers (eds), *Forensic Medicine of the Lower Extremity*. Totowa, NJ: Humana Press, pp. 3–15.

Poza-Carrion, C., T. Suslow, and S. Lindow. 2013. Resident bacteria on leaves enhance survival of immigrant cells of Salmonella enterica. *Phytopathology* 103: 341–351.

Price, G.M. 1974. Protein metabolism by the salivary glands and other organs of the larva of the blowfly, *Calliphora erythrocephala*. *Journal of Insect Physiology* 20: 329–347.

Price, G.M. 1975. Lipase activity in third instar larvae of the blowfly, *Calliphora erythrocephala*. *Insect Biochemistry* 5: 53–60.

Prinkkila, M.L. and I. Hanski. 1995. Complex competitive interactions in 4 species of *Lucilia* blowflies. *Ecological Entomology* 20: 261–272.

Pyke, G.H. 1984. Optimal foraging theory—A critical review. *Annual Review of Ecology and Systematics* 15: 523–575.

Reed Jr, H.B. 1958. A study of dog carcass communities in Tennessee, with special reference to the insects. *American Midland Naturalist* 59: 213–245.

Richard, R. and R.R. Gerrish. 1983. The first confirmed field case of myiasis produced by *Chrysomya* sp. (Diptera: Calliphoridae) in the continental United States. *Journal of Medical Entomology* 20: 685–685.

Richards, C.S., C.C. Rowlinson, L. Cuttiford, R. Grimsley, and M.J.R. Hall. 2013. Decomposed liver has a significantly adverse affect on the development rate of the blowfly *Calliphora vicina*. *International Journal of Legal Medicine* 127: 259–262.

Riddiford, L.M. 2008. Juvenile hormone action: A 2007 perspective. *Journal of Insect Physiology* 54: 895–901.

Rivers, D.B., C. Thompson, and R. Brogan. 2011. Physiological trade-offs of forming maggot masses by necrophagous flies on vertebrate carrion. *Bulletin of Entomological Research* 101: 599–611.

Rivers, D.B., J.A. Yoder, A.J. Jajack, and A.E. Rosselot. 2013. Water balance characteristics of pupae developing in different size maggot masses from six species of forensically important flies. *Journal of Insect Physiology* 59: 552–559.

Rohlfs, M. and T.S. Hoffmeister. 2004. Spatial aggregation across ephemeral resource patches in insect communities: An adaptive response to natural enemies? *Oecologia* 140: 654–661.

Rosati, J.Y. and S.L. Vanlaerhoven. 2007. New record of *Chrysomya rufifacies* (Diptera: Calliphoridae) in Canada: Predicted range expansion and potential effects on native species. *Canadian Entomologist* 139: 670–677.

Rosenheim, J.A., G.E. Heimpel, and M. Mangel. 2000. Egg maturation, egg resorption and the costliness of transient egg limitation in insects. *Proceedings of the Royal Society of London. Series B: Biological Sciences* 267: 1565–1573.

Rozen, D.E., D.J.P. Engelmoer, and P.T. Smiseth. 2008. Antimicrobial strategies in burying beetles breeding on carrion. *Proceedings of the National Academy of Sciences of the United States of America* 105: 17890–17895.

Saunders, D.S., I. Wheeler, and A. Kerr. 1999. Survival and reproduction of small blow flies (*Calliphora vicina*; Diptera: Calliphoridae) produced in severely overcrowded short-day larval cultures. *European Journal of Entomology* 96: 19–22.

Scheirs, J. and L. De Bruyn. 2002. Integrating optimal foraging and optimal oviposition theory in plant–insect research. *Oikos* 96: 187–191.

Schimel, J., T.C. Balser, and M. Wallenstein. 2007. Microbial stress–response physiology and its implications for ecosystem function. *Ecology* 88: 1386–1394.

Schlacher, T.A., S. Strydom, and R.M. Connolly. 2013. Multiple scavengers respond rapidly to pulsed carrion resources at the land–ocean interface. *Acta Oecologica* 48: 7–12.

Schmitz, O.J., J.L. Cohon, K.D. Rothley, and A.P. Beckerman. 1997. Reconciling variability and optimal behaviour using multiple criteria in optimization models. *Evolutionary Ecology* 12: 73–94.

Schoenly, K. 1992. A statistical analysis of successional patterns in carrion-arthropod assemblages: Implications for forensic entomology and determination of the postmortem interval. *Journal of Forensic Sciences* 37: 1489.

Schoofs, A., S. Niederegger, and R. Spiess. 2009. From behavior to fictive feeding: Anatomy, innervation and activation pattern of pharyngeal muscles of *Calliphora vicina* 3rd instar larvae. *Journal of Insect Physiology* 55: 218–230.

Schoofs, A. and R. Spiess. 2007. Anatomical and functional characterisation of the stomatogastric nervous system of blowfly (*Calliphora vicina*) larvae. *Journal of Insect Physiology* 53: 349–360.

Sebastian-Gonzalez, E., J.A. Sanchez-Zapata, J.A. Donazar, N. Selva, A. Cortes-Avizanda, F. Hiraldo, M. Blazquez, F. Botella, and M. Moleon. 2013. Interactive effects of obligate scavengers and scavenger community richness on lagomorph carcass consumption patterns. *Ibis* 155: 881–885.

Sekirov, I., S.L. Russell, L.C.M. Antunes, and B.B. Finlay. 2010. Gut microbiota in health and disease. *Physiological Reviews* 90: 859–904.

Sherman, R.A., M.J.R. Hall, and S. Thomas. 2000. Medicinal maggots: An ancient remedy for some contemporary afflictions. *Annual Review of Entomology* 45: 55–81.

Sherratt, T.N., D.M. Wilkinson, and R.S. Bain. 2006. Why fruits rot, seeds mold and meat spoils: A reappraisal. *Ecological Modelling* 192: 618–626.

Shiao, S.F. and T.C. Yeh. 2008. Larval competition of *Chrysomya megacephala* and *Chrysomya rufifacies* (Diptera: Calliphoridae): Behavior and ecological studies of two blow fly species of forensic significance. *Journal of Medical Entomology* 45: 785–799.

Shingleton, A.W., W.A. Frankino, T. Flatt, H.F. Nijhout, and D.J. Emlen. 2007. Size and shape: The developmental regulation of static allometry in insects. *Bioessays* 29: 536–548.

Shorrocks, B., W. Atkinson, and P. Charlesworth. 1979. Competition on a divided and ephemeral resource. *Journal of Animal Ecology* 48: 899–908.

Simmons, T., R.E. Adlam, and C. Moffatt. 2010a. Debugging decomposition data—Comparative taphonomic studies and the influence of insects and carcass size on decomposition rate. *Journal of Forensic Sciences* 55: 8–13.

Simmons, T., P.A. Cross, R.E. Adlam, and C. Moffatt. 2010b. The influence of insects on decomposition rate in buried and surface remains. *Journal of Forensic Sciences* 55: 889–892.

Singer, M.C., D. Vasco, C. Parmesan, C.D. Thomas, and D. Ng. 1992. Distinguishing between preference and motivation in food choice—An example from insect oviposition. *Animal Behaviour* 44: 463–471.

Smith, K.G.V. 1986. *A Manual of Forensic Entomology*. London: Trustees of the British Museum (Natural History).

Smith, C.R. and A.R. Baco. 2003. Ecology of whale falls at the deep-sea floor. *Oceanography and Marine Biology: An Annual Review* 41: 311–354.

Smith, K.E. and R. Wall. 1997. Asymmetric competition between larvae of the blowflies *Calliphora vicina* and *Lucilia sericata* in carrion. *Ecological Entomology* 22: 468–474.

Stafford, K. and D. Bay. 1987. Dispersion pattern and association of house fly, *Musca domestica* (Diptera: Muscidae), larvae and both sexes of *Macrocheles muscaedomesticae* (Acari: Macrochelidae) in response to poultry manure moisture, temperature, and accumulation. *Environmental Entomology* 16: 159–164.

Stokes, K.L., S.L. Forbes, L.A. Benninger, D.O. Carter, and M. Tibbett. 2009. Decomposition studies using animal models in contrasting environments: Evidence from temporal changes in soil chemistry and microbial activity. In: Ritz, K., L. Dawson, and D. Miller (eds), *Criminal and Environmental Soil Forensics*. The Netherlands: Springer, pp. 357–377.

Tabor, K.L., C.C. Brewster, and R.D. Fell. 2004. Analysis of the successional patterns of insects on carrion in Southwest Virginia. *Journal of Medical Entomology* 41: 785–795.

Tack, A.J.M., S. Gripenberg, and T. Roslin. 2012. Cross-kingdom interactions matter: Fungal-mediated interactions structure an insect community on oak. *Ecology Letters* 15: 177–185.

Tarone, A.M. and D.R. Foran. 2006. Components of developmental plasticity in a Michigan population of *Lucilia sericata* (Diptera: Calliphoridae). *Journal of Medical Entomology* 43: 1023–1033.

Tarone, A.M. and D.R. Foran. 2008. Generalized additive models and *Lucilia sericata* growth: Assessing confidence intervals and error rates in forensic entomology. *Journal of Forensic Sciences* 53: 942–948.

Tarone, A.M., C.J. Picard, C. Spiegelman, and D.R. Foran. 2011. Population and temperature effects on *Lucilia sericata* (Diptera: Calliphoridae) body size and minimum development time. *Journal of Medical Entomology* 48: 1062–1068.

Tarone, A.M., J.K. Tomberlin, J.S. Johnston, J.R. Wallace, M.E. Benbow, R. Mohr, E.B. Mondor, and S.L. Vanlaerhoven. 2014. Reply: A correspondence from a maturing discipline. *Journal of Medical Entomology* 51: 490–492.

The Human Microbiome Consortium. 2012. A framework for human microbiome research. *Nature* 486: 215–221.

Theis, K.R., T.M. Schmidt, and K.E. Holekamp. 2012. Evidence for a bacterial mechanism for group-specific social odors among hyenas. *Scientific Reports* 2: 615.

Theis, K.R., A. Venkataraman, J.A. Dycus, K.D. Koonter, E.N. Schmitt-Matzen, A.P. Wagner, K.E. Holekamp, and T.M. Schmidt. 2013. Symbiotic bacteria appear to mediate hyena social odors. *Proceedings of the National Academy of Sciences* 110: 19832–19837.

Tomberlin, J.K. and P.H. Adler. 1998. Seasonal colonization and decomposition of rat carrion in water and on land in an open field in South Carolina. *Journal of Medical Entomology* 35: 704–709.

Tomberlin, J.K. and M.E. Benbow. 2015. *Forensic Entomology: International Dimensions and Frontiers*. Boca Raton, FL: CRC Press.

Tomberlin, J.K., M.E. Benbow, A.M. Tarone, and R.M. Mohr. 2011b. Basic research in evolution and ecology enhances forensics. *Trends in Ecology & Evolution* 26: 53–55.

Tomberlin, J.K., T.L. Crippen, A.M. Tarone, B. Singh, K. Adams, Y.H. Rezenom, M.E. Benbow et al. 2012. Interkingdom responses of flies to bacteria mediated by fly physiology and bacterial quorum sensing. *Animal Behaviour* 84: 1449–1456.

Tomberlin, J.K., R. Mohr, M.E. Benbow, A.M. Tarone, and S. Vanlaerhoven. 2011a. A roadmap for bridging basic and applied research in forensic entomology. *Annual Review of Entomology* 56: 401–421.

Towne, E.G. 2000. Prairie vegetation and soil nutrient responses to ungulate carcasses. *Oecologia* 122: 232–239.

Tuomisto, S., P.J. Karhunen, and T. Pessi. 2013. Time-dependent post mortem changes in the composition of intestinal bacteria using real-time quantitative PCR. *Gut Pathogens* 5: 35.

Turnbaugh, P.J., R.E. Ley, M. Hamady, C.M. Fraser-Liggett, R. Knight, and J.I. Gordon. 2007. The Human Microbiome Project. *Nature* 449: 804–810.

Ullyett, G. 1950. Competition for food and allied phenomena in sheep-blowfly populations. *Philosophical Transactions of the Royal Society of London. Series B, Biological Sciences* 234: 77–174.

Urban, D.L., R.V. Oneill, and H.H. Shugart. 1987. Landscape ecology. *Bioscience* 37: 119–127.

Valachova, I., J. Bohova, Z. Palosova, P. Takac, M. Kozanek, and J. Majtan. 2013. Expression of lucifensin in *Lucilia sericata* medicinal maggots in infected environments. *Cell and Tissue Research* 353: 165–171.

Van Der Plas, M.J., G.N. Jukema, S.-W. Wai, H.C. Dogterom-Ballering, E.L. Lagendijk, C. Van Gulpen, J.T. Van Dissel, G.V. Bloemberg, and P.H. Nibbering. 2008. Maggot excretions/secretions are differentially effective against biofilms of *Staphylococcus aureus* and *Pseudomonas aeruginosa*. *Journal of Antimicrobial Chemotherapy* 61: 117–122.

Vass, A.A. 2011. The elusive universal post-mortem interval formula. *Forensic Science International* 204: 34–40.

Vass, A.A. 2012. Odor mortis. *Forensic Science International* 222: 234–241.

Vass, A.A., R.R. Smith, C.V. Thompson, M.N. Burnett, N. Dulgerian, and B.A. Eckenrode. 2008. Odor analysis of decomposing buried human remains. *Journal of Forensic Sciences* 53: 384–391.

Vass, A.A., R.R. Smith, C.V. Thompson, M.N. Burnett, D.A. Wolf, J.A. Synstelien, N. Dulgerian, and B.A. Eckenrode. 2004. Decompositional odor analysis database. *Journal of Forensic Sciences* 49: 760–769.

Vieira, F.G., A. Sánchez-Gracia, and J. Rozas. 2007. Comparative genomic analysis of the odorant-binding protein family in 12 *Drosophila* genomes: Purifying selection and birth-and-death evolution. *Genome Biol* 8: R235.

Villet, M.H. 2011. African carrion ecosystems and their insect communities in relation to forensic entomology. *Pest Technology* 5: 1–15.

Vinson, S.B. 1998. The general host selection behavior of parasitoid Hymenoptera and a comparison of initial strategies utilized by larvaphagous and oophagous species. *Biological Control* 11: 79–96.

Visser, J. 1988. Host–plant finding by insects: Orientation, sensory input and search patterns. *Journal of Insect Physiology* 34: 259–268.

Vogt, R.G., G.D. Prestwich, and M.R. Lerner. 1991. Odorant-binding-protein subfamilies associate with distinct classes of olfactory receptor neurons in insects. *Journal of Neurobiology* 22: 74–84.

Vogt, W. and T. Woodburn. 1980. The influence of temperature and moisture on the survival and duration of the egg stage of the Australian sheep blowfly, *Lucilia cuprina* (Wiedemann) (Diptera: Calliphoridae). *Bulletin of Entomological Research* 70: 665–671.

Wajnberg, E. 2006. Time allocation strategies in insect parasitoids: From ultimate predictions to proximate behavioral mechanisms. *Behavioral Ecology and Sociobiology* 60: 589–611.

Wall, R. and K.E. Smith. 1997. The potential for control of the blowfly *Lucilia sericata* using odour-baited targets. *Medical and Veterinary Entomology* 11: 335–341.

Wall, R. and M.L. Warnes. 1994. Responses of the sheep blowfly *Lucilia sericata* to carrion odor and carbon dioxide. *Entomologia Experimentalis Et Applicata* 73: 239–246.

Wallis, D.I. 1962. Olfactory stimuli and oviposition in blowfly, *Phormia regina* Meigen. *Journal of Experimental Biology* 39: 603–615.

Walter, G.H. 2005. *Insect Pest Management and Ecological Research*. New York, NY: Cambridge University Press.

Ward, S.A. 1987. Optimal habitat selection in time-limited dispersers. *American Naturalist* 129: 568–579.

Watson, E.J. and C.E. Carlton. 2003. Spring succession of necrophilous insects on wildlife carcasses in Louisiana. *Journal of Medical Entomology* 40: 338–347.

Watson, E.J. and C.E. Carlton. 2005. Insect succession and decomposition of wildlife carcasses during fall and winter in Louisiana. *Journal of Medical Entomology* 42: 193–203.

Wells, J.D. 2014. To the editor: Misstatements concerning forensic entomology practice in recent publications. *Journal of Medical Entomology* 51: 489–490.

Wells, J.D. and B. Greenberg. 1992a. Interaction between *Chrysomya rufifacies* and *Cochliomyia macellaria* (Diptera, Calliphoridae)—The possible consequences of an invasion. *Bulletin of Entomological Research* 82: 133–137.

Wells, J.D. and B. Greenberg. 1992b. Laboratory interaction between introduced *Chrysomya rufifacies* and native *Cochliomyia macellaria* (Diptera: Calliphoridae). *Environmental Entomology* 21: 641–645.

Wells, J.D. and B. Greenberg. 1994. Effect of the red imported fire and (Hymenoptera: Formicidae) and carcass type on the daily occurrence of postfeeding carrion fly larvae (Diptera, Calliphoridae, Sarcophagidae). *Journal of Medical Entomology* 31: 171–174.

Wells, J.D. and J. King. 2001. Incidence of precocious egg development in flies of forensic importance (Calliphoridae). *Pan-Pacific Entomologist* 77: 235–239.

Wells, J.D. and L.R. Lamotte. 2010. Estimating the postmortem interval. In: Byrd, J.H. and J.L. Castner (eds), *Forensic Entomology: The Utility of Arthropods in Legal Investigations*. Boca Raton, FL: CRC Press.

Wilson, E.E. and E.M. Wolkovich. 2011. Scavenging: How carnivores and carrion structure communities. *Trends in Ecology & Evolution* 26: 129–135.

Woodcock, L., D. Gennard, and P. Eady. 2013. Egg laying preferences and larval performance in *Dermestes maculatus*. *Entomologia Experimentalis Et Applicata* 148: 188–195.

Woodcock, B.A., A.D. Watt, and S.R. Leather. 2002. Aggregation, habitat quality and coexistence: A case study on carrion fly communities in slug cadavers. *Journal of Animal Ecology* 71: 131–140.

Yang, L.H. 2004. Periodical cicadas as resource pulses in North American forests. *Science* 306: 1565–1567.

Yang, L.H. 2006. Interactions between a detrital resource pulse and a detritivore community. *Oecologia* 147: 522–532.

Yang, L.H., J.L. Bastow, K.O. Spence, and A.N. Wright. 2008. What can we learn from resource pulses? *Ecology* 89: 621–634.

Young, A.R., E.N.T. Meeusen, and V.M. Bowles. 1996. Characterization of ES products involved in wound initiation by *Lucilia cuprina* larvae. *International Journal for Parasitology* 26: 245–252.

Yovanovitch, G.P. 1888. *Entomologie Appliquée à la Médicine Légale*. Paris, France: Ollier-Henry.

Zheng, L., T.L. Crippen, L. Holmes, B. Singh, M.L. Pimsler, M.E. Benbow, A.M. Tarone, S. Dowd, Z. Yu, and S.L. Vanlaerhoven. 2013. Bacteria mediate oviposition by the black soldier fly, *Hermetia illucens* (L.) (Diptera: Stratiomyidae). *Scientific Reports* 3: 2563.

9
Chemical Ecology of Vertebrate Carrion

Jonathan A. Cammack, Meaghan L. Pimsler, Tawni L. Crippen, and Jeffery K. Tomberlin

CONTENTS

9.1 Introduction

As many chapters in this book have discussed, vertebrate carrion is a nutrient-rich, ephemeral resource that is utilized by many different organisms, ranging from vertebrate and invertebrate scavengers to microbes (Janzen 1977; Early and Goff 1986; Braack 1987). The organisms that consume carrion play an important ecological role, as decomposition is vital to ecosystem function (Carter et al. 2007). Without these scavengers, vital elements (especially carbon, sulfur, nitrogen, and phosphorus) that are initially trapped in primary producers (plants) and subsequently concentrated in animal tissues would not be released back into the ecosystem in a timely manner (Putman 1978; Stevenson and Cole 1999; Parmenter and MacMahon 2009).

The goal of this chapter is to explore the chemical ecology of the terrestrial carrion system. Chemical ecology encompasses the range of molecules that organisms produce or encounter, and what interactions these molecules have with other organisms (Shivik 2006). This chapter will focus on the important chemical mechanisms occurring during the process of carrion decomposition. The chapter begins with a discussion of the various volatile organic compounds (VOCs), or semiochemicals, released during vertebrate carrion decomposition, followed by a review of the various sources of these compounds. The chapter next examines the behaviors exhibited by organisms in response to these semiochemicals (VOCs that convey information within or between species) and the role these behaviors play in a larger ecological context. Finally, an introduction is provided on the methods for collection and analysis that are appropriate for answering questions about these chemical compounds in an ecological context.

9.1.1 Biogeochemical Cycles

Carbon (C) is commonly referred to as "the building block of life," and indeed, all known organisms use carbon as the basis for their organic structure (Normile 2009). This characterization of C is predicated on the availability of four valence electrons for the formation of strong ionic bonds. Carbon is generally present in the atmosphere as carbon dioxide (CO_2), and in the process of photosynthesis, plants use energy derived from sunlight to break the double bonds between oxygen and carbon to eventually form glucose (Prentice et al. 2001). Glucose makes the C biologically available for the construction of many other molecules, including fats, nucleotides, and various amino acids (Montagnes et al. 1994). Herbivores and omnivores acquire C through the consumption of plant material, and predators move the carbon up through the food web. Many metabolic processes utilized by animals result in the release of CO_2 into the atmosphere through respiration, but much of the consumed C also remains in the tissues (Normile 2009). However, once an animal dies, microbes and other scavengers break down its complex carbon-containing molecules into simpler building blocks, that both microbes and larger scavengers (including vertebrates and invertebrates) utilize and then release back into the atmosphere as waste CO_2, thus perpetuating the C cycle (Stevenson and Cole 1999; Carter et al. 2007).

Sulfur (S), an important part of many proteins, is also considered a vital nutrient for life. It is found normally in the atmosphere in very small concentrations in the form of sulfur dioxide (SO_2) (Langer and Rohde 1991). In the S cycle, SO_2 reacts with water, oxygen, nitrogen oxide gases, and other chemicals in the atmosphere to form various acidic compounds such as sulfate ions (SO_4^{2-}) that rain down on the earth (Kellogg et al. 1972); plants absorb sulfur through their roots in the form of dissolved SO_4^{2-}. As with carbon, this sequestered S moves through the food chain, first into consumers of plants and then into higher trophic levels. The S is generally incorporated into amino acid R groups, and their S—S bonds are important for secondary and tertiary protein structure (Jacob et al. 2003; Lill and Mühlenhoff 2005). The highest concentration of sulfur-containing molecules is in protein and enzyme-rich animal tissues. As with C, the decomposition of animal and plant tissues releases S back into the environment (Parmenter and MacMahon 2009) in the form of waste hydrogen sulfide (H_2S) and SO_4^{2-} (Jørgensen 1977; Stevenson and Cole 1999).

Nitrogen (N) is present in vast quantities in the environment; 70% of the atmosphere is made of nitrogen gas (N_2) (Bernhard 2012). However, the element in the form of N_2 is not biologically available to most organisms, as a great deal of energy is required to break the triple bond between the two N molecules. The N cycle, therefore, relies heavily on soil-based microorganisms that produce nitrogen-fixing enzymes, which convert atmospheric N_2 into ammonium (NH_4^+), nitrites (NO_2^-), and nitrates (NO_3^-) (Wagner 2012). Some plants, such as legumes, have a symbiotic relationship with nitrogen-fixing bacteria and have a high concentration of these microorganisms in special root nodules (Cocking 2003). Nitrogen is vital for the production of amino and nucleic acids, and without enough N, plants cannot grow (Montagnes et al. 1994). Similar to the primary elements previously discussed, animals consuming plant tissue acquire N, and the element moves through the food web into higher trophic levels through predation and parasitism. Many animals produce large quantities of nitrogenous waste in their feces, and the type of nitrogenous waste excreted by the kidneys (vertebrates) or malpighian tubules (insects) can be an indicator of water stress (Wright 1995). Many aquatic organisms produce ammonia (NH_3), terrestrial organisms often produce the nitrogen-based compounds urea or uric acid, and many desert-dwelling species of arthropods produce nearly solid nitrogenous waste in the form of xanthine or guanine (Yokota and Shoemaker 1981; Wright 1995). As with S, much of the sequestered N remains in the proteins or genetic material of the organism, and once an animal dies, microorganisms break down large organic molecules (Parmenter and MacMahon 2009), eventually releasing excess N as NH_4^+, NO_2^-, and NO_3^-. While N is in high demand, and does not remain in the environment in these forms for long. Denitrifying bacteria reduce nitrites and nitrates and release N back into the atmosphere as N_2, thereby continuing the N cycle (Sprent 1987; Stevenson and Cole 1999).

The phosphorus (P) cycle is distinct from others, as it does not have an atmospheric component, and occurs on a much more protracted temporal scale (Stevenson and Cole 1999). Phosphorus is present in high concentration in rocks and soils and leaches out during erosion, generally forming dissolved phosphate (PO_4^{2-}) (Cross and Schlesinger 1995). This molecule is in high demand and quickly absorbed by plants (Cross and Schlesinger 1995). Phosphorus is a vital component of DNA and RNA (Watson and Crick 1953), the phospholipid bilayer of cells (Nagle and Tristram-Nagle 2000), and the "energy" molecule of life, adenosine triphosphate (Mildvan 1979). As with the other elements discussed, P moves through the food web through consumption, and through the action of decomposers, sequestered P is metabolized out of larger organic molecules and eventually released back into the environment as PO_4^{2-} (Parmenter and MacMahon 2009). While PO_4^{2-} has a short half-life in terrestrial and aquatic environments, there are times when an organism dies in a place where decomposers are unable to access them, such as in oceans (Stevenson and Cole 1999). Over a time scale of millions of years, the phosphate-rich marine sediment solidifies, and later geologic action causes this rock layer to uplift as mountains, making the rock susceptible to erosion and releasing PO_4^{2-} back into the environment.

The biological necessity of these elements coupled with their relative inaccessibility in the environment has contributed to the evolution of physiologies and behaviors crucial for locating these resources. As previously mentioned, some plants have symbiotic relationships with N-fixing microbes that allow them to utilize atmospheric N (Cocking 2003). However, with the exception of some marine systems, animals to not appear to have such obligate relationships with microbes (Dubilier et al. 2008). Therefore, animals have evolved a suite of chemosensory capabilities that allow for the location and detection of chemicals at both a long (olfaction) and short range (gustation).

9.1.2 Senses for Detecting Carrion

Given the limited availability of various nutrients and the ephemeral nature of carrion, competition for this resource is predicted to be intense. Therefore, there is likely strong selective pressure for sarcosaprophages ("rotting meat-eating organisms") to evolve strategies to quickly locate carrion, avoid or dominate intraspecific or interspecific interactions, and increase their portion of nutrient uptake (Janzen 1977; Shivik 2006). Vision is an important component of carrion location for species such as in the black vulture *Coragyps atratus* Le Maout (Cathartiformes: Cathartidae), which responds strongly to colors associated with decomposing animal remains (Buckley 1996). Carrion-mimicking flowers have taken advantage of these preexisting sensory preferences to enhance their attraction to pollinating insects and have evolved to exhibit carrion-like brown, purple, and dark-red colors, and many are also thermogenic (Meve and Liede 1994). See Chapter 16 for a more in-depth review of the ecology and biology of these plants.

Olfaction is likely to be the most important long-distance cue in locating carrion, and therefore obligate macroscavengers should be under strong selective pressure for sensitive olfaction (Shivik 2006). Indeed, turkey vultures *Cathartes aura* (L.) (Cathartiformes: Cathartidae) are well known for their keen sense of smell (Owre and Northington 1961), and the relative size of their olfactory bulbs in comparison to the rest of their brain is second only to that of kiwis, *Apteryx australis* Shaw and Nodder (Apterygiformes: Apterygidae) (Bang and Cobb 1968). Some scavenging snakes and carrion-breeding flies also rely on olfaction for the detection and location of resources (Jojola-Elverum et al. 2001; LeBlanc and Logan 2010). The brown tree snake, *Boiga irregularis* Merrem (Squamata: Colubridae) uses chemical cues for locating carrion (Shivik and Clark 1997); similarly, odors from carrion are important in regulating the attraction of female blow flies (Diptera: Calliphoridae) (Wall and Warnes 1994). Although large numbers of volatiles have been identified from carrion, specific classes or individual compounds may be regulating the behavior. Therefore, identifying the classes of VOCs present is crucial in chemical ecology research.

9.2 Classes of Volatiles

Researchers have used multiple classification schemes to describe and delineate the 500 (and growing!) volatile chemicals that have been collected and detected from vertebrate carrion in terrestrial environments (Vass et al. 2008; Dekeirsschieter et al. 2009; Stadler 2013; Forbes and Perrault 2014). As this chapter is meant to be an overview, and researchers are still converging on a common classification scheme, this chapter will remain cautiously rudimentary. The following list is far from complete, and interested readers are encouraged to refer to the growing body of literature on the subject.

9.2.1 Hydrocarbons

Hydrocarbons, the subject of organic chemistry, are composed entirely of hydrogen and carbon molecules. Most hydrocarbons are the product of organic processes, and a majority of those on earth are found in crude oil. They exist as gases, liquids, and solids, and readily bond with themselves in a process known as "catenation," producing complex, high molecular weight compounds.

Saturated hydrocarbons are the simplest form, composed entirely of single bonds and saturated with hydrogen. Unsaturated hydrocarbons have at least one double or triple bond between two carbon atoms (alkenes and alkynes, respectively). These can be found as a single linear molecule with or without one or more branches, or even possess a carbon ring. Hydrocarbons with a saturated carbon ring are known as "cycloalkanes," while those with a ring with alternating single and double bonds are called "aromatic hydrocarbons" or "arenes." These are known as "aromatic" due to a specific chemical property called "aromaticity," which is related to the stability and geometry of a molecule and not to any olfactory property (Hofmann 1856; Schleyer 2001).

These compounds are associated with the decomposition process particularly during the active decay and skeletal stages (Forbes and Perrault 2014). However, unlike nitrogenous compounds (discussed below) which are more commonly associated with air samples from vertebrate remains, these compounds are dominant in the soil (Forbes and Perrault 2014). These compounds could also originate from the exoskeletons of insects commonly associated with vertebrate carrion (Paczkowski and Schütz 2011).

9.2.2 Oxygen-Containing Compounds

Oxygen-containing compounds released from vertebrate carrion are diverse and include acids and acid esters, ethers, aldehydes, and ketones (Stadler 2013). Those identified from carrion vary between experiments, possibly due to differences in substrate availability, biochemistry, and the interaction between the species of carrion and scavengers (Herbert and Shewan 1976; Dekeirsschieter et al. 2009; Kalinová et al. 2009; Frederickx et al. 2012a; Stadler 2013).

The Brønsted–Lowry definition of an "acid" is a compound that can donate a proton (H^+) to another compound (Ebbing and Gammon 2012), and many organic processes result in the production of acids. Carboxylic acids, those that contain a carboxyl group ($R-CO_2H$) are common products of decomposition. By common naming convention, the compound is named by selecting the longest carbon chain and then replacing the "-e" in the parent alkane and replacing it with an "-oic acid" suffix (Moss et al. 1995). For example, formic acid, a carboxylic acid formed from methane and one commonly associated with ants is referred to as methanoic acid under IUPAC convetion (Löfqvist 1976). A diversity of carboxylic acids are volatile, and in the presence of oxygen, the metabolic activities of bacteria such as acetic acid bacteria, can metabolize alcohols into acids. Such metabolizing bacteria exist on decomposing carrion; for example, several carboxylic acids are produced by *Corynebacteria* spp. commonly found in human sweat (Natsch et al. 2006). Thus, many of these acids likely play a role in the detection of carrion by scavengers. In human decomposition, alcohols are detected from the decomposition of blood, muscle, body fat attached to skin, fat, and bone (Hoffman et al. 2009). Interestingly, qualitative differences were found in the composition of volatile alcohols among the different tissue samples analyzed by Hoffman et al. (2009): adipocere, some bone, and teeth produced the fewest alcohols, primarily 1-pentanol, while muscle, some bone, blood, fat, and body fat attached to skin produced a more diverse array of alcohols, including pentanols, hexanols, and octanols.

An ester is a compound derived from an acid, in which at least one hydroxyl group (—OH) is replaced with an alkoxy group (—O—alkyl). These chemicals often result in a "fruity aroma" during fermentation and decomposition, most likely from the combined action of esterification (formation of esters from alcohols) and alcoholysis (a reaction where fatty acyl groups are directly transferred to alcohols) (Bullard et al. 1978; Liu et al. 2004). In human decomposition, esters and acid esters appear to be primarily derived from the decomposition of fat and fat by-products such as adipocere (Hoffman et al. 2009). See Chapter 2 for more information on this process.

Aldehydes and ketones are two types of molecules that contain a carbonyl group (C=O) (Clayden 2001). Aldehydes are organic compounds that contain a formyl group (R—CHO), a hydrogen-containing carbonyl group; many of these compounds are components used in fragrant cosmetic products such as lotions and perfumes (Rastogi et al. 1996; Clayden 2001). Ketones are larger molecules in which the carbonyl group is a bridge between two functional R groups, which may or may not have the same identity (Clayden 2001). Aldehyde and ketone products are frequently the result of aerobic decomposition of fat, muscle, and bone, but can also be detected as part of the odors released from living humans (Degreeff and Furton 2011).

9.2.3 Nitrogenous Compounds

Knowledge of the diversity and function of VOCs within some of these classes is limited, but nitrogenous compounds have been described in great detail and are known to serve as mechanisms regulating the behaviors of animals seeking out these resources for food, shelter, or the location of mates. As the name implies, nitrogenous compounds are those that contain nitrogen.

Cadaverine and putrescine are well-known by-products associated with the decomposition of vertebrate carrion (Vass et al. 2002, 2008) by microbes (Schulz and Dickschat 2007; LeBlanc 2008). These compounds are similar in structure and are produced through the breakdown of protein and, as their names imply, are foul-smelling. Indole is another by-product of protein degradation that is commonly encountered with vertebrate decomposition and most likely produced by associated bacteria (Frederickx et al. 2012b; Chaudhury et al. 2014). However, unlike putrescine and cadaverine, indole is an aromatic heterocyclic compound.

As reviewed in Chapter 20, some of these molecules serve as quorum-sensing compounds for bacteria and play an integral role in regulating insect attraction and colonization of carrion. Putrescine, the product of the *speB* gene, represses *speA:lacZ* and restores swarming, a quorum-sensing response of the *Proteus mirabilis* mutant PM437 (Sturgill and Rather 2004). Similarly, indole serves as an extracellular signal in *Escherichia coli* (Wang et al. 2001). Both of these bacteria are often associated with vertebrate carrion and can serve important roles in the ecology (e.g., nutrition [Zurek et al. 2000] or production of antimicrobials [Erdmann 1987]) of the insects that consume carrion.

9.2.4 Sulfur-Containing Compounds

Sulfur has been an important part of human history for thousands of years and was likely isolated from volcanic soils until isolation and purification techniques were refined (Meyer 1976). Several well-known pungent odors, including that of garlic, onions, and rotten eggs are due to sulfur-containing compounds such as thiols/mercaptans, pyruvic acid, and H_2S, respectively (Schwimmer and Weston 1961; Bentley 2006). Many malodorous compounds released from living humans, especially those anaerobic products that are significant contributors to halitosis, contain sulfur (Schmidt et al. 1978).

The sulfur compounds detected from decomposition are generally sulfides. Dimethyl disulfide (DMDS) is one of the most prominent compounds released from decomposing human remains (Statheropoulos et al. 2005, 2007) and is also found in headspace analysis of blood clots, testicles, skin, adipocere, and bone (Hoffman et al. 2009). In Australia, *Lucilia cuprina* (Wiedemann) (Diptera: Calliphoridae) exhibits dose-dependent chemotaxis in response to DMDS and ethane thiol (Morris et al. 1998), though more recent work suggests that these insects respond to dimethyl trisulfide (DMTS) instead (Zito et al. 2014). Regardless, it is clear that sulfur-containing compounds are important, persistent, bioactive constituents of the volatile blend released during the process of vertebrate carrion decomposition. However, research in chemical ecology relies upon more than a catalog of the VOCs; identifying the source of the compounds is also crucial.

9.3 Sources of Volatiles

While the list of volatiles described above is far from exhaustive, more progress has been made in cataloguing the VOCs released than in determining their origin. Initial work in biochemistry provided a foundation, as researchers identified substrates and products of enzymes isolated from various tissues. More recent technological advancements in microbiology, protein isolation, and sequencing, and chemical analysis have streamlined the ability of researchers to investigate the source, and not just the identity, of VOCs associated with carrion. There are two major categories identified and discussed in this chapter: the carrion itself, and the organisms, especially insects and microbes, that use the carrion as a nutrient source.

9.3.1 Carrion

There is limited research on the VOCs produced endogenously during the autolysis of carrion. Some tissues, such as liver, stomach, and intestines, contain a higher concentration and diversity of proteolytic and metabolic enzymes than other tissues (e.g., muscle and skin) (LaRosa et al. 1972; Anderson et al. 2002). Studies on the autolysis of animal tissues after death are complicated by the inherent difficulty in removing microbes without affecting enzymes in the tissue (Beatty and Collins 1939). Some progress has been made in identifying organic compounds produced by autolytic enzymes, but most of these are not volatile (Zender et al. 1958; Aksnes 1988). However, beef muscle contains an esterase that functions in the production of the volatile butyric acid (Matlack and Tucker 1940), enzymes in the leg muscles of guinea pigs release volatile sulfide compounds (Osborne 1928), and some of the enzymes in fish muscle release unsaturated fatty acids that rapidly oxidize and produce a rancid flavor and odor (Mukundan et al. 1986). Studies on the biochemistry of autolysis are limited, and current research tends to focus on bacterial action (Gram and Huss 1996), as microbes clearly are the principal players in the breakdown of carrion (Herbert and Shewan 1976; Gram and Huss 1996; Hansen et al. 1996).

9.3.2 Insects

The chemical ecology of insects has been a fertile ground for research (Bell and Cardé 1984). Pheromones mediate everything from aggregation and mate location to learning in insects, and this research has been largely focused on herbivores, eusocial insects, and parasitoids, due to their agrarian significance (Mayer and McLaughlin 1990; Landolt 1997; Van Dyke Parunak et al. 2005). However, limited work in the carrion system has focused on insects as the producers of VOCs rather than bacteria. Here, the chemical ecology of blow flies (Diptera: Calliphoridae) is discussed throughout the different stages of the life cycle.

   The mating behavior of blow flies is difficult to study in the field, though there is evidence of a sex pheromone produced by males of the primary screwworm, *Cochliomyia hominivorax* (Coquerel) (Diptera: Calliphoridae) (Mackley and Broce 1981). In comparison, work in *L. cuprina* based on findings in *Musca domestica* L. (Diptera: Muscidae) and *Drosophila melanogaster* Meigen (Diptera: Drosophilidae) suggests that females produce a chemical that stimulates courtship behaviors in males (Rogoff et al. 1964; Bartell et al. 1969; Shorey and Bartell 1970). In contrast, work with another calliphorid, *Protophormia terraenovae* Robineau-Desvoidy, was unable to differentiate between the possibility of a male-secreted aphrodisiac or a female-secreted mounting stimulant (Parker 1968). Results from this work also suggested that aggregation around a protein source might be the mechanism by which males locate receptive females for mating. Therefore, it is unclear at present whether sex pheromone production is common in this family, which sex produces the chemicals, or the exact behavior elicited by these chemicals.

There is limited evidence of the role the presence of eggs themselves may play in the location and acceptance of oviposition sites, as bacteria present on the eggs are known to play a major role in this process (Emmens and Murray 1982; Chaudhury et al. 2010; Brundage et al. 2014). Far less work has been done to elucidate the VOCs released by immature calliphorids, but differences between species and temporal shifts in cuticular hydrocarbon (nonvolatile compounds) profiles of immatures have been identified (Roux et al. 2008; Drijfhout 2010). While there are clear differences in profiles between species, life stages, male and females, and even puparia of increasing age (Ye et al. 2007; Roux et al. 2008; Thomas and Simmons 2008; Xu et al. 2014), it is unclear whether these are detectable by insects at a distance.

Thus far, only a single study is known to have examined the VOCs released by immature insects (Frederickx et al. 2012b). Twenty of the compounds released from immatures of the *Calliphora vicina* Robineau-Desvoidy have also been detected from carrion and include hydrocarbons, alcohols, acids, and acid esters, and various sulfur-containing compounds. Though the insects were washed with distilled water, they were not raised under axenic conditions; therefore, a microbiological origin of these VOCs cannot be excluded. More work is clearly needed to catalogue and investigate the metabolic origins and behavioral impact of volatiles secreted by insects and other animals that utilize carrion.

9.3.3 Microbial Metabolism and VOCs

The bacterial community found in and on living animals varies greatly and is influenced by genetics, species, and diet. Upon the death of an animal, there ensues a complex progression of time-dependent fluctuations in a highly diverse assembly of bacterial species inhabiting the carrion (see Chapter 8). To fuel their own metabolism, bacteria break down organic matter into inorganic substances. These metabolic reactions result in the production of microbial volatile organic compounds (MVOCs), many of which are used as semiochemicals. MVOC-producing bacteria can be found in all environments: aqueous, terrestrial, subterranean, and atmospheric (Stotzky and Schenck 1976; Wenke et al. 2010). Shultz and Dickschat (2007) estimated that between 50 and 80% of bacteria produce volatiles and defined almost 350 compounds from the bacteria they investigated. A single microbial species can produce an impressive array of MVOCs; 32 have been identified from *Staphylococcus aureus* and 37 from *Pseudomonas aeruginosa* (Filipiak et al. 2012).

MVOCs are low molecular weight, lipophilic, carbon-based compounds produced during primary and secondary microbial metabolic processes. Primary metabolism results in the production of vital cellular components, while compounds produced during secondary metabolism are of more ecological importance. MVOCs from secondary metabolism encompass a vast assortment of compounds (see Section 10.2), dependent on the available substrate, and the microbial species utilizing the resource (Filipiak et al. 2012). Some of the compounds function to coordinate group action (quorum sensing) that produces biofilms or defense compounds, such as the production of virulence factors or toxins that affect larger hosts (Anand and Griffiths 2003; Burkepile et al. 2004; de Kievit 2009).

Many MVOCs can be produced within minutes of death, and due to their high vapor pressure, volatilize relatively easily. Such qualities make these substances readily available for other organisms to detect and exploit for their own benefit (Stotzky and Schenck 1976; Lowery et al. 2008; Tomberlin et al. 2012; Davis et al. 2013). For example, plants such as the orchid *Satyrium pumilum* exploit this signaling interaction by mimicking the scent of carrion compounds to attract carrion flies to serve as pollinators (van der Niet et al. 2011). See Chapter 16 for additional examples of mimicking.

This interkingdom (or interdomain) communication can direct behaviors that govern attraction and repulsion of various fauna during interspecies competition for the carrion resource (Easton and Feir 1991; Sperandio et al. 2003; Rumbaugh 2004; Ma et al. 2012; Davis et al. 2013). For example, methionine is broken down by *Proteus* spp. (Hayward et al. 1977) and other bacteria (Khoga et al. 2002) to form DMDS, a chemical cue used by adult blow flies to locate resources (Frederickx et al. 2012a,b; Liu 2014). The degradation of the amino acid tryptophan by *Bacteroides*, *Lactobacillus*, *Clostridium*, *Bifidobacterium*, and *Peptostreptococcus* yields indole, among other secondary metabolites (Smith and Macfarlane 1996). Like DMDS, indole is a known attractant for necrophagous flies, including *Fannia canicularis* (L.), *Muscina stabulans* (Fallen), and *M. domestica* (L.) (Diptera: Muscidae) and is a known quorum-sensing molecule affecting biofilm formation in bacteria (Hwang et al. 1978; Wang et al. 2001;

Urech et al. 2004; Hu et al. 2010). On the other hand, enzymes produced by competing microbes on carrion can disrupt these quorum-sensing pathways, breakdown biofilms, and kill bacteria. This has been well studied in fly larvae (Yamada and Natori 1994; Bexfield et al. 2004, 2008, 2010; Zhang and Dong 2004; Gonzáles and Keshavan 2006; Jones and Wall 2008; van der Plas et al. 2008, 2010; Cazander et al. 2009; Ceřovský et al. 2011).

Such mechanisms of interkingdom communication would afford species advantages when engaged in competition for limited and ephemeral resources, and likely benefit the survival and proliferation of the responding organism (Stotzky and Schenck 1976; Vining 1990; Kai et al. 2009). A recent study shows that some species of bacteria associated with *Lucilia* spp. (Diptera: Calliphoridae) can be horizontally and trans-generationally maintained, and are also shared between flies and their environment (Singh et al. 2015). A lack of this valuable communication may lower the ability of a species to locate a resource necessary for survival and reproduction. Therefore, natural selection will act on recipients of these cues for acute perception mechanisms from a distance, as this would be beneficial for locating and utilizing associated resources that are limited in the environment.

The structure of the microbial community on carrion and the array of MVOCs generated are highly interconnected with the biotype and environmental conditions in which the carrion is decomposing (Campobasso et al. 2001; Dekeirsschieter et al. 2009). Temperature, the availability of nutrients, oxygen, humidity levels, as well as the content of fat, carbohydrates, and protein of the carrion influences the structure of the microbial community and the metabolic pathways utilized (Korpi et al. 2009). Even the growth phase of specific bacteria species and the progression of decomposition of carrion impacts the quantity and diversity of the MVOCs released (Dekeirsschieter et al. 2009; Kai et al. 2009). Additionally, the carrion species may influence the quantity and type of MVOCs produced. For example, biogenic amines (e.g., cadaverine and putrescine) are commonly associated with decomposition and correspond to microbial growth on many vertebrate carcasses (Statheropoulos et al. 2005; Dekeirsschieter et al. 2009; Paczkowski and Schütz 2011).

MVOCs may provide important cues related to temporal aspects of habitat or resource. Arthropods appear to recognize MVOC profiles that differ over the course of decomposition and use them to determine whether a resource is suitable for their needs. As decomposition progresses, the nutrients available to microbes shift and those microbes possessing the pathways necessary to utilize the changing resource remain, while those lacking this ability either disperse, become dormant, or perish. This temporal aspect to the decomposition process manifests in a fluctuating microbial community over the progression of decay, resulting in an equally dynamic MVOC release dependent on the structure of the microbial community present (Dekeirsschieter et al. 2009). On a fresh carcass, bacteria find themselves in a nutrient-rich environment and likely prioritize metabolic activity and pathways to enhance growth and facilitate reproduction (Vining 1990). However, as decay progresses and nutrients are depleted, microbes adapt to changing resource conditions by utilizing alternative metabolic pathways resulting in a different MVOC profile. For example, after depletion of oxygen from the carcass, facultative and obligatory anaerobic bacteria metabolize substrates using organic carbon fermentation pathways instead of aerobic oxygen pathways. This fermentation produces hydrogen, lactate, succinate, butyrate, and ethanol. Many anaerobes, such as *Eubacterium* spp., *Roseburia* spp., *Butyrivibrio* spp., *Faecalibacterium* spp., and *Clostridium* spp. produce butyrate from butyric acid through their fermentation of sugars and starches (Louis and Flint 2009). These bacteria are often found inhabiting the gut of animals and the soil. Both butyric acid and the various butyrate esters produced are strongly aromatic; though butyric acid is known to have an unpleasant rancid smell by human detection, the odor of butyrate esters are considered to be more pleasant (Uchida et al. 2000). Responses to these short-chain fatty acids have also been found in other species, such as a broad range of arthropods. Intracellular measurements of olfactory mosquitoes neurons measured output suggested inhibitory effects from butyric acid (Meijerink and van Loon 1999). While butyric acid at high concentrations acted as an oviposition attractant for the gravid fruit fly *Dacus tryoni* (Froggatt) (Diptera: Tephritidae) (Eisemann and Rice 1992), it caused inhibitory, as well as excitatory responses from the cabbage butterfly, *Pieris brassicae* L. (Lepidoptera: Pieridae) (Den Otter et al. 1980).

The burying beetles, *Nicrophorus vespillo* (Linnaeus) and *N. vespilloides* Herbst (Coleoptera: Silphidae), respond to sulfur-containing MVOCs commonly emitted by fresh mouse carrion (Kalinová et al. 2009). The metabolism of amino acids into sulfur-containing MVOCs can occur in several bacteria taxa inhabiting carcasses, such as species of *Brevibacterium, Corynebacterium, Micrococcus, Staphylococcus, Arthrobacter*, and lactic acid bacteria (Schulz and Dickschat 2007). The hide beetle, *Dermestes maculatus* (De Geer) (Coleoptera: Dermestidae), is most attracted to a carcass during the postbloat decomposition processes, which coincides with the production of significant amounts of benzyl butyrate (von Hoermann et al. 2011). Benzyl butyrate can be produced during anaerobic fermentation by the enteric bacterium *Clostridium butyricum* (Popoff 1984; Zhang et al. 2009a,b). MVOCs from a mix of *Enterobaceriaceae* spp. isolated from wounds infested by *Cochliomyia hominivorax* (Coquerel) (Diptera: Calliphoridae) attracted gravid females for oviposition after 48 and 72 h of growth to a greater extent than after 24 or 96 h (Chaudhury et al. 2010). In addition, the effects of secondary metabolic products of bacteria on other species, such as insects, can differ depending on the developmental stage of the recipient organism (Korpi et al. 2009). The age and nutritional status of *Lucilia sericata* (Meigen) (Diptera: Calliphoridae) influences the blow fly response to bacterial volatiles, such as isobutyl amine and phenylethyl alcohol, when detecting an oviposition site (Tomberlin et al. 2012).

Semiochemicals produced from microbial decomposition mediate mosquito (Diptera: Culicidae) oviposition (Clements 1999), while facultative ectoparasites (i.e., blow flies) are preferentially attracted to and oviposit in animal wounds colonized by *P. aeruginosa* (Kingsford and Raadsma 1997; Jackson et al. 2002). House flies avoid unsuitable resources once semiochemicals produced by fungi on the resource are detected. Lam et al. (2010) used antenna-stimulatory analysis of volatiles produced by *Phoma* and *Rhizopus* fungi to identify likely compounds and subsequently demonstrated that DMTS and 2-phenylethanol deterred oviposition by house flies. In plant–microbe–nematode interactions, parasitic nematodes vector bacteria to a plant resulting in the release of plant-induced volatiles that attract more nematodes (Horiuchi et al. 2005; Bais et al. 2008). These scenarios involve a mix of volatiles produced by bacterial species. Such olfaction in the responsive invertebrates (e.g., insects, nematodes) appears to be dependent on specific ratios of a volatile chemical working in concert in a milieu rather than a single species-specific compound (Bruce et al. 2005). To enhance attraction of a target species, blends of volatile chemical attractants must be used. At present, synthetic blends that attract blow flies are not as effective as carrion derived MVOCs (Aak et al. 2010). Such optimization of synthetic chemicals is perhaps a much more daunting task than using blends produced from naturally-occurring sources, for example, a specific bacterial or fungal species. An example is quorum-sensing molecules produced by the bacterium *P. mirabilis*, which play a significant role in regulating fly attraction and colonization of a resource, or the bacteria present on con-generic eggs that mediate fly oviposition behavior (Ma et al. 2012; Tomberlin et al. 2012).

Fungi are also decomposers of animal materials and utilize many of the same metabolic pathways as bacteria during the process of decomposition, particularly in soil (Figure 9.1). The compound 1-octen-3-ol is produced by many fungi during lipid degradation (Bennett et al. 2012) and has also been collected

FIGURE 9.1 Unknown fungal growth on human remains decomposing in the Edwards Plateau region of Texas. (Photo courtesy of the Forensic Anthropology Center at Texas State University. All rights reserved.)

from decomposing pigs (Statheropoulos et al. 2011). DMDS, another volatile associated with carrion, is produced by *Phoma*, *Rhizopus*, and *Fusarium* sp. found on chicken feces and has negative effects on oviposition activity by house flies (Lam et al. 2010). Thus, fungi produce a multitude of aromatic secondary metabolites; most of which are polyketide oligomers (Vining 1990) that have a range of bioactivities as opposed to a single function (Keller et al. 2005; Crawford et al. 2009). A review by Pelaez (2004) found more than 1500 fungal metabolites with biological activity. Those activities included antibacterial, antifungal, and a variety of insecticidal activity, such as those of the fungus *Beauveria bassiana* that causes white muscardine disease in insects (Zimmermann 2007). Thus, some fungi influence the occurrence of other microbial species and consequently the insects utilizing the carrion during the process of decomposition, although clarity of these interactions requires more research.

9.4 Roles of Volatiles

In many cases, VOCs serve as a mechanism that initiates or inhibits sequences of the arrival and colonization patterns of animals that utilize carrion as a source of nutrition for themselves, their offspring, or to locate a mate (see Chapter 10). Smell is a key sense used by many animals, including insects, to detect, locate, and utilize a resource, whether as a host, prey, or mate (Cardé and Willis 2008). As in the case of vertebrate carrion, each behavior is a potential outcome depending on the animal of interest (e.g., turkey vulture or blow fly).

The compounds released from vertebrate carrion are diverse and can be present in complex odor bouquets (LeBlanc 2008; Kasper et al. 2012). Determining which compound(s) an animal uses to locate a carrion source can be challenging, as in many cases, the response is the result of multiple compounds being present. Thus, research in this area can be painstakingly slow as an investigator will need to describe the bouquet produced by the carrion source at the time the animal demonstrates a response (e.g., physiologically or behaviorally) and then determine which components are responsible for the response. Furthermore, once the primary compounds responsible for the response have been determined, the investigator is charged with determining the concentration of each compound responsible for the behavioral response as the behavior most likely operates within a given range and exposure outside this range does not elicit a response. However, a major hurdle for investigators is to determine the relationship between concentration-dependent responses and background odors. For example, *Chrysonotomyia ruforum* Krausse (Hymenoptera: Eulophidae) is an egg parasitoid of the sawfly *Diprion pini* L. (Hymenoptera: Diprionidae), and the response of the adult to the sesquiterpene (E)-β-farnesene, which is produced in significantly greater quantities by oviposition-induced pine (previously already colonized by the sawfly), was only exhibited when presented in combination with the background odor from an egg-free pine twig (Mumm and Hilker 2005).

As previously discussed, compound concentration and mixtures influence the behavioral response of animals utilizing carrion as a resource. Compounds produced at a specific nanogram level are known to elicit electroantennogram (EAG) responses of arthropods attracted to a resource (Rains et al. 2004). Slight variations above or below this level could be the difference between the animal responding or not. In the case of carrion and the arthropods that utilize it as a resource, these VOCs and their associated concentrations serve as signals of the presence of conspecifics; however, the presence of conspecifics can be beneficial or detrimental. For example, the house fly, which can colonize vertebrate carrion, is very specific to when eggs are laid in the presence of conspecific eggs; this response is due to the VOCs emitted by microbes associated with the conspecific eggs. Individuals lay eggs in the presence of freshly laid eggs over those aged 24 h (Lam et al. 2007). Oviposition correlated with the population dynamics of the bacterium *Klebsiella oxytoca* associated with older eggs. Further investigation determined this behavior (timing of oviposition) to be highly significant for offspring survival; oviposition in the presence of eggs older than 24 h (when *K. oxytoca* counts were high) resulted in reduced survivorship of offspring, as the larvae resulting from the older eggs would consume the newly laid eggs (Lam et al. 2007).

VOCs emitted from vertebrate carrion serve many roles. However, determining these roles is a challenge as the odor plume produced can be composed of many compounds at various concentrations (Verhulst et al. 2010) and be by-products of the carrion or the metabolic processes of associated microbes. Consequently, microbial responses to change in the carrion resource have a horizontal and vertical cascade effect on higher organisms that might use them to navigate through an environment to locate the carrion source. Furthermore, VOCs that might be a by-product of one organism can afford an animal an advantage in one circumstance (e.g., avoid competition with heterospecifics—allomone), but could be detrimental in other scenarios, such as the same compound could attract a predator or parasite (kairomone).

9.4.1 Ecological Interactions

As evident through reading many of the chapters within this text, locating resources is a primary driver behind organisms, both single and multicellular, associating with vertebrate carrion. In many instances, VOCs produced by the carrion or associated animals serve as a mechanism regulating the attraction, colonization, and utilization of the carrion or associated animals present on the carrion (e.g., adult staphylinid beetles visiting carrion to consume fly larvae). The exact ecological means of these interactions are highly diverse; some animals use the carrion as food, whereas others locate mates or prey associated with the carrion. The semiochemicals driving these ecological interactions can be divided into two groups: pheromones and allelochemicals; pheromones are those that result in intraspecific interactions, while allelochemicals result in intraspecific interactions. These two groups can be further broken down depending on the effect on the organism emitting or receiving the compound. For more information on semiochemicals, see LeBlanc and Logan (2010).

9.4.1.1 *Consumption*

The most obvious use of vertebrate carrion is as a source of food (Cossé and Baker 1996; Chaudhury et al. 2002, 2010; Aak et al. 2010). As discussed in other chapters, animals ranging from microbes to insects to vertebrates consume carrion, and in many instances, insects and vertebrates detect and locate these resources through VOCs such as DMDS, phenol, and indole, at specific ratios (Chaudhury et al. 2014) or individually (Frederickx et al. 2012a,b). In the postbloat phases of decomposition, benzyl butyrate is prevalent and attracts the carrion beetle *D. maculatus* (von Hoermann et al. 2011, 2012). Most likely, these responses are related to mate location and resource quality for consumption and rearing of offspring.

Many vertebrates (e.g., hyenas, red foxes, raptors) also consume carrion (DeVault et al. 2003). In some instances, these species also rely partially or totally on VOCs to detect and locate carrion (Smith and Paselk 1986; DeVault et al. 2003). The brown treesnake, *B. irregularis*, feeds on mouse carrion. Laboratory and field tests indicate this snake detects and responds to VOCs produced by bacteria associated with mouse carrion (Jojola-Elverum et al. 2001). In this system, researchers have attempted to utilize this information to develop a synthetic lure that can be used to trap this invasive species.

Many of these scavengers feed on fresh, rather than highly decomposed vertebrate carrion (DeVault et al. 2003). Thus, longitudinal research comparing the VOC profiles associated with various stages of carrion decomposition and the level of response of vertebrates and invertebrates would be of interest for identifying the key odors responsible for their attraction and consumption of carrion as well as understanding the evolutionary underpinnings for these preferences.

9.4.1.2 *Predator–Prey and Intra-/Interspecific Interactions*

Predator–prey interactions mediated by VOCs have been examined in many systems but little is known about the role of VOCS regulating the behaviors of animals predating (or avoiding predation) on others associated with carrion. Brundage (2012) examined the role that microbes, and indirectly the associated VOCs, associated with the facultative predator *Chrysomya rufifacies* (Macquart) (Diptera: Calliphoridae) and prey *Cochliomyia macellaria* (Fabricius) (Diptera: Calliphoridae) play in regulating heterospecific and conspecific interactions. *Cochliomyia macellaria* showed little to no preference for eggs or associated microbes of either species. However, *C. rufifacies* was significantly more attracted to *C. macellaria* (prey) eggs and associated microbes than conspecifics, indicating that the VOCs could be a mechanism regulating the ability of this predator to locate appropriate resources, while also avoiding competition with conspecifics. Furthermore, she suggested that the lack of response of *C. macellaria* to the predator could be due to a lack of selection to avoid the predator or the possibility of the microbes and their associated VOCs were serving as a camouflage for the predator. Thus, the prey colonizes a resource without detecting the predator, and the predator is presented with additional resources to supplement their development if the carrion is depleted (Brundage 2012).

Flores (2013) examined the responses of adult *C. macellaria* to resources infested for 1 h with either third instars of conspecifics or *C. rufifacies*. He determined that conspecifics of *C. macellaria* were repellent to adults, while heterospecifics were attractive (e.g., aggregation and oviposition occurred in the presence of heterospecifics). While this seemed counter-intuitive, he suspected that conspecifics at that stage released VOCs that indicated the resource was poor, due to common utilization of resources, while the predatory larvae, which do not feed on eggs or first instars of *C. macellaria*, represented no danger or a reduction of resources needed for their development. These data provide some insights to intra- and interspecific interactions for the resource rather than a predator–prey interaction.

Intra- and interspecific competition is discussed in Chapter 11. However, a brief review of VOCs as a mechanism regulating these interactions is provided here. The role of VOCs in these interactions is best depicted through succession studies where assemblages of arthropod species on carrion change temporally (Barton et al. 2013; Pechal et al. 2014; Perez et al. 2014). These shifts in community structure are governed by many factors associated with carrion, including the VOCs that initially attract arthropods to the carrion source. VOCs play an important role in regulating attraction of conspecifics (e.g., silphid adults) and heterospecifics (e.g., blow flies). These resources have been instrumental in the selection of mechanisms for detecting and locating resources quickly as they play an important role in food and mate location (von Hoermann et al. 2011, 2012, 2013). However, this area of research is still in its infancy, and additional research is needed to decipher and identify the complex VOC profiles associated with carrion, how they shift over time, what is responsible for their production, and determine which VOCs are responsible for the behavioral shifts observed by the various animals attracted to, and utilizing, the carrion source. As discussed in Chapter 20 and in this chapter, microbes associated with the carrion as well as the animals consuming it produce many of these VOCs.

9.4.1.3 Parasitoid–Host Interactions

Tachinaephagus zealandicus Ashmead (Hymenoptera: Encyrtidae), a parasitoid of dipteran larvae associated with vertebrate carrion, exhibits specific preferences to VOCs associated with carrion. Attraction was reported to be greatest when both the host *Calliphora albifrontalis* Malloch (Diptera: Calliphoridae) and host resource (sheep liver) were present, in comparison to liver of differing ages which had not been fed on by host larvae, or decomposed liver that had the host larvae removed (Voss et al. 2009). Furthermore, much like the dermestid beetle discussed in Sections 9.3.3 and 9.4.1.1, attraction was stage specific, with female wasps not attracted to fresh liver (Voss et al. 2009). The same has been determined for other decomposing material, such as plant matter. Boone et al. (2008) determined that VOCs produced by microbial symbionts associated with the bark beetle *Ips pini* Say (Coleoptera: Curculionidae) were utilized by parasitic wasps (Hymenoptera: Pteromalidae) to locate the beetle larvae in decomposing trees. These represent a few examples of VOCs regulating trophic interactions, specifically between parasitoids and hosts. Obviously, additional research is needed to better understand trophic interactions, and their degree of connectedness and relevance as related to network analysis and predicting ecosystem robustness.

9.5 Collection and Analysis of VOCs

The identification and the biological activity of VOCs associated with carrion and carrion-feeding animals can be determined through a number of techniques. A typical workflow for chemical ecology research consists of collection of the volatiles from the source, followed by identification of the compounds through mass spectrometry. Next would be to identify the role of the collected chemicals, through bioassays with the organisms of interest. In some instances, especially with insects, an intermediate step would include exposing an insect antenna (see discussion of electrophysiological techniques below) to the compounds at the same time they are being identified. Doing so allows the research to simultaneously identify and determine which compounds elicit a physiological response and can help narrow down which compounds in the blend are good candidates to investigate for biological activity. For each step of the process, a variety of techniques, equipment, and supplies are available to conduct the research. In this section, an overview of the methods and supplies that have been used to study the chemical ecology of decomposition are reviewed. For a comprehensive review of chemical ecology methods, see Millar and Haynes (1998).

9.5.1 Sampling Methods

9.5.1.1 Headspace Analysis

A chamber for trapping the volatiles produced by the source of interest and preventing contamination of the air being sampled is an essential part of the volatile collection system (Agelopoulos et al. 1999). The air within the chamber (known as headspace) is sampled by drawing air through a sorbent trap inserted into a port in the chamber. The collection of volatiles from this trapped air is known as headspace analysis and is a well-established method in decomposition research. LeBlanc (2008) collected volatiles from decomposing swine carcasses (*Sus scrofa* L.), with two types of air-entrainment devices. The first was a mylar thermal blanket supported with arched stainless steel wires; the edges of the blanket were held down to the ground with sand bags to minimize the collection of air from outside the "chamber." This set up was difficult to maintain because the blanket was not heavy enough to withstand the windy conditions that occurred during the experiment. A rectangular stainless steel chamber ($1.5 \times 1.0 \times 1.0$ m) was therefore constructed. The weight of the chamber allowed for an adequate seal with the ground and minimized contamination from external air. Forbes and Perrault (2014) also used stainless steel chambers ($1.3 \times 0.9 \times 0.6$ m) to collect volatiles from decomposing swine carcasses. An example of a similar chamber is shown in Figure 9.2. Statheropoulos et al. (2011) sampled decomposing pigs that were enclosed in polyethylene body bags, and von Hoermann et al. (2011) enclosed piglets in commercially available plastic oven bags. Regardless of the type of chamber used or precautions taken to avoid contamination from outside air, blank samples should always be collected without the carrion present to determine whether any volatiles are associated with the collection system or environment.

FIGURE 9.2 Researchers in the field standing in front of a stainless steel air entrainment device ($220 \times 91 \times 62$ cm) used to sample VOCs above carrion. (Photo by J.L. Pechal. All rights reserved.)

9.5.1.2 Sorbent Traps

A number of materials on which to collect volatiles are available and can be broken into two main groups: those requiring a solvent for recovery of the trapped volatiles (solvent desorption [SD]), and those that rely on rapid heating (thermal desorption [TD]) (Harper, 2000). Currently, there is no gold standard for trap choice. The choice should be based on the types of volatiles being collected and the method of sample introduction into the instrument analyzing the volatiles (e.g., gas chromatography–mass spectrometry [GC–MS], gas chromatography/time-of-flight–mass spectrometry [GC/TOF-MS], etc.).

LeBlanc (2008) utilized traps of both desorption methods to collect volatiles emitted by decomposing swine carcasses. Tenax® tubes (TD) were used to take short-term samples (30 min) during the day, and Porapak™ (SD) tubes were used to collect samples for four hours during the night. Further, Hayesep® Q (SD) has also been used to collect volatiles from swine carcasses (A.M. Bucci, personal communication). Trapping volatiles with more than one sorbent material inside the tube is also possible. Forbes and Perrault (2014) used tubes containing both Tenax® TA and Carbograph 5TD™, both of which are analyzed by TD. Tubes containing Tenax® TA and Carbopack™ X (also TD) were used by Statheropoulos et al. (2011) to collect volatiles from swine carcasses; von Hoermann et al. (2011) used tubes containing Porapak Q, followed by SD. Solid-phase microextraction (SPME) fibers (TD) were used by Hoffman et al. (2009) to collect volatiles from various human tissue types and by Kalinová et al. (2009) to collect volatiles from freshly euthanized mice. Numerous other sorbent materials are available for trapping volatiles; see Harper (2000) for a review of available sorbents and the types of compounds trapped by each.

The air flow rate and duration of sample collection also vary across studies, from 100 mL/min for 10 min (Statheropoulos et al. 2011; Forbes and Perrault 2014), 100 mL/min for 4 h (von Hoermann et al. 2011), to 1400 mL/min for 30 min or 4 h (LeBlanc 2008). Although each of the above studies used GC–MS to identify compounds collected in headspace samples, numerous other methods and instruments are available for identifying compounds (see Millar and Haynes 1998 for a review of instruments and techniques).

9.5.1.3 Electrophysiological Methods

Hundreds of compounds make up the odor plume associated with carrion (Statheropoulos et al. 2011; Forbes and Perrault 2014), but identification on a GC–MS or other instrument does not indicate that an isolated VOC is detectable by an organism. The EAG technique was developed to measure electrophysiological responses of insect antennae to VOCs (Schneider 1957). In this technique, an excised insect antenna is connected to electrodes, a voltage amplifier, and oscilloscope, and when exposed to a physiologically active VOC, results in a voltage deflection visible on the oscilloscope (Bjostad 1998). Moorhouse et al. (1969) greatly increased the power of the EAG technique by coupling with a GC system, and using the EAG as a detector (conveniently known as electroantennogram detection [EAD]). In the GC-EAD system, the volatiles going into the GC column are split, such that the sample reaches the GC detector and the insect antenna at the same time, simultaneously recording a chromatogram and EAG response. LeBlanc (2008) conducted GC-EAD analyses on volatiles extracted from swine carrion and found four compounds that elicited a response in the blow fly *Calliphora vomitoria* L.: propyl butyrate, DMDS, DMTS, and dimethyl tetrasulfide. GC-EAG studies on *D. maculatus* revealed 18 compounds that elicited an electrophysiological response, 13 of which could be identified by mass spectrometry: ethyl butyrate, isoamyl butyrate, isobutyl hexanoate, hexyl butyrate, acetic acid, 1-octen-3-ol, proprionic acid, isobutyric acid, butyric acid, hexanoic acid, benzyl butyrate, 2-phenylethanol, and 2-phenylethyl butyrate. Kalinová et al. (2009) used GC-EAD to determine whether VOCs collected from dead mice were attractive to the beetles *N. vespilloides* and *N. vespillo*; the beetles responded to methanethiol, methyl thiolacetate, dimethyl sulfide, DMDS, and DMTS. Although this technique is quite sophisticated and allows for faster screening of compounds from headspace that the insects respond to, an EAG response is not a measure of the type of behavioral response elicited by detection of the compound.

9.5.2 Bioassays

The odor plume associated with a decomposing animal contains a mixture of many chemical compounds (Statheropoulos et al. 2011; Forbes and Perrault 2014), all of which do not necessarily have a biological function for the animals that consume carrion. Therefore, further testing is needed to more confidently determine which compounds in the complex blend are biologically relevant. Although many bioassays for chemical ecology research have been developed, few have been used to study the behavior of animals that feed on carrion. For a review of bioassay methods, see Haynes and Millar (1998).

9.5.2.1 Bioassays with Insects

Behavior bioassays involving insects can be broken into two general classes, those using moving air, and those using still air (Baker and Cardé 1984), both have been used to study responses of insects to volatiles emitted from decomposing organic matter. These types of studies are conducted in devices known as olfactometers and wind tunnels (see Chapter 10 for a detailed discussion).

Depending on design, olfactometers used in moving-air bioassays can allow for the testing of one or multiple compounds. A dual-choice (Y-tube) olfactometer (Figure 9.3) has been used to test the attractiveness of freshly thawed and 24-hour-old beef liver to the blow flies *C. macellaria* and *C. rufifacies* (Brundage, 2012). A Y-tube olfactometer was also used by Kalinová et al. (2009) to test attractiveness of compounds identified as physiologically active via GC-EAD to the beetles *N. vespilloides* and *N. vespillo*. Although DMDS elicited an EAG response, the compound was not attractive to *N. vespilloides*; adults oriented upwind, but did not travel toward the source. In these types of olfactometers, organisms are presented with two odor sources, and once released, the organism moves upwind (usually via walking) and chooses a branch to enter (Hare, 1998). Von Hoermann et al. (2011) tested attraction of *D. maculatus* to volatiles isolated from pig carrion in a double-Y olfactometer.

A wind tunnel (see Figure 9.4, for example) was used to test the attractiveness of volatiles isolated from pig manure, many of which have also been isolated from carrion, to house flies (Cossé and Baker 1996). In this device, air flows across the compound of interest, and if attracted, the organism travels upwind (usually via flight) towards the source (Hare 1998). Although house flies were attracted to each of the 10 compounds tested, none elicited the same level of response as seen with pig manure, the positive control (Cossé and Baker 1996).

A cube olfactometer has been used in conducting still-air bioassays of attraction in carrion-feeding insects (see Ma et al. 2012 for a description of the olfactometer). Flores (2013) used this system to study the nonconsumptive effects previously discussed in Section 9.4.1.2 (see Chapter 20 for an image).

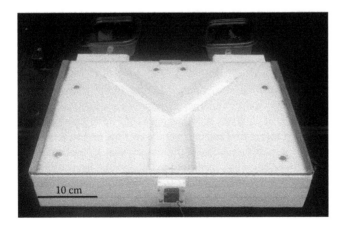

FIGURE 9.3 A dual-choice (Y-tube) olfactometer used for testing attraction or repellency (via walking) of insects to VOCs. (Photo by J.A. Cammack. All rights reserved.)

FIGURE 9.4 An example of a wind tunnel used for testing attraction of flying insects to VOCs, intake (a) exhaust (b) sides. (Photo by J.K. Tomberlin. All rights reserved.)

He released adults of *C. macellaria* into the cube and recorded attraction to different odor sources; all paired combinations of a blank, beef liver, third instars of *C. rufifacies* feeding on beef liver, and third instars of *C. macellaria* were tested. The attraction of *L. sericata* Meigen to volatiles produced by the bacteria *P. mirabilis* has also been tested in a cube olfactometer (Ma et al. 2012; Tomberlin et al. 2012).

9.5.2.2 Bioassays with Vertebrates

Studies to evaluate the role in olfaction in carrion location by turkey vultures have been conducted by obscuring the view of the carrion source, such as covering white-tailed deer (*Odocoileus virginianus* Zimmerman) with hay (McShea et al. 2000) or woodchuck (*Marmota monax* L.) with soil (Smith et al. 2002). Although the birds were able to find and uncover the carrion, the attractant was not identified. The olfactory sensitivity of *C. aura* to different concentrations of volatiles associated with carrion (ethanethiol, methylamine, and butanoic acid) was measured by monitoring heart rate via electrocardiography (Smith and Paselk 1986). An enclosure (0.63 × 0.33 × 0.30 m) was placed over the birds, and the volatiles introduced through a port near the head. Changes in heart rate in response to exposure to odorants have been recorded in numerous bird species (Wenzel and Sieck 1972).

Snakes do not typically consume carrion in the wild; however, in Guam, brown tree snakes, are attracted to decomposing mice and quartered mice at the same rate as to live mice (Shivik and Clark 1997). Laboratory studies on the behavior of brown tree snakes have indicated that directional probing of the head is a good indicator of attraction to olfactory cues (Shivik 1998). However, baiting in the field with compounds emitted from carrion (e.g., cadaverine, dimethylamine, DMDS, dimethyl sulfide, ethanethiol, trimethylamine, and putrescine), either singly or as blends, was not as efficient at attracting snakes as were live mice (Shivik and Clark 1999). This difficulty in identifying effective chemical baits for attraction of these snakes illustrates the importance of a systematic research approach to identification of bioactive volatiles in particular species, regardless of the system of study or goals of the research.

9.6 Conclusion

This limited view of chemical ecology is far from exhaustive, as this field encompasses all of the chemicals that mediate interactions between organisms, both living and dead (Eisner and Meinwald 1995). Entire books have been written about pheromones mediating sexual interactions, chemicals of biological origin affecting marine ecosystems, and chemicals that have been important to the evolution of species within specific niches (Wyatt 2003; McClintock and Baker 2001; Wajnberg and Colazza 2013). Chemical ecology can be studied on nearly any scale, from genetic pathways leading to chemical production, to ecosystem-level effects of specific chemical compounds or blends (Dicke and Takken 2006). Initial work in vertebrate decomposition has begun and the understanding of the VOCs that mediate and affect interactions between species associated with carrion will give a solid foundation for the abundant and fertile research that will inevitably follow.

REFERENCES

Aak, A., G.K. Knudsen, and A. Soleng. 2010. Wind tunnel behavioural response and field trapping of the blowfly *Calliphora vicina*. *Medical and Veterinary Entomology* 24: 250–257.

Agelopoulos, N.G., A.M. Hooper, S.P. Maniar, J.A. Pickett, and L.J. Wadhams. 1999. A novel approach for isolation of volatile chemicals released by individual leaves of a plant in situ. *Journal of Chemical Ecology* 25: 1411–1425.

Aksnes, A. 1988. Location of enzymes responsible for autolysis in bulk-stored capelin (*Mallotus villosus*). *Journal of the Science of Food and Agriculture* 44: 263–271.

Anand, S.K. and M.W. Griffiths. 2003. Quorum sensing and expression of virulence in *Escherichia coli* O157:H7. *International Journal of Food Microbiology* 85: 1–9.

Anderson, P.M., M.A. Broderius, K.C. Fong, K.N. Tsui, S.F. Chew, and Y.K. Ip. 2002. Glutamine synthetase expression in liver, muscle, stomach and intestine of *Bostrichthys sinensis* in response to exposure to a high exogenous ammonia concentration. *Journal of Experimental Biology* 205: 2053–2065.

Bais, H.P., C.D. Broeckling, and J.M. Vivanco. 2008. Root exudates modulate plant-microbe interactions in the rhizosphere. In: Karlovsky, P. (ed.), *Secondary Metabolites in Soil Ecology*, Vol. 14. Berlin, Heidelberg: Springer, pp. 241–252.

Baker, T.C. and R.T. Cardé. 1984. Techniques for behavioral bioassays. In: Hummel, H.E. and T.A. Miller (eds), *Techniques in Pheremone Research*. New York: Springer-Verlag, pp. 45–73.

Bang, B.G. and S. Cobb. 1968. The size of the olfactory bulb in 108 species of birds. *The Auk* 85: 55–61.

Barton, P.S., S.A. Cunningham, B.C.T. Macdonald, S. McIntyre, D.B. Lindenmayer, and A.D. Manning. 2013. Species traits predict assemblage dynamics at ephemeral resource patches created by carrion. *PLoS ONE* 8: e53961.

Bartell, R.J., H.H. Shorey, and L. Barton Browne. 1969. Pheromonal stimulation of the sexual activity of males of the sheep blowfly *Lucilia cuprina* (Calliphoridae) by the female. *Animal Behaviour* 17: 576–585.

Beatty, S.A. and V.K. Collins. 1939. Studies of fish spoilage: VI. The breakdown of carbohydrates, proteins and amino acids during spoilage of cod muscle press juice. *Journal of the Fisheries Board of Canada* 4: 412–423.

Bell, W.J. and R.T. Cardé. 1984. *Chemical Ecology of Insects*. London, UK: Chapman & Hall.

Bennett, J.W., R. Hung, S. Lee, and S. Padhi. 2012. Fungal and bacterial volatile organic compounds: An overview and their role as ecological signaling agents. In: Esser, K. (ed.), *The Mycota*, Vol. 9. Berlin: Springer-Verlag, pp. 373–393.

Bentley, R. 2006. The nose as a stereochemist. Enantiomers and odor. *Chemical Reviews* 106: 4099–4112.

Bernhard, A. 2012. The nitrogen cycle: Processes, players, and human impact. *Nature Education Knowledge* 3: 25.

Bexfield, A., A.E. Bond, C. Morgan, J. Wagstaff, R.P. Newton, N.A. Ratcliffe, E. Dudley, and Y. Nigam. 2010. Amino acid derivatives from *Lucilia sericata* excretions/secretions may contribute to the beneficial effects of maggot therapy via increased angiogenesis. *The British Journal of Dermatology* 162: 554–562.

Bexfield, A., A.E. Bond, E.C. Roberts, E. Dudley, Y. Nigam, S. Thomas, R.P. Newton, and N.A. Ratcliffe. 2008. The antibacterial activity against MRSA strains and other bacteria of a <500 Da fraction from maggot excretions/secretions of *Lucilia sericata* (Diptera: Calliphoridae). *Microbes and Infection* 10: 325–333.

Bexfield, A., Y. Nigam, S. Thomas, and N.A. Ratcliffe. 2004. Detection and partial characterisation of two antibacterial factors from the excretions/secretions of the medicinal maggot *Lucilia sericata* and their activity against methicillin-resistant *Staphylococcus aureus* (MRSA). *Microbes and Infection* 6: 1297–1304.

Boone, C.K., D.L. Six, Y. Zheng, and K.F. Raffa. 2008. Parasitoids and dipteran predators exploit volatiles from microbial symbionts to locate bark beetles. *Environmental Entomology* 37: 150–161.

Bjostad, L.B. 1998. Electrophysiological methods. In: Millar, J. and K. Haynes (eds), *Methods in Chemical Ecology, Vol. 1. Chemical Methods*. New York: Springer Science+Business Media.

Braack, L.E.O. 1987. Community dynamics of carrion-attendant arthropods in tropical African woodland. *Oecologia* 72: 402–409.

Bruce, T.J.A., L.J. Wadhams, and C.M. Woodcock. 2005. Insect host location: A volatile situation. *Trends in Plant Science* 10: 269–274.

Brundage, A.L. 2012. Fitness effects of colonization time of *Chrysomya rufifacies* and *Cochliomyia macellaria* and their response to intra- and inter-specific eggs and egg-associated microbes. PhD Dissertation, Texas A&M University, College Station, TX.

Brundage, A.L., M.E. Benbow, and J.K. Tomberlin. 2014. Priority effects on the life-history traits of two carrion blow fly (Diptera, Calliphoridae) species. *Ecological Entomology* 39: 539–547.

Buckley, N.J. 1996. Food finding and the influence of information, local enhancement, and communal roosting on foraging success of North American vultures. *The Auk* 113: 473–488.

Bullard, R.W., T.J. Leiker, J.E. Peterson, and S.R. Kilburn. 1978. Volatile components of fermented egg, an animal attractant and repellent. *Journal of Agricultural and Food Chemistry* 26: 155–159.

Burkepile, D.E., J.D. Parker, C.B. Woodson, H.J. Mills, J. Kubanek, P.A. Sobecky, and M.E. Hay. 2004. Chemically mediated competition between microbes and animals: Microbes as consumers in food webs. *Ecology* 87: 2821–2831.

Campobasso, C.P., G.D. Vella, and F. Introna. 2001. Factors affecting decomposition and Diptera colonization. *Forensic Science International* 120: 18–27.

Cardé, R.T. and M.A. Willis. 2008. Navigational strategies used by insects to find distant, wind-borne sources of odor. *Journal of Chemical Ecology* 34: 854–866.

Carter, D.O., D. Yellowlees, and M. Tibbett. 2007. Cadaver decomposition in terrestrial ecosystems. *Naturwissenschaften* 94: 12–24.

Cazander, G., K.E. van Veen, A.T. Bernards, and G.N. Jukema. 2009. Do maggots have an influence on bacterial growth? A study on the susceptibility of strains of six different bacterial species to maggots of *Lucilia sericata* and their excretions/secretions. *Journal of Tissue Viability* 18: 80–87.

Ceřovský, V., J. Slaninová, V. Fučik, L. Monincová, L. Bednárová, P. Maloň, and J. Stokrová. 2011. Lucifensin, a novel insect defensin of medicinal maggots: Synthesis and structural study. *ChemBioChem* 12: 1352–1361.

Chaudhury, M.F., S.R. Skoda, A. Sagel, and J.B. Welch. 2010. Volatiles emitted from eight would-isolated bacteria differentially attracted gravid screwworms (Diptera: Calliphoridae) to oviposit. *Journal of Medical Entomology* 47: 349–354.

Chaudhury, M.F., J.B. Welch, and L.A. Alvarez. 2002. Response of fertile and sterile screwworm (Diptera: Calliphoridae) flies to bovine blood inoculated with bacteria originating from screwworm infested animal wounds. *Journal of Medical Entomology* 39: 130–134.

Chaudhury, M.F., J.J. Zhu, A. Sagel, H. Chen, and S.R. Skoda. 2014. Volatiles from waste larval rearing media attract gravid screwworm flies (Diptera: Calliphoridae) to oviposit. *Journal of Medical Entomology* 51: 591–595.

Clayden, J. 2001. *Organic Chemistry*. Oxford, New York: Oxford University Press.

Clements, A. 1999. *The Biology of Mosquitoes, Vol. 2: Sensory Reception and Behaviour*. Wallingford, UK: CAB International.

Cocking, E.C. 2003. Endophytic colonization of roots by nitrogen-fixing bacteria. *Plant and Soil* 252: 169–175.

Cossé, A.A. and T.C. Baker. 1996. House flies and pig manure volatiles: Wind tunnel behavioral studies and electrophysiological evaluations. *Journal of Agricultural Entomology* 13: 301–317.

Crawford, J.M., T.P. Korman, J.W. Labonte, A.L. Vagstad, E.A. Hill, O. Kamari-Bidkorpeh, S.-C. Tsai, and C.A. Townsend. 2009. Structural basis for biosynthetic programming of fungal aromatic polyketide cyclization. *Nature* 461: 1139–1144.

Cross, A.F. and W.H. Schlesinger. 1995. A literature review and evaluation of the Hedley fractionation: Applications to the biogeochemical cycle of soil phosphorus in natural ecosystems. *Geoderma* 64: 197–214.

Davis, T.S., T.C. Crippen, R.W. Hofstetter, and J.K. Tomberlin. 2013. Microbial volatiles as arthropod semio-chemicals. *Journal of Chemical Ecology* 39: 840–859.

DeGreeff, L. and K. Furton. 2011. Collection and identification of human remains volatiles by non-contact, dynamic airflow sampling and SPME-GC/MS using various sorbent materials. *Analytical and Bioanalytical Chemistry* 401: 1295–1307.

Dekeirsschieter, J., F.J. Verheggen, M. Gohy, F. Hubrecht, L. Bourguignon, G. Lognay, and E. Haubruge. 2009. Cadaveric volatile organic compounds released by decaying pig carcasses (*Sus domesticus* L.) in different biotopes. *Forensic Science International* 189: 46–53.

de Kievit, T.R. 2009. Quorum sensing in *Pseudomonas aeruginosa* biofilms. *Environmental Microbiology* 11: 279–288.

Den Otter, C.J., M. Behan, and F.W. Maes. 1980. Single cell responses in female *Pieris brassicae* (Lepidoptera: Pieridae). *Journal of Insect Physiology* 26: 465–472.

DeVault, T.L., O.E. Rhodes Jr, and J.A. Shivik. 2003. Scavenging by vertebrates: Behavioral, ecological, and evolutionary perspectives on an important energy transfer pathway in terrestrial ecosystems. *Oikos* 102: 225–234.

Dicke, M. and W. Takken. 2006. *Chemical Ecology. From Genes to Ecosystem*, Vol. 16. Dordrecht, the Netherlands: Springer.

Drijfhout, F.P. 2010. Cuticular hydrocarbons: A new tool in forensic entomology? In: Amendt, J., M.L. Goff, C.P. Campobasso, and M. Grassberger (eds), *Current Concepts in Forensic Entomology*. Dordrecht, the Netherlands: Springer, pp. 179–203.

Dubilier, N., C. Bergin, and C. Lott. 2008. Symbiotic diversity in marine animals: The art of harnessing chemosynthesis. *Nature Reviews. Microbiology* 6: 725–740.

Early, M. and L.M. Goff. 1986. Arthropod succession patterns in exposed carrion on the island of O'ahu, Hawaiian Islands, USA. *Journal of Medical Entomology* 32: 520–531.

Easton, C. and D. Feir. 1991. Factors affecting the oviposition of *Phaenicia sericata* (Meigen) (Diptera: Calliphoridae). *Journal of the Kansas Entomological Society* 64: 287–294.

Ebbing, D. and S.D. Gammon. 2012. *General Chemistry*, 10th edition. Kentucky: Cengage Learning.

Eisemann, C.H. and M.J. Rice. 1992. Attractants for the gravid Queensland fruit fly *Dacus tryoni. Entomologia Experimentalis et Applicata* 62: 125–130.

Eisner, T. and Meinwald, J. 1995. *Chemical Ecology: The Chemistry of Biotic Interaction*, Vol. 92. Washington, D.C.: National Academy Press.

Emmens, R.L. and M.D. Murray. 1982. The role of bacterial odours in oviposition by *Lucilia cuprina* (Wiedemann) (Diptera: Calliphoridae), the Australian sheep blowfly. *Bulletin of Entomological Research* 72: 367–375.

Erdmann, G.R. 1987. Antibacterial action of myiasis-causing flies. *Parasitology Today* 3: 214–216.

Filipiak, W., A. Sponring, M.M. Baur, A. Filipiak, C. Ager, H. Wiesenhofer, M. Nagl, J. Troppmair, and A. Amann. 2012. Molecular analysis of volatile metabolites released specifically by *Staphylococcus aureus* and *Pseudomonas aeruginosa. BMC Microbiology* 12: 113–129.

Flores, M. 2013. Life-history traits of *Chrysomya rufifacies* (Macquart) (Diptera: Calliphoridae) and its associated non-consumptive effects on *Cochliomyia macellaria* (Fabricius) (Diptera: Calliphoridae) behavior and development. PhD Dissertation, Texas A&M University, College Station, TX.

Forbes, S.L. and K.A. Perrault. 2014. Decomposition odour profiling in the air and soil surrounding vertebrate carrion. *PLoS ONE* 9: e95107.

Frederickx, C., J. Dekeirsschieter, Y. Brostaux, J.P. Wathelet, F.J. Verheggen, and E. Haubruge. 2012b. Volatile organic compounds released by blowfly larvae and pupae: New perspectives in forensic entomology. *Forensic Science International* 219: 215–220.

Frederickx, C., J. Dekeirsschieter, F.J. Verheggen, and E. Haubruge. 2012a. Responses of *Lucilia sericata* Meigen (Diptera: Calliphoridae) to cadaveric volatile organic compounds. *Journal of Forensic Sciences* 57: 386–390.

Gonzáles, J.E. and N.D. Keshavan. 2006. Messing with bacterial quorum sensing. *Microbiology and Molecular Biology Reviews* 70: 859–875.

Gram, L. and H.H. Huss. 1996. Microbiological spoilage of fish and fish products. *International Journal of Food Microbiology* 33: 121–137.

Hansen, L.T., T. Gill, S.D. Røntved, and H.H. Huss. 1996. Importance of autolysis and microbiological activity on quality of cold-smoked salmon. *Food Research International* 29: 181–188.

Hare, J.D. 1998. Bioassay methods with terrestrial invertebrates.. In: Haynes, K. and J. Millar (eds), *Methods in Chemical Ecology, Vol. 2. Bioassay Methods*. New York: Springer Science+Business Media, pp. 212–270.

Harper, M. 2000. Sorbent trapping of volatile organic compounds from air. *Journal of Chromatography A* 885: 129–151.

Haynes, K. and J. Millar (eds). 1998. *Methods in Chemical Ecology, Vol. 2. Bioassay Methods*. New York: Springer Science+Business Media, 406 pp.

Hayward, N.J., T.H. Jeavons, and A.J. Nicholson. 1977. Methyl mercaptan and dimethyl disulfide production from methionine by *Proteus* species detected by head-space gas–liquid chromatography. *Journal of Clinical Microbiology* 6: 187–194.

Herbert, R.A. and J.M. Shewan. 1976. Roles played by bacterial and autolytic enzymes in the production of volatile sulphides in spoiling North Sea cod (*Gadus morhua*). *Journal of the Science of Food and Agriculture* 27: 89–94.

Hoffman, E.M., A.M. Curran, N. Dulgerian, R.A. Stockham, and B.A. Eckenrode. 2009. Characterization of the volatile organic compounds present in the headspace of decomposing human remains. *Forensic Science International* 186: 6–13.

Hofmann, A.W. 1856. On insolinic acid. *Proceedings of the Royal Society of London* 8: 1–3.

Horiuchi, J.-I., B. Prithiviraj, H.P. Bais, B.A. Kimball, and J.M. Vivanco. 2005. Soil nematodes mediate positive interactions between legume plants and rhizobium bacteria. *Planta* 222: 848–857.

Hu, M., C. Zhang, Y. Mu, Q. Shen, and Y. Feng. 2010. Indole affects biofilm formation in bacteria. *Indian Journal of Microbiology* 50: 362–368.

Hwang, Y.-S., M.S. Mulla, and H.A. Axelrod. 1978. Attractants of synanthropic flies: Ethanol as attractant for *Fannia cannicularis* and other pest flies in poultry ranches. *Journal of Chemical Ecology* 4: 463–470.

Jackson, T.A., J.F. Pearson, S.D. Young, J. Armstrong, and M. O'Callaghan. 2002. Abundance and distribution of microbial populations in sheep fleece. *New Zealand Journal of Agricultural Research* 45: 49–55.

Jacob, C., G.I. Giles, N.M. Giles, and H. Sies. 2003. Sulfur and selenium: The role of oxidation state in protein structure and function. *Angewandte Chemie International Edition* 42: 4742–4758.

Janzen, D.H. 1977. Why fruits rot, seeds mold, and meat spoils. *The American Naturalist* 111: 691–713.

Jojola-Elverum, S.M., J.A. Shivik, and L. Clark. 2001. Importance of bacterial decomposition and carrion substrate to foraging brown treesnakes. *Journal of Chemical Ecology* 27: 1315–1331.

Jones, G. and R. Wall. 2008. Maggot-therapy in veterinary medicine. *Research in Veterinary Science* 85: 394–398.

Jørgensen, B.B. 1977. The sulfur cycle of a coastal marine sediment (Limfjorden, Denmark). *Limnology and Oceanography* 22: 814–832.

Kai, M., M. Haustein, F. Molina, A. Petri, B. Scholz, and B. Piechulla. 2009. Bacterial volatiles and their action potential. *Applied Microbiology and Biotechnology* 81: 1001–1012.

Kalinová, B., H. Podskalská, J. Růžička, and M. Hoskovec. 2009. Irresistible bouquet of death—How are burying beetles (Coleoptera: Silphidae: Nicrophorus) attracted by carcasses. *Naturwissenschaften* 96: 889–899.

Kasper, J., R. Mumm, and J. Ruther. 2012. The composition of carcass volatile profiles in relation to storage time and climate conditions. *Forensic Science International* 223: 64–71.

Keller, N.P., G. Turner, and J.W. Bennett. 2005. Fungal secondary metabolism—From biochemistry to genomics. *Nature Reviews—Microbiology* 3: 937–947.

Kellogg, W.W., R.D. Cadle, E.R. Allen, A.L. Lazrus, and E.A. Martell. 1972. The sulfur cycle. *Science* 175: 587–596.

Khoga, J.M., E. Tóth, K. Márialigeti, and J. Borossay. 2002. Fly-attracting volatiles produced by *Rhodococcus fascians* and *Mycobacterium aurum* isolated from myiatic lesions of sheep. *Journal of Microbiological Methods* 48: 281–287.

Kingsford, N.M. and H.W. Raadsma. 1997. The occurrence of *Pseudomonas aeruginosa* in fleece washings from sheep affected and unaffected with fleece rot. *Veterinary Microbiology* 54: 275–285.

Korpi, A., J. Järnberg, and A.-L. Pasanen. 2009. Microbial volatile organic compounds. *Critical Reviews in Toxicology* 39: 139–193.

Lam, K., D. Babor, B. Duthie, E.-M. Babor, M. Moore, and G. Gries. 2007. Proliferating bacterial symbionts on house fly eggs affect oviposition behaviour of adult flies. *Animal Behaviour* 74: 81–92.

Lam, K., M. Tsang, A. Labrie, R. Gries, and G. Gries. 2010. Semiochemical-mediated oviposition avoidance by female house flies, *Musca domestica*, on animal feces colonized with harmful fungi. *Journal of Chemical Ecology* 36: 141–147.

Landolt, P.J. 1997. Sex attractant and aggregation pheromones of male phytophagous insects. *American Entomologist* 43: 12–22.

Langer, J. and H. Rodhe. 1991. A global three-dimensional model of the tropospheric sulfur cycle. *Journal of Atmospheric Chemistry* 13: 225–263.

LaRosa, J.C., R.I. Levy, H.G. Windmueller, and D.S. Fredrickson. 1972. Comparison of the triglyceride lipase of liver, adipose tissue, and postheparin plasma. *Journal of Lipid Research* 13: 356–363.

LeBlanc, H.N. 2008. Olfactory stimuli associated with the different stages of vertebrate decomposition and their role in the attraction of the blowfly *Calliphora vomitoria* (Diptera: Calliphoridae) to carcasses. PhD Dissertation, The University of Derby, Derby, UK.

LeBlanc, H.N. and J.G. Logan. 2010. Exploiting insect olfaction in forensic entomology. In: Amendt, J., M.L. Goff, C.P. Campobasso, and M. Grassberger (eds), *Current Concepts in Forensic Entomology*. The Netherlands: Springer, pp. 205–221.

Lill, R. and U. Mühlenhoff. 2005. Iron–sulfur–protein biogenesis in eukaryotes. *Trends in Biochemical Sciences* 30: 133–141.

Liu, W. 2014. Chemical and nutritional ecology of *Lucilia sericata* (Meigen) (Diptera: Calliphoridae) as related to volatile organic compounds and associated essential amino acids. PhD Dissertation, Texas A&M University, College Station, TX.

Liu, S.-Q., R. Holland, and V.L. Crow. 2004. Esters and their biosynthesis in fermented dairy products: A review. *International Dairy Journal* 14: 923–945.

Löfqvist, J. 1976. Formic acid and saturated hydrocarbons as alarm pheromones for the ant *Formica rufa*. *Journal of Insect Physiology* 22: 1331–1346.

Louis, P. and H.J. Flint. 2009. Diversity, metabolism and microbial ecology of butyrate-producing bacteria from the human large intestine. *FEMS Microbiology Letters* 294: 1–8.

Lowery, C.A., T.J. Dickerson, and K.D. Janda. 2008. Interspecies and interkingdom communication mediated by bacterial quorum sensing. *Chemical Society Reviews* 37: 1337–1346.

Ma, Q., A. Fonseca, W. Liu, A.T. Fields, M.L. Pimsler, A.F. Spindola, A.M. Tarone, T.L. Crippen, J.K. Tomberlin, and T.K. Wood. 2012. *Proteus mirabilis* interkingdom swarming signals attract blow flies. *The International Society of Microbial Ecology Journal* 6: 1356–1366.

Mackley, J.W. and A.B. Broce. 1981. Evidence of a female sex recognition pheromone in the screwworm fly. *Environmental Entomology* 10: 406–408.

Matlack, M.B. and I.W. Tucker. 1940. An esterase from muscular tissue. *The Journal of Biological Chemistry* 132: 663–667.

Mayer, M.S. and J.R. McLaughlin. 1990. *Handbook of Insect Pheromones and Sex Attractants*. Boca Raton, FL: CRC Press.

McClintock, J.B. and B.J. Baker. 2001. *Marine Chemical Ecology*. Boca Raton, FL: CRC Press.

McShea, W.J., E.G. Reese, T.W. Small, and P.J. Weldon. 2000. An experiment on the ability of free-ranging turkey vultures (*Cathartes aura*) to locate carrion by chemical cues. *Chemoecology* 10: 49–50.

Meijerink, J. and J.J.A. van Loon. 1999. Sensitivities of antennal olfactory neurons of the malaria mosquito, *Anopheles gambiae*, to carboxylic acids. *Journal of Insect Physiology* 45: 365–373.

Meve, U. and S. Liede. 1994. Floral biology and pollination in stapeliads—New results and a literature review. *Plant Systematics and Evolution* 192: 99–116.

Meyer, B. 1976. Elemental sulfur. *Chemical Reviews* 76: 367–388.

Mildvan, A.S. 1979. The role of metals in enzyme-catalyzed substitutions at each of the phosphorus atoms in ATP. *Advances in Enzymology and Related Areas of Molecular Biology* 49: 103–126.

Millar, J. and K. Haynes (eds). 1998. *Methods in Chemical Ecology, Vol. 1. Chemical Methods*. New York: Springer Science+ Business Media, 390 pp.

Montagnes, D.J.S., J.A. Berges, P.J. Harrison, and F.J.R. Taylor. 1994. Estimating carbon, nitrogen, protein, and chlorophyll *a* from volume in marine phytoplankton. *Limnology and Oceanography* 39: 1044–1060.

Moorhouse, J.E., R. Yeadon, P.S. Beevor, and B.F. Nesbitt. 1969. Method for use in studies of insect chemical communication. *Nature* 223: 1174–1175.

Morris, M.C., A.D. Woolhouse, B. Rabel, and M.A. Joyce. 1998. Orientation stimulants from substances attractive to *Lucilia cuprina* (Diptera, Calliphoridae). *Australian Journal of Experimental Agriculture* 38: 461–468.

Moss, G.P., P.A.S. Smith, and D. Tavernier. 1995. Glossary of class names of organic compounds and reactivity intermediates based on structure. *Pure and Applied Chemistry* 67: 1307–1375.

Mukundan, M.K., P.D. Antony, and M.R. Nair. 1986. A review on autolysis in fish. *Fisheries Research* 4: 259–269.

Mumm, R. and M. Hilker. 2005. The significance of background odour for an egg parasitoid to detect plants with host eggs. *Chemical Senses* 30: 337–343.

Nagle, J.F. and S. Tristram-Nagle. 2000. Lipid bilayer structure. *Current Opinion in Structural Biology* 10: 474–480.

Natsch, A., S. Derrer, F. Flachsmann, and J. Schmid. 2006. A broad diversity of volatile carboxylic acids, released by a bacterial aminoacylase from axilla secretions, as candidate molecules for the determination of human-body odor type. *Chemistry & Biodiversity* 3: 1–20.

Normile, D. 2009. Round and round: A guide to the carbon cycle. *Science* 325: 1642–1643.

Osborne, W.A. 1928. A note on volatile sulphide from muscle. *The Biochemical Journal* 22: 1312.

Owre, O.T. and P.O. Northington. 1961. Indication of the sense of smell in the turkey vulture, *Cathartes aura* (Linnaeus), from feeding tests. *American Midland Naturalist* 66: 200–205.

Paczkowski, S. and S. Schütz. 2011. Post-mortem volatiles of vertebrate tissue. *Applied Microbiology and Biotechnology* 91: 917–935.

Parker, G. 1968. The sexual behaviour of the blowfly, *Protophormia terrae-novae* R.-D. *Behaviour* 32: 291–308.

Parmenter, R.R. and J.A. MacMahon. 2009. Carrion decomposition and nutrient cycling in a semiarid shrub-steppe ecosystem. *Ecological Monographs* 79: 637–661.

Pechal, J.L., M.E. Benbow, T.L. Crippen, A.M. Tarone, and J.K. Tomberlin. 2014. Delayed access alters necrophagous insect community assembly and carrion decomposition. *Ecosphere* 5: 45.

Pelaez, F. 2004. Biological activities of fungal metabolites. In: An, Z. (ed.), *Handbook of Industrial Mycology*. Boca Raton, FL: CRC Press.

Perez, A.E., N.H. Haskell, and J.D. Wells. 2014. Evaluating the utility of hexapod species for calculating a confidence interval about a succession based postmortem interval estimate. *Forensic Science International* 241: 91–95.

Popoff, M.R. 1984. Selective medium for isolation of *Clostridium butyricum* from human feces. *Journal of Clinical Microbiology* 20: 417–420.

Prentice, I.C., G.D. Farquhar, M.J.R. Fasham, M.L. Goulden, M. Heimann, V.J. Jaramillo, H.S. Kheshgi, C. LeQuéré, R.J. Scholes, and D.W.R. Wallace. 2001. The carbon cycle and atmospheric carbon dioxide. In: Houghton, J.T., Y. Ding, D.J. Griggs, M. Noguer, P.J. van der Linden, X. Dai, K. Maskell, and C.A. Johnson (eds), *Climate Change 2001: The Scientific Basis Contributions of Working Group I to the Third Assessment Report of the Intergovernmental Panel on Climate Change*. Cambridge, UK: Cambridge University Press, pp. 185–237.

Putman, R.J. 1978. Flow of energy and organic matter from a carcase during decomposition; Decomposition of small mammal carrion in temperate systems 2. *Oikos* 31: 58–68.

Rains, G.C., J.K. Tomberlin, M. D'Alessandro, and W.J. Lewis. 2004. Limits of volatile chemical detection of a parasitoid, *Microplitis croceipes*, and an electronic nose: A comparative study. *Transactions of the ASAE* 47: 2145–2152.

Rastogi, S.C., J.D. Johansen, and T. Menné. 1996. Natural ingredients based cosmetics. Content of selected fragrance sensitizers. *Contact Dermatitis* 34: 423–426.

Rogoff, W.M., A.D. Beltz, J. Johnsen, and F. Plapp. 1964. A sex pheromone in the housefly, *Musca domestica* L. *Journal of Insect Physiology* 10: 239–246.

Roux, O., C. Gers, and L. Legal. 2008. Ontogenetic study of three Calliphoridae of forensic importance through cuticular hydrocarbon analysis. *Medical and Veterinary Entomology* 22: 309–317.

Rumbaugh, K.P. 2004. The language of bacteria ... and just about everything else. *The Scientist* 18: 26–27.

Schleyer, P.v.R. 2001. Introduction: Aromaticity. *Chemical Reviews* 101: 1115–1117.

Schmidt, N.F., S.R. Missan, W.J. Tarbet, and A.D. Cooper. 1978. The correlation between organoleptic mouth-odor ratings and levels of volatile sulfur compounds. *Oral Surgery, Oral Medicine, Oral Pathology* 45: 560–567.

Schneider, D. 1957. Elektrophysiologische untersuchungen von chemo-und mechanorezeptoren der antenne des seidenspinners *Bombyx mori* L. *Zeitschrift für Vergleichende Physiologie* 40: 8–41.

Schulz, S. and J. Dickschat. 2007. Bacterial volatiles: The smell of small organisms. *Natural Product Reports* 24: 814–842.

Schwimmer, S. and W.J. Weston. 1961. Onion flavor and odor, enzymatic development of pyruvic acid in onion as a measure of pungency. *Journal of Agricultural and Food Chemistry* 9: 301–304.

Shivik, J.A. 1998. Brown tree snake response to visual and olfactory cues. *The Journal of Wildlife Management* 62: 105–111.

Shivik, J.A. 2006. Are vultures birds, and do snakes have venom, because of macro- and microscavenger conflict? *BioScience* 56: 819–823.

Shivik, J.A. and L. Clark. 1997. Carrion seeking in brown tree snakes: Importance of olfactory and visual cues. *The Journal of Experimental Zoology* 279: 549–553.

Shivik, J.A. and L. Clark. 1999. Development of attractants for brown tree snakes. In: Johnston, R.E., D. Müller-Schwarze, and P.W. Sorensen (eds), *Advances in Chemical Signals in Vertebrates*. New York: Plenum Press, pp. 649–654.

Shorey, H.H. and R.J. Bartell. 1970. Role of a volatile female sex pheromone in stimulating male courtship behaviour in *Drosophila melanogaster*. *Animal Behaviour* 18: 159–164.

Singh, B., T.L. Crippen, L. Zheng, A.T. Fields, Z. Yu, Q. Ma, T.K. Wood et al. A metagenomic assessment of the bacteria associated with *Lucilia sericata* and *Lucilia cuprina* (Diptera: Calliphoridae). *Applied Microbiology and Biotechnology* 99: 869–883.

Smith, H.R., R.M. DeGraaf, and R.S. Miller. 2002. Exhumation of food by turkey vulture. *Journal of Raptor Research* 36: 144–145.

Smith, E.A. and G.T. Macfarlane. 1996. Enumeration of human colonic bacteria producing phenolic and indolic compounds: Effects of pH, carbohydrate availability and retention time on dissimilatory aromatic amino acid metabolism. *Journal of Applied Bacteriology* 81: 288–302.

Smith, S.A. and R.A. Paselk. 1986. Olfactory sensitivity of the turkey vulture (*Cathartes aura*) to three carrion-associated odorants. *The Auk* 103: 586–592.

Sperandio, V., A.G. Torres, B. Jarvis, J.P. Nataro, and J.B. Kaper. 2003. Bacteria–host communication: The language of hormones. *PNAS* 100: 8951–8956.

Sprent, J.I. 1987. *The Ecology of the Nitrogen Cycle*. Cambridge, UK: Cambridge University Press.

Stadler, S. 2013. Analysis of the volatile organic compounds produced by the decomposition of pig carcasses and human remains. PhD Dissertation, University of Ontario Institute of Technology, Ontario, Canada.

Statheropoulos, M., A. Agapiou, C. Spiliopoulou, G. Pallis, and E. Sianos. 2007. Environmental aspects of VOCs evolved in the early stages of human decomposition. *Science of the Total Environment* 385: 221–227.

Statheropoulos, M., A. Agapiou, E. Zorba, K. Mikedi, S. Karma, G.C. Pallis, C. Elipopoulos, and C. Spiliopoulou. 2011. Combined chemical and optical methods for monitoring the early decay stages of surrogate human models. *Forensic Science International* 210: 154–163.

Statheropoulos, M., C. Spiliopoulou, and A. Agapiou. 2005. A study of volatile organic compounds evolved from the decaying human body. *Forensic Science International* 153: 147–155.

Stevenson, F.J. and M.A. Cole. 1999. *Cycles of Soils: Carbon, Nitrogen, Phosphorus, Sulfur, Micronutrients*. Canada: John Wiley & Sons.

Stotzky, G. and S. Schenck. 1976. Volatile organic compounds and microorganisms. *Critical Reviews in Microbiology* 4: 333–382.

Sturgill, G. and P.N. Rather. 2004. Evidence that putrescine acts as an extracellular signal required for swarming in *Proteus mirabilis*. *Molecular Microbiology* 51: 437–446.

Thomas, M.L. and L.W. Simmons. 2008. Sexual dimorphism in cuticular hydrocarbons of the Australian field cricket *Teleogryllus oceanicus* (Orthoptera: Gryllidae). *Journal of Insect Physiology* 54: 1081–1089.

Tomberlin, J.K., T.L. Crippen, A.M. Tarone, B. Singh, K. Adams, Y.H. Rezenom et al. 2012. Interkingdom responses of flies to bacteria mediated by fly physiology and bacterial quorum sensing. *Animal Behavior* 84: 1449–1456.

Uchida, N., Y.K. Takahashi, M. Tanifuji, and K. Mori. 2000. Odor maps in the mammalian olfactory bulb: Domain organization and odorant structural features. *Nature Neuroscience* 3: 1035–1043.

Urech, R., P.E. Green, M.J. Rice, G.W. Brown, F. Duncalfe, and P. Webb. 2004. Composition of chemical attractants affects trap catches of the Australian sheep blowfly, *Lucilia cuprina*, and other blowflies. *Journal of Chemical Ecology* 30: 851–866.

van der Niet, T., D.M. Hansen, and S.D. Johnson. 2011. Carrion mimicry in a South African orchid: Flowers attract a narrow subset of the fly assemblage on animal carcasses. *Annals of Botany* 107: 981–992.

van der Plas, M.J., C. Dambrot, H.C. Dogterom-Ballering, S. Kruithof, J.T. van Dissel, and P.H. Nibbering. 2010. Combinations of maggot excretions/secretions and antibiotics are effective against *Staphylococcus aureus* biofilms and the bacteria derived therefrom. *The Journal of Antimicrobial Chemotherapy* 65: 917–923.

van der Plas, M.J., G.N. Jukema, S.W. Wai, H.C. Dogterom-Ballering, E.L. Lagendijk, C. van Gulpen, J.T. van Dissel, G.V. Bloemberg, and P.H. Nibbering. 2008. Maggot excretions/secretions are differentially effective against biofilms of *Staphylococcus aureus* and *Pseudomonas aeruginosa*. *The Journal of Antimicrobial Chemotherapy* 61: 117–122.

Van Dyke Parunak, H., S.A. Brueckner, R. Matthews, and J. Sauter. 2005. Pheromone learning for self-organizing agents. *Systems, Man and Cybernetics, Part A: Systems and Humans, IEEE Transactions on* 35: 316–326.

Vass, A.A., S.A. Barshick, G. Sega, J. Caton, J.T. Skeen, J.C. Love, and J.A. Synstelien. 2002. Decomposition chemistry of human remains: A new methodology for determining the postmortem interval. *Journal of Forensic Sciences* 47: 542–553.

Vass, A.A., R.R. Smith, C.V. Thompson, M.N. Burnett, N. Dulgerian, and B.A. Eckenrode. 2008. Odor analysis of decomposing buried human remains. *Journal of Forensic Sciences* 53: 384–391.

Verhulst, N.O., W. Takken, M. Dicke, G. Schraa, and R.C. Smallegange. 2010. Chemical ecology of interactions between human skin microbiota and mosquitoes. *FEMS Microbiology Ecology* 74: 1–9.

Vining, L.C. 1990. Functions of secondary metabolites. *Annual Reviews of Microbiology* 44: 395–427.

von Hoermann, C., J. Ruther, and M. Ayasse. 2012. The attraction of virgin female hide beetles (*Dermestes maculatus*) to cadavers by a combination of decomposition odour and male sex pheromones. *Frontiers in Zoology* 9: 18.

von Hoermann, C., J. Ruther, S. Reibe, B. Madea, and M. Ayasse. 2011. The importance of carcass volatiles as attractants for the hide beetle *Dermestes maculatus* (De Geer). *Forensic Science International* 212: 173–179.

von Hoermann, C., S. Steiger, J.K. Müller, and M. Ayasse. 2013. Too fresh is unattractive! The attraction of newly emerged *Nicrophorus vespilloides* females to odour bouquets of large cadavers at various stages of decomposition. *PLoS ONE* 8: e58524.

Voss, S.C., H. Spafford, and I.R. Dadour. 2009. Host location and behavioral response patterns of the parasitoid, *Tachinaephagus zealandicus* Ashmead (Hymenoptera: Encyrtidae), to host and host-habitat odours. *Ecological Entomology* 34: 204–213.

Wagner, S.C. 2012. Biological nitrogen fixation. *Nature Education Knowledge* 3: 15.

Wajnberg, E. and S. Colazza. 2013. *Chemical Ecology of Insect Parasitoids.* Chichester, UK: John Wiley & Sons.

Wall, R. and M. Warnes. 1994. Responses of the sheep blowfly *Lucilia sericata* to carion odour and carbon dioxide. *Entomologia Experimentalis et Applicata* 73: 239–246.

Wang, D., X. Ding, and P.N. Rather. 2001. Indole can act as an extracellular signal in *Escherichia coli*. *Journal of Bacteriology* 183: 4210–4216.

Watson, J.D. and F.H.C. Crick. 1953. Molecular structure of nucleic acids: A structure for deoxyribose nucleic acid. *Nature* 171: 737–738.

Wenke, K., M. Kai, and B. Piechulla. 2010. Belowground volatiles facilitate interactions between plant roots and soil organisms. *Planta* 231: 499–506.

Wenzel, B.M. and M. Sieck. 1972. Olfactory perception and bulbar electrical activity in several avian species. *Physiology and Behavior* 9: 287–293.

Wright, P.A. 1995. Nitrogen excretion: Three end products, many physiological roles. *The Journal of Experimental Biology* 198: 273–281.

Wyatt, T.D. 2003. *Pheromones and Animal Behaviour: Communication by Smell and Taste.* Cambridge, UK: Cambridge University Press.

Xu, H., G.-Y. Ye, Y. Xu, C. Hu, and G.-H. Zhu. 2014. Age-dependent changes in cuticular hydrocarbons of larvae in *Aldrichina grahami* (Aldrich) (Diptera: Calliphoridae). *Forensic Science International* 242: 236–241.

Yamada, K. and S. Natori. 1994. Characterization of the antimicrobial peptide derived from sapecin B, an antibacterial protein of *Sarcophaga peregrina* (flesh fly). *The Biochemical Journal* 298: 623–628.

Ye, G., K. Li, J. Zhu, G. Zhu, and C. Hu. 2007. Cuticular hydrocarbon composition in pupal exuviae for taxonomic differentiation of six necrophagous flies. *Journal of Medical Entomology* 44: 450–456.

Yokota, S.D. and V.H. Shoemaker. 1981. Xanthine excretion in a desert scorpion, *Paruroctonus mesaensis*. *Journal of Comparative Physiology* 142: 423–428.

Zender, R., C. Lataste-Dorolle, R.A. Collet, P. Rowinski, and R.F. Mouton. 1958. Aseptic autolysis of muscle: Biochemical and microscopic modifications occurring in rabbit and lamb muscle during aseptic and anaerobic storage. *Journal of Food Science* 23: 305–326.

Zhang, L.-H. and Y.-H. Dong. 2004. Quorum sensing and signal interference: Diverse implications. *Molecular Microbiology* 53: 1563–1571.

Zhang, C.H., Y.J. Ma, F.X. Yang, W. Liu, and Y.D. Zhang. 2009b. Optimization of medium composition for butyric acid production by *Clostridium thermobutyricum* using response surface methodology. *Bioresource Technology* 100: 4284–4288.

Zhang, C., H. Yang, F. Yang, and Y. Ma. 2009a. Current progress on butyric acid production by fermentation. *Current Microbiology* 59: 656–663.

Zimmermann, G. 2007. Review on safety of the entomopathogenic fungi *Beauveria bassiana* and *Beauveria brongniartii*. *Biocontrol Science and Technology* 17: 553–596.

Zito, P., M. Sajeva, A. Raspi, and S. Dötterl. 2014. Dimethyl disulfide and dimethyl trisulfide: So similar yet so different in evoking biological responses in saprophilous flies. *Chemoecology* 24: 1–7.

Zurek, L., C. Schal, and D.W. Watson. 2000. Diversity and contribution of the intestinal bacterial community to the development of *Musca domestica* (Diptera: Muscidae) larvae. *Journal of Medical Entomology* 37: 924–928.

10

Vertebrate Carrion as a Model for Conducting Behavior Research

Jeffery K. Tomberlin, Michelle R. Sanford,
Meaghan L. Pimsler, and Sherah L. VanLaerhoven

CONTENTS

10.1 Introduction

Death is not the end of life. At a macroscopic level, a dead animal would seem to be an end or stopping point, as the creature has stopped moving, breathing, playing, and mating—all of the outward signs of life are gone. The carrion is no longer a viable host for the parasites, pathogens, or symbionts that are dependent on the living animal host for sustenance, and so these organisms either die or disperse to other living individuals.

Though death has enveloped the animal, life persists on, or in, what is now the remaining carrion. For those that are patient, and maybe macabre enough to watch, carrion is teeming with life—through a different community of interconnected living organisms. Some species (e.g., Calliphoridae) are visible to the naked eye (Fuller 1934; DeVault et al. 2011), whereas others (e.g., bacteria) are only known to be present through the use of microscopic or by molecular means (Barnes et al. 2010; Pechal 2012; Hyde et al. 2013; Metcalf et al. 2013; Pechal et al. 2013). These new inhabitants represent species that directly depend on this resource for food (Payne et al. 1968), shelter (Haefner et al. 2004), hunting grounds (Wells and Greenberg 1992, 1994), and possibly as a place to locate a mate (Thomas 1993).

The diversity of animals utilizing carrion is extensive. Insects and other arthropods may use carrion directly as a habitat or a source of nutrition (Payne 1965) (see Chapter 8 for more discussion on carrion community ecology) or indirectly by consuming other animals that make use of the resource (Baumgartner 1993). However, invertebrates are not the only utilizers of carrion; vertebrates (Margalida et al. 2008; Wilson and Wolkovich 2011) (see Chapters 6 and 21) and microbes (Janzen 1977; Burkepile et al. 2006) (see Chapters 3 and 18–20) also compete for carrion resources.

Vertebrate carrion represents an ideal model for studying behavior. Carrion is a specific resource base that is ubiquitous worldwide under a wide array of abiotic and biotic conditions. Furthermore, carrion can be easily modified and manipulated under controlled conditions, allowing researchers to address basic questions with broader ramifications to the study of behavior including evolutionary implications, associated variability, and significance to the animal utilizing the resource. And, as previously discussed, the diversity of life associated with carrion along with their ecological interactions is immense and allows for researchers to be selective as to which model organisms they will use to study general behavior common to other systems (e.g., plant–herbivore).

While the purpose of this chapter is on behavior, it is important for the reader to recognize what it does not cover. This chapter does not provide a complete anthology of behavioral ecology as there are a number of insightful texts available on the subject (e.g., Westneat and Fox 2010; Moore and Breed 2012). Additionally, this chapter is not an attempt to summarize behavior across all Classes of animals associated with carrion. Such an attempt would be naïve to say the least, as the study of behavior of a single Class, Order, or even species of animal associated with vertebrate carrion could, in some cases, constitute its own separate book.

This chapter is a general overview of the field of behavior within the context of Tinbergen's four questions as related to a vertebrate carrion model. Such an approach allows for the utilization of a well-established framework (Tinbergen 1963; Bateson and Laland 2013) to explore some aspects of behavioral ecology of animals associated with vertebrate carrion and demonstrates the usefulness of carrion as a model. The majority of the examples in this chapter are about insects, though vertebrates are also represented. This approach provides the reader with some context of behaviors exhibited by those organisms utilizing carrion that could translate to a greater understanding of the role of vertebrate carrion within a given ecosystem and its associated ecological complexities. An additional goal is to inspire more rigorous behavioral research of animals associated with vertebrate carrion within the broader field of decomposition ecology.

10.1.1 Historical Context

Nikolaas Tinbergen is recognized as a pioneer in the field of behavioral ecology and ethology. He shared the 1973 Nobel Prize in Physiology or Medicine with Konrad Lorenz and Karl von Frisch for their discoveries regarding behaviors within the realm of individual and social groups of animals (Dewsbury 2003). This was, in part, due to their work elucidating the honey bee, *Apis mellifera* L. (Hymenoptera: Apidae) waggle dance (Dyer 2002).

Tinbergen is also known for developing a framework that has been successfully used by researchers to study animal behavior (Tinbergen 1963). His framework (Figure 10.1) built on the three basic questions posed by J.S. Huxley as the "three major problems of biology": (1) causation (mechanism), (2) survival value (function, adaptation, or current utility as a less confusing term [Bateson and Laland 2013]), and (3) evolution (phylogeny), to which he added (4) development (ontogeny) (Tinbergen 1963).

It is an extraordinary time to conduct behavior research. The diversity of tools available to modern researchers is both astounding and daunting, as they allow us to truly appreciate the level of complexity within each category of behavioral research. With advancements in technology, researchers are now able to ask questions in studies ranging from those that use meta-genomics to those that require geographic information systems (e.g., satellite input) or a combination of the two in order to study the behavior of an animal from a local to global scale.

Today's research is truly multidisciplinary in many cases. However, these novel methods also increase complexity and give rise to major challenges, such as the sheer volume of data gathered from a study, and once data are gathered, having the computing infrastructure to appropriately analyze the data. Concurrently, researchers must either be trained to analyze these datasets or, at a minimum, locate collaborators that can do so. Like in nature, what is a challenge for one individual or field is an opportunity for another. Consequently, the integration of computer science with the biological sciences is now a fruitful area of research that has advanced the process of data acquisition and analysis. However, without an appreciation of the framework presented by Tinbergen more than 50 years ago, it would not be possible to fully understand the complexities of behavior nor how new technologies might allow researchers to

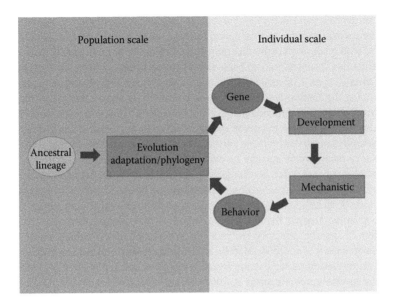

FIGURE 10.1 Schematic representation of Tinbergen's four questions.

address specific questions. Therefore, this chapter will present the four components of his framework with discussions of the proximate and then ultimate levels of behavior as associated with animals frequently encountered on vertebrate carrion.

10.2 Proximate Questions

The proximate level of the study of behavior involves understanding how a specific behavior is actually accomplished. This approach is associated with explaining the physiological mechanisms (Tinbergen's first question) regulating the response of the animal. Examples of these questions predominately encompass aspects of physiology: What is the morphology of the structures utilized to detect and respond to a given stimulus? What structures are used to produce a behavior? Furthermore, what are the biological mechanisms of these structures that allow for such a behavior to be demonstrated? Such studies can examine morphological structure and innervation as well as hormonal and gene-level mechanisms.

10.2.1 Causation (Mechanisms)

Historically, the fields of morphology and physiology served as cornerstones of behavioral research. Such avenues have allowed scientists to answer fundamental questions as to how behaviors occur without dependence on much technology (e.g., requiring only pen and paper for illustrating gross anatomy of an animal). The characteristics that constitute a morphological structure exist along a vast continuum from the organismal level, consisting of visible structures (e.g., hairs, eyes, ears), to the molecular level (e.g., DNA, RNA). Developing research questions in this area requires defining the specific question with regard to the mechanism, or structure, under examination to ensure that the focus of the research remains on the particular behavior in question.

The behaviors of animals associated with carrion, such as the Turkey Vulture, *Cathartes aura* L. (Accipitriformes: Cathartidae) (Smith and Paselk 1986), or burying beetles, *Nicrophorus vespilloides* Herbst (Coleoptera: Silphidae) (Rozen et al. 2008) have been studied in detail. However, debate with regard to the mechanisms regulating their behaviors persists, with new data supporting different paradigms. For example, the debate over the relative importance of vision versus olfaction in resource location in the Turkey Vulture has been occurring for the last 140+ years (Smith and Paselk 1986). In the past, many of these debates were due to technological limitations preventing in-depth examination of the mechanisms regulating the behaviors exhibited by these animals. In many instances, these behaviors were studied in the wild or through gross anatomical studies of the animal. Initial work relied on gross observations of structures essential for the behaviors of interest that could be viewed and manipulated without the aid of microscopes. Consequently, the conclusions drawn were based on broad generalities of the animals comprising the carrion-feeding community. Today, it is known that the primary senses used to locate such carrion are species-dependent and sweeping generalizations are inappropriate.

Researchers today have been able to conduct studies that integrate genetic and physiological components to explain behaviors observed by these animals. Such projects are able to characterize and explain the design of olfactory receptors, which can be essential for detecting vertebrate carrion, their layout in the epithelium, and how they are wired neurologically (Steiger et al. 2009). More recent work has focused on examining these behaviors as an extension of the genes encoding for olfactory receptors in order to explain the sensitivity, or lack thereof, exhibited by many of these animals (Steiger et al. 2009). Detailed work has been done in the vinegar fly, *Drosophilia melanogaster* Meigen (Diptera: Drosophilidae), to determine the effect of specific olfactory receptors or importance of individual neurons on avoidance and attraction behaviors (Suh et al. 2004; Marella et al. 2006; Schlief and Wilson 2007).

Defecation and regurgitation on the carrion resource are two common behaviors exhibited by associated vertebrates and invertebrates (Rozen et al. 2008). From the perspective of a vertebrate ecologist, the following was used to describe the regurgitation behavior of vultures (Cook 1997):

> In certain social circles, human males defend their alcoholic beverages from poachers, or at least claim to, by declaring that they have spit in their beer. This mildly repulsive suggestion is usually offered more for its contribution to the male bonding ritual than for any real resource protection. Vultures, on the other hand, especially in nutrient poor areas like deserts, take resource defense a bit more seriously, and are frequently apt to regurgitate their stomach contents on an assailant (Wagner 1980). Putrefied to begin with, added rarefaction of the carrion in the bird's stomach makes the stomach contents an effective defense against predators as well as a deterrent to many, particularly mammalian, competitors for a food source (many mammals rely on smell to find food).

The author of the preceding quote points out that behaviors can have multiple utilities in a given animal. Both regurgitation and defecation could be essential for consumers of carrion to be successful. Vultures, as well as flies (Diptera) and beetles (Coleoptera), are examples of animals that exhibit these behaviors in association with carrion.

Defecation is a common mechanism used by arthropods to introduce microbes into a given environment and this inoculation could result in greater larval survivorship. Adult vinegar flies introduce fungi to larval food substrates. These fungi break down plant tissues and release nutrients used by larvae and volatiles that recruit adults to oviposition sites (Ort et al. 2012). These data suggest that these microbes are used in selection for host preference (Ort et al. 2012) and potentially speciation. Furthermore, the digestive tract of *Drosophila* has been well described, thus providing greater understanding of the mechanisms that allow beneficial bacteria to proliferate while pathogenic bacteria are suppressed (Buchon et al. 2013). These beneficial bacteria are introduced into new habitats through regurgitation and defecation; larvae consume these bacteria, which then result in intestinal stimulation, turnover, and new tissue growth (Buchon et al. 2013).

Regurgitation produced by animals that consume carrion can have multiple roles depending on the animal. Vultures use regurgitation to feed their offspring, but also as a defense mechanism against predators (Fergus 2003). As pointed out previously with flies, regurgitation could reduce competition for a given resource (Cook 1997). In *Drosophila*, regurgitation can serve as a nuptial gift resulting in mate selection, whereas defecation inoculates an environment with fungi that breakdown plant tissue-releasing nutrients increase larval survivorship (Ort et al. 2012). The metabolic activities of the fungi released from adult regurgitation also release volatiles that recruit other adult *Drosophila*-seeking oviposition sites. Similarly, blow flies (Calliphoridae) also release bacteria into their environment (Barnes et al. 2010); however, it is not known at this time whether such behavior results in greater larval survivorship. Furthermore, it is unknown whether carrion-feeding animals also use regurgitation to provide a nuptial gift.

The beauty of *Drosophila* is that not only is it associated with ephemeral resources, albeit plant-based decomposition for the most part, but also that it is a well-established model organism through which behaviors observed in other organisms may be explained. However, other invertebrate models that commonly occur on vertebrate carrion, such as blow flies, can also serve as nontraditional models to validate or refute concepts that have been developed with the *Drosophila* model in terms of broader evolutionary implications.

Comparatively, mechanisms regulating defecation behaviors of insects that colonize vertebrate carrion appear to be better studied than with vultures and other associated vertebrates that utilize carrion resources. This research is best exemplified with *D. melanogaster* (Kaplan and Trout III 1968; Yoshihara and Ito 2012) and blow flies. Clearly, these well-studied organisms can utilize a wide range of nutritional resources (high protein:low carbohydrates in carrion to low protein:high carbohydrates in cactus). Furthermore, the morphology of the alimentary canal of flies, such as the blow fly adult and immature, has been previously described (Boonsriwong et al. 2011; Boonsriwong et al. 2012), allowing for researchers to determine where microbes reside within the arthropod and how they survive. Additionally, the mechanisms governing the physiological functioning of the digestive tract of the blow fly have been described. Synthesized, this information provides insights into the role of the foregut and hindgut (Hobson 1931), as well as what is produced during regurgitation and defecation (Hobson 1932; Evans 1956). Moreover, as discussed later in this chapter, hormones serve as mechanisms regulating these behaviors.

Future research is needed to further understand those mechanisms regulating expression of regurgitation and defecation. Previous research has demonstrated that these behaviors are partially explained by the microbiomes (i.e., fungi and bacteria in *Drosophila* and blow flies, respectively) within the host (Ezenwa et al. 2012). In at least some instances, these microbiomes regulate behavior through the production of quorum-sensing compounds or compounds that regulate quorum-sensing behavior, indicating a level of interkingdom communication (Ma et al. 2012) (see Chapter 20).

Much is known about the genetic regulation of behaviors exhibited by *Drosophila* (Powell 1997; Sokolowski 2001; Turner et al. 2014) that can potentially be used to explain the same behaviors observed in other arthropods. For example, initial work in *Drosophila* suggested that some variation in locomotory behavior in larvae was due to two alleles, rover and sitter, at the *foraging* gene locus (Sokolowski et al. 1984; Sokolowski 1985a,b). Follow-up work in other orders of insects, most notably work with bees and ants (Hymenoptera), suggested that this gene affected locomotory behavior in a wide range of arthropods (Lucas et al. 2010; Ingram et al. 2011; Tobback et al. 2011; Tarès et al. 2013). However, recent work using genome-wide association mapping seems to refute the initial results in *Drosophila* that variation in this behavior was due to allelic differences in a single gene (Turner et al. 2014).

Furthermore, research with other model organisms, such as the nematode *Caenorhabditis elegans* (Maupas) (Rhabditida: Rhabditidae), allows for parallel studies of these behaviors across phyla to be conducted, including examination of the mechanisms regulating defecation (Iwasaki and Thomas 1997). Thus, researchers interested in carrion-frequenting animals should strive to be cognizant of data produced in other models outside the realm of decomposition ecology.

10.2.2 Development (Ontogeny)

Another important aspect of Tinbergen's framework, and the unique question that he added to the baseline three questions posed by J.S. Huxley for studying animal ethology, was the ontogeny, or development, of behaviors (Tinbergen 1963). Initially this topic was limited in application as it was overshadowed by the development of other disciplines, such as sociobiology; however, questions related to development and resulting manifestations of behavior provided a platform to address questions of interest in broader research fields including evolutionary biology (Stamps 2003). Consequently, research in this area has steadily increased over the course of the last decade. A prime example, as pointed out by Stamps (2003), is the tremendous amount of interest in phenotypic plasticity; generally defined as the ability of a single genotype to generate multiple phenotypes, it is often the result of the interaction of genetics and environment. Phenotypic plasticity has been demonstrated with carrion flies, specifically blow flies, with regard to larval development rate as related to temperature (Gallagher et al. 2010; Owings et al. 2014). Larvae from fly populations from different regions of the United States exhibit significantly different responses to shifts in temperatures, including reduced development times from larva to pupa relative to other populations (Tarone and Foran 2008; Gallagher et al. 2010; Owings et al. 2014). These responses can result in different-sized adults, which in turn could influence subsequent behaviors including, but not limited to, flight, mate acquisition, and reproductive choices.

Ontogeny in an ethological context refers to the change or development of behaviors over time within an individual, or as Tinbergen (1963) phrased it, the "change in behavior machinery during development." These alterations can be influenced by internal factors such as gene expression and hormone titers, or external factors such as social interactions and ecological processes. Specific factors that are known to affect behavioral ontogeny in necrophagous animals are sex, age, temperature, density, and nutritional status (Wilson et al. 1984; Trumbo et al. 1995; Scott 1998; Moe et al. 2002; Hoback et al. 2004; Scott and Panaitof 2004; Trumbo and Robinson 2008; Boncoraglio and Kilner 2012).

Behavioral changes in insects are often tied to discrete developmental stages. Flies are holometabolous insects that go through four distinct life history stages: egg, larva, pupa, and adult (Truman and Riddiford 1999). Two of the stages, egg and pupa, are sedentary, whereas the larva and adult are not as restricted. Consequently, the morphology of the two active stages (larva and adult) is drastically different from that of the egg and pupa (Figure 10.2). Due to these complex changes in the body shape, the types of behaviors an individual is capable of will change over the course of its development. For example, the wings of adult blow flies allow them to disperse long distances, reported to be as far as approximately 4 km away from their site of origin, or nearly 540,000 times the length of the individual (Greenberg and Bornsetein 1964; Whitworth 2006). In comparison, larval blow flies are poor dispersers due to their lack of wing and locomotory appendages.

An interesting similarity between the adult and larval stages of blow flies is that both are known to deposit oral and anal secretions (regurgitation and defecation) on the carcass (Bexfield et al. 2004). For the larvae, these "excretions/secretions" as they are called, are known to function in extra-oral digestion and have important antimicrobial and antifungal properties (Vistnes et al. 1981; Anderson 1982; Cazander et al. 2009; van der Plas et al. 2010; Jiang et al. 2012). Adult flies have often been observed to remain stationary approximately 1 h after feeding on either a sugar or protein source and engage in a

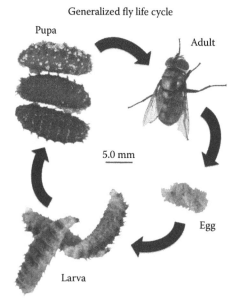

Generalized fly life cycle

Pupa

Adult

5.0 mm

Egg

Larva

FIGURE 10.2 (**See color insert.**) Life cycle of a fly.

behavior known as bubbling, in which the liquid contents of the crop are regurgitated and held at the end of the proboscis for between 5 and 10 min before being reingested (Stoffolano et al. 2008; Larson and Stoffolano 2011). The utility of this behavior is not yet understood, but it is clear that there is some sexual dimorphism in frequency and duration of bubbling events (Stoffolano et al. 2008). Furthermore, adult flies are known to act as mechanical vectors of pathogens (Greenberg 1973) and will often leave behind fecal and regurgitation artifacts (Benecke and Barksdale 2003; Durdle et al. 2013), yet the frequency, timing, and content of these "fly specks" has only been the subject of limited study.

Changes in the circulating titers of hormone can cause quick and sometimes long-lasting behavioral shifts in insects. This has been studied extensively in burying beetles. Adult males and females are known to be attracted to carcasses by the production of sulfur-containing volatile organic compounds and use their antennae not only for chemosensory input, but also to detect and follow the direction of flowing air (Heinzel and Böhm 1989; Kalinova et al. 2009). Females searching for a suitable oviposition site generally do not have fully matured ovaries and in most cases mating occurs on the carcass (Trumbo et al. 1995). Upon encountering a suitable carcass, juvenile hormone (JH) titers in females rise quickly, which stimulates the onset of carcass preparation behaviors (such as burying and skinning) as well as ovary maturation, whether or not a male is present. In comparison, male JH titers are much lower when a suitable mate is not present, and although single males will bury the carcass, they also release sex pheromones that are attractive to females (Scott and Panaitof 2004; Trumbo and Robinson 2008). These examples demonstrate shifts in behaviors over the lifetime of the insect and the significance of understanding the underlying mechanism and the fitness effects of the exhibited behaviors.

Insects are not the only carrion-utilizing organisms to exhibit changes in behavior during different developmental periods. Turkey Vultures exhibit different defensive behaviors dependent upon their age (Vogel Jr 1950). Adult Turkey Vultures are known to regurgitate and play dead when disturbed by humans, even going so far as to remain "expired" when being handled. The smell of the vultures and the death feign have led some individuals to try to convince a researcher that a vulture that was incubating her eggs was "deader 'n a doornail … Smelled like it'd been dead a couple days" (Mossman 1976). Although defensive behaviors in the young do include regurgitation, nestlings employ aggressive subterfuge in an attempt at intimidation of the intruder. They stomp, hiss, and even feign attack in a behavior described as a "scare jump" (Mossman 1976). The mechanism leading to these distinct defensive behaviors along a temporal gradient is not well understood, nor is it known when in the development of the shift occurs.

These are just a few examples of the ontological ethology of carrion-utilizing animals. Many researchers in the field focus on the ecology, physiology, or life history of these organisms rather than their behavior, and there is still much to discover related to the causes and mechanisms of behavioral shifts in carrion associated animals. Furthermore, many researchers who have done studies on behavior in these organisms often focus on the ultimate, rather than proximate, causes of the behaviors. They are frequently more interested in the adaptive value of different behaviors or comparing behavior between different species.

10.3 Ultimate Questions

As previously discussed, behaviors can be broadly classified into the categories of proximate and ultimate. Proximate questions in ethology are broadly interested in the "what" of behavior—"what is the behavior?", "what morphological adaptations allow an organism to exhibit that behavior?", or "what causes behavioral differences between individuals or over time?". In contrast, ultimate questions are more broadly interested in the "why" of behavior—"why does this behavior exist (i.e., what is its adaptive significance)?" or "why are there differences or similarities in this behavior between closely related species?". Another distinction is that ultimate questions are focused on the evolutionary view of a behavioral trait. As with proximate mechanisms, Tinbergen further divided ultimate mechanisms into two subtopics: these are function (or adaptation) and evolution (or phylogeny) (Tinbergen 1963). However, as stated initially in the chapter, function can be a confusing term depending on the context in which it is used. Therefore, for the purposes of this chapter, the term current utility (what is the purpose of the behavior) is used to describe this aspect of behavior instead of function, which has many definitions and thus can be confusing (Bateson and Laland 2013).

10.3.1 Current Utility (Function)

Why do organisms behave in specific ways, and what are the benefits of such behavior? From the ultimate perspective, a simple answer to these questions is that organisms behave in specific ways because these are the behaviors that allowed their predecessors to survive and successfully reproduce in their environment. Such success allows for the propagation of the genetic material that leads to those behaviors. However, environments are not constant; thus, behavior that is beneficial to an organism in a current environment may not be as beneficial to its future offspring. Therefore, flexibility in behavioral responses provides an organism the ability to adapt to environmental changes. For blow flies utilizing carrion, the female choice of where and when to oviposit eggs has vital implications for the survival of her offspring and her ability to successfully pass on her genetic material. The presence of other fly species that may be competitors for prime oviposition locations that reduce egg desiccation and ensure increased larval survival should select for a female to adopt a more aggressive strategy in the timing of oviposition and seek to oviposit faster than her competitors. In contrast, the ability to accept a less desirable oviposition site in the presence of a predatory species can also be beneficial. Creating spatial distance from a predator on the carrion resource reduces predation pressures on her offspring, which is another potential adaptation as demonstrated by Gião and Godoy (2007). Yet another adaptation within the range of behavior exhibited by blow flies is quickly ovipositing upon arriving at a resource or at sites near a facilitating species. Larvae of the competing species produce extra-oral secretions that may assist in her offsprings' ability to ingest nutrients and allow for faster development or larger, fitter adults. Ultimately, females are expected to make oviposition choices that result in higher offspring survival and more fecund offspring.

Due to the behavioral choices made by females, blow fly larvae must adapt to the environmental conditions they find themselves in, which may include competitors, predators, and variable resource quality even within a single carrion resource. In the presence of larvae of the facultative predator *Chrysomya rufifacies* Macquart (Diptera: Calliphoridae), other blow fly species such as *Phormia regina* (Meigen) and *Lucilia sericata* (Meigen) develop faster and leave the carrion resource sooner to pupate; thus, limiting the acquisition of additional food resources which results in greater survivorship but smaller adults that produce fewer eggs (Reid 2012). The same was observed for *Cochliomyia macellaria* Fabricius (Diptera: Calliphoridae) interacting with *C. rufifacies*. Third-instar *C. macellaria* larvae develop faster in the presence of the excretion/secretions of *C. rufifacies* (Flores 2013). This altered developmental response could reduce mortality stemming from competition. Furthermore, predation has been observed in interactions between *C. rufifacies* and other blow fly species, such as *C. megacephala* (Fabr.) (Shiao and Yeh 2008) and *C. macellaria* (Brundage et al. 2014), and altered development rates may permit escape from predation as well. Yet this is simply a snapshot of why blow flies behave and adapt their behavior in different ways in response to other species within the carrion habitat.

So, why does the behavior of an organism change and adapt in the absence of other species? Aside from these biotic interactions, organisms are subject to abiotic factors that can determine whether or not a species will survive. For poikilotherms, organisms such as insects that are unable to generate their own body heat, development rate is dependent on the ambient temperature. Insects colonizing carrion resources may not seek out carrion that is found in thermal environments that are too hot or too cold to successfully support oviposition and subsequent development of offspring. This was demonstrated with the current and predicted range expansion of *C. rufifacies* based on its thermal tolerances (Rosati and VanLaerhoven 2007). Humidity can also be an important factor determining oviposition site choice. Black soldier flies, *Hermetia illucens* L. (Diptera: Stratiomyidae), which utilize carrion, as well as other decomposing organic materials such as plant or animal waste, as larval resources (Nguyen et al. 2013) are heavily impacted by humidity. Successful egg hatch with this species is reduced as humidity declines from 70 to 25% (Holmes et al 2012). Therefore, the behavioral strategy of *H. illucens* preferentially laying eggs in cracks or crevices, or in multiple spherical layers in the absence of cracks or crevices is likely a behavior that minimizes desiccation and increases offspring survival.

The physical structure of the environment may also play an important role in the behavioral choices made by carrion-inhabiting organisms, leading them to adopt behaviors that result in increased survivorship through development. For example, the choice of pupation substrate by larvae that crawl away from the carrion resource when they are completing their development can impact their ability to successfully emerge as an adult (Lewis and Benbow 2011; SLV personal observation). The importance of this to some of these species' behaviors can be observed in extended wandering behavior that prolongs their development as they search for a suitable pupation site when no pupation media is available (such as on highly compacted ground or rock). Extended wandering uses up stored nutritional resources needed to survive pupariaton and develop into an adult, as the larvae are no longer feeding at this stage (Mohr 2012). This behavioral adaptation is well known in blow flies, but has also been observed in black soldier flies. For *H. illucens*, the presence of any pupation substrate results in a shorter pupation interval and greater survivorship than when no pupation substrate is present (Holmes et al. 2013).

As demonstrated by these examples, there is a range of behavioral adaptations that organisms exploiting carrion may choose and each choice has consequences for the survival and fitness of any one individual. Yet, only those advantageous behaviors that are heritable can persist in a population, thus leading to Tinbergen's final question—evolution.

10.3.2 Evolution (Phylogeny)

The fourth of Tinbergen's questions is that of ultimate causes or the "why" of behavioral ecology. It refers to the phylogeny of the organisms resulting from ancestral environments and evolutionary constraints. Animals associated with carrion represent an excellent system for examining the relationships among species associated with the intense competitive environment of an ephemeral resource.

Although many insect groups are associated with carrion, the beetles of the family Silphidae are an especially well-studied group from which to investigate questions related to ultimate causes of behavior. Carrion, or burying, beetles illustrate phylogenetic relationships that suggest evidence of strong competition and niche partitioning. Figure 10.3 illustrates the phylogenetic relationships of some species in the family. At more basal levels, the subfamily division represents a dichotomy in habitat preference with Silphinae preferring larger carrion where their larvae are predatory, and Nicrophorinae preferring small carrion, where biparental care of the larvae confers significant competitive advantages (Scott 1998).

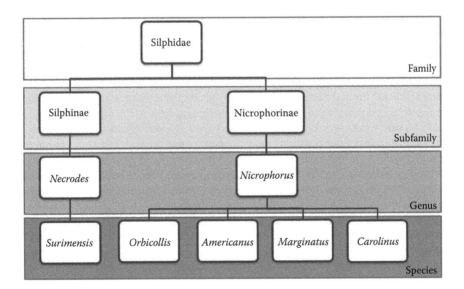

FIGURE 10.3 The phylogenetic relationships among several carrion beetles in the family Silphidae. (With permission from Springer Science+Business Media: *Journal of Chemical Ecology*, Differences among antimicrobial properties of carrion beetle secretions reflect phylogeny and ecology, 30, 2004, 719–729, Hoback, W.W. et al.)

Some of the behavioral characteristics of burying beetles that make them so successful are biparental care, the process of burying small carrion and slowing decomposing the resource through the deposition of oral and/or anal secretions (Scott 1998). In a test of the antibacterial properties of the oral and anal secretions of several of the burying beetles, Hoback et al. (2004) found that the presence and activity of these compounds is reflected in the beetle's phylogeny. The authors of the study examined beetles in both subfamilies, one species of the subfamily Silphinae, including *Necrodes surinamensis* (Fabricius), and seven species in the subfamily Nicrophorinae in the genus *Nicrophorus*, including those displayed in Figure 10.3. The single member of the Silphinae that produces secretions, *N. surinamensis*, produces only anal secretions which have antibacterial properties. This is very different from the species in the Nicrophorinae subfamily that exhibit biparental care; five of the seven *Nicrophorus* spp. examined produced oral secretions with antibacterial properties. These five species are known to bury small carrion and care for their larvae. The two species that did not produce oral secretions with antibacterial properties either produced anal secretions (*N. orbicollis*) or lacked antibacterial secretions completely (*N. pustulatus*). In the case of *N. pustulatus*, there is reason to suspect that this may reflect a possible trade-off, because this species is a known brood parasite (Hoback et al. 2004). In a follow-up experiment on these same species, Jacques et al. (2009) found that the antibacterial properties of *N. carolinus* appeared to be diet- and temperature-dependent, as this species had greater antimicrobial activity in its oral secretions at cooler (4°C) rather than warmer temperatures (10–30°C). *Nicrophorus marginatus* was the opposite with greater antimicrobial activity in the oral secretions of individuals maintained at the greater temperatures. These data suggest that there is flexibility within these beetles to adapt to changing environments and partition available resources.

Another option for mitigating competition between burying beetle species for ephemeral carrion resources is through temporal partitioning of activity patterns. Wilson et al. (1984) observed diurnal and seasonal partitioning of *Nicrophorus* sp. at two sites in Michigan, USA. The authors used suspended pitfall traps and observed at multiple times during the day to determine the diurnal activity pattern of burying beetles and found that several of the species had very different activity patterns. *Nicrophorus tomentosus* Weber was a day flier, while *Nicrophorus defodiens* Mannerheim was a crepuscular flier that did not overlap in activity with the nocturnally active species observed in the study. The authors also used marked pairs of beetles placed in the field with a whole mouse carcass and explored whether these beetles maintained control of their carcass or lost it to another species. They found that daily air temperature altered the competitive outcome between these two species. Seasonal activity partitioning among the species was also observed. Wilson et al. (1984) reported that *N. sayi* was the predominate taxon in the spring, while *N. orbicollis* was a summer active species. Seasonal activity partitioning also occurs in other species and appears to differ based on geographic location. In North Carolina, *Nicrophorus* spp. were only active during spring and fall though not during the middle of summer, with *N. orbicollis* active in the spring and *N. tomentosus* activity beginning in the early fall (Trumbo 1990).

In addition to the strong competitive pressure by other carrion beetle species, burying beetles compete with microbes and other invertebrates for carrion-derived nutrients. Rozen et al. (2008) demonstrated the negative fitness effects of bacterial degradation of carrion on the burying beetle *N. vespilloides*. Old (7 days after thaw) mouse carcasses, presumably exposed to a longer period of microbial activity, resulted in smaller larvae that migrated for pupation later than those reared on fresh carcasses. The authors also observed that parental care through increased feedings and offspring tending helped to compensate for the degraded nutritional quality of the carrion. In addition, female beetles preferred fresh carcasses for laying their eggs and rearing larvae. Other competitors for carrion include the carrion-feeding flies (Scott 1994) and ants (Scott et al. 1987), which appear to alter beetle behavior and limit the geographic distribution of the carrion beetles. Early studies suggested that phoretic mites were important in controlling fly eggs but beetles themselves will also feed on larval flies and eggs to protect their carrion resource (Wilson and Knollenberg 1987).

Competition for vertebrate carrion also comes from vertebrate scavengers. Old and New world vultures are the only known obligate vertebrate carrion scavengers, as most vertebrate predators merely facultatively scavenge (DeVault et al. 2003). In addition to morphological and physiological adaptations for feeding almost exclusively on carrion, these birds have behavioral adaptations that allow them to compete for this ephemeral resource (Wink 1995). These two types of carrion-feeding raptors are thought to be an example of convergent evolution for the specialized food resource that carrion represents (Wink 1995), suggesting that strong competition and similar evolutionary constraints also exist among vertebrates specialized for carrion feeding. As in the burying beetles, behavioral variation is evident in the phylogeny of the group.

In North America, the behavioral and physiological adaptations of the overlapping distributions of the Turkey Vulture and the Black Vulture, *Coragyps atratus* (Bechstein) (Accipitriformes: Cathartidae) appears to reflect intense competition between the species. The Turkey Vulture is known for its highly developed sense of olfaction used for detection of carrion. In fact, the brain of the Turkey Vulture is characterized by enlarged olfactory lobes (Tinbergen's first question). Its direct competitor is the Black Vulture, which relies heavily on its vision (Tinbergen's first question) rather than olfaction in locating carrion. It possesses comparatively reduced olfactory lobes when compared to the Turkey Vulture. Where these two species co-occur, Black Vultures are observed to follow Turkey Vultures to carrion and force them off with aggressive behavior (Buckley 1996). The aggressive behavior of Black Vultures has also been observed in birds that attack young calves and small mammals (Humphrey et al. 2002). Efforts to scare the vultures from carrion are often not sufficient to chase them from an area, as they are also very efficient at learning. The vultures learn that threats such as loud noises and decoys are not a real threat and revert back to their aggressive behaviors in a relatively short time period (Level 2013).

As can be observed in the above examples, ancestral environments and factors (e.g., intense competition) combined with variation in behavior and differential fitness have shaped, and continue to shape, the life histories of the species observed in the present. Strong competition for carrion as an ephemeral resource is a major force in shaping the behavior of obligate carrion feeding organisms. Invertebrate specialists associated with carrion have short generation times and are easily manipulated for experimentation and thus provide valuable opportunities to investigate ultimate questions in behavioral ecology. Intense competition amongst the few vertebrate species specialized for carrion feeding also shapes the life histories and behavioral flexibility of these species groups.

10.4 Methods for Studying Behavior

There are many different approaches for conducting behavior studies in association with animals that utilize vertebrate carrion. For the purposes of this chapter, a few techniques have been highlighted. Of course, the method selected is heavily reliant on the question(s) being asked by the researcher. For those interested in molecular methods useful for explaining a behavior, such material has been reviewed in other chapters (Chapters 3, 17, and 19) and thus will only be briefly reviewed here.

To state the obvious, identifying the behavior is critical. Thus, the simplest tools available for researchers with an interest in behavior research are their abilities to hear, see, and smell. The ability to observe the behavior with one's own senses, maybe with the assistance of a microscope, binoculars, or video or sound recorder, primarily serves as the beginning point of any fundamental behavior research. Typically, once the behavior is observed in the field, studies are then moved to a controlled environment such as a laboratory when possible. Initial studies focusing on the variation in the behavior exhibited by an individual, group, or population and the impact of abiotic and biotic conditions on the frequency, intensity, and results of the behavior are critical for appreciating the plasticity and ecological ramifications impacting the ability of an individual to replicate.

Regardless of the method selected, it is vital that researchers remain aware of the impact of manipulations on behaviors as they relate to the real world. Behaviors observed in the laboratory or controlled conditions in the field only potentially give insights into its significance, as results could be a result of the experimental conditions. Typically, field experiments, in combination with laboratory experiments, are most informative of the true nature and significance of the behavior.

Technology used to examine behavior has significantly advanced. With the aid of geographic information systems, researchers are now able to examine "larger-scale" behaviors, such as dispersal and migration. Computer programs such as Observer™ allow for controlled laboratory experiments where multiple individuals can be monitored in unison and analyzed with an "unbiased" mathematical algorithm. Such programs allow "group" behavior to be examined in detail. However, these programs are not solely restricted to the laboratory. Other models have been developed to assist with researcher group behavior outside of the laboratory (Strandburg-Peshkin et al. 2013). The rapid technological and theoretical advancements in DNA, RNA, and protein sequencing and bioinformatics also represent tools researchers are turning to in order to investigate mechanism, ontology, and evolution on a molecular scale (Carter et al. 2001; Ceriani et al. 2002; Anholt et al. 2003; Lamy et al. 2009).

Simple systems have been developed for monitoring decision-making of animals in the laboratory. Olfactometers (see Chapter 9) can be used to determine individual, or group, responses to individual or mixed ratios of resources. These devices can range from very simple (two choices) to highly complex (64 choices) and allow researchers to measure responses such as attraction or repellence. These devices can also be used to examine the impact of physiological state (e.g., male vs. female, mated vs. nonmated, and age response) on the responses of individuals to given stimuli. Other devices, such as a wind tunnel (Cardé and Willis 2008) and a Kramer sphere (Von Hoermann et al. 2013), can also be used to measure such behavioral responses. Researchers can also use techniques such as electroantennograms that can be used to determine the physiological responses that are related to the manifestation, or not, of select behaviors.

These methods represent just a few of the many different tools available for individuals interested in conducting behavior research. Each technique has benefits and limitations in deciphering the significance of a given behavior to the proliferation of a given species. However, because so many methods are available, maybe the greatest challenge of a researcher is selecting the one most appropriate to address the question being asked.

10.5 Challenges and Advantages of Studying Behavior with Carrion Model

The use of carrion as a model system for studying animal behavior can be ideal in that it is a well-defined ecosystem unit, yet challenging in that it is such a temporally restricted unit. As a model for forensic entomology, carrion has major advantages in the development of both succession models and development data that can be applied to human decomposition in the estimation of insect colonization and postmortem intervals. The behavior of insects associated with decomposing animal models (i.e., swine carcasses) has been found to be very similar to that of human bodies during decomposition (Schoenly et al. 2007). Human bodies are much more restricted in their use for decomposition studies for ethical and public health reasons, and therefore the use of animal carrion models represents an excellent surrogate. Another advantage of carrion as a model for the study of the behavior of animals associated with ecosystem succession is the speed at which succession can be observed. Carrion is an ephemeral resource that can be observed in a short period of time (in the order of weeks) (Early and Goff 1986) as opposed to other systems which may take long periods of time (e.g., community recovery after wildfire, which can take years or decades; Connell and Slatyer 1977).

The rapid progression of succession on carrion may also be one of its biggest challenges. The pace at which succession progresses on carrion is so rapid that the behavior of the animals associated with it can be difficult to observe. As soon as an animal dies, the bacterial community in the host gut changes roles from commensal to consumer. This process kicks off the series of events that lead to the skeletonization of the carrion, with help from insects and vertebrates. The speed of carrion decomposition is perhaps one source of selection pressure for the development of such adaptations as the highly developed olfaction observed in many carrion specialists and the behaviors associated with reducing bacterial competition through secretions also seen in carrion specialists (Wilson and Wolkovich 2011).

The rapid pattern of carrion decay also lends itself well to manipulation. The small size and rapid decay of many carrion animal models lend themselves well to controlled replication for rigorous statistical analyses and experimental design (Michaud et al. 2012). The carcasses can be placed in various scenarios that can aid in the evaluation of environmental variables as well as in the evaluation of questions associated with applied aspects of decomposition, as in forensic entomology. For these reasons, carrion animal decomposition models are highly utilized to evaluate behaviors and patterns of the animals associated with the decomposition of tissue in the study of both basic and applied ecological questions.

10.6 Conclusions

Behaviors exhibited by animals that compete for carrion are quite complex. However, many of these behaviors are not unique to the carrion system as they are found throughout the animal kingdom; what is unique is the animal community associated with carrion. This distinctive group of organisms provides an excellent model system to conduct comparative studies with other animals associated with disparate environments. Furthermore, animals, vertebrate, and invertebrate provide excellent models for testing hypotheses associated with the evolution of behavior and its significance to ecosystem function. Indeed, vertebrate carrion might be less appealing than other systems, but it does represent a resource that can be easily manipulated and studied within the greater context of biota.

REFERENCES

Anderson, O.D. 1982. Enzyme activities in the larval secretion of *Calliphora erythrocephala*. *Comparative Biochemistry and Physiology B* 72: 569–575.

Anholt, R.R., C.L. Dilda, S. Chang, J.-J. Fanara, N.H. Kulkarni, I. Ganguly, S.M. Rollmann, K.P. Kamdar, and T.F. Mackay. 2003. The genetic architecture of odor-guided behavior in *Drosophila*: Epistasis and the transcriptome. *Nature Genetics* 35: 180–184.

Barnes, K.M., D.E. Gennard, and R.A. Dixon. 2010. An assessment of the antibacterial activity in larval excretion/secretion of four species of insects recorded in association with corpses, using *Lucilia sericata* Meigen as the marker species. *Bulletin of Entomological Research* 100: 635–640.

Bateson, P. and K.N. Laland. 2013. Tinbergen's four questions: An appreciation and an update. *Trends in Ecology and Evolution* 28: 712–718.

Baumgartner, D.L. 1993. Review of *Chrysomya rufifacies* (Diptera: Calliphoridae). *Journal of Medical Entomology* 30: 338–352.

Benecke, M. and L. Barksdale. 2003. Distinction of bloodstain patterns from fly artifacts. *Forensic Science International* 137: 152–159.

Bexfield, A., Y. Nigam, S. Thomas, and N.A. Ratcliffe. 2004. Detection and partial characterisation of two antibacterial factors from the excretions/secretions of the medicinal maggot *Lucilia sericata* and their activity against methicillin-resistant *Staphylococus aureus*. *Microbes and Infection* 6: 1297–1304.

Boncoraglio, G. and R.M. Kilner. 2012. Female burying beetles benefit from male desertion: Sexual conflict and counter-adaptation over parental investment. *PLoS ONE* 7: e31713.

Boonsriwong, W., K. Sukontason, T. Chaiwong, U. Chaisri, R.C. Vogtsberger, and K. Sukontason. 2012. Alimentary canal of the adult blow fly, *Chrysomya megacephala* (F.) (Diptera: Calliphoridae). Part I: Ultrastructure of salivary glands. *Journal of Parasitology Research* 2012: 7.

Boonsriwong, W., K. Sukontason, R.C. Vogtsberger, and K.L. Sukontason. 2011. Alimentary canal of the blow fly *Chrysomya megacephala* (F.) (Diptera: Calliphoridae): An emphasis on dissection and morphometry. *Journal of Vector Ecology* 36: 2–10.

Brundage, A.L., M.E. Benbow, and J.K. Tomberlin. 2014. Priority effects on the life-history traits of two carrion blow fly (Diptera, Calliphoridae) species. *Ecological Entomology* 39: 539–547.

Buchon, N., N.A. Broderick, and B. Lemaitre. 2013. Gut homeostasis in a microbial world: Insights from *Drosophila melanogaster*. *Nature Reviews Microbiology* 11: 615–626.

Buckley, N.J. 1996. Food finding and the influence of information, local enhancement, and communal roosting on foraging success of North American vultures. *The Auk* 113: 473–488.

Burkepile, D.E., J.D. Parker, C.B. Woodson, H.J. Mills, J. Kubanek, P.A. Sobecky, and M.E. Hay. 2006. Chemically mediated competition between microbes and animals: Microbes as consumers in food webs. *Ecology* 87: 2821–2831.

Cardé, R. and M. Willis. 2008. Navigational strategies used by insects to find distant, wind-borne sources of odor. *Journal of Chemical Ecology* 34: 854–866.

Carter, T.A., J.A. Del Rio, J.A. Greenhall, M.L. Latronica, D.J. Lockhart, and C. Barlow. 2001. Chipping away at complex behavior: Transcriptome/phenotype correlations in the mouse brain. *Physiology & Behavior* 73: 849–857.

Cazander, G., K.E.B. van Veen, A.T. Bernards, and G.N. Jukema. 2009. Do maggots have an influence on bacterial growth? A study on the susceptibility of strains of six different bacterial species to maggots of *Lucilia sericata* and their excretions/secretions. *Journal of Tissue Viability* 18: 80–87.

Ceriani, M.F., J.B. Hogenesch, M. Yanovsky, S. Panda, M. Straume, and S.A. Kay. 2002. Genome-wide expression analysis in *Drosophila* reveals genes controlling circadian behavior. *The Journal of Neuroscience* 22: 9305–9319.

Connell, J.H. and R.O. Slatyer. 1977. Mechanisms of succession in natural communities and their role in community stability and organization. *The American Naturalist* 111: 1119–1144.

Cook, M.E. 1997. *Avian Desert Predators*. Berlin, Heidelberg: Springer.

DeVault, T.L., Z.H. Olson, J.C. Beasley, and J.O.E. Rhodes. 2011. Mesopredators dominate competition for carrion inan agricultural landscape. *Basic and Applied Ecology* 12: 268–274.

DeVault, T.L., J.O.E. Rhodes, and J.A. Shivik. 2003. Scavenging by vertebrates: Behavioral, ecological, and evolutionary perspectives on an important energy transfer pathway in terrestrial ecosystems. *Oikos* 102: 225–234.

Dewsbury, D.A. 2003. The 1973 Nobel Prize for physiology or medicine: Recognition for behavioral science? *American Psychologist* 58: 747–752.

Durdle, A., R.A. van Oorschot, and R.J. Mitchell. 2013. The morphology of fecal and regurgitation artifacts deposted by the blow fly *Lucilia curpina* fed a diet of human blood. *Journal of Forensic Sciences* 58: 897–903.

Dyer, F.C. 2002. The biology of the dance language. *Annual Review of Entomology* 47: 917–949.

Early, M. and L. M. Goff. 1986. Arthropod succession patterns in exposed carrion on the island of O'ahu, Hawaiian Islands, USA. *Journal of Medical Entomology* 32: 520–531.

Evans, W.A.L. 1956. Studies on the digestive enzymes of the blowfly *Calliphora erythrocephala*: I. The carbohydrases. *Experimental Parasitology* 5: 191–206.

Ezenwa, V.O., N.M. Gerardo, D.W. Inouye, M. Medina, and J.B. Xavier. 2012. Animal behavior and the microbiome. *Science* 338: 198–199.

Fergus, C. 2003. *Wildlife of Virginia and Maryland and Washington, D.C.* Mechanicsburg, PA: Stackpole Books.

Flores, M. 2013. Life-history traits of *Chrysomya rufifacies* (Macquart) (Diptera: Calliphoridae) and its associated non-consumptive effects on *Cochliomyia macellaria* (Fabricius) (Diptera: Calliphorirdae) behavior and development. PhD Dissertation, Texas A&M University.

Fuller, M.E. 1934. The insect inhabitants of carrion: A study in animal ecology. *Council for Scientific and Industrial Research* 82: 1–63.

Gallagher, M.B., S. Sandhu, and R. Kimsey. 2010. Variation in development time for geographically distinct populations of the common green bottle fly, *Lucilia sericata* (Meigen). *Journal of Forensic Sciences* 55: 438–442.

Gião, J. and W. Godoy. 2007. Ovipositional behavior in predator and prey blowflies. *Journal of Insect Behavior* 20: 77–86.

Greenberg, B. 1973. *Flies and Disease. Biology and Disease Transmission,* Vol II. Englewood Cliffs, NJ: Princeton University Press.

Greenberg, B. and A.A. Bornsetein. 1964. Fly dispersion from a rural Mexican slaughterhouse. *American Journal of Tropical Medicine and Hygiene* 13: 881–886.

Haefner, J.N., J.R. Wallace, and R.W. Merritt. 2004. Pig decomposition in lotic aquatic systems: The potential use of algal growth in establishing a postmortem submersion interval (PMSI). *Journal of Forensic Sciences* 49: 1–7.

Heinzel, H.-G. and H. Böhm. 1989. The wind-orientation of walking carrion beetles. *Journal of Comparative Physiology A* 164: 775–786.

Hoback, W.W., A. Bishop, J. Kroemer, J. Scalzitti, and J. Shaffer. 2004. Differences among antimicrobial properties of carrion beetle secretions reflect phylogeny and ecology. *Journal of Chemical Ecology* 30: 719–729.

Hobson, R.P. 1931. Studies on the nutrition of blow-fly larvae: I. Structure and function of the alimentary tract. *Journal of Experimental Biology* 8: 109–123.

Hobson, R.P. 1932. Studies on the nutrition of blow-fly larvae: III. The liquefaction of muscle. *Journal of Experimental Biology* 9: 359–365.

Holmes, L.A., S.L. Vanlaerhoven, and J.K. Tomberlin. 2012. Relative humidity effects on the life history of *Hermetia illucens* (Diptera: Stratiomyidae). *Environmental Entomology* 41: 971–978.

Holmes, L.A., S.L. Vanlaerhoven, and J.K. Tomberlin. 2013. Substrate effects on pupation and adult emergence of *Hermetia illucens* (Diptera: Stratiomyidae). *Environmental Entomology* 42: 370–374.

Humphrey, J.S., E.A. Tillman, and M.L. Avery. 2002. Vulture–cattle interactions at a central Florida ranch. In: Timms, R.M. and W.P. Gorenzel (eds), *Proceedings of the Twentieth Vertebrate Pest Conference*, University of California, Davis, pp. 122–125.

Hyde, E.R., D.P. Haarmann, A.M. Lynne, S.R. Bucheli, and J.F. Petrosino. 2013. The living dead: Bacterial community structure of a cadaver at the onset and end of the bloat stage of decomposition. *PLoS ONE* 8: e77733.

Ingram, K.K., L. Kleeman, and S. Peteru. 2011. Differential regulation of the *foraging* gene associated with task behaviors in harvester ants. *BMC Ecology* 11: 9.

Iwasaki, K. and J.H. Thomas. 1997. Genetics in rhythm. *Trends in Genetics* 13: 111–115.

Jacques, B.J., S. Akahane, M. Abe, W. Middleton, W.W. Hoback, and J.J. Shaffer. 2009. Temperature and food availability differentially affect the production of antimicrobial compounds in oral secretions produced by two species of burying beetle. *Journal of Chemical Ecology* 35: 871–877.

Janzen, D.H. 1977. Why fruits rot, seeds mold, and meat spoils. *American Naturalist* 111: 691–713.

Jiang, K.C., X.J. Sun, W. Wang, L. Liu, Y. Cai., Y.C. Chen, N. Luo, J.H. Yu, D.Y. Cai, and A.P. Wang. 2012. Excretion/sections from bacteria-pretreated maggots are more effective against *Pseudomonas aeruginosa* biofilms. *PLoS ONE* 7: e49815.

Kalinova, B., H. Podskalska, J. Ruzicka, and M. Hoskovec. 2009. Irresistible bouquet of death—How are burying beetles (Coleoptera: Silphidae: *Nicrophorus*) attracted by carcasses. *Naturwissenschaften* 96: 889–899.

Kaplan, W.D. and W.E. Trout III. 1968. The behavior of four neurological mutants of *Drosophila*. *Genetics* 61: 399–409.

Lamy, E., G. da Costa, R. Santos, F. Capela e Silva, J. Potes, A. Pereira, A.V. Coelho, and E. Sales Baptista. 2009. Sheep and goat saliva proteome analysis: A useful tool for ingestive behavior research? *Physiology & Behavior* 98: 393–401.

Larson, K. and J.G.J. Stoffolano. 2011. Effect of high and low concentrations of sugar solutions fed to adult male, *Phormia regina* (Diptera: Calliphoridae), on "bubbling" behavior. *Annals of the Entomological Society of America* 104: 1399–1403.

Level, A. 2013. *Vultures and Livestock*. AgNIC Wildlife Damage Management. Accessed Date: 03/04/2014. (http://lib.colostate.edu/research/agnic/vultures-livestock.html).

Lewis, A.J. and M.E. Benbow. 2011. When entomological evidence crawls away: *Phormia regina* en masse larval dispersal. *Journal of Medical Entomology* 48: 1112–1119.

Lucas, C., R. Kornfein, M. Chakaborty-Chatterjee, J. Schonfeld, N. Geva, M.B. Sokolowski, and A. Ayali. 2010. The locust *foraging* gene. *Insect Biochemistry and Physiology* 74: 52–66.

Ma, Q., A. Fonseca, W. Liu, A.T. Fields, M.L. Pimsler, A.F. Spindola, A.M. Tarone, T.L. Crippen, J.K. Tomberlin, and T.K. Wood. 2012. *Proteus mirabilis* interkingdom swarming signals attract blow flies. *International Society of Microbial Ecology Journal* 6: 1356–1366.

Marella, S., W. Fischle, P. Kong, S. Asgarian, E. Rueckert, and K. Scott. 2006. Imaging taste responses in the fly brain reveals a functional map of taste category and behavior. *Neuron* 49: 285–295.

Margalida, A., J.J. Negro, and I. Galván. 2008. Melanin-based color variation in the bearded vulture suggests a thermoregulatory function. *Comparative Biochemistry and Physiology Part A: Molecular & Integrative Physiology* 149: 87–91.

Metcalf, J.L., L. Wegener Parfrey, A. Gonzalez, C.L. Lauber, D. Knights, G. Ackermann, G.C. Humphrey et al. 2013. A microbial clock provides an accurate estimate of the postmortem interval in a mouse model system. *eLife* 2: e01104.

Michaud J.-P., K.G. Schoenly, and G. Moreau. 2012. Sampling flies or sampling flaws? Experimental design and inference strength in forensic entomology. *Journal of Medical Entomology* 49: 1–10.

Moe, S.J., N.C. Stenseth, and R.H. Smith. 2002. Density dependence in blowfly populations: Experimental evaluation of non-parametric time-series modelling. *Oikos* 98: 523–533.

Mohr, R.M. 2012. Female blow fly (Diptera: Calliphoridae) arrival patterns and consequences for larval development on ephemeral resources. PhD Dissertation, Texas A&M University, College Station, TX.

Moore, M.D. and J. Breed (eds.). 2012. *Animla Behavior*. San Diego, CA: Academic Press.

Mossman, M. 1976. Turkey vultures in the Baraboo Hills, Sauk County, Wisconsin. *Passenger Pigeon* 38: 93–99.

Nguyen, T.T.X., J.K. Tomberlin, and S.L. VanLaerhoven. 2013. Influences of resources on *Hermetia illucens* (Diptera: Stratiomyidae) larval development. *Journal of Medical Entomology* 50: 898–906.

Ort, B.S., R.M. Bantay, N.A. Pantoja, and P.M. O'Grady. 2012. Fungal diversity associated with Hawaiian *Drosophila* host plants. *PLoS ONE* 7: e40550.

Owings, C., C. Spiegelman, A. Tarone, and J. Tomberlin. 2014. Developmental variation among *Cochliomyia macellaria* Fabricius (Diptera: Calliphoridae) populations from three ecoregions of Texas, USA. *International Journal of Legal Medicine* 128: 709–717.

Payne, J.A. 1965. A summer carrion study of the baby pig *Sus scrofa* Linnaeus. *Ecology* 46: 592–602.

Payne, J.A., E.W. King, and G. Beinhart. 1968. Arthropod succession and decomposition of buried pigs. *Nature* 219: 1180–1181.

Pechal, J.L. 2012. The importance of microbial and primary colonizer interactions on an ephemeral resource. PhD Dissertation, Texas A&M University, College Station, TX.

Pechal, J.L., T.L. Crippen, M.E. Benbow, A.M. Tarone, S. Dowd, and J.K. Tomberlin. 2013. The potential use of bacterial community succession in forensics as described by high throughput metagenomic sequencing. *International Journal of Legal Medicine* 128: 193–205.

Powell, J.R. 1997. *Progress and Prospects in Evolutionary Biology: The* Drosophila *Model*. New York: Oxford University Press.

Reid, C.L.M. 2012. The role of community composition and basal resources in a carrion community. M.S. Thesis, University of Windsor.

Rosati, J.Y. and S.L. VanLaerhoven. 2007. New record of *Chrysomya rufifacies* (Diptera: Calliphoridae) in Canada: Predicted range expansion and potential effects on native species. *The Canadian Entomologist* 139: 670–677.

Rozen, D.E., D.J.P. Engelmoer, and P.T. Smiseth. 2008. Antimicrobial strategies in burying beetles breeding on carrion. *Proceedings of the National Academy of Sciences* 105: 17890–17895.

Schlief, M.L. and R.I. Wilson. 2007. Olfactory processing and behavior downstream from highly selective receptor neurons. *Nature Neuroscience* 10: 623–630.

Schoenly, K., N.H. Haskell, R.D. Hall, and J.R. Gbur. 2007. Comparative performance and complementarity of four sampling methods and arthropod preference tests from human and porcine remains at the Forensic Anthropology Center in Knoxville, Tennessee. *Journal of Medical Entomology* 44: 881–894.

Scott, M.P. 1994. Competition with flies promotes communal breeding in the burying beetle, Competition with flies promotes communal breeding in the burying beetle, *Nicrophorus tomentosus*. *Behavioral Ecology and Sociobiology* 34: 367–373.

Scott, M.P. 1998. The ecology and behavior of burying beetles. *Annual Review of Entomology* 43: 595–618.

Scott, M.P. and S.C. Panaitof. 2004. Social stimuli affect juvenile hormone during breeding in biparental burying beetles (Silphidae: *Nicrophorus*). *Hormones and Behavior* 45: 159–167.

Scott, M.P., J.F.A. Traniello, and I.A. Fetherston. 1987. Competition for prey between ants and burying beetles (*Nicrophorus* spp.): Differences between northern and southern temperate sites. *Psyche: A Journal of Entomology* 94: 325–332.

Shiao, S.F. and T.C. Yeh. 2008. Larval competition of *Chrysomya megacephala* and *Chrysomya rufifacies* (Diptera: Calliphoridae): Behavior and ecological studies of two blow fly species of forensic significance. *Journal of Medical Entomology* 45: 785–799.

Smith, S.A. and R.A. Paselk. 1986. Olfactory sensitivity of the Turkey vulture (*Cathartes aura*) to three carrion-associated odorants. *The Auk* 103: 586–592.

Sokolowski, M.B. 1985a. Genetics and ecology of *Drosophila melanogaster* larval foraging and pupation behaviour. *Journal of Insect Physiology* 31: 857–864.

Sokolowski, M.B. 1985b. Genetic aspects to differences in foraging behavior. *Behavioral and Brain Sciences* 8: 348–349.

Sokolowski, M.B. 2001. *Drosophila*: Genetics meets behaviour. *Nature Reviews Genetics* 2: 879–890.

Sokolowski, M.B., C. Kent, and J. Wong. 1984. *Drosophila* larval foraging behaviour: Developmental stages. *Animal Behaviour* 32: 645–651.

Stamps, J. 2003. Behavioural processes affecting development: Tinbergen's fourth question comes to stage. *Animal Behaviour* 66: 1–13.

Steiger, S., V. Kuryshev, M. Stensmyr, B. Kempenaers, and J. Mueller. 2009. A comparison of reptilian and avian olfactory receptor gene repertoires: Species-specific expansion of group γ genes in birds. *BMC Genomics* 10: 446.

Stoffolano, J.G., A. Acaron, and M. Conway. 2008. "Bubbling" or droplet regurgitation in both sexed of adult *Phormia regina* (Diptera:Calliphoridae) fed various concentrations of sugar and protein solutions. *Annals of the Entomological Society of America* 101: 964–970.

Strandburg-Peshkin, A., C.R. Twomey, N.W.F. Bode, A.B. Kao, Y. Katz, C.C. Ioannou, S.B. Rosenthal et al. 2013. Visual sensory networks and effective information transfer in animal groups. *Current Biology* 23: R709–R711.

Suh, G.S.B., A.M. Wong, A.C. Hergarden, J.W. Wang, A.F. Simon, S. Benzer, R. Axel, and D.J. Anderson. 2004. A single population of olfactory sensory neurons mediates an innate avoidance behaviour in *Drosophila*. *Nature* 431: 854–859.

Tarès, S., L. Arthaud, M. Amichot, and A. Robichon. 2013. Environmental exploraion and colonization behavior of the pea aphid associated with expression of the *foraging* gene. *PLoS ONE* 8: e65104.

Tarone, A.M. and D.R. Foran. 2008. Generalized additive models and *Lucilia sericata* growth: Assessing confidence intervals and error rates in forensic entomology. *Journal of Forensic Sciences* 53: 942–949.

Thomas, D.B. 1993. Behavioral aspects of screwworm ecology. *Journal of the Kansas Entomological Society* 66: 13–30.

Tinbergen, N. 1963. On aims and methods of ethology. *Zeitschrift für Tierpsychologie* 20: 410–433.

Tobback, J., V. Mommaerts, H.P. Vandersmissen, G. Smagghe, and R. Huybrechts. 2011. Age- and task-dependent *foraging* gene expression in the bumblebee *Bombus terrestris*. *Insect Biochemistry and Physiology* 76: 30–42.

Truman, J.W. and L.M. Riddiford. 1999. The origins of insect metamorphosis. *Nature* 401: 447–452.

Trumbo, S.T. 1990. Reproductive success, phenology and biogeography of burying beetles (Silphidae: *Nicrophorus* spp.). *American Midland Naturalist* 124: 1–11.

Trumbo, S.T., D.W. Borst, and G.E. Robinson. 1995. Rapid elevation of juvenile hormone titer during behavioral assessment of the breeding resource by the burying beetle, *Nicrophorous orbicollis*. *Journal of Insect Physiology* 41: 535–543.

Trumbo, S.T. and G.E. Robinson. 2008. Social and nonsocial stimuli and juvenile hormone titer in a male burying beetle, *Nicrophorus orbicollis*. *Journal of Insect Physiology* 54: 630–635.

Turner, T., C.C. Giauque, D.R. Schrider, and A.D. Kern. 2014. Genome-wide association of foraging behavior in *Drosophila melanogaster* fails to support large-effect alleles at the foraging gene. *bioRxiv*.

van der Plas, M.J., C. Dambrot, H.C. Dogterom-Ballering, S. Kruithof, J.T. van Dissel, and P.H. Nibbering. 2010. Combinations of maggot excretions/secretions and antibiotics are effective against *Staphylococcus aureus* biofilms and the bacteria derived therefrom. *Journal of Antimicrobial Chemotherapy* 65: 917–923.

Vistnes, L., R. Lee, and G. Ksander. 1981. Proteolytic activity of blowfly larvae secretions in experimental burns. *Surgery* 90: 835–841.

Vogel Jr, H.H. 1950. Observations on social behavior in Turkey vultures. *The Auk* 67: 210–216.

Von Hoermann, C., S. Steiger, J.K. Müller, and M. Ayasse. 2013. Too fresh is unattractive: The attraction of newly emerged *Nicrophorus vespilloides* females to odour bouquets of large cadavers at various stages of decomposition. *PLoS ONE* 8: e58524.

Wagner, F.H. 1980. *Wildlife of the Desert*. New York: Abrams.

Wells, J.D. and B. Greenberg. 1992. Interaction between *Chrysomya rufifacies* and *Cochliomyia macellaria* (Diptera, Calliphoridae)—The possible consequences of an invasion. *Bulletin of Entomological Research* 82: 133–137.

Wells, J.D. and B. Greenberg. 1994. Effect of the red imported fire ant (Hymenoptera: Formicidae) and carcass type on the daily occurrence of postfeeding carrion-fly larvae (Diptera: Calliphoridae, Sarcophagidae). *Journal of Medical Entomology* 31: 171–174.

Westneat, D.F. and C.W. Fox. 2010. *Evolutionary Behavioral Ecology*. New York: Oxford Press.

Whitworth, T. 2006. Keys to the genera and species of blow flies (Diptera: Calliphoridae) of America north of Mexico. *Proceedings of the Entomological Society of Washington* 108: 689–725.

Wilson, D.S. and W.G. Knollenberg. 1987. Adaptive indirect effects: The fitness of burying beetles with and without their phoretic mites. *Evolutionary Ecology* 1: 139–159.

Wilson, D.S., W.G. Knollenberg, and J. Fudge. 1984. Species packing and temperature dependent competition among burying beetles (Silphidae, *Nicrophorus*). *Ecological Entomology* 9: 205–216.

Wilson, E.E. and E.M. Wolkovich. 2011. Scavenging: How carnivores and carrion structure communities. *Trends in Ecology and Evolution* 26: 129–135.

Wink, M. 1995. Phylogeny of old and new world vultures (Aves: Accipitridae and Cathartidae) inferred from nucleotide sequences of the mitochondrial cytochrome b gene. *Zeritschrift fur Naturforsch* C.50: 868–882.

Yoshihara, M. and K. Ito. 2012. Acute genetic manipulation of neuronal activity for the functional dissection of neural circuits—A dream come true for the pioneers of behavioral genetics. *Journal of Neurogenetics* 26: 43–52.

11

Modeling Species Interactions within Carrion Food Webs

Sherah L. VanLaerhoven

CONTENTS

11.1 Introduction

In the span of less than a day, a once living animal becomes the resource upon which numerous organisms colonize and utilize to rear their offspring. With favorable abiotic conditions, such as warm temperatures and no precipitation, this carrion resource can be colonized, within minutes after death, by those species that are well adapted to locating such patchy resources. Other species arrive and within a few days, the carrion resource may host tens to hundreds of species and thousands of individuals, developing into a rich, albeit ephemeral, community. Yet how is it similar species are able to utilize the same resources within the carrion simultaneously, whereas other species are excluded? What factors determine who is, and is not, successful? How do these dynamics change when some species are present or excluded? These are some of the questions at the heart of community ecology, and carrion has the advantage of providing the means of testing these questions using multiple replicates and easily manipulated communities that assemble and disperse well within the lifetime of the researcher (see Chapter 7), making them invaluable as model systems.

So who is successful in colonizing carrion resources? Obviously, individuals have to be able to find the resource, and those who have evolved to recognize and rapidly respond to cues associated specifically with animal death should be the earliest colonizers (see Chapter 2). Larger carrion resources should support more species than smaller ones, as is the generalized pattern in ecology (Arrhenius 1921). Given that carrion resources can occur anywhere in the landscape and are not predictable at local temporal or spatial scales, the number of species supported by carrion resources should be determined by the theory of island biogeography (MacArthur and Wilson 1967), whereby the closer carrion resources are to each other, as well as the larger they are, the more species they should support.

Based on island biogeography theory (MacArthur and Wilson 1967), larger carrion resources should have more microhabitats and support larger populations with less risk of extinction prior to successfully reaching reproductive maturity. Thus, any context, such as clothing or fur, that increases the potential microhabitats associated with carrion resources should also have the potential to increase species diversity. Species diversity within a carrion community is a function of immigration to the carrion resource and emigration or extinction. Immigration is a function of dispersal and success-ful colonization, and for the purposes of this chapter, it will be assumed that species can get to the resource (for a discussion of community assembly on carrion, see VanLaerhoven 2010 and Chapter 8 of this book), and instead, we will focus on a species, ability to successfully colonize as it is influenced by species interactions. Extinction of species from a carrion resource may be due to random chance events, emigration to new resources at the completion of a life stage, or negative species interactions such as predation or competition. As we compare across increased carrion resource size and as more species arrive to colonize a resource, rates of local extinction events should accelerate, as there are potentially more species present at a given time to go extinct, smaller populations are more suscep-tible to extinction, and there is a higher probability of species interactions driving extinction from that carrion resource. Thus, at some point, the rate of species immigration becomes equal to the rate of extinction on the carrion resource "island." However, given the ephemeral nature of carrion, there is no true equilibrium from this point onward, and local extinction increases faster than species immi-gration until the resource is fully recycled into the surrounding ecosystem. Yet, this switching point in species diversity should be faster on smaller carrion resources than larger ones and on carrion found closer together than those further apart.

Although species diversity on a carrion resource is a function of immigration and extinction (emigra-tion), the population of each individual species is also a function of the birth and death rates. As it is rare for more than the initial colonizing generation of a species (excluding microbes) to be supported within the timeframe of a single carrion resource (Beaver 1977), it is assumed that the per capita birth rate of the F_2 generation is zero within the patch, requiring dispersal and emigration of the F_1 generation to a new carrion patch. Therefore, the population dynamics of interacting species on carrion occurs at the metapopulation scale, not simply individual local carrion communities. Within each local carrion com-munity, resources are limited and finite, thus the growth of populations is limited by the number and size of carrion resource patches available across the scale of the metapopulation.

As population density increases, the per capita growth rate declines due to negative interactions between individuals and depletion of resources. This negative feedback is the basis for density-dependent population regulation, which is described by the logistic equation for population growth (Verhulst 1838):

$$\frac{dN}{dt} = rN\left(\frac{K - N}{K}\right)$$

where
  r = per capita population growth rate,
  N = population density,
  K = carrying capacity of the environment.

This model assumes there is no immigration/emigration, no time lags, and no individual age or size effects on reproduction (i.e., all individuals within the population are equal). Although accumulation of wastes is often cited as a major factor maintaining population densities (Bedhomme et al. 2005), in a carrion resource context, this may not be a density control since the F_1 generation of the initial colonizing flies is required to disperse to a new patch for reproduction. Expanding this idea of density dependence to the community level, if a carrion resource can only support a fixed number or biomass of individuals, the increase in the abundance of one species must result in the decrease in the abun-dance of other species within the same trophic level as a result of compensatory dynamics, also called zero-sum dynamics (Hubbell 2001; Gonzalez and Loreau 2009). Zero-sum dynamics assumes that individuals within the same trophic level are utilizing the same resources within carrion, resulting in a community-level carrying capacity. Although individuals interact, it is the effect of these species interactions at the population level over time that result in the dynamics that structure the carrion communities we observe. As such, we will further explore species interactions from an individual and population level.

11.2 Species Interactions

Species interactions fall into six theoretical types, which may be positive, neutral, or negative for either of the interacting species (Figure 11.1; Haskell 1947).

These interactions may occur within the same trophic level, between trophic levels, and may be between individuals of the same species (intraspecific) or between individuals of different species (interspecific). Although it is possible that all types of interaction could occur between species utilizing carrion resources, this chapter will focus mainly on two interactions that are expected to be the most common and strongest at structuring carrion communities: competition, and predation.

11.2.1 Competition

Assuming density-dependent population regulation, if an individual encounters increasing density of other individuals, potential outcomes are to decrease its own growth rate and take longer to acquire carrion resources, trade off smaller adult size for successfully reaching reproductive maturity, deposit fewer offspring, or deposit offspring of lower quality. In this scenario, mortality of individuals increases with increasing density as a product of competition. If the individuals are of two different species, then interspecific competition is defined by the reduction in per capita growth of one species due to the shared use of the carrion resource by the second species. The mechanisms by which this competition occurs may take different forms in a carrion resource context but there are two main types. In resource competition (or exploitative competition), one species consumes the carrion resource, making this limited resource less available to other species for consumption, such as blow fly (Diptera: Calliphoridae) larvae feeding on muscle tissue, making it unavailable for flesh fly (Diptera: Sarcophagidae) larvae (Denno and Cothran 1976).

In contrast, interference competition (or contest competition) is when one species reduces access of other species to the limited carrion resource through direct or indirect means. This may take the form of monopolizing physical space for oviposition, such as the blow fly *Lucilia sericata* (Meigen) (Diptera: Calliphoridae) laying its eggs in the eyes, nose, and mouth of carrion, leaving less optimal locations for other flies. Also called preemptive competition, it restricts other species from utilizing those preferred locations. Another form of interference competition is when one species literally grows over another, taking up the space and resource, which may occur with microbial colonies (Hibbing et al. 2010). In chemical competition, the interference is by toxic chemicals that drive the other species away, such as the antimicrobial compounds secreted by blow flies (Cerovsky et al. 2010) that change the microbial community on carrion (see Chapters 3 and 18 through 20). Territorial competition, such as fights and mate guarding by dung beetles (Coleoptera: Scarabaeidae) (Sato and Hiramatsu 1993), is a behavioral interaction whereby one species defends specific physical space, mates, or resources with aggressive displays, excluding other species. The final major type of interference competition is encounter competition which occurs when predators or parasitoids run into each other and cause one of them to stop foraging on prey/hosts, diverting time away from acquiring resources; however, as this blurs the lines with predatory

		Species 1	
	−	0	+
−	Competition	Amensalism	Predation/parasitism
Species 2 **0**	Amensalism	Neutralism	Commensalism
+	Predation/parasitism	Commensalism	Mutualism

FIGURE 11.1 Pairwise species interactions based on positive, neutral, or negative effects on each species. (From Haskell, E.F. 1947. *Transactions of the NY Academy of Sciences* 9:186–196.)

interactions, we will discuss this later in the context of intraguild predation. Although we have defined the different types of competition, it is the outcomes of these competitive interactions on populations that may determine carrion community structure.

Building on the logistic equation for population growth, the Lotka–Volterra (L–V model) competition model (Lotka 1925; Volterra 1926) provides a starting point to describing the outcomes of interspecific competition on population growth of two species.

$$\frac{dN_1}{dt} = r_1 N_1 \left(\frac{k_1 - N_1 - \alpha N_2}{k_1} \right)$$

$$\frac{dN_2}{dt} = r_2 N_2 \left(\frac{k_2 - N_2 - \beta N_1}{k_2} \right)$$

where
   r = per capita population growth rate of species 1 or 2,
   N = population density of species 1 or 2,
   k = carrying capacity of the environment for species 1 or 2,
   α = competition coefficient with per capita effect of species 2 on species 1,
   β = competition coefficient with per capita effect of species 1 on species 2.

If there is no competition between the two species, then α and β are equal to zero and both populations grow as described by the logistic growth model. The L–V model assumes that the effect of one species on the other is linear, carrying capacities are constant, and the environment is stable. Although this model is not realistic for carrion communities at local patch levels, it may be more informative if the metapopulation of multiple patches is considered. Most importantly, the L–V model makes four mutually exclusive predictions regarding the outcomes of interspecific competition, which can be modeled using zero net growth isoclines (Gotelli 2008, Mittelbach 2012). Species 1 per capita growth rate will be zero when its density is equal to its carrying capacity, and it will also be zero when the density of species 2 is equivalent to the carrying capacity of species 1, representing the combination of densities of species 1 and 2 that is the zero net growth isocline for species 1 (Figure 11.2). At densities above the isocline, populations of species 1 are declining, whereas at densities below the isocline, populations of species 1 are growing. Because the per capita density dependence is linear, the zero growth isocline is linear.

Thus, by plotting the zero net growth isoclines for both species and based on the initial starting population levels, the outcomes for interspecific competition can be predicted (Figure 11.2). In scenarios A and B where the isoclines do not cross, the species whose isocline is above the other species will competitively exclude the other. Only when the isoclines cross as illustrated in scenarios C and D will there be the possibility for coexistence, yet this coexistence may be stable or unstable. Depending on the orientation of the isoclines, there are two possibilities such that either all starting population levels (indicated in red letters) have paths that will stabilize at the equilibrium point (scenario C), or if the starting density of the populations falls within two of the possible four quadrants that point away from equilibrium, then one species will be competitively excluded by the other (scenario D). Alternatively, there are two quadrants in scenario D where the starting population densities of species 1 and 2 will result in an unstable temporary coexistence. Therefore, the initial density of the two species determines the outcome of competition, with the more abundant species likely to prevail. This scenario is also known as founder control (Mittelbach 2012) and may explain the conflicting results reported with coexistence or competitive exclusion of blow flies on carrion (i.e., Hanski 1987; Wells and Greenberg 1994a). Each time a new carrion resource patch becomes available and is colonized, it resets the initial population densities, allowing for different outcomes of competition at local scales. This changes the coexistence dynamics within the wider metapopulation that is a sum of the outcomes of the individual populations within individual communities on each resource patch. According to Chesson (2000), when intraspecific competition is greater

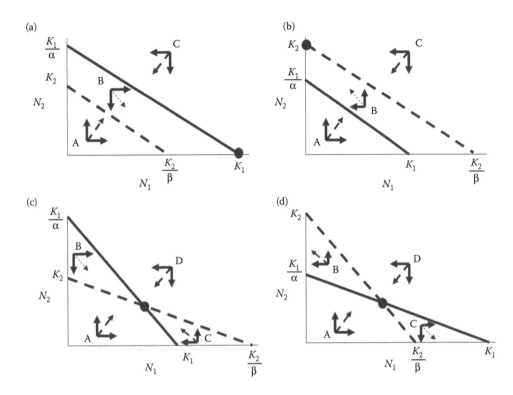

FIGURE 11.2 Outcomes of interspecific competition using zero percent growth isoclines. Each letter indicates a potential starting population density for species 1 and 2. Arrows indicate the direction of population dynamics due to competition and the black circle is the population densities after interspecific competition. (a) Competitive exclusion—species 1 wins, (b) competitive exclusion—species 2 wins, (c) stable equilibrium coexistence, and (d) unstable nonequilibrium coexistence. (Modified from Mittelbach, G.G. 2012. after Gotelli, N.J. 2008. *A Primer of Ecology*, 4th Edition. Sinauer Associates, Sunderland, MA.)

than interspecific competition, then two species coexist in the wider metacommunity and the L–V model can be expanded from relative competition coefficients to absolute competition coefficients.

However, far more than two species colonize a carrion resource patch and the L–V model can be extended into a community matrix (Levins 1968). Yet this model assumes that species effects are additive, and competitive interactions between single pairs of species are unaffected by other species in the community. This assumption is unrealistic for natural communities due to the presence of nonadditive or higher-order interactions such as trait-mediated interactions between predators and prey (Mittelbach 2012). Additionally, the L–V model does not explicitly include resources, yet carrion-associated communities are explicitly defined by their use of the resource, and exploitative competition is likely to be a driving force in structuring these communities. Thus, consumer–resource competition models may provide further insights into the criteria for species coexistence within communities on carrion resource patches.

Tilman (1982) developed consumer–resource models for two competing species utilizing the same essential resource. He demonstrated that at constant population densities, both species cannot coexist on one limiting resource and that whichever species can maintain its population density utilizing less of the resource will competitively exclude the second species. This definition is analogous to Gause's (1934) competitive exclusion principle. Tilman (1982) further expanded this model to multiple competing species utilizing two essential resources that allowed for four mutually exclusive outcomes of competition, similar to that predicted by the L–V model. The amount of an essential resource required to maintain a species population density (i.e., zero net growth isocline) compared to the availability of that resource can be graphed for two essential resources and compared to the requirements of a second species for those two resources to maintain the second species' population density. Coexistence between two species requires a trade-off in their ability to compete for the two resources, demonstrated by their zero net growth isoclines crossing. This coexistence is stable only if the resource that is consumed most by each species is the one that most limits its growth. Alternatively, if each species preferentially consumes the resource that most limits the growth of the other species, then the coexistence is unstable. Similar to the results of the L–V model, the starting supply of resources will influence the outcome of competition and lead to variability in competitive exclusion. With multiple species and spatial heterogeneity in the supply of resources such as the situation with carrion resources of different sizes and types that are unpredictable over the landscape, this promotes multiple consumer species coexistence. These multiple species may occur spatially separated as one- or two-species pairs across the available resource patches (Tilman 2007).

Thus, consumer–resource models predict that temporal or spatial heterogeneity can promote species coexistence in carrion resource communities, which are unpredictable both spatially and temporally. However, this coexistence requires trade-offs in the abilities of competing species to utilize carrion resources. In dung beetle guilds of the afro-tropics, telecroprids (rollers) are able to efficiently utilize buffalo dung resources during the heat of the day in the savannah, whereas endocroprids (dwellers) are dominant at night in the grassland river valley and paracroprids (tunnelers) are able to utilize these dung resources in multiple habitats during the night (Krell et al. 2003), suggesting that spatial and temporal heterogeneity with trade-offs in the different guilds' ability to utilize the dung resources has promoted the coexistence of all three guilds. Although not stated in this exact context, Hanski (1987) tested a form of the consumer–resource model utilizing multiple carrion-breeding flies. He found that the starting relative population sizes of the flies (influenced by season in his study) impacted the outcomes of competition, as would be expected by the predictions of the L–V model. Additionally, resource spatial heterogeneity influenced the outcomes of competition, as suggested by predictions of consumer–resource models, yet it is unclear if the mechanism was due to trade-offs in how the different species utilized the resource as only one overall type of resource was manipulated (i.e., fresh cow liver of different sizes with total amount remaining constant).

Another form of spatial heterogeneity that should impact the coexistence and outcome of competition is the spatial distribution of individuals, rather than resources. Spatial aggregation has been proposed as a mechanism to enable competitors to coexist in a metapopulation of habitat patches (Atkinson and Shorrocks 1981). Coexistence requires that the competing species have independently aggregated distributions across local resource patches throughout a landscape of resource islands that comprise the wider metapopulation, as this magnifies intraspecific competition instead of interspecific competition (Atkinson and Shorrocks 1981; Ives 1991). Hanski's (1987) study was designed around this concept, yet whether it is the spatial aggregation of individuals on resource patches or it is a potential trade-off in how the resource was utilized that resulted in coexistence or extinction of species in his experiment was unclear. Kneidel (1985) also studied spatial resource patchiness and demonstrated that increased spatial resource patchiness resulted in less interspecific competition and increased coexistence of two species of carrion-breeding flies, *Fannia howardi* Malloch (Diptera: Muscidae) and *Megaselia scalaris* Loew (Diptera: Phoridae). In that study, species aggregated their eggs at higher resource patchiness, thereby increasing intraspecific competition at higher resource patchiness. A recent review by Fiene et al. (2014) further discussed aggregation on carrion resources, reanalyzed *Chrysomya rufifacies* Macquart and *Cochliomyia macellaria* (F.) (Diptera: Calliphoridae) emergence data by Wells and Greenberg (1992) and adult *L. sericata*, *L. illustris* Meigen, and *Phormia regina* Meigen (Diptera: Calliphoridae) trapping data from my own lab (VanLaerhoven, Svilans and Yamashita, unpublished data), and demonstrated the presence of aggregation at both a spatial and potentially temporal scale. Fiene et al. (2014) suggest that spatial and temporal aggregation is a common mechanism structuring competitive interactions in these communities on both scales, the increase in intraspecific competition instead of intraspecific competition, and resulting occurrence of refugia from interspecific competitors should increase coexistence on patchy resources. Tying this idea together with previous discussion regarding the temporal availability of resources provides the foundation to further explore outcomes at these scales.

As predicted by consumer resource models, temporal heterogeneity, not just spatial, influences competitive interaction outcomes at local population and metapopulation scales. The lottery model of Chesson and Warner (1981) combines the temporal availability of resources with the temporal fluctuation in population dynamics of competing species. Environmental fluctuation prevents competitive exclusion and extinction at the metapopulation scale of either species that utilizes temporally unpredictable resources. Instead, the initial species, then followed by the other species, is favored by environmental stochasticity. Specifically, whichever species has more offspring available when an ephemeral resource becomes available has an increased probability of establishing a local population to utilize it. As both the occurrence of carrion resource availability and the timing of which species has a greater population at that moment is unpredictable, successfully finding and colonizing a carrion resource is like winning a lottery. Shorrocks and Bingley (1994) demonstrated this using two species of flies on decaying mushrooms. When the two species arrive together, *Drosophila phalerata* Meigen (Diptera: Drosophilidae) outcompetes *D. subobscura* Collin (Diptera: Drosophilidae); however, when one species arrives a few days before the other, both are able to colonize, but the later-arriving species has a longer development time, lower survival rate, smaller adult size, and, subsequently, lower fitness. The lesser competitor, *D. subobscura*, is a fugitive species and persists in the metapopulation by discovering new mushrooms first. In another example, the blow fly *Co. macellaria* has increased fecundity and survival when it colonizes carrion resources at least 2 days prior to the invasive blow fly *C. rufifacies* (Brundage et al. 2014). This competitive advantage of arriving first, known as a priority effect, has been demonstrated for insects colonizing carrion resources (e.g., Hanski and Kuusela 1977; Kneidel 1983; Hanski 1987). As a full discussion of priority effects, arrival time, dispersal, and spatial and temporal scale effects in community assembly on carrion have been discussed elsewhere (VanLaerhoven 2010), the majority of discussion here will be restricted to species interactions after successful colonization of carrion resources.

Interactions between arthropods on carrion may occur at the level of adults or larvae, both of which may utilize the resource for food or shelter. But outside manipulative laboratory experiments, how is interspecific competition detected and quantified on carrion? In natural communities, the outcomes of interspecific competition can be inferred by niche shifts, resource partitioning, and character displacement/convergence in the presence or absence of competitors (Mittelbach 2012). Certainly, different types and sizes of carrion may be colonized by different flies (e.g., Denno and Cothran 1975; Kneidel 1984; Ives 1991), and these observed patterns have been proposed as mechanisms of reducing niche overlap in carrion communities (Denno and Cothran 1975; Hanski and Kuusela 1980), yet this has not been empirically tested in the presence and absence of competitors in many instances. Certainly, exclusion of arthropods from carrion for different periods of time results in a different community of arthropods when they are provided access to the resource but whether this is due to competitive interactions, other species interactions, or differential attraction due to changing cues that species utilize to locate appropriate resources is not yet known (Pechal et al. 2014). There is increasing evidence that temporal and spatial resource partitioning occurs between competitors, with examples of different arrival times to carrion, presence/absence of species within different seasons, on different types/sizes of carrion, and even within single carrion resources as females should compete for oviposition locations on carrion that provide their offspring with the greatest survivorship. Yet, the question remains as to what degree is this driven by competitive interactions?

Other mechanisms such as physiological constraints, dispersal constraints, spatial or temporal heterogeneity, indirect interactions, or interactions with higher tropic levels may also play a role in explaining the patterns we have observed in carrion communities (VanLaerhoven 2010). Yet, the complexity of multiple mechanisms structuring carrion communities does not negate the empirical laboratory measurements of reduced survivorship, smaller adult size. and lower fitness of numerous blow fly species reared in the presence of competing blow fly species (e.g., Shiao and Yeh 2008). As has been demonstrated, competition is likely an important interaction structuring carrion communities, but as these communities contain more than two trophic levels, the discussion is expanded to other species interactions that may also play important roles.

11.2.2 Predation

Moving up a trophic level, another form of consumer–resource interaction that is likely to play an important role in structuring carrion communities is that of predation, with species attracted to the carrion resource "island" not because of the carrion itself, but because of the prey available. Feeding rates of predators is a fundamental aspect of this interaction, and the number of prey consumed are expected to increase with prey density on a per predator, per unit time basis (Turchin 2003). Yet, this rate is not likely to increase indefinitely, as a predator's feeding rate should be limited by the amount of time required to capture and consume each prey (its handling time) and satiation of the predator.

This relationship between prey handling time, satiation, prey density, and predator feeding rate was classified into three types of predator functional responses by Holling (1959). Type I functional response assumes that a predator's feeding rate increases linearly with prey density, with no handling time and no predator satiation. More realistic, type II functional response assumes that a predator's feeding rate is limited by handling time and satiation when prey density increases. This response has been demonstrated for facultative predators such as the blow fly *Chrysomya albiceps* (Wiedemann) (Diptera: Calliphoridae) on conspecifics (Faria et al. 2004). Type III functional response is sigmoidal, as it assumes that predator's feeding rate is initially slow at low prey densities, increases when prey density increases, and then slows again due to handling time and satiation. Using a type II functional response, Rosenzweig and MacArthur (1963) developed a more realistic predator–prey model than originally proposed by Lotka (1925) and Volterra (1926):

$$\frac{dR}{dt} = rR\left(1 - \frac{R}{K}\right) - \frac{aRP}{1 + ahR}$$

$$\frac{dP}{dt} = \frac{faRP}{1 + ahR} - qP$$

where
  R = prey population density,
  P = predator population density,
  a = predator per capita attack rate,
  h = handling time,
  q = predator's per capita mortality rate,
  f = predator's efficiency at turning prey consumed into new predators.

Under this model, an increase in prey density supports an increase in the number of predators; however, as the prey population approaches its carrying capacity, the environmental limits (such as the amount of resources) suppress growth of the prey population. Thus, there are conditions that will allow the predator and prey populations to reach a stable equilibrium or limit cycle. Yet, if either population is disturbed by some other factor, the predator, prey, or both may go extinct within the local population. As has already been discussed, insects colonizing carrion rarely develop more than one population on a single carrion resource, this predator–prey interaction plays out over numerous local populations within a larger metapopulation context as each local population of prey goes extinct from the resource as the resource is used up and prey disperse or die. Thus, determining the role of predators in prey dynamics on carrion is more complex than simply local patch interactions. The efficiency of the predator on multiple patches must be considered to better describe and understand the interaction.

Predators that are highly efficient at turning consumed prey into more predators maintain prey populations at low densities relative to the carrying capacity for the prey population. Additional carrion resources potentially would increase the prey population, reducing predation; yet, Rosenzweig (1971) recognized that exactly the opposite might happen if a prey's food supply is increased. In a paradox of enrichment, additional resources may allow the predator population to increase to the point that it drives its prey to extinction at the local patch level.

However, predators make choices and may forage on more than one species of prey, depending on whether specific prey types are required or may be substituted by other resources in the predator's diet. Within carrion communities, predators are typically generalists on more than one prey species, including intraguild prey for predators that feed on other predators or omnivores that feed on both prey and the carrion resource. Parasitoids such as *Nasonia vitripennis* (Walk) (Hymenoptera: Pteromalidae) attack pupae of numerous fly species but certainly exhibit preferences (Cornell and Pimentel 1978). Predators should make foraging decisions based on the probabilities that they will encounter, attack, and success-fully capture and consume prey of different types (Mittelbach 2012). In addition to population densities, these probabilities will depend on prey size, mobility of the prey, escape behavior, prey defenses, and activity patterns of the predator and prey. Thus, the preference for different prey types can be expressed using the Manly–Chesson index (Manly 1974; Chesson 1978).

$$\alpha_i = \frac{d_i/N_i}{\sum_{j=1}^{k} (d_j/N_j)}$$

where
$i = 1, 2, \ldots, k$,
k = number of prey types,
d_i = number of prey of type i in the predator's diet,
N_i = number of prey of type i in the environment.

A predator should seek to optimize the amount of energy it receives from its prey, while expending the least amount of energy on searching and handling prey. This idea of optimal diet or optimal forag-ing makes the predictions that predators should prefer the most profitable prey, yet when the density of this high-value prey is scarce, predators should switch to more low-value prey so as to not expend much energy searching for prey (Charnov 1976; Stephens and Krebs 1986). Yet, only those prey whose profit-ability is greater than the energetic cost of pursuing, attacking, and consuming that prey type compared to other available prey types that should be included in a predator's diet. Thus, according to optimal diet theory, a prey type is either always eaten or never eaten upon encounter, and the encounter rate that determines whether or not a prey type is included in the diet breadth of a predator. This model assumes that the predator has perfect knowledge of all types of available prey, but natural systems rarely operate in this manner. With a few rare exceptions, such as the red imported fire ant, *Solanopsis invicta* Buren (Hymenoptera: Formicidae), predation on fly eggs, and larvae (Wells and Greenberg 1994b), to what degree direct predatory interactions structure the species diversity and dynamics of carrion communities is not known.

Holt (1977) demonstrated that noncompeting prey species can have negative indirect effects on each other through a shared predator. An increase in the abundance of one prey species can result in an increase in predator abundance that then increases predation on the second prey species. Without exam-ining the next trophic level and the effect of predation, concluding that competition between the two prey species is responsible for the observed changes in the prey populations would be expected. Whichever prey species is better able to maintain its population density in the presence of the shared predator will exclude other prey in this apparent competition. The mechanism by which the better "competitor" prey species maintains its population could be due to a lower vulnerability to predation, a high intrinsic rate of increase, or both. If the predator exhibits prey switching to the most abundant prey, prey have refugia away from predation. Alternatively, the predator's population growth may be limited by factors other than prey availability, and in case the intensity of apparent competition is lessened. This was demon-strated with the parasitoid *N. vitripennis* in a three host system, attacking pupae of the house fly *Musca domestica* L. (Diptera: Muscidae), and the blow flies *P. regina* Meigen (Diptera: Calliphoridae) and *L. sericata* (Meigen) (Diptera: Calliphoridae) (Cornell and Pimentel 1978). The parasitoid exhibited pref-erences for whichever species it had previous experience with and by attacking particular species, was able to reverse competitive outcomes and change the dynamics between the three host species, thereby demonstrating an apparent competition interaction. However, *N. vitripennis* also switched to the most abundant prey at high densities of parasitoids, lessening the impact of the apparent competition.

In discussing predator–prey interactions in terms of population dynamics of predators and prey, the discussion has been restricted to consumptive effects of predators on prey. Within a prey species, predators can act as agents of selection by removing individuals of different age classes or specific traits. Between prey species, predators can change species diversity, through changes in abundance and richness. However, there are considerable nonconsumptive effects predators may also have on their prey which have the potential to change the structure of communities on carrion.

In one type of nonconsumptive effect, prey may shift their habitat preference or activity levels to avoid predators. In the presence of the facultative predator *C. rufifacies*, other blow flies may avoid laying eggs and may leave the carrion resource sooner (Rosati and VanLaerhoven 2007). For organisms that undergo habitat changes or metamorphosis during their development, they should switch habitats or undergo metamorphosis when growth in one stage declines below the potential growth in a second stage (Werner and Gilliam 1984), as long as mortality is equal in both stages. However, if mortality is higher in one habitat/stage, then this shifts when an organism should change habitats/stages as it should reduce the time spend in the stage or habitat that has the higher mortality. These types of changes can also result in changes in prey morphology or life history traits as predators may act as agents of selection. For example, if other blow fly species develop faster and leave carrion resources earlier in the presence of *C. rufifacies* than when it is not present, resulting in a smaller body size of prey species that survive *C. rufifacies*, then over time, this may result in an overall change in the adult size and developmental rate of the prey species. The potential for these types of changes was demonstrated with *Co. macellaria* in the presence of *C. rufifacies*, as *Co. macellaria* had lower fitness and survivorship when *C. rufifacies* colonized within 2 days after *Co. macellaria* (Brundage et al. 2014). Yet, not all interactions between species on carrion resources are negative.

11.2.3 Mutualism and Facilitation

Mutualism is a common species interaction with positive interactions between both species involved. In one example, *L. sericata* may utilize cues released by microbes to determine the suitability of carrion resources and may facilitate transport of microbes between carrion resources (Tomberlin et al. 2012). Thus, both microbes and blow flies benefit. However, facilitation (or commensalism), where at least one species benefits and the other species is not affected, is likely even more common in carrion communities by the very nature of the ephemeral resource. Bronstein (2009) defined facilitation as "an interaction in which the presence of one species alters the environment in a way that enhances growth, survival, or reproduction of a second, neighbouring species." Connell and Slatyer (1977) defined facilitation as early colonists, making the environment more suitable for later colonists and is one potential mechanism for succession. These species have been called ecosystem engineers (Jones et al. 1994). Certainly, blow flies modify the carrion resource making it less suitable for themselves and more suitable for later colonists. At shorter timescales such as within hours of a single day, research in my own laboratory has demonstrated that *L. sericata* facilitates the colonization of *P. regina* and *C. rufifacies* as both blow flies oviposit faster in the presence of *L. sericata* eggs than they do on their own (Rosati and VanLaerhoven, unpublished data). Whether this *L. sericata* cue is mediated by a mutualistic interaction with microbes has yet to be fully explored.

Modeling mutualistic interactions in terms of Lotka–Volterra by changing the interspecific effect sign from negative to positive is possible (Mittelbach 2012). This approach results in a stable coexistence over time only if a positive interspecific mutualistic effect is balanced by a negative intraspecific effect in a density-dependent manner that may be due to environmental fluctuation between carrion resource "islands" (see lottery model), or predation or competition of other life stages or with other species in the food web. Thus, to fully understand mutualistic interactions, it is vital to understand the balance between this positive interaction and negative interactions in other parts of the food web. None of these interactions occur in isolation.

11.3 Food Web Interactions

Until now, species interactions in isolation have been considered, yet carrion communities are food webs with multiple tropic levels and numerous interactions that make up complex networks (see also Chapter

8). Functional food webs, or interaction webs, are usually based on species removal experiments and indicate the strength of species interactions. These can be measured in different ways. The raw difference is used to show the absolute treatment effects, whereas the community importance quantifies the effect of the species based on its relative abundance (Berlow et al. 1999):

$$\text{Raw difference} = \frac{N - D}{Y}$$

$$\text{Community importance} = \frac{N - D}{N_{\text{Py}}}$$

$$\text{Dynamic index} = \frac{\ln(N - D)}{Yt}$$

where
   N = prey abundance in the presence of predators,
   D = prey abundance in the absence of predators,
   Y = predator abundance,
   Py = proportional abundance of the predator,
   t = time.

The dynamic index is analogous to the L–V predator–prey model interaction coefficients (Berlow et al. 1999). These three indices of interaction strength can be calculated for direct interactions from experimental manipulation experiments and do not include indirect effects or compensatory dynamics due to density-dependent feedback. Alternatively, interaction strength can be measured from observational data (Bascompte et al. 2005).

$$\text{Interaction strength} = \frac{(Q/B)_j \times \text{DC}_{ij}}{B_i}$$

where
   Q/B = number of times predator j population eats its own weight/day,
   DC_{ij} = prey i proportion in predator j's diet,
   B_i = prey i biomass.

Not all interactions are equal in strength or in their effect on the web as a whole. It is expected that most food webs have many weak interactions and only a few strong interactions. Some species have disproportionally large effects on the food web relative to their abundance, called keystone species, but no species have yet been identified from carrion communities that fulfill this definition. The possible exception is the role of fire ants in removing/delaying colonization of carrion by blow flies (Wells and Greenberg 1994b).

Certainly, this variability in interaction strength can be observed in carrion resource communities. Therefore, what do the consumer–resource models predict when we expand interactions from species pairs into the complexity of food webs? Firstly, increases in the availability or size of carrion resources should increase abundance of the top trophic levels and alternating tropic levels, but not the intervening levels (Mittelbach 2012). Second, increases in the abundance or size of carrion resources should increase the number of trophic levels within the local food web, similar to the species diversity. Finally, any decrease in population abundance at the top trophic level should alternatively increase and decrease the abundance of the levels below in a trophic cascade.

To date, only simple food chain manipulation experiments have been conducted with carrion resource communities. In order to explore these experimentally with carrion resources, one would have to take a metapopulation approach, manipulating multiple carrion resource patches to see the changes in population abundance within different trophic levels over time. It would be easy to assume that resource limitation drives all the dynamics for carrion communities; however, predators may have an important role in determining structure of these communities as well.

In contrast to resource- or predator-driven models, network models may provide different predictions at the food-web level and metacommunity level for carrion food-web dynamics. Natural biological systems are complex and when moving from ideas of a carrion resource—consumer—predator interactions to the natural ecological system of numerous species interacting in a food web that may span multiple resources and abiotic conditions over space and time is better explained by network theory. Network theory of food webs and community stability suggests that factors that affect species with the most connections to other members or species that represent energy bottlenecks are most likely to have the greatest impacts on metacommunity stability (Proulx et al. 2005). Yet, we cannot understand the important processes structuring carrion communities simply by studying interactions at a local resource patch level as "the theory of complex systems states that each of the different organizational levels reveal 'emergent properties' that can't be predicted by adding up the properties of the next-lower level" (Leuschner 2013). Emergent properties of food-web structure suggest that it is the distribution of different interaction strengths throughout the food web, rather than the number of links or particular strengths of the links that may be most important for stability (Pascual and Dunne 2006).

But is the important currency for carrion communities energy flow or population numbers? In this chapter, the focus has been on individuals and populations, but the structure of the community and the understanding of species interactions changes depending on the currency considered. Further insights will be gained by considering the bioenergetics of these interactions and considering emergent properties of the ecological network (Pascual and Dunne 2006).

11.4 Conclusions

Are arthropod communities associated with carrion structured by the bottom-up influence of the resource itself that is unpredictable spatially, temporally, and ephemeral once found? There is no doubt that the size and distribution of carrion resources influences the community. Yet, what is the role of predation in structuring the community of species utilizing the carrion resource? Or, is it the interactions between competitors that are the driving mechanisms of these communities, resulting in compensatory or zero-sum dynamics with a community carrying capacity? Houlahan et al. (2007) tested 41 communities at different spatial scales for evidence of compensatory dynamics and found that species abundance was driven more by abiotic factors than species interactions. Certainly, species interactions play an important role in structuring the species diversity and abundance within ephemeral communities such as carrion, but the relative importance of these different direct and indirect interactions in relation to abiotic mechanisms has yet to be answered. Rather than a review of all previous species interaction studies on carrion, this chapter highlighted different species interaction models and provided testable predictions. By identifying potential interactions occurring at local and metapopulation scales that may influence population dynamics of species on ephemeral resources, this chapter identified tools to measure species interactions and the carrion food web, thereby pushing forward our understanding of ephemeral-resource use communities beyond pattern documentation and instead into mechanisms that structure these fascinating communities.

REFERENCES

Arrhenius, O. 1921. Species and area. *Journal of Ecology* 9: 95–99.

Atkinson, W.D. and B. Shorrocks. 1981. Competition on a divided and ephemeral resource: A simulation model. *Journal of Animal Ecology* 50: 461–471.

Bascompte, J., C.J. Melian, and E. Sala. 2005. Interaction strength combinations and the overfishing of a marine food web. *Proceedings of the National Academy of Sciences of the United States of America* 102: 5443–5447.

Beaver, R.A. 1977. Non-equilibrium island communities: Diptera breeding in dead snails. *Journal of Animal Ecology* 46: 783–798.

Bedhomme, S., P. Agnew, C. Sidobre, and Y. Michalakis. 2005. Pollution by conspecifics as a component of intraspecific competition among *Aedes aegypti* larvae. *Ecological Entomology* 30: 1–7.

Berlow, E.L., S.A. Navarrete, C.J. Briggs, M.E. Power, and B.A. Menge. 1999. Quantifying variation in the strengths of species interactions. *Ecology* 80: 2206–2224.

Bronstein, J.L. 2009. The evolution of facilitation and mutualism. *Journal of Ecology* 97: 1160–1170.

Brundage, A., M.E. Benbow, and J.K. Tomberlin. 2014. Priority effects on the life history traits of two carrion blow fly (Diptera: Calliphoridae) species. *Ecological Entomology* 39: 539–547.

Cerovsky, V., J. Zdarek, V. Fucik, L. Monincova, Z. Voburka, and R. Bem. 2010. Lucifensin, the long-sought antimicrobial factor of medicinal maggots of the blowfly *Lucilia sericata*. *Cellular and Molecular Life Sciences* 67: 455–466.

Charnov, E.L. 1976. Optimal foraging: Attack strategy of a mantid. *American Naturalist* 110: 141–151.

Chesson, J. 1978. Measuring preference in selective predation. *Ecology* 59: 211–215.

Chesson, J. 2000. General theory of competitive coexistence in spatially-varying environments. *Theoretical Population Biology* 58: 211–237.

Chesson, P.L. and R.R. Warner. 1981. Environmental variability promotes coexistence in lottery competitive systems. *American Naturalist* 117: 923–943.

Connell, J.H. and R.O. Slatyer. 1977. Mechanisms of succession in natural communities and their role in community stability and organization. *American Naturalist* 111: 1119–1144.

Cornell, H. and D. Pimentel. 1978. Switching in the parasitoid *Nasonia vitripennis* and its effects on host competition. *Ecology* 59: 297–308.

Denno, R.F. and W.R. Cothran. 1975. Niche relationships of a guild of necrophagous flies. *Annals of the Entomological Society of America* 68: 741–754.

Denno, R.F. and W.R. Cothran. 1976. Competitive interactions and ecological strategies of Sarcophagid and Calliphorid flies inhabiting rabbit carrion. *Annals of the Entomological Society of America* 69: 109–113.

Faria, L.D.B., L.A. Trinca, and W.A.C. Godoy. 2004. Cannibalistic behavior and functional response of *Chrysomya albiceps* (Diptera: Calliphoridae). *Journal of Insect Behavior* 17: 251–261.

Fiene, J.G., G.A. Sword, S.L. VanLaerhoven, and A.M. Tarone. 2014. The role of spatial aggregation in forensic entomology. *Journal of Medical Entomology* 51: 1–9.

Gause, G.F. 1934. *The Struggle for Existence*. Williams and Wilkins, Baltimore, MD.

Gonzalez, A. and M. Loreau 2009. The causes and consequences of compensatory dynamics in ecological communities. *Annual Review of Ecology, Evolution and Systematics* 40: 393–414.

Gotelli, N.J. 2008. *A Primer of Ecology*, Fourth Edition. Sinauer Associates, Sunderland, MA.

Hanski, I. 1987. Carrion fly community dynamics: Patchiness, seasonality and coexistence. *Ecological Entomology* 12: 257–266.

Hanski, I. and S. Kuusela. 1977. An experiment on competition and diversity in the carrion fly community. *Annales Entomologicae Fennicae* 43: 108–115.

Hanski, I. and S. Kuusela. 1980. The structure of carrion fly communities: Differences in breeding seasons. *Annales Zoologici Fennicae* 17: 185–190.

Haskell, E.F. 1947. The natural classification of societies. *Transactions of the New York Academy of Sciences* 9: 186–196.

Hibbing, M.E, C. Fuqua, M.R. Parsek, and S.B. Peterson. 2010. Bacterial competition: Surviving and thriving in the microbial jungle. *Nature Reviews Microbiology* 8: 15–25.

Holling, C.S. 1959. The components of predation as revealed by a study of small mammal predation on the European pine sawfly. *The Canadian Entomologist* 91: 293–320.

Holt, R.D. 1977. Predation, apparent competition, and structure of prey communities. *Theoretical Population Biology* 12: 197–229.

Houlahan, J.E., D.J. Currie, K. Cottenie, G.S. Cumming, S.K.M. Ernest, C.S. Findlay, S.D. Fuhlendorf et al. 2007. Compensatory dynamics are rare in natural ecological communities. *Proceedings of the National Academy of Sciences of the United States of America* 104: 3273–3277.

Hubbell, S.P. 2001. *The Unified Neutral Theory of Biodiversity and Biogeography*. Princeton University Press, Princeton, NJ.

Ives, A.R. 1991. Aggregation and coexistence in a carrion fly community. *Ecological Monographs* 61: 75–94.

Jones, C.G., J.H. Lawton, and M. Shachak 1994. Organisms as ecosystem engineers. *Oikos* 69: 373–386.

Kneidel, K.A. 1983. Fugitive species and priority during colonization in carrion-breeding Diptera communities. *Ecological Entomology* 8: 163–169.

Kneidel, K.A. 1984. Influence of carcass taxon and size on species composition of carrion-breeding Diptera. *American Midland Naturalist* 111: 57–63.

Kneidel, K.A. 1985. Patchiness, aggregation, and the coexistence of competitors for ephemeral resources. *Ecological Entomology* 10: 441–448.

Krell, F.T., S. Krell-Westerwalbesloh, I. Weiß, P. Eggleton, and K.E. Linsenmair. 2003. Spatial separation of Afrotropical dung beetle guilds: A trade-off between competitive superiority and energetic constraints (Coleoptera: Scarabaeidae). *Ecography* 26: 210–222.

Leuschner, C. 2013. Vegetation and ecosystem. In: *Vegetation Ecology*, 2nd Edition. van der Maarel, E. and J. Franklin (eds). John Wiley & Sons Ltd, Oxford, UK.

Levins, R. 1968. *Evolution in Changing Environments*. Princeton University Press, Princeton, NJ.

Lotka, A.J. 1925. *Elements of Physical Biology*. Williams and Wilkins, Baltimore, MD.

MacArthur, R.H. and E.O. Wilson 1967. *The Theory of Island Biogeography*. Princeton University Press, Princeton, NJ.

Manly, B. 1974. A model for certain types of selection experiments. *Biometrics* 30: 281–294.

Mittelbach, G.G. 2012. *Community Ecology*. Sinauer Associates, Sunderland, MA.

Pascual, M. and J.A. Dunne. 2006. From small to large ecological networks in a dynamic world. In: *Ecological Networks Linking Structure to Dynamics in Food Webs*. Pascual, M. and J.A. Dunne (eds). Oxford University Press, New York, NY.

Pechal, J.L., M.E. Benbow, T.L. Crippen, A.M. Tarone, and J.K. Tomberlin. 2014. Delayed insect access alters carrion decomposition and necrophagous insect community assembly. *Ecosphere* 5: 45.

Proulx, S.R., D.E.L. Promislow, and P.C. Phillips. 2005. Network thinking in ecology and evolution. *Trends in Ecology and Evolution* 20: 345–353.

Rosati, J.Y. and S.L. VanLaerhoven. 2007. New record of *Chrysomya rufifacies* (Diptera: Calliphoridae) in Canada: Predicted range expansion and potential effects on native species. *The Canadian Entomologist* 139: 670–677.

Rosenzwieg, M.L. 1971. Paradox of enrichment: Destabilization of exploitation ecosystems in ecological time. *Science* 171: 385–387.

Rosenzweig, M.L. and R.H. MacArthur. 1963. Graphical representation and stability conditions of predator–prey interactions. *American Naturalist* 97: 209–223.

Sato, H. and K. Hiramatsu. 1993. Mating behaviour and sexual selection in the African ball-rolling scarab *Khepher platynotus* (Bates) (Coleoptera: Scarabaeidae). *Journal of Natural History* 27: 657–668.

Shiao, S. and T. Yeh. 2008. Larval competition of *Chrysomya megacephala* and *Chrysomya rufifacies* (Diptera: Calliphoridae): Behavior and ecological studies of two blow fly species of forensic significance. *Journal of Medical Entomology* 45: 785–799.

Shorrocks, B. and M. Bingley. 1994. Priority effects and species coexistence: Experiments with fungal-breeding *Drosophila*. *Journal of Animal Ecology* 63: 799–806.

Stephens, D.W. and J.R. Krebs. 1986. *Foraging Theory*. Princeton University Press, Princeton, NJ.

Tilman, D. 1982. *Resource Competition and Community Structure*. Princeton University Press, Princeton, NJ.

Tilman, D. 2007. Interspecific competition and multispecies coexistence. In: *Theoretical Ecology: Principles and Applications*, 3rd Edition. May, R.M. and A. McLean (eds). Oxford University Press, Oxford.

Tomberlin, J.K., T.L. Crippen, A.M. Tarone, B. Singh, K. Adams, Y.H. Rezenom, B.M. Eric et al. 2012. Interkingdom responses of flies to bacteria mediated by fly physiology and bacterial quorum sensing. *Animal Behavior* 84: 1449–1456.

Turchin, P. 2003. *Complex Population Dynamics: A Theoretical/Empirical Synthesis*. Princeton University Press, Princeton, NJ.

VanLaerhoven, S.L. 2010. Ecological theory and its application in forensic entomology. In: *Forensic Entomology: The Utility of Arthropods in Legal Investigations*, 2nd edition. Byrd, J.H. and J.L. Castner (eds.). CRC Press, Boca Raton, FL.

Verhulst, P.F. 1838. Notice sur la loi que la population suit dans son accroissement. *Correlative Mathematics and Physics* 10: 113–121.

Volterra, V. 1926. Variations and fluctuations in the numbers of individuals in animal species living together (reprinted in 1931. In: Chapman, R.N. *Animal Ecology*, McGraw-Hill, New York).

Wells, J.D. and B. Greenberg. 1992. Laboratory interaction between introduced *Chrysomya rufifacies* and native *Cochliomyia macellaria* (Diptera: Calliphoridae). *Environmental Entomology* 21: 640–645.

Wells, J.D. and B. Greenberg. 1994a. Resource use by an introduced and native carrion flies. *Oecologia* 99: 181–187.

Wells, J.D. and B. Greenberg. 1994b. Effect of the red imported fire ant (Hymenoptera: Formicidae) and carcass type on the daily occurrence of post feeding carrion-flylarvae (Diptera: Calliphoridae, Sarcophagidae). *Journal of Medical Entomology* 31: 171–174.

Werner, E.E. and J.F. Gilliam. 1984. The ontogenetic niche and species interactions in size-structured populations. *Annual Review of Ecology, Evolution and Systematics* 15: 393–425.

12

Aquatic Vertebrate Carrion Decomposition

John R. Wallace

CONTENTS

12.1 Introduction: Background and Chapter Aims

Most of the globe is covered by ocean, where animals may die far from where they lived. Large carcasses like those of whales may sink into the cold, dark depths. Salmon may live most of their lives in the ocean, yet they come inland to die and be deposited in fresh water, and the major effects of their recycling are on land, not in the oceans where they lived ....

—Bernd Heinrich (2012)

Among the Earth ecosystems, aquatic habitats are comprised of three types: (1) freshwater including standing waters (lentic), for example, natural container habitats such as tree holes or phytotelmata (bromeliads), ponds and lakes, as well as moving water systems (lotic), for example, streams and rivers; (2) transitional communities such as estuaries or embayments and other wetlands, for example, temporary ponds; and (3) marine systems that include shorelines, inland saline lakes, and open ocean. While this list of aquatic systems is not exclusive and their diversity may appear to prohibit their discussion in terms of carrion decomposition in one chapter, the aim of this chapter is to demonstrate that although there are many processes that occur in these disparate aquatic habitats, there is fundamental unity between them and carrion decomposition.

The treatise of decomposition ecology throughout most ecology textbooks today has focused primarily on the role of carrion ecology in terrestrial ecosystem functioning (Swift et al. 1979). As Heinrich (2012) stated in the quote above, carrion decomposition in aqueous environments may follow similar patterns to what has been observed in terrestrial ecosystems, but there are major differences in these processes that are fundamentally constrained to aquatic ecosystems. What is learned about such differences can be gathered from comparisons of lentic and lotic freshwater systems as well as the largest of aquatic systems, oceans, which appear to exhibit biogeochemical processes of both lentic and lotic water bodies. Thus, with water bodies of all size and type, it is a series of complex physical, chemical, and biological interactions that allow biogeochemical cycles and biotic communities to function (Schindler 1991a), which in turn ultimately influences the role of decomposition in these systems.

The specific aims of this chapter are to discuss the various physical and chemical influences on decomposition and demonstrate how these abiotic factors are interconnected with the mechanisms involved with nutrient cycling—the *raison d'être*, or defining purpose of decomposition. The relevance of community diversity and organization across different aquatic systems will by synthesized in order to foster a better understanding of aquatic ecosystem functioning and how this information is applied to fields outside of traditional ecological contexts. Finally, general insights into aquatic carrion decomposition from comparative analyses among aquatic ecosystems will be discussed to highlight challenges for future investigations.

12.2 Special Physical/Chemical Properties and Decomposition

In general, a variety of physical and chemical attributes influence and sustain ecological systems (EPA 2008). While decomposition in aquatic systems may be mediated through biological mechanisms, for example, microbial—macroinvertebrate trophic interactions, it is a suite of physical-chemical parameters that influence the dominant pathway of decomposition in aquatic habitats (Merritt and Wallace 2010) (Table 12.1). Several of these important parameters are discussed below in detail.

12.2.1 Temperature

Aquatic organisms are exposed to a thermal diversity that varies both spatially and temporally in all types of aquatic systems, and such variation can be influenced both naturally and from anthropogenic alteration (Ward and Stanford 1982). In lotic systems, such as streams and rivers, both geographical (e.g., latitude, altitude, and topographical elements) and meteorological factors generally explain the thermal diversity patterns observed in these systems (Smith and Lavis 1974). For example, air temperatures can have significant effects on streams, ponds, and littoral zones of lakes (Macan and Maudsley 1966; Brown 1969; Smith and Lavis 1974; Dale and Gillespie 1977). However, morphometry (i.e., the size and shape) as well as continentality are major determinants of thermal regimes in lentic systems from lakes to oceans (Hutchinson 1957).

TABLE 12.1

Selected Physical/Chemical Parameters Affecting Decomposition That Directly or Indirectly Affect Abiotic and Biotic Factors Important to Aquatic Carrion Decomposition

Parameter	Effects
Temperature	Dissolved oxygen
Gas exchange	Influences organismal physiology from bacteria to animals
Hydrology	Current regimes, depth, sea levels, and distribution of organic matter and nutrients
Nutrients	Nutrient cycling, transformation, and distribution influencing community structure and function
Other physical and chemical variables (e.g., pH, salinity, light)	Species distributions and community structure and function

Thermal stratification, the vertical layering of water temperature into three regions within lakes and oceans, has profound effects on physical, chemical, and biological properties in these systems (Schindler 1991a). These disparate regions of these systems are the following: (1) the epilimnion—shallowest and warmest; (2) metalimnion or thermocline—an intermediate layer exhibiting a rapid decline in temperature with depth; and (3) the hypolimnion—the deepest and cold, dark regions below the thermocline (Schindler 1991a). Although the epilimnion represents a region where nutrients are fixed via photosynthesis and where allochthonous organic material necessary for biological communities is produced, the hypolimnion is the zone where the products of decomposition accumulate (Schindler 1991a). Temperatures in the hypolimnion may be near 4°C—the maximum density of water and temperature that requires special adaptations by organisms to survive and metabolize these organic matter products of decomposition (Schindler 1991a).

Temperature is a major factor affecting the life histories of those aquatic organisms intimately involved in all stages of the decomposition of organic material from aquatic bacteria (White et al. 1991; Pomeroy and Wiebe 2001), invertebrates (Sweeney and Vannote 1978; Ward and Stanford 1982; Burkepile et al. 2006) to freshwater and marine fish (Pepin 1991). For example, when the half-life of carrion was plotted against temperature in a variety of freshwater systems from streams to lakes, there was a negative correlation between carrion half-life and increasing temperature; this was presumably due to the positive relationship between temperature and most bacterial and scavenger metabolic rates (Chidami and Amyot 2008).

Therefore, it should be no surprise that the rate of decomposition of organic matter in lotic and lentic systems tends to increase as a function of temperature, regardless of the system or seasonal fluctuations in temperate water bodies or in more stable climates such as in tropical seas. The effect of temperature on poikilothermic organisms involved with decomposition activities is considered a dominant factor in aquatic organic matter processing (Webster and Benfield 1986; Sorg et al. 1997; Merritt and Wallace 2010; Tank et al. 2010).

12.2.2 Gas Exchange

Dissolved oxygen and temperature are two of the fundamental variables that affect lake and pond ecology (Addy and Green 1997). Dissolved oxygen is the amount of oxygen in solution and can fluctuate daily depending on the temperature, salinity, altitude, groundwater inflow, and anthropogenic activities or accidents (Addy and Green 1997). For example, greater amounts of oxygen can be held in solution in cold water than warmer water, with variations ranging from diel differences, that is, 24 h period, to seasonal fluctuations (Minnesota Pollution Control Agency 2009).

Most aquatic organisms are intrinsically linked to respiring oxygen in solution; their survival and subsequent role in decomposition activities require oxygen. Because oxygen solubility in water is negatively correlated with temperature, certain groups of aquatic invertebrates that play substantial roles as shredding of carrion tissue may be excluded from such low oxygen environments (Merritt and Wallace 2010). Thus, depending on the oxygen concentration in a specific habitat, the faunal community available to colonize carrion may be quite different (Hobischak 1997). During high rates of decomposition, such as that which occurs after an algal bloom, dissolved oxygen levels can crash and cause massive mortality of higher organisms involved in the decomposition process in both lotic and lentic systems (Osmond et al. 1995). Further, while deep-ocean benthic habitats are diverse, dissolved oxygen concentrations are either very low or essentially absent. This is a problem for most marine organisms, indicating that decomposition at depths below 2000 m via bacteria is either nonexistent or very slow (Tunnicliffe et al. 2003; Heinrich 2012).

12.2.3 Hydrology

Flow regime in lotic aquatic systems is intimately linked to five hydrologic characteristics: (1) the magnitude or monthly discharge; (2) frequency of high- or low-pulse episodic flows; (3) the duration of either meteorological or anthropogenically generated inputs; (4) their timing or whether they are predictable; and (5) the rate of change or degree of flashiness (Poff et al. 1997). In lotic systems, these hydrologic characteristics are also mediated by instream and bank factors such as channel width, sinuosity, and habitat heterogeneity. In lentic systems, the hydroperiod (or length of time these systems are inundated) is a function of not only those characteristics that influence lotic systems, but also the variability of seasonal precipitation patterns and topographic microrelief variation in variables such as elevation gradients, degree of isolation, presence or absence of inlets or outlets, and, at times, water management (Stevenson and Childers 2004; Bauder 2005).

Normal hydrologic flux allows for the exchange of nutrients, detritus, and passage of aquatic life between systems (Egglishaw 1972; Osmond et al. 1995). However, changes in this flux (e.g., the frequency, duration, and timing of the hydroperiod) may impact species composition, fish spawning and migration, and, ultimately, the food chain support of the wetland and associated downstream systems (Osmond et al. 1995). Alterations in both lotic and lentic system hydrology have negatively impacted the biological diversity to such a degree that ecological integrity has been compromised and such impacts have serious repercussions on the rate of vertebrate carrion decomposition in these systems (Sakaris 2013).

12.2.4 Nutrient Limitations

To set the stage for aquatic carrion decomposition in regard to nutrient availability, it is important to understand how nutrients and other chemical constituents in aquatic systems impact the foundation of any food web via decomposing plant matter.

The fate of nutrients and compounds released from decomposing plant material in aquatic ecosystems may be exported into a soluble or particulate form (streams), incorporated into the soil (wetland) by microbial decomposers or eventually biotically transformed and released to the atmosphere (Almazon and Boyd 1978; Andersen 1978; Brock et al. 1985a,b; Ruppel et al. 2004; Debusk and Reddy 2005). The availability of phosphorous in freshwater ecosystems such as streams and some wetlands is considered to be a limiting nutrient for plant growth (Steinman and Mulholland 2006). In these types of aquatic ecosystems, specifically northern latitude arctic and subarctic waters with salmon subsidies (Figure 12.1), as well as subtropical wetlands (e.g., the Everglades Florida, USA) a fish detritus pathway for energy transfer may possibly compensate for seasonal variations in the availability of phosphorus and essentially facilitate the necessary nutrient cycling to maintain primary production in these systems and overall ecosystem integrity (Stevenson and Childers 2004).

12.2.5 Other Chemical Factors

Other water chemistry factors, such as pH and salinity, may influence aquatic carrion decomposition indirectly by impacting the diversity of organisms that initiate the process, specifically primary and

FIGURE 12.1   Salmon subsidies in streams of Alaska. (Photo credit: Emily Campbell.)

secondary decomposers such as bacteria, fungi, and invertebrates, thereby affecting ecosystem functioning in general. Unless impacted from anthropogenic sources, the normal pH for freshwater ecosystems ranges from 6.5 to 8.5 (Berezina 1999), whereas for saltwater, the normal pH is from 7.8 to 8.4 (Paletta 1999). However, acidification of streams and rivers through anthropogenic activities has been occurring on a global scale and this process has been shown to negatively impact a number of aquatic micro- and macroorganisms essential in the decomposition of aquatic organic material including carrion (Schindler 1991b).

A microorganism pathway consisting of fungi and bacteria (Anderson and Sedell 1979) drives the breakdown of dead plant or animal matter in streams. Because few terrestrial fungi can survive in an aqueous environment, aquatic hyphomycete fungi are estimated to be more important than bacteria in the decomposition process of organic matter (Bärlocher and Kendrick 1974); however, many of these fungi are incapable of growing in water with a pH of 5.0. Interestingly, for many bacteria essential to the polysaccharide biofilm matrix that colonizes leaf matter, a pH above 5.0 is necessary for bacterial growth and survival (Richardson 1995). This limitation that pH has on hyphomycete fungal and bacterial growth has been shown to have a ripple effect on aquatic organisms such as amphipods that have been documented as integral players in the fragmentation of vertebrate carrion in streams and oceans (Anderson and Hobischak 2002; Merritt and Wallace 2010).

Acid stress on aquatic systems, such as those that might be created naturally or by anthropogenic impact, for example, acid mine drainage, is a major factor limiting processing rates of detritus on substrates that rely on extracellular enzymes for decomposition (e.g., lignocellulose) (Haines 1981; Schoenberg et al. 1990). In systems such as the Okefenokee Swamp (Georgia, USA), and other lake systems, low pH has been responsible for a slower breakdown rate for lignocellulose that has contributed to the peat formation in the Okefenokee Swamp relative to comparable salt marshes (Clymo 1983; Benner et al. 1985). While the effects of pH on plant tissue may regulate the microorganisms capable of carrying out decomposition, depending on the aquatic system, the effect on animal tissue has been rarely examined.

Postmortem changes due to the pH in aquatic systems during decomposition have been primarily focused on human remains (Haglund and Sorg 1997, 2002). The extreme pH in the digestive tracts of predator animals can be observed on bone and the remainder of partially digested remains eaten by predator species, for example, black tipped reef sharks (Papastamatiou et al. 2007). Controlled laboratory studies on bovine tissue and bone have shown that pH levels of 4 (a pH level that would be encountered in acidified streams) removed and dissolved soft tissue over a 3-week period, but had little to no effect on bone even after 1-year exposure (Christensen and Myers 2011). Because of the negative impacts of low pH on microbial decomposers, it is assumed that decomposition rates of carrion in the absence of scavengers would be slowed as well.

Salinity represents the total concentrations of sodium and chloride and other cations and ions in aquatic systems and is an integral part of the biochemistry of both terrestrial and aquatic ecosystems (Dunlop et al. 2005). Studies specifically focused on wildlife decomposition are lacking, but what we do understand is from field experiments utilizing domestic swine carcasses to estimate a postmortem submersion interval (PMSI) for forensic purposes (Haefner et al. 2004; Zimmerman and Wallace 2008). In terms of differences in decomposition rates between a stream (Haefner et al. 2004) and a brackish pond (Zimmerman and Wallace 2008), researchers have found that swine carcasses required almost twice the number of degree days (physiological thermal units) to decompose in more brackish waters (typically higher in salinity) than riffle habitats in streams.

The impacts of salinity are known to result in significant shifts in the integrity and diversity of aquatic ecosystems (Dunlop et al. 2005). Salinity increases may cause toxic effects on the physiologies of freshwater taxa, thereby resulting in a change in diversity (Nielsen et al. 2003), which may significantly impact decomposition (Gessner et al. 2010). Factors such as salinity along a freshwater gradient will create divergent habitat conditions, thereby altering evolutionary trajectories for aquatic organisms (Gessner et al. 2010). The time required to progress through the various stages of decomposition in aquatic systems may be as much as two times longer in the freshwater systems that have been impacted to be more brackish compared to unimpacted freshwater systems (Haefner et al. 2004; Zimmerman and Wallace 2008).

12.3 Mechanisms of Energy Transfer via Decomposition

Decomposition processes of plant and animal tissue in aquatic systems involves the dissipation of energy stored in these forms of organic matter (Allan 1995). The pathways and processes involved in the decomposition of plant matter whether from allochthonous (from outside of aquatic system) or autochthonous (from within an aquatic system) sources has been thoroughly examined in streams, rivers, lakes, and oceans (Cummins et al. 1973, 1989; Cummins 1974; Minshall 1978; Vannote et al. 1980; Wallace et al. 1982; Peterson et al. 1985; Parmenter and Lamarra 1991). Recently, increased attention has focused on vertebrate and invertebrate carrion decomposition as an important resource subsidy and pathway for nutrient cycling in freshwater lotic and lentic systems as well as marine ecosystems (Parmenter and Lamarra 1991; Chidami and Amyot 2008).

12.3.1 Pulse Perturbations and Nutrient Cycling

Resource subsidies can create tight linkages among ecosystems (Polis et al. 1997). For example, leaf litter subsidies that fall into aquatic habitats, insect emergences (e.g., mayfly spinner fall, periodic mass cicada emergences), and parasite-mediated insect subsidies to streams (e.g., nematamorph-infected crickets that are behaviorally hijacked by these endoparasites and are driven into streams in mass) all can have pronounced effects on aquatic systems in terms of nutrient cycling and effects on macroinvertebrate community structure and overall ecosystem functioning (Kaushik and Hynes 1971; Meyer and Johnson 1983; Wallace et al. 1997; Nakano et al. 1999; Baxter et al. 2004, 2005; Sato et al. 2011, 2012). Examples of the biological and climatological drivers of resource pulses in aquatic systems (e.g., anadromous fish reproduction such as salmon and insect emergences), the type of aquatic system (e.g., streams, ponds, lakes and oceans), and the quality and frequency of such pulses are described in a review by Nowlin et al. (2008).

Because of energetic and stoichiometry differences between terrestrial and aquatic systems, primary producers (i.e., plants, algae) may respond differently to pulses of limiting abiotic resources such as nutrients (Carpenter et al. 1998; Wold and Hershey 1999; Symmank 2009; Nowlin et al. 2008). Therefore, primary producer biomass response to such pulses may lead to greater herbivore reproductive success due to more efficient energy transfer and nutrient cycling in aquatic systems (Fairchild et al. 1985; Nowlin et al. 2008). Hence, from an ecosystem perspective, the rate at which aquatic food webs might respond to resource pulses whether they are regular events such as leaf litter inputs to streams or ponds (Wallace et al. 1997) or unpredictable events such as whale falls (Smith and Baco 2003) may depend on a variety of factors including, but not limited to the following: (1) the nature and quality of the pulse; (2) carrion size structure; and (3) the relative energy flow and nutrient cycling through decomposer and herbivore channels within the food web (Nowlin et al. 2008).

In some regions, productivity in freshwater systems is limited by the availability of nitrogen (N) or phosphorous (P), especially in the Pacific Northwest (Ashley and Slaney 1997; Chaloner and Wipfli 2002). Carrion throughput in these systems can vary in scale and source of the carrion contribution to a given system, for example, when a single deer drowns in a river compared to thousands of salmon deposited in and along hundreds of stream systems regionally or a whale fall in the deep ocean. The influence on nutrient dynamics or nutrient budgets in any of these systems begins as soon as the carrion enters the system through the excretion of nitrogenous compounds resulting from protein catabolism (Schuldt and Hershey 1995; Kline et al. 1997). For example, winter mortality of waterfowl on migratory routes may provide a storage pool of nutrients in certain habitats, and due to a slower decay rate for such carcasses over winter months, nutrient return to the marsh ecosystems may be extended (Parmenter and Lamarra 1991). The annual pulse of nutrients from a given run of spawning anadromous fish, such as Pacific salmon (*Oncorhynchus* sp.), is analogous to an upstream pump of nitrogen and carbon into these headwater systems (Hicks et al. 2005). The literature refers to this delivery of N and carbon (C) in the form of adult salmon biomass from marine environments as marine-derived nutrients (MDNs; Chaloner and Wipfli 2002). This pulse of marine-derived nitrogen can be a significant source of available N, C, and P to the trophic structure of streams that experience MDNs in the form of salmon transport and deposition during spawning (Kline et al. 1993) (Figure 12.2). Therefore, MDN can be considered an

FIGURE 12.2 Postspawning, decomposing chum salmon (*Oncorhynchus keta*) in Montana Creek of the Susitna River drainage. (Photo credit: Suzanne Yocom.)

important nutrient subsidy to nutrient-impoverished coastal freshwater ecosystems, one in which decomposing salmon carcasses may elevate stream productivity (Polis et al. 1997; Wipfli 1997). However, recent research has also indicated that the physical disturbance caused by the migrating salmon can mediate such increases in in-stream productivity as well (Tiegs et al. 2008, 2009; Lessard et al. 2009; Campbell et al. 2011).

The salmon delivery of marine-derived nitrogen and carbon to streams has been traced into freshwater and terrestrial organisms and has been empirically shown to be a resource subsidy critical for maintaining both freshwater and riparian productivity (Schuldt and Hershey 1995; Kline et al. 1997; Chaloner et al. 2002). In general, invertebrates in marine and freshwater systems are important mediators of the fragmentation of organic matter, nutrient cycling, and energy flow within food webs (Cummins 1973; Britton and Morton 1994). Several studies have observed freshwater invertebrate colonization on salmon and other carrion subsidies (Keiper et al. 1997; Kline et al. 1997; Haefner et al. 2004). Invertebrate colonization and subsequent feeding on submerged or floating carrion increases the growth and standing stock of macroinvertebrates, inherently increasing the secondary production in such streams (Chaloner and Wipfli 2002; Lessard and Merritt 2006). The transfer of MDNs to higher trophic levels such as predatory fish may be facilitated by such colonization (Chaloner et al. 2002). The end result of stream macroinvertebrate utilization of salmon carcasses as a food resource may enhance the retention and incorporation of MDNs into stream food webs (Chaloner and Wipfli 2002; Hocking et al. 2009).

Salmon subsidies also influence terrestrial scavenger and predator populations of the adjacent forests of salmon-bearing riverine systems (Hocking et al. 2009) (see Chapter 22). MDNs from salmon carcasses are distributed from streams and lakes to adjoining terrestrial ecosystems through foraging activities by wildlife. The carcasses transported to riparian areas are typically consumed by terrestrial scavengers, such as bears, birds, and saprophagic insects (Hocking and Reimchen 2006). Thus, MDNs provide a linkage of nutrient cycling between aquatic and terrestrial ecosystems (Hocking et al. 2009).

Carrion that falls into freshwater lakes or marine systems can provide protracted perturbation events that release carbon and other nutrients over considerable temporal and spatial scales (Yamamoto et al. 2008; Frost et al. 2012). Because P can be a limiting nutrient in aquatic ecosystems and nutrient budgets are substantially tied to fish, a significant portion of the total fish P is held in bones and scales. Parmenter and Lamarra (1991) proposed the role of vertebrate wildlife carrion with respect to the cycling of P may be that they serve as a P sink rather than a recyclable storage pool of nutrients. However, the journey of a dead fish or other wildlife carrion typically begins with an initial sinking and then a rise to the surface followed by a fall toward the sediments. In the end, the carrion sources may experience other changes, leading to different decomposition outcomes (Haefner et al. 2004; Chidami and Amyot 2008; Zimmerman and Wallace 2008). Studies on the nutrient release from fish debris in Pennsylvania, USA, and Wisconsin lakes, USA, indicated that the vertical flux of P to littoral sediments is substantial and may be greater than allochthonous P inputs (Kitchell et al. 1975; Niriagu 1983). Fish carcasses in the littoral zone may be degraded within days via terrestrial and aquatic invertebrate and sometimes vertebrate activity (Chidami and Amyot 2008). Conversely, fish carcasses sinking into pelagic zones may show a slower decomposition rate due to the thermocline in deeper waters and may take months before the mineralized P is available to re-enter the food chain. This difference in carcass half-life determines how soon carcass-bound P will be available for biomass production and reincorporated into aquatic food webs (Chidami and Amyot 2008).

Pelagic scyphozoan (*Chrysaroa quinquecirrha* Desor.) medusae occur in masses that may last for weeks to months (Sexton et al. 2010). While medusae are comprised primarily of water, they do assimilate and release nutrients such as C, N, and phosphate (PO_4) through excretory and secretory processes (Frost et al. 2012). When medusae densities reach appreciable levels, these releases create a press perturbation (Glasby and Underwood 1996) that may support phytoplankton and bacterial production. However, this response is from live medusae, but what happens when a medusae bloom dies as proposed by Frost et al. (2012)?

Observations on scyphozoan mass falls indicate that the medusae initiate decomposition as they sink and large densities of medusae carry significant biomass to the sea floor (Miyake et al. 2002). Over a short period (days), this mass deposition of medusae carcasses yield increased nutrients via bacterial decomposition and reduced oxygen levels as evidence by hydrogen sulfide production (Billett et al. 2006). These observations support experimentally simulated *C. quinquecirrha* scyphozoan pulse perturbations by Frost et al. (2012). They demonstrate the release of carbon and other nutrients through bacterial decomposition, and certain bacteria drove decomposition in these simulated conditions associated with scyphozoan carrion falls (Frost et al. 2012). These results are valuable for predicting the magnitude and duration of such pulse perturbations in nature and how such perturbations can be modeled to explain nutrient translocation from living and nonliving gelatinous zooplankton into microbial food webs (Frost et al. 2012).

Most of the benthic regions of the world's oceans are nutrient-limited, with a very slow supply cycling to these profound depths from surface waters. Just as in freshwater streams, the input of organic detritus plays an essential role in the structure and function of nutrient cycling and ecosystem dynamics in marine systems. The death of large whales (30–160 ton adult body weight) and the subsequent fall of these large carcasses into deep-sea depths also subsidize deep-ocean benthic zones with massive pulses of labile organic matter (Smith and Baco 2003) (also see Chapter 22).

Considering whale biomass is approximately 5% of organic carbon, the flux of whale fall subsidies of particulate organic carbon (POC) has been estimated to be less than 0.5% of seafloor POC (Smith 2006). However, whales do not sink as one large object, but are scavenged and sink as large, organic-rich lumps that can be distributed over areas of ≈50 m² or greater. Therefore, the sediments underneath these whale falls have been estimated to sustain in a single pulse, equivalent to approximately 2000 years of background POC flux at abyssal depths (Smith and Baco 2003). The work by Smith and Baco (2003) has conservatively estimated that with North Pacific gray whale (*Eschrichtius robustus*, Lilljeborg) ranges, an annual pulse of whale fall events may deposit bits of carcass with a nearest-neighbor proximity of <16 km. Scattered descending pulses of carbon as whale fall upon sea floors essentially creates high-quality or organic-rich "islands" in a nutrient depauperate benthic system that could last for extended periods of time (Stockton and DeLaca 1982). Certainly, within these habitats, the whale carrion detritus pool can be directly tied to key ecological services such as nutrient cycling within and nearby these 'carrion-island' ecosystems.

12.3.2 Quality of Carrion Detritus

The structure and function of the community ecology of organic detritus (both allochthonous and autochthonous) and the interconnectedness of detrital linkages among terrestrial and aquatic food webs have been well studied (Moore et al. 2004). Despite this attention, many studies treat detritus as a single resource pool and ignore detrital quality and the extreme variation between plant and carrion organic material (Wilson and Wolkovich 2011). For example, in aquatic and terrestrial ecosystems, plant detritus is considered by community ecologists to be low quality and slow to decay (Odum and Biever 1984; Cebrian and Lartigue 2004). Therefore, detritus can be considered as multiple resource pools with a gradient of quality from lowest for plant tissue to highest for animal tissue (Moore et al. 2004; Wilson and Wolkovich 2011).

Carrion decomposition via microbes, invertebrates, and vertebrates generally occurs rapidly because of intra- and interkingdom competition for such a high-quality food resource (Burkepile et al. 2006). The quality of carrion may influence nutrient cycling as well as overall food web structure and stability in several ways. For example, one important pathway may be through consumption efficiency, that is, if consuming carrion increases assimilation efficiencies, then the retention of carrion-derived nutrients would be greater in the consumer versus the detrital pool (Wilson and Wolkovich 2011). In addition, greater assimilation efficiencies would regulate consumer population dynamics, thereby controlling prey populations and ultimately stabilize food webs (Wilson and Wolkovich 2011). Certainly, the consumption of vertebrate carrion floating in aquatic systems by terrestrial arthropod taxa expedites the ephemeral status of carrion at the surface of these systems and facilitates the transport of nutrients from the surface to the benthos (Smith and Baco 2003; Berry and Nygard unpublished data).

12.4 Community Organization and Dynamics

An important distinction between terrestrial and aquatic ecosystems in the cycling of carrion resources from both spatial and temporal perspectives is two-fold (Stergiou and Browman 2005). First, the processing of carrion is directly related to the specific properties of water and air, such as narrow temperature ranges, relative density of water, and the inherent three-dimensional nature of aquatic systems that can influence the movement of vertebrate carrion laterally in the water column via wave action, upwelling to the surface, or by sinking to the benthos (Beasley et al. 2012). This relationship may cause the rate of carrion decomposition to be slower in water than on land in some instances, whereas under different circumstances, it may increase the rate. For example, water movement associated with sinking, moving laterally through current and wave action or floating due to bacterial activity resulting in gas accumulation in the gut, facilitates carrion across the aquatic landscape, thus, increasing exposure to scavengers (Britton and Moore 1994; Merritt and Wallace 2010; Beasley et al. 2012). Second, carrion indicator species are ecologically characterized in terrestrial systems according to having coevolved to feed on carrion and their trophic relationship with decomposing carcasses (Merritt and Wallace 2010), whereas in freshwater and marine ecosystems, there is a dichotomy in carrion decomposition with regard to the fauna that utilize it. For example, in freshwater systems, there does not appear to be an evolved "normal" carrion community from a taxonomic standpoint (Fenoglio et al. 2014). Rather, the carrion arthropod fauna in freshwater may be better considered using a functional feeding group (FFG) approach. This approach classifies aquatic invertebrates based on sets of morphological and behavioral adaptations evolved to utilize their basic nutritional resources; in this case, a carcass or algal and fungal matter growing on that carcass (Cummins and Klug 1979; Minshall et al. 1985; Merritt and Cummins 2006; Merritt and Wallace 2010). However, in marine systems, the invertebrates that utilize carrion are considered to be more scavenger-oriented in their ecology (Fenoglio et al. 2014); the community dynamics of organisms that utilize carrion in natural forms (e.g., whale falls; Smith and Baco 2003) or artificial supplementation (e.g., pig carcasses used for research; Anderson 2012) that follow a pattern of organismal scavenger succession as the carcass proceeds through various stages of decomposition.

12.4.1 Succession: From Microbes to Scavengers

In order to understand how the biota changes during carrion decomposition, it is critical to understand the stages of decomposition that an animal carcass progresses. Therefore, the stages of decomposition for carrion or corpses in terrestrial settings have been principally characterized in the forensic literature (Payne 1965; Smith 1986; Anderson and VanLaerhoven 1996). Since the Payne and King (1972) treatise of pig decomposition described in six stages in water, more recent work has modified this description by reducing the number of stages as well as the duration of each stage, the carcass condition, and whether aquatic arthropods are present or absent in freshwater, estuarine, or marine habitats (Vance et al. 1995; Minikawa 1997; Tomberlin and Adler 1998; Hobischak and Anderson 2002; Haefner et al. 2004). Due to the influence of terrestrial insects on floating carcasses and the difficulty distinguishing between the two stages of decay as described by Payne and King (1972), the number of stages were reduced from six to

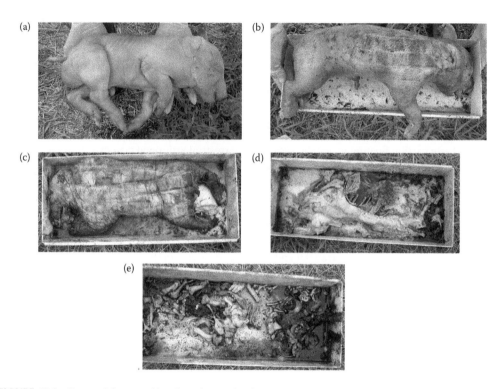

FIGURE 12.3 Stages of decomposition for swine carrion in a Delaware salt marsh, Smyrna, Delaware: (a) Fresh stag, (b) early floating, (c) early floating decay, (d) advanced floating decay, and (e) sunken remains. (Photo credit: Kate Zimmerman.)

five to include: (1) submerged fresh, (2) early floating, (3) early floating decay, (4) advanced floating decay, and (5) sunken remains (Merritt and Wallace 2010) (Figure 12.3 and Table 12.2). The "biotic participants" involved in these stages range from a microbial matrix to a suite of macroinvertebrate processors and vertebrate scavengers (Hobischak and Anderson 2002; Smith and Baco 2003; Merritt and Wallace 2010).

Microbial biofilms in aquatic systems exhibit a progression of species succession as well as biomass differences on a variety of organic (e.g., wood, leaves) and inorganic substrates (e.g., rocks or sand)

TABLE 12.2

Physical Descriptions of the Decompositional Stages from Submerged Pig Carrion in South Central Pennsylvania Streams

Stage of Decomposition	Physical Description of Carcass
(1) Submerged fresh	Fresh; no outward signs of decomposition; still sunken; stage ends when body floats to surface
(2) Early floating	Bloated; floating on surface of water; cage indentations on carcass as it presses against the top of the cage; algal growth evident
(3) Early floating decay	Minor decay becoming apparent; sloughing of flesh; loss of muscle mass or "thinning" of hind limbs; eyes and soft tissues becoming disarticulated; head and legs remains intact; identity of carcass as being that of a pig still evident
(4) Advanced floating decay	Major deterioration visible; ribs and skull exposed; breaks in and loss of bones, including skull; leg bones gone; carcass identity becoming indistinguishable as a result of major appendage and skull loss; stage ends as remains sink
(5) Sunken remains	Remains sunken to bottom of cage; any skin takes on a "soup-like" consistency; stage ended arbitrarily with mostly small pieces of bones remaining

Source: Adapted from Byrd, J.L. and Castner, J.H. (eds). 2010. *Forensic Entomology.* Boca Raton, FL: Taylor & Francis.

(Giller and Malmqvist 1998). The challenges for researchers studying microbial assemblages primarily focus on abundance and succession of microbial taxa as well as functional and metabolic changes, or how antibiotics affect these communities. However, little published work exists, examining microbial assemblages on carrion in aquatic systems (Burkepile et al. 2006; Rozen et al. 2008). The effects of carrion resource pulses on bacterial and fungal diversity, as well as primary production on upper trophic levels may be a function of their magnitude and frequency within aquatic systems (Nowlin et al. 2008). Recently, Benbow et al. (in press) determined that there were similar shifts in bacterial community composition during both summer and winter decomposition trials of swine carcasses within a temperate stream (Figure 12.4). During the summer trial, Bacteroidetes, Firmicutes, and Proteobacteria were the three predominant phyla of the communities. The mean relative abundance was variable in Bacteroidetes during the first (6.3%), second (1.7%), third (2.7%), and fourth (10.2%) weeks of decomposition; Firmicutes increased in mean relative abundance at weekly intervals, increasing from 2.3% during the first week to 23.6 to 36.3 to 58% by the fourth and final week; whereas Proteobacteria demonstrated an inverse trend and decreased in mean relative abundance from 91.3% in week 1, 74.5% in week 2, 60.4% in week 3, down to 30.5% in week four (Figure 12.4a).

During the winter trial, bacterial communities were dominated by Firmicutes and Proteobacteria and demonstrated similar trends over an increased temporal scale of decomposition (7 weeks) with an increasing mean relative abundance of Firmicutes and a decreasing mean relative abundance of Proteobacteria. The Proteobacteria mean relative abundance declined from 81.9 to 38.2% from the first to seventh week in biweekly intervals (Figure 12.4b). These data offer some of the first descriptions of microbial succession on aquatic carrion; however, additional work is needed to go beyond the phylum level to evaluate family- or genera-level carrion microbiome succession. Nevertheless, carrion in aquatic habitats can have profound influence on microbial communities.

Little free oxygen and low temperatures in abyssal ocean depths could result in bacterial decay of carrion being very slow or nonexistent (Beasley et al. 2012; Heinrich 2012). While the role of microbiome effects on decomposing carrion has yet to be clearly elucidated in either freshwater or marine ecosystems, future studies should begin to examine how the combination of autotrophic and heterotrophic microbes in biofilms on carrion function to form the base of the stream or oceanic food webs and ecosystem processes and how carrion effects the microbial communities below these resources during and after decomposition.

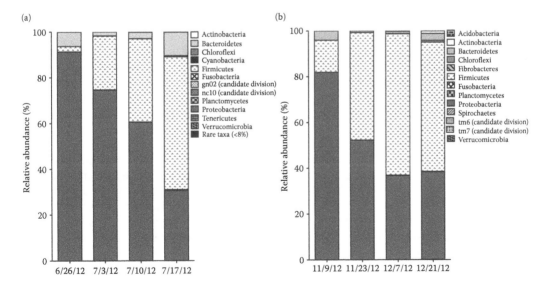

FIGURE 12.4 Seasonal ((a) summer; (b) winter) comparison of the bacterial community composition on implanted pig carcasses in a south central Pennsylvania second-order stream ($n = 3$ for summer trial; $n = 2$ for winter trial).

Because aquatic arthropods in freshwaters have evolved a vast array of morphological, physiological, and behavioral adaptations, they have been able to occupy virtually all bodies of water including specialized habitats such as hot/cold springs, intertidal pools, temporary/vernal ponds, phytotelmata, saline lakes, and marine intertidal zones (Merritt and Wallace 2010). Because of this high variability in habitats, the FFG approach may be a useful technique to monitor carrion decomposition in freshwater and estuarine systems through the documentation of FFGs based on changes in the nutritional resources of a decomposing carcass over time. Several studies utilizing pig carcasses in streams, freshwater ponds, and salt marshes, as well as a few homicide cases, have demonstrated the use of FFGs for this purpose (Chaloner et al. 2002; Wallace et al. 2008; Merritt and Wallace 2010).

The succession fauna on carrion in freshwater systems often consists primarily of crayfish (Decapoda: Cambaridae) and other omnivores, whereas shredders and some aquatic predators also engage in scavenging by removing small bits of tissue from the carcass (Wallace et al. 2008; Merritt and Wallace 2010). Shredders play a more important role in pool habitats than riffles, but for the most part occur on carrion throughout all stages of decomposition. Collector filterers, such as the larvae of net-spinning caddisflies and black flies (Diptera: Simuliidae) are predominate in riffle habitats (Merritt and Wallace 2010). Interestingly, scraper mayflies (Ephemeroptera: Heptageniidae) were reported to be present during the submerged fresh to floating decay stages presumably grazing on the microbial matrix that included a substantial algal component (Haefner et al. 2004; Zimmerman and Wallace 2008). The primary issue in associating aquatic invertebrate taxa with a particular stage in carrion decomposition is that it has been thought that aquatic invertebrates have not evolved as purely sarcophagous feeders (Haskell et al. 1989; Fenoglio et al. 2014). Therefore, a myriad of studies have demonstrated that a wide variety of taxa may appear at random or in no predictable succession sequence in aquatic carrion decomposition.

For example, Barrios and Wolff (2011) found collector-gatherer taxa such as hydrophilid beetles (Coleoptera: Hydrophilidae), leptophlebiid mayflies (Ephemeroptera: Leptophlebiidae), and talitrid amphipods (Crustacea: Amphipoda), collector-filterer taxa such as chironomid midges (Diptera: Chironomidae), black flies (Diptera: Simuliidae), elmid beetles (Coleoptera: Elmidae), and polycentropodid caddisflies (Trichoptera: Polycentropodidae), and predator species such as dragonflies and damselflies (Order: Odonata), true bugs (Order: Hemiptera), and diving beetles (Coleoptera: Dytiscidae) to be abundant throughout succession on pig carcasses in Colombia. The predators *Dytiscus marginicollis* LeConte and *Colymbetes incognatus* Zimmerman (Coleoptera: Dytiscidae) were attracted to fish carrion in a minnow trap in a small lake in the mountains of north-central Colorado, USA (DeJong, pers. obs.). Many aquatic arthropods simply use submerged carrion opportunistically as a substrate for attachment (Hirudinea) or for a convenient feeding location downstream of the shredders (collector-gatherers such as *Chironomus*, collector-filterers such as Simuliidae and Hydropsychidae, and predators feeding on dislodged carrion fauna) (Merritt and Wallace 2010).

Crustaceans are the most common omnivorous taxa in aquatic carrion communities found in both freshwater and marine habitats. Commercial and recreational lobstermen, crabbers, and crayfish collectors frequently make use of the attraction of these crustaceans to decaying carrion by placing pieces of decayed meat in their traps. Two to three crayfish can have a significant effect on decomposition by opening up carrion and expediting the access of larger, vertebrate scavengers (Wallace, pers. obs.). Decomposition of chicken and rat carcasses in minnow traps in Slaughterhouse Gulch, a small, urban stream in metropolitan Denver, Colorado, USA, required only a third of the time when two crayfish were present in comparison to the time required when crayfish were completely excluded (DeJong, pers. obs.). Even some microcrustaceans appear to be attracted to carrion: statistically larger numbers of the copepods *Paracyclops fimbriatus* (Fischer) and *Nitocrella* sp. on mouse carcasses buried in the hyporheic zone of the South Platte River in Denver, Colorado, USA, were found compared to background communities (DeJong, pers. obs.). Nevertheless, it appears that even crustaceans may have preferences for different kinds of carrion, as Groenewold and Fonds (2000) demonstrated using beam-trawl fishery discards in the North Sea.

Empirical work on saprophagic aspects of carrion decomposition in marine environments is lacking (Anderson and Hobischak 2002; Smith and Baco 2003). Much of the knowledge gained in these areas has been through implantation studies, that is, placing either pig carcasses into shallow and deep regions of Bays and Sounds (Anderson and Hobischak 2004) or whale carcasses into deep-ocean environments (Smith and Baco 2003; Dahlgren et al. 2004). Contrary to the longitudinal migrations of salmon moving from marine systems up into headwater freshwater systems for spawning purposes, and the subsequent die-off in streams and rivers contributing a pulse of carrion in these systems, whale carcasses travel vertically in the water column coming from the top layer of the ocean, an area driven by photosynthesis, and are deposited in nutrient-poor and environmentally challenging habitats of deep-sea floors. Therefore, carrion breakdown or decomposition in these environments is primarily due to scavenger activity.

Carrion decomposition on swine carcasses intentionally placed in shallow depths from 7 to 15 m in Horseshoe Bay, British Columbia, Canada, exhibited four primary stages of decomposition: fresh, bloat, active decay, and remains and are described more completely in terms of abiotic factors influencing duration of each stage in Anderson and Hobischak (2002). Mollusks, such as whelks and tritons, were observed to initially scrape carcasses and produce access holes into swine carcasses (Anderson and Hobischak 2002). Macroorganism scavengers such as large crustaceans (e.g., shrimp, lobsters, crabs, and some amphipods and isopods) and echinoderms (e.g., Crinoidea, Asteroidea, Ophiuroidea, Echinoidea, and Holothuroidea) were more abundant on carcasses at a deeper depth (15 m) than those maintained at a shallow depth (7 m) (Anderson and Hobischak 2002). In terms of vertebrate scavengers between these two depths, only fish taxa such as gobies (Perciformes: Gobiidae), sculpins (Scorpaeniformes: Cottoidea), and lingcod (Scorpaeniformes: Hexagrammmidae) were observed early and late in decomposition, respectively (Anderson and Hobischak 2002).

Seasonal differences associated with depth were also observed in these studies with respect to scavenger abundance; however, these differences were inconsistent and interpreted to be associated with methodological issues (Anderson and Hobischak 2002). In a similar study conducted in Howe Sound, near Vancouver, Canada, the primary faunal species involved in a shallow decomposition study were whelks, wrinkled Amphissa urchins (*Strongylocentrotus droebachiensis* Müller), and sunflower sea stars (*Pycnopodia helianthoides* Brandt). Seafloor geology (i.e., substrate type) had a more significant effect on the rate of decomposition by scavengers as compared to depth as well as on the diversity of faunal scavengers involved (Anderson and Hobischak 2004).

One of the more recent findings in marine carrion decomposition deals with the diversity and specialization of scavenger fauna associated with sunken whale carcasses or whale falls (Witte 1999; Smith and Baco 2003; Dahlgren et al. 2006). Previous studies on deep-sea scavengers suggested that fresh whale falls would exhibit four overlapping stages of ecological succession: a mobile-scavenger stage; enrichment-opportunist stage; sulfophilic stage; and a reef stage (Bennett et al. 1994; Smith et al. 1998). Abyssal community succession is discussed in detail in Chapter 22.

Interesting ecological details that have not been addressed in previous carrion decomposition studies in shallow or deep-sea (abyssal) marine systems are the trophic relationship differences observed among freshwater, shallow marine, and abyssal marine organisms as well as microbial presence and duration of time required for carrion decomposition. In freshwater systems, microbial succession and influence on decomposition occurs early in the fresh stage and extends to the active decay stage of decomposition (Benbow et al. in review). Whereas, in abyssal environments, sulfophilic bacteria do not appear until large vertebrate and macroinvertebrate (e.g., amphipods and crustaceans) activity has removed most or all of the flesh, thus colonizing bone surfaces (Smith and Baco 2003).

Invertebrate scavengers in freshwater systems are represented by both shredder species (e.g., caddisflies and amphipods) and scavenger taxa such as crayfish. These scavengers may in fact perform the same task as small invertebrate taxa (e.g., amphipods, mollusks, and echinoderms) in shallow marine systems or large vertebrate fish taxa such as hagfish (Myxiniformes: Myxinidae), sleeper sharks (Squaliformes: Sominiosidae), and grenadiers (Gadiformes: Macrouridae) in abyssal environments. Carrion in freshwater systems are exposed to a suite of abiotic influences, such as current regimes which impact the decomposition process, and are analogous to wave action on carcasses that sink in littoral areas of shorelines (Smith 2009). Certainly, a common thread among all three aquatic systems is that, ultimately, carrion may serve as a substrate upon which other noncarrion-dependent taxa such as suspension feeders, for example, freshwater collector-filterer insects and sabellid, chaetopterid, and serpulid polychaetes (Phylum: Annelida, Class: Polychaeta) can exploit (Smith and Baco 2003; Merritt and Wallace 2010).

12.4.2 Diversity Meets Food-Web Complexity

Changes in biodiversity can affect fundamental ecosystem processes, such as organic matter decomposition (Gartner and Cardon 2004; Hättenschwiler and Gasser 2005). The rate of leaf litter decomposition is based on leaf chemistry and microbial presence (Cornwell et al. 2008), effects of biodiversity on leaf litter resources (Gartner and Cardon 2004; Kominoski et al. 2007), and consumer-level manipulations of microorganisms and invertebrates (Bärlocher and Corkum 2003; Dangles and Malmqvist 2004; Dang et al. 2005). Similar processes that favor such rapid exploitation of leaf litter should be applicable to high nutrient quality resources such as carrion in aquatic systems (Rader and Richardson 1994). This applicability would be especially true for those systems with nutrient limitation because of habitat (e.g., deep ocean depths) or location, such as flowing waters in certain geographical regions that may be either N or P limited as a function soil type (e.g., Pacific Northwest or southwestern USA) (Dillon and Kirchner 1975). Nutrient limitation can also occur during periods of low-stream flow discharge when denitrification may reduce the supply of N (Lohman et al. 1991), or when there are pulses of P enrichment that may render N secondarily limiting (Winterbourne 1990).

Carrion may play a critical role in augmenting horizontal diversity (within-trophic level) to support more complex food webs. Interestingly, with large (whale) or small (swine) carcass implantation studies, it has also been documented that larger invertebrate or vertebrate scavengers facilitated a combination of both horizontal and vertical diversity (across-trophic level) in deep-sea habitats (Anderson and Hobischak 2002; Smith and Baco 2003). Because trophic directional effects on carrion decomposition has been less studied, it is difficult to determine whether bottom-up or top-down effects on carrion decomposition are consistent between freshwater lotic (running water) or lentic (standing water) systems compared to oceanic shallow or deep-sea habitats.

The concept of habitat islands has been applied to whale fall events in both shallow and deep-water depths to serve as a model for chemosynthetic ecosystems (Dahlgren et al. 2006). In fact, more than 400 species have been associated with whale falls (Smith and Baco 2003) and there is a high degree of endemism, especially among annelids associated with these carcasses (Dahlgren et al. 2004). Communities of fossilized chemoautotrophs dating back to the Oligocene ($\approx$30 mya) have been found on fossilized whale remains from the bathyal northeast Pacific (Goedert et al. 1995). Therefore, chemoautotrophic faunal assemblages and their feeding relationships with whale carrion have existed for tens of millions of years (Smith and Baco 2003). These "stepping stones" of a diverse assemblage of sulfophilic taxa closely tied to the decomposition of concentrated bone lipids scattered about these "resource islands" may represent a dispersal mechanism for those taxa (Smith and Baco 2003).

The loss of biodiversity affecting carrion decomposition has been shown in recent experiments where there has been a tendency to reduce the consumptive power and conversion of carrion biomass and inherently impact ecosystem functioning when reducing species diversity within a given trophic level (Cardinale et al. 2006; Srivastava et al. 2009). However, these studies have unfortunately focused on a single trophic level; the more important relationship would examine the role of multiple trophic levels, or trophic structure, in the control of carrion biomass distribution among food webs (Srivastava and Vellend 2005; Srivastava et al. 2009).

12.5 Applications and Broader Implications

Key questions in terms of microbial, invertebrate, and vertebrate consumption of carrion in aquatic systems are by what mechanism is this process initiated and why is it assumed that specialists have not evolved in aquatic systems (Fenoglio et al. 2014). Without a doubt, as Beasley et al. (2012) suggested, the decomposition of carrion in terrestrial systems is integral to many ecological processes from secondary production and the preservation of food web stability (Wilson and Wolkovich 2011) to diminishing the spread of disease either through natural scavenging of carcasses or human intervention via composting (Jennelle et al. 2009; Xu et al. 2009). However, the role of carrion decomposition in aquatic ecosystems in terms of anthropogenic perturbations has yet to be elucidated in terms of any cascading effects on ecosystem functioning, food-web stability, or disease spread (for a detailed treatment of anthropogenic effects on scavenging through shifts in carrion frequency and distribution, see Chapter 6).

12.5.1 Ecological Impacts

The spatial and temporal dynamics of carrion subsidies in concert with the competitive hierarchy among organisms that exploit these resources are integral drivers in terrestrial system decomposition processes that are subsequently linked through evolutionary processes to obligate necrophages (Beasley et al. 2012; Mondor et al. 2012). The transfer of nutrients and carbon compounds from carrion either by passive means, such as leaching, or through active measures, such as detritivore feeding, may influence how efficiently necrophagic populations are able to compete and exploit the ephemeral nature of carrion subsidies, thereby impacting the rate of decomposition and ultimately ecosystem functioning (Gessner et al. 2010). Therefore, in systems such as freshwater and marine environments, certain organisms are critical in the movement of such nutrients; however, what is less understood is how environmental perturbations such as climate influence this rather tenuous balance between facultative decomposers (e.g., invertebrates and vertebrates) and obligate decomposers (e.g., bacteria with maintaining food-web stability).

Freshwater systems that retain salmon add valuable nutrients to not only these systems, but also to adjoining riparian buffer habitats (Hocking et al. 2009). This reverse nutrient subsidy not only impacts the freshwater organisms where salmon are deposited, but also terrestrial insects and large, mammalian fauna such as brown bears (*Ursus arctos* Linnaeus) (Henfield and Naiman 2006; Hocking and Reimchen 2006; Hocking et al. 2006). If water temperature increases related to climate change influence the presence of these invertebrate carrion-feeders, then processing and nutrient transfer of such subsidies may be altered in complex ways. Concomitantly, if water levels decrease and water temperatures increase with impacts on salmon fry survivorship, then future subsidies to these systems are in danger of serious decline. Such a reduction in salmon carrion subsidy to these breeding waters could result in negative impact in riparian biodiversity by limiting species dependent on a carrion food supply (Wilmers et al. 2003a,b). Conversely, in marine abyssal ecosystems, climate change impacts are not as regulating as they are on land or perhaps in shallow, freshwater systems (Beasley et al. 2012).

Without large or small carrion-feeding organisms to cycle nutrient pulses received from salmon subsidies, decomposition rates in aquatic systems will be slower, hence creating nutrient-enriched loads that would generate excess algal growth and eutrophication in these systems (Chislock et al. 2013). From an ecosystem perspective, hypereutrophic aquatic systems have reduced water quality due to limited light and reduced energy-transfer efficiency in food webs, thus changing aquatic community structure (Chislock et al. 2013). Aquatic systems severely impacted by nutrient-enriching events such as cattle grazing have shown increased abundance of *Tubifex tubifex*, the primary oligochaete host associated with the salmonid whirling disease parasite, *Myxobolus cerebralis*, thus reducing wild trout populations (Zendt and Bergersen 2000). Therefore, a reduction in salmon subsidy processing that potentially leads to nutrient enrichment and direct increases in whirling disease may in fact further impact salmon and trout fry further reducing future carrion subsidies.

12.5.2 Forensic Investigation Applications

Bridging the intimate relationship between carrion and terrestrial insects with criminal and civil investigations has been facilitated by studies of their life histories and basic ecology (Keiper and Casamata 2001). Terrestrial insects and their arthropod relatives play a diverse role in such investigations, and the expanding body of knowledge over the past decade has elucidated the effects of temperature on development, behavioral, and dispersal activities, thus providing a firm foundation that is crucial to their application in death scene investigations. To a degree, similar applications of benthic ecology to forensic investigations are becoming more frequent in estimating a PMSI, that is, the time period from corpse submersion in an aquatic ecosystem to the moment of discovery (Keiper and Casamata 2001). There are several candidates of evidence that can be used to estimate a PMSI: (1) algal growth and succession (Haefner et al. 2004; Zimmerman and Wallace 2008), (2) adipocere formation (Yan et al. 2001); (3) larval insect activity (Haskell et al. 1989; Hobischak 1997; Wallace and Merritt 2008; Myskowiak et al. 2010); and (4) bacterial communities (Benbow et al. in press). However, there is a paucity of information for using aquatic organisms in PMSI estimates, most probably because there are few evolved necrophagous aquatic insects (Chaloner et al. 2002), and there are few studies that have examined aquatic invertebrate colonization of carrion or corpses (Merritt and Wallace 2010).

Research efforts addressing the use of aquatic invertebrates to estimate a PMSI have expanded since early studies that focused on developing sampling devices to improve collections of aquatic insect evidence (Vance et al. 1995). Recent statistical modeling research has focused on the modification of unique multivariate approaches (e.g., Wagner's parsimony method typically used in cladistic phylogenetic analyses) to classify different crime scene characteristics and compare them to the collected invertebrates from corpses in order to isolate pertinent bio-indicators and improve accuracy in PMSI estimates (Myskowiak et al. 2010). Recent empirical work has included the identification of indicator taxa (e.g., caddisflies and crayfish) in actual case work that may provide important life history connections to PMSI estimates. And the application of FFG classification approaches used in detritus processing (as described earlier in this chapter) may be used to develop the necessary associations between aquatic insect succession and the stage of decomposition for statistical models (Wallace et al. 2008; Merritt and Wallace 2010).

The concept of different stages of decomposition of carrion originating from terrestrial ecological work with a forensic focus (Smith 1986; Anderson and VanLaerhoven 1996), but was modified by Hobischak and Anderson (2002), Haefner et al. (2004), and Zimmerman and Wallace (2008) for carrion in lotic and lentic aquatic ecosystems. Using swine implantation studies in Pennsylvania, USA, streams, the five stages of decomposition that have been described include: (1) submerged fresh, (2) early floating, (3) early floating decay, (4) advanced floating decay, and (5) sunken remains (Merritt and Wallace 2010) (Table 12.2).

As suggested by Keiper and Casamata (2001), macroinvertebrate assemblages associated with each stage of aquatic decomposition (Table 12.2) as described by Schultenover and Wallace (unpublished data, 1997) indicated that collector-gatherer and -filterers such as the net-spinning caddisfly (Trichoptera: Hydropsychidae), black fly larvae (Diptera: Simuliidae), and midge larvae (Diptera: Chironomidae) were most abundant on carcasses throughout decomposition (Figure 12.5).

Similar results were found by Keiper et al. (1997) on rat (*Rattus rattus*) carcasses in stream riffle and pool habitats in Ohio, USA, documenting a diverse assemblage of the collector-gatherer midge family, Chironomidae. This has also been reported by Chaloner et al. (2002) on salmon carrion. Interestingly,

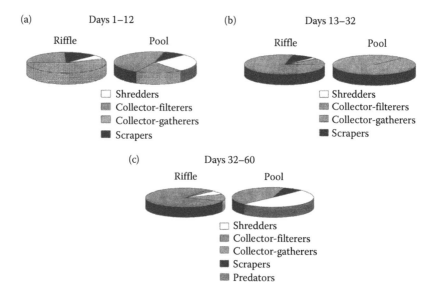

FIGURE 12.5 Percent composition and comparative changes in macroinvertebrate FFGs colonizing pigs at five different stages of decomposition ((a) Submerged fresh—early floating; (b) early floating—advanced decay; (c) advanced decay—sunken remains) in riffle and pool habitats in an Indiana stream. (Adapted from Byrd, J.L. and Castner, J.H. (eds). 2010. *Forensic Entomology*. Boca Raton, FL: Taylor & Francis.)

many scraper invertebrates (e.g., Heptageniidae mayfly and Glossosomatidae caddisfly larvae) were observed feeding on the algae accumulating on a decomposing pig. These observations prompted a new line of enquiry by Keiper and Casamata (2001) who examined algal succession on carrion for use in estimating a PMSI.

While the exact role of algal colonization on carrion in the decomposition process has not been elucidated, the excess nutrients leaching from a carcass in aquatic systems does provide a substrate upon which ecological concepts of primary production can be examined from a PMSI perspective. Other microbial communities, such as bacteria, may also respond in similar way to carrion in the aquatic habitat. Chlorophyll *a* production and taxonomic diversity analyses of algal colonization may provide an important quantitative connection between carrion decomposition in fresh and brackish water systems and forensic investigations (Haefner et al. 2004; Zimmerman and Wallace 2008). In Pennsylvania, USA, streams, Haefner et al. (2004) found that not only did chlorophyll *a* concentrations differ between pig carrion and ceramic tiles, but they also increased significantly over time; the authors also provided a regression model that could be used to estimate the amount of chlorophyll *a* on a human body for any given time period up to approximately 1 month. Concomitantly, Zimmerman and Wallace (2008) found that algal diversity differed between pig carcasses and tiles in brackish salt marshes and that taxonomic diversity was correlated to specific stages of decomposition. Applying general benthic ecological methodologies to understand primary production in aquatic systems has demonstrated promise for use in the world of criminal justice and forensic science.

12.6 Conclusions and Perspectives

A common theme emerging from comparing carrion decomposition across freshwater, estuarine, and marine ecosystems is that the abiotic effects on decomposition, the underlying mechanisms governing energy transfer and nutrient cycling, and the effects of diversity and succession on community organization are fundamentally similar across all three major types of aquatic ecosystems. In part, this similarity may be attributed to the intimate selection pressures that have shaped the evolutionary paths of microbial—invertebrate—vertebrate interactions in processing decomposing plant material. Support for this contention may be explained by the fact that in many instances, the occurrence of carrion (e.g., small-scale carrion deposition events) is unpredictable and results in intense competition for these high-quality ephemeral resources (Tomberlin et al. 2012). Therefore, arthropods or vertebrates that utilize carrion resources in aquatic systems may have evolved highly evolved chemosensory systems that expedite the location, colonization, and utilization of these resources similar to certain terrestrial arthropods (Tomberlin et al. 2012; Fenoglio et al. 2014).

However, to what extent do patterns emerge from freshwater and marine system comparisons that more generally describe energy transfer, diversity, and food-web complexity associated with aquatic carrion? The answer may lie in the recent underpinnings of a new paradigm involving the role of microbial communication in terrestrial decomposition ecology (Tomberlin et al. 2011). Potentially, microbial populations in aquatic systems may initiate the release of volatile organic compounds (see Chapter 20) that serve as small signaling compounds that facilitate "cross-talk" communication (i.e., quorum sensing) within and among bacterial species that may attract shredder invertebrates and possibly vertebrate scavengers (Hughes and Sperandio 2008; Tomberlin et al. 2012). Recently, several studies have documented this phenomenon occurring with saprophagic Diptera that use carrion remains on land (Ma et al. 2012; Tomberlin et al. 2012). The implication in terms of nutrient transfer, community structure, and biodiversity of carcass–microbe–secondary consumer–tertiary consumer interactions would serve as a new paradigm for understanding decomposition ecology not just in terrestrial systems as proposed by Tomberlin et al. (2011), but for aquatic systems as well. Certainly, such experiments would be particularly enlightening regarding interkingdom communication between microbes and higher forms of carrion decomposers in aquatic systems.

Anthropogenic impacts (e.g., climate change) on aquatic systems represent significant threats not only to ecosystem functioning in general, but also to the stability of trophic relationships of carrion decomposition (Wilson and Wolkovich 2011). While it is arguable that aquatic system response to these impacts may vary, it is understood that decomposition rates for carrion double for every 10°C increase in temperature in terrestrial systems (Parmenter and MacMahon 2009), and while this has not been studied in aquatic habitats, it is assumed that rates will increase in aquatic systems accordingly. However, a greater threat to ecosystem functioning would be the potential loss of carrion species in aquatic systems as global average temperatures rise. For terrestrial systems, it is predicted that there will be a reduction in carrion availability, especially in the arctic regions (Fuglei et al. 2003). However, how will mid-latitude stream systems respond to such temperature increases? If there are negative impacts on salmon egg and fry survivorship, there could be substantial top-down repercussions leading to a reduction of oceanic feeding salmon.

Smith et al. (2008) have stated that current climate change models project that temperature shifts will negatively impact surface primary production by as much as 50% of organic material in marine systems, and such a decline would drastically reduce available resources for abyssal ecosystems and subsequently constrain scavenger populations, with negative impacts on the biodiversity of these deep-sea environments. It has only been in the last 10–15 years where there has been increased research interest into the role of MDNs in freshwater systems and the importance of whale carrion in abyssal ecosystems. The dearth of research into carrion-dependent ecosystem response is alarming, but one that can be corrected in order to more fully understand the ecosystem processes influenced by imminent environmental change associated with natural or anthropogenic sources such as global climate change, pollution, or overfishing (Smith and Baco 2003).

The connection between aquatic carrion decomposition with forensic science is relatively recent; however, three landmark cases in the court of law have highlighted arthropod taxa that are key carrion decomposers in aquatic systems; cases where black fly larvae, caddisfly larvae, and crayfish were important forms of evidence (Merritt and Wallace 2010). These cases demonstrate how a deeper understanding of aquatic carrion decomposition has profound applications.

Tomberlin et al. (2011) eloquently stated that perhaps the identification and pursuit of novel areas of research could lead to a better understanding of the key components that govern the evolutionary ecology of carrion and how it relates to ecosystem structure and function and how this increase in knowledge could inform the forensic sciences. Therein lies the lynchpin to understanding aquatic carrion decomposition, it is this empirical roadmap that is needed to build a better bridge between the vast foundation of basic and applied terrestrial decomposition ecological research and connect with what is most inextricably linked with our watery worlds.

Acknowledgments

I would like to acknowledge Ms. Rebecca McCabe and Don Brown in the acquisition of key resources to write this chapter. The contributions of important images by Ms. Suzanne Yocom and Ms. Emily Campbell are greatly appreciated. I would like to express my appreciation to Dr. Rich Merritt for comments in the writing process and to the editors for their patience.

REFERENCES

Addy, K. and L. Green. 1997. Dissolved oxygen and Temperature. Cooperative Extension Fact Sheet 96-3. Kingston, RI: University of Rhode Island. 5 pp.

Allan, J.D. 1995. *Stream Ecology—Structure and Function of Running Waters*. London: Chapman & Hall, 388 pp.

Anderson, G.S. and L.S. Bell. 2014. Deep coastal marine taphonomy: Investigation into carcass decomposition in the Saanich Inlet, British Columbia using a baited camera. *PLoS ONE* 9(10): e110710.

Almazon, G. and C.E. Boyd. 1978. Effects of nitrogen levels on rates of oxygen consumption during decay of aquatic plants. *Aquatic Botany* 5: 119–126.

Andersen, F.O. 1978. Effects of nutrient level on the decomposition of *Phragmites communis* Trin. *Archiv für Hydrobiologie* 84: 42–54.

Anderson, G.S. and N.R. Hobischak. 2002. Determination of time since death for humans discovered in salt-water using aquatic organism succession and decomposition rates. Canadian Police Research Centre, Ottawa, Ontario. 76 pp. TR-09–2002.

Anderson, G.S. and N.R. Hobischak. 2004. Decomposition of carrion in the marine environment in British Columbia, Canada. *International Journal of Legal Medicine* 11(4): 206–209.

Anderson, N.H. and J.R. Sedell. 1979. Detritus processing by macroinvertebrates in stream ecosystems. *Annual Review of Entomology* 24: 531–577.

Anderson G.S. and S.L. VanLaerhoven. 1996. Initial studies on insect succession on carrion in southwestern British Columbia. *Journal of Forensic Science* 41: 617–625.

Ashley, K.I. and P.A. Slaney. 1997. Accelerating recovery of stream and pond productivity by low-level nutrient enrichment. In: Slaney, P.A. and D. Zaldokas (eds), *Fish Habitat Rehabilitation Procedures*. Watershed Restoration Technical Circular No. 9, pp. 239–262. Vancouver, BC: British Columbia Ministry of Environment, Lands and Parks and Ministry of Forests.

Bärlocher, F. and M. Corkum. 2003. Nutrient enrichment overwhelms diversity effects in leaf decomposition by stream fungi. *Oikos* 101: 247–252.

Bärlocher, F. and B. Kendrick. 1974. Dynamics of the fungal population on leaves in a stream. *Ecology* 11(1): 761–791.

Barrios, M. and M. Wolff. 2011. Initial study of arthropods succession and pig carrion decomposition in two freshwater ecosystems in the Colombian Andes. *Forensic Science International* 212: 164–172.

Bauder, E.T. 2005. The effects of an unpredictable precipitation regime on vernal pool hydrology. *Freshwater Biology* 50: 2129–2135.

Baxter, C.V., K.D. Fausch, M. Murakami, and P.L. Chapman. 2004. Fish invasion restructures stream and forest food webs by interrupting reciprocal prey subsidies. *Ecology* 85: 2656–2663.

Baxter, C.V., K.D. Fausch, and W.C. Saunders. 2005. Tangled webs: Reciprocal flows of invertebrate prey link streams and riparian zones. *Freshwater Biology* 50(2): 201–220.

Beasley, J.C., Z.H. Olson, and T.L. DeVault. 2012. Carrion cycling in food webs: Comparisons among terrestrial and marine ecosystems. *Oikos* 121: 1021–1026.

Benbow, M.E., J.L. Pechal, J.M. Lang, R.E. Erb, and J.R. Wallace. 2015. The potential of high-throughput metagnomic sequencing of aquatic bacterial communities to estimate the postmortem submersion interval. *Journal of Forensic Science* (in press).

Benner, R., M.A. Moran, and R.E. Hodsos. 1985. Effects of pH and plant source on lignocellulose biodegradation rates in two wetland ecosystems, the Okefenokee Swamp and a Georgia salt marsh. *Limnology and Oceanography* 30: 489–499.

Bennett, B.A., C.R. Smith, B. Glaser, and H.L. Maybaum. 1994. Faunal community structure of a chemoautotrophic assemblage on whale bones in the deep northeast Pacific Ocean. *Marine Ecology* 138: 803–817.

Berezina, N.A. 1999. Influence of ambient pH on freshwater invertebrates under experimental conditions. *Russian Journal of Ecology* 32(5): 343–351.

Billett, D.S., B.J. Bett, C.L.O. Jacobs, I.P. Rouse, and B.D. Wigham. 2006. Mass deposition of jellyfish in the deep Arabian Sea. *Limnology and Oceanography* 51: 2077–2083.

Britton, J.C. and B. Morton. 1994. Marine carrion and scavengers. *Oceanography and Marine Biology: An Annual Review* 32: 369–434.

Brock, T.C.M., J.J. Boon, and B.G.P. Paffen. 1985a. The effects of the season and of the water chemistry on the decomposition of *Nymphaea alba* L.: Weight loss and pyrolysis spectrometry of the particulate matter. *Aquatic Botany* 21: 197–229.

Brock, T.C.M., M.J.H. De Lyon, E.M.J. Van Laar, and E.M.M. Van Loon. 1985b. Field studies on the breakdown of *Nuphar lutea* (L.) SM. (Nymphaeaceae), and a comparison of three mathematical models for organic weight loss. *Aquatic Botany* 21: 1–22.

Brown, W. 1969. Predicting temperatures of small streams. *Water Resources Research* 5: 68–75.

Burkepile, D.E., J.D. Parker, C.B. Woodson, H.J. Mills, J. Kubanek, P.A. Sobecky, and M.E Hay. 2006. Chemically mediated competition between microbes and animals: Microbes as consumers in food webs. *Ecology* 87: 2821–2831.

Byrd, J.L. and Castner, J.H. (eds). 2010. *Forensic Entomology*. Boca Raton, FL: Taylor & Francis.

Campbell, E.Y., M.E. Benbow, S.D. Tiegs, J.P. Hudson, G.A. Lamberti, and R.W. Merritt. 2011. Timber harvest intensifies spawning-salmon disturbance of macroinvertebrates in southeast Alaska streams. *Journal of North American Benthological Society* 301(1): 49–59.

Cardinale, B.J., D.S. Srivastava, J.E. Duffy, J.P. Wright, A.L. Downing, M. Sankaran, and C. Jouseau. 2006. Effects of biodiversity on the functioning of trophic groups and ecosystems. *Nature* 443: 989–992.

Carpenter, S., N.F. Caraco, D.L. Correll, R.W. Howarth, A.N. Sharpley, and V.H. Smith. 1998. Nonpoint pollution of surface waters with phosphorus and nitrogen. *Issues in Ecology* 3: 1–12.

Cebrian, J. and J. Lartigue. 2004. Patterns of herbivory and decomposition in aquatic and terrestrial ecosystems. *Ecological Monographs* 74: 237–259.

Chaloner, D.T. and M.S. Wipfli. 2002. Influence of decomposing Pacific Salmon on macroinvertebrate growth and standing stock in southeastern Alaska streams. *Journal of the North American Benthological Society* 21(3): 430–442.

Chaloner, D.T., M.S. Wipfli, and J.P. Caquette. 2002. Mass loss and macroinvertebrate colonisation of Pacific salmon carcasses in south-eastern Alaskan streams. *Freshwater Biology* 47: 263–273.

Chidami, S. and M. Amyot. 2008. Fish decomposition in boreal lakes and biogeochemical implications. *Limnology and Oceanography* 53(5): 1988–1996.

Chislock, M.F., E. Doster, R.A. Zitomer, and A.E. Wilson. 2013. Eutrophication: Causes, consequences, and controls in aquatic ecosystems. *Nature Education Knowledge* 4(4): 10.

Christensen, A.M. and S.W. Myers. 2011. Macroscopic observations of the effects of varying freshwater pH on bone. *Journal of Forensic Sciences* 56: 475–479.

Clymo, R.S. 1983. Peat. In: Gore, A.J.P. (ed.), *Mires: Swamp, Bog, Fen and Moor*. Amsterdam: Elsevier, pp. 159–224.

Cornwell, W.K., J.H.C. Cornelissen, K. Amatangelo, E. Dorrepaal, V.T. Eviner, O. Godoy, H.E. Hobbie, B. Hoorens, and H. Kurokawa. 2008. Plant species traits are the predominant control on litter decomposition rates within biomes worldwide. *Ecology Letters* 11: 1065–1071.

Cummins, K.W. 1973. Trophic relations of aquatic insects. *Annual Review of Entomology* 18: 183–206.

Cummins, K.W. 1974. Structure and function of stream ecosystems. *Bioscience* 24: 631–641.

Cummins, K.W. and M.J. Klug. 1979. Feeding ecology of stream invertebrates. *Annual Review of Ecological Systematics* 10: 147–172.

Cummins, K.W., R.C. Peterson, F.O. Howard, J.C. Wuycheck, and V.I. Holt. 1973. The utilization of leaf litter by stream detritivores. *Ecology* 54: 336–345.

Cummins, K.W., M.A. Wilzbach, D.M. Gates, J.B. Perry, and W.B. Taliaferro. 1989. Shredders and riparian vegetation. *Bioscience* 39: 24–30.

Dahlgren, T.G., A.G. Glover, and A. Baco. 2004. Fauna of whale falls: Systematics and ecology of a new polychaete (Annelida: Chrysopetialidae) from the deep Pacific Ocean. *Deep-Sea Research I* 51: 1873–1887.

Dahlgren, T.G., H. Wiklund, B. Kallstrom, T. Lundalv, C.R. Smith, and A.G. Glover. 2006. A shallow-water whale-fall experiment in the North Atlantic. *Cahiers de Biologie Marine* 47(4): 385–389.

Dale, H.M. and T. Gillespie. 1977. Diurnal fluctuations of temperature near the bottom of shallow water bodies as affected by solar radiation, bottom color and water circulation. *Hydrobiologia* 55: 87–92.

Dang, C.K. et al. 2005. Magnitude and variability of process rates in fungal diversity litter decomposition relationships. *Ecology Letters* 8: 1129–1137.

Dangles, O. and B. Malmqvist. 2004. Species richness–decomposition relationships depend on species dominance. *Ecology Letters* 7: 395–402.

Debusk, W.F. and K.R. Reddy. 2005. Litter decomposition and nutrient dynamics in a phosphorus enriched everglades marsh. *Biogeochemistry* 75: 217–240.

Dillon, P.J. and W.B. Kirchner. 1975. The effects of geology and land use on the export of phosphorous from watersheds. *Water Research* 9: 135–148.

Duffy, J.E., B.J. Cardinale, K.E. France, P.B. McIntyre, E. Thébault, and M. Loreau. 2007. The functional role of biodiversity in ecosystems: Incorporating trophic complexity. *Ecology Letters* 10: 522–538.

Dunlop, J., G. McGregor, and N. Horrigan. 2005. Potential impacts of salinity and turbidity in riverine ecosystems. National Action Plan for Salinity and Water Quality. WQ06 Technical Report. QNRM05523, ISBN 1741720788.

Egglishaw, H.J. 1972. An experimental study of the breakdown of cellulose in fast-flowing streams. *Memorie Dell'Istituto Italiano Di* (Supplement) 29: 23–27.

EPA. 2008. What are the trends in the critical physical and chemical attributes of the nation's ecological systems? 2008 Report on the Environment. pp. 31–45. http://www.epa.gov/ncea/roe/docs/roe_final/roe_final_eco_chap6_physical.pdf.

Estes, J.A., D.P. DeMaster, R.L. Brownell Jr., D.F. Doak, and T.M. Williams (eds). 2006. *Whales, Whaling and Ocean Ecosystems*. Berkeley, CA, USA: University of California Press, pp. 286–301.

Fairchild, G.W., R.L. Lowe, and W.B. Richardson. 1985. Algal periphyton growth on nutrient-diffusing substrates: An *in situ* bioassay. *Ecology* 66(2): 465–472.

Fenoglio, S.R., R.W. Merritt, and K.W. Cummins. 2014. Why do no specialized necrophagous species exist among aquatic insects? *Freshwater Science* 33(3): 711–715.

Frost, J.R., C.A. Jacoby, T.K. Frazer, and A.R. Zimmerman. 2012. Pulse perturbations from bacterial decomposition of *Chrysaora quinquechrrha* (Scyphozoa: Pelagidae). *Hydrobiologia* 690: 247–256.

Fuglei, E. et al. 2003. Local variation in arctic fox abundance on Svalbard, Norway. *Polar Biology* 26: 93–98.

Gartner, T.B. and Z.G. Cardon. 2004. Decomposition dynamics in mixed-species leaf litter. *Oikos* 104: 230–246.

Gessner, M.O., C.M. Swann, C.K. Dang, B.G. McKie, R.D. Bardgett, D.H. Wall, and S. Hattenschwiler. 2010. Diversity meets decomposition. *Trends in Ecology and Evolution* 25(6): 372–380.

Giller, P.S. and B. Malmqvist. 1998. *The Biology of Streams and Rivers*. New York: Oxford University Press, 296 pp.

Glasby, T.M. and A.J. Underwood. 1996. Sampling to differentiate between pulse and press perturbations. *Environmental Monitoring and Assessment* 42: 241–252.

Goedert, J.L., R.L. Squires, and L.G. Barnes. 1995 Paleoecology of whale-fall habitats from deep-water Oligocene rocks, Olympic Peninsula, Washington State. *Paleogeography, Palaeoclimatology, Paleoecology*, 118: 151–158.

Groenewold, S. and M. Fonds. 2000. Effects on benthic scavengers of discards and damaged benthos produced by the beam-trawl fishery in the southern North Sea. *ICES Journal of Marine Science* 57: 1395–1406.

Haefner, J.N., J.R. Wallace, and R.W. Merritt. 2004. Pig decomposition in lotic aquatic systems: The potential use of algal growth in establishing a postmortem submersion interval (PMSI). *Journal of Forensic Science* 49(2): 330–336.

Haglund, W.D. and M.H. Sorg. (eds). 1997. *Forensic Taphonomy: The Postmortem Fate of Human Remains*. Boca Raton, FL: CRC Press, 672 pp.

Haglund, W.D. and M.H. Sorg (eds). 2002. Human remains in water environments. In: *Advances in Forensic Taphonomy: Method, Theory, and Archaeological Perspectives*. Boca Raton, FL: CRC Press, pp. 202–218.

Haines, T.A. 1981. Acidic precipitation and its consequences for aquatic ecosystems: A review. *Transactions of the American Fisheries* 110: 669–707.

Haskell, N.H., D.G. McShaffrey, D.H. Hawley, R.E. Williams, and J.E. Pless. 1989. Use of aquatic insects in determining submersion interval. *Journal of Forensic Science* 34(3): 622–632.

Hättenschwiler, S. and P. Gasser. 2005. Soil animals alter plant litter diversity effects on decomposition. *Proceedings of the National Academy of Sciences* 102: 1519–1524.

Heinrich, B. 2012. *Life Everlasting: The Animal Way of Death*. Boston, MA: Houghton Mifflin Harcourt Publishing, 235 pp.

Henfield, J.M. and R.J. Naiman. 2006. Keystone interactions: Salmon and bear in riparian forests of Alaska. *Ecosystems* 9: 167–180.

Hicks, B.J., M.S. Wipfli, D.W. Lang, and M.E. Lang. 2005. Marine-derived nitrogen and carbon in freshwater-riparian food webs of the Copper River Delta, southcentral Alaska. *Oecologia* 144: 558–569.

Hobischak, N.R. 1997. Freshwater invertebrate succession and decomposition studies on carrion in British Columbia. Master's thesis in pest management, Simon Fraser University, Burnaby, British Columbia, Canada.

Hobischak, N.R. and G.S. Anderson. 2002. Time of submergence using aquatic invertebrate succession and decompositional changes. *Journal of Forensic Science* 47: 142–151.

Hocking, M.D. and T.E. Reimchen. 2006. Consumption and distribution of salmon (*Onchorrhyncus* spp.) nutrients and energy by terrestrial flies. *Canadian Journal of Fisheries Aquatic Sciences* 63: 2076–2086.

Hocking, M.D., R.A. Ring, and T.E. Reimchen. 2006. Burying beetle *Nicrophorus investigator* reproduction on Pacific salmon carcasses. *Ecological Entomology* 31: 5–12.

Hocking, M.D., R.A. Ring, and T.E. Reimchen. 2009. The ecology of terrestrial invertebrates on Pacific salmon carcasses. *Ecological Research* 24: 1091–1100.

Hughes, D.T. and V. Sperandio. 2008. Interkingdom signalling: Communication between bacteria and their hosts. *National Review Microbiology* 6(2): 111–120.

Hutchinson, G.E. 1957. *A Treatise on Limnology*. Vol. I. New York: Wiley, 1015 pp.

Jennelle, C.S., M.D. Samuel, C.A. Nolden, and E.A. Berkley. 2009. Deer carcass decomposition and potential scavenger exposure to chronic wasting disease. *The Journal of Wildlife Management* 73: 655–662.

Kaushik, N.K. and H.B.N. Hynes. 1971. The fate of the dead leaves that fall into streams. *Archiv für Hydrobiologie* 68: 465–515.

Keiper, J.B. and D.A. Casamata. 2001. Benthic organisms as forensic indicators. *Journal of the North American Benthological Society* 20(2): 311–324.

Keiper, J.B., E.G. Chapman, and B.A. Foote. 1997. Midge larvae (Diptera: Chironomidae) as indicators of postmortem submersion interval of carcasses in a woodland stream: A preliminary report. *Journal of Forensic Science* 42: 1074–1079.

Kitchell, J.F., J.F. Koonce, and P.S. Tennis. 1975. Phosphorous flux through fishes. *Internationalen Vereinigung für Theoretische und Angewandte Limnologie* 19: 2478–2484.

Kline Jr, T.C. et al. 1993. Recycling of elements transported upstream by runs of pacific salmon: I. $\delta^{15}N$ and $\delta^{13}C$ evidence in the Kvichak River watershed, Bristol Bay, Southwestern Alaska. *Canadian Journal of Fisheries and Aquatic Sciences.* 50: 2350–2365.

Kline Jr, T.C., J.J. Goering, O.A. Mathison, and P.H. Poe. 1997. The effect of salmon carcasses on Alaskan freshwaters. In: Milner, A.H. and M.W. Oswood (eds), *Freshwaters of Alaska: Ecological Synthesis.* New York: Springer-Verlag, pp. 179–204.

Kominoski, J.S. et al. 2007. Nonadditive effects of litter species diversity on breakdown dynamics in a detritus-based stream. *Ecology* 88: 1167–1176.

Lessard, J.L. and R.W. Merritt. 2006. Influence of marine-derived nutrients from spawning salmon on aquatic insect communities in southeast Alaskan streams. *Oikos* 113: 334–343.

Lessard, J.L., R.W. Merritt, and M.B. Berg. 2009. Investigating the effect of marine-derived nutrients from spawning salmon on macroinvertebrate secondary production in southeast Alaskan streams. *Journal of North American Benthological Society* 28: 683–693.

Lohman, K., J.R. Jones, and C. Baysinger-Daniels. 1991. Experimental evidence for nitrogen limitation in a northern Ozark stream. *Journal of the North American Benthological Society* 10(1): 14–23.

Ma, Q., A. Fonseca, W. Liu, A.T. Fields, M.L. Pimsler, A.F. Spindola, A.M. Tarone, T.L. Crippen, J.K. Tomberlin, and T.K. Wood. 2012. *Proteus mirabilis* interkingdom swarming signals attract blow flies. *ISME Journal* 6(7): 1356–1366.

Macan, T.T. and Maudsley, R. 1966. The temperature of a moorland fishpond. *Hydrobiologia* 27: 1–22.

Merritt, R.W. and K.W. Cummins. 2006. Trophic relationships. In: Hauer, F.R. and G.A. Lamberti (eds), *Methods in Stream Ecology,* 2nd ed. San Diego, CA: Academic Press, pp. 585–609.

Merritt, R.W. and J.R. Wallace. 2010. The role of aquatic insects in forensic investigations. In: Byrd, J.H. and J.L. Castner (eds), *Forensic Entomology: The Utility of Arthropods in Legal Investigations.* Boca Raton, FL: CRC Press, 681 pp.

Meyer, J.L. and C. Johnson. 1983. The influence of elevated nitrate concentration on rate of leaf decomposition in a stream. *Freshwater Biology* 13: 177–183.

Minikawa, N. 1997. The dynamics of aquatic insect communities associated with salmon spawning. Ph.D. Dissertation. University of Washington, Seattle, WA.

Minnesota Pollution Control Agency. 2009. Low dissolved oxygen in water causes, impact on aquatic life—An Overview. *Water Quality/Impaired Waters* 3.24. 12 pp.

Minshall, G.W. 1978. Autotrophy in stream ecosystems. *BioScience* 28(12): 767–771.

Minshall, G.W., K.W. Cummins, R.C. Peterson, C.E. Cushing, D.A. Bruns, J.R. Sedell, and R.L. Vannote. 1985. Developments in stream ecosystem theory. *Canadian Journal of Fisheries and Aquatic Sciences* 42: 1045–1055.

Miyake, H.D., J. Lindsay, J.C. Hunt, and T. Hamatsu. 2002. Scyphomedusae *Aurelia limbata* (Brandt, 1938) found in deep waters off Kushiro, Hokkaido, northern Japan. *Plankton Biology and Ecology* 49: 44–46.

Mondor, E.B., M.N. Tremblay, J.K. Tomberlin, M.E. Benbow, A.M. Tarone, and T.L. Crippen. 2012. The ecology of carrion decomposition. *Nature Education Knowledge* 3(10): 21.

Moore, J.C. et al. 2004. Detritus, trophic dynamics and biodiversity. *Ecology Letters* 7: 584–600.

Myskowiak, J.B., G. Masselot, L. Fanton, and Y. Schuliar. 2010. Freshwater invertebrates and Wagner's parsimony method (WPM): Tools for the submersion time estimation of a cadaver found in a natural aquatic environment. Description of a sampling protocol. *La Revue de Médecine Légale* 1: 47–60.

Nakano, S., H. Miyasaka, and N. Kuhara, 1999. Terrestrial–aquatic linkages: Riparian arthropod inputs alter trophic cascades in a stream food web. *Ecology* 80: 2435–2441.

Nielsen, D.L., M.A. Brock, G.N. Rees, and D.S. Baldwin. 2003. Effects of increasing salinity on freshwater ecosystems in Australia. *Australian Journal of Botany* 51: 655–665.

Nowlin, W.H., M.J. Vanni, and L.H. Yang. 2008. Comparing resource pulses in aquatic and terrestrial ecosystems. *Ecology* 89(3): 647–659.

Nriagu, L.X. 1983. Rapid decomposition of fish bones in Lake Erie sediments. *Hydrobiologia* 106: 217–222.

Odum, E.P. and L.J. Biever. 1984. Resource quality, mutualism and energy partitioning in food chains. *American Naturalist* 124: 360–376.

Osmond, D.L., D.E. Line, J.A. Gale, R.W. Gannon, C.B. Knott, K.A. Bartenhagen, M.H. Turner et al. 1995. WATERSHEDS: Water, soil and hydro-environmental decision support system. http://h2osparc.wq.ncsu.edu.

Paletta, M.S. 1999. *The New Marine Aquarium: Step-By-Step Setup & Stocking Guide*. Neptune City, NJ: TFH Publications.

Papastamatiou, Y., S. Purkis, and K. Holland. 2007. The response of gastric pH and motility to fasting and feeding in free-swimming blacktip reef sharks, *Carcharhinus melanopterus*. *Journal of Experimental Marine Biology and Ecology* 345: 129–140.

Parmenter, R.R. and V.A. Lamarra. 1991. Nutrient cycling in a freshwater marsh: The decomposition of fish and waterfowl carrion. *Limnology and Oceanography* 36(5): 976–987.

Parmenter, R.R. and J.A. MacMahon. 2009. Carrion decomposition and nutrient cycling in a semiarid shrub-steppe ecosystem. *Ecological Monographs* 79: 637–661.

Payne, J.A. 1965. A summer carrion study of the baby pig, *Sus scrofa*. *Ecology* 46: 592–602.

Payne, J.A. and E.W. King. 1972. Insect succession and decomposition of pig carcasses in water. *Journal of the Georgia Entomologist Society* 7(3): 153–162.

Pepin, P. 1991. Effect of temperature and size on development, mortality and survival rates of the pelagic early life history stages of marine fish. *Canadian Journal of Fisheries and Aquatic Sciences* 48: 503–518.

Peterson, B.J., J.E. Hobbie, A.E. Hershey, M.A. Lock, T.E. Ford, J.R. Vestral, V.L. McKinley et al. 1985. Transformation of a tundra river from heterotrophy to autotrophy by addition of phosphorous. *Science* 229: 1383–1386.

Poff, N.L, J.D. Allan, M.B. Bain, J.R. Karr, K.L. Prestegaard, B.D. Richter, R.E. Sparks, and J.C. Stromberg. 1997. The natural flow regime: A paradigm for river conservation and restoration. *BioScience* 47: 769–784.

Polis, G.A., W.B. Anderson, and R.D. Holt. 1997. Toward an integration of landscape and food web ecology: The dynamics of spatially subsidized food webs. *Annual Review of Ecology and Systematics* 28: 289–316.

Pomeroy, L.R. and W.J. Wiebe. 2001. Temperature and substrate as interactive limiting factors for marine heterotrophic bacteria. *Aquatic Microbial Ecology* 23: 187–204.

Rader R.B. and C.J. Richardson. 1994. Response of macroinvertebrates and small fish to nutrient enrichment in the Northern Everglades. *Wetlands* 14: 134–143.

Richardson, C.J. 1995. Wetlands ecology. In: Nierenberg, W.A. (ed.), *Encyclopedia of Environmental Biology*, Vol. 3. San Diego, CA: Academic Press, pp. 535–550.

Rozen, D.E., D.J.P. Engelmoer, and P.T. Smiseth. 2008. Antimicrobial strategies in burying beetles breeding on carrion. *Proceedings of the National Academy of Sciences* 105: 17890–17895.

Ruppel, R.E., K.E. Setty, and M. Wu. 2004. Decomposition rates of *Typha* spp. in northern freshwater wetlands over a stream–marsh–peatland gradient. *Scientia Disciputorum* 1: 26–37.

Sakaris, P. 2013. A review of the effects of hydrologic alteration on fisheries and biodiversity and the management and conservation of natural resources in regulated river systems. In: Bradley, P.M. (ed.), *Environmental Sciences: Current Perspectives in Contaminant Hydrology and Water Resources Sustainability*. Rijeka, Croatia: In Tech Publishing.

Sato, T., T. Egusa, K. Fukushima, T. Oda, N. Ohte, N. Tokuchi, K. Watanabe, M. Kanaiwa, I. Murakami, and K.D. Lafferty, 2012. Nematomorph parasites indirectly alter the food web and ecosystem function of streams through behavioral manipulation of their cricket hosts. *Ecology Letters* 15: 786–793.

Sato, T., K. Watanabe, N. Tokuchi, H. Kamauchi, Y. Harada, and K.D. Lafferty. 2011. A nematomorph parasite explains variation in terrestrial subsidies to trout streams in Japan. *Oikos* 120: 1595–1599.

Schindler, D.W. 1991a. Lakes and oceans as functional wholes. In: Barnes, R.S.K. and K.H. Mann (eds), *Fundamentals of Aquatic Ecology*. Oxford, UK: Blackwell Science Publishers, 270 pp.

Schindler, D.W. 1991b. Aquatic ecosystems and global ecology. In: Barnes, R.S.K. and K.H. Mann (eds), *Fundamentals of Aquatic Ecology*. Oxford, UK: Blackwell Science Publishers, 270 pp.

Schoenberg, S.A., R. Benner, A. Armstrong, P. Sobecky, and R.E. Hudson. 1990. Effects of acid stress on aerobic decomposition of algal and aquatic macrophyte detritus: Direct comparison in a radiocarbon assay. *Applied and Environmental Microbiology* 56(1): 237–244.

Schuldt, J.A. and A.E. Hershey. 1995. Effect of salmon carcass decomposition of Lake Superior tributary streams. *Journal of the North American Benthological Society* 14: 259–268.

Sexton, M.A., R.R. Hood, J. Sarkodee-adoo, and A.M. Liss. 2010. Response of *Chrysaora quinquecirrha* medusae to low temperature. *Hydrobiologia* 645: 125–133.

Smith, K.G.V. 1986. *A Manual of Forensic Entomology*. Ithaca, NY: Comstock Publishing Associates, Cornell University Press, 205 pp.

Smith, C.R. 2006. Bigger is better: The role of whales as detritus in marine ecosystems. In: Estes, J.A., P. DeMaster, D.F. Doak, T.M. Williams, R.L. Bownell Jr. (eds), *Whales, Whaling and Ocean Ecosystems*, Univ. Calif Press, Berkeley, CA, pp. 286–300.

Smith, C.R. 2009. Organic falls on the ocean floor. In: Gillespie, R. and D. Clague (eds), *The Encyclopedia of Islands*. University of California Press, pp. 700–702.

Smith, C.R. and A.R. Baco. 2003. Ecology of whale falls at the deep-sea floor. *Oceanography and Marine Biology: An Annual Review* 41: 311–354.

Smith, K. and M.E. Lavis. 1974. Environmental influences on the temperature of a small upland stream. *Oikos* 26: 228–236.

Smith, C.R., F.C. De Leo, A.F. Bernandino, A.K. Sweetman, and P.M. Arbizu. 2008. Abyssal food limitation, ecosystem structure, and climate change. *Trends in Ecology and Evolution* 23(9): 518–528.

Smith, C.R., L.S. Mullineaux, and L.A. Levin. 1998. Deep-sea biodiversity: A compilation of recent advances in honor of Robert R. Hessler. *Deep-Sea Research* 45: 1–12.

Sorg, M.H., J.H. Dearborn, E.I. Monahan, H.F. Ryan, K.G. Sweeney, and E. David. 1997. Forensic taphonomy in marine contexts. In: Haglund, W.D. and M.H. Sorg (eds), *Forensic Taphonomy: The Postmortem Fate of Human Remains*. Boca Raton, FL: CRC Press, pp. 567–604.

Srivastava, D.S., B.J. Cardinale, A.L. Downing, J.E. Duffy, C. Jouseau, M. Sankaran, and J.P. Wright. 2009. Diversity has stronger top-down than bottom-up effects on decomposition. *Ecology* 90(4): 1073–1083.

Srivastava, D.S. and M. Vellend. 2005. Biodiversity-ecosystem function research: Is it relevant to conservation? *Annual Review Ecology, Evolution and Systematics* 36: 267–294.

Steinman, A.D. and P.J. Mulholland. 2006. Phosphorus limitation, uptake, and turnover in benthic stream algae. In: Hauer, F.R. and G.A. Lamberti (eds), *Methods in Stream Ecology*. Burlington, MA: Elsevier, pp. 187–212.

Stergiou, K.I. and H. Browman. 2005. Bridging the gap between aquatic and terrestrial ecosystems. *Marine Ecology Progress Series* 304: 271–307.

Stevenson, C. and D.L. Childers. 2004. Hydroperiod and seasonal effects on fish decomposition in an oligotrophic Everglades marsh. *Wetlands* 24(3): 529–537.

Stockton, W.L. and T.E. DeLaca. 1982. Food falls in the deep-sea: Occurrence, quality and significance. *Deep-Sea Research I* 29: 157–169.

Sweeney, B.W. and R.L. Vannote. 1978. Size variation and the distribution of hemimetabolous aquatic insects: Two thermal equilibrium hypotheses. *Science* 200: 444–446.

Swift, M.J., O.W. Heal, and J.M. Anderson 1979. *Decomposition in Terrestrial Ecosystems*. Oxford, UK: Blackwell Scientific, 509 pp.

Symmank, S. 2009. Nutrient limitation of periphyton in western Wisconsin streams during winter baseflow months. *UW-L Journal of Undergraduate Research* XII: 1–3.

Tank, J.L., E.J. Rosi-Marshall, N.A. Griffiths, S.A. Entrekin, and M.L. Stephen. 2010. A review of allochthonous organic matter dynamics and metabolism in streams. *Journal of the North American Benthological Society* 29(1): 118–146.

Tiegs, S.D., P.O. Akinwole, and M.O. Gessner. 2009. Litter decomposition across multiple spatial scales in stream networks. *Oecologia* 161: 343–351.

Tiegs, S.D., D.T. Chaloner, P. Levi, J. Ruegg, J.L. Tank, and G.A. Lamberti. 2008. Timber harvest transforms ecological roles of salmon in southeast Alaska rain forest streams. *Ecological Applications* 18: 4–11.

Tomberlin, J.K. and P.H. Adler. 1998. Seasonal colonization and decomposition of rat carrion in water and on land in an open field in South Carolina. *Journal of Medical Entomology* 35: 704–709.

Tomberlin, J.K., T.L. Crippen, A.M. Tarone, B. Singh, K. Adams, Y.H. Rezenom, M.E. Benbow et al. 2012. Interkingdom responses of flies to bacteria mediated by fly physiology and bacterial quorum sensing. *Animal Behavior* 30: 1–8.

Tunnicliffe, V, S.K. Juniper, and M. Sibuet. 2003. Reducing environments of the deep-seafloor. In: Tyler, P.A. (ed.), *Ecosystems of the Deep Oceans*. Amsterdam: Elsevier, pp. 81–110.

Tomberlin, J.K., R. Mohr, M.E. Benbow, A.M. Tarone, and S. VanLaerhoven. 2011. A roadmap for bridging basic and applied research in forensic entomology. *Annual Review of Entomology* 56: 401–421.

Vance, G.M., J.K. VanDyk, and W.A. Rowley. 1995. A device for sampling aquatic insects associated with carrion in water. *Journal of Forensic Sciences* 40: 479–482.

Vannote, R.L., G.W. Minshall, K.W. Cummins, J.R. Sedell, and C.E. Cushing. 1980. The river continuum concept. *Canadian Journal of Fisheries and Aquatic Sciences* 37: 130–137.

Wallace, J.B., S.L. Eggert, J.L. Meyer, and J.R. Webster. 1997. Multiple trophic levels of a forest stream linked to terrestrial litter inputs. *Science* 277: 102.

Wallace, J.R. and R.W. Merritt. 2008. The use of aquatic insect evidence in criminal investigations. In: Haskell, N.H. and R.E. Williams (eds), *Entomology & Death: A Procedural Guide*, 2nd ed. Clemson, SC: Forensic Entomology Partners, 216 pp.

Wallace, J.R., R.W. Merritt, R. Kimbirauskas, M.E. Benbow, and M. McIntosh. 2008. Caddisfly cases assist homicide case: Determining a postmortem submersion interval (PMSI) using aquatic insects. *Journal of Forensic Science* 53(1): 1–3.

Wallace, J.B., J.R. Webster, and T.F. Cuffney. 1982. Stream detritus dynamics: Regulation by invertebrate consumers. *Oecologia* 53: 197–200.

Ward, J.V. and J.A. Stanford. 1982. Thermal responses in the evolutionary ecology of aquatic insects. *Annual Review of Entomology* 27: 97–117.

Webster, J.R. and E.F. Benfield. 1986. Vascular plant breakdown in freshwater systems. *Ann. Rev. Ecol. Syst.* 17: 567–594.

White, P.A., J. Kaliff, J.B. Rasmussen, and J.M. Gasol. 1991. The effect of temperature and algal biomass on bacterial production and specific growth rate in freshwater and marine habitats. *Microbial Ecology* 21: 99–118.

Wilmers, C.C. et al. 2003a. Resource dispersion and consumer dominance: Scavenging at wolf and hunter-killed carcasses in Greater Yellowstone, USA. *Ecology Letters* 6: 996–1003.

Wilmers, C.C. et al. 2003b. Trophic facilitation by introduced top predators: Grey wolf subsidies to scavengers in Yellowstone National Park. *Journal of Animal Ecology* 72: 909–916.

Wilson, E.E. and E.M. Wolkovich. 2011. Scavenging: How carnivores and scavenging structure communities. *Trends in Ecology and Evolution* 26: 129–135.

Winterbourne, M.J. 1990. Interactions among nutrients algae and invertebrates in a New Zealand mountain stream. *Freshwater Biology* 23: 463–474.

Wipfli, M.S. 1997. Terrestrial invertebrates as salmonid prey and nitrogen sources in streams: Contrasting old-growth and younger growth riparian forests in southeastern Alaska, USA. *Canadian Journal of Fisheries and Aquatic Sciences* 54: 1259–1269.

Witte, U. 1999. Consumption of large carcasses by scavenger assemblages in the deep Arabian Sea: Observations by baited camera. *Marine Ecology Progress Series* 183: 139–147.

Wold, A.P. and A.E. Hershey. 1999. Spatial and temporal variability of nutrient limitation in 6 north shore tributaries to Lake Superior. *Journal of the North American Benthological Society* 18(1): 2–14.

Xu, W., T. Reuter, G.D. Inglis, F.J. Larney, T.W. Alexander, J. Guan, K. Stanford, Y. Xu, and T.A. McAlllister. 2009. A biosecure composting system for disposal of cattle carcasses and manure following infectious disease outbreak. *Journal of Environmental Quality* 38(2): 437–450.

Yamamoto, J., M. Hirose, T. Ohtani, K. Sugimoto, K. Hirase, N. Shimamoto, T. Shimura, N. Honda, Y. Fujimori, and T. Mukai. 2008. Transportation of organic matter to the sea floor by carrion falls of the giant jellyfish *Nemopilema nomurai* in the Sea of Japan. *Marine Biology* 153: 311–317.

Yan, F., R. McNally, E.J. Kontanis, and O.A. Sadik. 2001. Preliminary quantitative investigation of postmortem adipocere formation. *Journal of Forensic Science* 46: 609–614.

Zendt, J.S. and E.P. Bergersen. 2000. Distribution and abundance of the aquatic oligochaete host *Tubifex turbifex* for the salmonid whirling disease parasite *Myxobolus cerebralis* in the Upper Colorado River Basin. *North American Journal of Fisheries Management* 20: 502–512.

Zimmerman, K.A. and J.R. Wallace. 2008. The potential to determine a postmortem submersion interval based on algal/diatom diversity on decomposing mammalian carcasses in brackish ponds in Delaware. *Journal of Forensic Science* 53(4): 935–941.

13

The Role of Carrion in Ecosystems

Philip S. Barton

CONTENTS

13.1 Introduction

Carrion, in the broadest sense, is nonliving animal tissue. It is the product of the death of living animals and can be the result of many natural processes, such as disease, starvation, or old age. Many animals are depredated when alive, but unconsumed remains can also be considered carrion. Carrion decomposition involves many interactions between organisms and their environment. These interactions determine the way carrion-derived energy and nutrients are recycled through microbes, animals, and plants and this has implications for the structure and functioning of whole ecosystems. The purpose of this chapter is to give an overview of the broader ecological context of carrion and the way it can affect many different biotic components of ecosystems.

This chapter is divided into four main parts. First, a summary is given of carrion as a distinct resource pool in ecosystems, separate from plants, and a description is given of the features of carrion that produce its disproportionate effects on biota relative to other forms of detritus. Second, the range of effects of carrion on different components of ecosystems is illustrated by case studies from different and contrasting ecosystems. Third, approaches to studying the spatial and temporal effects of carrion are given. Fourth, current gaps in our knowledge of the role of carrion in landscapes and ecosystems are described, and some priority areas are identified for future research. These different sections provide an outline of the range of effects of carrion on the biotic components of ecosystems and points to several of the most important carrion ecology papers to date. This gives a detailed background to our current understanding of carrion in ecosystems and highlights several research areas in need of further development.

13.2 Carrion Is a Unique Resource in Ecosystems

13.2.1 Detritus-Based Food Web

All animals, regardless of trophic position, will ultimately re-enter the carrion pool in an ecosystem upon their death, unless first consumed by a predator. Thus, the ecology of carrion is a central unifying concept that links all animals to the cycling of energy and nutrients in their ecosystem. The fate of carrion-derived energy and nutrients in an ecosystem is a function of how they are recycling through different organisms and the environment. This, in turn, determines how carrion affects the functioning of ecosystems and their associated biota.

The understanding of the role of carrion in ecosystems is quite poor relative to that of plant detritus (Moore et al. 2004). Indeed, few studies have attempted to even quantify the relative contribution of plant and animal biomass to terrestrial ecosystems (Barton et al. 2013a). Although the role of carrion in ecosystem nutrient and energy flow is poorly understood, it can be conceptualized in a similar way to plant detritus. Figure 13.1 summarizes the flow of energy and nutrients in a simplified carrion-centered food web. Environmental variables, such as temperature and moisture, have strong controlling effects on rates of carrion decay (Carter et al. 2007, 2010; Parmenter and Macmahon 2009) and are especially critical to the rate of nutrient cycling via microbial decomposers. However, the dispersal of nutrients away from carrion is largely driven by the activity of arthropod and vertebrate detritivores and scavengers and their predators (Payne et al. 1968; Putman 1978a; Braack 1987; Devault et al. 2003; Schmitz et al. 2008; Parmenter and Macmahon 2009). Carrion nutrients are also channeled through below-ground pathways by bacteria and fungi, which sequentially degrade the large and complex organic molecules comprising animal tissue (Carter et al. 2008; Macdonald et al. 2014). Some loss of energy occurs through the

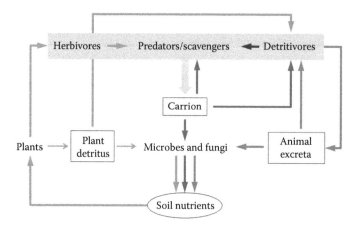

FIGURE 13.1 **(See color insert.)** Simplified pathways of energy and nutrient flow in a carrion-centered detrital food web. The grey box represents all living animals and their potential contribution to the carrion pool in an ecosystem. Brown arrows depict the flow of energy and material from carrion resources through vertebrate scavengers and invertebrate detritivores and the channeling of carrion nutrients through microbes and fungi into the soil nutrient pool. Blue arrows show the transfer of energy from live animals to excreta into detritivores and microbes. The green arrows depict the flow of nutrients via plants, starting with the uptake of nutrients from soil and their flow through herbivores and their predators. Most plant biomass is not consumed by animals and gives the largest contribution to the detrital resource pool, which is also channeled through invertebrate detritivores and microbes. (Reproduced from Barton, P.S. et al. 2013a. *Oecologia* 171: 761–772. With permission.)

release of gases, such as carbon dioxide, from a carcass (Putman 1978b). The degradation of proteins by microbes releases key macronutrients such as nitrogen and phosphorus and make them available for uptake by plants (Towne 2000). These recycled carrion nutrients are incorporated into plant tissues, which may then be consumed by herbivores (Yang 2008) and their predators (Moore et al. 2004).

13.2.2 Distinguishing Features of Carrion

Two key properties of carrion make it a resource very different from plant detritus and thus produce very different effects on ecosystems and their biota. First, carrion is nutrient-rich, whereas plant litter is typically nutrient-poor, with low concentrations of key macronutrients such as nitrogen (Swift et al. 1979; Moore et al. 2004). For example, different kinds of detritus and their respective nutrient content are given by Carter et al. (2007), who showed nitrogen content of carrion being up to five times higher, and moisture up to 10 times higher, than several kinds of plant litter. Second, carrion decomposes at much faster rates than plant litter, perhaps up to three orders of magnitude faster (Parmenter and Macmahon 2009). These two key qualities of accelerated temporal dynamics and high nutrient concentration make carrion a quite distinct component of the detritus pool in ecosystems. It also makes carrion a valuable and highly sought-after resource by various consumers.

13.3 Effects of Carrion on Different Kinds of Ecosystems

Research quantifying the effects of carrion on whole ecosystems is rare, but there are many studies that provide details on the localized effects of carrion. This includes studies on different ecosystem components, including soil, plants, and animals. Below is an overview of various studies that describe the effects of carrion in a range of ecosystems. These examples serve to highlight the findings of carrion ecology studies in some contrasting environments. This information is intended to give the reader a starting point for further exploring the role of carrion in specific ecosystems.

13.3.1 Grassland Ecosystems

Grasslands are characterized by a near-continuous layer of grasses and herbs (Coupland 1992). Typically, most pristine natural grasslands are dominated by perennial species, but many disturbed grasslands have a higher proportion of annual species (Coupland 1992). Grassland plant species can generally respond faster to nutrient changes than the larger shrub and tree species present in forests. This has enabled research on the response of plant functional types to carrion, such as the differential response of graminoids versus forbs and annuals versus perennials. One of the first examples of this research was by Towne (2000), who showed a gradient in soil nitrogen levels away from bison carcasses in a Prairie grassland, Kansas, USA. The gradient in soil nitrogen led to variable responses among annual and perennial forbs and grasses. Recent work by Barton et al. (2013b) used kangaroo (*Macropus giganteus*) carcasses to examine grassland plant responses in a modified grassy woodland ecosystem. They showed elevated plant species richness at carcasses was driven by the annual grasses and that traits such as specific leaf area could facilitate the rapid colonization of carcasses by annual grasses compared with perennial forbs (Figure 13.2). This same experiment also revealed new details on the flux of proteins in soil beneath carcasses and how changes in amino acids might affect the dynamics of microbial communities and plant nutrient uptake (Macdonald et al. 2014). Further, they calculated that a kangaroo carcass of 30 kg adds the equivalent of 4 kg of nitrogen per square meter of surface soil beneath each carcass. This is a quantity greater than that delivered by atmospheric nitrogen deposition or microbial fixation over 1000 years, or equivalent to 100 years of fertilization in many agricultural systems (Macdonald et al. 2014).

13.3.2 Forest Ecosystems

Forest ecosystems are characterized by occurrence of trees that form a canopy overstorey, with various levels of midstorey and understorey vegetation, and occupy moderate to high rainfall zones from tropical

FIGURE 13.2   (See color insert.) An example of how ecosystem context mediates localized responses of plant communities 1 year after addition of kangaroo carcasses. (a) Very little recolonization has occurred in grassland dominated by nutrient-sensitive species, whereas (b) recolonization has progressed significantly in grassland a few kilometers away but dominated by exotic nutrient-tolerant species after the same period of time. (Reproduced from Barton, P.S. et al. 2013b. *PLoS ONE* 8: e53961. With permission.)

to temperate latitudes (Ovington 1983; Röhrig and Ulrich 1991). Several studies on carrion in forest ecosystems have demonstrated both localized and large-scale effects on soil nutrients, microbes, and plants. For example, Yang (2004) quantified the large-scale effects of the mass emergence of cicadas (*Magicicada* spp.) on soil and plants in the deciduous forests of the eastern United States. Yang (2004) showed that the decomposition of cicadas on the forest floor led to an increase in soil microbial biomass and nitrogen availability, thus generating a bottom-up cascade through plants, as well as their potential herbivores (Yang 2008). This emphasis of the study on the effects of insect carrion, rather than that of large vertebrates, highlights the significant and influential role of large numbers of small animals in the functioning of ecosystems.

Several other studies have documented the localized effects of large carcasses on soil and plants in forest ecosystems. Bump et al. (2009b) documented the competitive advantage given to some plants by the increased nutrient availability associated with the decomposition of white-tailed deer carcasses (*Odocoileus virginianus*). This shift in competitive dynamics among plants able to respond to dramatic changes in nutrient levels provides evidence of carrion being important for the maintenance of biodiversity. This effect was attributed to the large effect of carcasses on the key soil macronutrients N, P, and K, which are often limiting resources in forest ecosystems. The use of a stable nitrogen isotope ($\delta^{15}N$) by Bump et al. (2009b) also confirmed the nutrient-cycling pathway from carcass to soil to plant. Melis et al. (2007) in the Białowieza Forest, Poland, documented the temporal variation in soil nutrient changes after the addition of bison carcasses. They found that N had a faster turnover compared to other macronutrients in this temperate forest and that Na and K levels were elevated 20 m distant from the carcass after 4 and 5 years, respectively. However, Melis et al. (2007) cautioned that the more heterogeneous structure of forests, compared with grasslands or tundra, could result in carrion effects being more variable and difficult to detect.

13.3.3 Arid Ecosystems

Arid ecosystems cover approximately one-third of terrestrial land area (Collins et al. 2008) and are characterized by either very low or highly variable rainfall (Evenari et al. 1985). Arid ecosystems can also experience a wide range of daily temperatures, from freezing to very hot, and soils generally have low fertility and organic matter content (Evenari et al. 1985). These features mean that arid ecosystems have lower productivity compared with more mesic ecosystems (Austin et al. 2004; Huxman et al. 2004).

Because of its low productivity, the presence of carrion in arid ecosystems may therefore have a dispro-portionate role in nutrient budgets, and in supporting biodiversity, although this has not been examined in detail. One of the few quantitative studies of carrion-derived nutrients in arid ecosystems has suggested vertebrate carcasses may constitute only a small fraction of the total aboveground biomass, perhaps less than 1%, in a semiarid shrub–steppe ecosystem in Wyoming, USA (Parmenter and Macmahon 2009). Despite this apparently trivial quantity at landscape scales, they argue that localized nutrient cycling at individual carcasses is significant, thus having a nontrivial effect on nutrient patchiness.

Many of the world's arid and semiarid rangelands, including low-rainfall grassland and savanna eco-systems, contain large herds of grazing herbivores harvested for human consumption. The harvesting of both wild herbivores and livestock from rangelands has been identified as a potential negative impact on these low-nutrient ecosystems through the export of scarce nutrients (Read and Wilson 2004). The impact of harvesting of herbivores from rangelands, and other arid and semiarid ecosystems, may be mitigated by the leaving of animal remains to supplement the diet of diverse scavenger communities in these ecosystems (Read and Wilson 2004), as well as the predators whose diet is also supplemented greatly by carrion (Wilson and Wolkovich 2011) (also see Chapter 6).

13.3.4 Island and Cave Ecosystems

Small oceanic islands (<10 km^2) and caves are defined by their limited, and often absent, primary pro-ductivity (Polis and Hurd 1995; Schneider et al. 2011). For these islands, this might be because they are relatively young with no soil for plants to thrive, or so small that saline maritime conditions create a hos-tile growing environment. For cave systems, the absence of light limits plant growth. In both situations, allochthonous resources underpin ecosystem productivity and functioning at higher trophic levels, and this is typically in the form of detritus, including carrion.

Pioneering work by Polis and Hurd (1995) on islands in the Gulf of California demonstrated the sig-nificance of carrion (and other forms of detritus) to island ecosystem productivity. One of the ways they measured this was quantifying spider densities on a range of islands of different size. Given the low pre-cipitation and primary productivity, detritus-based inputs were shown to be critical to these ecosystems, including supporting high abundances of invertebrates and their spider predators. Further, it was shown that smaller islands had higher densities of spiders than larger islands due to their greater perimeter to area ratio and thus higher receipt of carrion resources per unit area (Polis and Hurd 1995). This finding revealed a causal relationship between marine-derived carrion and terrestrial food-web dynamics of islands. This work has inspired further research on the effect of movement of nutrients across ecosystems (Polis et al. 1997; Colombini and Chelazzi 2003; Marczak et al. 2007).

Cave ecosystems are also sensitive to external resource subsidies. In an experiment by Schneider et al. (2011), a number of caves were cleaned of their naturally occurring detritus. Carrion and leaf detritus were added to these caves and the carrion-associated arthropods were monitored over time. A key find-ing from this study was that arthropod communities changed dramatically in abundance after carrion input and also changed in composition, thus revealing the extent to which cave invertebrate communities are dependent upon outside resource subsidies (Schneider et al. 2011).

13.3.5 Riparian Ecosystems

Riparian ecosystems provide an interesting case study for carrion ecology because they represent the interface between two distinct environments: terrestrial and aquatic. Aquatic decomposition pro-cesses are described in Chapter 12, and here the focus will be on the terrestrial component of the riparian environment. Much of the literature on carrion in riparian ecosystems has focused on the effects of anadromous salmon (*Oncorhynchus* spp.) in North America (Helfield and Naiman 2006; Tiegs et al. 2011). Each year, several species of salmon undertake spawning runs up streams and riv-ers from their ocean-feeding grounds in the Atlantic and Pacific Oceans. One of the consequences of this is the mass senescence and death of enormous numbers of fish in the upper reaches of inland waterways, and the delivery of carrion into the adjacent riparian ecosystems (Figure 13.3; Helfield and Naiman 2006; Tiegs et al. 2011).

FIGURE 13.3   (See color insert.) The annual spawning run of salmon (*Oncorhynchus* spp.) from the ocean into the upper reaches of inland rivers (a) and streams can deliver large quantities of marine-derived nutrients to riparian ecosystems (b). (Photo credits: M.E. Benbow.)

Mass salmon die-off can have profound and far-reaching impacts on whole catchments. For example, Hocking and Reynolds (2011) found that plant biomass and diversity were altered in catchments where salmon migration occurred. This occurs though the uptake of salmon-derived nitrogen by plants (Hocking and Reynolds 2012). Other studies have identified the role of salmon carrion in top-predator diets, such as brown bear (*Ursus arctos*) (Hilderbrand et al. 1999) and bald eagle (*Haliaeetus leucocephalus*) (Jackman and Hunt 2007). These predators and scavengers perform the final role of transferring energy and nutrients that were originally captured from marine ecosystems, channeled through aquatic ecosystems, and then delivered to terrestrial riparian ecosystems. The death of migrating salmon, and their consumption by vertebrate predators and scavengers, provides an excellent example of how carrion can have a significant influence on ecosystem structure and function. Additional details of carrion importance to fisheries and wildlife management can be found in Chapter 22, and the role of vertebrate scavengers in carrion cycling in Chapter 6.

13.3.6 Polar Ecosystems

Polar ecosystems that support life year-round occur in high-latitude regions of the northern and southern hemispheres and are dominated by simple tundra or barren ground cover (Wielgolaski 1997). Polar ecosystems not under perpetual snow and ice are more expansive in the northern hemisphere, as only some peninsulas in the Antarctic and islands of the sub-Antarctic remain ice-free year round (Wielgolaski 1997).

Polar ecosystems are very low-resource environments, largely due to the immobility of nutrients in frozen and cold soils (Chapin 1983; Wielgolaski 1997). The addition of nutrients to soil from a large carcass can have significant effects on nearby plant growth and quality, but its effects are also long-lasting. A study by Danell et al. (2002) showed a steep gradient in plant nitrogen concentrations at 1–2 m distant from the carcasses of muskox in the northern Canadian arctic tundra. This effect was detected at a carcass estimated to be greater than 10 years old (Figure 13.4). Thus, the long-term effect of nitrogen enrichment in the soil near carcasses in this nutrient-limited and featureless environment was to create small islands of plant microcommunities. These islands could further attract or trap plant seeds or litter, thereby having additional ecosystem effects (Danell et al. 2002).

The absence of vegetation growth and associated animal herbivores for much of the year presents a significant challenge to many of the predator animals that persist all year round in polar environments. Scavenging of carrion by many predators is a common strategy to supplement their diet during the winter months. The few studies that have examined carrion in polar ecosystems have focused on the population dynamics of vertebrate scavengers. A range of vertebrate species, such as red foxes (*Vulpes vulpes*), arctic foxes (*Vulpes lagopus*), ermine (*Mustela erminea*), ravens (*Corvus corax*), and gulls (*Larus hyperboreus*), are known to feed at the carcasses of dead animals in Canadian arctic tundra. The percentage

FIGURE 13.4   The 10-year-old remains of a muskox in the Canadian arctic. The grass at this carcass is noticeably taller than the surrounding tundra vegetation. (Reproduced from Danell, K., D. Berteaux, and K.A. Brathen. 2002. *Arctic* 55: 389–392. With permission.)

of their diet that comprises carrion has been estimated to be 5%–15% (Krebs et al. 2003). Research has shown a high degree of dependence of some scavenger species on the remains left unconsumed by large predators such as polar bears (Roth 2003). For example, ringed seal (*Phoca hispida*) carcasses left by polar bears are a critical resource to arctic foxes and can help to maintain populations through harsh winters (Roth 2003). This may have important implications for the population dynamics of smaller prey species, such as collared lemmings (*Dicrostonyx richardsoni*), which may suffer as a result of high populations of foxes (Roth 2003). The subsidization of carrion resources provided by large predators to smaller scavengers might therefore have far-reaching effects in polar ecosystems.

13.3.7 Deep-Sea Ecosystems

Deep-sea and abyssal seafloor ecosystems (depths >1000 m) cover 50% of the earth's surface and form the largest habitat on earth (Tyler 2003). The role of carrion in marine ecosystems is in some ways very different from terrestrial ecosystems (Beasley et al. 2012). For example, in deep-sea ecosystems, heterotrophic organisms dominate rather than autotrophs, with many species dependent on the influx of detritus that falls from surface waters (Britton and Morton 1994). Thus, changes in detritus input can have significant effects on the abundance of deep-sea organisms, such as abyssal fish and crustacean populations (Smith and Baco 2003; Drazen et al. 2012; Amon et al. 2013).

The largest parcels of detritus that occur in deep-sea marine ecosystems are carcasses of great whales, which are thought to play unusual roles in deep-sea food webs (Smith and Demopolous 2003). The enormous size of a great whale carcass (up to 160 ton) presents a fantastically rich nutrient resource, yet the population and production rates of whales are relatively low compared with other marine organisms. Thus, whale carcasses are extremely rare when averaged across the vastness of oceans, but also represent one of the richest sources of concentrated detritus available on the seafloor.

The rarity of whales in the context of the vastness of oceans means that their contribution to total energy and nutrient flux is likely to be quite small, even in nutrient-poor abyssal deep-sea ecosystems (Smith and Demopolous 2003). One of the few quantitative estimates of the contribution of carrion to ecosystem nutrient budgets has been done for whale carcasses in the deep sea. Smith and Baco (2003) performed calculations suggesting that the estimated 69,000 great whale deaths per year, averaged across an ocean floor of 3.6×10^8 km^2, would result in an input of 3.8×10^{-4} g of organic carbon per square meter per year (other estimates can be found in Smith and Demopolous 2003). This represents approximately 0.1% of the background input of organic carbon to the deep-sea floor (Smith and Demopolous 2003). In terms of absolute nutrient input, therefore, whale carcasses cannot be considered unusual in deep-sea ecosystems. Rather, it is the highly concentrated nutrients embodied in a whale carcass relative to background levels that makes whale carcasses unusual.

Whale carcasses support an extraordinary range of animals, both in terms of the number of species and individuals found at carcasses. The numbers and kinds of species present at a whale carcass were first documented off the coast of California in 1987 (Smith et al. 1989), with most subsequent studies also from this area (Smith and Demopolous 2003; Lundsten et al. 2010). These studies have generally identified four different successional stages of species assemblages visiting decomposing whale carcasses: (i) mobile scavenger, (ii) enrichment-opportunist, (iii) sulfophilic, and (iv) a reef. The diversity of species associated with each of these stages can vary substantially. For example, 38 species have been identified from the mobile scavenger stage, with most being generalist scavengers (Smith and Baco 2003). The highest densities of organisms are thought to occur during the enrichment-opportunist stage. For example, Smith and Baco (2003) reported extraordinary densities of 20,000–45,000 individuals per square meter, and included various polychaetes, gastropods, and crustaceans. The sulfophilic stage of decomposition is also very species rich, with 185 associated species reported by Smith and Baco (2003).

Whale carcasses in the deep sea can be difficult to find, and an understanding of their contribution to biodiversity in these ecosystems remains poor. Clear areas for future research have been alluded to by several researchers and include the study of species assemblages at whale carcasses beyond the Californian coast and in the different oceans of the world (Smith and Demopolous 2003; Amon et al. 2013). An experimental approach should be employed, with carcasses and their associated assemblages monitored simultaneously across different oceans and at different depths. A further area for research is the role of large predators, such as killer whales (*Orcinus orca*), in generating surface carrion that might subsequently contribute to deep sea carrion. Prey such as gray whales (*Eschrichtius robustus*) off the coast of Alaska is often only partially processed by killer whales, with the remains washing ashore or sinking to the sea floor (Barrett-Lennard et al. 2011). The extent to which this occurs in open water and over abyssal plains is largely unknown.

13.4 Approaches to Understanding the Effects of Carrion on Ecosystems

In this section, an overview is given for some of the different approaches used to examine the effects of carrion on ecosystems. These effects have been separated into spatial and temporal effects. A synthesis of both spatial and temporal approaches is described to provide a framework for future research on the role of carrion within ecosystems.

13.4.1 Spatial Dynamics of Carrion: Heterogeneity and Resource Subsidies

The input of carrion can vary spatially within and across ecosystems and is a result of the death of individual animals due to predation or disease, or the die-off of animals undergoing mass migration events. Various human activities can also influence the spatial input of carrion, such as hunting or culling. These spatial processes can affect ecosystems at both small and large spatial scales, each discussed in turn below.

Carrion is a spatially patchy resource, with each carcass forming a "cadaver decomposition island" (sensu Carter et al. 2007). This has the effect of creating a spatially distinct hotspot of biological and chemical activity that becomes more pronounced for larger vertebrate animals. The small-scale spatial effects of carrion are commonly documented but are rarely placed in the context of whole ecosystems. For example, studies have shown the depth of soil nutrient changes under carcasses (Coe 1978; Parmenter and Macmahon 2009), and the distance at which plants still respond to elevated nutrients (Danell et al. 2002; Melis et al. 2007). This "gradient approach" of taking samples or measurements at certain distances from carcasses has revealed the extent of the localized effects of individual carcasses and the significant nutrient and biodiversity "islands" created by large carcasses (Carter et al. 2007; Barton et al. 2013a). Only a few studies have been designed specifically to quantify the small-scale effects of carrion in ecosystems and have used a case–control design (Bump et al. 2009a; Barton et al. 2013b; Macdonald et al. 2014). These studies compare measurements taken at carcasses with those taken a small distance away from a carcass, and then couple this with high levels of spatial replication across a landscape. These studies have revealed the additive effects of carcass "islands" on ecosystems and demonstrated high levels of heterogeneity in soil nutrients (Bump et al. 2009a; Macdonald et al. 2014), plants (Bump et al. 2009b), and arthropods (Barton et al. 2013b).

At larger scales, the spatial movement of carrion resources within and across ecosystems can subsidize the existing available resources at destination sites. This process fits within the "spatial subsidies" concept of ecology, which applies more broadly to nutrient supplementation to an ecosystem or ecological community (Polis et al. 1997). This concept is exemplified by isolated island ecosystems that have limited autochthonous biomass production. In these instances, carrion from nesting seabirds, or bird and seal carcasses washed up from sea, supplement the resource base for the island food web (Polis and Hurd 1995). Quantifying the effects of nutrient subsidies is difficult at whole-of-ecosystem scales, but can be achieved through standardized surveys of different ecosystem components. These measurements can then be scaled-up after multiplying by factors relevant to ecosystem size, such as island area and perimeter (Polis and Hurd 1995, 1996). The spatial movement of carrion can also occur through the migration of live animals. A notable example is the migration of salmon up streams as part of their annual spawning runs (Hocking and Reynolds 2011). This migration and die-off results in the deposit of significant volumes of salmon biomass into the streams and riparian zones, with animals such as bears subsequently moving many carcasses further inland and into the broader ecosystem (Helfield and Naiman 2006).

13.4.2 Temporal Dynamics Carrion: Succession and Resource Pulses

The temporal dynamics of carrion in ecosystems occurs at both small and large scales. At small scales, the processes occurring during the different phases of decomposition of individual carcasses (see, e.g., Mégnin 1894; Bornemissza 1957; Carter et al. 2007) provide the underlying mechanisms through which carcasses affect their surroundings and associated organisms over time. This has been discussed in detail in other chapters of this book. At larger scales, however, carrion can sometimes become available in regular "pulses" in ecosystems. This can lead to a widespread and dramatic increase in resource and nutrient availability and have important effects on whole ecosystems. One of the best documented examples of this comes from North America and includes the mass emergence of periodical cicadas (*Magicicada* spp.) in deciduous forests in the east of the United States. For most of their life, the immature cicadas live underground but emerge every 13 or 17 years depending on the species. Adults climb trees, call to one another to find a mate, lay their eggs, and then die, all in only a few days or weeks. This results in the widespread delivery of large amounts of concentrated nutrients onto the soil surface. Although many are undoubtedly scavenged by various birds and small animals, the vast majority decompose at the soil surface. Some of the documented consequences of this includes a change in microbial biomass in the soil and a widespread spike in nitrogen levels in forest plants (Yang 2004). These effects can ramify through the ecosystem in many ways, such as producing higher quality foliage that is targeted by insect herbivores (Yang 2008).

13.4.3 A Framework for Research on Carrion in Landscapes and Ecosystems

Efforts to synthesize a general understanding of the dynamics of carrion and its effect on ecosystems have focused on spatial and temporal properties. This has led to carrion being conceptualized as discrete patches of concentrated energy and nutrients that are spread across the landscape (Doube 1987). Such a perspective, encapsulated by the term "ephemeral resource patch" has been further developed in greater detail by Finn (2001), with additional detail from Carter et al. (2007). This single idea captures many of the concepts underpinning succession and aggregation theory. For example, the rapid change in carrion quality during the process of decomposition drives the succession of species through time (Schoenly and Reid 1987; Moura et al. 2005; Matuszewski et al. 2010), and the rare occurrence of carrion produces intense inter- and intraspecific competition for a limited resource through space (Kneidel 1984; Ives 1991; Kouki and Hanski 1995). Thus, the ephemeral resource patch concept provides a general framework for how carrion resources can affect the spatiotemporal dynamics of species populations or communities, and ecological processes. This also highlights key conceptual parallels between carrion resources and other spatially discrete and ephemeral resources, such as dung pads (Doube 1987), fungal fruiting bodies (Heard 1998), and plant fruits (Sevenster and Vanalphen 1996).

Barton et al. (2013a) described a framework to facilitate future carrion ecology research that builds on the ephemeral resource patch concept and associated theory of temporal succession and spatial aggregation. This framework is useful because it links the localized effects of carrion to large-scale effects on landscapes. There are many similarities, for example, between the change in plant biomass and spike in soil nutrients at an individual carcass (Towne 2000; Melis et al. 2007), and the change in plants and soil nutrients across an entire landscape due to the mass input of animal carrion (Yang 2004; Bump et al. 2009a; Hocking and Reynolds 2011). In both scenarios, carrion generates patchiness in biodiversity and ecological processes, but it occurs at very different spatial scales. The landscape-scale analogue to the small-scale ephemeral resource patch concept is the idea of "landscape fluidity." This idea was proposed as a way of viewing the ebb and flow of organisms in landscapes through time (Manning et al. 2009). It is described as a tool to conceptualize the dynamics of landscapes and how processes such as changes in the connectivity of habitat will affect the movement and persistence of organisms. Most notably, with particular relevance to carrion ecology, the concept encapsulates the continuity of species presence both within sites and between sites across a given landscape through time. It is this idea of continuity and overlap in carrion patches that has yet to be fully developed in the carrion ecology literature.

Merging the concepts of landscape fluidity and ephemeral resource patches provides a framework for testing hypotheses about the role of carrion in maintaining continuity in biodiversity and ecological processes across multiple spatial and temporal scales within ecosystems (Figure 13.5). For example, it is widely documented that the decomposition process drives temporal succession in carrion assemblages (Figure 13.5a). This means that two carcasses at different stages of decay host more species than two carcasses at the same stage of decay. Similarly, different habitat types host distinct carrion consumer species (Figure 13.5b), and two carcasses in different habitats might be expected to support more species than two carcasses in the same habitat. Thus, an important question emerging from this framework is: What spatiotemporal arrangement of carrion patches promotes the persistence of the highest diversity

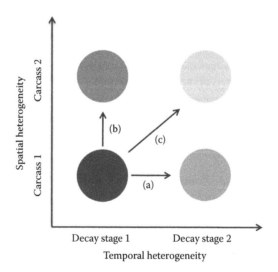

FIGURE 13.5 The ephemeral resource patch concept allows for the scaling-up of localized effects of carrion to better understand the consequences for ecosystem heterogeneity. This framework builds on widely documented links between (a) the temporal change in structure and nutrient content of carrion and assemblage succession and (b) the spatial differences in habitat affinities and stochastic colonization. This information can be used to test hypotheses about (c) what spatiotemporal arrangements of carrion promote heterogeneity in biodiversity and nutrient cycling within ecosystems. (Reproduced from Barton, P.S. et al. 2013a. *Oecologia* 171: 761–772. With permission.)

of carrion consumers within a landscape (Figure 13.5c)? Answering this question requires the explicit linking of local carrion community dynamics with landscape-scale carrion resource dynamics. Barton et al. (2013a) suggest that simulation models might be used to extrapolate alternate scenarios of carrion recourse availability based on population abundance, recruitment, and mortality data (Wilmers and Getz 2004). This might be combined with data on assemblage composition dynamics observed across different kinds of carcasses (Schoenly and Reid 1987; Parmenter and Macmahon 2009). Further development of this into a more formal mathematical framework would require the integration of succession theory with aggregation and coexistence theory (Ives 1991; Schoenly 1992). A fully developed research agenda that combines this framework with existing succession models and simulations of carrion availability would be able to quantitatively predict how differing scenarios of carrion input, density, and decay stage might lead to differing biodiversity and nutrient cycling outcomes for whole ecosystems.

13.5 Future Research Directions

The major gap in our knowledge of the role of carrion in ecosystems can be formed as a question: Is carrion important to ecosystems beyond its localized effects? Many studies have now documented the different ways carcasses can affect the chemical properties of soil and the abundance and diversity of microbes, plants, and animals (Barton et al. 2013a). A research agenda for the role of carrion in ecosystems must move beyond localized studies and address three broad priorities: (i) carrion resources, (ii) emergent effects of community interactions, and (iii) global change and carrion management (Table 13.1). Examination of the spatial and temporal dimensions of each of these research priorities might borrow from the previously described framework (see Figure 13.5), but will also require new conceptual advancements and integration of different subdisciplines of ecosystem ecology, community ecology, landscape ecology, and biogeochemistry.

TABLE 13.1

A List of Knowledge Gaps in Carrion Ecology, Including Example Questions and Some of Their Spatial and Temporal Dimensions

Knowledge Gap	Example Questions	Spatial and Temporal Dimensions
Quantifying carrion resources	What is the contribution of carrion to ecosystem biomass? What is the contribution of differently sized animal carcasses to the carrion resource pool? Does the contribution of carrion to nutrient budgets vary among ecosystems?	What is the spatial density and turnover of differently sized carcasses within an ecosystem, such as insects versus mice? When are the main factors affecting temporal inputs of carrion? Do carrion nutrients recycle more slowly in low-productivity ecosystems?
Community interactions	How are carrion resources partitioned among the carrion community? Is there functional compensation among carrion feeding species?	How do flies and scavenging carnivores disperse carrion nutrients through the landscape? Do different arrival and colonization dynamics lead to different rates of nutrient cycling? Does the absence of species alter the dispersal or turnover rates of carrion nutrients?
Global change and carrion	What anthropogenic factors are changing carrion supply and distribution, and its decomposition and recycling?	How does hunting or culling affect the spatial movement of carrion nutrients? How might carrion be managed to promote spatial and temporal heterogeneity in ecosystems?

Source: Reproduced from Barton, P.S. et al. 2013a. *Oecologia* 171: 761–772. With permission.

13.5.1 Quantifying Carrion Resources

There are very few studies that have attempted to describe a complete nutrient budget accounting for the different forms of both plant and animal detritus in an ecosystem. This means that there is little quantitative knowledge of the contribution of various forms of carrion to nutrient cycling. In terms of absolute quantity, there is evidence that carrion might only contribute to a small fraction of the total nutrient and energy budget in some ecosystems (Parmenter and Macmahon 2009), yet in other ecosystems it is clear that carrion is a critical resource (Polis and Hurd 1995). Across all of earth's ecosystems, the importance of carrion as a resource ranges from trivial to crucial.

An increased understanding of the role of carrion in ecosystems can only be achieved through further quantitative studies on carrion biomass and nutrients in all its forms. For example, what is the net contribution of differently sized animal carcasses to the total carrion resource pool? Answering this question requires an understanding of the composition and abundance dynamics of the animal communities that will contribute to the carrion pool. Focusing first on vertebrate carrion, it is clear that some species contribute disproportionately to carrion input in terms of overall biomass. For example, sea birds on islands may be the major source of carrion input (Polis and Hurd 1996), whereas a bison (*Bison bison*) may be only one of several species of ungulate contributing to the carrion pool in a grassland or forest ecosystem (Towne 2000; Melis et al. 2007). Further, the input of vertebrate carrion can vary widely from such massive (but rare) quantities of millions of kilograms of carrion per square kilometer, which was recorded during the mass starvation of elephants (*Loxodonta africana*) in Kenya (Coe 1978), to thousands of kilograms of moose (*Alces alces*) carrion per square kilometer in Michigan, USA (Bump et al. 2009a). An average of 5000 kg of carrion per square kilometer can be introduced into a bison grazing ecosystem each year, representing a substantial input of at least 180 kg N/km^2 (Carter et al. 2007). Hundreds of kilograms of mammal biomass per square kilometer have been estimated in a neotropical forest (Houston 1985). These various carrion inputs may add up to much more than the 1% of organic matter in some terrestrial ecosystems suggested by researchers to date (Carter et al. 2007; Parmenter and Macmahon 2009).

It is widely understood that invertebrate biomass exceeds that of vertebrates in many ecosystems. This is exemplified in some tropical systems, where 90% of the estimated 200 kg ha^{-1} of aboveground animal biomass consists of invertebrates (Wilson 1987). Soil invertebrate biomass may be even higher in some ecosystems, ranging from 51 kg ha^{-1} in cropped land, to 205 kg ha^{-1} in temperate forests, and 732 kg ha^{-1} in pastures (Lavelle et al. 1997). Invertebrates also have shorter life spans and are recycled through ecosystems at much faster rates than vertebrates. They also are comparatively more uniformly distributed than vertebrates. This has different ecological implications. Within the root system of a single individual plant, soil invertebrate deaths and decomposition will provide a continuous source of nutrients throughout a growing season. However, recent research has also identified the indirect effect insect carrion on plant functioning by demonstrating how fear of predation by spiders changed insect herbivore stoichiometry (Hawlena et al. 2012). This change in insect nutrient content did not alter carrion decomposition rates, but rather decreased rates of subsequent plant litter decomposition. In contrast, the nutrient gain for the same plant from a decomposing vertebrate carcass is a lower probability "hit or miss" scenario, but many orders of magnitude greater.

Animal biomass in terrestrial ecosystems will always be tiny compared with plant biomass (Fahey et al. 2005; Parmenter and Macmahon 2009). Nevertheless, an understanding of the relative biomass of the different kinds of carrion from both vertebrates and invertebrates, and their stoichiometry and input rates, will provide for a more objective approach and give a fuller picture of the way different kinds of carrion are distributed and partitioned within ecosystems.

13.5.2 Emergent Effects of Community Interactions

Who gets to a carcass first will affect the way it is consumed and how the nutrients and energy will be recycled and distributed into the ecosystem. The action of microbes, invertebrates, and vertebrates together facilitate the decomposition and dispersion of carrion nutrients through an ecosystem (Carter et al. 2007; Parmenter and Macmahon 2009; Beasley et al. 2012). Yet, how these different components of the carrion community interact is poorly understood (Olson et al. 2012; Tomberlin et al. 2012). Further, there is little known about how the interactions between carrion and its consumers relates to ecosystem functioning (Tomberlin et al. 2011; Wilson and Wolkovich 2011). How are carrion resources partitioned among the carrion community? The competitive interactions between microbial decomposers and vertebrate carrion consumers have been known for some time (Janzen 1977; Devault et al. 2003). However, it is only more recently that examples have been documented for the competitive exclusion of arthropods by bacteria (Burkepile et al. 2006; Tomberlin et al. 2012) and bacteria by arthropods (Hoback et al. 2004; Rozen et al. 2008). This competitive dynamic might have significant implications for the fate of carrion nutrients in ecosystems. This is because nutrients can be dispersed large distances by insects, whereas they are converted in the soil via bacteria and fungi at the site of a carcass. This highlights the lack of studies that have compared communities of multiple taxa such as microbes, arthropods, birds, and carnivores (but see Read and Wilson 2004; Selva and Fortuna 2007; Parmenter and Macmahon 2009; Tomberlin et al. 2012).

Changes to land use and habitat also may affect the composition of carrion communities (Klein 1989; Devault et al. 2011). When this happens, some consumers of carrion may be lost. This in turn may affect the process of carrion decomposition (Klein 1989). Where and when is there functional redundancy among carrion consumers? It is already known that many vertebrates will opportunistically scavenge on carrion (Devault et al. 2003), and there is evidence that some intraguild compensation occurs when dominant scavengers are removed (Olson et al. 2012). Similarly, several species of blow fly (Diptera: Calliphoridae) co-occur at carcasses (Lang et al. 2006), and inter- and intraspecific competition may affect carrion consumption rates. However, far less is known about redundancy among other arthropod or microbial components of carrion food webs (Pechal et al. 2013). An increased understanding of how carrion resources are partitioned among carrion communities, and the spatial and temporal overlap of consumers, will allow for a greater appreciation of the consequences of species loss to carrion decomposition processes and the implications for ecosystem functioning.

13.5.3 Global Change and Carrion Management

The loss of top predators through hunting or extirpation (Strong and Frank 2010; Estes et al. 2011) is a leading cause of changes to the supply and distribution of carrion in ecosystems (Wilmers and Post 2006; Wilson and Wolkovich 2011). Top predators, such as the wolf (*Canis lupis*) in parts of Europe and North America, various carnivores in Africa, and the dingo (*Canis lupis dingo*) in Australia, play an important role in the trophic regulation of ecosystems (Glen et al. 2007; Beschta and Ripple 2009; Estes et al. 2011). This occurs through the consumption and control of herbivore populations and provision of carrion "left-overs" to smaller scavengers (Wilmers et al. 2003; Wilson and Wolkovich 2011; Ripple et al. 2013). When these top predators are lost, major pathways of energy and nutrient flow are also lost, with implications for the functioning of ecosystems (Strong and Frank 2010; Estes et al. 2011; Ripple et al. 2013). Attempts to reintroduce top predators, such as wolves, can therefore have beneficial effects on other scavengers (Wilmers et al. 2003), as well as soil and plants (Bump et al. 2009a). Further research is required to quantify the consequences of both the absence of top predators and the restorative effects of predator reintroduction (also see Chapters 6 and 22).

Another cause of altered carrion distribution and input in ecosystems is changes in populations of large vertebrate herbivores, which may in part be due to the loss of predators. Large herbivores play an important role in the redistribution of nutrients across landscapes though their consumption of plant biomass and heterogeneous deposition of dung and carrion (Doughty et al. 2013). In recent times, some species of large herbivore have become over-abundant in different parts of the world (Côté et al. 2004; Barton et al. 2011). A commonly applied management response to this problem is hunting or culling, with carcasses often buried, taken off-site for disposal or consumption by humans. When this occurs, nutrients become unavailable to animal and microbial consumers and are not recycled through food webs and the immediate ecosystem. This might have significant consequences for high-order food-web dynamics driven by large carnivores and scavengers (Devault et al. 2003; Wilson and Wolkovich 2011), as well as the redistribution of nutrients across landscapes (Doughty et al. 2013). Where carrion resources have been reduced, alternative sources may need to be supplied to provide spatial and temporal continuity in resources and allow the persistence of carrion-dependent species in a given landscape, such as has occurred for some species of endangered vulture in the Pyrenees mountains of Europe (Margalida et al. 2009, 2011). Conversely, excess carrion from high densities of livestock, for example (see Chapter 23), might subsidize the diet of some vertebrate scavenger species and this may also have negative implications for biodiversity (Ripple et al. 2013). Managing carrion in ecosystems might be one of the most effective short-term approaches to maintaining biodiversity associated, directly or indirectly, with this important nutrient cycling process. Of course, there are likely to be several impediments to the management of carrion in landscapes, including cultural, legal, and economic factors (Dupont et al. 2012). Recognizing them in the first place, however, may help to mitigate some of these challenges.

13.6 Conclusions

There are many studies that have examined the effects of carrion on soils, plants, and animals *in* different ecosystems, but few that have quantified the effects of carrion *on* ecosystems. The few studies that have achieved this, however, have produced surprising and valuable data on the magnitude of the effects, both small and large (Yang 2004; Parmenter and Macmahon 2009; Hocking and Reynolds 2011). This has generated a new understanding of the potentially critical role of carrion in ecosystems. Nevertheless, there remains much to be done to increase the understanding of how the localized phenomena driven by the decomposition of individual carcasses ramifies through ecosystems and the larger-scale maintenance of biodiversity and cycling of energy and nutrients. Advancing this knowledge can be achieved by tackling questions about the supply and distribution of all kinds of animal carrion, and the interactions and linkages among different biotic components of ecosystems associated with carrion. The question of whether carrion is important beyond localized scales will ultimately be answered by performing more comprehensive studies on carrion nutrient budget, and by quantifying experimentally how much diversity and heterogeneity is added to communities through the dynamics of multiple carrion patches across whole landscapes. Such knowledge will serve to enhance the understanding and potential management of carrion in ecosystems.

REFERENCES

Amon, D.J., A.G. Glover, H. Wiklund, L. Marsh, K. Linse, A.D. Rogers, and J.T. Copley. 2013. The discovery of a natural whale fall in the Antarctic deep sea. *Deep-Sea Research II* 92: 87–96.

Austin, A.T., L. Yahdjian, J.M. Stark, J. Belnap, A. Porporato, U. Norton, D.A. Ravetta, and S.M. Schaeffer. 2004. Water pulses and biogeochemical cycles in arid and semiarid ecosystems. *Oecologia* 141: 221–235.

Barrett-Lennard, L.G., C.O. Matkin, J.W. Durban, E.L. Saulitis, and D. Ellifrit. 2011. Predation on gray whales and prolonged feeding on submerged carcasses by transient killer whales at Unimak Island, Alaska. *Marine Ecology Progress Series* 421: 229–241.

Barton, P.S., S.A. Cunningham, D.B. Lindenmayer, and A.D. Manning. 2013a. The role of carrion in maintaining biodiversity and ecological processes in terrestrial ecosystems. *Oecologia* 171: 761–772.

Barton, P.S., S.A. Cunningham, B.C.T. Macdonald, S. Mcintyre, D.B. Lindenmayer, and A.D. Manning. 2013b. Species traits predict assemblage dynamics at ephemeral resource patches created by carrion. *PLoS ONE* 8: e53961.

Barton, P.S., A.D. Manning, H. Gibb, J.T. Wood, D.B. Lindenmayer, and S.A. Cunningham. 2011. Experimental reduction of native vertebrate grazing and addition of logs benefit beetle diversity at multiple scales. *Journal of Applied Ecology* 48: 943–951.

Beasley, J.C., Z.H. Olson, and T.L. Devault. 2012. Carrion cycling in food webs: Comparisons among terrestrial and marine ecosystems. *Oikos* 121: 1021–1026.

Beschta, R.L. and W.J. Ripple. 2009. Large predators and trophic cascades in terrestrial ecosystems of the western United States. *Biological Conservation* 142: 2401–2414.

Bornemissza, G.F. 1957. An analysis of arthropod succession in carrion and the effect of its decomposition on the soil fauna. *Australian Journal of Zoology* 5: 1–12.

Braack, L.E.O. 1987. Community dynamics of carrion-attendant arthropods in tropical African woodland. *Oecologia* 72: 402–409.

Britton, J.C. and B. Morton. 1994. Marine carrion and scavengers. *Oceanography and Marine Biology* 32: 369–434.

Bump, J.K., R.O. Peterson, and J.A. Vucetich. 2009a. Wolves modulate soil nutrient heterogeneity and foliar nitrogen by configuring the distribution of ungulate carcasses. *Ecology* 90: 3159–3167.

Bump, J.K., C.R. Webster, J.A. Vucetich, R.O. Peterson, J.M. Shields, and M.D. Powers. 2009b. Ungulate carcasses perforate ecological filters and create biogeochemical hotspots in forest herbaceous layers allowing trees a competitive advantage. *Ecosystems* 12: 996–1007.

Burkepile, D.E., J.D. Parker, C.B. Woodson, H.J. Mills, J. Kubanek, P.A. Sobecky, and M.E. Hay. 2006. Chemically mediated competition between microbes and animals: Microbes as consumers in food webs. *Ecology* 87: 2821–2831.

Carter, D.O., D. Yellowlees, and M. Tibbett. 2007. Cadaver decomposition in terrestrial ecosystems. *Naturwissenschaften* 94: 12–24.

Carter, D.O., D. Yellowlees, and M. Tibbett. 2008. Temperature affects microbial decomposition of cadavers (rattus rattus) in contrasting soils. *Applied Soil Ecology* 40: 129–137.

Carter, D.O., D. Yellowlees, and M. Tibbett. 2010. Moisture can be the dominant environmental parameter governing cadaver decomposition in soil. *Forensic Science International* 200: 60–66.

Chapin, F.S. 1983. Direct and indirect effects of temperature on arctic plants. *Polar Biology* 2: 47–52.

Coe, M. 1978. The decomposition of elephant carcasses in the Tsavo (east) National Park, Kenya. *Journal of Arid Environments* 1: 71–86.

Collins, S.L., R.L. Sinsabaugh, C. Crenshaw, L. Green, A. Porras-Alfaro, M. Stursova, and L.H. Zeglin. 2008. Pulse dynamics and microbial processes in aridland ecosystems. *Journal of Ecology* 96: 413–420.

Colombini, I. and L. Chelazzi. 2003. Influence of marine allochthonous input on sandy beach communities. *Oceanography and Marine Biology* 41: 115–159.

Côté, S.D., T.P. Rooney, J.P. Tremblay, C. Dussault, and D.M. Waller. 2004. Ecological impacts of deer over-abundance. *Annual Review of Ecology Evolution and Systematics* 35: 113–147.

Coupland, R.T. 1992. *Natural Grasslands: Introduction and Western Hemisphere.* Amsterdam: Elsevier Science Publishers.

Danell, K., D. Berteaux, and K.A. Brathen. 2002. Effect of muskox carcasses on nitrogen concentration in tundra vegetation. *Arctic* 55: 389–392.

Devault, T.L., Z.H. Olson, J.C. Beasley, and O.E. Rhodes. 2011. Mesopredators dominate competition for carrion in an agricultural landscape. *Basic and Applied Ecology* 12: 268–274.

Devault, T.L., O.E. Rhodes, and J.A. Shivik. 2003. Scavenging by vertebrates: Behavioral, ecological, and evolutionary perspectives on an important energy transfer pathway in terrestrial ecosystems. *Oikos* 102: 225–234.

Doube, B.M. 1987. Spatial and temporal organization in communities associated with dung pads and carcasses. In: Gee, J.H.R. and P.S. Giller (eds), *Organization of Communities Past and Present.* Oxford: Blackwell Scientific Publications, pp. 255–280.

Doughty, C.E., A. Wolf, and Y. Malhi. 2013. The legacy of the Pleistocene megafauna extinctions on nutrient availability in amazonia. *Nature Geoscience* 6: 761–764.

Drazen, J.C., D.M. Bailey, H.A. Ruhl, and K.L. Smith. 2012. The role of carrion supply in the abundance of deep-water fish off California. *PLoS ONE* 7: e49332.

Dupont, H., J.B. Mihoub, S. Bobbe, and F. Sarrazin. 2012. Modelling carcass disposal practices: Implications for the management of an ecological service provided by vultures. *Journal of Applied Ecology* 49: 404–411.

Estes, J.A., J. Terborgh, J.S. Brashares, M.E. Power, J. Berger, W.J. Bond, S.R. Carpenter et al. 2011. Trophic downgrading of planet earth. *Science* 333: 301–306.

Evenari, M., I. Noy-Meir, and D.W. Goodall. 1985. *Hot Deserts and Arid Shrublands.* Amsterdam: Elsevier Science Publishers.

Fahey, T.J., T.G. Siccama, C.T. Driscoll, G.E. Likens, J. Campbell, C.E. Johnson, J.J. Battles et al. 2005. The biogeochemistry of carbon at Hubbard Brook. *Biogeochemistry* 75: 109–176.

Finn, J.A. 2001. Ephemeral resource patches as model systems for diversity-function experiments. *Oikos* 92: 363–366.

Glen, A.S., C.R. Dickman, M.E. Soule, and B.G. Mackey. 2007. Evaluating the role of the dingo as a trophic regulator in Australian ecosystems. *Austral Ecology* 32: 492–501.

Hawlena, D., M.S. Strickland, M.A. Bradford, and O.J. Schmitz. 2012. Fear of predation slows plant–litter decomposition. *Science* 336: 1434–1438.

Heard, S.B. 1998. Resource patch density and larval aggregation in mushroom-breeding flies. *Oikos* 81: 187–195.

Helfield, J.M. and R.J. Naiman. 2006. Keystone interactions: Salmon and bear in riparian forests of Alaska. *Ecosystems* 9: 167–180.

Hilderbrand, G.V., S.G. Jenkins, C.C. Schwartz, T.A. Hanley, and C.T. Robbins. 1999. Effect of seasonal differences in dietary meat intake on changes in body mass and composition in wild and captive brown bears. *Canadian Journal of Zoology-Revue Canadienne De Zoologie* 77: 1623–1630.

Hoback, W.W., A.A. Bishop, J. Kroemer, J. Scalzitti, and J.J. Shaffer. 2004. Differences among antimicrobial properties of carrion beetle secretions reflect phylogeny and ecology. *Journal of Chemical Ecology* 30: 719–729.

Hocking, M.D. and J.D. Reynolds. 2011. Impacts of salmon on riparian plant diversity. *Science* 331: 1609–1612.

Hocking, M.D. and J.D. Reynolds. 2012. Nitrogen uptake by plants subsidized by Pacific salmon carcasses: A hierarchical experiment. *Canadian Journal of Forest Research-Revue Canadienne De Recherche Forestiere* 42: 908–917.

Houston, D.C. 1985. Evolutionary ecology of Afrotropical and Neotropical vultures in forests. In: Foster, M. (ed.), *Neotropical Ornithology*. Washington, DC: American Ornithologists' Union, pp. 856–864.

Huxman, T.E., K.A. Snyder, D. Tissue, A.J. Leffler, K. Ogle, W.T. Pockman, D.R. Sandquist, D.L. Potts, and S. Schwinning. 2004. Precipitation pulses and carbon fluxes in semiarid and arid ecosystems. *Oecologia* 141: 254–268.

Ives, A.R. 1991. Aggregation and coexistence in a carrion fly community. *Ecological Monographs* 61: 75–94.

Jackman, R.E. and W.G. Hunt. 2007. Bald eagle foraging and reservoir management in northern California. *Journal of Raptor Research* 41: 202–211.

Janzen, D.H. 1977. Why fruits rot, seeds mold, and meat spoils. *American Naturalist* 111: 691–713.

Klein, B.C. 1989. Effects of forest fragmentation on dung and carrion beetle communities in central Amazonia. *Ecology* 70: 1715–1725.

Kneidel, K.A. 1984. Competition and disturbance in communities of carrion breeding Diptera. *Journal of Animal Ecology* 53: 849–865.

Kouki, J. and I. Hanski. 1995. Population aggregation facilitates coexistence of many competing carrion fly species. *Oikos* 72: 223–227.

Krebs, C.J., K. Danell, A. Angerbjorn, J. Agrell, D. Berteaux, K.A. Brathen, O. Danell et al. 2003. Terrestrial trophic dynamics in the Canadian Arctic. *Canadian Journal of Zoology-Revue Canadienne De Zoologie* 81: 827–843.

Lang, M.D., G.R. Allen, and B.J. Horton. 2006. Blowfly succession from possum (*Trichosurus vulpecula*) carrion in a sheep-farming zone. *Medical and Veterinary Entomology* 20: 445–452.

Lavelle, P., D. Bignell, M. Lepage, V. Wolters, P. Roger, P. Ineson, O.W. Heal, and S. Dhillion. 1997. Soil function in a changing world: The role of invertebrate ecosystem engineers. *European Journal of Soil Biology* 33: 159–193.

Lundsten, L., K.L. Schlining, K. Frasier, S.B. Johnson, L.A. Kuhnz, J.B.J. Harvey, G. Clague, and R.C. Vrijenhoek. 2010. Time-series analysis of six whale-fall communities in Monterey Canyon, California, USA. *Deep-Sea Research Part I-Oceanographic Research Papers* 57: 1573–1584.

Macdonald, B.C.T., M. Farrell, S. Tuomi, P.S. Barton, S.A. Cunningham, and A.D. Manning. 2014. Carrion decomposition causes large and lasting effects on soil amino acid and peptide flux. *Soil Biology and Biochemistry* 69: 132–140.

Manning, A.D., J. Fischer, A. Felton, B. Newell, W. Steffen, and D.B. Lindenmayer. 2009. Landscape fluidity—A unifying perspective for understanding and adapting to global change. *Journal of Biogeography* 36: 193–199.

Marczak, L.B., R.M. Thompson, and J.S. Richardson. 2007. Meta-analysis: Trophic level, habitat, and productivity shape the food web effects of resource subsidies. *Ecology* 88: 140–148.

Margalida, A., J. Bertran, and R. Heredia. 2009. Diet and food preferences of the endangered bearded vulture *Gypaetus barbatus*: A basis for their conservation. *Ibis* 151: 235–243.

Margalida, A., M.A. Colomer, and D. Sanuy. 2011. Can wild ungulate carcasses provide enough biomass to maintain avian scavenger populations? An empirical assessment using a bio-inspired computational model. *PLoS ONE* 6: e20248.

Matuszewski, S., D. Bajerlein, S. Konwerski, and K. Szpila. 2010. Insect succession and carrion decomposition in selected forests of central Europe. Part 2: Composition and residency patterns of carrion fauna. *Forensic Science International* 195: 42–51.

Mégnin, P. 1894. La faune des cadavres application de l'entomologie à la médecine légale. *Encylopédie Scientifique des Aide-Mémoire*, Paris.

Melis, C., N. Selva, I. Teurlings, C. Skarpe, J.D.C. Linnell, and R. Andersen. 2007. Soil and vegetation nutrient response to bison carcasses in Bialeowieza primeval forest, Poland. *Ecological Research* 22: 807–813.

Moore, J.C., E.L. Berlow, D.C. Coleman, P.C. De Ruiter, Q. Dong, A. Hastings, N.C. Johnson et al. 2004. Detritus, trophic dynamics and biodiversity. *Ecology Letters* 7: 584–600.

Moura, A.O., E.L.D. Monteiro-Filho, and C.J.B. De Carvalho. 2005. Heterotrophic succession in carrion arthropod assemblages. *Brazilian Archives of Biology and Technology* 48: 477–486.

Olson, Z.H., J.C. Beasley, T.L. Devault, and O.E. Rhodes. 2012. Scavenger community response to the removal of a dominant scavenger. *Oikos* 121: 77–84.

Ovington, J.D. 1983. *Temperate Broad-Leaved Evergreen Forests*. Amsterdam: Elsevier Science Publishers.

Parmenter, R.R. and J.A. Macmahon. 2009. Carrion decomposition and nutrient cycling in a semiarid shrub-steppe ecosystem. *Ecological Monographs* 79: 637–661.

Payne, J.A., E.W. King, and G. Beinhart. 1968. Arthropod succession and decomposition of buried pigs. *Nature* 219: 1180–1181.

Pechal, J.L., A.J. Lewis, J.K. Tomberlin, and E. Benbow. 2013. Microbial community functional change during vertebrate carrion decomposition. *PLoS ONE* 8(11): e79035.

Polis, G.A., W.B. Anderson, and R.D. Holt. 1997. Toward an integration of landscape and food web ecology: The dynamics of spatially subsidized food webs. *Annual Review of Ecology and Systematics* 28: 289–316.

Polis, G.A. and S.D. Hurd. 1995. Extraordinarily high spider densities on islands: Flow of energy from the marine to terrestrial food webs and the absence of predation. *Proceedings of the National Academy of Sciences of the United States of America* 92: 4382–4386.

Polis, G.A. and S.D. Hurd. 1996. Linking marine and terrestrial food webs: Allochthonous input from the ocean supports high secondary productivity on small islands and coastal land communities. *American Naturalist* 147: 396–423.

Putman, R.J. 1978a. Flow of energy and organic matter from a carcass during decomposition. 2. Decomposition of small mammal carrion in temperate systems. *Oikos* 31: 58–68.

Putman, R.J. 1978b. Patterns of carbon dioxide evolution from decaying carrion. 1. Decomposition of small mammal carrion in temperate systems. *Oikos* 31: 47–57.

Read, J.L. and D. Wilson. 2004. Scavengers and detritivores of kangaroo harvest offcuts in arid Australia. *Wildlife Research* 31: 51–56.

Ripple, W.J., A.J. Wirsing, C.C. Wilmers, and M. Letnic. 2013. Widespread mesopredator effects after wolf extirpation. *Biological Conservation* 160: 70–79.

Röhrig, E. and B. Ulrich. 1991. *Temperate Deciduous Forests*. Amsterdam: Elsevier Science Publishers.

Roth, J.D. 2003. Variability in marine resources affects arctic fox population dynamics. *Journal of Animal Ecology* 72: 668–676.

Rozen, D.E., D.J.P. Engelmoer, and P.T. Smiseth. 2008. Antimicrobial strategies in burying beetles breeding on carrion. *Proceedings of the National Academy of Sciences of the United States of America* 105: 17890–17895.

Schmitz, O.J., H.P. Jones, and B.T. Barton. 2008. Scavengers. In: Jorgensen, S.E. and B. Fath (eds), *Encyclopedia of Ecology*. Amsterdam, the Netherlands: Elsevier, pp. 3160–3164.

Schneider, K., M.C. Christman, and W.F. Fagan. 2011. The influence of resource subsidies on cave invertebrates: Results from an ecosystem-level manipulation experiment. *Ecology* 92: 765–776.

Schoenly, K. 1992. A statistical analysis of successional patterns in carrion arthropod assemblages: Implications for forensic entomology and determination of the postmortem interval. *Journal of Forensic Sciences* 37: 1489–1513.

Schoenly, K. and W. Reid. 1987. Dynamics of heterotrophic succession in carrion arthropod assemblages: Discrete series or a continuum of change? *Oecologia* 73: 192–202.

Selva, N. and M.A. Fortuna. 2007. The nested structure of a scavenger community. *Proceedings of the Royal Society B-Biological Sciences* 274: 1101–1108.

Sevenster, J.G. and J.J.M. Vanalphen. 1996. Aggregation and coexistence. 2. A neotropical *Drosophila* community. *Journal of Animal Ecology* 65: 308–324.

Smith, C.R. and A.R. Baco. 2003. Ecology of whale falls at the deep-sea floor. *Oceanography and Marine Biology* 41: 311–354.

Smith, C.R. and A.W.J. Demopolous. 2003. The deep Pacific ocean floor. In: Tyler, P.A. (ed.), *Ecosystems of the Deep Ocean*. Amsterdam: Elsevier, pp. 179–218.

Smith, C.R., H. Kukert, R.A. Wheatcroft, P.A. Jumars, and J.W. Deming. 1989. Vent fauna on whale remains. *Nature* 34: 27–28.

Strong, D.R. and K.T. Frank. 2010. Human involvement in food webs. *Annual Review of Environment and Resources* 35: 1–24.

Swift, M.J., O.W. Heal, and J.M. Anderson. 1979. *Decomposition in Terrestrial Ecosystems*. Oxford: Blackwell Scientific Publications.

Tiegs, S.D., P.S. Levi, J. Rueegg, D.T. Chaloner, J.L. Tank, and G.A. Lamberti. 2011. Ecological effects of live salmon exceed those of carcasses during an annual spawning migration. *Ecosystems* 14: 598–614.

Tomberlin, J.K., M.E. Benbow, A.M. Tarone, and R.M. Mohr. 2011. Basic research in evolution and ecology enhances forensics. *Trends in Ecology & Evolution* 26: 53–55.

Tomberlin, J.K., T.L. Crippen, A.M. Tarone, B. Singh, K. Adams, Y.H. Rezenom, M.E. Benbow et al. 2012. Interkingdom responses of flies to bacteria mediated by fly physiology and bacterial quorum sensing. *Animal Behaviour* 84: 1449–1456.

Towne, E.G. 2000. Prairie vegetation and soil nutrient responses to ungulate carcasses. *Oecologia* 122: 232–239.

Tyler, P.A. 2003. *Ecosystems of the Deep Oceans*. Amsterdam: Elsevier Science.

Wielgolaski, F.E. 1997. *Polar and Arctic Tundra*. Amsterdam: Elsevier Science Publishers.

Wilmers, C.C., R.L. Crabtree, D.W. Smith, K.M. Murphy, and W.M. Getz. 2003. Trophic facilitation by introduced top predators: Grey wolf subsidies to scavengers in Yellowstone National Park. *Journal of Animal Ecology* 72: 909–916.

Wilmers, C.C. and W.M. Getz. 2004. Simulating the effects of wolf–elk population dynamics on resource flow to scavengers. *Ecological Modelling* 177: 193–208.

Wilmers, C.C. and E. Post. 2006. Predicting the influence of wolf-provided carrion on scavenger community dynamics under climate change scenarios. *Global Change Biology* 12: 403–409.

Wilson, E.O. 1987. The little things that run the world (the importance and conservation of invertebrates). *Conservation Biology* 1: 344–346.

Wilson, E.E. and E.M. Wolkovich. 2011. Scavenging: How carnivores and carrion structure communities. *Trends in Ecology & Evolution* 26: 129–135.

Yang, L.H. 2004. Periodical cicadas as resource pulses in North American forests. *Science* 306: 1565–1567.

Yang, L.H. 2008. Pulses of dead periodical cicadas increase herbivory of American bellflowers. *Ecology* 89: 1497–1502.

Section III

Evolutionary Ecology of Carrion Decomposition

14

Ecological Genetics

Aaron M. Tarone

CONTENTS

14.1 Introduction to Ecological Genetics

The famous twentieth-century biologist Theodosius Dobzhansky, one of the scientists credited with ushering in a "new synthesis" of evolutionary biology that incorporated genetics into the Darwinian framework, wrote a famous essay noting, "nothing in biology makes sense, except in the light of evolution" (Dobzhansky 1973). Yet, although the significance of this essay is obvious in the context of evolutionary biology, the importance of ecology to evolution, outlined in the same essay, is frequently overlooked. One of Dobzhansky's principal arguments in the essay was that evolution occurs in response to the myriad environments available to organisms, so that diverse forms of life evolved to exploit them. This concept, connecting evolution and ecology, permeates the work of Dozhansky and his colleagues that pushed the new synthesis of evolutionary biology forward into the twentieth century. One of the hallmarks of the new synthesis was the origin of population and quantitative genetics, which are centered on two principles: (1) the demographic processes experienced by a population and (2) the selective pressures exerted on the members of a population. Clearly, ecology plays an important role in understanding population dynamics and selective pressures. For this reason, evolutionary and ecological studies are frequently linked: If you want to understand the ecology of a system, you must understand the evolutionary pressures in that system, and similarly if you want to study the evolution of an organism, you must understand its ecology.

One classical example in evolutionary ecology comes from the work of Dobzhansky himself. He studied how genetic variation is distributed in space. To do this, he decided to study wild *Drosophila* (Diptera: Drosophilidae) populations, typically *Drosophila pseudoobscura* (Frolova and Astaurov). This species is distributed across numerous ecological zones in North America, from Canada to Mexico. In his work, Dobzhansky studied the frequencies of chromosomal inversions across the various ecological zones. Through this work, he documented one of the first examples connecting a genetic factor (types of chromosomes), evolution (the differential distribution of those chromosomes), and ecology (the different seasons or regions of the world exerting evolutionary pressure on those chromosomes, altering their distributions). He did this by observing how these chromosomal inversions were distributed across time (Wright and Dobzhansky 1946; Dobzhansky 1948a; Powell 1997) and space (Dobzhansky and Epling 1944; Dobzhansky 1948b; Powell 1997), finding that certain chromosomes appeared to have temporal and/or spatial biases throughout North America. Some chromosomes increased in frequency during the summer or in different parts of the U.S. desert southwest. This finding spurred a great deal of interest in the connection between genetics and ecology. Today, the connections between chromosomal inversions and adaptations are still being dissected by a variety of researchers interested in topics spanning a range of biological diversity from how gut bacteria survive in their hosts to genetic changes associated with global warming to speciation (Anderson et al. 1991; Rodríguez-Trelles and Rodríguez 1998; Komano 1999; Krinos et al. 2001; Noor et al. 2001; Rieseberg 2001; Navarro et al. 2003; Kennington et al. 2006; Kirkpatrick 2010).

Yet, this subject is just a small subsection of the field of ecological genetics that has risen from the work of this extraordinary scientist and his contemporaries. Now research in this area permeates biology, and the field is rapidly advancing, as genomic tools and technologies have generated interest in such studies with nonmodel organisms. This chapter will review basic knowledge in the field of ecological genetics and relate it to concepts that are relevant to carrion ecology. This chapter is not meant to be exhaustive, as a book could be written on any of the topics in this and subsequent related chapters, but is meant to provide an entry point into considering the process of carrion decomposition from an ecological genetic perspective. Additional information on related topics can be found in several published works (Barker et al. 1990; Roff 1992, 2002; Powell 1997; Lynch and Walsh 1998; Schlichting and Pigliucci 1998; Fox et al. 2001; Conner and Hartl 2004; DeWitt and Scheiner 2004; Whitman and Ananthakrishnan 2009; Westneat and Fox 2010) and additional details in the carrion system can be found in subsequent chapters.

14.2 Evolution

14.2.1 Introduction

A generally accepted definition of evolution is the change in frequency of a phenotypic or genetic variant within a group of organisms over multiple generations (Conner and Hartl 2004; Ridley 2004; Futuyma 2009). Typically, there are two accepted ways by which a phenotypic or genetic variant can change in frequency over time: Either through selection or genetic drift. Selection happens as described by Darwin (1859), where in each generation advantageous traits within a species lead to differential success of the members of the species that possess those traits. Over time, and across generations, this differential in success leads to an increase in frequencies of the genetic factors that contribute to those traits (Figure 14.1). This process is easily conceptualized and has been utilized by humanity for thousands of years in the process of domesticating plants and animals. The development of insecticide (Roush and McKenzie 1987) and antibiotic resistance (Neu 1992) is also easily demonstrated genetic shifts directly due to the selective pressures of insecticides and antibiotics, respectively. However, genetic variants can also change over time due to stochastic factors, typically referred to as genetic drift (Malecot 1944; Kerr and Wright 1954a,b; Kimura 1954; Wright and Kerr 1954). For instance, a meteor that crashes down on a population, indiscriminately destroying 90% of the population, has not exerted a selective effect on the population, but may have randomly eliminated many genetic variants from the population. Less spectacular examples of drift occur regularly and are typically associated with small (effective) population sizes. Small population sizes result in a higher probability that allele frequencies will change over

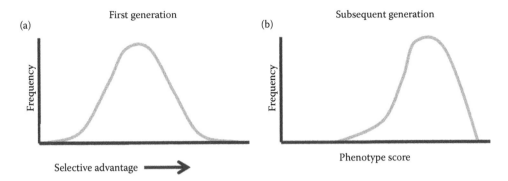

FIGURE 14.1 An example of a selection response in a population. The distribution in panel (a) represents the distribution of a fitness-related phenotype in an initial generation. In this scenario, the selective advantage goes to the individuals with phenotypes at the right side of the phenotype scale. This advantage means that individuals with the advantage will produce offspring more often than others in the population. They will also bequeath their genes to more individuals of subsequent generations. This differential reproductive advantage will result in a change in the genetic composition of the population, such that the average phenotype of the population will shift in the direction of the selective advantage as seen in panel (b). This is an example of directional selection, where selection favors one side of a distribution. However, there are other types of selection that can favor the mean phenotype, extreme phenotypes, or can favor traits in a frequency-dependent manner. Each of these types of selection results in a different type of change to the phenotypic and genetic distribution of a population. Details on each of these selection responses can be found in the evolution texts cited at the beginning of this chapter.

time due to chance and can sometimes act against selection. For instance, imagine a population with 10 individuals where there are five males and five females. Imagine also that the environment will support only one successful mating pair. Evolutionarily "optimal" genetic variants carried by four of the five females have a small but appreciable probability of not being part of the next generation. In such an instance, those variants will be lost from the population despite their evolutionary advantage. However, in a population where four, or all five, pairs can be supported, there will be a decreased probability of losing those same genetic variants from the population. This same sampling process can lead to stochastic shifts in the frequencies of selectively neutral alleles. The combination of selection and drift has been observed in artificial selection experiments of varying effective population sizes, where smaller effective population sizes result in greater variance around the same average selection response (Buri 1956; Rich et al. 1979) (Figure 14.2). A classic example of this concept was demonstrated empirically in *Tribolium castaneum* (Herbst) (Coleoptera: Tenebrionidae), where small effective population sizes were shown to result in the same trajectory of selection for a wild-type allele in the laboratory, but with much greater variance, so that it was possible for unselected alleles to fix in some experimental populations and selectively advantageous alleles to be lost—something that did not happen in populations with larger effective population sizes (Rich et al. 1979). Thus, any phenotypic or genetic change over time can be considered an evolutionary change, and these changes can be due to two, sometimes competing, forces.

Obviously, time is a major factor in the evolutionary process. The study of genetic change over large timescales is known as macroevolutionary biology. This area of research is largely not covered in this chapter, but can be considered in the context of carrion ecology by comparing and contrasting the varying strategies of microbes, insects, and vertebrates that utilize carrion. For instance, avian and mammalian scavengers (noted in Chapters 6 and 21) use different strategies for identifying and utilizing carrion, which is ultimately due to their genetic differences (the most obvious of which is the widespread ability of one group to fly). At an even larger timescale of divergence, bacteria utilize carrion in a completely different way than the vertebrates. For instance, bacteria possess biochemical capabilities for degrading remains that do not exist in vertebrates.

Of primary concern in this chapter is microevolutionary theory, which deals with relatively short evolutionary timescales. Principal questions in microevolution research are targeted at understanding how differences can arise among populations of the same species, how the early stages of speciation occur, and these questions are intimately associated with the ecology of the organisms in question. These questions are frequently addressed by studying the population (Chapter 17) or quantitative (Chapter 15) genetics of

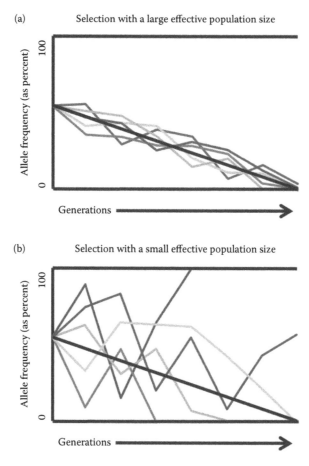

FIGURE 14.2 A theoretical example of the interplay between selection and drift. In this example, there is selection against a hypothetical allele in five populations. The black line represents the theoretical selection response, whereas the other lines represent the actual responses of each of the replicate populations experiencing selection against the target allele. When the effective population size is high (a), there is little fluctuation in allele frequencies or deviation from the theoretical trend in allele frequency. However, when effective population size is low (b), drift impacts the selection response by increasing variation around the theoretical response and can even lead to fixation of the deleterious allele as demonstrated by the line reaching 100% frequency in one population.

target populations and/or species. This chapter will provide a general basis for understanding subsequent chapters and provide case examples of studies in ecological genetics (defined here as the study of how organisms adapt to their environments) that have yielded information relevant to the carrion system.

One of the most basic tools an evolutionary biologist can employ when studying a population in an attempt to determine whether it is differentiated from others is the Hardy–Weinberg equation (Hardy 1908; Weinberg 1908) (Chapter 15). This classic biological formula describes the expected frequencies of genetic variants within a population, given that several key assumptions are met. These assumptions are that the population: (1) exhibits random mating, (2) has a large effective population size, (3) has no mutation, (4) has no recombination, and (5) has no selection. Very frequently, these assumptions are violated; however, when they are violated and the genotypes do not differ significantly from expected frequencies, they are considered unaffected by the violations. When the Hardy–Weinberg expected frequencies are not observed, it is then important to evaluate which of the violated assumptions has led to the deviation from expectations that are creating an opportunity for evolution to occur. There are numerous biochemical and statistical tests for evaluating deviations from Hardy–Weinberg equilibrium, and other similar measures of divergence among populations, but they all lead to one of two answers: either drift or selection is the principal driving force for evolution of a genetic locus in the population.

14.2.2 Fitness

So how do these forces described above, selection and drift, connect to ecology? They connect through evolutionary fitness, a concept used to describe how an organism survives, mates, and produces offspring in its environment (Mousseau and Roff 1987; Roff and Mousseau 1987; Conner and Hartl 2004; Ridley 2004; Futuyma 2009; Orr 2009). If an allele is under selection, it must contribute to fitness somehow or be linked to such an allele. If not, then that allele may still speak to the population dynamics driving genetic shifts in the genome and act in opposition to or in conjunction with selection. Often population dynamics are affected by the ecology of the system, thus it is not surprising that fitness is the driving concept underpinning many ecological genetic studies. The field devotes a considerable amount of effort to understanding how an organism survives and reproduces in its environment, and survival is tied intimately to evolutionary fitness as a prerequisite for reproduction.

Interestingly, fitness is an inherently fuzzy concept, discussions of which can be found in several good texts (Roff 2002; Conner and Hartl 2004; Ridley 2004; Futuyma 2009; Orr 2009; Westneat and Fox 2010). All definitions share a common theme of success in the wild, which is the ability to produce more viable descendants than competitors. How does one study the relative performance of one individual in the population compared to another? Typically, studies are directed toward phenotypes that are generally believed to improve survival, such as fecundity, predator avoidance, growth and survival rates, and the ability to find resources (Roff 1992, 2002; Conner and Hartl 2004; Westneat and Fox 2010). Chapter 15 covers some extremely relevant examples of fitness traits in the dung system, which shares many similarities with the carrion system.

In any biological system, it can be challenging to evaluate and determine which specific factors contribute most to fitness. Different traits can be more tied to fitness for some genotypes than others or are more important in some environments than others. For example, in some biological systems, reproduction may be the biggest limitation on fitness such that reproduction-related traits like body size, mating success, and egg production may be the most important factors to study if one is interested in evolutionary fitness. However, in other species, the ability to avoid predation may be the most important component of fitness, meaning that studies of traits like behavior, chemical defenses, and physical traits that enable avoidance would be the most important to study. It may even be possible that there are environments where the major components of fitness may differ, for instance, in years with abundant nutrient availability, reproduction traits may contribute most to fitness, but in contrasting years, other physiological traits such as osmotic stress resistance or dispersal to new habitats may contribute more to fitness. This logic is easily transferred to the carrion system. For instance, for some carrion-feeding species, fitness could be affected by the ability to find a resource, whereas for others, fitness may be affected more by the ability to defend or dominate a resource once it is found. Similarly, there is a considerable difference in the ability of carrion-feeding animals to manage the microbial communities associated with remains (Chapter 21), suggesting that a strong immune system is a large component of evolutionary fitness for some, but not all, carrion-feeding animals. Thus, while it is easy to imagine that a certain trait could impact fitness in the carrion system, it is often difficult to ascertain which traits are most critical to fitness in a particular population, condition, location, and time. This is potentially a rich area of enquiry of the carrion system.

14.2.3 Common Genetic Studies

One interesting feature of ecologically relevant fitness-associated traits is that they are very frequently complex in nature. This means that they typically exist in a population as a continuum of states, not as a few discrete states. Such phenotypes are often the result of numerous genetic and environmental inputs that culminate in the ultimate expression of the trait. Consequently, quantitative genetic (see Chapter 15) principles are frequently used to dissect the genetic and environmental underpinnings of ecologically relevant phenotypes. There are several types of experiments that are extremely important to understanding the biology of fitness-related traits: Common garden, analyses of relatives, gene mapping, and selection experiments. Each of these can be done to evaluate various aspects of ecologically important traits and are important to understand.

Common garden experiments are centered on the evaluation of natural genotypes raised together in a common environment (or garden). Such experiments produce what is known as a reaction norm, which can be used to evaluate the general degree to which a trait is influenced by genotype, the environment, and their interaction. Reaction norms are essentially a graphical representation of Equation 15.1, which can be evaluated statistically using analysis of variance and similar statistics. These studies provide a basis for comparing the relative performance of specific genetic groups when they are raised in their different environments, providing information regarding their evolutionary trajectories. One important feature of a reaction norm is whether there is an interaction between genotype and the environment. A genotype–environment interaction in a trait that influences fitness is considered to be indicative of local adaptation (Schlichting 1986; Scheiner 1993; Via 1993; Conner and Hartl 2004). Several carrion-feeding flies (Diptera) have been studied in this capacity and have exhibited genetic differences among populations and genotype–temperature interactions (Gallagher et al. 2010; Tarone et al. 2011; Owings et al. 2014). There are several instances where plasticity and genotype environment interactions are expected to evolve (see Schlichting 1986; Scheiner 1993; Via 1993; Conner and Hartl 2004; DeWitt and Scheiner 2004; Whitman and Ananthakrishnan 2009), and this evolution is expected to impact evolutionary processes (Pfennig and McGee 2010; Pfennig et al. 2010). Plasticity in carrion flies is demonstrated frequently, but only at a superficial level compared to other systems, though theory indicates that migration between sub-populations (see Chapters 15 and 17 for more details on this in carrion-feeding insects) would increase the probability of evolved plasticity (Kingsolver et al. 2002).

One way of evaluating whether an ecologically important trait is due to genetic factors is through a comparison of the phenotypes of relatives. Logically, this makes sense, as the similarities of numerous traits exhibited by relatives can be appreciated. These studies can be approached in a variety of ways, but often include analyses of siblings (both full and half), monozygotic versus dizygotic twins, and parent–offspring comparisons (Lynch and Walsh 1998; Conner and Hartl 2004). The logic of all of these studies is to dissect the degree to which relatives of differing genetic distances (such as a full versus half siblings, which share roughly half versus a quarter of their genetic material, respectively) also share phenotypes. If relatedness explains the majority of the variation in the trait, then the trait is mostly affected by genetics. If relatedness explains little phenotypic variation, then the trait is under the influence of the environment. Typically, the genetic contribution to a trait is estimated by calculating heritability (Chapter 15), where high heritability means a high genetic contribution to the trait. Heritability estimates can be useful in understanding the genetic architecture of a fitness-related trait (Mousseau and Roff 1987; Roff and Mousseau 1987), though it is a population- and environment-specific measure that can easily be misinterpreted (Visscher et al. 2008).

Once a trait is determined to be fully or partially explained by genetics, dissecting the nature of this genetic connection and determining the extent to which each genetic variant may contribute to the trait of interest is necessary. One interesting example of a genetic adaptation can be found in the case of the *abnormal abdomen* phenotype in *D. mercatorum* (Patterson and Wheeler) (Templeton et al. 1985, 1993; DeSalle and Templeton 1986; Hollocher et al. 1992; Hollocher and Templeton 1994). These cactophilic *Drosophila* live on a volcanic mountain in Hawaii, USA, where ecological challenges differ depending on humidity and thermal shifts along a cline (or ecological gradient) in elevation. At high elevations, the environment is essentially a cloud forest; at low elevations, the environment is much drier and hotter, leading to a shift in life-history strategies. Lower elevation flies place a high priority on reproduction at the expense of longevity. Interestingly, the populations of this species exhibit a high degree of phenotypic divergence, which manifests visibly as an abdominal phenotype when combined with a controlled genetic background in the laboratory. The genotype has other life-history consequences as well in wild genetic backgrounds, including differences in development time, longevity, and fecundity. Ultimately, this genetic adaptation, which is a response to two different ecological challenges, has been mapped to the presence of R1 transposable element sequences inserted into the X-linked 28S rDNA gene. This insertion results in the abdominal phenotype found in the laboratory crosses. The *abnormal abdomen* case is a simple example of a genetic difference that explains phenotypic divergence between relatively close (separated by less than one kilometer) populations of an organism that experience different ecological challenges. Interestingly, this adaptation has repeated itself in another cactophilic fly, *D. hydei* (Sturtevant) (Templeton et al. 1989). Populations of *D. hydei* exhibit the same adaptations to the ecological challenges encountered along the cline and involve the same rDNA gene. However, the detail of the genetic mechanism leading to the adaptation is species-specific.

In many cases, the genetic bases of ecological adaptations are more complex than the *abnormal abdomen* example. Complicating the problem is the fact that interactions with the environment will occur and are both critical and difficult to document and understand. Dissection of the genetics of complex traits can be done through a few approaches (Lynch and Walsh 1998; Mackay 2001; Conner and Hartl 2004; Mackay 2004), described below and in Chapter 15. In quantitative trait locus mapping, panels of recombinant genotypes from a limited number of genetic backgrounds are used to statistically associate genotypes with phenotypes. With a large enough density of genetic markers, it is possible to identify a region or regions of chromosomes statistically associated with a particular trait and, in some cases, even specific genes or alleles associated with a trait of interest. Owing to the complex nature of these traits, false-positive results are possible and require confirmation of genotype–phenotype linkages. In a process called association mapping, a similar approach can also be taken with wild populations, where individuals or inbred lines derived from wild populations are individually genotyped and phenotyped. This approach will allow for a greater sampling of genotypes than a study derived from quantitative trait mapping experiments, making the results more relevant to wild populations. However, association mapping will still suffer from possible false-positive associations and the potential for low statistical power without a large panel of target genotypes. In addition, if the environment is not controlled in such an experiment, for instance, by just sampling wild-caught individuals, false-negatives may arise due to the variable effects of the environment on the trait of interest. The quantitative genetic and association studies mentioned in Chapter 15 reflect these types of strategies for dissecting the genetics of complex traits. Although these kinds of experiments suffer from limitations associated with statistical power and the genetic architectures of the traits of interest, there has been considerable success in implementing these approaches to find genes responsible for variation in a number of fitness-related traits of model organisms (e.g., Wayne et al. 2001; Kopp et al. 2003; Bickel et al. 2009; Atwell et al. 2010; Brachi et al. 2010; Hohenlohe et al. 2010a,b; Nemri et al. 2010; Keurentjes et al. 2011; Strange et al. 2011; Mackay et al. 2012; Bastide et al. 2013; Huang et al. 2014).

Functional approaches can also be taken to identify genes that may influence a phenotype of ecological interest (Anholt et al. 2001; Ayroles et al. 2009; Harbison et al. 2009). Global gene expression studies are regularly used to evaluate the expression of mRNA in different treatment groups (genetic or environmental) relevant to the survival of an organism in the wild (Coffman et al. 2005; Ayroles et al. 2009; Reed et al. 2014). In these cases, genes that are differentially expressed can be considered connected to the process of interest, though it may not always be clear whether the genes are upstream or downstream of the trait of interest. Clusters and classes of genes associated with a trait can also be identified. With the advance of next-generation sequencing, genomic studies of this sort are much easier to conduct, allowing for an identification of genes that are associated with a genetic or environmental response. Several genomic studies have been conducted on arthropods in the carrion system (Sze et al. 2012; Zhang et al. 2013) and these types of studies are likely to increase in numbers as the field develops better tools.

Another means of evaluating a genetic contribution to an ecologically important trait is a selection experiment (Figure 14.3; Conner and Hartl 2004, Chapter 15). As noted above, these studies rely on the same breeding principles that agriculture has employed for centuries in the domestication of crops or livestock, taking advantage of the selection process identified in natural populations by Darwin. In this manner, over generations of experimentation, determining the degree to which wild populations harbor genetic variation for a particular trait and the strength with which selection imparts change in the phenotype is possible. Using the breeder's equation (see Chapter 15 for details), these experiments can yield an estimate of realized heritability. Other experiments, like parent–offspring, sibling, and twin analyses described above can also be used to estimate heritability. In any of these experiments, it must be remembered that heritability scores are population- and environment-specific values that provide some useful but limited information regarding the genetic contribution to a trait. Selection experiments can also be combined with genotyping experiments to identify specific genetic targets of selection for fitness components, as has been done for longevity and body size in *D. melanogaster* (Meigen) (Burke et al. 2010; Turner et al. 2011).

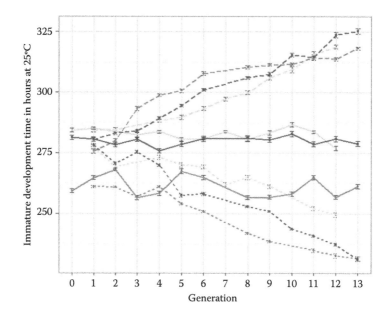

FIGURE 14.3   A selection experiment on immature development time in the carrion-feeding insect *Cochliomyia macellaria*. In this experiment, three laboratory populations were established from three distinct Texas populations of this species derived from different ecoregions (represented by different colored lines). Each population was then split into three subpopulations that were each exposed to separate experimental regimes: Control, Fast, and Slow selection. Mean development times and standard errors are reported for each group. Control populations were exposed to no selection on development time and are represented by the solid lines. In the Fast selection regime, represented by the short dashed lines, the fastest developing flies were used to found each subsequent generation. Similarly, in the Slow selection regime, represented by the long dashed lines, each generation was founded by the slowest developers. Note that the original populations exhibit mean immature development times that are within ~1 day of each other. However, after 12–13 generations of selection, there is ~4 days difference in development times between Slow and Fast groups. Note also that the selection response is imbalanced, with a stronger response to Slow selection in the same timeframe.

By employing genetic approaches, it is possible to determine the degree and nature of genetic impacts on ecologically relevant traits in any organism, including those that utilize carrion. Currently, these sorts of studies are rare in carrion ecology; however, there are numerous relevant examples of ecological genetic research that can help shed light on the potential genetic features of a carrion ecology system. This chapter presents concepts from general biology research that is perceived as highly relevant to the carrion system. These research vignettes are meant to inspire deeper consideration of ecological genetic processes in carrion ecologists and to encourage a nontraditional, nonapplied perspective on carrion ecology.

14.3 Ecological Genetic Research in Action

14.3.1 Classical Examples in Evolutionary Ecology Research

Dobzhansky's own research model—the *Drosophila* model—has been an important system for discovery in evolutionary ecology (for a good review, see Powell 1997). In addition, the Drosophilidae share evolutionary and developmental similarities to carrion-feeding flies such as the Calliphoridae and Sarcophagidae and many dung-feeding flies discussed in Chapter 15, as all of these flies are Cyclorhapphan Diptera (Yeates and Wiegmann 2005). Accordingly, if the *Drosophila* model can be considered a powerful model for human medical research, it is definitely a good model for beginning to understand the evolutionary ecology of carrion-feeding insects, especially when juxtaposed against the biology of dung flies in Chapter 15.

One evolutionary topic where the *Drosophila* model has contributed considerably is in the study of local adaptation, a topic relevant to any organism utilizing carrion. As noted above, much of this kind of research was facilitated by Dobzhansky and his contemporaries, through the initial discovery that chromosomal inversions segregated nonrandomly in time and space. Much of this work centered on observations that *Drosophila* inversions and phenotypes often segregate along ecological clines, though this is only part of the story that is currently understood by *Drosophila* evolutionary biologists.

D. melanogaster is an old world fly, originally from Africa (David and Capy 1988). However, with the colonial expansion of Europeans, this species was distributed to every continent except Antarctica relatively recently (on an evolutionary timescale). This species has established similar latitudinal clines in body size and other body size-related traits on each continent, suggesting that, in this species, increasing body size contributes to fitness in cooler environments (James et al. 1997; Zwaan et al. 2000; Bochdanovits and De Jong 2003; De Jong and Bochdanovits 2003). These latitudinal clines have been studied extensively, revealing that populations on each continent likely followed their own evolutionary trajectories, which resulted in the same basic phenotypic outcomes among the continents, though they result from different genetic changes.

One obvious genetic candidate that may influence this cline is chromosomal inversions like those studied by Dobzhansky. Reduced recombination among these chromosomal rearrangements allows co-adapted suites of alleles to co-exist on a chromosome (Figure 14.4). Accordingly, theory predicts that such co-adapted alleles would lead to temporal or spatial changes in inversion frequencies such as those seen in several *Drosophila* species, as specific chromosomes will possess clusters of alleles that are adaptive for specific environments (Neiman and Linksvayer 2006; Kirkpatrick 2010). However, even as

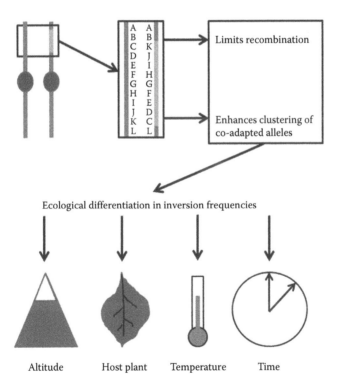

FIGURE 14.4 Ecological differentiation through changes in inversion frequency. Inversions in chromosomes result in changes in the organization of genes on homologous chromosomes in a manner that inhibits recombination. The inhibition of recombination allows for the clustering of coadapted alleles of genes within the inverted region to cosegregate. This cosegregation enables the accumulation of these chromosomes in certain ecological zones (such as altitudes, host plants, or temperatures) or at certain times.

inverted chromosomes are expected to segregate adaptively, they also clearly do not explain all clinal patterns of variation in phenotypes of adaptive significance (Turner et al. 2008; Kolaczkowski et al. 2011; Reinhardt et al. 2014).

There are a number of other traits such as alcohol tolerance, pigmentation, desiccation resistance, starvation resistance, and metabolism that have been identified as exhibiting clinally variable patterns, and which have in some instances repeated across continents, all suggesting selection as the driving force behind the evolution of these phenotypes (Gockel et al. 2001; Hallas et al. 2002; Kennington et al. 2003a,b; Rion and Kawecki 2007; Turner et al. 2008; Kolaczkowski et al. 2011; Fabian et al. 2012). One good example of this principle is pigmentation. In *D. melanogaster*, there are clines in pigmentation, which have been extensively studied. In this and other *Drosophila* species, alleles of a number of genes whose mutants exhibit pigmentation phenotypes have been associated with variation in wild pigmentation levels (Wittkopp et al. 2002a,b, 2003a,b, 2009, 2011; Jeong et al. 2008; Arnoult et al. 2013; Bastide et al. 2013, 2014). The most correlated factor associated with this pattern in African populations is ultraviolet light intensity, suggesting that one role for melanization is the protection of wild populations from this environmental stressor (Bastide et al. 2014), though this result is not consistent with correlates of pigmentation on other continents.

Studies of local adaptation are not limited to studies of latitudinal clines. There are numerous examples in the *Drosophila* literature of local adaptation to numerous ecological factors. For instance, in India, different populations of *D. melanogaster*, *D. repleta* (Wollaston), and *D. immigrans* (Sturtevant) collected from ordinary and alcohol-rich areas have been shown to exhibit different levels of tolerance to alcohol and actetate (Karan et al. 1999). Studies of *D. pachea* (Patterson and Wheeler) and *D. mojavensis* (Patterson and Crow) and several other cactophilic species living in the Baja Peninsula and Sonoran Desert demonstrate different *alcohol dehydrogenase* and inversion allele frequencies, which are associated with survival on different host plants (Ward et al. 1974; Ward 1975; Brussard et al. 1978; Etges et al. 1995; Pfeiler and Markow 2001). Elevation clines have also been demonstrated in the Sierra Nevada mountains of California, USA, for both *D. pseudoobscura* and *D. persimilis* (Dobzhansky and Epling 1944) inversion frequencies (Dobzhansky 1948a).

Clinal variation is not restricted to the *Drosophila* system either. The apple maggot *Rhagoletis pomonella* (Walsh) (Diptera: Tephritidae) is a model for speciation in which different populations of the species are associated with different hosts (Feder et al. 1990a,b; Powell et al. 2014). These different hosts produce fruit asynchronously in the wild, leading to evolution of differences among races of the species in eclosion timing and diapause induction. Host-associated races of these flies are genetically distinct even though these populations have very few private alleles (Xie et al. 2008). The species also exhibits a latitudinal cline in development time with some evidence for a role of chromosomal inversions in development time variation (Feder et al. 2003a,b). When development times of one race were studied from northern Mexico to Minnesota, USA, there was a clear cline in immature degree day requirements, with more northern flies completing development in fewer degree days. Another such example can be found in the *Anopheles gambiae* (Giles) (Diptera: Culicidae) species complex, which exhibits a longitudinal pattern in inversion frequency that follows a cline in humidity found between the tropical forests of Africa to the dry deserts of Arabia (Coluzzi et al. 2002; Fouet et al. 2012; Ayala et al. 2013). Ecological divergence is associated with chromosomal inversions in numerous species.

Recently, Keller et al. (2013) conducted a comprehensive review of altitudinal adaptations in a number of animal species and study sites. They identified evidence of local altitudinal adaptations in several *Drosophila*, amphibian, beetle (Coleoptera), butterfly (Lepidoptera), and grasshopper (Orthoptera) species. These adaptations manifested in significant shifts in a number of phenotypes including wing size, cold tolerance, mass, and development time. In addition, they identified numerous genetic variants in candidate genes that had differentiated among populations of various vertebrate and invertebrate species, which they attributed to selection pressures associated with temperature, oxygen, and parasites. The most obvious of these candidate loci were hemoglobin alleles, which would be expected to vary in function as oxygen levels are expected to differ across altitudes. Clearly, adaptations to various ecological zones can be identified along with genetic factors that contribute to local adaptations to those zones.

The ecological genetics community is replete with examples of local adaptations. The goal of this section was to provide some basic examples of how environmental and genetic variation can result in local adaptations. Local selection can structure variation among populations in phenotypes, chromosomes, or alleles of specific genes. Ultimately, carrion-feeding organisms will face similar challenges to their survival from their ecosystems as the organisms discussed above. A more detailed study of carrion systems in this capacity will enhance our understanding of how evolution operates on the organisms that feed upon decomposing remains, which will inform applied and basic understanding of how decomposition of carrion impacts ecology.

14.3.2 Plasticity

Studies in evolutionary ecology have contributed considerably to our understanding of fitness and can provide a basis for understanding evolutionary principles associated with local adaptation in any system, including carrion ecology. Whether local adaptation is connected to chromosomal inversions or not, the overriding principle of local adaptation is that local environments lead to differential selection regimes that select for distinct phenotypic optima, resulting in different genetic patterns. But that raises the question of what traits lead to success, or differential success, across ecosystems? In evaluating responses to the environment, it is critical to consider the concept of plasticity.

One common feature of local adaptation is the concept of phenotypic plasticity, where different environments lead to different phenotypic outcomes from the same genotype (Conner and Hartl 2004; DeWitt and Scheiner 2004; Whitman and Ananthakrishnan 2009; Westneat and Fox 2010). An example of this concept is the temperature-size rule, which is manifested in the vast majority of poikilotherms. Organisms following this rule, such as *Drosophila*, generally produce larger body sizes when developing in cooler environments (David et al. 1994, 1997, 2011; Petavy et al. 1997a,b; Karan et al. 1998a). This response mirrors the genetic variation along clines observed for *Drosophila* populations where different populations evolving in different thermal regimes achieved different body sizes over time (Huey et al. 2000; Gilchrist and Huey 2004; Gilchrist et al. 2004). Both of these phenomena demonstrate a common relationship between organismal size and temperature. Further, there are complex interactions between body size and development time in insects, which also interact with temperature (Gotthard et al. 1994; Nylin and Gotthard 1998; Davidowitz and Nijhout 2004; Gotthard 2004; Davidowitz et al. 2012).

Some of the best examples of phenotypic plasticity come from insects that exhibit phase polyphenism, where environmental effects result in not just a change in one trait, but a syndrome of organismal reorganization. The pea aphid *Acyrthosiphon pisum* (Harris) (Hemiptera: Aphididae) exhibits interesting polyphenisms related to dispersal and reproduction (reviewed in Srinivasan and Brisson 2012). In pea aphids, asexual-phase females are wingless and produce many clones on their host plant. This can be a highly effective strategy (as any gardener can confirm), until the population overruns its resource or predation increases. Those conditions trigger the production of winged individuals, which involve upregulation of some genes known in *Drosophila* to impact wing development (Brisson 2010; Brisson et al. 2010). Interestingly, male wing production is not a function of population density, but rather is restricted to individuals with certain alleles of the *aphicarus* locus, where some genotypes produce winged males and others produce unwinged males (Braendle et al. 2006). This locus also appears to be connected to the female polyphenism. Relatively independent of wing polyphenism are the environmental responses to temperature and photoperiod that trigger a change from asexual to sexual reproduction that produces eggs that will overwinter. Emerging from these eggs will be the asexuals that begin the next phase of the reproductive cycle. The aphid system has and will continue to contribute to our understanding of phenotypic plasticity, as it offers several striking examples to study.

Polyphenism is not restricted to aphids. Environmental cues are also responsible for the manifestation of the gregarious phases of numerous locusts (Orthoptera: Acrididae) (Sword and Simpson 2000; Sword 2003; Gray et al. 2009) and Mormon crickets (Orthoptera: Tettigoniidae) (Lorch et al. 2005; Sword et al. 2005; Simpson et al. 2006; Srygley et al. 2009). In these cases, the environmental triggers are population density and nutrition. High population densities lead to the development of populations that exhibit "plague locust" style behaviors, physiologies, and morphologies that can have dire consequences for human societies wherever these insects occur. Interestingly, a review of locust polyphenism implicates numerous independent events in the evolution of such behaviors, events that are triggered by different environments across taxa (Song 2011).

In the cricket *Gryllus firmus* (Scudder) (Orthoptera: Gryllidae), nutritional variation results in a trade-off between dispersal and reproduction. Individuals experiencing high nutritional quality invest in reproductive success at the cost of dispersal capabilities, while individuals experiencing low nutritional quality invest less in reproduction and more in dispersal (Fairbairn and Roff 1990; Zera and Brink 2000; Zera 2005). In all of these cases, it is clear that the same genotype can produce different traits or trait combinations, depending on environmental exposure, which can have drastic impacts on organismal survival and fitness.

Genotype interacts with the environment, which in some cases is expected to produce local adaptation (Kingsolver et al. 2002; Pfennig and Murphy 2002; Conner and Hartl 2004; DeWitt and Scheiner 2004; Pfennig et al. 2006; Pfennig and Martin 2010). Plasticity and genotype–environment interactions are expected to evolve under a number of conditions that are beyond the scope of this chapter but are discussed further in Kingsolver et al. (2002), DeWitt and Scheiner (2004), Whitman and Ananthakrishnan (2009), Pfennig et al. (2010), and Westneat and Fox (2010). Once again, the pea aphid serves as a model for understanding genotype–environment interactions. In this case, there are aphid lineages that are very successful on some plants but unsuccessful on others (e.g., successful on clover, but are not very successful on alfalfa, and vice versa; Via 1991a,b; Hawthorne and Via 2001; Ferrari et al. 2008) due to possession of different types of obligate symbionts such as those from the genus *Buchnera* (Prosser and Douglas 1991, 1992; Douglas and Prosser 1992; Prosser et al. 1992; Baumann et al. 1995; Rouhbakhsh et al. 1996; Douglas 1998; Sandstrom et al. 2000; Moran and Degnan 2006; Ferrari et al. 2012). These symbionts provide different essential amino acids, resulting in differential survival on host plants that reflects the amino acid contents of the hosts.

In *Drosophila*, a classic example of genotype–environment interactions is due to genetic variation at the *foraging* locus (Debelle et al. 1987, 1989; Debelle and Sokolowski 1987; Partridge and Sgro 1998; Sokolowski 1998, 2001) (Figure 14.5). This locus controls what are known as "rover" and "sitter" genotypes. In the absence of food, larvae of both genotypes will search for additional resources. However, in the presence of ample larval nutritional resources, the "sitter" maggots are content to feed, whereas "rover" genotypes continue to quest for additional resources. These alleles are pleiotropic, impacting

The *foraging* locus of *Drosophila melanogaster*

"rover" alleles

- Long foraging path in the absence of food
- Long foraging path in the presence of food

- Forms puparia a relatively greater distance from food

- Better short-term and poorer long-term memory

"sitter" alleles

- Long foraging path in the absence of food
- Short foraging path in the presence of food

- Forms puparia a relatively shorter distance from the food

- Better long-term and poorer short-term memory

FIGURE 14.5 Pleiotropic impacts of the foraging locus on *D. melanogaster* biology. The foraging locus has "rover" and "sitter" alleles. The genotype of an individual at this locus can dictate whether it continues to forage in good nutritional conditions, will pupate close to its food, and whether it has good long-term or short-term memory. Genetic variation at loci such as this can have drastic impacts on the biology of carrion-feeding organisms as the traits affected would impact how an organism manages trade-offs associated with limited nutrition and how it avoids predation.

larval feeding, larval migration, adult postfeeding dispersal, and memory. The maintenance of both alleles in wild populations, with such a divergent environmental response to food availability, is suggestive of different habitats that exhibit varying degrees of resource availability. The existence of "rovers" and "sitters" suggest that in some instances fitness is optimized by staying put on a useable resource, whereas in other instances it is optimized by continuing to forage in anticipation of the depletion of the current resource. It is then not a surprise that genotype–environment interactions appear to be very common in nutritional biology.

In a study of numerous inbred genotypes of *D. melanogaster* raised on a range of diet types, the primary factor explaining phenotypic shifts with respect to diet was the interaction between genotype and environment (Reed et al. 2010). In this study, phenotypes of some strains were impacted more by fat and calories in the diet, while others were not. Gene expression differences have been connected to these differential responses (Reed et al. 2014). In *D. melanogaster*, a recent study of genetic variation across populations demonstrated latitudinal variation in 65% of 127 metabolism-associated genetic variants tested, implicating genes in the upper glycolytic pathway, its pentose shunt, glycerol shuttling, and glutamate/pyruvate metabolism as locally adaptive along a latitudinal cline (Lavington et al. 2014). Thus, variation in diet is expected to differentially impact members of a species and is likely in some cases, to impact evolutionary differentiation across space.

Since carrion is a nutritionally heterogeneous resource, such dynamics are likely important to the evolution of carrion-feeding organisms. For instance, the chemical compositions of soils under decomposing remains differ based on the removal of organs due to autopsy (Aitkenhead-Peterson et al. 2012). Such variation is likely to impact the biology of carrion-feeding organisms. For instance, carrion-feeding blow flies can exhibit diet-dependent phenotypic responses when fed different tissues (Kaneshrajah and Turner 2004; Clark et al. 2006; Ireland and Turner 2006; Boatright and Tomberlin 2010; Flores et al. 2014). These flies will grow at different rates when they are fed different tissues (e.g., heart, liver, etc.). They will also develop differently if there are fed the same tissue derived from different animals and when fed the same tissue from the same organism when it is allowed to desiccate at different rates. Recent studies also indicate that the immature development rates of some blow flies may respond more to variation in food source, whereas others appear more sensitive to temperature (Boatright and Tomberlin 2010; Flores et al. 2014). Therefore, nutritionally derived genotype–environment interactions are likely to occur in the carrion system.

As noted above, carrion-feeding organisms must also deal with thermal challenges (Denlinger 1972; Henrich and Denlinger 1982; Turner and Howard 1992; Slone and Gruner 2007; Rinehart et al. 2010; Charabidze et al. 2011; Rivers and Dahlem 2013; Johnson and Wallman 2014). There is much that can be learned about thermal responses in the *Drosophila* system as the literature is replete with studies of potential thermal adaptations (Burnell et al. 1991; James et al. 1997; Hercus et al. 2000; Hoffmann et al. 2003; Magiafoglou and Hoffmann 2003; Loeschcke and Hoffmann 2007; Overgaard et al. 2011a,b, 2014; Kristensen et al. 2012; Telonis-Scott et al. 2013; Parkash et al. 2014; Sillero et al. 2014). As one example, Gibert et al. (2004) evaluated populations of *D. simulans* and *D. melanogaster*, two sister species with similar geographical distributions that are most successful at different thermal exposures. While they both exhibit similar types of responses to thermal environments, they differ in the degree to which they respond to temperature. It is also clear that a variety of phenotypes from each species (and populations within the species) exhibit distinct thermal responses. These data are suggestive of the two species occupying different thermal niches that partially explain their differential success and coexistence. Other experiments have been conducted with various comparisons of conspecific and heterospecific populations, finding extensive evidence of temperature–genotype interactions in several *Drosophila* species (Gibert et al. 1996; Morin et al. 1996, 1997, 1999; David et al. 1997, 2004; Vieira et al. 2000; Norry and Loeschcke 2002a,b; Ayrinhac et al. 2004; Trotta et al. 2006; Cooper et al. 2012; Ghosh et al. 2013).

All organisms must manage a host of environmental challenges to survive in their environments. Genetics and the environment, as well as their interactions, can impact fitness through a variety of mechanisms. Ecological genetic studies allow researchers to dissect the evolutionary and ecological challenges these organisms face. This approach has been successfully used in a number of biological systems to dissect local and global adaptations and can inform studies of carrion biology.

14.4 Ecological Genetic Studies Most Relevant to the Carrion System

14.4.1 Introduction

Carrion does not decompose in a vacuum—it decomposes in an ecosystem (Tomberlin et al. 2011). That ecosystem is comprised of numerous biotic and abiotic components that will impact the evolution of the organisms that utilize carrion as a resource. There are relatively few studies of carrion that specifically address the evolutionary ecology of the system raising the question: What are typical evolutionary challenges that might be encountered by a carrion-feeding organism?

To answer this question, it is useful to evaluate the typical conditions of carrion itself. When an organism dies, the process of decomposition that results in the ultimate recycling of its nutrients into the ecosystem is initiated. This means that through autolysis and putrefaction, the remains will begin to break down from their complex structures of biological origin (carbohydrates, proteins, nucleic acids, etc.) into simpler compounds derived from their core elements (carbon, nitrogen, sulfur, phosphorus, etc.) such as butyric acid, putrescine, dimethyl disulfide, and ammonia. The details of this process can be found in Evans (1963), Forbes et al. (2002, 2003), Vass et al. (2002), Dent et al. (2004), Statheropoulos et al. (2005), Carter et al. (2007), Benninger et al. (2008), Aitkenhead-Peterson et al. (2012), and Rivers and Dahlem (2013). In addition, animals are organisms that are comprised mainly of water, which is maintained in the living animal in part by skin or cuticle (in an arthropod). In terrestrial systems, decomposition of carrion is associated with water loss through the various orifices in the body and perforations of skin. Water loss and moisture are important factors during decomposition (Payne 1965; Carter et al. 2007, 2010). Thus, during terrestrial decomposition, biotic and abiotic, chemical and physical decomposition of remains typically co-occur with dehydration of remains.

Essentially, at its core decomposition is a multifaceted chemical reaction that reduces complex once-living tissues into simpler components. Like all chemical reactions, there are several obvious conditions that will impact the progress of the reaction. For instance, temperature—considered above as an environmental factor important to local adaptation—is a key feature of any chemical reaction and can drastically impact the rate at which a chemical reaction progresses. Therefore, it should not be surprising that carrion will decompose at a slower rate in cooler temperatures. Furthermore, many chemical reactions are impacted by the availability of water and/or oxygen. Some ecosystems or local climates may impact the availability of water or oxygen, again altering the trajectory of decomposition. For example, mummies are often formed in dry, hot environments. For these reasons, an understanding of thermal biology from both a molecular and ecological perspective can aid in understanding the pressures that carrion-feeding organisms may encounter.

Numerous different biological components are associated with decomposing remains. Microbes sensitive to different environmental insults drive putrefaction of carrion, and certain communities can push the trajectory of decomposition toward different end points (mummification, saponification, etc.) depending on conditions (see Chapters 2, 3, 18, and 19). Further, microbes produce various by-products that are toxic to different other consumers of carrion. As an example, have you ever known someone to walk by a dead animal and seriously say "I'm hungry?" Did you ever wonder why that is? One answer to that question is that we have adapted to detecting and avoiding the microbial spoiling of foods that would be harmful to us. This hypothesis was raised by Janzen (1977), who posited that microbes spoil food in part to avoid competition with larger consumers. Recently, evidence of such a competition in a carrion system has been demonstrated for consumers of fish carrion, where microbial communities deterred feeding by other carrion consumers (Burkepile et al. 2006). A perusal of Chapter 21 seems to support this hypothesis as well, with different carrion consumers, which rely on carrion to varying degrees, evolving varying degrees of resistance to microbes. Of the mammals that feed on carrion in Africa, only the hyena, renowned for its tough stomach, will eat remains in practically any stage of decomposition.

Hyenas are not the only carrion consumers that can manage microbe-laden foods. One form of medical research is oriented toward the biology of interactions between carrion consumers and microbes (Sherman et al. 2000; Sherman 2003; Cazander et al. 2010, 2013). The research is centered on the fact that blow flies feed on remains when they are wet and covered in bacteria; and a better understanding of how blow flies interact with bacteria would help in dissecting how blow flies are effective in debriding wounds in a practice called maggot therapy. As it turns out, several insect taxa that utilize carrion are known for their antibiotic properties (Solter et al. 1989; Hoback et al. 2004; Rozen et al. 2008; Arce et al. 2012; Cazander et al. 2013; Thompson et al. 2013). Taken together, these examples indicate that if an organism evolves to use carrion as a primary food source, it will require a good immune system, and presumably the ability of blow flies to debride wounds is a consequence of that immune system.

Decomposing remains are also a potentially toxic resource independent of microbial toxins. It is well known in food-web ecology that toxins will bioaccumulate in organisms that occupy higher levels of the food web (Ahel et al. 1993; Cabana and Rasmussen 1994; Cabana et al. 1994; Morel et al. 1998; van der Oost et al. 2003; Wang and Wong 2003). Interestingly, while apex predators are often considered at the top of a food web, there are many organisms that will eat those apex predators–when they die. On some level, this means that carrion consumers are even more susceptible to bioaccumulation. As an example, California Condors were endangered, in part, due to the ingestion of toxic gunshot encountered in carrion (Cade 2007; Rideout et al. 2012). Toxins are similarly implicated in the deaths of many other carrion-feeding vertebrates (Fisher et al. 2006; Jung et al. 2009; Ogada et al. 2012). A natural example of this process can be found in insects feeding on fish runs (see Chapters 12 and 22). The fish accumulate mercury as oceanic predators. Then, en masse, they swim up inland streams to spawn before dying. This represents a massive influx of carrion into the ecosystem and with it—mercury (Sarica et al. 2005a,b). Carrion-feeding insects use these remains as a resource and are demonstrated to accumulate mercury as they develop. However, in one fly feeding on mercury-contaminated fish, the lowest levels of mercury is found in eggs, with higher levels found in each successive developmental stage. Interestingly, after eclosion, these same flies release a meconium and mercury levels in the adult drops (Nuorteva and Hasanen 1972; Nuorteva and Nuorteva 1982). For a predator of blow fly larvae, this accumulated mercury can be lethal (Nuorteva and Nuorteva 1982). Similar drops in toxin levels have been observed in another carrion-feeding fly with cadmium (Aoki and Suzuki 1984; Aoki et al. 1984). Avoidance of bioaccumulation appears to be toxin-specific (Ferhat et al. 2010) and represents a striking example of how a carrion consumer may encounter and avoid the effects of a toxin in its environment.

As the above examples demonstrate, carrion consumers face a number of abiotic and biotic challenges. Evolutionary ecology research has addressed many of these challenges from a basic biological perspective that could inform the numerous applications of carrion ecology. An understanding of what is known about other systems is instructive in evaluating the challenges faced by carrion feeding organisms and the potential solutions that they may employ to engage in their useful and unappealing lifestyles.

14.4.2 Abiotic Components of the Environment

The previous section has established and outlined the basic importance of several abiotic components of carrion systems that are important to the fitness of carrion-feeding organisms. This section will explore more detailed aspects of what is known in these systems, especially as they are related to these systems. This section is intended to demonstrate the specific molecular and genetic adaptations organisms are known to employ to survive these abiotic challenges.

14.4.2.1 Temperature

Temperature is a common environmental stressor that all organisms must manage which often dictates species distributions and can impact numerous organismal traits, including life-history traits. Accordingly, there has been a great deal of research interest paid to the ecology of thermal biology (for reviews on this topic, see Cossins and Bowler 1987; Angilletta 2009; Tattersall et al. 2012). Temperature is also a commonly acknowledged factor influencing the progression of decomposition of remains. Naturally formed mummies are examples of this phenomenon, where extreme temperatures and low humidity results in very slow rates of decomposition. Under more typical temperate conditions, it is well known that temperature increases the rate of decomposition, such that forensic anthropologists use thermal summation models to predict how long human remains have been at a death scene (Megyesi et al. 2005). Temperature is therefore expected to impact all members of the necrobiome, from the microbial to the vertebrate components.

At the microbial scale, there is a great deal of concern regarding the effects of temperature on changes in microbially mediated functions, most specifically related to the release of carbon dioxide from soil (Russell 1984; Suutari and Laakso 1994; Bradford et al. 2008; Allison et al. 2010; Wallenstein and Hall 2012; Bradford 2013) which is essentially a decomposition process. Though much consideration is centered on whether climate change is expected to impact the release of carbon from soils, the concepts considered in this debate are useful for consideration in a carrion decomposition context and are relevant to prokaryotic and eukaryotic cells. Essentially, all organismal function requires enzymes. In order for these enzymes to perform their functions, they must be able to effectively interact with their target substrates. This process is greatly impacted by thermal dynamics as a consequence of the basic physics of chemical reactions. Accordingly, there is a trade-off between enzymatic function in low and high temperatures, where enzymes that function in relatively higher temperatures must be relatively structurally rigid and enzymes required to function in relatively lower temperatures must be relatively loose or flexible (Somero 1978, 1995; Zavodszky et al. 1998). This dynamic results in the trade-off organisms face with respect to temperature, as enzymatic structures cannot be both rigid and flexible. As a result, there are known amino acid biases that are expected in cold-adapted versus warm-adapted organisms, as some amino acids allow more rotation or rigidity of polypeptides than others (Vieille and Zeikus 1996, 2001; Demirjian et al. 2001; Fields 2001). For instance, the simple structure of glycine allows for a considerable degree of rotational flexibility in a polypeptide, which is generally favored in enzymes adapted to cooler environments.

In addition, prokaryotic and eukaryotic cells produce "heat shock proteins" and other metabolites that are generally responsible for stabilizing protein structures in stressful environments (Kusukawa and Yura 1988; Li et al. 1992; Jakob et al. 1993; Feder et al. 1997; Krebs and Feder 1997; Feder and Hofmann 1999; Tattersall et al. 2012). In some of these studies, variation in heat shock protein expression has been linked to thermal adaptation. Cell membranes also must differ across a range of thermal (and other stressful) environments, with different components required at different temperatures to achieve ideal membrane function (Wodtke 1981; Hazel et al. 1991, 1992; Hazel 1995; Konings et al. 2002; Michaud and Denlinger 2006). The trade-off in enzyme and cellular stability that is in part mediated by amino acid composition, additional structural effects in proteins, maintenance of protein folding, and temperature-related membrane composition results in thermal specialist species, which are highly efficient at specific temperatures, and thermal generalist species that can function at relatively broader temperature ranges but will be less efficient across that range of temperatures when compared to a specialist in its preferred environment.

Much of the current consideration of thermal microbiology in the literature that is most relevant to carrion ecology is centered on the specifics of the microbial response to thermal shifts in the soil as previously mentioned, but for the purposes of carrion ecology it is instructive to note that alleles of genes expected to function well in decomposition at low temperatures are not likely to function well at high temperatures and vice versa. Thus, the microbial taxa and enzymes employed in carrion decomposition are also likely to differentially evolve across thermal environments.

Thermal adaptations are not limited to bacteria. There are a host of adaptations—physiological, behavioral, and structural—that are associated with specific thermal environments in eukaryotes (Angilletta 2009; Tattersall et al. 2012). While eukaryotes are not exempt from the limitations of proteins as noted for microbes, they exhibit additional adaptations to thermal variation. Carrion-feeding flies demonstrate some good examples of thermal challenges faced when feeding on carrion (these are reviewed in Byrd and Castner 2009; Rivers and Dahlem 2013). These flies do not regulate their temperatures metabolically. However, their larval aggregates create heat, allowing for the possibility of behavioral management of temperature (Cianci and Sheldon 1990; Catts 1992; Huntington et al. 2007; Slone and Gruner 2007; Charabidze et al. 2011; Rivers et al. 2011). In larval aggregations, these masses can create enough heat to raise temperatures above ambient (and potentially up to lethal) temperatures. They must also survive in the cold, thus they express heat shock proteins and express anti-freeze proteins or other metabolites to limit the damaging effects of stressful low temperatures (Lee et al. 1987a,b;

Joplin et al. 1990; Tachibana et al. 2005), which is a feature shared with other cold-tolerant insects (Ring 1982). In addition, they can use environmental cues to survive the winter through diapause, which requires a number of metabolic adjustments that are well known in carrion-feeding *Sarcophaga* (Diptera: Sarcophagidae) species (Fraenkel and Hsiao 1968; Denlinger 1972; Adedokun and Denlinger 1985; Michaud and Denlinger 2006, 2007; Fraenkel and Hsiao 1968; Michaud and Denlinger 2007). It should then not be surprising that some carrion-feeding flies are associated with distinct locations or seasons. For example, some *Calliphora* species are associated with cooler climates, seasons, and conditions depending on the geographic location (Greenberg and Tantawi 1993; Faucherre et al. 1999). In addition, recent experiments indicate the potential for local adaptation of blow flies, with respect to temperature, as there are indications of temperature–genotype interactions (Tarone et al. 2011; Owings et al. 2014).

There are other adaptations that are generally associated with extreme thermal environments (these are well reviewed in Tattersall et al. 2012). Warmer climates require adaptations for releasing excess heat. These can include mechanisms for altering body temperature (e.g., thermoregulatory thermogenesis or hyperthermia), enhancing evaporative heat loss (e.g., sweating or panting), or changing locations to enhance heat loss through conductance of heat (e.g., lying in the shade or a pond). These behaviors can result in alterations in water requirements or preferences in location, which can conceivably influence patterns in carrion consumption by carrion-feeding organisms.

Colder environments require a different set of adaptations. These can include mechanisms for maintaining heat (e.g., insulation via feathers, fur, or fat), slowing metabolism (e.g., hibernation or diapause), production of antifreeze metabolites (e.g., glycogol or chaotropic salts), or heat production (e.g., shivering). Many of these adaptations are energetically expensive. For example, storage of lipids to survive hibernation or diapause and/or to produce a fatty layer of insulation requires a considerable energetic input. Given that carrion is a high-nutrient resource, it is quite possible that the energetic requirements of cold adaptations impact the value of carrion to cold-adapted organisms or populations.

14.4.2.2 Desiccation Resistance

Just as different ecosystems vary with respect to temperature, there are environments that vary in the amount of water available to the biotic component of the ecosystem. Given the importance of water to life and many biological functions (including some thermal adaptations noted above), this ecological challenge can impact evolution. Desiccation resistance has been studied in the *Drosophila* system (Karan et al. 1998b; Hoffmann and Harshman 1999). This can theoretically happen by increasing water content, decreasing rates of water loss, or increasing tolerance to water loss (Bazinet et al. 2010). In *D. melanogaster*, rates of water loss appear to be a result of decreasing water loss as a function of alterations to cuticular permeability. For instance, cuticular hydrocarbon profiles result in longer chains of hydrocarbons after 100 generations of selection for desiccation resistance (Gibbs et al. 1997). Thus, cuticular hydrocarbon profiles have been demonstrated to vary across regions in *Drosophila* species and that some of these alterations are associated with desiccation resistance (Hoffmann et al. 2003; Frentiu and Chenoweth 2010). However, in other research, hemolymph volume and correlated shifts in carbohydrates may also be important for *Drosophila* desiccation resistance (Folk and Bradley 2001; Folk et al. 2001; Archer et al. 2007). Thus, desiccation responses do not appear to follow a single strategy and can be selected for as a consequence of selection for other traits like longevity (Rose 1989). Currently, little is known about the specifics of the genetic responses studied, but a genetic study of desiccation resistant lines was able to link resistance to the first and third chromosomes of *D. melanogaster* (Telonis-Scott et al. 2006) and some mapping and functional genetic experiments have begun to approach this question (Sinclair et al. 2007; Foley and Telonis-Scott 2011; Telonis-Scott et al. 2012). Interestingly, the trait appears to negatively correlate with starvation resistance, making it difficult to understand which phenotype is the ultimate target of selection (Karan et al. 1998b). Given that carrion is a potential source of water, it is very likely that it is an important source of water for carrion-feeding organisms in arid environments. The carrion system is thus poised to contribute to the unresolved issues apparent in the desiccation resistance literature.

14.4.3 Biotic Components of the Environment

The previous section considered adaptations organisms typically use to survive abiotic challenges in their environments. This section will similarly consider the adaptations organisms use to deal with biotic challenges in their environments. Several of these factors have been discussed above, but here they are considered in more detail as they relate to the challenges encountered by carrion-feeding organisms.

14.4.3.1 Density Effects

One common feature of carrion systems that is obvious with a quick review of this book is variability in density, both of the availability of a resource and in the availability of consumers of a carrion resource. Typically, the high nutrient value of carrion can lead it to result in high densities of consumers competing for that resource. Density effects are a great example of a biologically derived environmental impact on the evolution of a system. From a very broad perspective, there is a considerable amount of research devoted to density-dependent factors in an ecosystem that limit demographic expansions of all sorts of species. However, more targeted examples of this are useful to illustrate how this may affect carrion biology.

One of the more obvious concepts relevant to carrion biology is that of the Allee effect (Allee 1927; Allee and Bowen 1932; Courchamp et al. 1999; Stephens and Sutherland 1999; Stephens et al. 1999; Dennis 2002; Kramer et al. 2009). This effect is represented by a correlation between population size and fitness of organisms in low to intermediate densities (at some point, high densities are always expected to decrease fitness). Typically, the effect is considered to manifest such that small groups of individuals fare better than individuals would on their own. For example, Allee invoked this mechanism in explaining resistance to desiccation in isopods (Isopoda) (Allee 1927). Aggregation (nonrandom clustering of individuals) is often studied in ecology, and its relevance to carrion ecology has been addressed in several instances (Prinkkila and Hanski 1995; Rivers et al. 2011; Barton et al. 2013; Boulay et al. 2013; Brodie et al. 2015; Fiene et al. 2014).

Aggregation may be important in carrion biology for several reasons. First, it may help to dilute toxins. For goldfish, Allee suggested that aggregation limited toxic effects of colloidal silver, as the fish grew more rapidly when there were more individuals in the tank (Allee and Bowen 1932). Second, it may impact predator–prey interactions as the ability to avoid predators varies depending on the densities of prey and predators. Third, it may improve the ability of some organisms to feed on carrion and compete with con- or hetero-specifics. As an example, many larvae of the flies that utilize carrion engage in extra-oral digestion. Their excretions and secretions have biochemical properties and can moisten the carrion in large numbers, whereas in smaller numbers larvae are unable to keep their food moist, and drier food treatments have been shown to slow the development of a carrion-feeding fly (Tarone and Foran 2006). This effect of dry food is not surprising, as maceration by larval mouth parts and the enzymes excreted and secreted onto food will be limited by a dry resource.

Hydra effects (Abrams and Matsuda 2005), which are related to Allee effects, are also often connected to density variation in biological systems. Similar to the mythological Hydra, that grew back two heads when one was cut off, biological Hydra effects result in a non-intuitive population increase as a result of mortality increases. Such effects are expected in predator–prey systems (Sieber and Hilker 2012). However, there are other systems that may demonstrate Hydra effects or the potential for Hydra effects. There are several hypotheses for the features of systems that are expected to demonstrate these types of response, including population fluctuations, species interactions, temporal dependence of mortality effects, or mortality associated changes in resource exploitation (Abrams 2009). One of the earliest examples in support of Hydra effects is in a carrion-associated species. Nicholson (1950, 1954, 1957) demonstrated in a series of papers that the carrion-feeding blow fly *Lucilia cuprina* (Wiedemann) exhibits density-dependent compensatory responses. In his experiments, populations of *L. cuprina* were exposed to various environmental conditions. He observed that, as adult mortality increased, the highest survivorship of immature flies occurred, since high densities of adults resulted in high competition among larvae for resources. As adult populations began to die off, he observed the greatest peaks on production of immatures, resulting in an inverse relationship between immature and adult densities and repeatable oscillations in the representations of various life-history stages. In general, he concluded that removal of any life-history stage from the experiments resulted in a rapid replacement of that stage, suggesting compensatory responses of the flies to alterations in densities. The concept of compensation was expanded more recently to another member of the genus, *Lucilia sericata* (Meigen), in a study of cadmium toxin dependent compensation (Moe et al. 2002). Recently, the effects of copper toxicity on the blow fly *Protophormia terraenovae* (Robineau-Desvoidy) have been evaluated with respect to its interaction with temperature (Polkki et al. 2014), demonstrating the potential for studying numerous aspects of compensatory responses. There is currently a renewed interest in these types of systems, focused primarily on identifying examples that support or refute specific hypotheses about how Hydra effects evolve or operate. The carrion system appears to harbor some species that could provide further clarification on the evolution and dynamics of Hydra effects, especially as they relate to toxin and thermal biology.

The impact of density effects on the evolution of life-history traits is also well studied. The *Drosophila* model has provided some valuable insights into how population densities impact the evolution of population stability, stress resistance, energy storage, foraging behavior, and other components of fitness (Frahm and Kojima 1966; Guo et al. 1991; Mueller et al. 1991a,b, 1993; Graves and Mueller 1993; Santos et al. 1997; Sokolowski et al. 1997; Fitzpatrick et al. 2007; Shingleton et al. 2009; Dey et al. 2012; Mueller and Cabral 2012; Vijendravarma et al. 2012a,b). Typically, experiments evaluating the effects of population density have manipulated either adult or larval density in selection experiments and then compared the life-history traits of populations exposed to different densities. One interesting result from these experiments is the increase in "rover" alleles of the *foraging* locus when population densities are high and increases in "sitter" alleles when densities are low (Sokolowski et al. 1997; Fitzpatrick et al. 2007; Vijendravarma et al. 2012b). Interestingly, there is an observation that some density-associated trait correlations decay over time, suggesting differences in long-term and short-term evolution in high-density populations (Phelan et al. 2003). While it is worth noting that there are limitations to such studies (Burke and Rose 2009), generally there are indications that selection for survival in different densities impacts numerous fitness-related traits in a complex manner. Given that the carrion system is rife with density variation, it is logically a system that could contribute to our understanding of the evolutionary ecology of density-dependent fitness effects.

14.4.3.2 Toxin Resistance

As noted previously, microbes have been hypothesized to compete with eukaryotes for carrion and other decomposing resources (Janzen 1977). There is some evidence to support this contention, as microbe-laden fish are consumed less by scavengers than those that are not inoculated with microbes (Burkepile et al. 2006). Recent studies of the microbial communities associated with carrion would suggest that there is ample opportunity for such competition to occur (Hyde et al. 2013; Metcalf et al. 2013; Can et al. 2014; Finley et al. 2015; Pechal et al. 2014). Other ecological interactions, such as microbial facilitation of certain consumers, are also likely. This contention regarding microbial facilitation or deterrence of consumers appears intuitively. Most people have encountered an unpalatable food overrun by microbial communities, which is not a response unique to humans but is common to African carrion consumers as discussed in Chapter 21 and reflects one of several strategies for avoiding deleterious effects of toxic foods (Glendinning 2007). One of the major means by which potential consumers are deterred from any resource is through chemical defense (e.g., plants that produce toxic compounds to prevent herbivory). This is a critical feature of any biological system where one organism may consume another and consumers must develop strategies for utilizing chemically defended resources (reviewed in Glendinning 2007). The toxicity of carrion has already been addressed here in the context of bioaccumulation of toxic compounds, where certain organisms that consume carrion are at risk from such processes, while others are more resistant to the risk. However, another potential source of toxins derives from the microbes of the system as well.

A very classical example of this concept can be found in the *Drosophila* literature. These flies rely on yeasts in their food as a source of nutrients, including amino acids, which are uncommon in the rotting plant material they use as a resource (Delcourt and Guyenot 1910; Baumberger 1919). Yeasts are known to produce chemical by-products of decomposition, including alcohols through fermentation. *Drosophila* use *alcohol dehydrogenase* to help in degrading this toxic compound in their food. As with any chemical reaction, this is a temperature-dependent reaction, thus there is no expectation that: (1) alcohols will be encountered at the same rates in foods in different thermal environments or that (2) alleles that function well in colder environments are equally good at processing alcohol in warmer environments (Vigue and Johnson 1973; Vigue and Sofer 1974; Vigue et al. 1982; Barker et al. 1990; Heinstra 1993; Powell 1997). One of the oldest examples of ecological genetic differentiation in *Drosophila* is of phenotypic variation in alcohol tolerance and of alleles of *alcohol dehydrogenase* that segregate along latitudinal clines or under differential selective pressures (Oakeshott et al. 1982a,b, 1985; Cohan and Graf 1985; Cohan and Hoffmann 1986; Geer et al. 1989, 1993; Barker et al. 1990; Berry and Kreitman 1993; Heinstra 1993; Powell 1997). Alleles of this locus are also demonstrated to have differential distributions across ecological zones, including regions with different host plants (Brussard et al. 1978) and locations that are close to wineries (McKechnie and Geer 1993).

Yeasts are not just a source of toxins for *Drosophila* species either. Work with cactophilic species, which require yeasts in cactus rots to colonize toxic plants, are known to perform better on their hosts when a greater diversity of yeast species are present in the rot (Starmer and Aberdeen 1990). This has been attributed, in part, to the ability of these yeasts to detoxify foods for the flies (Barker et al. 1982; Fogleman et al. 1982; Starmer 1982; Starmer et al. 1986; Starmer and Fogleman 1986). Thus, microbes associated with *Drosophila* food are both a source of toxins and detoxification.

While the *Drosophila* system differs from carrion in the types of microbes involved, other components of these systems are comparable. Both involve microbes that can be a source of toxicity. Any organism using carrion will need to overcome the toxins associated with microbes growing on carrion. In addition, these microbes may also be capable of detoxifying the remains. However, the carrion system is replete with amino acids, which are relatively rare in *Drosophila* foods, meaning that the specific toxins produced by carrion will differ greatly from that system and will provide a useful point of comparison in evaluating these microbial challenges to utilizing a rotting resource.

14.4.3.3 Tradeoffs Associated with Pathogen Resistance

The Hyena example in Chapter 21 highlights how some carrion consumers can dominate a resource by investing heavily in immune functions. But that example also raises the question: Why have other similar organisms not adopted the Hyena strategy to manage microbial populations associated with carrion? The answer to that question likely lies in the fact that immunity is connected to a number of other trade-offs with other fitness components, making it difficult to be highly immune and fecund and long-lived (Sheldon and Verhulst 1996; Verhulst et al. 1997; Adamo et al. 2001; McKean and Nunney 2001, 2008; Wikelski and Ricklefs 2001; Zera and Harshman 2001; DeVeale et al. 2004; Fedorka et al. 2004; Jacot et al. 2004; Ahtiainen et al. 2005; Ardia 2005; Bauer et al. 2006; Harshman and Zera 2007; May 2007; Lazzaro et al. 2008; Lazzaro and Little 2009; van der Most et al. 2011).

One of the most important components of fitness is the ability to reproduce. There is a clear connection between sexual reproduction and immunity (Sheldon and Verhulst 1996; Verhulst et al. 1997; Adamo et al. 2001; McKean and Nunney 2001; Westendorp et al. 2001; Ahtiainen et al. 2005; Ardia 2005; Lazzaro et al. 2008; van der Most et al. 2011). For instance, in mammals, maternal immune responses can result in miscarriages (Moffett and Loke 2006a,b; Arck et al. 2008; Robertson 2010). In *D. melanogaster*, the act of sex itself has been demonstrated to depress immune function such that mating will result in the inability to fight an infection that a virgin female from the same strain would be able to overcome (Short and Lazzaro 2010; Short et al. 2012). The details of this effect appear to lie in changes in female physiology associated with mating itself, in response to, and independent of, the influence of male ejaculate. Male immunity has also been shown to be reduced by increased mating (McKean and Nunney 2001). Overall, it is clear that reproduction and immunity are sometimes incompatible strategies.

Immunity is also connected to a trade-off in longevity (DeVeale et al. 2004; Jacot et al. 2004; May 2007), though this seems to be less studied than other trade-offs. In general, antimicrobial responses are costly to employ and stressful on eukaryotic systems. These systems have evolved to recognize infection before employing costly antimicrobial measures. In fact, aging is associated with shifts from adaptive (conditionally induced) to innate (constitutively induced) immune responses (DeVeale et al. 2004). Thus, it is not surprising that a genotype that regularly initiates antimicrobial defenses may also exhibit a shorter lifespan. The carrion system is ideal for evaluating this interaction between immunity and longevity, as the system includes consumers of carrion that employ a range of immune system activities. The longevities of members of the guild across the range of immune activities as compared to their non-carrion feeding relatives could provide valuable information regarding the relationships between these traits in high nutrient conditions.

There is also a known trade-off between longevity and reproduction where high reproductive output typically negatively correlates with longevity (Rose 1989; Kaitala 1991; Djawdan et al. 1996; Westendorp and Kirkwood 1998; Westendorp et al. 2001; Partridge et al. 2005; Bauer et al. 2006; Harshman and Zera 2007; Flatt et al. 2008; Toivonen and Partridge 2009; Kenyon 2010). This trade-off is well understood, but complex, with a number of hypotheses developed from different perspectives to explain why the trade-off exists (Westendorp et al. 2001; Harshman and Zera 2007; Lee et al. 2008; Toivonen and Partridge 2009; Kenyon 2010). One explanation from nutritional biology suggests that high-protein diets optimize reproduction, whereas low-protein diets optimize longevity (Lee et al. 2008; Maklakov et al. 2008). Another interesting aspect of the trade-off is studied in social Hymenoptera, where longer-lived castes are also the reproductives. Interestingly, vitellogenins (major components of the egg contents) have been linked to this differential in reproduction and survival, and vitellogenins have been implicated in increased resistance to oxidative stress and increase immunity (Amdam et al. 2004; Seehuus et al. 2006; Corona et al. 2007).

In *D. melanogaster*, variation in yolk protein expression has been linked to longevity as well (Tarone et al. 2012). Females producing more yolk protein than males and live longer than males. However, within a sex, long life is associated with expression of less yolk protein such that females and males expressing the least yolk protein live the longest (Tarone et al. 2012). The yolk proteins are distinct from but evolutionarily convergent with vitellogenins (Brandt et al. 2005; Tufail and Takeda 2008, 2009) and have similar impacts on reproductive success (Bownes et al. 1991; Terashima and Bownes 2004). Taken together, the *D. melanogaster* sexes appear to respond to yolk protein expression in a manner similar to Hymenopteran castes, where more yolk protein is associated with longer life. However, their within-sex responses of longevity to yolk protein levels are counter to this pattern and remain be explained. Clearly, there is a complex connection between reproduction, immunity, and longevity with many open questions regarding the exact nature of the interactions among these fitness components. Carrion biology, with a potentially large and diverse microbial community and a distinct nutritional profile, is a potentially fertile testing ground for many of the hypotheses related to these tradeoffs.

14.5 Conclusion

The goal for this chapter was to introduce the reader to a general and nonexhaustive set of concepts considered in ecological genetics with the intention of stimulating a consideration and appreciation of these and similar concepts in future studies of carrion biology. All organisms encounter specific challenges to surviving in any environment and those that feed on carrion are not exceptions. The field of carrion biology is relatively understudied in this context, even though a basic understanding of the evolutionary principles important to the system will be critical for the applications of carrion biology discussed at the end of this book. Fortunately, dung biology has been studied from this perspective for many years and the next chapter will cover these topics from the perspective of that system, as interpreted by a leader in that field. Chapter 16 will present issues related to the evolution of plant manipulation of dung and carrion insects, which exhibits some generalizable features. Chapter 17 will evaluate what is known about carrion-consuming insects and serves as a starting point for future studies of the genetics and evolution of carrion insects. The final chapters consider the microbial and vertebrate aspects of the evolutionary ecology and ecological genetics of carrion biology.

Acknowledgments

I would like to thank Texas A&M University, the Department of Entomology, our publisher, and my co-editors for the support that enabled the birth of this chapter and the entire book. The data for Figure 14.3 was funded by grant number 2012-DN-BX-K024 from the National Institute of Justice to Aaron Tarone, Christine Picard, and Sing-Hoi Sze. Points of view in this document are mine and do not necessarily represent the official position or policies of the U.S. Department of Justice or U.S. Government. I offer a special thanks to my fellow editors and all the authors who were willing to contribute chapters to this book, especially this section. Thanks to my colleagues at the North American Forensic Entomology Association for helping to push the boundaries of research in carrion biology. I also want to thank Bob Kimsey at UC Davis for introducing me to the carrion system, Sergey Nuzhdin at USC for introducing me to *Drosophila* evolutionary ecology, and J. Spencer Johnston at Texas A&M for providing a continued valuable perspective on these topics as well as commentary on the chapter. Finally, and most importantly, I want to thank Lauren Kalns for her continued love and support, especially during the arduous process of editing a book at all hours of the day, and for her contribution to our own personal experiment in evolutionary fitness.

REFERENCES

Abrams, P.A. 2009. When does greater mortality increase population size? The long history and diverse mechanisms underlying the hydra effect. *Ecology Letters* 12: 462–474.

Abrams, P.A. and H. Matsuda. 2005. The effect of adaptive change in the prey on the dynamics of an exploited predator population. *Canadian Journal of Fisheries and Aquatic Sciences* 62: 758–766.

Adamo, S.A., M. Jensen, and M. Younger. 2001. Changes in lifetime immunocompetence in male and female *Gryllus texensis* (formerly *G. integer*): Trade-offs between immunity and reproduction. *Animal Behaviour* 62: 417–425.

Adedokun, T.A. and D.L. Denlinger. 1985. Metabolic reserves associated with pupal diapause in the flesh fly, *Sarcophaga crassipalpis*. *Journal of Insect Physiology* 31: 229–233.

Ahel, M., J. Mcevoy, and W. Giger. 1993. Bioaccumulation of the lipophilic metabolites of nonionic surfactants in fresh-water organisms. *Environmental Pollution* 79: 243–248.

Ahtiainen, J.J., R.V. Alatalo, R. Kortet, and M.J. Rantala. 2005. A trade-off between sexual signalling and immune function in a natural population of the drumming wolf spider *Hygrolycosa rubrofasciata*. *Journal of Evolutionary Biology* 18: 985–991.

Aitkenhead-Peterson, J.A., C.G. Owings, M.B. Alexander, N. Larison, and J.A. Bytheway. 2012. Mapping the lateral extent of human cadaver decomposition with soil chemistry. *Forensic Science International* 216: 127–134.

Allee, W.C. 1927. Animal aggregations. *The Quarterly Review of Biology* 2: 367–398.

Allee, W.C. and E.S. Bowen. 1932. Studies in animal aggregations: Mass protection against colloidal silver among goldfishes. *Journal of Experimental Zoology* 61: 185–207.

Allison, S.D., M.D. Wallenstein, and M.A. Bradford. 2010. Soil-carbon response to warming dependent on microbial physiology. *Nature Geoscience* 3: 336–340.

Amdam, G.V., Z.L.P. Simoes, A. Hagen, K. Norberg, K. Schroder, O. Mikkelsen, T.B.L. Kirkwood, and S.W. Omholt. 2004. Hormonal control of the yolk precursor vitellogenin regulates immune function and longevity in honeybees. *Experimental Gerontology* 39: 767–773.

Anderson, W.W., J. Arnold, D.G. Baldwin, A.T. Beckenbach, C.J. Brown, S.H. Bryant, J.A. Coyne, L.G. Harshman, W.B. Heed, and D.E. Jeffery. 1991. Four decades of inversion polymorphism in *Drosophila pseudoobscura*. *Proceedings of the National Academy of Sciences* 88: 10367–10371.

Angilletta, M.J. 2009. *Thermal Adaptation: A Theoretical and Empirical Synthesis*. Oxford University Press, Oxford.

Anholt, R.R.H., J.J. Fanara, G.M. Fedorowicz, I. Ganguly, N.H. Kulkarni, T.F.C. Mackay, and S.M. Rollmann. 2001. Functional genomics of odor-guided behavior in *Drosophila melanogaster*. *Chemical Senses* 26: 215–221.

Aoki, Y. and K.T. Suzuki. 1984. Excretion of cadmium and change in the relative ratio of iso-cadmium-binding proteins during metamorphosis of fleshfly *Sarcophaga peregrina*. *Comparative Biochemistry and Physiology Part C* 78: 315–317.

Aoki, Y., K.T. Suzuki, and K. Kubota. 1984. Accumulation of cadmium and induction of its binding-protein in the digestive-tract of fleshfly *Sarcophaga peregrina* larvae. *Comparative Biochemistry and Physiology Part C* 77: 279–282.

Arce, A.N., P.R. Johnston, P.T. Smiseth, and D.E. Rozen. 2012. Mechanisms and fitness effects of antibacterial defences in a carrion beetle. *Journal of Evolutionary Biology* 25: 930–937.

Archer, M.A., T.J. Bradley, L.D. Mueller, and M.R. Rose. 2007. Using experimental evolution to study the physiological mechanisms of desiccation resistance in *Drosophila melanogaster*. *Physiological and Biochemical Zoology* 80: 386–398.

Arck, P.C., M. Rueckel, M. Rose, J. Szekeres-Bartho, A.J. Douglas, M. Pritsch, S.M. Blois et al. 2008. Early risk factors for miscarriage: A prospective cohort study in pregnant women. *Reproductive BioMedicine Online* 17: 101–113.

Ardia, D.R. 2005. Tree swallows trade off immune function and reproductive effort differently across their range. *Ecology* 86: 2040–2046.

Arnoult, L., K.F. Su, D. Manoel, C. Minervino, J. Magrina, N. Gompel, and B. Prud'homme. 2013. Emergence and diversification of fly pigmentation through evolution of a gene regulatory module. *Science* 339: 1423–1426.

Atwell, S., Y.S. Huang, B.J. Vilhjalmsson, G. Willems, M. Horton, Y. Li, D. Meng et al. 2010. Genome-wide association study of 107 phenotypes in *Arabidopsis thaliana* inbred lines. *Nature* 465: 627–631.

Ayala, D., R.F. Guerrero, and M. Kirkpatrick. 2013. Reproductive isolation and local adaptation quantified for a chromosome inversion in a malaria mosquito. *Evolution* 67: 946–958.

Ayrinhac, A., V. Debat, P. Gibert, A.G. Kister, H. Legout, B. Moreteau, R. Vergilino, and J.R. David. 2004. Cold adaptation in geographical populations of *Drosophila melanogaster*: Phenotypic plasticity is more important than genetic variability. *Functional Ecology* 18: 700–706.

Ayroles, J.F., M.A. Carbone, E.A. Stone, K.W. Jordan, R.F. Lyman, M.M. Magwire, S.M. Rollmann, L.H. Duncan, F. Lawrence, R.R. Anholt, and T.F.C. Mackay. 2009. Systems genetics of complex traits in *Drosophila melanogaster*. *Nature Genetics* 41: 299–307.

Barker, J.S.F., W.T. Starmer, and R.J. MacIntyre. 1990. *Ecological and Evolutionary Genetics of Drosophila*. Plenum Press, New York.

Barker, J.S.F., W.T. Starmer, and US/Australia Co-operative Science Program. 1982. *Ecological Genetics and Evolution: The Cactus–Yeast–Drosophila Model System*. Academic Press, Sydney.

Barton, P.S., S.A. Cunningham, D.B. Lindenmayer, and A.D. Manning. 2013. The role of carrion in maintaining biodiversity and ecological processes in terrestrial ecosystems. *Oecologia* 171: 761–772.

Bastide, H., A. Betancourt, V. Nolte, R. Tobler, P. Stobe, A. Futschik, and C. Schlotterer. 2013. A genome-wide, fine-scale map of natural pigmentation variation in *Drosophila melanogaster*. *PLoS Genetics* 9: e1003534.

Bastide, H., A. Yassin, E.J. Johanning, and J.E. Pool. 2014. Pigmentation in *Drosophila melanogaster* reaches its maximum in Ethiopia and correlates most strongly with ultra-violet radiation in sub-Saharan Africa. *BMC Evolutionary Biology* 14: 179.

Bauer, M., J.D. Katzenberger, A.C. Hamm, M. Bonaus, I. Zinke, J. Jaekel, and M.J. Pankratz. 2006. Purine and folate metabolism as a potential target of sex-specific nutrient allocation in *Drosophila* and its implication for lifespan-reproduction tradeoff. *Physiological Genomics* 25: 393–404.

Baumann, P., L. Baumann, C.Y. Lai, D. Roubakhsh, N.A. Moran, and M.A. Clark. 1995. Genetics, physiology, and evolutionary relationships of the genus *Buchnera*—Intracellular symbionts of aphids. *Annual Review of Microbiology* 49: 55–94.

Baumberger, J.P. 1919. A nutritional study of insects, with special reference to microorganisms and their sub-strata. *Journal of Experimental Zoology* 28: 1–81.

Bazinet, A.L., K.E. Marshall, H.A. MacMillan, C.M. Williams, and B.J. Sinclair. 2010. Rapid changes in desiccation resistance in *Drosophila melanogaster* are facilitated by changes in cuticular permeability. *Journal of Insect Physiology* 56: 2006–2012.

Benninger, L.A., D.O. Carter, and S.L. Forbes. 2008. The biochemical alteration of soil beneath a decomposing carcass. *Forensic Science International* 180: 70–75.

Berry, A. and M. Kreitman. 1993. Molecular analysis of an allozyme cline *alcohol dehydrogenase* in *Drosophila melanogaster* on the East Coast of North America. *Genetics* 134: 869–893.

Bickel, R.D., W.S. Schackwitz, L.A. Pennacchio, S.V. Nuzhdin, and A. Kopp. 2009. Contrasting patterns of sequence evolution at the functionally redundant *bric a brac* paralogs in *Drosophila melanogaster*. *Journal of Molecular Evolution* 69: 194–202.

Boatright, S.A. and J.K. Tomberlin. 2010. Effects of temperature and tissue type on the development of *Cochliomyia macellaria* (Diptera: Calliphoridae). *Journal of Medical Entomology* 47: 917–923.

Bochdanovits, Z. and G. De Jong. 2003. Temperature dependent larval resource allocation shaping adult body size in *Drosophila melanogaster*. *Journal of Evolutionary Biology* 16: 1159–1167.

Boulay, J., C. Devigne, D. Gosset, and D. Charabidze. 2013. Evidence of active aggregation behaviour in *Lucilia sericata* larvae and possible implication of a conspecific mark. *Animal Behaviour* 85: 1191–1197.

Bownes, M., K. Lineruth, and D. Mauchline. 1991. Egg-production and fertility in *Drosophila* depend upon the number of *Yolk protein* gene copies. *Molecular Genetics and Genomics* 228: 324–327.

Brachi, B., N. Faure, M. Horton, E. Flahauw, A. Vazquez, M. Nordborg, J. Bergelson, J. Cuguen, and F. Roux. 2010. Linkage and association mapping of *Arabidopsis thaliana* flowering time in nature. *PLoS Genetics* 6(5): e1000940.

Bradford, M.A. 2013. Thermal adaptation of decomposer communities in warming soils. *Frontiers in Microbiology* 4: 333.

Bradford, M.A., C.A. Davies, S.D. Frey, T.R. Maddox, J.M. Melillo, J.E. Mohan, J.F. Reynolds, K.K. Treseder, and M.D. Wallenstein. 2008. Thermal adaptation of soil microbial respiration to elevated temperature. *Ecology Letters* 11: 1316–1327.

Braendle, C., G.K. Davis, J.A. Brisson, and D.L. Stern. 2006. Wing dimorphism in aphids. *Heredity* 97: 192–199.

Brandt, B.W., B.J. Zwaan, M. Beekman, R.G.J. Westendorp, and P.E. Slagboom. 2005. Shuttling between species for pathways of lifespan regulation: A central role for the vitellogenin gene family? *Bioessays* 27: 339–346.

Brisson, J.A. 2010. Aphid wing dimorphisms: Linking environmental and genetic control of trait variation. *Philosophical Transactions of the Royal Society of London. Series B Biological Sciences* 365: 605–616.

Brisson, J.A., A. Ishikawa, and T. Miura. 2010. Wing development genes of the pea aphid and differential gene expression between winged and unwinged morphs. *Insect Molecular Biology* 19(Suppl 2): 63–73.

Brodie, B.S., W.H.L. Wong, S. VanLaerhoven, and G. Gries. 2015. Is aggregated oviposition by the blow flies *Lucilia sericata* and *Phormia regina* (Diptera: Calliphoridae) really pheromone-mediated? *Insect Science* In Press.

Brussard, P.F., R.W. Allard, and Society for the Study of Evolution. 1978. *Ecological Genetics: The Interface.* Springer-Verlag, New York.

Buri, P. 1956. Gene-frequency in small populations of mutant *Drosophila*. *Evolution* 10: 367–402.

Burke, M.K., J.P. Dunham, P. Shahrestani, K.R. Thornton, M.R. Rose, and A.D. Long. 2010. Genome-wide analysis of a long-term evolution experiment with *Drosophila*. *Nature* 467: 587–590.

Burke, M.K. and M.R. Rose. 2009. Experimental evolution with *Drosophila*. *American Journal of Physiology–Regulatory, Integrative and Comparative Physiology* I 296: R1847–R1854.

Burkepile, D.E., J.D. Parker, C.B. Woodson, H.J. Mills, J. Kubanek, P.A. Sobecky, and M.E. Hay. 2006. Chemically mediated competition between microbes and animals: Microbes as consumers in food webs. *Ecology* 87: 2821–2831.

Burnell, A.M., C. Reaper, and J. Doherty. 1991. The effect of acclimation temperature on enzyme-activity in *Drosophila melanogaster*. *Comparative Biochemistry and Physiology Part B* 98: 609–614.

Byrd, J.H. and J.L. Castner. 2009. *Forensic Entomology: The Utility of Arthropods in Legal Investigations*, 2nd edition. Taylor & Francis, Boca Raton, Florida.

Cabana, G. and J.B. Rasmussen. 1994. Modeling food-chain structure and contaminant bioaccumulation using stable nitrogen isotopes. *Nature* 372: 255–257.

Cabana, G., A. Tremblay, J. Kalff, and J.B. Rasmussen. 1994. Pelagic food-chain structure in Ontario lakes—A determinant of mercury levels in lake trout *Salvelinus namaycush*. *Canadian Journal of Fisheries and Aquatic Sciences* 51: 381–389.

Cade, T.J. 2007. Exposure of California condors to lead from spent ammunition. *Journal of Wildlife Management* 71: 2125–2133.

Can, I., G.T. Javan, A.E. Pozhitkov, and P.A. Noble. 2014. Distinctive thanatomicrobiome signatures found in the blood and internal organs of humans. *Journal of Microbiological Methods* 106: 1–7.

Carter, D.O., D. Yellowlees, and M. Tibbett. 2007. Cadaver decomposition in terrestrial ecosystems. *Naturwissenschaften* 94: 12–24.

Carter, D.O., D. Yellowlees, and M. Tibbett. 2010. Moisture can be the dominant environmental parameter governing cadaver decomposition in soil. *Forensic Science International* 200: 60–66.

Catts, E.P. 1992. Problems in estimating the postmortem interval in death investigations. *Journal of Agricultural Entomology* 9: 245–255.

Cazander, G., D.I. Pritchard, Y. Nigam, W. Jung, and P.H. Nibbering. 2013. Multiple actions *of Lucilia sericata* larvae in hard-to-heal wounds. *Bioessays* 35: 1083–1092.

Cazander, G., M.C. van de Veerdonk, C.M.J.E. Vandenbroucke-Grauls, M.W.J. Schreurs, and G.N. Jukema. 2010. Maggot excretions inhibit biofilm formation on biomaterials. *Clinical Orthopaedics and Related Research* 468: 2789–2796.

Charabidze, D., B. Bourel, and D. Gosset. 2011. Larval-mass effect: Characterisation of heat emission by necrophageous blowflies (Diptera: Calliphoridae) larval aggregates. *Forensic Science International* 211: 61–66.

Cianci, T. and J. Sheldon. 1990. Endothermic generation by blow fly larvae *Phormia regina* developing in pig carcasses. *Bulletin of the Society for Vector Ecology* 15: 33–40.

Clark, K., L. Evans, and R. Wall. 2006. Growth rates of the blowfly, *Lucilia sericata*, on different body tissues. *Forensic Science International* 156: 145–149.

Coffman, C.J., M.L. Wayne, S.V. Nuzhdin, L.A. Higgins, and L.M. McIntyre. 2005. Identification of co-regulated transcripts affecting male body size in *Drosophila*. *Genome Biology* 6: R53.

Cohan, F.M. and J.D. Graf. 1985. Latitudinal cline in *Drosophila melanogaster* for knockdown resistance to ethanol fumes and for rates of response to selection for further resistance. *Evolution* 39: 278–293.

Cohan, F.M. and A.A. Hoffmann. 1986. Genetic divergence under uniform selection II. Different responses to selection for knockdown resistance to ethanol among *Drosophila melanogaster* populations and their replicate lines. *Genetics* 114: 145–164.

Coluzzi, M., A. Sabatini, A. della Torre, M.A. Di Deco, and V. Petrarca. 2002. A polytene chromosome analysis of the *Anopheles gambiae* species complex. *Science* 298: 1415–1418.

Conner, J.K. and D.L. Hartl. 2004. *A Primer of Ecological Genetics*. Sinauer Associates, Sunderland, MA.

Cooper, B.S., J.M. Tharp, I.I. Jernberg, and M.J. Angilletta. 2012. Developmental plasticity of thermal tolerances in temperate and subtropical populations of *Drosophila melanogaster*. *Journal of Thermal Biology* 37: 211–216.

Corona, M., R.A. Velarde, S. Remolina, A. Moran-Lauter, Y. Wang, K.A. Hughes, and G.E. Robinson. 2007. Vitellogenin, juvenile hormone, insulin signaling, and queen honey bee longevity. *Proceedings of the National Academy of Sciences of the United States of America* 104: 7128–7133.

Cossins, A.R. and K. Bowler. 1987. *Temperature Biology of Animals*. Chapman & Hall, London.

Courchamp, F., T. Clutton-Brock, and B. Grenfell. 1999. Inverse density dependence and the Allee effect. *Trends in Ecology and Evolution* 14: 405–410.

Darwin, C. 1859. *On the Origin of Species by Means of Natural Selection*. J. Murray, London.

David, J.R., R. Allemand, P. Capy, M. Chakir, P. Gibert, G. Petavy, and B. Moreteau. 2004. Comparative life histories and ecophysiology of *Drosophila melanogaster* and *D. simulans*. *Genetica* 120: 151–163.

David, J.R. and P. Capy. 1988. Genetic variation of *Drosophila melanogaster* natural populations. *Trends in Genetics* 4: 106–111.

David, J.R., P. Gibert, E. Gravot, G. Petavy, J.P. Morin, D. Karan, and B. Moreteau. 1997. Phenotypic plasticity and developmental temperature in *Drosophila*: Analysis and significance of reaction norms of morphometrical traits. *Journal of Thermal Biology* 22: 441–451.

David, J.R., B. Moreteau, J.P. Gauthier, G. Petavy, A. Stockel, and A.G. Imasheva. 1994. Reaction norms of size characters in relation to growth temperature in *Drosophila melanogaster*—An isofemale lines analysis. *Genetics Selection Evolution* 26: 229–251.

David, J.R., A. Yassin, J.C. Moreteau, H. Legout, and B. Moreteau. 2011. Thermal phenotypic plasticity of body size in *Drosophila melanogaster*: Sexual dimorphism and genetic correlations. *Journal of Genetics* 90: 295–302.

Davidowitz, G. and H.F. Nijhout. 2004. The physiological basis of reaction norms: The interaction among growth rate, the duration of growth and body size. *Integrative and Comparative Biology* 44: 443–449.

Davidowitz, G., H.F. Nijhout, and D.A. Roff. 2012. Predicting the response to simultaneous selection: Genetic architecture and physiological constraints. *Evolution* 66: 2916–2928.

Debelle, J.S., A.J. Hilliker, and M.B. Sokolowski. 1987. Genetic localization of the rover sitter larval foraging polymorphism in *Drosophila melanogaster*. *Behavioral Genetics* 17: 620.

Debelle, J.S., A.J. Hilliker, and M.B. Sokolowski. 1989. Genetic localization of *foraging (for)*—A major gene for larval behavior in *Drosophila melanogaster*. *Genetics* 123: 157–163.

Debelle, J.S. and M.B. Sokolowski. 1987. Heredity of rover sitter—Alternative foraging strategies of *Drosophila melanogaster* larvae. *Heredity* 59: 73–83.

De Jong, G. and Z. Bochdanovits. 2003. Latitudinal clines in *Drosophila melanogaster*: Body size, allozyme frequencies, inversion frequencies, and the insulin-signalling pathway. *Journal of Genetics* 82: 207–223.

Delcourt, A. and E. Guyenot. 1910. The possibility of studying certain Diptera in a defined environment. *Comptes Rendus de l'Académie des Sciences* 151: 255–257.

Demirjian, D.C., F. Moris-Varas, and C.S. Cassidy. 2001. Enzymes from extremophiles. *Current Opinion in Chemical Biology* 5: 144–151.

Denlinger, D.L. 1972. Seasonal phenology of diapause in flesh fly *Sarcophaga bullata* (Diptera: Sarcophagidae). *Annals of the Entomological Society of America* 65: 410–414.

Dennis, B. 2002. Allee effects in stochastic populations. *Oikos* 96: 389–401.

Dent, B.B., S.L. Forbes, and B.H. Stuart. 2004. Review of human decomposition processes in soil. *Environmental Geology* 45: 576–585.

DeSalle, R. and A.R. Templeton. 1986. The molecular through ecological genetics of *abnormal abdomen*. III. Tissue-specific differential replication of ribosomal genes modulates the *abnormal abdomen* phenotype in *Drosophila mercatorum*. *Genetics* 112: 877–886.

DeVeale, B., T. Brummel, and L. Seroude. 2004. Immunity and aging: The enemy within? *Aging Cell* 3: 195–208.

DeWitt, T.J. and S.M. Scheiner. 2004. *Phenotypic Plasticity: Functional and Conceptual Approaches*, Oxford University Press, Oxford, New York.

Dey, S., J. Bose, and A. Joshi. 2012. Adaptation to larval crowding in *Drosophila ananassae* leads to the evolution of population stability. *Ecology and Evolution* 2: 941–951.

Djawdan, M., T.T. Sugiyama, L.K. Schlaeger, T.J. Bradley, and M.R. Rose. 1996. Metabolic aspects of the trade-off between fecundity and longevity in *Drosophila melanogaster*. *Physiological Zoology* 69: 1176–1195.

Dobzhansky, T. 1948a. Genetics of natural populations: Altitudinal and seasonal changes produced by natural selection in certain populations of *Drosophila pseudoobscura* and *Drosophila persimilis*. *Genetics* 33: 158–176.

Dobzhansky, T. 1948b. Genetics of natural populations: Experiments on chromosomes of *Drosophila pseudoobscura* from different geographic regions. *Genetics* 33: 588–602.

Dobzhansky, T. 1973. Nothing in biology makes sense except in the light of evolution. *The American Biology Teacher* 35: 125–129.

Dobzhansky, T. and C. Epling. 1944. *Contributions to the Genetics, Taxonomy, and Ecology of Drosophila pseudoobscura and its Relatives*. Washington, DC: Carnegie Institute of Washington.

Douglas, A.E. 1998. Host benefit and the evolution of specialization in symbiosis. *Heredity* 81: 599–603.

Douglas, A.E. and W.A. Prosser. 1992. Synthesis of the essential amino acid tryptophan in the pea aphid (*Acyrthosiphon pisum*) symbiosis. *Journal of Insect Physiology* 38: 565–568.

Etges, W.J., W.E. Johnson, G.A. Duncan, G. Huckins, and W.B. Heed. 1995. Ecological genetics of cactophilic *Drosophila*: Inversion polymorphism in *Drosophila mojavensis* and *Drosophila pachea*. In: R. Robichaux (ed.), *Ecology and Conservation of the Sonoran Desert*. University of Arizona Press, Tucson, AZ.

Evans, W.E.D. 1963. *The Chemistry of Death*. Thomas, Springfield, IL.

Fabian, D.K., M. Kapun, V. Nolte, R. Kofler, P.S. Schmidt, C. Schlotterer, and T. Flatt. 2012. Genome-wide patterns of latitudinal differentiation among populations of *Drosophila melanogaster* from North America. *Molecular Ecology* 21: 4748–4769.

Fairbairn, D.J. and D.A. Roff. 1990. Genetic correlations among traits determining migratory tendency in the sand cricket, *Gryllus firmus*. *Evolution* 44: 1787–1795.

Faucherre, J., D. Cherix, and C. Wyss. 1999. Behavior of *Calliphora vicina* (Diptera: Calliphoridae) under extreme conditions. *Journal of Insect Behavior* 12: 687–690.

Feder, J.L., S.H. Berlocher, J.B. Roethele, H. Dambroski, J.J. Smith, W.L. Perry, V. Gavrilovic, K.E. Filchak, J. Rull, and M. Aluja. 2003b. Allopatric genetic origins for sympatric host–plant shifts and race formation in *Rhagoletis*. *Proceedings of the National Academy of Sciences* 100: 10314–10319.

Feder, M.E., N. Blair, and H. Figueras. 1997. Natural thermal stress and heat-shock protein expression in *Drosophila* larvae and pupae. *Functional Ecology* 11: 90–100.

Feder, J.L., C.A. Chilcote, and G.L. Bush. 1990a. The geographic pattern of genetic differentiation between host associated populations of *Rhagoletis pomonella* (Diptera: Tephritidae) in the eastern United States and Canada. *Evolution* 44: 570–594.

Feder, J.L., C.A. Chilcote, and G.L. Bush. 1990b. Regional, local and microgeographic allele frequency variation between apple and hawthorn populations of *Rhagoletis pomonella* in western Michigan. *Evolution* 44: 595–608.

Feder, M.E. and G.E. Hofmann. 1999. Heat-shock proteins, molecular chaperones, and the stress response: Evolutionary and ecological physiology. *Annual Review of Physiology* 61: 243–282.

Feder, J.L., F.B. Roethele, K. Filchak, J. Niedbalski, and J. Romero-Severson. 2003a. Evidence for inversion polymorphism related to sympatric host race formation in the apple maggot fly, *Rhagoletis pomonella*. *Genetics* 163: 939–953.

Fedorka, K.M., M. Zuk, and T.A. Mousseau. 2004. Immune suppression and the cost of reproduction in the ground cricket, *Allonemobius socius*. *Evolution* 58: 2478–2485.

Ferhat, A., K.A. Yavuz, and F. Kose. 2010. Effects of some toxic heavy metals on larval growth rates of *Calliphora vicina* (Diptera: Calliphoridae) and estimation of PMI. *Fresenius Environmental Bulletin* 19: 1064–1073.

Ferrari, J., S. Via, and H.C.J. Godfray. 2008. Population differentiation and genetic variation in performance on eight hosts in the pea aphid complex. *Evolution* 62: 2508–2524.

Ferrari, J., J.A. West, S. Via, and H.C.J. Godfray. 2012. Population genetic structure and secondary symbionts in host-associated populations of the pea aphid complex. *Evolution* 66: 375–390.

Fields, P.A. 2001. Review: Protein function at thermal extremes: Balancing stability and flexibility. *Comparative Biochemistry and Physiology A: Molecular and Integrative Physiology* 129: 417–431.

Fiene, J.G., G.A. Sword, S.L. Vanlaerhoven, and A.M. Tarone. 2014. The role of spatial aggregation in forensic entomology. *Journal of Medical Entomology* 51: 1–9.

Finley, S.J., M.E. Benbow, and G.T. Javan. 2015. Microbial communities associated with human decomposition and their potential use as postmortem clocks. *International Journal of Legal Medicine* 129: 623–632.

Fisher, I.J., D.J. Pain, and V.G. Thomas. 2006. A review of lead poisoning from ammunition sources in terrestrial birds. *Biological Conservation* 131: 421–432.

Fitzpatrick, M.J., E. Feder, L. Rowe, and M.B. Sokolowski. 2007. Maintaining a behaviour polymorphism by frequency-dependent selection on a single gene. *Nature* 447: U210–U215.

Flatt, T., K.J. Min, C. D'Alterio, E. Villa-Cuesta, J. Cumbers, R. Lehmann, D.L. Jones, and M. Tatar. 2008. *Drosophila* germ-line modulation of insulin signaling and lifespan. *Proceedings of the National Academy of Sciences* 105: 6368–6373.

Flores, M., M. Longnecker, and J.K. Tomberlin. 2014. Effects of temperature and tissue type on *Chrysomya rufifacies* (Diptera: Calliphoridae) (Macquart) development. *Forensic Science International* 245C: 24–29.

Fogleman, J.C., W.T. Starmer, and W.B. Heed. 1982. Comparisons of yeast florae from natural substrates and larval guts of southwestern *Drosophila*. *Oecologia* 52: 187–191.

Foley, B.R. and M. Telonis-Scott. 2011. Quantitative genetic analysis suggests causal association between cuticular hydrocarbon composition and desiccation survival in *Drosophila melanogaster*. *Heredity* 106: 68–77.

Folk, D.G. and T.J. Bradley. 2001. Ion regulation and water balance in *Drosophila melanogaster* selected for enhanced desiccation-tolerance. *American Zoologist* 41: 1445.

Folk, D.G., C. Han, and T.J. Bradley. 2001. Water acquisition and partitioning in *Drosophila melanogaster*: Effects of selection for desiccation resistance. *Journal of Experimental Biology* 204: 3323–3331.

Forbes, S.L., J. Keegan, B.H. Stuart, and B.B. Dent. 2003. A gas chromatography–mass spectrometry method for the detection of adipocere in grave soils. *European Journal of Lipid Science and Technology* 105: 761–768.

Forbes, S.L., B.H. Stuart, and B.B. Dent. 2002. The identification of adipocere in grave soils. *Forensic Science International* 127: 225–230.

Fouet, C., E. Gray, N.J. Besansky, and C. Costantini. 2012. Adaptation to aridity in the malaria mosquito *Anopheles gambiae*: Chromosomal inversion polymorphism and body size influence resistance to desiccation. *PLoS ONE* 7: e34841.

Fox, C.W., D.A. Roff, and D.J. Fairbairn. 2001. *Evolutionary Ecology: Concepts and Case Studies*. Oxford University Press, Oxford, New York.

Fraenkel, G. and C. Hsiao. 1968. Morphological and endocrinological aspects of pupal diapause in a fleshfly *Sarcophaga argyrostoma*. *Journal of Insect Physiology* 14: 707–718.

Frahm, R.R. and K.I. Kojima. 1966. Comparison of selection responses on body weight under divergent larval density conditions in *Drosophila pseudoobscura*. *Genetics* 54: 625–637.

Frentiu, F.D. and S.F. Chenoweth. 2010. Clines in cuticular hydrocarbons in two *Drosophila* species with independent population histories. *Evolution* 64: 1784–1794.

Futuyma, D.J. 2009. *Evolution*, 2nd edition. Sinauer Associates, Sunderland, MA.

Gallagher, M.B., S. Sandhu, and R. Kimsey. 2010. Variation in developmental time for geographically distinct populations of the common green bottle fly, *Lucilia sericata* (Meigen). *Journal of Forensic Sciences* 55: 438–442.

Geer, B.W., L.K. Dybas, and L.J. Shanner. 1989. *Alcohol dehydrogenase* and ethanol tolerance at the cellular level in *Drosophila melanogaster*. *Journal of Experimental Zoology* 250: 22–39.

Geer, B.W., P.W.H. Heinstra, and S.W. McKechnie. 1993. The biological basis of ethanol tolerance in *Drosophila*. *Comparative Biochemistry and Physiology Part B* 105: 203–229.

Ghosh, S.M., N.D. Testa, and A.W. Shingleton. 2013. Temperature-size rule is mediated by thermal plasticity of critical size in *Drosophila melanogaster*. *Proceedings of the Royal Society B: Biological Sciences* 280: 20130174.

Gibbs, A.G., A.K. Chippindale, and M.R. Rose. 1997. Physiological mechanisms of evolved desiccation resistance in *Drosophila melanogaster*. *Journal of Experimental Biology* 200: 1821–1832.

Gibert, P., P. Capy, A. Imasheva, B. Moreteau, J.P. Morin, G. Petavy, and J.R. David. 2004. Comparative analysis of morphological traits among *Drosophila melanogaster* and *D. simulans*: Genetic variability, clines and phenotypic plasticity. *Contemporary Issues in Genetics* and *Evolution* 11:165–179.

Gibert, P., B. Moreteau, J.C. Moreteau, and J.R. David. 1996. Growth temperature and adult pigmentation in two *Drosophila* sibling species: An adaptive convergence of reaction norms in sympatric populations? *Evolution* 50: 2346–2353.

Gilchrist, G.W. and R.B. Huey. 2004. Plastic and genetic variation in wing loading as a function of temperature within and among parallel clines in *Drosophila subobscura*. *Integrative and Comparative Biology* 44: 461–470.

Gilchrist, G.W., R.B. Huey, J. Balanya, M. Pascual, and L. Serra. 2004. A time series of evolution in action: A latitudinal cline in wing size in South American *Drosophila subobscura*. *Evolution* 58: 768–780.

Glendinning, J.I. 2007. How do predators cope with chemically defended foods? *The Biological Bulletin* 213: 252–266.

Gockel, J., W.J. Kennington, A. Hoffmann, D.B. Goldstein, and L. Partridge. 2001. Nonclinality of molecular variation implicates selection in maintaining a morphological cline *of Drosophila melanogaster*. *Genetics* 158: 319–323.

Gotthard, K. 2004. Growth strategies and optimal body size in temperate *Pararginii butterflies*. *Integrative and Comparative Biology* 44: 471–479.

Gotthard, K., S. Nylin, and C. Wiklund. 1994. Adaptive variation in growth rate: Life-history costs and consequences in the speckled wood butterfly, *Pararge aegeria*. *Oecologia* 99: 281–289.

Graves, J.L. and L.D. Mueller. 1993. Population density effects on longevity. *Genetica* 91: 99–109.

Gray, L.J., G.A. Sword, M.L. Anstey, F.J. Clissold, and S.J. Simpson. 2009. Behavioural phase polyphenism in the Australian plague locust *Chortoicetes terminifera*. *Biology Letters* 5: 306–309.

Greenberg, B. and T.I. Tantawi. 1993. Different developmental strategies in two boreal blow flies (Diptera: Calliphoridae). *Journal of Medical Entomology* 30: 481–484.

Guo, P.Z., L.D. Mueller, and F.J. Ayala. 1991. Evolution of behavior by density-dependent natural selection. *Proceedings of the National Academy of Sciences* 88: 10905–10906.

Hallas, R., M. Schiffer, and A.A. Hoffmann. 2002. Clinal variation in *Drosophila serrata* for stress resistance and body size. *Genetical Research* 79: 141–148.

Harbison, S.T., M. Carbone, J.F. Ayroles, E.A. Stone, R.F. Lyman, and T.F.C. Mackay. 2009. Co-regulated transcriptional networks contribute to natural genetic variation in *Drosophila* sleep. *Sleep* 32: A399–A400.

Hardy, G.H. 1908. Mendelian proportions in a mixed population. *Science* 28: 49–50.

Harshman, L.G. and A.J. Zera. 2007. The cost of reproduction: The devil in the details. *Trends in Ecology and Evolution* 22: 80–86.

Hawthorne, D.J. and S. Via. 2001. Genetic linkage of ecological specialization and reproductive isolation in pea aphids. *Nature* 412: 904–907.

Hazel, J.R. 1995. Thermal adaptation in biological membranes: Is homeoviscous adaptation the explanation. *Annual Review of Physiology* 57: 19–42.

Hazel, J.R., S.J. Mckinley, and E.E. Williams. 1992. Thermal adaptation in biological membranes: Interacting effects of temperature and pH. *Journal of Comparative Physiology B-Biochemical Systemic and Environmental Physiology* 162: 593–601.

Hazel, J.R., E.E. Williams, R. Livermore, and N. Mozingo. 1991. Thermal adaptation in biological membranes: Functional significance of changes in phospholipid molecular species composition. *Lipids* 26: 277–282.

Heinstra, P.W.H. 1993. Evolutionary genetics of the *Drosophila alcohol dehydrogenase* gene–enzyme system. *Genetica* 92: 1–22.

Henrich, V.C. and D.L. Denlinger. 1982. Selection for late pupariation affects diapause incidence and duration in the flesh fly *Sarcophaga bullata*. *Physiological Entomology* 7: 407–411.

Hercus, M.J., D. Berrigan, M.W. Blows, A. Magiafoglou, and A.A. Hoffmann. 2000. Resistance to temperature extremes between and within life cycle stages in *Drosophila serrata*, *D. birchii* and their hybrids: Intraspecific and interspecific comparisons. *Biological Journal of the Linnean Society* 71: 403–416.

Hoback, W.W., A.A. Bishop, J. Kroemer, J. Scalzitti, and J.J. Shaffer. 2004. Differences among antimicrobial properties of carrion beetle secretions reflect phylogeny and ecology. *Journal of Chemical Ecology* 30: 719–729.

Hoffmann, A.A. and L.G. Harshman. 1999. Desiccation and starvation resistance in *Drosophila*: Patterns of variation at the species, population and intrapopulation levels. *Heredity* 83: 637–643.

Hoffmann, A.A., J.G. Sorensen, and V. Loeschcke. 2003. Adaptation of *Drosophila* to temperature extremes: Bringing together quantitative and molecular approaches. *Journal of Thermal Biology* 28: 175–216.

Hohenlohe, P.A., S. Bassham, P.D. Etter, N. Stiffler, E.A. Johnson, and W.A. Cresko. 2010b. Population genomics of parallel adaptation in three spine stickleback using sequenced RAD tags. *PLoS Genetics* 6(2): e1000862.

Hohenlohe, P.A., P.C. Phillips, and W.A. Cresko. 2010a. Using population genomics to detect selection in natural populations: Key concepts and methodological considerations. *International Journal of Plant Sciences* 171: 1059–1071.

Hollocher, H. and A.R. Templeton. 1994. The molecular through ecological genetics of *abnormal abdomen* in *Drosophila mercatorum*. VI. The nonneutrality of the Y chromosome rDNA polymorphism. *Genetics* 136: 1373–1384.

Hollocher, H., A.R. Templeton, R. DeSalle, and J.S. Johnston. 1992. The molecular through ecological genetics of *abnormal abdomen*. IV. Components of genetic variation in a natural population of *Drosophila mercatorum*. *Genetics* 130: 355–366.

Huang, W., A. Massouras, Y. Inoue, J. Peiffer, M. Ramia, A.M. Tarone, L. Turlapati et al. 2014. Natural variation in genome architecture among 205 *Drosophila melanogaster* Genetic Reference Panel lines. *Genome Research* 24: 1193–1208.

Huey, R.B., G.W. Gilchrist, M.L. Carlson, D. Berrigan, and L. Serra. 2000. Rapid evolution of a geographic cline in size in an introduced fly. *Science* 287: 308–309.

Huntington, T.E., L.G. Higley, and F.P. Baxendale. 2007. Maggot development during morgue storage and its effect on estimating the post-mortem interval. *Journal of Forensic Sciences* 52: 453–458.

Hyde, E.R., D.P. Haarmann, A.M. Lynne, S.R. Bucheli, and J.F. Petrosino. 2013. The living dead: Bacterial community structure of a cadaver at the onset and end of the bloat stage of decomposition. *PLoS ONE* 8(10): e77733.

Ireland, S. and B. Turner. 2006. The effects of larval crowding and food type on the size and development of the blowfly *Calliphora vomitoria*. *Forensic Science International* 159: 175–181.

Jacot, A., H. Scheuber, and M.W.G. Brinkhof. 2004. Costs of an induced immune response on sexual display and longevity in field crickets. *Evolution* 58: 2280–2286.

Jakob, U., M. Gaestel, K. Engel, and J. Buchner. 1993. Small heat-shock proteins are molecular chaperones. *Journal of Biological Chemistry* 268: 1517–1520.

James, A.C., R.B. Azevedo, and L. Partridge. 1997. Genetic and environmental responses to temperature of *Drosophila melanogaster* from a latitudinal cline. *Genetics* 146: 881–890.

Janzen, D.H. 1977. Why fruits rot, seeds mold, and meat spoils. *American Naturalist* 111: 691–713.

Jeong, S., M. Rebeiz, P. Andolfatto, T. Werner, J. True, and S.B. Carroll. 2008. The evolution of gene regulation underlies a morphological difference between two *Drosophila* sister species. *Cell* 132: 783–793.

Johnson, A.P. and J.F. Wallman. 2014. Infrared imaging as a non-invasive tool for documenting maggot mass temperatures. *Australian Journal of Forensic Sciences* 46: 73–79.

Joplin, K.H., G.D. Yocum, and D.L. Denlinger. 1990. Diapause specific proteins expressed by the brain during the pupal diapause of the flesh fly *Sarcophaga crassipalpis*. *Journal of Insect Physiology* 36: 775–779, 781–783.

Jung, K., Y. Kim, L. Hang, and J.T. Kim. 2009. *Aspergillus fumigatus* infection in two wild Eurasian black vultures (*Aegypius monachus* Linnaeus) with carbofuran insecticide poisoning: A case report. *The Veterinary Journal* 179: 307–312.

Kaitala, A. 1991. Phenotypic plasticity in reproductive behavior of waterstriders: Trade-offs between reproduction and longevity during food stress. *Functional Ecology* 5: 12–18.

Kaneshrajah, G. and B. Turner. 2004. *Calliphora vicina* larvae grow at different rates on different body tissues. *International Journal of Legal Medicine* 118: 242–244.

Karan, D., N. Dahiya, A.K. Munjal, P. Gibert, B. Moreteau, R. Parkash, and J.R. David. 1998b. Desiccation and starvation tolerance of adult *Drosophila*: Opposite latitudinal clines in natural populations of three different species. *Evolution* 52: 825–831.

Karan, D., J.P. Morin, B. Moreteau, and J.R. David. 1998a. Body size and developmental temperature in *Drosophila melanogaster*: Analysis of body weight reaction norm. *Journal of Thermal Biology* 23: 301–309.

Karan, D., R. Parkash, and J.R. David. 1999. Microspatial genetic differentiation for tolerance and utilization of various alcohols and acetic acid in *Drosophila* species from India. *Genetica* 105: 249–258.

Keller, I., J.M. Alexander, R. Holderegger, and P.J. Edwards. 2013. Widespread phenotypic and genetic divergence along altitudinal gradients in animals. *Journal of Evolutionary Biology* 26: 2527–2543.

Kennington, W.J., J. Gockel, and L. Partridge. 2003a. Testing for asymmetrical gene flow in a *Drosophila melanogaster* body-size cline. *Genetics* 165: 667–673.

Kennington, W.J., J.R. Killeen, D.B. Goldstein, and L. Partridge. 2003b. Rapid laboratory evolution of adult wing area in *Drosophila melanogaster* in response to humidity. *Evolution* 57: 932–936.

Kennington, W.J., L. Partridge, and A.A. Hoffmann. 2006. Patterns of diversity and linkage disequilibrium within the cosmopolitan inversion *In(3R)Payne* in *Drosophila melanogaster* are indicative of coadaptation. *Genetics* 172: 1655–1663.

Kenyon, C.J. 2010. The genetics of ageing. *Nature* 464: 504–512.

Kerr, W.E. and S. Wright. 1954a. Experimental studies of the distribution of gene frequencies in very small populations of *Drosophila melanogaster*. I. *Forked. Evolution* 8: 172–177.

Kerr, W.E. and S. Wright. 1954b. Experimental studies of the distribution of gene frequencies in very small populations of *Drosophila melanogaster*. III. *Aristapedia* and *Spineless. Evolution* 8: 293–302.

Keurentjes, J.J.B., G. Willems, F. van Eeuwijk, M. Nordborg, and M. Koornneef. 2011. A comparison of population types used for QTL mapping in *Arabidopsis thaliana*. *Plant Genetic Resources* 9: 185–188.

Kimura, M. 1954. Process leading to quasi-fixation of genes in natural populations due to random fluctuation of selection intensities. *Genetics* 39: 280–295.

Kingsolver, J.G., D.W. Pfennig, and M.R. Servedio. 2002. Migration, local adaptation and the evolution of plasticity. *Trends in Ecology and Evolution* 17: 540–541.

Kirkpatrick, M. 2010. How and why chromosome inversions evolve. *PLoS Biology* 8(9): e1000501.

Kolaczkowski, B., A.D. Kern, A.K. Holloway, and D.J. Begun. 2011. Genomic differentiation between temperate and tropical Australian populations of *Drosophila melanogaster*. *Genetics* 187: 245–260.

Komano, T. 1999. SHUFFLONS: Multiple inversion systems and integrons. *Annual Review of Genetics* 33: 171–191.

Konings, W.N., S.V. Albers, S. Koning, and A.J.M. Driessen. 2002. The cell membrane plays a crucial role in survival of bacteria and archaea in extreme environments. *Antonie Van Leeuwenhoek International Journal of General and Molecular Microbiology* 81: 61–72.

Kopp, A., R.M. Graze, S. Xu, S.B. Carroll, and S.V. Nuzhdin. 2003. Quantitative trait loci responsible for variation in sexually dimorphic traits in *Drosophila melanogaster*. *Genetics* 163: 771–787.

Kramer, A.M., B. Dennis, A.M. Liebhold, and J.M. Drake. 2009. The evidence for Allee effects. *Population Ecology* 51: 341–354.

Krebs, R.A. and M.E. Feder. 1997. Natural variation in the expression of the heat-shock protein *Hsp70* in a population of *Drosophila melanogaster* and its correlation with tolerance of ecologically relevant thermal stress. *Evolution* 51: 173–179.

Krinos, C.M., M.J. Coyne, K.G. Weinacht, A.O. Tzianabos, D.L. Kasper, and L.E. Comstock. 2001. Extensive surface diversity of a commensal microorganism by multiple DNA inversions. *Nature* 414: 555–558.

Kristensen, T.N., J. Overgaard, A.A. Hoffmann, N.C. Nielsen, and A. Malmendal. 2012. Inconsistent effects of developmental temperature acclimation on low-temperature performance and metabolism in *Drosophila melanogaster*. *Evolutionary Ecology Research* 14: 821–837.

Kusukawa, N. and T. Yura. 1988. Heat shock protein *groE* of *Escherichia coli*: Key protective roles against thermal stress. *Genes and Development* 2: 874–882.

Lavington, E., R. Cogni, C. Kuczynski, S. Koury, E.L. Behrman, K.R. O'Brien, P.S. Schmidt, and W.F. Eanes. 2014. A small-system high-resolution study of metabolic adaptation in the central metabolic pathway to temperate climates in *Drosophila melanogaster*. *Molecular Biology and Evolution* 31: 2032–2041.

Lazzaro, B.P., H.A. Flores, J.G. Lorigan, and C.P. Yourth. 2008. Genotype-by-environment interactions and adaptation to local temperature affect immunity and fecundity in *Drosophila melanogaster*. *PLoS Pathogens* 4(3): e1000025.

Lazzaro, B.P. and T.J. Little. 2009. Immunity in a variable world. *Philosophical Transactions of the Royal Society B: Biological Sciences* 364: 15–26.

Lee, R.E., C.P. Chen, and D.L. Denlinger. 1987a. A rapid cold-hardening process in insects. *Science* 238: 1415–1417.

Lee, R.E., C.P. Chen, M.H. Meacham, and D.L. Denlinger. 1987b. Ontogenic patterns of cold-hardiness and glycerol production in *Sarcophaga crassipalpis*. *Journal of Insect Physiology* 33: 587–592.

Lee, K.P., S.J. Simpson, F.J. Clissold, R. Brooks, J.W. Ballard, P.W. Taylor, N. Soran, and D. Raubenheimer. 2008. Lifespan and reproduction in *Drosophila*: New insights from nutritional geometry. *Proceedings of the National Academy of Sciences* 105: 2498–2503.

Li, G.C., L.G. Li, R.Y. Liu, M. Rehman, and W.M.F. Lee. 1992. Heat-shock protein *Hsp70* protects cells from thermal stress even after deletion of its ATP-binding domain. *Proceedings of the National Academy of Sciences* 89: 2036–2040.

Loeschcke, V. and A.A. Hoffmann. 2007. Consequences of heat hardening on a field fitness component in *Drosophila* depend on environmental temperature. *American Naturalist* 169: 175–183.

Lorch, P.D., G.A. Sword, D.T. Gwynne, and G.L. Anderson. 2005. Radiotelemetry reveals differences in individual movement patterns between outbreak and non-outbreak Mormon cricket populations. *Ecological Entomology* 30: 548–555.

Lynch, M. and B. Walsh. 1998. *Genetics and Analysis of Quantitative Traits*. Sinauer Associates, Sunderland, MA.

Mackay, T.F.C. 2001. The genetic architecture of quantitative traits. *Annual Review of Genetics* 35: 303–339.

Mackay, T.F.C. 2004. The genetic architecture of quantitative traits: Lessons from *Drosophila*. *Current Opinion in Genetics and Development* 14: 253–257.

Mackay, T.F.C., S. Richards, E.A. Stone, A. Barbadilla, J.F. Ayroles, D. Zhu, S. Casillas et al. 2012. The *Drosophila melanogaster* Genetic Reference Panel. *Nature* 482: 173–178.

Magiafoglou, A. and A. Hoffmann. 2003. Thermal adaptation in *Drosophila serrata* under conditions linked to its southern border: Unexpected patterns from laboratory selection suggest limited evolutionary potential. *Journal of Genetics* 82: 179–189.

Maklakov, A.A., S.J. Simpson, F. Zajitschek, M.D. Hall, J. Dessmann, F. Clissold, D. Raubenheimer, R. Bonduriansky, and R.C. Brooks. 2008. Sex-specific fitness effects of nutrient intake on reproduction and lifespan. *Current Biology* 18: 1062–1066.

Malecot, G. 1944. A problem of probabilities in a series which poses genetics. *Comptes Rendus de l'Académie des Sciences* 219: 379–381.

May, R.C. 2007. Gender, immunity and the regulation of longevity. *Bioessays* 29: 795–802.

McKean, K.A. and L. Nunney. 2001. Increased sexual activity reduces male immune function in *Drosophila melanogaster*. *Proceedings of the National Academy of Sciences* 98: 7904–7909.

McKean, K.A. and L. Nunney. 2008. Sexual selection and immune function in *Drosophila melanogaster*. *Evolution* 62: 386–400.

McKechnie, S.W. and B.W. Geer. 1993. Micro-evolution in a wine cellar population: An historical perspective. *Genetica* 90: 201–215.

Megyesi, M.S., S.P. Nawrocki, and N.H. Haskell. 2005. Using accumulated degree-days to estimate the postmortem interval from decomposed human remains. *Journal of Forensic Sciences* 50: 618–626.

Metcalf, J.L., L.W. Parfrey, A. Gonzalez, C.L. Lauber, D. Knights, G. Ackermann, G.C. Humphrey et al. 2013. A microbial clock provides an accurate estimate of the postmortem interval in a mouse model system. *eLife* 2: e01104.

Michaud, M.R. and D.L. Denlinger. 2006. Oleic acid is elevated in cell membranes during rapid cold-hardening and pupal diapause in the flesh fly *Sarcophaga crassipalpis*. *Journal of Insect Physiology* 52: 1073–1082.

Michaud, M.R. and D.L. Denlinger. 2007. Shifts in the carbohydrate, polyol, and amino acid pools during rapid cold-hardening and diapause-associated cold-hardening in flesh flies *Sarcophaga crassipalpis*: A metabolomic comparison. *Journal of Comparative Physiology B-Biochemical Systemic and Environmental Physiology* 177: 753–763.

Moe, S.J., N.C. Stenseth, and R.H. Smith. 2002. Density-dependent compensation in blowfly populations give indirectly positive effects of a toxicant. *Ecology* 83: 1597–1603.

Moffett, A. and C. Loke. 2006a. Implantation, embryo–maternal interactions, immunology and modulation of the uterine environment: A workshop report. *Placenta* 27: S54–S55.

Moffett, A. and C. Loke. 2006b. Immunology of placentation in eutherian mammals. *Nature Reviews Immunology* 6: 584–594.

Moran, N.A. and P.H. Degnan. 2006. Functional genomics of *Buchnera* and the ecology of aphid hosts. *Molecular Ecology* 15: 1251–1261.

Morel, F.M.M., A.M.L. Kraepiel, and M. Amyot. 1998. The chemical cycle and bioaccumulation of mercury. *Annual Review of Ecology and Systematics* 29: 543–566.

Morin, J.P., B. Moreteau, G. Petavy, and J.R. David. 1999. Divergence of reaction norms of size characters between tropical and temperate populations of *Drosophila melanogaster* and *D. simulans*. *Journal of Evolutionary Biology* 12: 329–339.

Morin, J.P., B. Moreteau, G. Petavy, A.G. Imasheva, and J.R. David. 1996. Body size and developmental temperature in *Drosophila simulans*: Comparison of reaction norms with sympatric *Drosophila melanogaster*. *Genetics Selection Evolution* 28: 415–436.

Morin, J.P., B. Moreteau, G. Petavy, R. Parkash, and J.R. David. 1997. Reaction norms of morphological traits in *Drosophila*: Adaptive shape changes in a stenotherm circumtropical species? *Evolution* 51: 1140–1148.

Mousseau, T.A. and D.A. Roff. 1987. Natural selection and the heritability of fitness components. *Heredity* 59: 181–197.

Mueller, L.D. and L.G. Cabral. 2012. Does phenotypic plasticity for adult size versus food level in *Drosophila melanogaster* evolve in response to adaptation to different rearing densities? *Evolution* 66: 263–271.

Mueller, L.D., F. Gonzalezcandelas, and V.F. Sweet. 1991a. Components of density-dependent population dynamics: Models and tests with *Drosophila*. *American Naturalist* 137: 457–475.

Mueller, L.D., J.L. Graves, and M.R. Rose. 1993. Interactions between density-dependent and age-specific selection in *Drosophila melanogaster*. *Functional Ecology* 7: 469–479.

Mueller, L.D., P.Z. Guo, and F.J. Ayala. 1991b. Density-dependent natural selection and trade-offs in life-history traits. *Science* 253: 433–435.

Navarro, A., N.H. Barton, and D. Houle. 2003. Accumulating postzygotic isolation genes in parapatry: A new twist on chromosomal speciation. *Evolution* 57: 447–459.

Neiman, M. and T.A. Linksvayer. 2006. The conversion of variance and the evolutionary potential of restricted recombination. *Heredity* 96: 111–121.

Nemri, A., S. Atwell, A.M. Tarone, Y.S. Huang, K. Zhao, D.J. Studholme, M. Nordborg, and J.D. Jones. 2010. Genome-wide survey of Arabidopsis natural variation in downy mildew resistance using combined association and linkage mapping. *Proceedings of the National Academy of Sciences* 107: 10302–10307.

Neu, H.C. 1992. The crisis in antibiotic resistance. *Science* 257: 1064–1073.

Nicholson, A.J. 1950. Population oscillations caused by competition for food. *Nature* 165: 476–477.

Nicholson, A.J. 1954. Compensatory reactions of populations to stresses and their evolutionary significance. *Australian Journal of Zoology* 2: 1–8.

Nicholson, A.J. 1957. The self-adjustment of populations to change. *Cold Spring Harbor Symposia on Quantitative Biology* 22: 153–173.

Noor, M.A.F., K.L. Grams, L.A. Bertucci, and J. Reiland. 2001. Chromosomal inversions and the reproductive isolation of species. *Proceedings of the National Academy of Sciences* 98: 12084–12088.

Norry, F.M. and V. Loeschcke. 2002a. Temperature-induced shifts in associations of longevity with body size in *Drosophila melanogaster*. *Evolution* 56: 299–306.

Norry, F.M. and V.R. Loeschcke. 2002b. Longevity and resistance to cold stress in cold-stress selected lines and their controls in *Drosophila melanogaster*. *Journal of Evolutionary Biology* 15: 775–783.

Nuorteva, P. and E. Hasanen. 1972. Transfer of mercury from fishes to sarcosaprophagous flies. *Annales Zoologici Fennici* 9: 23–27.

Nuorteva, P. and S.L. Nuorteva. 1982. The fate of mercury in sarcosaprophagous flies and in insects eating them. *Ambio* 11: 34–37.

Nylin, S. and K. Gotthard. 1998. Plasticity in life-history traits. *Annual Review of Entomology* 43: 63–83.

Oakeshott, J.G., F.M. Cohan, and J.B. Gibson. 1985. Ethanol tolerances of *Drosophila melanogaster* populations selected on different concentrations of ethanol supplemented media. *Theoretical and Applied Genetics* 69: 603–608.

Oakeshott, J.G., J.B. Gibson, P.R. Anderson, W.R. Knibb, D.G. Anderson, and G.K. Chambers. 1982a. *Alcohol dehydrogenase* and *Glycerol-3-phosphate dehydrogenase* clines in *Drosophila melanogaster* on different continents. *Evolution* 36: 86–96.

Oakeshott, J.G., T.W. May, J.B. Gibson, and D.A. Willcocks. 1982b. Resource partitioning in five domestic *Drosophila* species and its relationship to ethanol metabolism. *Australian Journal of Zoology* 30: 547–556.

Ogada, D.L., F. Keesing, and M.Z. Virani. 2012. Dropping dead: Causes and consequences of vulture population declines worldwide. *Year in Ecology and Conservation Biology* 1249: 57–71.

Orr, H.A. 2009. Fitness and its role in evolutionary genetics. *Nature Reviews Genetics* 10: 531–539.

Overgaard, J., A.A. Hoffmann, and T.N. Kristensen. 2011a. Assessing population and environmental effects on thermal resistance in *Drosophila melanogaster* using ecologically relevant assays. *Journal of Thermal Biology* 36: 409–416.

Overgaard, J., M.R. Kearney, and A.A. Hoffmann. 2014. Sensitivity to thermal extremes in Australian *Drosophila* implies similar impacts of climate change on the distribution of widespread and tropical species. *Global Change Biology* 20: 1738–1750.

Overgaard, J., T.N. Kristensen, K.A. Mitchell, and A.A. Hoffmann. 2011b. Thermal tolerance in widespread and tropical *Drosophila* species: Does phenotypic plasticity increase with latitude? *American Naturalist* 178: S80–S96.

Owings, C.G., C. Spiegelman, A.M. Tarone, and J.K. Tomberlin. 2014. Developmental variation among *Cochliomyia macellaria* Fabricius (Diptera: Calliphoridae) populations from three ecoregions of Texas, USA. *International Journal of Legal Medicine* 128: 709–717.

Parkash, R., C. Lambhod, and D. Singh. 2014. Thermal developmental plasticity affects body size and water conservation of *Drosophila nepalensis* from the Western Himalayas. *Bulletin of Entomological Research* 104: 504–516.

Partridge, L., D. Gems, and D.J. Withers. 2005. Sex and death: What is the connection? *Cell* 120: 461–472.

Partridge, L. and C.M. Sgro. 1998. Behavioural genetics: Molecular genetics meets feeding ecology. *Current Biology* 8: R23–R24.

Payne, J.A. 1965. A summer carrion study of the baby pig *Sus scrofa* Linnaeus. *Ecology* 46: 592–602.

Pechal, J.L., T.L. Crippen, M.E. Benbow, A.M. Tarone, S. Dowd, and J.K. Tomberlin. 2014. The potential use of bacterial community succession in forensics as described by high throughput metagenomic sequencing. *International Journal of Legal Medicine* 128: 193–205.

Petavy, G., J.R. David, and B. Moreteau. 1997a. What is the adaptive value of correlated temperature-induced variations of body size traits and flight parameters in *Drosophila* fruit flies? *Bulletin de la Societe Zoologique de France* 122: 13–20.

Petavy, G., J.P. Morin, B. Moreteau, and J.R. David. 1997b. Growth temperature and phenotypic plasticity in two *Drosophila* sibling species: Probable adaptive changes in flight capacities. *Journal of Evolutionary Biology* 10: 875–887.

Pfeiler, E. and T.A. Markow. 2001. Ecology and population genetics of Sonoran Desert *Drosophila*. *Molecular Ecology* 10: 1787–1791.

Pfennig, D.W. and R.A. Martin. 2010. Evolution of character displacement in spadefoot toads: Different proximate mechanisms in different species. *Evolution* 64: 2331–2341.

Pfennig, D.W. and M. McGee. 2010. Resource polyphenism increases species richness: A test of the hypothesis. *Philosophical Transactions of the Royal Society of London Series B, Biological Sciences* 365: 577–591.

Pfennig, D.W. and P.J. Murphy. 2002. How fluctuating competition and phenotypic plasticity mediate species divergence. *Evolution* 56: 1217–1228.

Pfennig, D.W., A.M. Rice, and R.A. Martin. 2006. Ecological opportunity and phenotypic plasticity interact to promote character displacement and species coexistence. *Ecology* 87: 769–779.

Pfennig, D.W., M.A. Wund, E.C. Snell-Rood, T. Cruickshank, C.D. Schlichting, and A.P. Moczek. 2010. Phenotypic plasticity's impacts on diversification and speciation. *Trends in Ecology and Evolution* 25: 459–467.

Phelan, J.P., M.A. Archer, K.A. Beckman, A.K. Chippindale, T.J. Nusbaum, and M.R. Rose. 2003. Breakdown in correlations during laboratory evolution. I. Comparative analyses of *Drosophila* populations. *Evolution* 57: 527–535.

Polkki, M., K. Kangassalo, and M.J. Rantala. 2014. Effects of interaction between temperature conditions and copper exposure on immune defense and other life-history traits of the blow fly *Protophormia terraenovae*. *Environmental Science and Technology* 48: 8793–8799.

Powell, J.R. 1997. *Progress and Prospects in Evolutionary Biology: The Drosophila Model*. Oxford University Press, New York.

Powell, T.H.Q., A.A. Forbes, G.R. Hood, and J.L. Feder. 2014. Ecological adaptation and reproductive isolation in sympatry: Genetic and phenotypic evidence for native host races *of Rhagoletis pomonella*. *Molecular Ecology* 23: 688–704.

Prinkkila, M.L. and I. Hanski. 1995. Complex competitive interactions in four species of *Lucilia* blowflies. *Ecological Entomology* 20: 261–272.

Prosser, W.A. and A.E. Douglas. 1991. The aposymbiotic aphid: An analysis of chlortetracycline-treated pea aphid *Acyrthosiphon pisum. Journal of Insect Physiology* 37: 713–719.

Prosser, W.A. and A.E. Douglas. 1992. A test of the hypotheses that nitrogen is upgraded and recycled in an aphid *Acyrthosiphon pisum* symbiosis. *Journal of Insect Physiology* 38: 93–99.

Prosser, W.A., S.J. Simpson, and A.E. Douglas. 1992. How an aphid *Acyrthosiphon pisum* symbiosis responds to variation in dietary nitrogen. *Journal of Insect Physiology* 38: 301–307.

Reed, L.K., K. Lee, Z. Zhang, L. Rashid, A. Poe, B. Hsieh, N. Deighton, N. Glassbrook, R. Bodmer, and G. Gibson. 2014. Systems genomics of metabolic phenotypes in wild-type *Drosophila melanogaster. Genetics* 197: 781–793.

Reed, L.K., S. Williams, M. Springston, J. Brown, K. Freeman, C.E. DesRoches, M.B. Sokolowski, and G. Gibson. 2010. Genotype-by-diet interactions drive metabolic phenotype variation in *Drosophila melanogaster. Genetics* 185: 1009–1019.

Reinhardt, J.A., B. Kolaczkowski, C.D. Jones, D.J. Begun, and A.D. Kern. 2014. Parallel geographic variation in *Drosophila melanogaster. Genetics* 197: 361–373.

Rich, S.S., A.E. Bell, and S.P. Wilson. 1979. Genetic drift in small populations of *Tribolium. Evolution* 33: 579–584.

Rideout, B.A., I. Stalis, R. Papendick, A. Pessier, B. Puschner, M.E. Finkelstein, D.R. Smith et al. 2012. Patterns of mortality in free-ranging California condors *Gymnogyps californianus. Journal of Wildlife Diseases* 48: 95–112.

Ridley, M. 2004. *Evolution*, 3rd edition. Blackwell Publishing, Malden, MA.

Rieseberg, L.H. 2001. Chromosomal rearrangements and speciation. *Trends in Ecology and Evolution* 16: 351–358.

Rinehart, J.P., R.M. Robich, and D.L. Denlinger. 2010. Isolation of diapause-regulated genes from the flesh fly *Sarcophaga crassipalpis* by suppressive subtractive hybridization. *Journal of Insect Physiology* 56: 603–609.

Ring, R.A. 1982. Freezing-tolerant insects with low supercooling points. *Comparative Biochemistry and Physiology A Physiology* 73: 605–612.

Rion, S. and T.J. Kawecki. 2007. Evolutionary biology of starvation resistance: What we have learned from *Drosophila. Journal of Evolutionary Biology* 20: 1655–1664.

Rivers, D.B. and G. Dahlem. 2013. *The Science of Forensic Entomology*. John Wiley & Sons, Hoboken, NJ.

Rivers, D.B., C. Thompson, and R. Brogan. 2011. Physiological trade-offs of forming maggot masses by necrophagous flies on vertebrate carrion. *Bulletin of Entomological Research* 101: 599–611.

Robertson, S.A. 2010. Immune regulation of conception and embryo implantation—All about quality control? *Journal of Reproductive Immunology* 85: 51–57.

Rodríguez-Trelles, F. and M. Rodríguez. 1998. Rapid micro-evolution and loss of chromosomal diversity in *Drosophila* in response to climate warming. *Evolutionary Ecology* 12: 829–838.

Roff, D.A. 1992. *The Evolution of Life Histories: Theory and Analysis*. Chapman & Hall, New York.

Roff, D.A. 2002. *Life History Evolution*. Sinauer Associates, Sunderland, MA.

Roff, D.A. and T.A. Mousseau. 1987. Quantitative genetics and fitness: Lessons from *Drosophila. Heredity* 58: 103–118.

Rose, M.R. 1989. Genetics of increased lifespan in *Drosophila. Bioessays* 11: 132–135.

Rouhbakhsh, D., C.Y. Lai, C.D. vonDohlen, M.A. Clark, L. Baumann, P. Baumann, N.A. Moran, and D.J. Voegtlin. 1996. The tryptophan biosynthetic pathway of aphid endosymbionts *Buchnera*: Genetics and evolution of plasmid-associated anthranilate synthase (*trpEG*) within the aphididae. *Journal of Molecular Evolution* 42: 414–421.

Roush, R.T. and J.A. McKenzie. 1987. Ecological genetics of insecticide and acaricide resistance. *Annual Review of Entomology* 32: 361–380.

Rozen, D.E., D.J.P. Engelmoer, and P.T. Smiseth. 2008. Antimicrobial strategies in burying beetles breeding on carrion. *Proceedings of the National Academy of Sciences* 105: 17890–17895.

Russell, N.J. 1984. Mechanisms of thermal adaptation in bacteria: Blueprints for survival. *Trends in Biochemical Sciences* 9: 108–112.

Sandstrom, J., A. Telang, and N.A. Moran. 2000. Nutritional enhancement of host plants by aphids—A comparison of three aphid species on grasses. *Journal of Insect Physiology* 46: 33–40.

Santos, M., D.J. Borash, A. Joshi, N. Bounlutay, and L.D. Mueller. 1997. Density-dependent natural selection in *Drosophila*: Evolution of growth rate and body size. *Evolution* 51: 420–432.

Sarica, J., M. Amyot, J. Bey, and L. Hare. 2005a. Fate of mercury accumulated by blowflies feeding on fish carcasses. *Environmental Toxicology and Chemistry* 24: 526–529.

Sarica, J., M. Amyot, L. Hare, P. Blanchfield, R.A. Bodaly, H. Hintelmann, and M. Lucotte. 2005b. Mercury transfer from fish carcasses to scavengers in boreal lakes: The use of stable isotopes of mercury. *Environmental Pollution* 134: 13–22.

Scheiner, S.M. 1993. Genetics and evolution of phenotypic plasticity. *Annual Review of Ecology and Systematics* 24: 35–68.

Schlichting, C.D. 1986. The evolution of phenotypic plasticity in plants. *Annual Review of Ecology and Systematics* 17: 667–693.

Schlichting, C.D. and M. Pigliucci. 1998. *Phenotypic Evolution: A Reaction Norm Perspective.* Sinauer Associates, Sunderland, MA.

Seehuus, S.C., K. Norberg, U. Gimsa, T. Krekling, and G.V. Amdam. 2006. Reproductive protein protects functionally sterile honey bee workers from oxidative stress. *Proceedings of the National Academy of Sciences* 103: 962–967.

Sheldon, B.C. and S. Verhulst. 1996. Ecological immunology: Costly parasite defences and trade-offs in evolutionary ecology. *Trends in Ecology and Evolution* 11: 317–321.

Sherman, R.A. 2003. Maggot therapy for treating diabetic foot ulcers unresponsive to conventional therapy. *Diabetes Care* 26: 446–451.

Sherman, R.A., M.J.R. Hall, and S. Thomas. 2000. Medicinal maggots: An ancient remedy for some contemporary afflictions. *Annual Review of Entomology* 45: 55–81.

Shingleton, A.W., C.M. Estep, M.V. Driscoll, and I. Dworkin. 2009. Many ways to be small: Different environmental regulators of size generate distinct scaling relationships in *Drosophila melanogaster*. *Philosophical Transactions of the Royal Society of London. Series B Biological Sciences* 276: 2625–2633.

Short, S.M. and B.P. Lazzaro. 2010. Female and male genetic contributions to post-mating immune defence in female *Drosophila melanogaster*. *Philosophical Transactions of the Royal Society of London Series B Biological Sciences* 277: 3649–3657.

Short, S.M., M.F. Wolfner, and B.P. Lazzaro. 2012. Female *Drosophila melanogaster* suffer reduced defense against infection due to seminal fluid components. *Journal of Insect Physiology* 58: 1192–1201.

Sieber, M. and F.M. Hilker. 2012. The hydra effect in predator–prey models. *Journal of Mathematical Biology* 64: 341–360.

Sillero, N., M. Reis, C.P. Vieira, J. Vieira, and R. Morales-Hojas. 2014. Niche evolution and thermal adaptation in the temperate species *Drosophila americana*. *Journal of Evolutionary Biology* 27: 1549–1561.

Simpson, S.J., G.A. Sword, P.D. Lorch, and I.D. Couzin. 2006. Cannibal crickets on a forced march for protein and salt. *Proceedings of the National Academy of Sciences* 103: 4152–4156.

Sinclair, B.J., A.G. Gibbs, and S.P. Roberts. 2007. Gene transcription during exposure to, and recovery from, cold and desiccation stress in *Drosophila melanogaster*. *Insect Molecular Biology* 16: 435–443.

Slone, D.H. and S.V. Gruner. 2007. Thermoregulation in larval aggregations of carrion-feeding blow flies (Diptera: Calliphoridae). *Journal of Medical Entomology* 44: 516–523.

Sokolowski, M.B. 1998. Genes for normal behavioral variation: Recent clues from flies and worms. *Neuron* 21: 463–466.

Sokolowski, M.B. 2001. Drosophila: Genetics meets behaviour. *Nature Reviews Genetics* 2: 879–890.

Sokolowski, M.B., H.S. Pereira, and K. Hughes. 1997. Evolution of foraging behavior in *Drosophila* by density-dependent selection. *Proceedings of the National Academy of Sciences* 94: 7373–7377.

Solter, L.F., B. Lustigman, and P. Shubeck. 1989. Survey of medically important true bacteria found associated with carrion beetles (Coleoptera: Silphidae). *Journal of Medical Entomology* 26: 354–359.

Somero, G.N. 1978. Temperature adaptation of enzymes: Biological optimization through structure–function compromises. *Annual Review of Ecology and Systematics* 9: 1–29.

Somero, G.N. 1995. Proteins and temperature. *Annual Review of Physiology* 57: 43–68.

Song, H. 2011. Density-dependent phase polyphenism in nonmodel locusts: A minireview. *Psyche* 2011: Article ID 741769.

Srinivasan, D.G. and J.A. Brisson. 2012. Aphids: A model for polyphenism and epigenetics. *Genetics Research International* 2012: 431531.

Srygley, R.B., P.D. Lorch, S.J. Simpson, and G.A. Sword. 2009. Immediate protein dietary effects on movement and the generalised immunocompetence of migrating Mormon crickets *Anabrus simplex* (Orthoptera: Tettigoniidae). *Ecological Entomology* 34: 663–668.

Starmer, W.T. 1982. Analysis of the community structure of yeasts associated with the decaying stems of cactus. I. *Stenocereus gummosus*. *Microbial Ecology* 8: 71–81.

Starmer, W.T. and V. Aberdeen. 1990. The nutritional importance of pure and mixed cultures of yeasts in the development of *Drosophila mulleri* larvae in *Opuntia* tissues and its relationship to host plant shifts. *Ecological and Evolutionary Genetics of Drosophila*. Springer, New York, NY, pp. 145–160.

Starmer, W.T., J.S.F. Barker, H.J. Phaff, and J.C. Fogleman. 1986. Adaptations of *Drosophila* and yeasts: Their interactions with the volatile 2-propanol in the cactus microorganism—*Drosophila* model system. *Australian Journal of Biological Sciences* 39: 69–77.

Starmer, W.T. and J.C. Fogleman. 1986. Coadaptation of *Drosophila* and yeasts in their natural habitat. *Journal of Chemical Ecology* 12: 1037–1055.

Statheropoulos, M., C. Spiliopouiou, and A. Agapiou. 2005. A study of volatile organic compounds evolved from the decaying human body. *Forensic Science International* 153: 147–155.

Stephens, P.A. and W.J. Sutherland. 1999. Consequences of the Allee effect for behaviour, ecology and conservation. *Trends in Ecology and Evolution* 14: 401–405.

Stephens, P.A., W.J. Sutherland, and R.P. Freckleton. 1999. What is the Allee effect? *Oikos* 87: 185–190.

Strange, A., P. Li, C. Lister, J. Anderson, N. Warthmann, C. Shindo, J. Irwin, M. Nordborg, and C. Dean. 2011. Major-effect alleles at relatively few loci underlie distinct vernalization and flowering variation in *Arabidopsis* accessions. *PLoS ONE* 6: e19949.

Suutari, M. and S. Laakso. 1994. Microbial fatty-acids and thermal adaptation. *Critical Reviews in Microbiology* 20: 285–328.

Sword, G.A. 2003. To be or not to be a locust? A comparative analysis of behavioral phase change in nymphs of *Schistocerca americana* and *S. gregaria*. *Journal of Insect Physiology* 49: 709–717.

Sword, G.A., P.D. Lorch, and D.T. Gwynne. 2005. Migratory bands give crickets protection. *Nature* 433: 703–703.

Sword, G.A. and S.J. Simpson. 2000. Is there an intraspecific role for density-dependent colour change in the desert locust? *Animal Behaviour* 59: 861–870.

Sze, S.H., J.P. Dunham, B. Carey, P.L. Chang, F. Li, R.M. Edman, C. Fjeldsted, M.J. Scott, S.V. Nuzhdin, and A.M. Tarone. 2012. A *de novo* transcriptome assembly of *Lucilia sericata* (Diptera: Calliphoridae) with predicted alternative splices, single nucleotide polymorphisms and transcript expression estimates. *Insect Molecular Biology* 21: 205–221.

Tachibana, S.I., H. Numata, and S.G. Goto. 2005. Gene expression of heat-shock proteins (*Hsp23*, *Hsp70* and *Hsp90*) during and after larval diapause in the blow fly *Lucilia sericata*. *Journal of Insect Physiology* 51: 641–647.

Tarone, A.M. and D.R. Foran. 2006. Components of developmental plasticity in a Michigan population of *Lucilia sericata* (Diptera: Calliphoridae). *Journal of Medical Entomology* 43: 1023–1033.

Tarone, A.M., L.M. McIntyre, L.G. Harshman, and S.V. Nuzhdin. 2012. Genetic variation in the Yolk protein expression network of *Drosophila melanogaster*: Sex-biased negative correlations with longevity. *Heredity* 109: 226–234.

Tarone, A.M., C.J. Picard, C. Spiegelman, and D.R. Foran. 2011. Population and temperature effects on *Lucilia sericata* (Diptera: Calliphoridae) body size and minimum development time. *Journal of Medical Entomology* 48: 1062–1068.

Tattersall, G.J., B.J. Sinclair, P.C. Withers, P.A. Fields, F. Seebacher, C.E. Cooper, and S.K. Maloney. 2012. Coping with thermal challenges: Physiological adaptations to environmental temperatures. *Comprehensive Physiology* 2: 2151–2202.

Telonis-Scott, M., M. Gane, S. DeGaris, C.M. Sgro, and A.A. Hoffmann. 2012. High resolution mapping of candidate alleles for desiccation resistance in *Drosophila melanogaster* under selection. *Molecular Biology and Evolution* 29: 1335–1351.

Telonis-Scott, M., K.M. Guthridge, and A.A. Hoffmann. 2006. A new set of laboratory-selected *Drosophila melanogaster* lines for the analysis of desiccation resistance: Response to selection, physiology and correlated responses. *Journal of Experimental Biology* 209: 1837–1847.

Telonis-Scott, M., B. van Heerwaarden, T.K. Johnson, A.A. Hoffmann, and C.M. Sgro. 2013. New levels of transcriptome complexity at upper thermal limits in wild *Drosophila* revealed by exon expression analysis. *Genetics* 195: 809–830.

Templeton, A.R., T.J. Crease, and F. Shah. 1985. The molecular through ecological genetics of *abnormal abdomen* in *Drosophila mercatorum*. I. Basic genetics. *Genetics* 111: 805–818.

Templeton, A.R., H. Hollocher, and J.S. Johnston. 1993. The molecular through ecological genetics of *abnormal abdomen* in *Drosophila mercatorum*. V. Female phenotypic expression on natural genetic backgrounds and in natural environments. *Genetics* 134: 475–485.

Templeton, A.R., H. Hollocher, S. Lawler, and J.S. Johnston. 1989. Natural selection and ribosomal DNA in *Drosophila*. *Genome* 31: 296–303.

Terashima, J. and M. Bownes. 2004. Translating available food into the number of eggs laid by *Drosphila melanogaster*. *Genetics* 167: 1711–1719.

Thompson, C.R., R.S. Brogan, L.Z. Scheifele, and D.B. Rivers. 2013. Bacterial interactions with necrophagous flies. *Annals of the Entomological Society of America* 106: 799–809.

Toivonen, J.M. and L. Partridge. 2009. Endocrine regulation of aging and reproduction in *Drosophila*. *Molecular and Cellular Endocrinology* 299: 39–50.

Tomberlin, J.K., M.E. Benbow, A.M. Tarone, and R.M. Mohr. 2011. Basic research in evolution and ecology enhances forensics. *Trends in Ecology and Evolution* 26: 53–55.

Trotta, V., F.C. Calboli, M. Ziosi, D. Guerra, M.C. Pezzoli, J.R. David, and S. Cavicchi. 2006. Thermal plasticity in *Drosophila melanogaster*: A comparison of geographic populations. *BMC Evolutionary Biology* 6: 67.

Tufail, M. and M. Takeda. 2008. Molecular characteristics of insect vitellogenins. *Journal of Insect Physiology* 54: 1447–1458.

Tufail, M. and M. Takeda. 2009. Insect vitellogenin/lipophorin receptors: Molecular structures, role in oogenesis, and regulatory mechanisms. *Journal of Insect Physiology* 55: 87–103.

Turner, B. and T. Howard. 1992. Metabolic heat-generation in dipteran larval aggregations: A consideration for forensic entomology. *Medical and Veterinary Entomology* 6: 179–181.

Turner, T.L., M.T. Levine, M.L. Eckert, and D.J. Begun. 2008. Genomic analysis of adaptive differentiation in *Drosophila melanogaster*. *Genetics* 179: 455–473.

Turner, T.L., A.D. Stewart, A.T. Fields, W.R. Rice, and A.M. Tarone. 2011. Population-based resequencing of experimentally evolved populations reveals the genetic basis of body size variation in *Drosophila melanogaster*. *PLoS Genetics* 7: e1001336.

van der Most, P.J., B. de Jong, H.K. Parmentier, and S. Verhulst. 2011. Trade-off between growth and immune function: A meta-analysis of selection experiments. *Functional Ecology* 25: 74–80.

van der Oost, R., J. Beyer, and N.P.E. Vermeulen. 2003. Fish bioaccumulation and biomarkers in environmental risk assessment: A review. *Environmental Toxicology and Pharmacology* 13: 57–149.

Vass, A.A., S.A. Barshick, G. Sega, J. Caton, J.T. Skeen, J.C. Love, and J.A. Synstelien. 2002. Decomposition chemistry of human remains: A new methodology for determining the postmortem interval. *Journal of Forensic Sciences* 47: 542–553.

Verhulst, S., J.M. Tinbergen, and S. Daan. 1997. Multiple breeding in the great tit: A trade-off between successive reproductive attempts? *Functional Ecology* 11: 714–722.

Via, S. 1991a. Specialized host plant performance of pea aphid clones is not altered by experience. *Ecology* 72: 1420–1427.

Via, S. 1991b. The genetic structure of host plant adaptation in a spatial patchwork: Demographic variability among reciprocally transplanted pea aphid clones. *Evolution* 45: 827–852.

Via, S. 1993. Adaptive phenotypic plasticity: Target or by-product of selection in a variable environment. *American Naturalist* 142: 352–365.

Vieille, C. and J.G. Zeikus. 1996. Thermozymes: Identifying molecular determinants of protein structural and functional stability. *Trends in Biotechnology* 14: 183–190.

Vieille, C. and G.J. Zeikus. 2001. Hyperthermophilic enzymes: Sources, uses, and molecular mechanisms for thermostability. *Microbiology and Molecular Biology R* 65: 1–43.

Vieira, C., E.G. Pasyukova, Z.B. Zeng, J.B. Hackett, R.F. Lyman, and T.F.C. Mackay. 2000. Genotype–environment interaction for quantitative trait loci affecting life span in *Drosophila melanogaster*. *Genetics* 154: 213–227.

Vigue, C.L. and F.M. Johnson. 1973. Isoenzyme variability in species of genus *Drosophila*. VI. Frequency– property–environment relationships of allelic *alcohol dehydrogenases* in *D. melanogaster*. *Biochemical Genetics* 9: 213–227.

Vigue, C.L. and W. Sofer. 1974. *Adhn5* temperature-sensitive mutant at *Adh* locus in *Drosophila*. *Biochemical Genetics* 11: 387–396.

Vigue, C.L., P.A. Weisgram, and E. Rosenthal. 1982. Selection at the *alcohol dehydrogenase* locus of *Drosophila melanogaster*: Effects of ethanol and temperature. *Biochemical Genetics* 20: 681–688.

Vijendravarma, R.K., S. Narasimha, and T.J. Kawecki. 2012a. Chronic malnutrition favours smaller critical size for metamorphosis initiation in *Drosophila melanogaster*. *Journal of Evolutionary Biology* 25: 288–292.

Vijendravarma, R.K., S. Narasimha, and T.J. Kawecki. 2012b. Evolution of foraging behaviour in response to chronic malnutrition in *Drosophila melanogaster*. *Philosophical Transactions of the Royal Society of London Series B Biological Sciences* 279: 3540–3546.

Visscher, P.M., W.G. Hill, and N.R. Wray. 2008. Heritability in the genomics era: Concepts and misconceptions. *Nature Reviews Genetics* 9: 255–266.

Wallenstein, M.D. and E.K. Hall. 2012. A trait-based framework for predicting when and where microbial adaptation to climate change will affect ecosystem functioning. *Biogeochemistry* 109: 35–47.

Wang, W.X. and R.S.K. Wong. 2003. Bioaccumulation kinetics and exposure pathways of inorganic mercury and methylmercury in a marine fish, the sweetlips *Plectorhinchus gibbosus*. *Marine Ecology Progress Series* 261: 257–268.

Ward, R.D. 1975. *Alcohol dehydrogenase* activity in *Drosophila melanogaster*: A quantitative character. *Genetical Research* 26: 81–93.

Ward, B.L., W.T. Starmer, J.S. Russell, and W.B. Heed. 1974. Correlation of climate and host plant morphology with a geographic gradient of an inversion polymorphism in *Drosophila pachea*. *Evolution* 28: 565–575.

Wayne, M.L., J.B. Hackett, C.L. Dilda, S.V. Nuzhdin, E.G. Pasyukova, and T.F.C. Mackay. 2001. Quantitative trait locus mapping of fitness-related traits in *Drosophila melanogaster*. *Genetical Research* 77: 107–116.

Weinberg, W. 1908. Über den Nachweis der Vererbung beim Menschen. *Jahreshefte des Vereins für vaterländische Naturkunde in Württemberg* 64: 368–382.

Westendorp, R.G.J. and T.B.L. Kirkwood. 1998. Human longevity at the cost of reproductive success. *Nature* 396: 743–746.

Westendorp, R.G.J., G.M. van Dunne, T.B.L. Kirkwood, F.M. Helmerhorst, and T.W.J. Huizinga. 2001. Optimizing human fertility and survival. *Nature Medicine* 7: 873.

Westneat, D.F. and C.W. Fox. 2010. *Evolutionary Behavioral Ecology*. Oxford University Press, Oxford, New York.

Whitman, D. and T.N. Ananthakrishnan. 2009. *Phenotypic Plasticity of Insects: Mechanisms and Consequences*. Science Publishers, Enfield, NH.

Wikelski, M. and R.E. Ricklefs. 2001. The physiology of life histories. *Trends in Ecology and Evolution* 16: 479–481.

Wittkopp, P.J., S.B. Carroll, and A. Kopp. 2003b. Evolution in black and white: Genetic control of pigment patterns in *Drosophila*. *Trends in Genetics* 19: 495–504.

Wittkopp, P.J., G. Smith-Winberry, L.L. Arnold, E.M. Thompson, A.M. Cooley, D.C. Yuan, Q. Song, and B.F. McAllister. 2011. Local adaptation for body color in *Drosophila americana*. *Heredity* 106: 592–602.

Wittkopp, P.J., E.E. Stewart, L.L. Arnold, A.H. Neidert, B.K. Haerum, E.M. Thompson, S. Akhras, G. Smith-Winberry, and L. Shefner. 2009. Intraspecific polymorphism to interspecific divergence: Genetics of pigmentation in *Drosophila*. *Science* 326: 540–544.

Wittkopp, P.J., J.R. True, and S.B. Carroll. 2002a. Reciprocal functions of the *Drosophila yellow* and *ebony* proteins in the development and evolution of pigment patterns. *Development* 129: 1849–1858.

Wittkopp, P.J., K. Vaccaro, and S.B. Carroll. 2002b. Evolution of *yellow* gene regulation and pigmentation in *Drosophila*. *Current Biology* 12: 1547–1556.

Wittkopp, P.J., B.L. Williams, J.E. Selegue, and S.B. Carroll. 2003a. *Drosophila* pigmentation evolution: Divergent genotypes underlying convergent phenotypes. *Proceedings of the National Academy of Sciences* 100: 1808–1813.

Wodtke, E. 1981. Temperature adaptation of biological membranes: Compensation of the molar activity of *Cytochrome C oxidase* in the mitochondrial energy transducing membrane during thermal acclimation of the carp *Cyprinus carpio* L. *Biochimica et Biophysica Acta* 640: 710–720.

Wright, S. and T. Dobzhansky. 1946. Genetics of natural populations: Experimental reproduction of some of the changes caused by natural selection in certain populations of *Drosophila pseudoobscura*. *Genetics* 31: 125–156.

Wright, S. and W.E. Kerr. 1954. Experimental studies of the distribution of gene frequencies in very small populations of *Drosophila melanogaster*. II. Bar. *Evolution* 8: 225–240.

Xie, X., A.P. Michel, D. Schwarz, J. Rull, S. Velez, A.A. Forbes, M. Aluja, and J.L. Feder. 2008. Radiation and divergence in the *Rhagoletis pomonella* species complex: Inferences from DNA sequence data. *Journal of Evolutionary Biology* 21: 900–913.

Yeates, D.K. and B.M. Wiegmann. 2005. *The Evolutionary Biology of Flies*. Columbia University Press, New York.

Zavodszky, P., J. Kardos, A. Svingor, and G.A. Petsko. 1998. Adjustment of conformational flexibility is a key event in the thermal adaptation of proteins. *Proceedings of the National Academy of Sciences* 95: 7406–7411.

Zera, A.J. 2005. Intermediary metabolism and life history trade-offs: Lipid metabolism in lines of the wing-polymorphic cricket *Gryllus firmus* selected for flight capability versus early age reproduction. *Integrative and Comparative Biology* 45: 511–524.

Zera, A.J. and T. Brink. 2000. Nutrient absorption and utilization by wing and flight muscle morphs of the cricket *Gryllus firmus*: Implications for the trade-off between flight capability and early reproduction. *Journal of Insect Physiology* 46: 1207–1218.

Zera, A.J. and L.G. Harshman. 2001. The physiology of life history trade-offs in animals. *Annual Review of Ecology and Systematics* 32: 95–126.

Zhang, M., H. Yu, Y. Yang, C. Song, X. Hu, and G. Zhang. 2013. Analysis of the transcriptome of blowfly *Chrysomya megacephala* (Fabricius) larvae in responses to different edible oils. *PLoS ONE* 8: e63168.

Zwaan, B.J., R.B. Azevedo, A.C. James, J. Van't Land, and L. Partridge. 2000. Cellular basis of wing size variation *in Drosophila melanogaster*: A comparison of latitudinal clines on two continents. *Heredity* 84: 338–347.

15

Quantitative Genetics of Life History Traits in Coprophagous and Necrophagous Insects

Wolf Blanckenhorn

CONTENTS

15.1 Introduction: Why Life History Evolution Is Important for Forensics

One species' refuse can be another species' food—or entire habitat, for that matter. This old and fundamental ecological wisdom forms the basis of this chapter, in fact this whole book. That necrophagous insects eating dead or decaying animals are involved in recycling carcasses in nature has been long known, and this knowledge has been applied to investigating medical and criminal cases of human death for at least 150 years (Mégnin 1894). Since then, forensic entomology has become a full-fledged science (Tomberlin et al. 2011a), meaning that its applications should be grounded in firm, testable, and predictable principles derived from basic science in the relevant fields of evolution, ecology, behavior, physiology, and genetics (Tomberlin et al. 2011b). This book, including this chapter, should aid in achieving this goal.

Forensically relevant insects, the majority of which are flies and beetles (Byrd and Castner 2010; Chapter 24), at least occasionally but more often obligatorily feed on carrion. In my article, I shall stray away a bit from carrion. Instead, I shall primarily discuss animal excrements, or "dung." The ecological analogies between *necrophagous* carrion and *coprophagous* dung decomposers are obvious: both typi-cally live, breed in, and feed on remains from (mainly large) vertebrates. Carrion or dung thus becomes the prime habitat for all those insects that are not merely occasional visitors. Consequently, their life cycle, or life history, revolves around those habitats. The *life history* of an organism can be defined as its particular schedule of birth, growth, reproduction, and death. It is a trait syndrome that is best understood as an adaptation, evolved by natural selection, of the organism to its environment. There is a rich body of literature on life history theory and experimentation (e.g., Roff 1992; Stearns 1992; Nylin and Gotthard 1998), which is highly relevant if we want to fully understand the evolution, ecology, and behavior of forensic insects. The standard life history traits (or components) are longevity (duration of life), egg or (more generally) propagule size and number, fecundity (number of offspring), fertility (insemination success), survival (or mortality) rates, growth rate, development time, body size, age and size at maturity, diapause (i.e., hibernation, aestivation, or dormancy), dispersal, and the like (Roff 1992). All these traits have heritable (genetic) components and they are interconnected, that is, they do not evolve indepen-dently from each other. It should be obvious that any organism cannot grow infinitely fast to a giant size and live forever to endlessly produce large offspring. Typically, the environment is limiting in terms of food and other resources, so life history variation "is constrained in large measure by trade-offs between traits" (Roff 1992). Thus, for instance, an organism can either produce many small or few large offspring (Fretwell 1972). The resulting life histories follow classic patterns along a continuum. Large organisms in stable environments often have slow life cycles and few offspring, providing extensive parental care (so-called *K*-selected), while small organisms in unstable environments have fast life cycles and many offspring without parental care (so-called *r*-selected). Some organisms only reproduce once (semelpar-ity; e.g. salmon or mayflies), whereas most others reproduce repeatedly over extended periods of their life (iteroparity). Other such patterns exist.

This chapter focuses on the evolutionary ecology of life history and associated traits of coprophagous, dung-breeding but also necrophagous, carrion-breeding insects, primarily flies and beetles. It seems that more experimental laboratory and field work in this realm has been carried out on the former than the latter, perhaps because there is a long history of trying to understand the life and ecosystem ser-vice of dung-breeding insects in the context of human agriculture and farming (Holter 1979; Hanski and Cambefort 1991; Skidmore 1991; Jochmann et al. 2011). I shall discuss similarities and differences between the two habitat or ecotypes to identify fruitful avenues to pursue in the field of carrion ecology, evolution, and genetics in light of a suggested new framework to bridge basic and applied research in forensic entomology (Tomberlin et al. 2011a,b). My main argument and take-home message will be that insect life histories are complex, highly plastic, and perpetually evolve in response to changing environ-ments. Thus, populations of the same species as a rule will be geographically differentiated. For exam-ple, the duration of development of a deer bot fly (*Cephenemyia* spp.) in northern Norway is likely not the same as that of the same species further south (even at similar environmental conditions), a fact that is certainly relevant for forensics. This is because northern flies are likely adapted to the shorter season in their cooler climate and may therefore have evolved faster development rates (cf. Nilssen 1997a,b). For the same reason, Norwegian bot flies may also have evolved different dispersal regimes, and they may be physiologically attracted to different host species (e.g., reindeer vs. red deer). Such evolved genetic popu-lation differences can best be uncovered experimentally under so-called "common garden" conditions in the laboratory (see Section 15.2.5.1), because differences evident in the field are strongly influenced by environmental variation between the local population sites and thus primarily reflect phenotypic plastic-ity, which may or may not be adaptive. Due to this interplay between genetic and environmental factors, this kind of life history work is generally grounded in the theory of *quantitative genetics*, of which phe-notypic plasticity is an essential part.

Here I first review the major and most basic tenets and methods of quantitative genetic theory (Falconer 1989; Roff 1997; Lynch and Walsh 1998) as they relate to and are important for understanding research on insect life histories. I shall then contrast coprophagous (dung) and necrophagous (carrion) insects in terms of their commonalities and differences, before discussing quantitative genetic laboratory and field experiments investigating life history traits of dung- and carrion-breeding insects. This will include both single-population studies of phenotypic plasticity in response to various environmental factors and studies comparing geographic, altitudinal, or latitudinal populations. Whenever appropriate, this chapter will also touch upon population genetic aspects of population comparisons. Focus will be on the above-mentioned lifehistory traits, though behavioral mechanisms relating to dispersal, host, and mate searching will also be considered (cf. Tomberlin et al. 2011b).

15.2 Quantitative Genetics in a Nutshell

15.2.1 Basic Variance Partitioning of Quantitative Traits and Definition of Phenotypic Plasticity

Molecular genetics studies the function of (typically) single genes, or at best gene cascades and simple gene networks and their resulting phenotype. It goes bottom up, from the gene to the phenotype, and typically assumes a simple one-to-one relationship between genotype and phenotype. As such, it is inherently reductionist in nature, attempting to understand the whole (the organism) by the sum of its parts (the genes). Whole-organism studies of evolution, ecology, and behavior instead focus on complex reproductive (e.g., timing of egg laying), behavioral (e.g., mate or habitat choice), or life history traits (e.g., body size and those other traits listed above), which are typically affected by multiple genes: these traits are quantitative (rather than qualitative) or *polygenic* (Falconer 1989). Quantitative genetics "is concerned with the inheritance of those differences between individuals that are of degree rather than of kind," writes Falconer (1989). While humans have been breeding animals and plants for thousands of years without knowledge of the underlying genes, the scientific theory of quantitative genetics has been derived only over the past 100+ years from this experience by the architects of the modern evolutionary synthesis Fisher, Haldane, and Wright in the context of Darwinian thinking (Falconer 1989). Following the classic approach of Mendel (1865) in discovering genetics, the approach goes top down, from the phenotype to the gene (analogous to what is called reverse genetics by molecular biologists). The approach is entirely statistical and thus phenomenological, and can be applied without knowing the particular genes actually influencing a quantitative trait such as body size. It merely assumes the existence of several (even many) underlying genes with likely varying degrees of effect. Ultimately, of course, it is, or should be, also the aim of quantitative genetics to entirely work out the genetic basis of the traits in question. So-called *quantitative trait locus* mapping can be used to uncover the number and location in the genome of the genes affecting, say, body size with the help of molecular markers such as microsatellites or amplified fragment length polymorphisms (Lynch and Walsh 1998). In so doing, this classical technique combines and hence synthesizes the top-down methods of quantitative genetics with the bottom-up methods of molecular biology to achieve a common goal. More recent genomic techniques, such as genome-wide association mapping, have developed such methods further (Atwell et al. 2010 for *Arabidopsis thaliana* and Mackay et al. 2012 for *Drosophila melanogaster*).

The main objective of quantitative genetics is to, at least, qualitatively, but ultimately quantitatively determine the relative contributions of genetic and environmental sources to phenotypic variation in a particular trait in a population. This is crucial because traits can only evolve if there is underlying heritable (genetic) variation. In the simplest case, the phenotypic variance V_p of a quantitative trait in a population, such as body size, can be partitioned into three components due to genetic (V_g), environmental (V_e) and genotype-by-environment effects ($V_{g \times e}$) (Falconer 1989; Roff 1997):

$$V_p = V_g + V_e + V_{g \times e} \tag{15.1}$$

This is illustrated in Figure 15.1, where genotype 1 is generally larger than genotype 2 in either environment; thus, there is genetic variation in body size in the population (V_g). In response to improved environmental (e.g., food) quality (on the *x*-axis), genotype 1 increases in size, an environmental effect (V_e); so, this genotype is phenotypically plastic. As this is not the case for genotype 2, there is genetic variation in the response of genotypes to environmental quality in the population ($V_{g \times e}$; Scheiner and Lyman 1989). Phenotypic plasticity consequently has two components: a purely environmental component (V_e), mediated here by the food environment, and a heritable component ($V_{g \times e}$) mediated by the different response to food level of the different genotypes (Via and Lande 1985; Stearns et al. 1991). Nonparallel lines (in

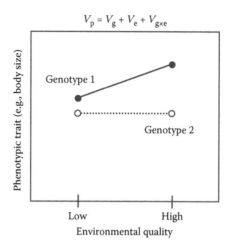

$$V_p = V_g + V_e + V_{g \times e}$$

FIGURE 15.1 Variance partitioning of a quantitative phenotypic trait (V_p), for example, body size, into three components due to genetic (V_g), environmental (V_e) and genotype-by-environment effects ($V_{g \times e}$). Genotype 1 is larger than genotype 2 in either environment, so there is genetic variation in body size within the population (= V_g). In response to improved environmental quality genotype 1 increases in size (= V_e), so it is phenotypically plastic. As this is not the case for genotype 2, there is genetic variation in the response of genotypes to environmental quality in the population (= $V_{g \times e}$). Phenotypic plasticity consequently has two components, a purely environmental (V_e) and a heritable ($V_{g \times e}$) component.

any figure such as Figure 15.1) indicate genetic variation in phenotypic plasticity, that is, genotype-by-environment interactions $V_{g \times e}$ (Via and Lande 1985). In general, relationships of phenotypic traits (on the y-axis), of genotypes or whole populations, in response to any variable environmental factor (on the x-axis), as depicted in Figure 15.1, are called *reaction norms*. These reaction norms need not be linear and can be complex (Stearns and Koella 1986; van Noordwijk and de Jong 1986). Thus, phenotypic plasticity occurs when a particular genotype assumes a different phenotype in a different environment (Schmalhausen 1949; Stearns 1989, 1992). Such plasticity is ubiquitous in plants and animals and affects almost all behavioral, physiological, morphological, and life history traits. Any variable environmental factor can produce plastic responses in organisms, the most prominent factors being food availability, predators, temperature, season length, photoperiod, and humidity, as will be discussed later.

It is important to note that there are two competing concepts about the genetics of phenotypic plasticity (Schlichting 1986; de Jong 1995; Via et al. 1995). One school of thought posits that the reaction norm itself (i.e., the slope of the line in Figure 15.1) can be regarded as a trait that is controlled by genes, which may therefore evolve somewhat independent of the trait (Via and Lande 1985; Via 1993a,b). The other school of thought believes that only the trait itself (i.e., the endpoints in Figure 15.1) is controlled by genes, perhaps by different genes and/or by additional regulatory genes in different environments, and thus evolves. In this view, the reaction norm (i.e., the line) linking the environment-specific phenotypic values is merely an epiphenomenon (Scheiner 1993a,b). In either case, phenotypic plasticity (the slope) has a heritable basis and *can* therefore evolve, directly or indirectly. It should also be clear that phenotypic plasticity can easily be reconciled with the concepts of gene expression and gene regulation, although their actual roles remain to be determined in any particular case.

15.2.2 Heritability Defined

From Equation 15.1, the central concept of the heritability h^2 of a quantitative trait can be easily derived, which is defined as the proportion (or percentage) of the total phenotypic variance V_p that is genetic:

$$h^2 (\text{broad sense}) = \frac{V_g}{V_p} \tag{15.2}$$

Theoretically, h^2 ranges between 0 and 1 (or 100%), though in practice estimates <0 and >1 can occur. Equation 15.2 defines the broad-sense heritability, as in addition to the evolutionarily relevant *additive genetic variance* V_a, the genetic variance V_g can also include *dominance variance* V_d, genetic variation that is due to dominance interactions of alleles at the same locus, and *epistatic variance* V_i, genetic variation that is due to interactions of alleles at different loci (i.e., $V_g = V_a + V_d + V_i$). These latter two variance components are not strictly passed on to the next generation (Falconer 1989; Roff 1997). Thus,

$$h^2 (\text{narrow sense}) = \frac{V_a}{V_p} \tag{15.3}$$

and h^2 (narrow sense) $\leq h^2$ (broad sense). The narrow sense heritability is the best available estimate of the heritable genetic component of any quantitative trait.

15.2.3 Genetic Correlation Defined

As previously mentioned, life history traits tend to be correlated. For example, to grow to a large size, an individual typically requires longer development and/or faster growth. Thus body size and development time will likely be positively correlated phenotypically. Analogous to the concept of genetic variance just explained, this phenotypic correlation (or covariance) also has a genetic component, called the genetic correlation (or covariance), resulting primarily from *pleiotropic* influences of several genes on a given phenotypic trait (Roff 1997). In statistics, the phenotypic correlation r_p of two traits x and y is defined as the phenotypic covariance standardized by the square-root of the product of the two individual variances:

$$r_p = \frac{\text{COV}_{pxy}}{\sqrt{V_{px} V_{py}}} \tag{15.4}$$

The phenotypic correlation again can be decomposed into genetic and environmental components, analogous to Equation 15.1 (but ignoring the g × e interaction here for simplicity), with only the covariances being additive (and *not* the correlation, being a ratio):

$$\text{COV}_{pxy} = \text{COV}_{gxy} + \text{COV}_{exy} \tag{15.5}$$

Thus, the genetic correlation is

$$r_g = \frac{\text{COV}_{gxy}}{\sqrt{V_{gx} V_{gy}}} \tag{15.6}$$

the environmental correlation being analogous, substituting p for g. For the strictly additive genetic effects (as above), again substitute a for g.

15.2.4 Coefficient of Relatedness Defined

The fundamental logic of quantitative genetic inference is that related individuals should resemble each other, because they share genes passed on by their ancestors that are identical by descent. Therefore, all empirical methods hinge on comparisons of genetically related individuals. The degree of resemblance predictably depends on the degree of relatedness. Clones, including identical twins, are 100% related, as they share all their genes; their coefficient of relatedness $r = 1$. Each offspring only shares 50% of its genes with each parent: $r = 0.5$. Full siblings again have $r = 0.5$, because at any locus they can either share both alleles, one allele or no alleles, with equal probability, so on average one of two alleles, or 50%, are shared. The coefficient of relatedness of grandparents and grand-offspring $r = 0.25$, as it is for half siblings; for cousins, it is $r = 0.125$, and so on.

15.2.5 Classic Methods for Establishing and Quantifying the Genetic Component (Heritability) or Genetic Covariance (Correlation) of Quantitative Traits

Based on the above logic of resemblance due to genes shared by decent, in what follows, I briefly characterize various classic methods used to estimate quantitative genetic variation and parameters: parent–offspring regression, sib analyses, analysis of any related individuals via animal models, iso-female line approaches, and artificial selection studies. For calculation details, please refer to the classic textbooks on quantitative genetics (Falconer 1989; Becker 1992; Roff 1997; Lynch and Walsh 1998).

15.2.5.1 Common Garden Experiments

Although the main objective of quantitative genetics is to quantify heritabilities and genetic correlations, we start with the simplest case of qualitative establishment of genetic differences between groups of individuals or populations by means of the so-called common garden experiments. Evolutionary ecologists are regularly interested in knowing whether, and in a next step to what degree, any differences between populations in quantitative traits are genetic or merely environmental. For example, are plants or animals living at high altitudes or latitudes smaller because of the harsh environmental conditions they face, or because they are genetically different (Franz 1979; Blanckenhorn 1997a; Chown and Klok 2003)? From Equation 15.1, it is intuitively obvious that if we raise offspring from different populations in any "common garden," meaning in the same laboratory or field environment, thus holding environmental variation V_e constant, any remaining differences between the populations must be genetic, that is, represent V_g or $V_{g \times e}$. Such experiments are most effectively presented in interaction plots such as Figure 15.1, which display population differentiation in quantitative traits (e.g., Figure 15.2) but not necessarily permit calculating heritabilities or genetic correlations unless relatives are directly compared, as outlined below.

15.2.5.2 Parent–Offspring Regression

As parents and offspring share 50% of their genes and consequently resemble each other, quantitative genetic parameters can be estimated by comparing their phenotypes (Falconer 1989; Roff 1997; Lynch and Walsh 1998). Thus, tall parents, of either sex, will produce tall offspring, on average, if body size is heritable (V_g). Of course, large size can also result from environmental causes, such as good nutrition (V_e). Nevertheless, when the average body size (or any other quantitative trait) of many families is compared in a least-squares regression plot of offspring size (on the y-axis) on parent size (on the x-axis), a positive correlation should result because larger parents will have larger offspring. The slope of the regression of the mid-offspring body size (measuring whichever trait), averaging over all offspring of both sexes, on mid-parent body size, averaging the two parents, will be equal to the heritability h^2 of the measured body size trait (Falconer 1989). As for any regression, these estimates come with a corresponding standard error.

From statistics, we know that the more scatter due to environmental sources there is, the shallower the slope and the lower the heritability. Therefore, it is important to standardize and to ideally equalize the environmental conditions under which the parents and offspring were raised, in the extreme and ideally requiring bringing up two generations in a controlled laboratory (if possible) or field environment. Nevertheless, it is possible to derive parent–offspring heritabilities from organisms that can be marked and tracked over generations in the field (Kruuk 2004), primarily larger vertebrates or plants, but not easily with small invertebrates. In this case, environmental conditions will surely vary unpredictably over generations, incurring extra environmental variance that is expected to dilute the field heritability estimate relative to any estimate derived under constant laboratory conditions. Thus, field heritabilities are expected to be lower on average (Weigensberg and Roff 1996; Blanckenhorn 2002). What should be obvious is that any heritability estimate will depend somewhat, and potentially strongly, on the particular environmental conditions under which it was estimated; it is hence population- and environment-specific and of limited value.

A final important point in this context is that traits can be sex-specific (e.g., egg size) or they can differ between the sexes, for example, when there is sexual dimorphism, which is common (Fairbairn et al. 2007), or when males bear particular secondary sexual traits. Computing a pooled mid-parent value is then not advised and statistically doubtful. In such cases, sex-specific (i.e., daughter–mother and son–father) regression estimates should be computed instead. As quantitative traits are biparentally inherited via the autosomes (unless the trait is sex-linked), sex-specific data only consider half the picture, and the slope estimates from daughter–mother and son–father regressions have to be doubled to obtain the sex-specific heritabilities.

15.2.5.3 Sib Analyses

Rather than parents and offspring, we can compare siblings. This has the advantage that we only have to consider or raise one generation, thus assuring that the offspring environment is roughly standardized. However, offspring of many families, the relevant unit of replication, have to be scored in order to arrive at a good estimate of the genetic variation for a particular population and environment. Individuals of the same family should resemble each other predictably relative to unrelated individuals from different families. There are various concrete analysis of variance designs depending on the organism. Typically, in animals with internal fertilization, a number of males are each mated to a number of females to produce families of full sibs (having the same father and mother) and half sibs (having the same father but different mothers). Two randomly picked offspring individuals can thus be either unrelated ($r = 0$), half sibs ($r = 0.25$), or full sibs ($r = 0.5$). By means of nested analysis of variance (Becker 1992), two heritability estimates can be obtained from such a design: a paternal, half sib estimate and a maternal, full sib estimate, corresponding to the above narrow sense (Equation 15.3) and broad sense (Equation 15.2) heritabilities, respectively. For statistical reasons according to the coefficients of relatedness, the maternal (full sib) component has to be doubled and the paternal (half sib) component quadrupled to yield the heritability (Falconer 1989; Becker 1992; Roff 1997). In plants and animals with external fertilization, such as fish or frogs, maternal half sibs in addition to paternal half sibs can be generated because sperm can be freely allocated to a well-defined proportion of a female's eggs. Thus, other sib designs bearing various historical names exist (e.g., North Carolina design, diallel cross, etc.; see Roff 1997, 49 ff).

Since males only contribute sperm and often offer no paternal care to the offspring, the corresponding paternal genetic variance component best reflects the pure, additive genetic variance V_a. In contrast, females typically contribute physiological substances nourishing the offspring with the egg and behavioral care, the so-called nongenetic *maternal effects* (e.g., Jann and Ward 1999). In addition, offspring generally grow up in a *common environment* such as a nest (in birds), a common household (in humans), or any common rearing container for insects. These latter effects contribute to the resemblance of siblings, thus inflating the estimate, even though they are, strictly, nongenetic but environmental (V_e). Finally, the maternal (full sib) component of half/full sib designs additionally contains *dominance variance*; the reasons for this are a bit more complicated to explain, so I refer to Falconer (1989), Roff (1997), or Lynch and Walsh (1998) for further reading. Consequently, the maternal variance component generally reflects the broad-sense genetic variance V_g and often includes some nongenetic variance due to maternal effects or common environment. Note that at least the maternal part of the parent–offspring heritability estimate described above in Section 15.2.5.2 can also contain maternal and common environment variance, but not dominance variance.

15.2.5.4 Generalized Animal Model Approaches

It should be obvious that the above designs based on resemblance among relatives with genes shared by decent can be generalized to include any defined relative such as grandparents, cousins, and so on. Whereas classic parent–offspring and sib designs have been very useful under controlled conditions in the laboratory and for systematic domesticated plant and animal breeding programs, the generalized approach has been particularly fruitful when applied to long-term data on field populations of larger vertebrates that can be individually identified, followed over time, scored for reproductive success, and for which pedigree information is consequently available over many generations (e.g., Merilä et al. 2001; Kruuk 2004; Postma et al. 2007; Wilson et al. 2007). A key advantage is that animal model programs can extract all relevant quantitative genetic estimates by statistically integrating over all family relationships defined by the pedigree for the whole population over many generations, thus arriving at representative genetic parameters for the natural field situation (Kruuk 2004). The approach can also be well integrated with classic population genetic methods (Postma et al. 2011). Unfortunately, this methodology does not work well with small invertebrates that cannot be tracked in the field, although animal model programs certainly serve equally well with insects and well substitute the above traditional statistical approaches.

15.2.5.5 Iso-Female Approaches

Although large vertebrates have the advantage that they can be tracked in the field, small insects can more easily be held in the laboratory under seminatural conditions approaching their field environment. Some organisms such as plants, few vertebrates, and quite a number of arthropods can be held as *clones* or *iso-female* lines, implying that inbreeding is not much of a problem for them. For example, clones of water fleas (*Daphnia* spp.) or aphids occur naturally (e.g., Ebert 1993; Rouchet and Vorburger 2012). Clones, just like identical human twins, share all their genes, so their coefficient of relatedness $r = 1$. Iso-female lines refer to the offspring of field-caught females being separately maintained in the laboratory as genetic lines over many generations. *Drosophila* spp. are the most prominent examples (Hoffmann et al. 2001; David et al. 2005), but the technique also works, for example, for sepsid dung flies (Berger et al. 2013), and I would suspect that at least some carrion-breeding flies or beetles can also be maintained in this way. Continuous brother–sister mating renders individuals more and more homozygous and genetically alike, such that their coefficient of relatedness with time becomes greater than $r = 0.5$ (corresponding to full sibs) and in the extreme approaches that of clones ($r \approx 1$). The main advantage of this approach is that the originally sampled genetic variance is preserved, even if some lines go extinct with time. When alternatively maintaining individuals in the laboratory in one (large) population cage, some or in the extreme one genotype will eventually become most abundant due to inadvertent laboratory selection, thus losing genetic variation. Quantitative genetic estimation methods for iso-female lines exist in the literature (Hoffmann et al. 2001; David et al. 2005), although they necessarily estimate broad-sense genetic parameters, that is, V_g rather than V_a. Iso-female approaches are also very useful for sequencing work, as they permit generating many genetically (almost) identical individuals and therefore sufficient genetic material bearing little unwanted intra-specific variation even for small sexual species.

15.2.5.6 Artificial Selection and the Breeder's Equation

As initially stated, the crucial importance of quantitative genetic estimates in the evolutionary context is that the rate of evolution of any particular trait is proportional to its genetic variation present in the population. In so many words, this is Fisher's so-called fundamental theorem of natural selection (Fisher 1930). That is, if we want to quantitatively predict the evolutionary response R of a trait to natural selection, for example, for breeding purposes, we need to know the strength of selection S and the amount of genetic variation present for the trait, that is, the heritability h^2, as

$$R = S \times h^2 \tag{15.7}$$

This relationship is also known as the breeder's equation, stating that the rate of evolution R is proportional to both the selection intensity S and the heritability of a trait h^2, which makes intuitive sense.

By reshuffling Equation 15.7, it follows that the realized heritability can also be estimated from artificial selection as $h^2 = R/S$ after the fact. For example, in one of our own experiments, we artificially selected for body size in the yellow dung fly (Teuschl et al. 2007) for 22 generations, with selection intensity $S \approx 1$ per generation, meaning that the mean body size of the selected individuals allowed to reproduce in every generation was one standard deviation unit greater (or smaller) than the base population mean. By this procedure, we increased (and decreased) the body size by ca. 10% overall, as evident in an equally linear response in both directions (see Figure 1 in Teuschl et al. 2007). From these responses, we could calculate the realized heritability of body size as $h^2 = 0.34$. Because selection was performed only on males, this number is twice the obtained slope of the regression through the selection response over the 22 generations of 0.17 (in absolute body size units per generation) = R/S. The realized heritability is of course the best available heritability estimate, as no assumptions are necessary about which components of variance are included. On the other hand, there is little point in estimating h^2 from artificial selection (which is equivalent to breeding), as typically the whole (and sole) reason for estimating h^2 is to predict the response to selection (and not vice versa).

15.3 Similarities and Differences of Dung and Carrion as a Habitat

Carrion or dung, stemming mainly from vertebrates, is the main habitat for necrophagous and coprophagous organisms, respectively, in which they live, feed, and reproduce. Their ecological similarities, summarized in Table 15.1, initially prompted this chapter, and a number of species knowingly use both resources (e.g., the generalist *Musca domestica*; see list in Byrd and Castner 2010). Patches of dung or dead animals, to use the technical generic term from foraging theory (Stephens and Krebs 1986), are highly ephemeral and unpredictable in nature. Carrion certainly has tremendous nutritional value for a multitude of predators, scavengers, parasites, and decomposers, and this holds also for dung, albeit probably to a lesser degree because large carcasses have no dung equivalent. Whereas the nutritional value of carrion is necessarily protein-rich, it is less clear what nutrients dung dwellers derive from the dung besides partially digested food of the defecator, and microbes or fungi growing in or on feces; however, the latter also abound on carcasses (Holter and Scholtz 2007).

Inter- and intraspecific competition for and in dung pats and animal carcasses will generally be strong because of their high nutritional value. As a consequence, these habitats harbor an entire interacting community of consumers, parasites, predators, and so on (Hanski and Cambefort 1991; Van Laerhoven 2010). For instance, the number of species of arthropods, bacteria, and fungi associated with a typical cowpat in north-central Europe has been estimated in the hundreds (Hammer 1941; Skidmore 1991; Jochmann et al. 2011). Each dung pat offers various ecological niches in space and time, the dung being drier on top but wetter at the bottom; some species prefer fresh dung, for example, the pioneers such as the yellow dung fly *Scathophaga stercoraria* (Diptera: Scathophagidae) or the black scavenger fly *Sepsis cynipsea* (Diptera: Sepsidae) (Parker 1970, 1972), while others come to the dung later when it is drier (Hammer 1941; Pont and Meier 2002). Some species are specialized on dung of particular grazers, whereas others are generalists (Pont and Meier 2002). Although the species richness of carrion and dung at any given site must be necessarily the same, the number of spatiotemporal niches in carrion is likely even greater than in dung, because each carcass is composed of various organs (or parts) with varying degrees of nutritional value, on which particular organisms might specialize. Thus, the diversity of necrophagous organisms can be expected to be even greater than the diversity of coprophagous organisms.

TABLE 15.1
Characteristic Similarities and Differences between Dung and Carrion

Characteristic	Dung	Carrion
Ephemerality	+	+
Unpredictability in time and space	+	+
Size range	+	++
Frequency of occurrence	++	(+)
Nutritional quality range	+	++
Nutritional quantity range	+	++
Diversity across species	+	+
Diversity of nutrients within patches	0	++
Diversity in niches within patches (dry/wet; hot/cool)	+	++
Diversity of community of consumers etc.	+	++
Anthropogenic influences on occurrence	++	+
Attraction mechanisms	+	+

Another important ecological difference is that dung, even of large grazers such as bison or moose, appears regularly over time in relatively small portions, whereas animal carcasses are rare but can be very large. The fact that carrion is removed quickly by vertebrate scavengers and predators (think of the African Savannah in Chapter 21) makes this habitat even rarer for the many necrophorous arthropods. Thus, the resource dung will be small in size, ephemeral, locally unpredictable at the small scale, but globally predictable, reliable, and frequent at the large scale. In contrast, animal carcasses, which are also ephemeral, tend to be larger but will likely be even more unpredictable and infrequent at both the local and global scales. Although there should be many more small carcasses of, for example, juvenile birds, voles, frogs, fish (in the aquatic realm), these will also be quickly removed by larger vertebrate scavengers and predators. Humans have exacerbated this pattern by wiping out many large top predators, and by regularly removing most carcasses of large (livestock) animals (including humans) as soon as they appear. At the same time, domesticated grazers such as cows, horses, sheep, and so on have become very common worldwide, and so has their dung. As a result, dung flies are and may have become even more common and widespread due to human activities. For example, the yellow dung fly *S. stercoraria* occurs throughout the northern hemisphere (Asia, Europe, North America) in places as far north as Spitzbergen (Blanckenhorn et al. 2010); nevertheless, based on neutral genetic marker populations, they are differentiated very little ($F_{st} < 4\%$; Demont et al. 2008). As populations seem to be more differentiated in quantitative traits such as body size and development time (i.e., $G_{st} > F_{st}$: Demont et al. 2008; see Section 15.4), the simplest explanation for this pattern is that population sizes are so huge that differentiation by genetic drift is negligible. The prediction would be that population sizes of necrophagous insects, especially those specialized on large carcasses, should be smaller and show greater neutral genetic differentiation because their habitat is and has become exceedingly rare (cf. Chapter 17). A population genetic study of the carrion beetle *Silpha perforata* (Coleoptera: Silphidae) showed considerable genetic differentiation in northern Japan (Hokkaido), especially when including offshore islands (Ito et al. 2010), although another study of the endangered (i.e., rare) burying beetle *Nicrophorus americanus* showed limited population differentiation in North America (Szalanski et al. 2000; see also Kozol et al. 1994). Generalists such as some muscid, sarcophagid, or calliphorid flies (Byrd and Castner 2010), some of which breed in flesh as well as dung (e.g., *M. domestica*), in contrast, should be widespread, common, and relatively undifferentiated in neutral markers (as found, e.g., for the widespread *Lucilia sericata*: Picard and Wells 2010; see also Gleeson 1995 for *Lucilia cuprina*).

Initial attraction to the habitat, primarily by odor, is likely similar for dung or carrion. All necrophagous and coprophagous organisms should be somehow adapted to perceiving the (for us) foul smell emitted so that they can home in efficiently. The actual volatile chemicals picked up by the various species attracted will certainly differ, if only by chance, be they produced by the carrion itself (Vass et al. 2002), the bacterial community (Janzen 1977), or other organisms using the remains (Watts et al. 1981), and this is a research field of chemical and (neuro)physiological ecology and behavior by itself not further discussed here (Dethier 1947; Visser 1986; Bruce et al. 2005; Schröder and Hilker 2008). Regardless of what the attractant is, the colonization of dung or carcasses can and should be studied from the behavioral perspective (Tomberlin et al. 2005, 2011b). This has been done for the above-mentioned common dung flies (Parker 1970; Blanckenhorn et al. 2000). For example, a simple null hypothesis derived from behavioral ecology and foraging theory is that the colonization of a habitat should be proportional to its size and nutrient value (Stephens and Krebs 1986), that is, the number of dung flies attracted to dung pats of different sizes should be ideal-free distributed (Blanckenhorn et al. 2000). The dung pat surface area (rather than linear diameter or dung volume) turned out to give the best fit, which suggests proportionality to the amount of odor emitted. Moreover, the effect was the same for males and females, resulting in equal operational sex ratios (i.e., the number of males per female) on all dung pats, according to the expected ideal free mate distribution. Thus, initial attraction to the habitat has immediate consequences for mating behavior and reproduction for all those species that also mate at the dung or carrion. Ultimately, this simple mechanism would also equalize the level of intraspecific larval competition among pats. Thus, colonization of the habitat can be studied by means of simple behavioral field experiments testing classic behavioral ecology theory.

Overall, the similarities between the two habitats, dung and carrion, far exceed the differences (Table 15.1), such that the two fields can certainly learn from each other. As discussed, major differences occur in the size, frequency, predictability, and the putative number of ecological niches of the habitat dung versus carrion, which should differentially and testably affect patterns of biodiversity and population genetic differentiation of necrophagous versus coprophagous insects.

15.4 Quantitative Genetic Life History Studies of Coprophagous and Necrophagous Insects

As pointed out in Section 15.1, the life histories of dung and carrion insects are highly plastic and very likely genetically differentiated geographically within species in response to various environmental factors. To motivate future research efforts with forensically relevant carrion-breeding insects in this direction, in what follows, I shall discuss quantitative genetic studies of various life history traits of, primarily, dung-breeding flies and beetles. Subsections will cover the diversity of life history traits and methods used (as outlined in Section 15.2.5), the extent and type of phenotypic plasticity exhibited, and studies of population differentiation within species. Unless pedigree information and breeding information can be obtained in the field (cf. Section 15.2.5.4), such work is typically performed on a necessarily arbitrary handful of "model" species that can be easily reared in the laboratory. These may or may not be representative, although the hope is that mechanisms and patterns can be generalized from them. As life history traits are interdependent and often assessed in conjunction, I shall further structure this section by the environmental variables potentially generating plasticity and differentiation, the most prominent being food availability, temperature, season length, photoperiod, predators, and humidity.

15.4.1 Heritability and Genetic Correlation Studies

If a species can be easily cultured in the laboratory, heritability studies are actually easy to conduct. Nonetheless, they will always be laborious, as high sample sizes are required to obtain reliable and statistically significant estimates (Becker 1992; Roff 1997). Also, as estimates are environment- and population-specific, the conditions have to be decided upon, the choice of which will necessarily be a bit arbitrary but should be guided by concrete research interests and questions. If multiple populations or environments are included, additional assessment of genotype-by-environment interactions ($V_{g \times e}$), reflecting the genetic component of phenotypic plasticity or population differentiation, becomes possible (cf. Section 15.4.2). When more than one trait is assessed per individual, which is strongly advised and efficient, genetic correlations can be computed.

As usual, quantitative estimates of life history and associated behavioral and physiological traits exist in the literature only for a handful of dung and carrion species, likely because these can be reared in the laboratory. The contexts naturally vary. Most estimates exist for the (in the northern hemisphere) cosmopolitan yellow dung fly *S. stercoraria*. In this species, the evolution, and hence the heritability, of body size and its relationship with juvenile development and growth is one central focus. Multiple estimates for these traits exist, generated by most of the methods described above; estimates vary but are overall consistent, with narrow sense $h^2 = 0–0.6$ and broad sense $h^2 = 0.6–1$ (Simmons and Ward 1991; Blanckenhorn 2002; Blanckenhorn and Hosken 2003; Blanckenhorn and Heyland 2004; Teuschl et al. 2007; Thüler et al. 2011). This range encompasses methodological (parent–offspring, sib analyses, artificial selection) as well as environmental variation (food, temperature, photoperiod, seasonality) affecting the estimates. It also includes estimates from sibships reared under seminatural conditions in the field (Blanckenhorn 1997b, 1998a), which is possible and more realistic in small insects, averaging across naturally variable environmental conditions. The genetic correlation between body size (hind tibia length) and development time, reflecting a trade-off because it takes time to get large, is moderate but positive, as expected ($r_g = 0.2–0.5$; Blanckenhorn 1998a; Blanckenhorn and Heyland 2004). The correlation is much lower than that of the ecologically similar *D. melanogaster* (Nunney 1996), presumably because yellow dung flies are well adapted to plastically alter their development, growth, and resulting body size depending on the quantity of the finite dung resource in which they grow up (Blanckenhorn 1998a, 1999). In contrast, the genetic correlations between the sexes for these two traits are expectedly high ($r_g(\text{sex}) = 0.57–0.78$) but clearly less than 1, allowing for independent evolution of both traits in males and females in this sexually dimorphic species (males larger; Blanckenhorn 2002; cf. Fairbairn et al. 2007). In general, body size, based on whichever morphological trait surrogate used (wing length, thorax length, leg length, etc.), is always an easy and very useful trait to measure as a baseline in any heritability assessment, if only because it affects most other life history, behavioral, and physiological traits (Blanckenhorn 2000a), and because it is a prominent surrogate for "condition," an important concept in behavioral ecology (Blanckenhorn and Hosken 2003).

Blanckenhorn (2002) also reports high $h^2 \approx 0.75$ for diapause in the yellow dung fly, a so-called threshold trait derived from presence/absence (1/0) data (Blanckenhorn 1998b), which, in general, requires special statistical correction (see Roff 1997 for details). The heritability of fecundity (clutch size) and egg size is lower ($h^2 = 0.1$–0.3), as can be expected for life history traits closely linked to fitness (Mousseau and Roff 1987), and the genetic correlation between these two traits, expected to be negative due to the presumed energetic trade-off, is also low $r_g = -0.05$ to 0.3; Blanckenhorn and Heyland 2004). Further quantitative genetic estimates in the yellow dung fly include behavioral, physiological, and morphological traits in the reproductive and sexual selection context such as copula duration (Mühlhäuser et al. 1996), sperm length (Ward 2000; Dobler and Hosken 2010), other internal reproductive structures (Simmons 2001; Thüler et al. 2011), phenoloxidase (a substance involved in the insect immune response: Schwarzenbach et al. 2005), lipid and glycogen reserves (Blanckenhorn and Hosken 2003), or fluctuating asymmetry of paired appendages (Blanckenhorn and Hosken 2003). Notably missing are traits relating to locating the resource and traits associated with predation.

Similar quantitative genetic estimates were generated for the black scavenger fly *S. cynipsea* in a sexual selection context (Blanckenhorn et al. 1998; Reusch and Blanckenhorn 1998; Mühlhäuser and Blanckenhorn 2004). Assessed traits include body size and fluctuating asymmetry (traits actually or potentially preferred by females), fecundity and egg size, as well as behavioral traits such as female shaking (indicating reluctance to mate) or male mating persistence, and correlations among them. Note that such genetic designs always permit calculation of the genetic correlation r_g as well as the phenotypic correlation r_p, the latter potentially being sufficient because it often corresponds to r_g (Roff 1995, 1996). Martin and Hosken (2003, 2004) allowed *S. cynipsea* to evolve experimentally under conditions of monogamy and polyandry (i.e., females copulating with multiple males, the natural situation generating conflict between the sexes), which produced several expected and unexpected correlated responses in traits such as longevity or fecundity, which in such a design are necessarily realized genetic correlations r_g. A similar study was conducted in the yellow dung fly by Hosken et al. (2001). A state-of-the-art genetic analysis using modern multivariate methods of quantitative genetic analysis and based on iso-female lines has been recently conducted by Berger et al. (2013) for development rate of the related *Sepsis punctum*. The housefly *M. domestica* (Diptera: Muscidae), a generalist species that is found in both dung and carrion, is also a good model species that can be reared in the laboratory and for which quantitative genetic estimates (on morphological traits) have been generated (Bryant and Meffert 1998).

The only dung beetle group for which quantitative genetic studies exist in the literature are horned beetles of the genus *Onthophagus* (Coleoptera: Scarabaeidae). Again, there is often a body size context, such as in Emlen's (1996) and Nijhout and Emlen's (1998) studies investigating the allometry of horn length in *O. acuminatus*. Another set of studies had a sexual selection and sperm competition context involving body size, horn expression, sperm length, and maternal brood provisioning in *O. taurus* and *O. sagittarius* (Hunt and Simmons 2002; Moczek et al. 2002; Simmons and Kotiaho 2007; Simmons et al. 2009; Watson and Simmons 2010, 2012; House and Simmons 2012).

In contrast to this work on dung breeders, the quantitative genetic literature on carrion organisms is rather sparse in comparison. The reasons for this are not obvious, as many necrophagous flies and beetles should be as easy to rear in the laboratory under controlled conditions. Interestingly, Cooper et al. (2002) estimated the heritability of autogeny (the ability to reproduce without adult protein feeding) in the blowfly *L. cuprina* (Diptera: Calliphoridae) by means of artificial selection. This trait points clearly to the more applied perspective in the forensic field, which is also evident in a quantitative genetic study investigating resistance of sheep to the same blowfly species (Smith et al. 2008). This and other calliphorid fly species also have been studied from a population genetic perspective (Gleeson 1995; Picard and Wells 2009, 2010), but again not as frequently as I would have expected based on their applied importance.

With close to 300 publications in the literature, one group of carrion insect stands out as an experimental model in the nonforensic context: burying beetles of the genus *Nicrophorus* (Coleoptera: Silphidae). Although there are also population genetic studies (e.g., Kozol et al. 1994; Szalanski et al. 2000), the vast majority of studies concern the behavioral ecology of social brood care of these large beetles that specialize on carcasses of small mammals such as mice or voles, research that apparently has a rich tradition in America, Europe, and Asia (Trumbo 1990; Eggert and Sakaluk 1995; Nagano and Suzuki 2007; Ikeda et al. 2011; Steiger et al. 2011). Nevertheless, only very few quantitative genetic studies investigating life history (body size, development time) and behavioral (parental care, offspring begging, etc.) traits exist (Rauter and Moore 2002a,b; Lock et al. 2004).

To summarize, minimum quantitative genetic estimates serve to prove whether and to what extent traits are subject to evolution; at their best, they serve to test particular mechanisms, often in the sexual selection context, facilitating or constraining the evolution of traits that can be and are often expected to be either negatively (by trade-offs) or positively (e.g., male traits and female preference for it) genetically associated (e.g., Simmons 2001; Blanckenhorn and Heyland 2004; Thüler et al. 2011). The discussion of whether cumbersome estimations of genetic parameters are truly necessary or even worthwhile for any particular trait, especially if the phenotypic equivalents are typically similar enough, has not been completely settled (Cheverud 1988; Roff 1995, 1996, 1997; Weigensberg and Roff 1996; Reusch and Blanckenhorn 1998) and may have reached a new level with the development of molecular genetic methods (Turner et al. 2011; Tarone et al. 2012).

15.4.2 Common Garden Studies Assessing Phenotypic Plasticity

Studies of phenotypic plasticity of animal life histories abound in the literature (Arendt 1997; Nylin and Gotthard 1998; Dmitriew 2011). Any study of any life history trait will, deliberately or inadvertently, include some aspect of phenotypic plasticity. Common garden experiments as described above are the best way to assess the quantitative effects of particular environmental variables on a trait under controlled conditions in the laboratory or the field. When combined with population comparisons (see Section 15.4.3) and/or comparisons of related individuals (see Section 15.4.1), such studies become effectively genetic and hence evolutionary.

Virtually all species exhibit phenotypic plasticity in most quantitative traits. For body size, arguably the best-investigated life history trait, ca. 30%–40% of variation is heritable (i.e., V_g: Mousseau and Roff 1987), the remainder being phenotypically plastic (i.e., V_e and V_{gxe}; cf. Figure 15.1). Body size variation is proximately produced by variation in growth rate and development time, as there are two ways to get large: individuals can either grow faster (i.e., increase growth rate) or grow for longer time (i.e., increase development time). These three traits are thus intimately, even mathematically, interconnected and traded off against each other; they can therefore not be investigated in isolation. Other important life history traits exhibiting strong phenotypic plasticity include fecundity, longevity, or offspring (egg) size, all of which tend to be influenced by body size (Blanckenhorn 2000a). In this section, I shall discuss common garden studies conducted with dung or carrion insects, investigating the response of these life history traits to major environmental variables such as food availability, temperature, season length, photoperiod, predators, and humidity, most of which produce well-known evolutionary patterns of phenotypic plasticity in animal life histories.

In dung or carrion insects, the food type, quality, or quantity presented to a set number of developing larvae can be easily manipulated (e.g., Blanckenhorn 1998a; Tarone and Foran 2006; Nagano and Suzuki 2007). Multiple individuals (siblings or not) may compete for their share of the resource, as in nature; alternatively, larvae may be reared individually, which is technically a bit trickier given their small size (Blanckenhorn et al. 2010). In dung organisms, we cannot easily separate effects of larval competition from those of food availability, because the (fluid) dung is the larval habitat and at the same time the food source. In either case, the per capita amount of food minimally necessary and that producing the maximal body size can be worked out experimentally (see Amano 1983 for yellow dung flies). Food restriction generally curtails growth and hence final body size (Berrigan and Charnov 1994), while at the same time prolonging, not changing, or even reducing development time (Blanckenhorn 1999). In the extreme, there is too little food so that all individuals will die before maturation (Blanckenhorn 1998a).

Any body size decrease in response to larval food limitation is an environmentally induced effect representing V_e in Equation 15.1. However, because full sib families were split among three food environments, Blanckenhorn's (1998a) experiment also revealed heritable genotype-by-environment interactions (V_{gxe}), as genotypes varied significantly in growth, development, and body size in response to food limitation. The resulting smaller-bodied individuals usually have low reproductive potential (fecundity, egg size, etc.: Jann and Ward 1999; Reim et al. 2006a), which is generally a positive function of body size in ectothermic (cold-blooded) organisms (Honek 1993; Blanckenhorn 2000a), and they may or may not live longer (Reim et al. 2006b, 2009).

The main conclusion from these studies is that ephemeral, rapidly disappearing habitats such as dung or carrion will likely be limiting depending on size, weather conditions, and the number of competitors present, all of which are unpredictable. Species inhabiting (and depleting) this habitat are likely adapted to this situation in that they should be prepared at any time to abbreviate their development, if necessary at the expense of large final body size, else they will die. Dung flies thus trade off faster development and smaller size against mortality, as they cannot switch to another patch and would certainly die if the pat were depleted before they pupate (Blanckenhorn 1998a). This life history response of maturing earlier but smaller is relatively rare and not predicted by standard life history models (Stearns and Koella 1986; Blanckenhorn 1999); it generally occurs in organisms that regularly (but unpredictably) face habitat depletion that they cannot escape, such as some tadpoles living in ephemeral ponds (Newman 1992), some beetles whose larvae develop in plant seeds (Møller et al. 1989; Fox et al. 1996), or some hymenopteran or dipteran parasitoids (Visser 1994). Whether this life history adaptation, which does contribute to intra- and interspecific variation and therefore should be of forensic relevance, also occurs in carrion feeders remains to be investigated systematically (cf. Tarone and Foran 2006); it would be expected primarily in systems where the resource is regularly limiting, that is, when the carcass is small relative to the number of insects feeding on it, as is the case, for example, for *Nicrophorus* (cf. Nagano and Suzuki 2007). One study by Ireland and Turner (2006) of the blowfly *Calliphora vomitoria* (Diptera: Calliphoridae) demonstrates production of undersized insects even after prolonged development time in crowded conditions, which is the standard response predicted by life history models (Stearns and Koella 1986) because individuals compensate food restriction by growing for longer time to meet a presumed minimal target size.

All biochemical and physiological processes in nature accelerate with temperature, an effect typically quantified as Q_{10}, the factor by which biochemical reactions speed up when temperature is increased by 10°C (Chown and Gaston 1999). Temperature therefore ultimately affects animal life histories; it can also be manipulated easily in experimental settings. From a life history point of view, temperature is an interesting exception to the rule that environments facilitating growth always produce larger body sizes, as high temperatures speed up growth but nevertheless result in smaller body sizes, a life history puzzle (Berrigan and Charnov 1994; Atkinson and Sibly 1997). Laboratory rearing studies confirm that a majority of ectotherms grow larger at lower temperatures, a pattern known as the temperature-size rule (Atkinson 1994; Angilletta 2009). This pattern is conceptually related but, strictly, separate from Bergmann's rule, the phenomenon that both endothermic and ectothermic animals tend to be larger in colder climates, particularly along latitudinal and altitudinal gradients (Bergmann 1847; Ray 1960; Blackburn et al. 1999). A unifying explanation for these two phenomena is still lacking (Atkinson 1994; Atkinson and Sibly 1997; Chown and Gaston 1999; Blanckenhorn and Demont 2004; Gaston et al. 2008; Stillwell 2010; Arendt 2011). These patterns occur among but also within species, making them relevant for forensics because geographic populations may be adapted to locally differing climatic conditions. Common garden experiments are a good method for investigating such patterns.

In agreement with the literature, dung flies of three species (*S. stercoraria*, *S. cynipsea*, and *S. punctum*) follow the temperature-size rule (Blanckenhorn 1997a; Berger et al. 2013): when grown in the laboratory at various constant temperatures, these flies have very much reduced development times and smaller body sizes at warmer temperatures (Figure 15.2). Matching these results, comparisons of nine laboratory-reared latitudinal *Sc. stercoraria* populations also show a slight genetic Bergmann cline in Europe (flies larger at higher latitudes: Blanckenhorn and Demont 2004; Demont et al. 2008; Scharf et al. 2010), but only at 24°C and not at 18°C, at the same time showing faster development towards

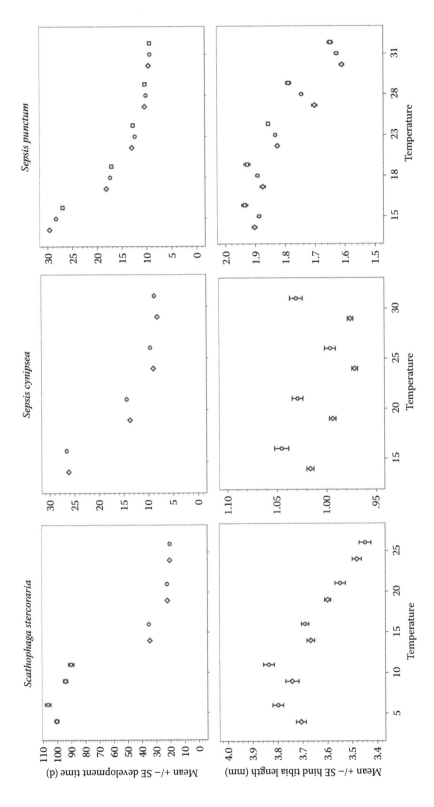

FIGURE 15.2 Development time (top panels) and body size (hind tibia length; bottom panels) of males only of (from left to right) the three dung fly species *S. stercoraria* (yellow dung fly), *S. cynipsea*, and *S. punctum* at various temperatures. For the left two species, low- (circles) and high- (diamonds) altitude populations are compared (Data from Blanckenhorn, W.U. 1997a. *Sepsis cynipsea. Oecologia* 109: 342–352.); in *S. punctum* southern (diamonds), central (circles), and northern European (squares) flies are being compared. (From Berger, D. et al. 2013. *Evolution* 67: 2385–2399.)

the north (Figure 15.3). Whether these body size patterns are adaptive is unclear and disputed. It could be that physiological constraints at the cellular level account for this temperature-mediated body size variation (Atkinson and Sibly 1997). Thus, differences in overall body size may merely reflect corresponding differences in cell size. Accordingly, egg (but not clutch) size (Blanckenhorn 2000b), wing cell and ommatidia size all decrease with increasing temperature in *Sc. stercoraria* (Blanckenhorn and Llaurens 2005), although sperm length follows the opposite pattern (Blanckenhorn and Hellriegel 2002). On the other hand, temperature seems to have little systematic effect on fecundity measures in this species (clutch size, total egg number produced; Blanckenhorn 1997a, 2000b; Blanckenhorn and Heyland 2004; Blanckenhorn and Henseler 2005), even though recent theory and empirical evidence implicates temperature-dependent changes in fecundity schedules as being crucial in evaluating the adaptive nature of temperature-size relationships (Berger et al. 2008; Arendt 2011).

Common garden temperature manipulations of course have been conducted in various carrion-breeding insects, as this represents the essence of estimating the time of death of the victim in forensics (e.g., Davies and Ratcliffe 1994; Byrd and Butler 1996, 1997; Nilssen 1997a; Grassberger and Reiter 2001; Donovan et al. 2006; Nabity et al. 2006; Hwang and Turner 2009; Voss et al. 2010; Tarone et al. 2011; Zuha et al. 2012). However, the contexts vary and seldom consider other life history traits such as body size; data are also sometimes presented in ways that do not allow easy extraction of the relevant information. It seems that with a bit of extra work, other traits could be additionally investigated in such experiments, which would tremendously increase the information content for forensic applications.

A third environmental factor potentially causing systematic life history variation is related to temperature but yet different: season length. A short growing season limits the time available for growth and development and consequently the final body size that can be attained at maturity, especially if development is long relative to season length (Roff 1980; Rowe and Ludwig 1991). As season length systematically decreases towards the poles, the resulting pattern of decreasing developmental periods and body size with increasing latitude has been aptly termed converse Bergmann rule (Park 1949; Mousseau 1997; Figure 15.3). Latitudinal changes in season length are generally accompanied and proximately mediated by photoperiod, which is a much more reliable indicator of latitudinal location and seasonal timing than is temperature, and to which organisms therefore often respond (Tauber et al. 1986; Danks 1987; Bradshaw and Holzapfel 2001). Photoperiod can also be manipulated in the laboratory, although separating effects on life histories of temperature and photoperiod, which correlate in nature, can be difficult in practice, requiring independent manipulation of the two factors; this can be cumbersome and limiting in terms of the necessary climate chambers. To simulate the natural situation, temperature and photoperiod can also be manipulated in conjunction (e.g., Blanckenhorn 1998b). The main life history trait to be investigated in response to photoperiod in seasonal environments is winter (sometimes summer) diapause or dormancy (Bradshaw and Holzapfel 2001). Diapause plasticity in response to photoperiod is expectedly strong in yellow dung flies (*Sc. steroraria*), which enter diapause at 12°C at much higher rates at short (8 h light) compared to long (16 h light) photoperiods, although the quantitative response expectedly differs among populations along a latitudinal gradient (Scharf et al. 2010). The same species also responds well to decreasing photoperiod/temperature conditions simulating the end of the season (Blanckenhorn 1998b), as does *Se. cynipsea* (Blanckenhorn 1998b). Common garden experiments manipulating photoperiod to investigate winter diapause have been conducted in some forensically relevant flies, often in combination with temperature manipulations (e.g., Saunders et al. 1986; Tachibana and Numata 2004a,b; Mello et al. 2012). It is important to stress that seasonal adaptations are trait syndromes (Tauber et al. 1986; Danks 1987; Nylin and Gotthard 1998), such that photoperiod manipulations will likely not only affect diapause, but with it also the other associated life history traits development time, growth rate, body size, and fecundity (Demont et al. 2008).

Predators can also mediate life history differentiation. Especially in aquatic habitats, the presence of predators (e.g., fish or dragonfly larvae) typically hamper the foraging activity and consequently ultimately growth, development, and body size of prey species (e.g., tadpoles or smaller fish: Fraser and Gilliam 1992; Werner and Anholt 1993). Insect examples are known from dragonflies (Martin et al. 1991) or mayflies (Dahl and Peckarsky 2003). The dung pat can be likened to a small albeit opaque pond that harbors an entire community of larval consumers, parasitoids, and predators (Hammer 1941; Hanski and Cambefort 1991; Skidmore 1991; Jochmann et al. 2011; Sladecek et al. 2013). Therefore, life history

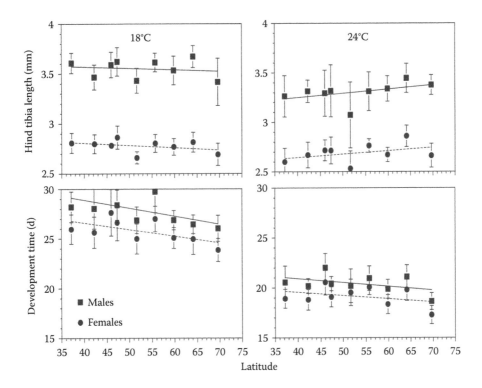

FIGURE 15.3 Population mean ± SE body size (hind tibia length; top panels) and development time (bottom panels) of male (squares) and female (circles) yellow dung flies for nine European populations from different latitudes (southern Spain to northern Norway) raised in the laboratory at common-garden 18°C (left) and 24°C (right). Development time displays a converse Bergmann cline (faster at higher latitudes), whereas body size displays a Bergmann cline (larger at higher latitudes) at 24°C and no (flat) cline at 18°C. Taken together, these results imply countergradient faster growth at higher latitude sites with shorter season length (Blanckenhorn et al. unpublished data).

adaptations to particular predators are expected in dung and carrion insects but remain to be investigated, largely due to practical difficulties with manipulating predation in such a system.

Humidity is a somewhat neglected environmental factor that nevertheless can also promote life history differentiation. Dung organisms indeed show preferences for particular humidity conditions, as some species prefer fresh, wet dung, whereas others older, dry dung (Hammer 1941; Pont and Meier 2002). When manipulating dung humidity in an experiment with yellow dung fly larvae (unpublished data), we expected that the drying of the dung would elicit earlier pupation at smaller size, analogous to dung limitation, because larvae can no longer feed on dry dung. Although development time was indeed prolonged when preventing dung drying, growth rate however significantly decreased, resulting in smaller final body sizes, contrary to expectation (Figure 15.4). Our interpretation of this result is that preventing dung drying by means of avoiding evaporation increases humidity to unnatural levels and thus produces maladaptive plastic responses of larvae. Apparently, the larvae prefer a dry place to pupate, which they searched for in the humidity treatments but could not find in the end expending considerable energy and time, which ultimately reduced final body size. Humidity of a carcass has been identified to be important for some carrion insects (Grassberger and Reiter 2001; Tarone and Foran 2006), so manipulations were more successful and in line with expectations.

To conclude this section, common garden manipulations in the laboratory, or even in the field under more natural conditions, are straightforward and powerful tools to systematically investigate presumed influences of key environmental variables on dung and carrion insect life histories. When combined with quantitative genetic studies (as described in Section 15.4.1) and population comparisons (discussed next in Section 15.4.3), the evolution of phenotypic plasticity and geographic differentiation can be studied.

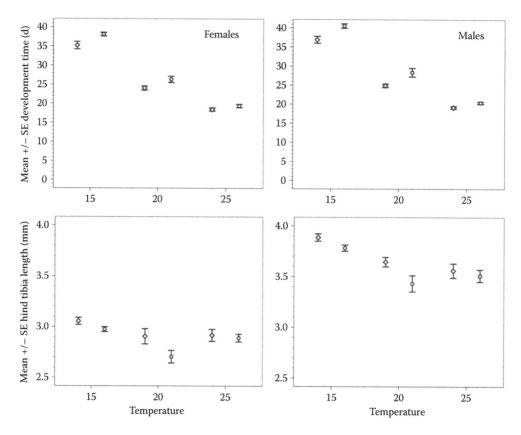

FIGURE 15.4 Development time (top panels) and body size (hind tibia length; bottom panels) of female (left) and male (right) yellow dung flies (*S. stercoraria*) at overabundant dung (>2 g per individual) and various temperatures under dry (diamonds) and humid (circles) conditions in the laboratory. Prevention of dung drying prolongs development and results in smaller adult flies (Blanckenhorn and Heyland unpublished data).

15.4.3 Population Comparisons

In general, when studying populations within species that are locally differentiated, it is not sufficient to compare field specimens, because such comparisons confound adaptive genetic and environmentally induced differences (Stillwell 2010). As outlined in Sections 15.2.5 and 15.4.2 genetic differences among populations are most effectively revealed by common garden experiments. Especially in widespread species, populations are likely to diverge in behavior and life history due to geographic variation in the previously discussed environmental variables such as food availability, temperature, photoperiod, predators, etc. A process that might ultimately lead to what has been termed ecological speciation (Lande and Kirkpatrick 1988; Schluter 1996; Rundle and Nosil 2005; Hendry et al. 2007). Once ecological niche differentiation has occurred, sexual selection often acts on mating behavior and associated traits to reinforce and accelerate speciation (Lande and Kirkpatrick 1988; Via 2009). Perhaps at least in part following human farming activities, a great number of dung and carrion insects have become very widespread, even if they were not widespread before. In this section, I shall discuss some population differentiation studies of dung and carrion insects.

Latitudinal body size variation following Bergmann or converse Bergmann clines is common in insects (Blanckenhorn and Demont 2004) and analogous patterns can be caused by altitude (Chown and Klok 2003). As already mentioned, yellow dung flies (*Sc. steroraira*) show a slight body size increase with latitude, indicating a Bergmann size cline in Europe (Blanckenhorn and Demont 2004; Demont et al. 2008; Scharf et al. 2010). This cline is genetic because it is expressed under common garden conditions in the laboratory, but also plastic because it differs as a function of rearing temperature (Figure 15.3). The corresponding development times actually decrease with latitude following a converse Bergmann effect, which is consistent with the idea that at higher latitudes, season length becomes limiting and populations consequently evolve faster growth rates. There, thus, is a mixed pattern indicating that Bergmann clines, due to temperature *per se* but of unclear adaptive value, and converse Bergmann clines, due to season length and clearly adaptive because they are predicted by optimality models (Roff 1980; Rowe and Ludwig 1991), are not necessarily mutually exclusive and may occur in conjunction, producing any intermediate pattern (Chown and Gaston 1999; Blanckenhorn and Demont 2004). That the cline itself is plastic complicates the picture and lends credence to the hypothesis that nonadaptive mechanistic artifacts at the cellular level may partially mediate such clines when assessed in the laboratory (van der Have and de Jong 1996; Walters and Hassall 2006). Yellow dung flies further show an inconsistent pattern of altitudinal differentiation, high-altitude populations being slightly smaller at high but slightly larger at low temperatures, with no differences in development time (Blanckenhorn 1997a; Figure 15.2). However, these differences are minute and thus probably of limited biological significance. In general, altitudinal differentiation along a mountain of course cannot be expected to be large, as the good flight capability of this mobile fly generates considerable dispersal and hence gene flow counteracting any local adaptation of populations over short distances of only few kilometers (Kraushaar et al. 2002).

Yellow dung fly (*Sc. stercoraria*) populations along a latitudinal cline in Europe also show strong differentiation in diapause response: northern populations enter winter diapause more readily at a given short light period/12°C temperature combination) than southern populations (Demont and Blanckenhorn 2008; Scharf et al. 2010). In contrast, the diapause response of pairs of altitudinal populations of yellow dung flies and the black scavenger fly *Se. cynipsea* are not locally differentiated, as can be expected based on their close proximity (Blanckenhorn 1998b).

Sepsid dung flies also show geographic population differentiation as revealed by common garden rearing. European *S. punctum* displays a converse Bergmann cline in development rate indicating seasonal time constraints in northern (Swedish) populations that is evident only at lower rearing temperatures, whereas southern (Italian) populations have slower development throughout (Berger et al. 2013; Figure 15.2). *S. punctum* also exhibits a Bergmann body size cline in Europe, but not in North America (Puniamoorthy et al. 2012a). However, the geographic differentiation of this species is complex as it also involves reproductive behavior and morphology (see below), so any conclusions are premature. Latitudinal population variation is currently investigated for the related *Se. cynipsea*, while altitudinal differentiation of this species follows a Bergmann pattern, body size, and development time being greater at high altitudes (Blanckenhorn 1997a; Figure 15.2).

I could not find any studies in the literature on carrion insects specifically focusing on latitudinal population differentiation. Gallagher et al. (2010) compared three geographic populations of *L. sericata* from North America, identifying different larval development times. Similarly, Tarone et al. (2011) compared three populations of the same species, finding the southernmost California flies being largest but taking longest to develop at high temperatures. Owings et al. (2014) recently compared three Texas populations of *Cochliomyia macellaria*, finding differences in a number of life history traits among the populations. Smith et al. (2000) studied altitudinal body size variation in the burying beetle *Nicrophorus investigator*, finding higher altitude beetles to be larger but not necessarily more fecund. Other studies assessing the distribution of carrion insects along an altitudinal gradient exist but report no life history traits (Fierros-Lopez and Navarrete-Heredia 2001; Baz et al. 2007).

A number of species of sepsid dung or black scavenger flies (Diptera: Sepsidae) are very widespread in the northern hemisphere, occurring in Europe, Asia, and North America, where they seem to be strongly differentiated and probably in the process of speciation. North American and European *Se. punctum* populations differ strongly in body size, associated male primary and secondary sexual traits (such as foreleg and genital morphology), and mating behavior. Although body size variation *per se* could be caused by thermal adaptation in response to climate (temperature and/or season length), as discussed above, sexual trait and behavioral differentiation is likely mediated by geographic variation in sexual selection, which might be a secondary consequence of some primary ecological differences among populations (Via 2009). North American *Se. punctum* males display a courtship dance, and large body size is not of great advantage in gaining access to females, whereas European males simply attack females and large male size is strongly advantageous. Mating rates are higher in Europe. As a result, European

Se. punctum are larger, with males being larger than females, whereas in North American populations, females are the larger sex (Puniamoorthy et al. 2012a,b; Dmitriew and Blanckenhorn 2012). Data are consistent with the hypothesis that the stronger sexual selection on male body size in Europe lead to a reversal in the sexual size dimorphism from female-biased to male-biased (cf. Fairbairn et al. 2007). Because European males have higher mating rates, they face increased sperm competition and consequently have to invest more in sperm production, so the allometry of testis-to-body size has evolved to be steeper. Thus pre-copulatory sexual selection on male courtship seems to be stronger in North America, whereas postcopulatory sexual selection on internal genital morphology seems to be stronger in Europe (Puniamoorthy et al. 2012b). There appears to be further differentiation in mating behavior between western and eastern North American populations, which has to be further explored (Puniamoorthy et al. 2012b). All these results were obtained by comparing various populations under seminatural conditions in a common garden laboratory, indicating that even behavioral differentiation in insects can so be studied.

15.5 Conclusions

In this chapter, I discussed quantitative genetic studies of life history traits of dung and carrion insects. Although dung and carrion as a habitat also differ in many ways, they share a great number of similarities. Research on the evolutionary ecology of life history traits is apparently much more developed in dung insects than in carrion insects. The aim here was to compare coprophagous and necrophagous insects to identify fruitful research avenues in carrion entomology in light of the new framework to bridge basic and applied research (Tomberlin et al. 2011a,b). There seems great potential in crossing fields: as a first step, carrion insect researchers may simply replicate some of the approaches used in dung flies and beetles. As initial guidance, I briefly summarized the necessary quantitative genetic methods. Clearly, further reading of standard genetic textbooks will be required at some point (Falconer 1989; Becker 1992; Roff 1997; Lynch and Walsh 1998), even if the material may appear difficult to the inexperienced. The hope is that at some stage dung insect researchers can learn from experiences and results derived from carrion insects, as at least some patterns are likely to be very different.

Acknowledgments

I thank Aaron Tarone for involving me in this book. I certainly learned a lot about carrion flies and beetles along the way, enriching my experience. I also thank the Swiss National Foundation, as well as the Zoological Museum and our new Institute of Evolutionary Biology and Environmental Studies of the University of Zurich for continuous support.

REFERENCES

Amano, K. 1983. Studies on the intraspecific competition in dung breeding flies. I. Effects of larval density on the yellow dung fly. *Japanese Journal of Sanitary Zoology* 34: 165–175.
Angilletta Jr, M.J. 2009. *Thermal Adaptation: A Theoretical and Empirical Synthesis*. Oxford: Oxford University Press.
Arendt, J.D. 1997. Adaptive intrinsic growth rates: And integration across taxa. *Quarterly Review of Biology* 72: 149–177.
Arendt, J.D. 2011. Size–fecundity relationships, growth trajectories and the temperature-size rule for ectotherms. *Evolution* 65: 43–51.
Atkinson, D. 1994. Temperature and organism size—A biological law for ectotherms? *Advances in Ecological Research* 25: 1–58.
Atkinson, D. and R.M. Sibly. 1997. Why are organisms usually bigger in colder environments? Making sense of a life history puzzle. *Trends in Ecology and Evolution* 12: 235–239.

Atwell, S., Y. Huang, B. Vilhjálmsson, G. Willems, M. Horton, Y. Li, D. Meng et al. 2010. Genome-wide association study of 107 phenotypes in a common set of *Arabidopsis thaliana* inbred lines. *Nature* 465: 627–631.

Baz, A., B. Cifrian, L.M. Diaz-Aranda, and D. Marin-Vega. 2007. The distribution of adult blow-flies (Diptera: Calliphoridae) along an altitudinal gradient in central Spain. *Annales de la Societe Entomologique de France* 43: 289–296.

Becker, W.A. 1992. *Manual of Quantitative Genetics*, 5th edition. Pullmann, WA: Academic Enterprises.

Berger, D., E. Postma, W.U. Blanckenhorn, and R.J. Walters. 2013. Quantitative genetic divergence and standing genetic (co)variance in thermal reaction norms along latitude. *Evolution* 67: 2385–2399.

Berger, D., R.J. Walters, and K. Gotthard. 2008. What limits insect fecundity? Body size- and temperature-dependent egg maturation and oviposition in a butterfly. *Functional Ecology* 22: 523–529.

Bergmann, C. 1847. Über die Verhältnisse der Wärmeökonomie der Thiere zu ihrer Grösse. *Göttinger Studien* 1: 595–708.

Berrigan, D. and E.L. Charnov. 1994. Reaction norms for age and size at maturity in response to temperature: A puzzle for life historians. *Oikos* 70: 474–478.

Blackburn, T.M., K.J. Gaston, and N. Loder. 1999. Geographic gradients in body size: A clarification of Bergmann's rule. *Diversity and Distributions* 5: 165–174.

Blanckenhorn, W.U. 1997a. Altitudinal life history variation in the dung flies *Scathophaga stercoraria* and *Sepsis cynipsea*. *Oecologia* 109: 342–352.

Blanckenhorn, W.U. 1997b. Effects of rearing temperature on growth, development and diapause in the yellow dung fly—Against all rules? *Oecologia* 111: 318–324.

Blanckenhorn, W.U. 1998a. Adaptive phenotypic plasticity in growth rate and diapause in the yellow dung fly. *Evolution* 52: 1394–1407.

Blanckenhorn, W.U. 1998b. Altitudinal differentiation in diapause response in two species of dung flies. *Ecological Entomology* 23: 1–8.

Blanckenhorn, W.U. 1999. Different growth responses to food shortage and temperature in three insect species with similar life histories. *Evolutionary Ecology* 13: 395–409.

Blanckenhorn, W.U. 2000a. The evolution of body size: What keeps organisms small? *Quarterly Review of Biology* 75: 385–407.

Blanckenhorn, W.U. 2000b. Temperature effects on egg size and their fitness consequences in the yellow dung fly. *Evolutionary Ecology* 14: 627–643.

Blanckenhorn, W.U. 2002. The consistency of heritability estimates in field and laboratory in the yellow dung fly. *Genetica* 114: 171–182.

Blanckenhorn, W.U. and M. Demont. 2004. Bergmann and converse Bergmann latitudinal clines in arthropods: Two ends of a continuum? *Integrative and Comparative Biology* 44: 413–424.

Blanckenhorn, W.U. and B. Hellriegel. 2002. Against Bergmann's rule: Fly sperm size increases with temperature. *Ecology Letters* 5: 7–10.

Blanckenhorn, W.U. and C. Henseler. 2005. Temperature-dependent ovariole and testis maturation in the yellow dung fly. *Entomologia Experimentalis et Applicata* 116: 159–165.

Blanckenhorn, W.U. and A. Heyland. 2004. The quantitative genetics of two life history trade-offs in the yellow dung fly in abundant and limited food environments. *Evolutionary Ecology* 18: 385–402.

Blanckenhorn, W.U. and D.J. Hosken. 2003. The heritability of three condition measures in the yellow dung fly. *Behavioral Ecology* 14: 612–618.

Blanckenhorn, W.U. and V. Llaurens. 2005. Effects of temperature on cell size and number in the yellow dung fly *Scathophaga stercoraria*. *Journal of Thermal Biology* 30: 213–219.

Blanckenhorn, W.U., C. Morf, and M. Reuter. 2000. Are dung flies ideal-free distributed at their mating and oviposition sites? *Behaviour* 137: 233–248.

Blanckenhorn, W.U., C. Mühlhäuser, and T. Reusch. 1998. Fluctuating asymmetry and sexual selection in the dung fly *Sepsis cynipsea*—Testing the good genes assumptions and predictions. *Journal of Evolutionary Biology* 11: 735–753.

Blanckenhorn, W.U., A.J. Pemberton, L.F. Bussière, J. Römbke, and K.D. Floate. 2010. Natural history and laboratory culture of the yellow dung fly, *Scathophaga stercoraria* L. (Diptera: Scathophagidae). *Journal of Insect Science* 10: 11.

Bradshaw, W.E. and C.M. Holzapfel. 2001. Genetic shift in photoperiodic response correlated with global warming. *Proceedings of the National Academy of Sciences of the United States of America* 98: 14509–14511.

Bruce, T.J.A., L.J. Wadhams, and C.M. Woodcock. 2005. Insect host location: A volatile situation. *Trends in Plant Sciences* 10: 269–274.

Bryant, E.H. and L.M. Meffert. 1998. Quantitative genetic estimates of morphometric variation in wild caught and laboratory reared houseflies. *Evolution* 52: 626–630.

Byrd, J.H. and J.F. Butler. 1996. Effects of temperature on *Cochliomyia macellaria* (Diptera: Calliphoridae) development. *Journal of Medical Entomology* 33: 901–905.

Byrd, J.H. and J.F. Butler. 1997. Effects of temperature on *Chrysomya rufifacies* (Diptera: Calliphoridae) development. *Journal of Medical Entomology* 34: 353–358.

Byrd, J.H. and J.L. Castner (eds). 2010. *Forensic Entomology: The Utility of Arthropods in Legal Investigations*, 2nd edition. Boca Raton, FL: CRC Press, 681 pp.

Cheverud, J.M. 1988. A comparison of genetic and phenotypic correlations. *Evolution* 42: 958–968.

Chown, S.L. and K.J. Gaston. 1999. Exploring links between physiology and ecology at macro-scales: The role of respiratory metabolism in insects. *Biological Reviews* 74: 87–120.

Chown, S.L. and C.J. Klok. 2003. Altitudinal body size clines: Latitudinal effects associated with changing seasonality. *Ecography* 26: 445–455.

Cooper, K., M. Burd, and K.S. Lefevere. 2002. Correlated response of autogeny to selection for adult starvation resistance in the blowfly *Lucilia cuprina*. *Heredity* 88: 35–38.

Dahl, J. and B.L. Peckarsky. 2003. Developmental responses to predation risk in morphologically defended mayflies. *Oecologia* 137: 188–194.

Danks, H.V. 1987. *Insect Dormancy: An Ecological Perspective*. Ottawa, Canada: Biological Survey of Canada.

David, J.R., P. Gibert, H. Legout, G. Petavy, P. Capy, and B. Moreteau. 2005. Isofemale lines in *Drosophila*: An empirical approach to quantitative trait analysis in natural populations. *Heredity* 94: 3–12.

Davies, L. and G.G. Ratcliffe. 1994. Development rates of some pre-adult stages in blowflies with reference to low temperatures. *Medical and Veterinary Entomology* 8: 245–254.

de Jong, G. 1995. Phenotypic plasticity as a product of selection in a variable environment. *American Naturalist* 145: 493–512.

Demont, M. and W.U. Blanckenhorn. 2008. Genetic differentiation in diapause response along a latitudinal cline in European yellow dung fly populations. *Ecological Entomology* 33: 197–201.

Demont, M., W.U. Blanckenhorn, D.J. Hosken, and T.W.J. Garner. 2008. Molecular and quantitative genetic differentiation across Europe in yellow dung flies. *Journal of Evolutionary Biology* 21: 1492–1503.

Dethier, V.G. 1947. *Chemical Insect Attractants and Repellents*. Philadelphia, PA: The Blakiston Co.

Dmitriew, C.M. 2011. The evolution of growth trajectories: What limits growth rate? *Biological Reviews* 86: 97–116.

Dmitriew, C.M. and W.U. Blanckenhorn. 2012. The role of sexual (selection and) conflict in mediating among-population variation in mating strategies and sexually dimorphic traits in the black scavenger fly *Sepsis punctum*. *PloS ONE* 7: e49511.

Dobler, R. and D.J. Hosken. 2010. Response to selection and realized heritability of sperm length in the yellow dung fly (*Scathophaga stercoraria*). *Heredity* 104: 61–66.

Donovan, S.E., M.J.R. Hall, B.D. Turner, and C.B. Moncrieff. 2006. Larval growth rates of the blowfly *Calliphora vicina* over a range of temperatures. *Medical and Veterinary Entomology* 20: 106–114.

Ebert, D. 1993. The trade-off between offspring size and number in *Daphnia magna*: The influence of genetic, environmental and maternal effects. *Archives of Hydrobiology* 4: 453–473.

Eggert, A.K. and S.K. Sakaluk. 1995. Female-coerced monogamy in burying beetles. *Behavioural Ecology and Sociobiology* 37: 147–153.

Emlen, D.J. 1996. Artificial selection on horn length–body size allometry in the horned beetle *Onthophagus acuminatus* (Coleoptera: Scarabaeidae). *Evolution* 50: 1219–1230.

Fairbairn, D.J., W.U. Blanckenhorn, and T. Székely (eds). 2007. *Sex, Size and Gender Roles: Evolutionary Studies of Sexual Size Dimorphism*. New York: Oxford University Press.

Falconer, D.S. 1989. *Introduction to Quantitative Genetics*, 3rd edition. Harlow, UK: Longman Scientific and Technical.

Fierros-Lopez, H.E. and J.L. Navarrete-Heredia. 2001. Altitudinal distribution and phenology of three species of carrion beetles (Coleoptera: Silphidae) from Nevado de Colima, Jalisco, Mexico. *Pan-Pacific Entomologist* 77: 45–46.

Fisher, R.A. 1930. *The Genetical Theory of Natural Selection*. Oxford: Clarendon Press.

Fox, C.W., J.D. Martin, M.S. Thakar, and T.A. Mousseau. 1996. Clutch size manipulations in two seed beetles: Consequences for progeny fitness. *Oecologia* 108: 88–94.

Franz, H. 1979. *Oekologie der Hochgebirge*. Stuttgart, Germany: Ulmer Verlag.

Fraser, D.F. and J.F. Gilliam. 1992. Nonlethal impacts of predator invasion: Facultative suppression of growth and reproduction. *Ecology* 73: 959–970.

Fretwell, S.D. 1972. *Populations in a Seasonal Environment*. Princeton, NJ: Princeton University Press.

Gallagher, M.B., S. Sandhu, and R. Kimsey. 2010. Variation in developmental time for geographically distinct populations of the common green bottle fly, *Lucilia sericata* (Meigen). *Journal of Forensic Science* 55: 438–442.

Gaston, K.J., S.L. Chown, and K.L. Evans. 2008. Ecogeographical rules: Elements of a synthesis. *Journal of Biogeography* 35: 483–500.

Gleeson, D.M. 1995. The effects on genetic variability following a recent colonization event: The Australian sheep blowfly *Lucilia cuprina* arrives in New Zealand. *Molecular Ecology* 4: 699–707.

Grassberger, M. and C. Reiter. 2001. Effect of temperature on *Lucilia sericata* (Diptera: Calliphoridae) development with special reference to the isomegalen- and isomorphen-diagram. *Forensic Science International* 120: 32–36.

Hammer, O. 1941. Biological and ecological investigations on flies associated with pasturing cattle and their excrement. *Videnskabelige Meddelelser fra Dansk Naturhistorisk Forening* 105: 140–393.

Hanski, I. and Y. Cambefort. 1991. *Dung Beetle Ecology*. Princeton, NJ: Princeton University Press.

Hendry, A.P., P. Nosil, and L.H. Rieseberg. 2007. The speed of ecological speciation. *Functional Ecology* 21: 455–464.

Hoffmann, A.A., R. Hallas, C. Sinclair, and P. Mitrovski. 2001. Levels of variation in stress resistance in *Drosophila* among strains, local populations, and geographic regions: Patterns for desiccation, starvation, cold resistance, and associated traits. *Evolution* 55: 1621–1630.

Holter, P. 1979. Effect of dung-beetles (*Aphodius* spp.) and earthworms on the disappearance of cattle dung. *Oikos* 32: 393–402.

Holter, P. and C.H. Scholtz. 2007. What do dung beetles eat? *Ecological Entomology* 32: 690–697.

Honek, A. 1993. Intraspecific variation in body size and fecundity in insects: A general relationship. *Oikos* 66: 483–492.

Hosken, D.J., T.W.J. Garner, and P.I. Ward. 2001. Sexual conflict selects for male and female reproductive characters. *Current Biology* 11: 489–493.

House, C.M. and L.W. Simmons. 2012. The genetics of primary and secondary sexual character trade-offs in a horned beetle. *Journal of Evolutionary Biology* 25: 1711–1717.

Hunt, J. and L.W. Simmons. 2002. The genetics of maternal care: Direct and indirect genetic effects on phenotype in the dung beetle *Onthophagus taurus*. *Proceedings of the National Academy of Sciences of the United States of America* 99: 6828–6832.

Hwang, C.C. and B.D. Turner. 2009. Small-scaled geographical variation in life history traits of the blowfly *Calliphora vicina* between rural and urban populations. *Entomological Experiments and Applications* 132: 218–224.

Ikeda, H., S. Shimano, and A. Yamagami. 2011. Differentiation in searching behavior for carcasses based on flight height differences in carrion beetles (Coleoptera: Silphidae). *Journal of Insect Behavior* 24: 167–174.

Ireland, S. and B. Turner. 2006. The effects of larval crowding and food type on the size and development of the blowfly *Calliphora vomitoria*. *Forensic Science International* 159: 175–181.

Ito, N., T. Katoh, N. Kobayashi, and H. Katakura. 2010. Effects of straits as dispersal barriers for the flightless roving carrion beetle *Silpha perforata* (Coleoptera: Silphidae). *Zoological Science* 27: 313–319.

Jann, P. and P.I. Ward. 1999. Maternal effects and their consequences for offspring fitness in the yellow dung fly. *Functional Ecology* 13: 51–58.

Janzen, D.H. 1977. Why fruits rot, seeds mold, and meat spoils. *American Naturalist* 111: 691–713.

Jochmann, R., W.U. Blanckenhorn, L.F. Bussière, C.E. Eirkson, J. Jensen, U. Kryger, J. Lahr et al. 2011. How to test nontarget effects of veterinary pharmaceutical residues in livestock dung in the field. *Integrated Environmental Assessment and Management* 7: 287–296.

Kozol, A.J., J.F.A. Traniello, and S.M. Williams. 1994. Genetic variation in the endangered burying beetle *Nicrophorus americanus* (Coleoptera: Silphidae). *Annals of the Entomological Society of America* 87: 928–935.

Kraushaar, U., J. Goudet, and W.U. Blanckenhorn. 2002. Geographic and altitudinal population genetic structure of two dung fly species with contrasting mobility and temperature preference. *Heredity* 89: 99–106.

Kruuk, L.E.B. 2004. Estimating genetic parameters in natural populations using the "animal model." *Philosophical Transactions of the Royal Society London B* 359: 873–890.

Lande, R. and M. Kirkpatrick. 1988. Ecological speciation by sexual selection. *Journal of Theoretical Biology* 133: 85–98.

Lock, J.E., P.T. Smiseth, and A.J. Moore. 2004. Selection, inheritance and the evolution of parent-offspring interactions. *American Naturalist* 164: 13–24.

Lynch, M. and B. Walsh. 1998. *Genetics and Analysis of Quantitative Traits.* Sunderland, MA: Sinauer Associates.

Mackay, T.F., S. Richards, E.A. Stone, A. Barbadilla, J.F. Ayroles, D. Zhu, S. Casillas et al. 2012. The *Drosophila melanogaster* Genetic Reference Panel. *Nature* 482: 173–178.

Martin, O.Y. and D.J. Hosken. 2003. Costs and benefits of evolving under experimentally enforced polyandry or monogamy. *Evolution* 57: 2765–2772.

Martin, O.Y. and D.J. Hosken. 2004. Reproductive consequences of population divergence through sexual conflict. *Current Biology* 14: 906–910.

Martin, T.H., D.M. Johnson, and R.D. Moore. 1991. Fish-mediated alternative life-history strategies in the dragonfly *Epitheca cynosura. Journal of the North American Benthological Society* 10: 271–279.

Mégnin, P. 1894. La faune des cadavres, application de l'entomologie à la médecine légale. In: *Encyclopédie Scientifique des Aide-Mémoire*, M. Leauté (ed.) Vol. 101B. Paris: Masson.

Mello, R.D., G.E.M. Borja, and M.M.D. Queiroz. 2012. How photoperiods affect the immature development of the forensically important blowfly species *Chrysomya albiceps* (Calliphoridae). *Parasitology Research* 111: 1067–1073.

Mendel, G. 1865. Versuche über Pflanzen-Hybriden. *Verhandlungen der Naturforschenden Gesellschaft Brünn* 4: 3–47.

Merilä, J., B.C. Sheldon, and L.E.B. Kruuk. 2001. Explaining stasis: Microevolutionary studies in natural populations. *Genetica* 112–113: 199–222.

Moczek, A.P., J. Hunt, D.J. Emlen, and L.W. Simmons. 2002. Threshold evolution in exotic populations of a polyphenic beetle. *Evolutionary Ecology Research* 4: 587–601.

Møller, H., R.H. Smith, and R.M. Sibly. 1989. Evolutionary demography of a bruchid beetle. II. Physiological manipulations. *Functional Ecology* 3: 683–691.

Mousseau, T.A. 1997. Ectotherms follow the converse Bergmann's rule. *Evolution* 51: 630–632.

Mousseau, T.A. and D.A. Roff. 1987. Natural selection and the heritability of fitness components. *Heredity* 59: 181–197.

Mühlhäuser, C. and W.U. Blanckenhorn. 2004. The quantitative genetics of sexual selection in the dung fly *Sepsis cynipsea. Behaviour* 141: 327–341.

Mühlhäuser, C., W.U. Blanckenhorn, and P.I. Ward. 1996. The genetic component of copula duration in the yellow dung fly, *Scathophaga stercoraria. Animal Behaviour* 51: 1401–1407.

Nabity, P.D., L.G. Higley, and T.M. Heng-Moss. 2006. Effects of temperature on development of *Phormia regina* (Diptera: Calliphoridae) and use of developmental data in determining time intervals in forensic entomology. *Journal of Medical Entomology* 43: 1276–1286.

Nagano, M. and S. Suzuki. 2007. Effects of carcass size and male presence on clutch size in *Nicrophorus quadripunctatus* (Coleoptera: Silphidae). *Entomological Science* 10: 245–248.

Newman, R.A. 1992. Adaptive plasticity in amphibian metamorphosis. *Bioscience* 42: 671–678.

Nijhout, H.F. and D.J. Emlen. 1998. Competition among body parts in the development and evolution of insect morphology. *Proceedings of the National Academy of Sciences of the United States of America* 95: 3685–3689.

Nilssen, A.C. 1997a. Effect of temperature on pupal development and eclosion dates in the reindeer oestrids *Hypoderma tarandi* and *Cephenemya trompe. Environmental Entomology* 26: 296–306.

Nilssen, A.C. 1997b. Factors affecting size, longevity and fecundity in the reindeer oestrid flies *Hypoderma tarandi* and *Cephenemya trompe. Ecological Entomology* 22: 294–304.

Nunney, L. 1996. The response to selection for fast larval development in *Drosophila melanogaster* and its effect on adult weight: An example of a fitness trade-off. *Evolution* 50: 1193–1204.

Nylin, S. and K. Gotthard. 1998. Plasticity in life history traits. *Annual Review of Entomology* 43: 63–83.

Owings, C.G., C. Spiegelman, A.M. Tarone, and J.K. Tomberlin. 2014. Developmental variation among *Cochliomyia macellaria* Fabricius (Diptera: Calliphoridae) populations from three ecoregions of Texas, USA. *International Journal of Legal Medicine* 128: 709–717.

Park, O. 1949. Application of the converse Bergmann principle to the carabid beetle, *Dicaelus purpuratus*. *Physiological Zoology* 22: 359–372.

Parker, G.A. 1970. The reproductive behaviour and the nature of sexual selection in *Scatophaga stercoraria* L. (Diptera: Scatophagidae) I. Diurnal and seasonal changes in population density around the site of mating and oviposition. *Journal of Animal Ecology* 39: 185–204.

Parker, G.A. 1972. Reproductive behaviour of *Sepsis cynipsea* (L) (Diptera: Sepsidae). I. A preliminary analysis of the reproductive strategy and its associated behaviour patterns. *Behaviour* 41: 172–206.

Picard, C.J. and J.D. Wells. 2009. Survey of the genetic diversity of *Phormia regina* (Diptera: Calliphoridae) using amplified fragment length polymorphisms. *Journal of Medical Entomology* 46: 664–670.

Picard, C.J. and J.D. Wells. 2010. The population genetic structure of North American *Lucilia sericata* (Diptera: Calliphoridae), and the utility of genetic assignment methods for reconstruction of postmortem corpse relocation. *Forensic Science International* 195: 63–67.

Pont, A.C. and R. Meier. 2002. The Sepsidae (Diptera) of Europe. *Fauna Entomologica Scandinavica* 37: 1–221.

Postma, E., F. Heinrich, U. Koller, R.J. Sardell, J.M. Reid, P. Arcese, and L.F. Keller. 2011. Disentangling the effect of genes, the environment and chance on sex ratio variation in a wild bird population. *Proceedings of the Royal Society London B* 278: 2996–3002.

Postma, E., J. Visser, and A. Van Noordwijk. 2007. Strong artificial selection in the wild results in predicted small evolutionary change. *Journal of Evolutionary Biology* 20: 1823–1832.

Puniamoorthy, N., W.U. Blanckenhorn, and M.A. Schäfer. 2012b. Differential investment in pre- versus post-copulatory sexual selection reinforces a cross-continental reversal of sexual size dimorphism in *Sepsis punctum* (Diptera: Sepsidae). *Journal of Evolutionary Biology* 25: 2253–2263.

Puniamoorthy, N., M.A. Schäfer, and W.U. Blanckenhorn. 2012a. Sexual selection accounts for geographic reversal of sexual size dimorphism in the dung fly, *Sepsis punctum* (Diptera: Sepsidae). *Evolution* 66: 2117–2126.

Rauter, C.M. and A.J. Moore. 2002a. Evolutionary importance of parental care performance, food resources, and direct and indirect genetic effects in a burying beetle. *Journal of Evolutionary Biology* 15: 407–417.

Rauter, C.M. and A.J. Moore. 2002b. Quantitative genetics of growth and development time in the burying beetle *Nicrophorus pustulatus* in the presence and absence of post-hatching paternal care. *Evolution* 56: 96–110.

Ray, C. 1960. The application of Bergmann's and Allen's rules to the poikilotherms. *Journal of Morphology* 106: 85–108.

Reim, C., C. Kaufmann, and W.U. Blanckenhorn. 2009. Size-dependent energetic costs of metamorphosis in the yellow dung fly *Scathophaga stercoraria*. *Evolutionary Ecology Research* 11: 1111–1130.

Reim, C., Y. Teuschl, and W.U. Blanckenhorn. 2006a. Size-dependent effects of larval and adult food availability on reproductive energy allocation in the yellow dung fly. *Functional Ecology* 20: 1012–1021.

Reim, C., Y. Teuschl, and W.U. Blanckenhorn. 2006b. Size-dependent effects of temperature and food stress on energy stores and survival in yellow dung flies (Diptera: Scathophagidae). *Evolutionary Ecology Research* 8: 1215–1234.

Reusch, T. and W.U. Blanckenhorn. 1998. Quantitative genetics of the dung fly *Sepsis cynipsea*: Cheverud's conjecture revisited. *Heredity* 81: 111–119.

Roff, D.A. 1980. Optimizing development time in a seasonal environment: The "ups and downs" of clinal variation. *Oecologia* 45: 202–208.

Roff, D.A. 1992. *The Evolution of Life Histories*. New York, NY: Chapman & Hall.

Roff, D.A. 1995. The estimation of genetic correlations from phenotypic correlations: A test of Cheverud's conjecture. *Heredity* 74: 481–490.

Roff, D.A. 1996. The evolution of genetic correlations: An analysis of patterns. *Evolution* 50: 1392–1403.

Roff, D.A. 1997. *Evolutionary Quantitative Genetics*. New York, NY: Chapman & Hall.

Rouchet, R. and C. Vorburger. 2012. Strong specificity in the interaction between parasitoids and symbiont-protected hosts. *Journal of Evolutionary Biology* 25: 2369–2375.

Rowe, L. and D. Ludwig. 1991. Size and timing of metamorphosis in complex life cycles: Time constraints and variation. *Ecology* 72: 413–427.

Rundle, H.D. and P. Nosil. 2005. Ecological speciation. *Ecology Letters* 8: 336–352.

Saunders, D.S., J.N. Macpherson, and K.D. Cairncross. 1986. Maternal and larval effects of photoperiod on the induction of larval diapause in two species of fly, *Calliphora vicina* and *Lucilia sericata*. *Experimental Biology* 46: 51–58.

Scharf, I., S.S. Bauerfeind, W.U. Blanckenhorn, and M.A. Schäfer. 2010. Effects of maternal and offspring environmental conditions on growth, development and diapause in latitudinal yellow dung fly populations. *Climate Research* 43: 115–125.

Scheiner, S.M. 1993a. Genetics and evolution of phenotypic plasticity. *Annual Review of Ecology and Systematics* 24: 35–68.

Scheiner, S.M. 1993b. Plasticity as a selectable trait—Reply. *American Naturalist* 142: 371–373.

Scheiner, S.M. and R.F. Lyman. 1989. The genetics of phenotypic plasticity. I. Heritability. *Journal of Evolutionary Biology* 2: 95–107.

Schlichting, C.D. 1986. The evolution of phenotypic plasticity in plants. *Annual Review of Ecology and Systematics* 17: 667–693.

Schluter, D. 1996. Ecological causes of adaptive radiation. *American Naturalist* 148: S40–S64.

Schmalhausen, I.I. 1949. *Factors of Evolution*. Philadelphia, PA: Blakiston.

Schröder, R. and M. Hilker. 2008. The relevance of background odor in resource location by insects: A behavioral approach. *BioScience* 58: 308–316.

Schwarzenbach, G.A., D.J. Hosken, and P.I. Ward. 2005. Sex and immunity in the yellow dung fly *Scathophaga stercoraria*. *Journal of Evolutionary Biology* 18: 455–463.

Simmons, L.W. 2001. *Sperm Competition and its Evolutionary Consequences in the Insects*. Princeton, NJ: Princeton University Press.

Simmons, L.W., C.M. House, J. Hunt, and F. Garcia-Gonzales. 2009. Evolutionary response to sexual selection in male genital morphology. *Current Biology* 19: 1442–1446.

Simmons, L.W. and J.S. Kotiaho. 2007. Quantitative genetic correlation between trait and preference supports a sexually selected sperm process in the beetle *Onthophagus taurus*. *Proceedings of the National Academy of Sciences of the United States of America* 104: 16604–16608.

Simmons, L.W. and P.I. Ward. 1991. The heritability of sexually dimorphic traits in the yellow dung fly *Scathophaga stercoraria* (L.). *Journal of Evolutionary Biology* 4: 593–601.

Skidmore, P. 1991. *Insects of the British Cow-dung Community*. Shrewsbury, UK: Field Studies Council Occasional Publication No. 21.

Sladecek, F.X.J., J. Hrcek, P. Klimes, and M. Konvicka. 2013. Interplay of succession and seasonality reflects resource utilization in an ephemeral habitat. *Acta Oecologica* 46: 17–24.

Smith, J.L., I.G. Coldiz, L.R. Piper, R.M. Sandeman, and S. Dominik. 2008. Genetic resistance to growth of *Lucilia cuprina* larvae in Merino sheep. *Australian Journal of Experimental Agriculture* 48: 12010–1216.

Smith, R.J., A. Hines, S. Richmond, M. Merrick, A. Drew, and R. Fargo. 2000. Altitudinal variation in body size and population density of *Nicrophorus investigator* (Coleoptera: Silphidae). *Environmental Entomology* 29: 290–298.

Stearns, S.C. 1989. The evolutionary significance of phenotypic plasticity. *Bioscience* 39: 436–445.

Stearns, S.C. 1992. *The Evolution of Life Histories*. Oxford, UK: Oxford University Press.

Stearns, S.C., G. de Jong, and R.A. Newman. 1991. The effects of phenotypic plasticity on genetic correlations. *Trends in Ecology and Evolution* 6: 122–126.

Stearns, S.C. and J. Koella. 1986. The evolution of phenotypic plasticity in life history traits: Predictions of reaction norms for age and size at maturity. *Evolution* 40: 893–914.

Steiger, S., W. Haberer, and J.K. Müller. 2011. Social environment determines degree of chemical signaling. *Biology Letters* 7: 822–824.

Stephens, D.W. and J.R. Krebs. 1986. *Foraging Theory*. Princeton, NJ: Princeton University Press.

Stillwell, R.C. 2010. Are latitudinal clines in body size adaptive? *Oikos* 119: 1387–1390.

Szalanski, A.L., D.S. Sikes, R. Bischof, and M. Fritz. 2000. Population genetics and phylogenetics of the endangered American burying beetle *Nicrophorus americanus* (Coleoptera: Silphidae). *Annals of the Entomological Society of America* 93: 589–594.

Tachibana, S.I. and H. Numata. 2004a. Parental and direct effects of photoperiod and temperature on the induction of larval diapause in the blow fly *Lucilia sericata*. *Physiological Entomology* 29: 39–44.

Tachibana, S.I. and H. Numata. 2004b. Effects of photoperiod and temperature on the termination in *Lucilia sericata*. *Zoological Science* 21: 197–202.

Tarone, A.M. and D.R. Foran. 2006. Components of developmental plasticity in a Michigan population of *Lucilia sericata* (Diptera: Calliphoridae). *Journal of Medical Entomology* 43: 1023–1033.

Tarone, A.M., L.M. McIntyre, L.G. Harshman, and S.V. Nuzhdin. 2012. Genetic variation in the Yolk protein expression network of *Drosophila melanogaster*: Sex-biased negative correlations with longevity. *Heredity* 109: 226–234.

Tarone, A.M., C.J. Picard, C. Spiegelman, and D.R. Foran. 2011. Population and temperature effects on *Lucilia sericata* (Diptera: Calliphoridae) body size and minimum development time. *Journal of Medical Entomology* 48: 1062–1068.

Tauber, M.J., C.A. Tauber, and S. Masaki. 1986. *Seasonal Adaptations of Insects*. Oxford, UK: Oxford University Press.

Teuschl, Y., C. Reim, and W.U. Blanckenhorn. 2007. Correlated responses to artificial body size selection in growth, development, phenotypic plasticity and juvenile viability in yellow dung flies. *Journal of Evolutionary Biology* 20: 87–103.

Thüler, K., L.F. Bussière, E. Postma, P.I. Ward, and W.U. Blanckenhorn. 2011. Genetic and environmental sources of covariance among internal reproductive traits in the yellow dung fly. *Journal of Evolutionary Biology* 24: 1477–1486.

Tomberlin, J.K., M.E. Benbow, A.M. Tarone, and R.M. Mohr. 2011b. Basic research in evolution and ecology enhances forensics. *Trends in Ecology and Evolution* 26: 53–55.

Tomberlin, J.K., R.M. Mohr, M.E. Benbow, A.M. Tarone, and S. VanLaerhoven. 2011a. A roadmap for bridging basic and applied research in forensic entomology. *Annual Review of Entomology* 56: 401–421.

Tomberlin, J.K., D.C. Sheppard, and J.A. Joyce. 2005. Black soldier fly (Diptera: Stratiomyidae) colonization of pig carrion in south Georgia. *Journal of Forensic Science* 50: 152–153.

Trumbo, S.T. 1990. Reproductive benefits of infanticide in a biparental burying beetle *Nicrophorus orbicollis*. *Behavioural Ecology and Sociobiology* 27: 269–273.

Turner, T.L., A.D. Stewart, A.T. Fields, W.R. Rice, and A.M. Tarone. 2011. Population-based resequencing of experimentally evolved populations reveals the genetic basis of body size variation in *Drosophila melanogaster*. *PloS Genetics* 7: e1001336.

van der Have, T.M. and de Jong, G. 1996. Adult size in ectotherms: Temperature effects on growth and differentiation. *Journal of Theoretical Biology* 18: 329–340.

Van Laerhoven, S.L. 2010. Ecological theory and its application in forensic entomology. In: Byrd, J.H. and J.L. Castner (eds), *Forensic Entomology: The Utility of Arthropods in Legal Investigations*. Boca Raton, FL: CRC Press, pp. 493–518.

van Noordwijk, A.J. and de Jong, G. 1986. Acquisition and allocation of resources: Their influence on variation in life history tactics. *American Naturalist* 128: 137–142.

Vass, A.A., S.A. Barshick, G. Sega, J. Caton, J.T. Skeen et al. 2002. Decomposition chemistry of human remains: A new methodology for determining the postmortem interval. *Journal of Forensic Science* 47: 542–553.

Via, S. 1993a. Adaptive phenotypic plasticity—Target or by-product of selection in a variable environment. *American Naturalist* 142: 352–365.

Via, S. 1993b. Regulatory genes and reaction norms. *American Naturalist* 142: 374–378.

Via, S. 2009. Natural selection in action during speciation. *Proceedings of the National Academy of Sciences of the United States of America* 106: 9939–9946.

Via, S., R. Gomulkiewicz, G. de Jong, S.M. Scheiner, C.D. Schlichting, and P.H. van Tienderen. 1995. Adaptive phenotypic plasticity—Consensus and controversy. *Trends in Ecology and Evolution* 10: 212–217.

Via, S. and R. Lande. 1985. Genotype–environment interaction and the evolution of phenotypic plasticity. *Evolution* 39: 505–522.

Visser, J.H. 1986. Host odor perception in phytophagous insects. *Annual Review of Entomology* 31: 121–144.

Visser, M.E. 1994. The importance of being large: The relationship between size and fitness in females of the parasitoid *Aphaereta minuta* (Hymenoptera: Braconidae). *Journal of Animal Ecology* 63: 963–978.

Voss, S.C., H. Spafford, and I.R. Dadour. 2010. Temperature-dependent development of the parasitoid *Tachinaephagus zealandicus* on five forensically important carrion fly species. *Medical and Veterinary Entomology* 24: 189–198.

Walters, R.J. and M. Hassall. 2006. The temperature-size rule: Does a general explanation exist after-all? *The American Naturalist* 167: 510–523.

Ward, P.I. 2000. Sperm length is heritable and sex-linked in the yellow dung fly (*Scathophaga stercoraria*). *Journal of Zoology* 251: 349–353.

Watson, N.L. and L.W. Simmons. 2010. Male and female secondary sexual traits show different patterns of quantitative genetic and environmental variation in the horned beetle *Onthophagus sagittatus*. *Journal of Evolutionary Biology* 23: 2397–2402.

Watson, N.L. and L.W. Simmons. 2012. Unraveling the effects of differential maternal allocation and male genetic quality on offspring viability in the dung beetle *Onthophagus sagittatus*. *Evolutionary Ecology* 26: 139–147.

Watts, J.E., G.C. Merritt, and B.S. Goodrich. 1981. The ovipositional response of the Australian sheep blowfly *Lucilia cuprina*, to fleece-rot odours. *Australian Veterinary Journal* 57: 450–454.

Weigensberg, I. and D.A. Roff. 1996. Natural heritabilities: Can they be reliably estimated in the laboratory? *Evolution* 50: 2149–2157.

Werner, E.E. and B.R. Anholt. 1993. Ecological consequences of the trade-off between growth and mortality rates mediated by foraging activity. *American Naturalist* 142: 242–272.

Wilson, A.J., J.M. Pemberton, J.G. Pilkington, T.H. Clutton-Brock, D.W. Coltman, and L.E.B. Kruuk. 2007. Quantitative genetics of growth and cryptic evolution of body size in an island population. *Evolutionary Ecology* 21: 337–356.

Zuha, R.M., T.A. Razak, N.W. Ahmad, and B. Omar. 2012. Interaction effects of temperature and food on the development of the forensically important fly *Megaselia scalaris* (Diptera: Phoridae). *Parasitology Research* 111: 2179–2187.

16

Carrion and Dung Mimicry in Plants

Andreas Jürgens and Adam Shuttleworth

CONTENTS

16.1 Introduction

Flowers are stereotypically viewed as symbols of beauty which have inspired poets for centuries. To the layperson, flowers are usually associated with bright colors and sweet fragrances. Of course flowers do not exist for the benefit of human observers, despite the reverence conferred upon them by poets such as Shakespeare and Goethe. In a biological context, flowers represent mechanisms which plants have evolved in order to exploit third-parties, mostly insects, as a means of transferring gametes between individuals and achieve cross-fertilization (Faegri and van der Pijl 1979).

Decomposing organic matter, such as carrion, dung, and rotting plant material, represent important resources which support a diverse assemblage of arthropods, ranging from opportunistic foragers to obligate saprophytes such as carrion flies (Diptera) and carcass beetles (Coleoptera). Many of these obligate saprophytes have evolved specialized life histories and rely on carrion or dung at various stages in their life cycle (see Chapter 4). The evolution of a group of arthropods obligately associated with carrion has in turn allowed for the evolution of a remarkable pollination system in which flowers have evolved various carrion-like traits in order to mimic decaying carcasses, thereby attracting insects associated with decaying organic matter for the purposes of pollination (Vereecken and McNeil 2010; Urru et al. 2011). These flowers are commonly categorized as carrion flowers or sapro(myio)philous flowers (following the tradition of pollination syndromes of Faegri and van der Pijl 1979). This terminology, however, can be misleading, as sapromyiophilous flowers are neither all imitating carrion nor are they all pollinated by flies (as myiophilous suggests) nor are they all flowers (the functional unit in the Araceae is an inflorescence, composed of many flowers). For practical reasons and for correctness, they will be referred to here as either oviposition site mimicry systems, in general (including associations with all types of detritus-feeding insects), or true carrion flowers, in particular (only associated with necrophagous insects). For the purposes of this contribution, the main focus will be on the more obviously carrion-mimicking species, but it is important to realize that these form part of a broader phenomenon of oviposition site mimicry by flowers (Vereecken and McNeil 2010; Urru et al. 2011).

The evolutionary success and high species diversity of flowering plants has often been explained as a result of coevolution with flower-visiting animals (Faegri and van der Pijl 1979). It is now widely accepted that coevolution has played an important role in shaping floral traits within the most common type of plant–pollinator interaction, where a plant attracts animals with rewards (e.g., nectar or pollen) in exchange for the transport of pollen (the plant's male gametes) between flowers to produce seed (see Johnson and Anderson 2010). However, the best examples of coevolution can be found within the nursery pollination systems, for example, the yucca–moth or the fig–wasp systems, where both partners not only benefit from the mutualistic relationship but are obligately dependent on their partners (Sakai 2002; Dötterl et al. 2006). In nursery pollination systems, pollinating insects use the flower (or inflorescence) as a brood site for their larvae and both partners benefit from the relationship. More recent studies, however, have reported that a great proportion of plant–pollinator relationships are asymmetrical in terms of benefit sharing (Jersáková et al. 2009; Vereecken and McNeil 2010). In these cases, plants are exploiting the innate and learned responses of animals to visual and/or olfactory signals without offering any reward (Schiestl et al. 1999; Jersáková et al. 2009; Vereecken and McNeil 2010). Carrion flowers and other oviposition site mimicry systems provide excellent examples of asymmetrical benefit relationships. These plants lure coprophagous and necrophagous insects to their flowers by olfactory and visual signals that resemble those of dung or carrion (Jürgens et al. 2006; Johnson and Jürgens 2010). In many of these systems, flower-visiting insects receive no reward from the plant, although this is not always the case and some species produce nectar (Meve and Liede 1994; Jürgens et al. 2006). In some instances, the features of the flower seem to be such a perfect mimic of carrion that flies find it irresistible to lay their eggs (Calliphoridae) or larvae (Sarcophagidae) on the flowers (Snow 1957; Bänziger 1996). However, the larvae are not provided with their usual brood site substrate and cannot survive on the flowers (Bänziger 1996).

Oviposition site mimicry systems have recently been used to study convergent evolution (the evolution of similar features in unrelated plant species) of flower scent in angiosperms (Jürgens et al. 2013). They represent an excellent model system for this because the volatile organic compounds (VOCs) emitted by the suspected models such as carrion and dung and the behavioral responses of different detritus-feeding insects to these models have been investigated in some detail (Dekeirsschieter et al. 2009). Carrion mimicry falls within a broader system of brood site mimicry (or oviposition site mimicry) in which flowers exploit the innate and learned cues that some insects use to find brood sites in order to attract these insects as pollinators (Urru et al. 2011). Brood site mimicry has evolved numerous times in angiosperms and involves the mimicry of various different brood site substrates including carrion, feces, and other animal waste products, and sometimes fermentation products (Urru et al. 2011; Jürgens et al. 2013). To this end, many flowers have evolved remarkably sophisticated methods of enticing insect visitors through the false promise of a particular brood site which the insects depend upon at some stage in their life history. These include the production of suites of VOCs usually associated with the respective brood sites (Jürgens et al. 2006) as well as various tactile and visual cues that complete the deception. Some have even gone so far as to evolve mechanisms of raising their internal temperatures to closely match the temperature changes of potential models as they decompose (Stensmyr et al. 2002; Angioy et al. 2004).

This chapter reviews the chemical basis for the behavior of detritus-feeding insects and provides a description of the mechanisms that plants have evolved to manipulate and exploit the innate and learned preferences of these insects. A historical perspective and review of carrion-mimicking flowers is initially presented, followed by a discussion of the *animal perspective* aimed at providing the reader with the basic background regarding the chemical ecology of detritus-feeding insects. However, the focus will be on aspects that are of relevance for understanding the ecology and evolution of carrion flowers. For a more detailed overview of chemical ecology, see Chapter 9. The animal perspective is followed by a discussion of the *plant perspective* which focuses on the chemical ecology of oviposition site mimicry systems. Finally, a discussion of the knowledge gaps and possibilities for interdisciplinary research approaches that could help to fill these gaps is given.

The primary focus of this chapter will be the truly carrion-mimicking flowers within the broad confines of brood site mimicry as a whole. However, it should be kept in mind that the boundary between carrion and other types of decaying organic matter (particularly feces) is often somewhat blurry and many of the insect groups concerned exhibit a slightly broader trophic niche than just carrion. This makes it difficult (if not meaningless!) to pigeonhole flowers that are associated with saprophilous insects into a purely carrion-mimicking mode of life. Although there are now several examples of very well-studied and understood carrion-mimicry systems, there is also a diversity of flowers which exhibit carrion-like traits but remain to be studied in detail. For the purposes of this chapter, flowers which either exhibit carrion-like traits (particularly odor, but also various visual and morphological traits) or, alternatively, have been shown to be visited (and ideally pollinated) by insects that are typically associated with carrion are considered to be carrion-mimicking flowers. The insects involved in these pollination systems are mostly flies (especially in the families Muscidae, Calliphoridae, and Sarcophagidae, but also several other carrion-feeding families) and beetles (particularly in the families Dermestidae, Silphidae, Scarabaeidae, Staphylinidae, Histeridae, and Hybosoridae).

16.2 Historical Perspective and Review of Carrion-Mimicry in Angiosperms

Flowers that exhibit carrion-like traits have been known to botanists since the seventeenth century, and at least two well-known South African carrion-mimicking stapeliads, *Orbea variegata* and *Stapelia hirsuta*, where apparently being cultivated in Europe by the mid- to late-1600s (White and Sloane 1937). The earliest record of *O. variegata* is from drawings made by Justus Heurnius in 1624, following a brief stop in Table Bay (White and Sloane 1937; Bruyns 2005). Heurnius' drawings were passed on to Johannes Bodaeus Stapelius in Amsterdam who included them, along with the first written description of *O. variegata* (under the name *Fritillaria crassa*) in his *Theophrasti eresii de historia plantarum* of 1644 (edited and published by his father Egbert Bodaeus Stapelius, following the death of Johannes in 1636) (Stapelius 1644; White and Sloane 1937). It is not clear when *S. hirsuta* was first collected, although it is first mentioned by Plukenet (1696) in his *Almagestum botanicum* (White and Sloane 1937). Both species were formally described to science by Linnaeus (1753) in his *Species Plantarum* (representing the beginning of plant nomenclature and taxonomy) although they had both already appeared in his earlier catalogue of the plants in George Clifford's personal collection, published as *Hortus Cliffortianus* (Linnaeus 1737).

By the early 1800s, a number of other carrion-mimicking plants had been discovered and described, including several more stapeliads (mainly from Africa, but also from Arabia, India, and southern Spain) as well as the two bizarre holoparasites (a completely parasitic plant that cannot exist without a host plant), *Hydnora africana* (Hydnoraceae; Figure 16.1c) from Africa (Thunberg 1775) and *Rafflesia arnoldi* (Rafflesiaceae; Figure 16.1a) from Sumatra (Brown 1822). *H. africana* was first collected and described by Carl Thunberg, an accomplished botanist and former student of Linnaeus, in 1775. However, Thunberg was apparently fooled by the weird morphology and lack of leaves, and initially described the plant as a fungus related to the genus *Hydnum* (Thunberg 1775; Bolin et al. 2011). *R. arnoldi*, now well known as the world's largest flower, was discovered in Sumatra in 1818 by James Arnold and Sir Stamford Raffles. The formal description of the plant was subsequently read before the Linnean Society in 1820 (and published in 1822) by Robert Brown, an influential nineteenth-century European botanist, who named the plant after its two codiscoverers.

Although many of the plants that are now recognized as carrion-mimics had been described by the mid-1800s, it is not clear when they were first recognized as mimicking decaying flesh. Linnaeus's original description of *S. hirsuta* in *Hortus Cliffortianus* included phrases such as "*flore pulchre fimbriato*" (flowers beautifully fringed), "*marginem interne undique crinibus longis rectis barbata*" (inner margin on all sides bearded with long straight hairs), and "*odor hircinus aphrodisiacus lascivus*" (a goat-like [or unwashed armpit-like], aphrodisiac, lewd smell). Linnaeus, however, was interested only in classifying flowers and had little interest in the functional significance of these peculiar floral traits (Vogel 1996).

It was only after Sprengel (1793) introduced the idea that floral traits must have a purpose that people began to consider the traits of carrion flowers in the light of their role in attracting carrion-associated

FIGURE 16.1 (See color insert.) Examples of carrion-mimicking flowers/inflorescences. (a) *Rafflesia arnoldi* (Rafflesiaceae), reproduction of the original illustration by Franz Bauer used in the type description by Robert Brown (1822). (b) *Aristolochia cymbifera* (Aristolochiaceae; Photo: R. Roth). (c) *Hydnora africana* (Hydnoraceae), reproduction of an illustration by Ferdinand Bauer used by Robert Brown (1845) in his description of the structure of *H. africana*. (d) *Duvalia polita* (Apocynaceae; Photo: A. Shuttleworth). (e) *Stapelia gigantea* (Apocynaceae; Photo: A. Shuttleworth). (f) *Helicodiceros muscivorus* (Araceae; Photo: M. Stensmyr).

insects. In the context of carrion flowers, Sprengel himself commented in his landmark book, *Das entdeckte Geheimniss der Natur im Bau und in der Befruchtung der Blumen* (The Secret of Nature in the Form and Fertilization of Flowers Discovered) published in 1793, that *S. hirsuta* "… stinks like carrion only in order to lure bottle and carrion flies to which the stench is highly agreeable, and seduce them to fertilize the flower" (translated and quoted by Vogel 1996). Sprengel's ideas regarding the supposed functions of floral traits were not widely accepted and remained somewhat obscure until after his death when his work was highlighted and praised by none other than Charles Darwin in the 1870s (Vogel 1996). Nonetheless, there are several references to the role of floral traits, particularly odor, in the attraction of carrion insects for pollination from the nineteenth century, suggesting that people had begun to recognize the link between looking and smelling like a carcass and attracting carcass-associated insects as pollinators.

This is particularly clear from Robert Brown's (1822) original description of *R. arnoldi*. In his paper, Brown (1822) reproduces a letter from Dr James Arnold, one of the plant's discoverers, to an "unknown friend" (Arnold died of malaria shortly after the discovery, and before news of the discovery reached Europe) in which he describes his first glimpse of the enormous flower: "When I first saw it a swarm of flies were hovering over the mouth of the nectary, and apparently laying their eggs in the substance of it. It had precisely the smell of tainted beef." Brown concluded from this statement that flies were essential for the plant's pollination and were attracted by the odor (Brown 1822). Interestingly, Robert Brown was one of the few people who appreciated the value of Sprengel's ideas about the functional significance of floral traits and had recommended Sprengel's book to Darwin (Vogel 1996). However, the importance of odor in attracting carrion flies was also recognized by other botanists at the time. The French botanist Henri Lecoq, for example, wrote in 1862: "les Stapelia … l'odeur cadavéreuse des fleurs, qui est telle que les mouches y déposent continuellement leurs oeufs, comme sur de la viande corrompue, n'a-telle pas pour but de faciliter la fécondation par le moyeu des insects" (in *Stapelia* … the cadaverous smell of flowers, which is such that the flies lay their eggs there continuously, as in rotten meat, is intended to facilitate fertilization by means of these insects).

Although clearly recognizing the similarity between rotting meat and the floral traits of these flowers, many botanists at this time favored a teleological interpretation of floral traits such as "stinking flowers." Sprengel, for example, although appreciating the functions of floral traits, interpreted them as proof of Divine Wisdom and viewed their functions purely as completing the interaction between insects and flowers (Vogel 1996). This interpretation of floral traits persisted until the late 1800s when the publication of Darwin's "On the Origin of Species" (1859) and, perhaps more importantly in the context of floral evolution, "On the various contrivances by which British and foreign orchids are fertilized by insects" (1862), introduced a new adaptive framework with which to interpret floral traits. This was particularly enlightening for understanding the existence of carrion flowers in which the adaptive significance of traits mimicking carcasses are immediately clear.

Interestingly, there is no clear record of Darwin himself having considered the evolution of carrion flowers, although several letters to Darwin mention *Stapelia* flowers. In one letter to Darwin dated November 9, 1867, F.E. Kitchener, a school teacher and amateur botanist, asked, "Has anyone yet investigated the fertilization of the *Stapelia* to see, whether the putrid smell may be regarded as a mimetic resemblance to carrion, which benefits the plant by attracting flies, under false pretences of its being a suitable place to lay their eggs?" (http://www.darwinproject.ac.uk, accessed April 23, 2013). Unfortunately, there is no record of a response from Darwin and his thoughts on carrion mimicry remain unknown. Intriguingly, Darwin was always sceptical of the existence of nonrewarding orchids and considered insects to be too clever to be outsmarted by deceptive plants (Darwin 1877). Darwin's views on the existence of orchids that produce no nectar but are still visited by insects is summarized in his statement that "we can hardly believe in so gigantic an imposture" (Darwin 1877). It could be argued that some of the carrion flowers are even greater imposters, and it would be interesting to know Darwin's thoughts on carrion-mimicking flowers which attract insects with a false promise of food or a brood site.

Regardless of how carrion-like floral traits were interpreted, the existence of a suite of flowers associated with carrion-feeding insects (mostly flies, but also some beetles) was well established by the end of the nineteenth century. This was perhaps most clearly recognized by the Italian botanist Federico Delpino who described observations of the fertilization of stapeliads by flies in an 1867 monograph (Delpino 1867; White and Sloane 1937). Delpino later devised a system of classifying plants based on their pollinators and included "sapromyiophilae (carrion fly flowers)" as one of his entomophilous (insect pollinated) groups (Delpino 1868–1875) (these would ultimately form the conceptual basis for the development of modern pollination syndrome theory; Waser 2006). Despite the name (*myio* is derived from the Greek *muia* meaning fly), this group included flowers visited by carrion beetles. Delpino's classification was expanded by Vogel (1954) to form his "myiophilous style" for which he outlined a suite of traits, including exposed nectar, fetid odor, and a glossy or dull, warty surface, often with cilia. A similar approach was followed by Faegri and van der Pijl (1979) for their outline of the "syndrome of sapromyophily," for which they described a similar suite of traits as Vogel (1954) but also including the frequent presence of filiform appendages, dark spots, and motile hairs.

Increasing botanical explorations since the 1900s have led to the recognition of a diversity of plants in various families that exhibit aspects of the sapromyiophilous syndrome normally associated with carrion mimicry (Table 16.1; Figure 16.1). It is now clear that carrion mimicry has evolved on numerous separate occasions in the angiosperms (Ollerton and Raguso 2006; Urru et al. 2011; Jürgens et al. 2013) and, with a slightly different function, in various fungi (Tuno 1998; Johnson and Jürgens 2010) and some of the "dung mosses" in the family Splachnaceae (Marino et al. 2009). Within the angiosperms, however, there are several groups in which the phenomenon is particularly well represented.

TABLE 16.1A

Summary of Carrion Mimicry in the Angiosperms

Family	Distribution	No. of Species (Genera)	Autotrophic or Holoparasitic	Thermogenesis	References
Annonaceae	Africa, North America	c. 15 (8)	Autotrophic	No	Goodrich and Raguso 2009; Goodrich 2012; Gottsberger et al. 2011
Apocynaceae	Africa, Arabia, Asia, Madagascar, Europe	c. 75 (15)	Autotrophic	No	Bhatnagar 1986; Coombs 2010; Coombs et al. 2011; Dyer 1983; Formisano et al. 2009; Geers, Shuttleworth and van der Niet, unpublished data; Herrera and Jürgens et al. 2006; Meve and Liede 1994; Meve et al. 2004; Nassar 2009; Pisciotta et al. 2011; Shuttleworth and Johnson 2009; Shuttleworth and Jürgens, unpublished data; Snow 1957; Zito et al. 2013.
Araceae	Africa, Asia, Mediterranean, North America	c. 29 (7)	Autotrophic	Yes	Angioy et al. 2004; Barthlott et al. 2009; Beath 1996; Borg-Karlson et al. 1994; Fujioka et al. 2012; Gibernau et al. 2005; Gibernau 2003; Kakishima et al. 2011; Kite and Hetterscheid 1997; Kite et al. 1998; Kite 2000; Seymour and Schultze-Motel 1999; Shirasu et al. 2010; Stensmyr et al. 2002; Stránský and Valterová 1999; Uemura et al. 1993; van der Pijl 1937
Aristolochiaceae	South America, North America, Russia	13 (1)	Autotrophic	Reported for some species	Berjano et al. 2009; Blanco 2002; Burgess et al. 2004; Johnson and Jürgens 2010; Nakonechnaya et al. 2008; Stotz and Gianoli 2013; Thien et al. 2000; Thien et al. 2009
Asparagaceae	Africa	c. 9 (1)	Autotrophic	No	Shuttleworth and Johnson 2010; Shuttleworth and Johnson, unpublished data
Asteraceae	South Africa	1 (1)	Autotrophic	No	Kite and Smith 1997
Colchicaceae	South Africa	3 (2)	Autotrophic	Not reported	Vogel 1954; Sivechurran, Jürgens, Shuttleworth and Johnson, unpublished data
Euphorbiaceae[a]	South Africa, Asia	4 (1)	Autotrophic	No	Bänziger et al. 2008; Johnson and Jürgens 2010; Vogel 1954
Hydnoraceae	Africa, Madagascar, Arabia	7 (1)	Holoparasitic	Low levels in some species	Bolin et al. 2009; Bolin et al. 2011; Burger et al. 1988; Marloth 1907; Seymour et al. 2009; Vogel 1954
Iridaceae	South Africa	6 (2)	Autotrophic	No	Goldblatt et al. 2009; Johnson and Jürgens 2010; Vogel 1954

(Continued)

TABLE 16.1A (*Continued*)

Summary of Carrion Mimicry in the Angiosperms

Family	Distribution	No. of Species (Genera)	Autotrophic or Holoparasitic	Thermogenesis	References
Liliaceae	Europe, Asia, North America	4 (1)	Autotrophic	Not reported	Anecdotal, see http://www.fritillariaicones.com/info/pollinators.html (accessed October 2013)
Lowiaceae[b]	Asia	1 (1)	Autotrophic	Not reported	Sakai and Inoue 1999
Malvaceae	Asia, North America	2 (1)	Autotrophic	Not reported	Atluri et al. 2004; Weber 2008
Orchidaceae	South America, South Africa, Reunion	4 (3)	Autotrophic	No	Humeau et al. 2011; van der Niet et al. 2011
Rafflesiaceae	Southeast Asia	c. 27 (3)	Holoparasitic	Reported for two species	Bänziger and Hansen 2000; Bänziger and Pape 2004; Bänziger 1991; Bänziger 1996; Bänziger 2001; Bänziger et al. 2000; Barcelona et al. 2009; Beaman et al. 1988; Bendikshy et al. 2010; Hidayati et al. 2000; Pape and Bänziger 2000; Patiño et al. 2000; Patiño et al. 2002
Rhamnaceae	South America, Asia	2 (2)	Autotrophic	No	Alves et al. 2005; Bänziger and Pape 2004
Solanaceae	South America	5 (1)	Autotrophic	Not reported	Moré et al. 2013; Vogel 1954; Moré, Raguso, Dötterl and Cocucci, unpublished data
Stemonaceae	Southeast Asia and tropical Australia	3 (1)	Autotrophic	No	Duyfjes 1991
Theophrastaceae	North America, Hispaniola	2 (2)	Autotrophic	Not reported	Knudsen and Ståhl 1994
Thismiaceae	South America	1 (1)	Holoparasitic	Not reported	Woodward et al. 2007
Taccaceae[c]	Asia	10 (1)	Autotrophic	Not reported	Faegri and van der Pijl 1979; Zhang et al. 2005

Note: The number of species for each family is estimated from published studies and personal observations. However, several of the families, particularly Araceae and Apocynaceae, undoubtedly contain considerably more carrion-mimicking species than have been listed here. Note that the family Apocynaceae also contains nonstapeliad carrion mimics, particularly in the genus *Brachystelma* (see Dyer 1983).

a Based on limited observation of flies visiting flowers, so should be treated with caution. Sweetly scented species may be generalists which are also visited by flies.

b This species may be a dung mimic, but the beetle pollinators were also attracted to carrion.

c Species in the genus *Tacca* have traditionally been considered sapromyiophilous based on their floral traits which fit with the syndrome (Faegri and van der Pijl 1979). However, a recent study showed that the flowers of one species are highly selfing and the only possible pollinators were stingless bees (see Zhang et al. 2005), so it remains unclear if these should be considered true carrion-mimicking flowers.

TABLE 16.1B

Summary of Carrion Mimicry in the Angiosperms Continued

Family	Known or Assumed Pollinators	Flower/ Inflorescence Type	Presence of Nectar	Odor	Colour and Tactile Cues	Ovi- or Larviposition on Flowers
Annonaceae	Flies and beetles	Open	Not reported	Foetid, fermented	Often maroon or yellow, sometimes with elongated textured petals	Not reported
Apocynaceae	Flies	Open	Usually	Diverse, but often carrion or feces	Yellow, brownish or reddish, often with contrasting markings on a pale background, fleshy textured lobes, often shiny, pubescent, papillate, often with vibratile clavate hairs	Sometimes
Araceae	Mostly beetles, sometimes flies	Trap	No	Carrion, gaseous	Often reddish or brownish, sometimes pubescent	Reported in one species
Aristolochiaceae	Flies	Trap	Sometimes	Foetid	Often deep reddish with paler markings and elongated "tails"	Sometimes
Asparagaceae	Flies	Open	Yes	Carrion or potatoes	Greenish-white, often with purple markings	No
Asteraceae	Flies	Open	Not reported	Foetid	Reddish	No
Colchicaceae	Flies	Open	Yes	Nauseating or faecal	White with dark spots or with modified petal-like leaves that are elongated and greenish speckled with reddish marks	Not reported
Euphorbiaceae[a]	Flies	Open	Not reported	Foetid or sweet	Greenish	No
Hydnoraceae	Mostly beetles, possibly also flies	Trap	No	Carrion	Brown, reddish-orange inside	Not reported
Iridaceae	Flies	Open	Yes	Carrion or pungent, but also sweet in two species	Contrasting brown and pale greenish, tepals crisped along margins, sometimes shiny	Not reported
Liliaceae	Flies	Open	Not reported	Foetid, dog feces	Contrasting brown and yellow speckles	Not reported
Lowiaceae[b]	Dung beetles	Open	No	Dung-like	Black with contrasting white lateral petals	Not reported
Malvaceae	Flies	Open	Not reported	Carrion, fetid	Brownish-reddish, sometimes with contrasting markings on a pale background, pubescent	Not reported
Orchidaceae	Flies	Open	No	Unpleasant or carrion	Dull green with reddish markings	Reported in one species
Rafflesiaceae	Flies	Open	No	Foetid, carrion	Often brownish with paler markings, sometimes pubescent	Sometimes
Rhamnaceae	Flies	Open	Not reported	Fecal for one species	Not reported	Not reported
Solanaceae	Flies	Open	No (but small amounts of a sugary solution are present)	Faeces, garlic or carrion	Dull reddish to black, pubescent	Yes
Stemonaceae	Flies	Open	Not reported	Unpleasant or carrion	Dull green with reddish markings	Not reported
Theophrastaceae	Flies	Open	Not reported	Not reported	Not reported	Reported in one species
Thismiaceae	Flies	Open	Not reported	Rotten fish	Dull yellow, shiny	Not reported
Taccaceae[c]	Selfing, but also stingless bees in one species	Open	Not reported	Musty	Maroon with filiform appendages	No

Note: See Table 16.1A for footnotes and references.

Perhaps one of the best known groups of carrion-mimicking flowers are the stem-succulent stapeliads of Africa, Europe, Arabia, and Asia. As already mentioned, in the context of carrion mimicry, stapeliads represent some of the earliest known examples of carrion flowers. However, a diversity of species have been described since the original descriptions of *S. hirsuta* and *O. variegata* by Linnaeus (1737, 1753) and the stapeliads now represent a group of between 300 and 400 species in 31–34 genera (depending on taxonomy). Morphological and molecular studies suggest that the stapeliads represent a monophyletic clade within the subtribe Stapeliinae in the tribe Ceropegieae (Apocynaceae: Asclepiadoideae) (Bruyns 2000, 2005; Meve and Liede 2002, 2004). Although widely distributed, stapeliads are typically associated with arid to semiarid habitats with two major centers of diversity, one in southwestern South Africa and Namibia and one in northeastern Africa (Kenya, Ethiopia, and Somalia) and Arabia (Bruyns 2000, 2005; Meve and Liede 2002, 2004).

Stapeliads, particularly in the genera *Stapelia* and *Orbea*, represent some of the archetypal carrion-mimicking flowers and they are well known for their foul odors and bizarre (sometimes grotesque) morphological adaptations which include fleshy, textured flowers covered in hairs, papillae, and vibratile clavate cilia (Figure 16.1d and e). The popular image of stapeliads, however, is biased toward the more derived genera such as *Stapelia* which usually have large, hairy, smelly flowers. In reality, the group comprises a diversity of different floral forms, colors, and sizes (see Bruyns 2005). Although many stapeliads are commonly known as "carrion flowers" (or "aasblomme" in Afrikaans) and often anecdotally assumed to be primarily fly-pollinated, there have been very few detailed studies of pollination in stapeliads (Meve and Liede 1994; Meve et al. 2004). Furthermore, much of the empirical evidence of fly-pollination is based on observations of cultivated plants outside their natural range (Meve and Liede 1994). Nonetheless, it appears that, aside from some of the smaller-flowered species which are pollinated by micro-Diptera, the majority of stapeliads are pollinated by carrion- or dung-associated flies, particularly in the families Calliphoridae, Muscidae, and Sarcophagidae (Snow 1957; Meve and Liede 1994, 2004; Meve 1997; Bruyns 2000, 2005; Herrera and Nassar 2009; Johnson and Jürgens 2010; Geers, Shuttleworth and van der Niet, unpublished data; Shuttleworth and Jürgens, unpublished data).

Flies are usually (although not always) attracted to stapeliad flowers by odors and collect pollinia on their mouthparts (or occasionally the legs) while probing the central gynostegium, a typical feature of most members of the Apocynaceae formed by the fusion of the stamens, styles, and stigmatic surfaces (Meve and Liede 1994; Meve 1997). Most stapeliad flowers produce nectar (Meve and Liede 1994, 2004; Bruyns 2005) which often serves as a reward for visiting flies. However, it is not clear whether nectar always functions as a reward, and it has also been suggested to play a role in enhancing the visual attractiveness of flowers by creating a shiny, reflective floral surface (Meve and Liede 1994). In flowers where the nectar is consumed by the flies it thus appears that, although often attracting flies with a dishonest signal (odors suggestive of the flies' brood site), the flowers are not deceptive in the sense of, for example, nonrewarding deceptive orchids (Meve and Liede 1994). Interestingly, flies are known to oviposit or larviposit on some stapeliad flowers (particularly members of the genus *Stapelia*), suggesting that in some cases there is an element of real deception which incurs a cost on the part of the pollinator. This is well illustrated by flowers of *Stapelia gigantea* (Figure 16.1e) which have often been observed to contain fly eggs and larvae (Snow 1957; Johnson and Jürgens 2010; Shuttleworth pers. obs.). Intriguingly, in a study of the pollinators of *S. gigantea* in its natural range, it was found that the blow fly pollinators and nonpollinating fly visitors, while initially attracted by the foul odor, exhibit clear foraging behavior and collect nectar from *S. gigantea* flowers (Geers, Shuttleworth and van der Niet, unpublished data). The interplay between cues governing ovi- or larviposition and those that govern foraging thus remain unclear and the levels of deception within stapeliads are difficult to assess.

It is also difficult to establish the frequency of true carrion mimicry within stapeliads. Many of the larger-flowered species have flowers strikingly reminiscent of carrion to the human observer (in terms of both odor and morphology). However, stapeliad flowers exhibit a range of different fetid odors. While many of these are similar to carrion to the human observer, there is some variation in the qualitative nature of these odors and a large number of stapeliads exhibit odors which are more reminiscent of excrement than carrion (Bruyns 2005). In a recent analysis of the chemical composition of the odors of 15 stapeliad species, Jürgens et al. (2006) identified several distinct chemotypes suggesting mimicry of different broodsite substrates, including carcasses/carrion, herbivore feces, and urine. This suggests that there are various different mimicry strategies within stapeliads and it remains unclear what proportion of stapeliads could be considered pure carrion mimics per se.

The Rafflesiaceae represents another angiosperm group in which carrion mimicry is particularly prevalent. The family comprises three genera (*Rafflesia*, *Sapria*, and *Rhizanthes*) distributed throughout southeast Asia from eastern India to the Philippines, and Java. The genus *Rafflesia* is perhaps best known for containing some of the world's largest individual flowers, including the already described *R. arnoldi* (Figure 16.1a) which can measure 1 m in diameter and weigh up to 7 kg (Davis et al. 2008; Bendiksby et al. 2010). All members of the Rafflesiaceae lack vegetative parts and are endophytic holoparasites (body plan reduced to tissue strands living entirely within the tissue of the host plant) of members of the genus *Tetrastigma* (Vitaceae), emerging only to flower during a brief period (Bendiksby et al. 2010). Visually, the members of the genus *Rafflesia* are somewhat convergent with some of the stapeliads and exhibit similar brownish coloring with paler markings. However, molecular studies suggest that the family Rafflesiaceae represents a monophyletic clade nested within the Euphorbiaceae (Davis et al. 2007; Wurdack and Davis 2009). In addition to having large flowers, members of the Rafflesiaceae are also well known for their foul odors (Bänziger 1991) and appear to be universally pollinated by carrion-associated flies (particularly Calliphoridae and Sarcophagidae) (Beaman et al. 1988; Bänziger 1991, 1996, 2001; Bänziger and Hansen 2000; Bänziger et al. 2000; Hidayati et al. 2000; Pape and Bänziger 2000; Bänziger and Pape 2004). Although nectar is offered as a reward in *Rhizanthes* (Bänziger 1996; Bänziger and Hansen 2000), it appears that most Rafflesiaceae are truly deceptive and offer no reward for their pollinators (Beaman et al. 1988). In addition, flies have been observed to oviposit extensively in the flowers of *Rhizanthes* species suggesting that the deception is often very effective (Bänziger 1996; Bänziger and Hansen 2000).

The arums (Family Araceae) represent another family which exhibits widespread brood-site mimicry and includes various carrion-mimicking species (Kite and Hetterscheid 1997; Kite et al. 1998; Stensmyr et al. 2002; Urru et al. 2011). The best known carrion flowers in the Araceae are the Titan Arum (*Amorphopallus titanum*) of the Sumatran rain forests (famous as the largest inflorescence in the world, with a height of up to height 3 m), and the well-studied Mediterranean Dead Horse Arum (*Helicodiceros muscivorus*; Figure 16.1f) (Kite 2000; Stensmyr et al. 2002; Seymour et al. 2003a; Angioy et al. 2004; Barthlott et al. 2009). However, carrion mimicry is widespread in the family and occurs in various other genera including *Dracunculus*, *Hydrosme*, *Pseudodracontium*, *Sauromatum*, and *Symplocarpus* (Uemura et al. 1993; Borg-Karlson et al. 1994; Kite and Hetterscheid 1997; Seymour and Schultze-Motel 1999; Stránský and Valterová 1999). As for other such systems, carrion mimicry in the Araceae is strongly characterized by the production of fetid odors particularly dominated by oligosulfides (Borg-Karlson et al. 1994; Kite and Hetterscheid 1997; Kite et al. 1998; Stránský and Valterová 1999; Kite 2000). However, visual and tactile cues are also apparent, and the flowers of carrion-mimicking Araceae are typically dark-brownish or purplish in color and are sometimes covered in hairs (see Figure 16.1f) (Kite et al. 1998; Stensmyr et al. 2002).

The pollinators of carrion-mimicking Araceae are mainly beetles (including members of the families Scarabaeidae, Silphidae, Histeridae, Staphylinidae, and Dermestidae; Beath 1996; Kite et al. 1998; Seymour and Schultze-Motel 1999), although flies are also often attracted and are the primary pollinators in some species (see Stensmyr et al. 2002). Pollinating insects are usually trapped and temporarily held captive inside a chamber at the base of the spathe (Kite et al. 1998). Although some species provide a food reward, most are nonrewarding (van der Pijl 1937; Young 1986; Diaz and Kite 2006). Oviposition inside flowers is uncommon but beetles have been observed to lay eggs in the inflorescences of *Amorphophallus variabilis* (van der Pijl 1937 cited in Beath 1996). Another interesting feature of the Araceae is the high frequency of thermogenesis (Gibernau et al. 2005; Barthlott et al. 2009; Kakishima et al. 2011 and references therein). This has been suggested to play a role in enhancing the volatilization and dispersion of compounds responsible for the foul odors (Seymour and Schultze-Motel 1999; Barthlott et al. 2009) and has also been experimentally shown to manipulate the behavior of floral visitors (Angioy et al. 2004).

Although the main groups of carrion-mimicking flowers are described above, there are also several other minor groups in which the strategy has evolved and is particularly well represented. The most notable of these minor radiations are the genera *Hydnora* and *Aristolochia*. The genus *Hydnora*, in the small holoparasitic family Hydnoraceae, contains about seven species occurring in Africa, Madagascar, and Arabia (Bolin et al. 2009, 2011; Nickrent et al. 2002). *Hydnora* flowers (Figure 16.1c) have a peculiar warty, fungus-like appearance and exhibit a foul carrion odor. They are pollinated mainly by carrion-beetles (particularly Dermestidae, but also Trogidae, Scarabaeidae, and Hybosoridae), but sometimes also attract flies which may contribute to pollination (Marloth 1907; Burger et al. 1988; Vogel 1954; Bolin et al. 2009, 2011 and references therein). Like the Araceae, *Hydnora* flowers hold their pollinators temporarily captive in a chamber at the base of the flower (Bolin et al. 2009). It is interesting to note that *Hydnora*, being holoparasitic trap flowers, combine aspects of two other major groups of carrion flowers, namely the parasitic life history of the Rafflesiaceae combined with the trap flowers of the Araceae. Trap flowers are also found in the large pantropical genus *Aristolochia* (Aristolochiaceae; Figure 16.1b) which also contains several carrion-mimicking species (Proctor et al. 1996; Burgess et al. 2004; Nakonechnaya et al. 2008; Berjano et al. 2009; Johnson and Jürgens 2010; Stotz and Gianoli 2013). The family Aristolochiaceae appears to be closely related to the Hydnoraceae, and trap flowers may thus be ancestral to these groups (Nickrent et al. 2002). It is also interesting that some of the carrion-mimicking aristolochias, particularly *A. grandiflora* and *A. gorgona*, are notable for the very large size of their flowers, a trend which appears repeatedly in carrion-mimicking systems (Blanco 2002; Burgess et al. 2004; Davis et al. 2008).

The carrion-mimicking flowers described above represent some of the major groups of plants associated with the phenomenon. However, carrion mimicry has evolved independently in various angiosperm lineages (Urru et al. 2011; Jürgens et al. 2013) and these systems offer useful models for the study of floral evolution. Modern advances in analytical techniques, particularly for examining odor chemistry, coupled with increasing studies of carcass decomposition (the physical and chemical breakdown of animal or plant tissue via proteolytic enzymes and micro-organisms) and associated insects now permit a considerably more detailed understanding of the phenomenon than could ever have been hoped for by the botanists who first described flowers that "stink like carrion."

16.3 VOCs Produced by Different Types of Decaying Material and Insect Olfaction

The ecology and evolution of trophic preferences in detritus-feeding insects are linked to their tactile, visual, and olfactory sensory systems, and the innate adaptive responses of these to food sources and oviposition sites. More generally, most animals depend on a diet consisting of at least one, but often all three, main macronutrients: protein, carbohydrate, and lipid. However, none of these macronutrients themselves can be identified by a characteristic color or specific odor, making them difficult to detect via vision or olfaction. Despite these difficulties, insects have evolved fine-tuned olfactory systems not only to detect and find, but also to evaluate the quantity and quality of potential food sources and their macronutrient content based on volatile chemicals that are emitted during the decomposition process (Carlsson and Hansson 2006). The evolution of a fine-tuned olfactory system was only possible because the emission of certain VOCs from decaying plant and animal matter is correlated with the macronutrient composition, and in turn with the type of food source (e.g., feces or carrion) and stage of decomposition.

It can be assumed that selection would favor individuals that show sensory preferences for a diet or an oviposition site that has the highest possible relative net gain in terms of macronutrients (see also Simpson and Raubenheimer 2009). Among the detritus-feeding insects, specialists and generalists are found (Cambefort 1991), and the evolution of terrestrial arthropods can be viewed from the perspective of their trophic preferences and how these are linked to macronutrient composition. The main food types represented by decomposing organic matter, namely plant matter (e.g., leaf litter, decomposing wood, wilting flowers, decomposing fruits, decomposing roots), animal feces, and carrion, correspond to the three general types of detritus-feeding insects, namely those that feed on decomposing plant material (phyto-saprophagous), mammal feces (coprophagous), or carcasses (necrophagous) (Cambefort 1991). However, the situation is often more complex and other factors besides macronutrient content, such as competition with other insects, the risk of being attacked by predators (Archer and Elgar 2003), or being infected by fungi (Lam et al. 2010), may also play a role in the evolution of trophic preferences.

The literature on detritus-feeding insects suggests that trophic preferences have evolved in response to both the type of food and its stage of decomposition (Hanski 1987a,b; Bänziger and Pape 2004). Carrion and feces differ in terms of both their nutrient composition and their physical and chemical properties (Hanski 1987a,b), and it has been shown that dung/carrion beetle and fly attraction differs depending on the type of feces or carrion (Bänziger and Pape 2004). Trophic preference in terms of protein content, food quantity (e.g., size of carcass), and arrival time at the food source have probably been critical selection factors for niche separation of detritus-feeding insects. Furthermore, insects have not only evolved sensory preferences for specific VOCs or combinations of VOCs that are good predictors for a given oviposition site and/or food source, but also to VOCs that allow them to find the source at the optimal stage of decay. In forensic research, the decomposition process of animal (and human) bodies has traditionally been divided into different stages (e.g., fresh, bloat, active decay, postdecay) until the tissue is either absorbed by the environment or transformed into relatively stable chemical products (e.g., dry remains) (see Dekeirsschieter et al. 2009). To illustrate this, a visualization of the potential for trophic niche separation of different detritus-feeding insects has been presented (Figure 16.2). It would be interesting to map onto this figure the different VOCs emitted from different types of flowers associated with phyto-saprophagous, coprophagous, and necrophagous insects, and to compare them with that emitted from decaying organic matter (plants, animals, herbivore and carnivore feces). However, data on the pollinators of flowers exhibiting oviposition site mimicry are still very limited.

In the last 20 years, several studies have shown that olfactory signals play a dominant role in the location of oviposition sites and food sources by detritus-feeding insects (e.g., Kalinová et al. 2009). Olfaction is a complex sense and involves detecting and processing many different molecules for odor recognition (Dobson 2006). Technological advances in the field of chemical ecology during the last two decades now allow for the identification of trace amounts of compounds that are emitted from degrading plant and animal matter (e.g., Paczkowski and Schütz 2011) and to test the physiological and behavioral responses of insects to these compounds (Stensmyr et al. 2002). The VOC composition of carrion or

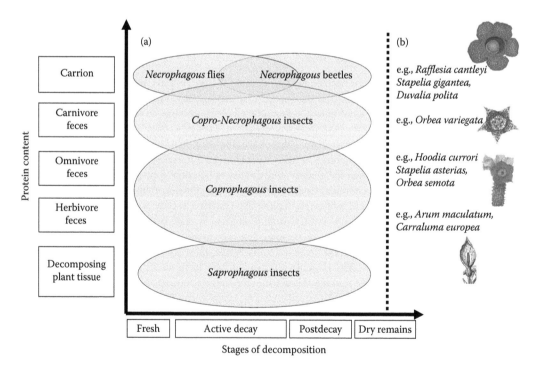

FIGURE 16.2   (a) Trophic niche matrix of different decomposition stages (*x*-axis) and main oviposition sites (different types of substrate) with different protein content (*y*-axis). (b) Examples of flowering plants which mimic different oviposition sites. It is possible that flowers not only resemble different oviposition sites, but also different stages. However, this hypothesis needs to be tested in future studies.

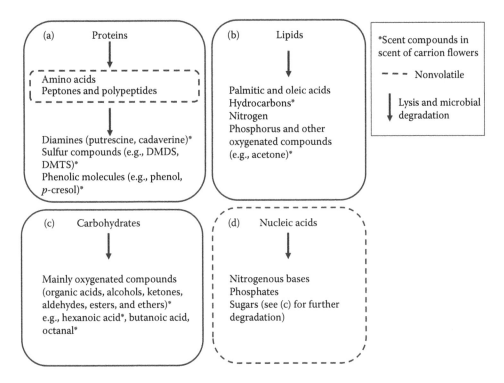

FIGURE 16.3 Some of the main decomposition by-products of carcass and their occurrence in scent samples of carrion flowers. (a) Proteins, (b) lipids, (c) carbohydrates, and (d) nucleic acids.

dung samples is often highly diverse and consists of many different chemical compound classes including sulfur compounds, nitrogen-containing compounds, monoterpenes, sesquiterpene, benzenoids, and aliphatic hydrocarbons (Statheropoulous et al. 2005; DeGreeff and Furton 2011). The same is true for VOC samples collected from the headspaces of flowers that attract detritus-feeding insects (Jürgens et al. 2006, 2013). Hundreds of compounds have been identified from the emissions of decaying plant and animal matter (Statheropoulous et al. 2005; DeGreeff and Furton 2011), and a single scent sample from a decaying animal carcass or flower may contain over 100 different compounds from various chemical compound classes (e.g., Skubatz et al. 1996; Jürgens et al. 2006, 2013). It has been shown that microorganisms play an important role in the decomposition process and are the main producers of VOCs (Dekeirsschieter et al. 2009; Paczkowski and Schütz 2011; Chapters 3 and 19). An overview of the main VOCs that are emitted as a result of the degradation process is given in Figure 16.3.

It has been reported that different types of tissue will produce different VOC profiles during decomposition. Decomposition of muscle tissue, for example, leads to proteins which are further broken down by proteolysis into proteases, peptones, polypeptides, and amino acids (Dekeirsschieter et al. 2009; Paczkowski and Schütz 2011). The volatile end products of proteolysis are compounds such as skatole (3-methylindole) and indole. Interestingly, indole has also been reported as a quorum-sensing molecule in bacteria, and it has been hypothesized that insects use these signals to evaluate the quality of the resource (Bansal et al. 2009; Tomberlin et al. 2012). Sulfur compounds are the end products of the decomposition of the sulfur-containing amino acids methionine and cysteine. Animal tissue contains a particularly high content of these two amino acids (Magee et al. 2000). Sulfur-containing compounds and nitrogen-containing compounds are often accompanied by aromatic hydrocarbons and alkanes (e.g., Dekeirsschieter et al. 2009). Dominant compound classes emitted from bones are alkanes, aromatic hydrocarbons, alcohols, and aldehydes (Dekeirsschieter et al. 2009). Decomposed animal fat samples often contain aldehydes, probably from the decomposition of fatty acids (DeGreeff and Furton 2011). Depending on whether the decomposition is in an aerobic or anaerobic environment, the unsaturated fatty acids are either oxidized to aldehydes and ketones or are saturated, respectively (DeGreeff and Furton 2011).

Significant differences but also considerable overlap in the scent composition of different decomposing objects has been reported. The odors of carnivore dung and carrion, for example, are dominated by similar sulfur-containing compounds (see Jürgens et al. 2013). In contrast, sulfur-containing compounds are rarely found in the odors of decomposing plant tissue or herbivore feces (see Jürgens et al. 2006, 2013). The differences stem from the higher content of the sulfur-containing amino acids (methionine and cysteine) in animal tissue compared to plant tissue (Magee et al. 2000). This explains, for example, why necrophagous insects are particularly sensitive to sulfur-containing compounds (e.g., dimethyldisulfide [DMDS] and dimethyl trisulfide [DMTS]) and are innately attracted to them: sulfur-containing compounds are indicators of a protein-rich food source and they are often emitted at an early stage of decomposition (Nilssen et al. 1996; Stensmyr et al. 2002; Kalinová et al. 2009).

16.4 The Plant Perspective

16.4.1 Chemical Ecology of Oviposition Site Mimicry Systems

One of the key features of oviposition site mimicry systems is their emission of VOCs identical to those found in scent samples of dung or carcasses (Jürgens et al. 2006; Johnson and Jürgens 2010). Furthermore, physiological and behavioral studies of the pollinators suggest that olfactory signals are the primary modality of pollinator attraction in these saprophilous pollination systems (e.g., Stensmyr et al. 2002). Comparative and case studies of the VOCs produced by the flowers have revealed that the floral scent is often dominated by oligosulfides (e.g., DMDS, DMTS), aliphatic acids, and *p*-cresol (e.g., Kite and Hetterscheid 1997; Kite et al. 1998; Jürgens et al. 2006, 2013; Johnson and Jürgens 2010; van der Niet et al. 2011). For some of these chemicals, like the oligosulfides, their function as pollinator attractants has been demonstrated in field experiments (Stensmyr et al. 2002; Shuttleworth and Johnson 2010; van der Niet et al. 2011). A recent comparative study by Jürgens et al. (2013) showed that the emission of sulfur-containing compounds has evolved independently in at least five different flowering plant families of the flowering plants. This is not surprising since oligosulfides such as DMDS and DMTS are among the most important compounds to attract necrophagous insects (Kalinová et al. 2009). For some other compounds, for example, *p*-cresol and indole, it has been established that they are potent lures to coprophagous flies (e.g., Kite et al. 1998).

Trophic specialization of detritus-feeding insects to different types of dung and/or carrion might have consequences for the evolution of flowers that mimic dung and/or carrion to attract these insects for pollination. Martín-Piera and Lobo (1996) pointed out that Scarabaeidae beetles have evolved specialized feeding on carrion and/or feces in tropical forest habitats of southeast Asia and South America (Halffter et al. 1992), whereas in African savannah habitats, where large carnivores and vultures feed on carcasses (Chapter 20), herbivore and omnivore dung is the main trophic niche of these beetles (Hanski and Combefory 1991). It is believed that the different feeding habits in detritus-feeding Scarabaeinae evolved from saprophagous ancestors about 65 million years ago, at the end of the dinosaurs and rise of the mammals (Martín-Piera and Lobo 1996 and references therein). The diversification and specialization of Scarabaeinae to coprophagy was then likely the result of an increase in mammal dung types during the Cenozoic, while necrophagy in the neotropical Scarabaeinae seems to be linked to reduced dung resources owing to mass extinctions of mammalian megafauna between two to five million years ago (Martín-Piera and Lobo 1996 and references therein). It would be interesting to compare this evolutionary scenario with the radiation of oviposition site mimicry systems in the angiosperms. However, pollinator data for plants exhibiting oviposition site mimicry systems are still very fragmentary and often only anecdotal, and more comparative studies in different regions are needed to analyze how complex multitrophic interactions may have shaped the evolution of oviposition site mimicry systems.

Dung and/or carrion, although a rich ephemeral resource, are limited in both spatial and temporal availability (Hanski 1987b; Carter et al. 2007). Thus, locating them from a distance is an important ability for insects feeding on such substrates. Because carrion-mimicking flowers often occur in small scattered populations, attracting pollinators from a distance probably plays an important role for them as well (Jürgens et al. 2006). For example, the reason for the low flower density in *Rafflesia, Rhizanthes,*

Sapria (southeast Asia), and *Hydnora* (Africa, Arabia, and Madagascar) might be that they are holo-parasitic plants without photosynthetic tissue and high asynchrony in the flowering time. Under these circumstances, however, it may be advantageous to adapt to pollination by insects that fly long distances to find food and/or oviposition sites. Carrion-seeking insects might therefore be easy to recruit as pollinators because they can locate carrion volatiles from a distance.

Another aspect of trophic specialization is that the composition of VOCs changes with the time elapsed since death and it is known from forensic and ecological studies that necrophagous insects will arrive and colonize a vertebrate carcass in a relatively predictable sequence (Schoenly and Reid 1987). It is likely that the arrival of insects is correlated with the emission of distinct VOC profiles for the different stages of decomposition. Thus, to attract specific pollinators, the VOC emission of flowers associated with carrion insects may not only reflect the olfactory preferences of insects to different types of decaying organic matter, but also to different stages of the decomposition process (Barkman et al. 2008; Dekeirsschieter et al. 2009).

16.5 Evolution of Carrion Mimicry Systems and Future Research

Carrion flowers have evolved independently in various plant lineages and are represented throughout the tropical and subtropical regions of the world (see Table 16.1 for examples of plant groups that exhibit carrion-like traits and/or are associated with carrion insects; Urru et al. 2011; Jürgens et al. 2013). The diversity and wide geographic distribution of carrion-mimicking flowers makes it difficult to speculate on the factors driving the evolution of carrion mimicry. However, there are several interesting correlates that may provide insights into the evolution of some of the major carrion-mimicking groups and may represent fruitful avenues for future research. These include the association of some of the major groups with either arid to semiarid regions (stapeliads, *Hydnora*) or tropical rainforests (Rafflesiaceae, Araceae); the frequent evolution of unusually large flowers or inflorescences in the carrion-mimicking members of a family (e.g., *Amorphophallus*, *Aristolochia*, *Rafflesia*, and *Stapelia*); the association of universal carrion-mimicry with a holoparasitic lifestyle in two of the major groups of carrion mimics (Rafflesiacae and *Hydnora*); and, finally, the association of carrion mimicry with trap flowers in three of the major groups of carrion mimics (Araceae, *Aristolochia*, and *Hydnora*).

The ecology and evolution of carrion flowers is closely linked with that of detritus-feeding insects and integrating information on the chemical ecology and behavior of coprophagous and necrophagous insects into pollination biology studies has been a fruitful approach (e.g., Stensmyr et al. 2002; van der Niet et al. 2011). However, the current picture of mechanisms that are responsible for pollinator attraction by carrion flowers, for example, scent emission, is still very static. So far, the comparison between VOCs of flowers of oviposition site mimics, and different types of decomposing organic matter are based on the idea that VOCs of flowers imitate different trophic models (*y*-axis in Figure 16.2). However, it is well established that the emission of VOCs from carrion changes significantly during the decomposition process (Dekeirsschieter et al. 2009). Furthermore, forensic research has shown that different insects colonize decomposing animal/human bodies in a relatively predictable sequence (Dekeirsschieter et al. 2009). It is therefore possible that different species of carrion flowers mimic different temporal stages of decomposition and are associated with different trophic specialists of the entomofaunal succession (Dekeirsschieter et al. 2009 and references therein) (*x*-axis in Figure 16.2). Blow flies, for example, are colonizers of early decomposition stages, whereas other insects, particularly carrion beetles, are late colonizers. It would be interesting to compare existing VOC data of flowers imitating carrion with that of different successional stages of animal carrion to analyze whether they not only represent specific trophic niches, but also different stages of the decomposition process. The following questions may provide a guideline for future research programmes that could be addressed within the next decade.

1. How does the presence of dung and carrion affect the fitness of oviposition site mimicking flowers?
2. What are the conditions that favor the evolution of carrion flowers?

3. Why is flower gigantism correlated with the carrion flower syndrome?
4. Do animals use universal scent signals to identify decomposing plant and animal matter and how are these sensory attractants exploited by other organisms?

16.5.1 How Does the Presence of Dung and Carrion Affect the Fitness of Oviposition Site Mimicking Flowers?

Data from forensic research and from studies interested in the ecology and evolution of detritus-feeding insects have helped to develop hypotheses regarding the ecological function and evolution of floral scent compounds in oviposition site mimicry systems. In other well-investigated pollination systems such as sexually deceptive orchids or moth-pollinated flowers, it is either difficult to study convergent evolution or difficult to establish an evolutionary scenario for the evolution of floral scent compounds. Moth-pollinated flowers, for example, produce strong sweet scent at night and they are probably the best investigated pollination system in terms of their scent composition and the role of scent in pollinator attraction (Knudsen and Tollsten 1993; Jürgens et al. 2002). Although they provide a classic example of convergent evolution of floral scent, it remains an enigma why plants have converged on certain compounds like phenylacetaldehyde or methylbenzoate that are part of the so-called moth floral scent syndrome (Knudsen and Tollsten 1993). In sexually deceptive orchids, where plants chemically and/or morphologically mimic a female insect, the problem is a different one. Here each species mimics a specific (different) model, the sex pheromone of a specific female, to attract a male. In such a scenario, convergence only occurs if plant species adapt to the same insect pheromone, but this is a very unlikely situation, because each insect species produces its own specific pheromone, making it difficult to study convergent evolution is such systems (but see Phillips et al. 2013).

Mimicry represents a classic example of evolution through natural selection (Vereecken and McNeil 2010). Most attention has focused on the role of visual signals, such as those deployed by palatable butterflies (Lepidoptera) to mimic the color patterns of unpalatable species in order to gain protection from predators (Hines et al. 2011). Floral mimicry in plants is gaining increasing attention (Schiestl and Johnson 2013). The best studied examples are the use of color in food deception mimicry (Jersáková et al. 2012; Newman et al. 2012) and floral scent signals in sexual mimicry (Schiestl et al. 1999). Oviposition site mimicry systems that exploit the innate and learned preferences of insects that feed on decomposing organic matter have also recently been used to study convergent evolution among different plant species and advergent evolution of these species to their potential models (advergent evolution: adaptive phenotypic resemblance to their models but not vice versa).

The essential criterion for evolutionary mimicry is that the species under consideration (the mimic) has evolved features (e.g., visual, olfactory) that are similar to that of another species (the model) in order to fool another organism (the dupe), which confuses the model with the mimic due to their similar features (Schiestl et al. 1999; Vereecken and McNeil 2010; Jersáková et al. 2012). This relationship either benefits only the mimic (as in *Batesian mimicry*) or both the mimic and the model (as in *Müllerian mimicry*). In carrion flowers, the features of the flowers do not resemble another living species; instead, they mimic the features of dung, carrion, or other models of decomposing organic matter to exploit the learned and innate preferences of detritus-feeding insects. It is widely accepted that the features of the mimic are the result of an adaptation process towards features of the model (see Jersáková et al. 2009). Although the resemblance between model and mimic is often used as the main criterion for mimicry, a more narrow definition of mimicry, based on the historical context of the original descriptions by Fritz Müller (1878) and Henry Walter Bates (1861), would include that the response of the operator has to be a learned (conditioned) response to the signals of the mimic (see Jersáková et al. 2009). Furthermore, in the original description of Müllerian and Batesian mimicry, the model was a species that co-occurred with the mimic. However, for oviposition site mimicry systems, very few studies have explored the co-occurrence of carrion flowers with their potential models, and operators, in the same habitat (but see Bänziger and Pape 2004; van der Niet et al. 2011). Thus, the importance of the co-occurring model (e.g., feces, carrion) for the evolution of the flowers remains unclear. It is also unclear whether learning by operators plays a role in the selection process of oviposition site mimicry systems or if mimicry systems exploit innate responses of operators. More generally, the question remains: how does the abundance of the model and mimic effect fitness of oviposition site mimicking flowers? Following the findings in Batesian mimicry systems (Jersáková et al. 2009), it is possible that negative frequency-dependent selection models would apply to carrion fly systems, meaning that the fitness of a phenotype increases as it becomes rarer.

16.5.2 What Are the Conditions That Favor the Evolution of Carrion Flowers?

As discussed previously, detritus-feeding insects play an important role in all ecosystems. It seems logical to assume that the occurrence of flowers that mimic dung or carrion is a reflection of the diversity and abundance of the multitrophic interactions between mammals and coprophagous/necrophagous insect communities in a given region. There are two scenarios in which the evolution of carrion flowers might be advantageous:

1. Since many detritus-feeding insects fly often relatively long distances to find food (e.g., Bishopp and Laake 1921; Hogsette and Ruff 1985), plants with low numbers of individuals per population and/or scattered distributions would benefit from adapting to pollination by such insects because they would provide long-distance pollen dispersal.
2. Flowering times of entomophilous plant communities are often synchronized and seasonal. However, detritus-feeding flies and beetles are still available as pollinators during times when other insect populations, that are normally associated with flowers, have collapsed or where flower-visitor communities are limited to certain guilds that are specialized to specific interactions.

One or both of these scenarios apply to the majority of oviposition site mimicry systems. The stapeliads and the genus *Hydnora*, for example, grow in desert and semidesert conditions which often exhibit highly seasonal flowering times with high levels of pollinator specialization. The parasitic plant groups that have oviposition site mimicry systems (see Table 16.1A) often have asynchronous flowering times with scattered individuals over wide areas.

16.5.3 Why Is Flower Gigantism Correlated with the Carrion Flower Syndrome?

Unusually large flowers are a striking feature of many carrion-mimicry systems (note that the functional unit in the Araceae is technically an inflorescence, but for ease of discussion they will be referred to as flowers; see discussion of terminology in Box 1 of Davis et al. 2008). Indeed, the largest flower in the world (*R. arnoldi*) and the largest inflorescence in the world (*Amorphophallus titanum*) are both carrion mimics (Brown 1822; Davis et al. 2008; Barthlott et al. 2009; Shirasu et al. 2010; Fujioka et al. 2012). Although not breaking records, other members of the Rafflesiaceae and Araceae also exhibit relatively large flowers. Moreover, large flowers are not limited to these two families, but appear sporadically in the carrion-mimicking members of other plant families, including the Apocynaceae (e.g., *S. gigantea* with a diameter of up to 40 cm) and Aristolochiaceae (e.g., *Aristolochia grandiflora* with a diameter of over 1 m if including the long perianth appendage) (Burgess et al. 2004; Bruyns 2005). The repeated appearance of large flowers in carrion-mimicry systems suggests that there is some selective advantage driving increases in floral size.

The traditional explanation for gigantism in some carrion flowers has been that increases in floral size are related to more effective long-distance attraction of pollinators (Davis et al. 2008). This could operate through both olfactory and visual cues as larger flowers could produce higher total volatile emissions and offer a greater target for possible visitors (Davis et al. 2008). Many carrion flowers occur in low density and are often widely separated in space or occur in poorly lit rainforests. These conditions would make it difficult for potential pollinators to locate and move between flowers and could conceivably result in selection for traits that enhance long-distance attraction. From a visual perspective, there is also evidence from behavioral experiments with necrophagous insects which indicates that for some insects, visual stimuli play a role in initiating landing on objects (Aak and Knudsen 2011).

An alternative explanation is that the increased size plays a role in increasing the attractiveness of flowers to a *specific* subset of pollinators that show preferences for larger carcasses (Barkman et al. 2008). It is known, for example, that Nicrophorinae (Silphidae, burying beetles) seem to prefer small carcasses such as birds and rodents, whereas Silphinae are typically found on large carcasses (Scott 1998). Likewise, van der Niet et al. (2011) demonstrated clear differences in the assemblage of flies visiting small versus large items of carrion, with sarcophagids preferring smaller carrion then calliphorids. Thus, evolution of increased floral size may conceivably be driven by the advantages of specialization to a subset of the available carcass-associated entomofauna (Kneidel 1984; van der Niet et al. 2011). Bigger flowers would represent more accurate mimics of larger carcasses (in terms of both total volatile emission and visual cues). If this were the case, it could be expected that the pollinators of large-flowered species would be closely correlated with insects that prefer larger carcasses (Kneidel 1984; Scott 1998). Testing this hypothesis requires detailed studies of the pollinators of particular large-flowered species in conjunction with studies of the insects associated with different carrion sizes in the same habitat (e.g., van der Niet et al. 2011).

Thermogenesis in some large flowers (particularly in the family Araceae) introduces an added complication to understanding the evolution of large flowers in carrion mimicry systems. Recent studies in the Araceae suggest that levels of temperature increase relative to ambient temperature are physically constrained by the volume of the inflorescence (Gibernau et al. 2005). Thus, in the Araceae, larger inflorescences can generate higher temperature differentials relative to ambient temperature. This suggests that increases in the size of Araceae inflorescences could result from selection for warmer flowers (Davis et al. 2008). Thermogenesis has also been demonstrated in *Rafflesia*, *Rhizanthes*, and *Hydnora* (Patiño et al. 2000, 2002; Seymour et al. 2009), although the morphological structure of these flowers may influence the relationship between size and potential heat differential relative to ambient as observed in the Araceae. In the context of carrion mimicry, thermogenesis has been suggested to play a role in volatilizing scent compounds (Meeuse and Raskin 1988; Patiño et al. 2000, 2002; Barthlott et al. 2009; Seymour et al. 2009), although there are other possible functions (Seymour et al. 2003b; Angioy et al. 2004). Whatever the role of thermogenesis, it is clear that attempts to understand the evolution of large flowers in carrion mimicry systems need to also consider the role of heat production by these flowers.

Ultimately, it may prove difficult to provide a single explanation for the evolution of large flowers in some carrion mimicry systems. Different habitats and insect communities are likely to apply different patterns of selection on floral size. Future studies need to examine the pollinators of large-flowered species in comparison to the insects attending large carcasses in the same habitat. Manipulative experiments exploring the relative contributions of scent emission and visual display in the attraction of pollinators would also enlighten our understanding of the mechanisms driving selection for large flowers.

16.5.4 Do Animals Use Universal Infochemicals to Identify Decomposing Plant and Animal Matter?

The question whether VOCs that are interpreted identically by all organisms (universal [scent] infochemicals) do exist is probably one of the most exciting evolutionary questions to be addressed by chemical ecologists (and neurobiologists) in the next decades (see Chapter 19). It is important to keep in mind that, in an evolutionary context, it is not an intrinsic feature of the chemical compound that produces the sensory response (e.g., sweet, stinky). It is the peripheral sensory apparatus and the brain that make the signal by evolving a useful link to a given VOC as part of a pattern recognition process. Although olfactory systems of vertebrates and invertebrates are very different, it can be hypothesized that both have evolved pattern recognition mechanisms in response to essentially similar macronutrient demands (namely protein, lipid, and carbohydrates). It is therefore likely that, based on the same evolutionary problem of finding macronutrients, invertebrates and vertebrates have evolved similar useful sensory categories for finding and distinguishing different types of food. This does not mean that a fly would perceive an odor that is sweet to a human also as sweet; it would mean that different brains probably have evolved similar pattern recognition systems to link specific VOCs to carbohydrate sources (e.g., in response to VOCs typically emitted from sources with carbohydrates such as flowers and fruits). The same would be true for oligosulfides (DMDS, DMTS), the breakdown products of amino acids that contain sulfur, that are useful protein indicators. In addition, DMDS, which is usually emitted during the early stages of decay of animal tissue, may also be used as an indicator of early decomposition. It is therefore not surprising that DMDS is one of the most commonly emitted scent compounds from angiosperm flowers that are mimicking carrion (Jürgens et al. 2013). Moreover, DMDS and other oligosulfides are also used in other sessile organisms, such as the dung moss *Tetraplodon mnioides* (Splachnaceae) and the stinkhorn fungi *Phallus impudicus* and *Clathrus archeri* (Phallaceae), as sensory attractants for spore dispersal by flies (Borg-Karlson et al. 1994; Marino et al. 2009; Johnson and Jürgens 2010).

In contrast to the many studies that have investigated VOC emissions of oviposition site mimicry systems, not much is known about the role of color and tactile features for oviposition site mimicry systems (see Jürgens et al. 2006, 2013). This is surprising because the red and dull colors and the peculiar hairs that these flowers often exhibit have interested botanists for a long time. The red color of the flowers are, however, also an enigma from a sensory viewpoint because color models of the flies visual system suggest that such colors are difficult for flies to perceive (Troje 1993; Chittka et al. 2001, but see Schaefer and Ruxton 2008). This also suggests that more studies are needed that integrate the visual, olfactory, and tactile signals into a model that can explain why flowers have evolved these features (e.g., Aak and Knudsen 2011).

Finally, floral traits of oviposition site mimicry systems have mainly be interpreted as adaptations to attract pollinators. More recently, Lev-Yadun et al. (2009) hypothesized that flowers with carrion scent might have an antiherbivore defensive function. This phenomenon has been reported for some insects. For example, feces mimicry has been reported as a visual defence in caterpillars and as an olfactory defence in lacewings (Eisner et al. 2005). It seems therefore plausible that plants could also evolve carrion-like features to reduce herbivore damage. However, as already stated by Lev-Yadun et al. (2009), more research is needed to test whether livestock and wild herbivores are deterred by the VOCs that are emitted by carrion flowers.

16.6 Conclusions

Convergent evolution is well documented in oviposition site mimicry systems (Johnson and Jürgens 2010; Jürgens et al. 2013). They are probably one the best investigated pollination system in terms of floral scent chemistry and its relevance for pollination and provide fascinating examples of how flowering plants manipulate the behavior of animals for pollen transport. The convergent evolution of sulfur-containing compounds in flowering plants that attract carrion flies and beetles as pollinators suggests that sulfur-containing VOCs are universal sensory attractants used by these insects to find carrion (Jürgens et al. 2013). Moreover, the high diversity of floral forms and chemical patterns suggests that different species of angiosperms exploit insects with different trophic niches. These trophic niches are determined by the protein content, size of the food source, and the stage of decomposition.

Integrating information from different research fields, such as sensory physiology, forensic entomology, and microbiology, has been a fruitful approach to better understand how carrion flowers attract their pollinators. The olfactory preferences of carrion insects are the product of a complex multitrophic decomposition process where bacteria play a central role. More recent work seems to suggest that carrion flies use quorum-sensing signals emitted by bacteria for detecting resources (Tomberlin et al. 2012). The same scent compounds are also used by carrion flowers to attract carrion insects for pollination (see Jürgens et al. 2013), and they might play a role as antiherbivore defence against vertebrates (Lev-Yadun et al. 2009). All this suggests that many of the VOCs used by carrion flowers represent universal infochemicals.

What can other fields learn from research on flowers that mimic carrion? Oviposition site mimicry systems have evolved multiple times independently in different plant families and in different regions of the world. This suggests that carrion and dung represent important trophic components of environments worldwide. The different ways in which these plants attract pollinators have been shaped by natural selection. Thus, the specialized floral features may provide clues to identify the key signals used by different carrion insects to find resources.

Acknowledgments

Adam Shuttleworth was supported by a University of KwaZulu-Natal postdoctoral fellowship. Andreas Jürgens was supported by the National Research Foundation (South Africa). We would like to thank Ashley Nicholas, Hugh Glen, Denis Brothers, and Florent Martos for assistance with Latin and French translations. We would also like to thank Marcus Stensmyr and Rogan Roth for the use of their photographs, and Timo van der Niet and Frank Geers for pointing out additional carrion-mimicking flowers. John Wiley & Sons Inc. are thanked for permission to reproduce images from Robert Brown's publications.

REFERENCES

Aak, A. and G.K. Knudsen. 2011. Sex differences in olfaction-mediated visual acuity in blowflies and its consequences for gender-specific trapping. *Entomologia Experimentalis et Applicata* 139: 25–34.

Alves, R.J.V., A.C. Pinto, A.V.M. da Costa, and C.M. Rezende. 2005. *Zizyphus mauritiana* Lam. (Rhamnaceae) and the chemical composition of its floral fecal odor. *Journal of the Brazilian Chemical Society* 16: 654–656.

Angioy, A.M., M.C. Stensmyr, I. Urru, M. Puliafito, I. Collu, and B.S. Hansson. 2004. Function of the heater: The dead horse arum revisited. *Proceedings of the Royal Society B (Suppl.)* 271: S13–S15.

Archer, M.S. and M.A. Elgar. 2003. Effects of decomposition on carcass attendance in a guild of carrion-breeding flies. *Medical and Veterinary Entomology* 17: 263–271.

Atluri, J.B., S.P.V. Ramana, and C.S. Reddi. 2004. Sexual system and pollination of *Sterculia foetida* Linn. *Beiträge zur Biologie der Pflanzen* 73: 223–242.

Bansal, T., T.K. Wood, and A. Jayaraman. 2009. The bacterial signal indole promotes epithelial cell barrier properties and attenuates inflammation. *Proceedings of the National Academy of Sciences, USA*, 107, 228e233.

Bänziger, H. 1991. Stench and fragrance: Unique pollination lure of Thailand's largest flower, *Rafflesia kerrii* Meijer. *Natural History Bulletin of the Siam Society* 39: 19–52.

Bänziger, H. 1996. Pollination of a flowering oddity: *Rhizanthes zippelii* (Blume) Spach (Rafflesiaceae). *Natural History Bulletin of the Siam Society* 44: 113–142.

Bänziger, H. 2001. Studies on the superlative deceiver: *Rhizanthes* Dumortier (Rafflesiaceae). *Bulletin of the British Ecological Society* 32: 36–39.

Bänziger, H. and B. Hansen. 2000. A new taxonomic revision of a deceptive flower, *Rhizanthes* Dumortier (Rafflesiaceae). *Natural History Bulletin of the Siam Society* 48: 117–143.

Bänziger, H., B. Hansen, and K. Kreetiyutanont. 2000. A new form of the Hermit's Spittoon, *Sapria himalayana* Griffith F. *albovinosa* Bänziger & Hansen F. Nov. (Rafflesiaceae), with notes on its ecology. *Natural History Bulletin of the Siam Society* 48: 213–219.

Bänziger, H. and T. Pape. 2004. Flowers, faeces and cadavers: Natural feeding and laying habits of flesh flies in Thailand (Diptera: Sarcophagidae, *Sarcophaga* spp.). *Journal of Natural History* 38: 1677–1694.

Bänziger, H., H. Sun, and Y.-B. Luo. 2008. Pollination of wild lady slipper orchids *Cypripedium yunnanense* and *C. flavum* (Orchidaceae) in south-west China: Why are there no hybrids? *Botanical Journal of the Linnean Society* 156: 51–64.

Barcelona, J.F., P.B. Pelser, D.S. Balete, and L.L. Co. 2009. Taxonomy, ecology, and conservation status of Philippine *Rafflesia* (Rafflesiaceae). *Blumea* 54: 77–93.

Barkman, T.J., M. Bendiksby, S.-H. Lim, K.M. Salleh, J. Nais, D. Madulid, and T. Schumacher. 2008. Accelerated rates of floral evolution at the upper size limit for flowers. *Current Biology* 18: 1508–1513.

Barthlott, W., J. Szarzynski, P. Vlek, W. Lobin, and N. Korotkova. 2009. A torch in the rain forest: Thermogenesis of the Titan arum (*Amorphophallus titanum*). *Plant Biology* 11: 499–505.

Bates, H.W. 1861. Contributions to an insect fauna of the Amazon valley. Lepidoptera: Heliconidae. *Transactions of the Linnean Society* 23: 495–566.

Beaman, R.S., P.J. Decker, and J.H. Beaman. 1988. Pollination of *Rafflesia* (Rafflesiaceae). *American Journal of Botany* 75: 1148–1162.

Beath, D.D.N. 1996. Pollination of *Amorphophallus johnsonii* (Araceae) by carrion beetles (*Phaeochrous amplus*) in a Ghanaian rain forest. *Journal of Tropical Ecology* 12: 409–418.

Bendiksby, M., T. Schumacher, G. Gussarova, J. Nais, K. Mat-Salleh, N. Sofiyanti, D. Madulid, S.A. Smith, and T. Barkman. 2010. Elucidating the evolutionary history of the Southeast Asian, holoparasitic, giant-flowered Rafflesiaceae: Pliocene vicariance, morphological convergence and character displacement. *Molecular Phylogenetics and Evolution* 57: 620–633.

Berjano, R., P.L. Ortiz, M. Arista, and S. Talavera. 2009. Pollinators, flowering phenology and floral longevity in two Mediterranean *Aristolochia* species, with a review of flower visitor records for the genus. *Plant Biology* 11: 6–16.

Bhatnagar, S. 1986. On insect adaptations for pollination in some asclepiads of central India. In: Kapil, R.P. (ed.), *Pollination Biology—An Analysis*. New Delhi: Inter-India Publications, pp. 37–57.

Bishopp, F.C. and E.W. Laake 1921. Dispersion of flies by flight. *Journal of Agricultural Research* 21: 729–766.

Blanco, M.A. 2002. *Aristolochia gorgona* (Aristolochiaceae), a new species with giant flowers from Costa Rica and Panama. *Brittonia* 54: 30–39.

Bolin, J.F., E. Maass, and L.J. Musselman. 2009. Pollination biology of *Hydnora africana* Thunb. (Hydnoraceae) in Namibia: Brood-site mimicry with insect imprisonment. *International Journal of Plant Sciences* 170: 157–163.

Bolin, J.F., E. Maass, and L.J. Musselman. 2011. A new species of *Hydnora* (Hydnoraceae) from southern Africa. *Systematic Botany* 36: 255–260.

Borg-Karlson, A.K., F.O. Englund, and C.R. Unelius. 1994. Dimethyl oligosulphides, major volatiles released from *Sauromatum guttatum* and *Phallus impudicus*. *Phytochemistry* 35: 321–323.

Brown, R. 1822. An account of a new genus of plants, named *Rafflesia*. *Transactions of the Linnean Society of London* 13: 201–234.

Brown, R. 1845. Description of the female flower and fruit of *Rafflesia Arnoldi*, with remarks on its affinities; and an illustration of the structure of *Hydnora africana*. *Transactions of the Linnean Society of London* 19: 221–247.

Bruyns, P.V. 2000. Phylogeny and biogeography of the stapeliads. *Plant Systematics and Evolution* 221: 199–226.

Bruyns, P.V. 2005. *Stapeliads of Southern Africa and Madagascar*, Vols. 1 and 2. Pretoria: Umdaus Press.

Burger, B.V., Z.M. Munro, and J.H. Visser. 1988. Determination of plant volatiles 1: Analysis of the insect-attracting allomone of the parasitic plant *Hydnora africana* using Grob-Habich activated charcoal traps. *Journal of High Resolution Chromatography* 11: 496–499.

Burgess, K.S., J. Singfield, V. Melendez, and P.G. Kevan. 2004. Pollination biology of *Aristolochia grandiflora* (Aristolochiaceae) in Veracruz, Mexico. *Annals of the Missouri Botanical Garden* 91: 346–356.

Cambefort, Y. 1991. *From Saprophagy to Coprophagy*, pp. 22–35. In: Hanski, I. and Y. Cambefort (eds), *Dung Beetle Ecology*. Princeton, NJ: Princeton University Press, 481 pp.

Carlsson, M.A. and B.S. Hansson. 2006. Detection and coding of flower volatiles in nectar-foraging insects. In: Dudareva, N. and E. Pichersky (eds), *Biology of Floral Scent*. Boca Raton, FL: CRC Press, pp. 243–261.

Carter, D.O., D. Yellowlees, and M. Tibbett. 2007. Cadaver decomposition in terrestrial ecosystems. *Naturwissenschaften* 94: 12–24.

Chittka L., J. Spaethe, A. Schmidt, and A. Hickelsberger. 2001. Adaptation, constraint, and chance in the evolution of flower color and pollinator color vision. In: Chittka, L. and J.D. Thomson (eds), *Cognitive Ecology of Pollination*. Cambridge, UK: Cambridge University Press, pp. 106–126.

Coombs, G. 2010. Ecology and degree of specialization of South African milkweeds with diverse pollination systems. Unpublished PhD Thesis, Rhodes University, Grahamstown.

Coombs, G., A.P. Dold, and C.I. Peter. 2011. Generalized fly-pollination in *Ceropegia ampliata* (Apocynaceae: Asclepiadoideae): The role of trapping hairs in pollen export and receipt. *Plant Systematics and Evolution* 296: 137–148.

Darwin, C. 1859. *On the Origin of Species by Means of Natural Selection, or the Preservation of Favoured Races in the Struggle for Life*. London: John Murray.

Darwin, C. 1862. *On the Various Contrivances by which British and Foreign Orchids are Fertilised by Insects*. London: John Murray.

Darwin, C. 1877. *The Various Contrivances by which Orchids are Fertilised by Insects*, 2nd edition. London: John Murray.

Davis C.C., P. Endress, and D.A. Baum. 2008. The evolution of floral gigantism. *Current Opinion in Plant Biology* 11: 49–57.

Davis, C.C., M. Latvis, D.L. Nickrent, K.J. Wurdack, and D.A. Baum. 2007. Floral gigantism in Rafflesiaceae. *Science* 315: 1812.

DeGreeff, L.E. and K.G. Furton. 2011. Collection and identification of human remains volatiles by non-contact, dynamic airflow sampling and SPME-GC/MS using various sorbent materials. *Analytical and Bioanalytical Chemistry* 401: 1295–1307.

Dekeirsschieter, J., F.J. Verheggen, M. Gohy, F. Hubrecht, L. Bourguignon, G. Lognay, and E. Haubruge. 2009. Cadaveric volatile organic compounds released by decaying pig carcasses (*Sus domesticus* L.) in different biotopes. *Forensic Science International* 189: 46–53.

Delpino, F. 1867. *Sugli Apparecchi Della Fecondazione Nelle Piante Antocarpee (Fanerogame).* Cellini: Firenze.

Delpino, F. 1868–1875. Ulteriori osservazione e considerazioni sulla dicogamia nel regno vegetale. *Atti della Societa Italiana di Scienze Naturale in Milano* 11: 265–332; 12: 21–141, 179–233.

Dobson, H.E.M. 2006. Relationship between floral fragrance composition and type of pollinator. In: Dudareva, N. and E. Pichersky (eds), *Biology of Floral Scent*. Boca Raton, FL: CRC Press, Taylor & Francis Group, pp. 147–198.

Dötterl, S., A. Jürgens, K. Seifert, T. Laube, B. Weißbecker, and S. Schütz. 2006. Nursery pollination by a moth in *Silene latifolia*: The role of odours in eliciting antennal and behavioural responses. *New Phytologist* 169: 707–718.

Diaz, A. and G.C. Kite. 2006. Why be a rewarding trap? The evolution of floral rewards in *Arum* (Araceae), a genus characterized by saprophilous pollination systems. *Biological Journal of Linnean Society* 88: 257–268.

Duyfjes, B.E.E. 1991. Stemonaceae and Pentastemonaceae: With miscellaneous notes on members of both families. *Blumea* 36: 239–252.

Dyer, R.A. 1983. *Ceropegia, Brachystelma and Riocreuxia in southern Africa.* Rotterdam: A.A. Balkema.

Eisner, T., M. Eisner, and M. Siegler. 2005. *Secret weapons. Defenses of Insects, Spiders, Scorpions, and Other Many-Legged Creatures.* Cambridge: Harvard University Press.

Faegri, K. and L. van der Pijl. 1979. *The Principles of Pollination Ecology*, 3rd edition. Oxford: Pergamon Press.

Formisano, C., F. Senatore, G. Della Porta, M. Scognamiglio, M. Bruno, A. Maggio, S. Rosselli, P. Zito, and M. Sajeva. 2009. Headspace volatile composition of the flowers of *Caralluma europaea* N.E.Br. (Apocynaceae). *Molecules* 14: 4597–4613.

Fujioka, K., M. Shirasu, Y. Manome, N. Ito, S. Kakishima, T. Minami, T. Tominaga et al. 2012. Objective display and discrimination of floral odors from *Amorphophallus titanum*, bloomed on different dates and at different locations, using an electronic nose. *Sensors* 12: 2152–2161.

Gibernau, M. 2003. Pollinators and visitors of aroid inflorescences. *Aroideana* 26: 73–91.

Gibernau, M., D. Barabe, M. Moisson, and A. Trombe. 2005. Physical constraints on temperature difference in some thermogenic aroid inflorescences. *Annals of Botany* 96: 117–125.

Goldblatt, P., P. Bernhardt, and J.C. Manning. 2009. Adaptive radiation of the putrid perianth: *Ferraria* (Iridaceae: Irideae) and its unusual pollinators. *Plant Systematics and Evolution* 278: 53–65.

Goodrich, K.R. 2012. Floral scent in Annonaceae. *Botanical Journal of the Linnean Society* 169: 262–279.

Goodrich, K.R. and R.A. Raguso. 2009. The olfactory component of floral display in *Asimina* and *Deeringothamnus* (Annonaceae). *New Phytologist* 183: 457–469.

Gottsberger, G., S. Meinke, and S. Porembski. 2011. First records of flower biology and pollination in African Annonaceae: *Isolona, Piptostigma, Uvariodendron, Monodora* and *Uvariopsis. Flora* 206: 498–510.

Halffter, G.M., E. Favila, and V. Halffter. 1992. A comparative study of the structure of the scarab guild in Mexican tropical rain forests and derived ecosystems. *Folia Entomológica Mexicana* 84: 131–156.

Hanski, I. 1987a. Nutritional ecology of dung- and carrion-feeding insects. In: Slansky Jr, F. Rodriguez, J.G. (eds), *Nutritional Ecology of Insects, Mites, and Spiders*. New York: John Wiley & Sons, pp. 837–884.

Hanski, I. 1987b. Carrion fly community dynamics: Patchiness, seasonality, and coexistence. *Ecological Entomology* 12: 257–266.

Hanski, I. and T. Cambefory. 1991. Resource partitioning. In: Hanski, I. and Y. Cambefort (eds), *Dung Beetle Ecology*. New Jersey: Princeton University Press, pp. 330–365.

Herrera, I. and J.M. Nassar. 2009. Reproductive and recruitment traits as indicators of the invasive potential of *Kalanchoe daigremontiana* (Crassulaceae) and *Stapelia gigantea* (Apocynaceae) in a Neotropical arid zone. *Journal of Arid Environment* 73: 978–986.

Hidayati, S.N., W. Meijer, J.M. Baskin, and J.L. Walck. 2000. A contribution to the life history of the rare Indonesian holoparasite *Rafflesia patma* (Rafflesiaceae). *Biotropica* 32: 408–414.

Hines, H.M., B.A. Counterman, R. Papa, P.A. de Moura, M.Z. Cardoso, M. Linares, J. Mallet et al. 2011. Wing patterning gene redefines the mimetic history of *Heliconius* butterflies. *Proceedings of the National Academy of Sciences USA* 108: 19666–19671.

Hogsette, J.A. and J.P. Ruff. 1985. Stable fly (Diptera: Muscidae) migration in northwest Florida. *Environmental Entomology* 14: 170–175.

Humeau, L., C. Micheneau, H. Jacquemyn, A. Gauvin-Bialecki, J. Fournel, and T. Pailler. 2011. Sapromyiophily in the native orchid, *Bulbophyllum variegatum*, on Réunion (Mascarene Archipelago, Indian Ocean). *Journal of Tropical Ecology* 27: 591–599.

Jersáková, J., S.D. Johnson, and A. Jürgens. 2009. Deceptive behaviour in plants II. Food deception by plants: From generalized systems to specialized floral mimicry. In: Baluška, F. (ed.), *Plant–Environment Interactions, Signalling and Communication in Plants, From Sensory Plant Biology to Active Plant Behaviour.* Berlin, Heidelberg: Springer-Verlag, pp. 223–246.

Jersáková, J., A. Jürgens, P. Šmilauer, and S.D. Johnson. 2012. The evolution of floral mimicry: Identifying traits that visually attract pollinators. *Functional Ecology* 26: 1381–1389.

Johnson, S.D. and B. Anderson. 2010. Coevolution between food-rewarding flowers and their pollinators. *Evolution: Education and Outreach.* 3: 32–39.

Johnson, S.D. and A. Jürgens. 2010. Convergent evolution of carrion and faecal scent mimicry in a stinkhorn fungus and fly-pollinated angiosperm flowers. *South African Journal of Botany* 76: 796–807.

Jürgens, A., S. Dötterl, and U. Meve. 2006. The chemical nature of fetid floral odours in stapeliads (Apocynaceae–Asclepiadoideae–Ceropegieae). *New Phytologist* 172: 452–468.

Jürgens, A., S.-L. Wee, A. Shuttleworth, and S.D. Johnson. 2013. Chemical mimicry of insect oviposition sites: A global analysis of convergence in angiosperms. *Ecology Letters* 16: 1157–1167.

Jürgens, A., T. Witt, and G. Gottsberger. 2002. Flower scent composition in night-flowering *Silene* species (Caryophyllaceae). *Biochemical Systematics and Ecology* 30: 383–397.

Kakishima, S., Y. Terajima, J. Murata, and H. Tsukaya. 2011. Infrared thermography and odour composition of the *Amorphophallus gigas* (Araceae) inflorescence: The cooling effect of the odorous liquid. *Plant Biology* 13: 502–507.

Kalinová, B., H. Podskalská, J. Růžička, and M. Hoskovec. 2009. Irresistible bouquet of death—How are burying beetles (Coleoptera: Silphidae: *Nicrophorus*) attracted by carcasses. *Naturwissenschaften* 96: 889–899.

Kite, G.C. 2000. Inflorescence odour of the foul-smelling aroid *Helicodiceros muscivorus. Kew Bulletin* 55: 237–240.

Kite, G.C. and W.L.A. Hetterscheid. 1997. Inflorescence odours of *Amorphophallus* and *Pseudodracontium* (Araceae). *Phytochemistry* 46: 71–75.

Kite, G.C., W.L.A. Hetterscheid, M.J. Lewis, P.C. Boyce, J. Ollerton, E. Cocklin, A. Diaz, and M.S.J. Simmonds. 1998. Inflorescence odours and pollinators of *Arum* and *Amorphophallus* (Araceae). In: Owens, S.J. and P.J. Rudall (eds), *Reproductive Biology.* London: Royal Botanic Gardens, pp. 295–315.

Kite, G.C. and S.A.L. Smith. 1997. Inflorescence odour of *Senecio articulatus*: Temporal variation in isovaleric acid levels. *Phytochemistry* 45: 1135–1138.

Kneidel, K.A. 1984. Influence of carcass taxon and size on species composition of carrion-breeding Diptera. *The American Midland Naturalist Journal* 111: 57–63.

Knudsen, J.T. and B. Ståhl. 1994. Floral odours in the Theophrastaceae. *Biochemical Systematics and Ecology* 22: 259–268.

Knudsen, J.T. and L. Tollsten.1993. Trends in floral scent chemistry in pollination syndromes: Floral scent composition in moth-pollinated taxa. *Botanical Journal of the Linnean Society* 113: 263–284.

Lam, K., M. Tsang, A. Labrie, R. Gries, and G. Gries. 2010. Semiochemical-mediated oviposition avoidance by female house flies, *Musca domestica*, on animal feces colonized with harmful fungi. *Journal of Chemical Ecology* 36: 141–147.

Lecoq, H. 1862. *De la fécondation naturelle et artificielle des végétaux f.t. de l'hybridation.* Paris: Librairie Agricole de la Maison Rustique.

Lev-Yadun, S., G. Ne'eman, and U. Shanas 2009. A sheep in wolf's clothing: Do carrion and dung odours of flowers not only attract pollinators but also deter herbivores? *BioEssays* 31: 84–88.

Linnaeus, C. 1737. *Hortus Cliffortianus*. Amsterdam: Privately published.

Linnaeus, C. 1753. *Species Plantarum*. Stockholm: Laurentius Salvius.

Magee, E.A., C.J. Richardson, R. Hughes, and J.H. Cummings. 2000. Contribution of dietary protein to sulphide production in the large intestine: An *in vitro* and a controlled feeding study in humans. *American Journal of Clinical Nutrition* 72: 1488–1494.

Marino, P., R. Raguso, and B. Goffinet. 2009. The ecology and evolution of fly dispersed dung mosses (Family Splachnaceae): Manipulating insect behaviour through odour and visual cues. *Symbiosis* 47: 61–76.

Marloth, R. 1907. Notes on the morphology and biology of *Hydnora africana* Thunb. *Transactions of the South African Philosophical Society* 16: 465–468.

Martín-Piera, F. and J.M. Lobo. 1996. A comparative discussion of trophic preferences in dung beetles communities. *Miscellània Zoològica* 19: 13–31.

Meeuse, B.J.D. and I. Raskin. 1988. Sexual reproduction in the arum lily family, with emphasis on thermogenicity. *Sexual Plant Reproduction* 1: 3–15.

Meve, U. 1997. *The Genus Duvalia (Stapelieae): Stem-Succulents between the Cape and Arabia*. New York: Springer-Verlag/Wien.

Meve, U., G. Jahnke, S. Liede, and F. Albers. 2004. Isolation mechanisms in the stapeliads (Apocynaceae–Asclepiadoideae–Ceropegieae). *Schumannia 4 Biodiversity & Ecology* 2: 107–126.

Meve, U. and S. Liede. 1994. Floral biology and pollination in stapeliads—New results and a literature review. *Plant Systematics and Evolution* 192: 99–116.

Meve, U. and S. Liede. 2002. A molecular phylogeny and generic rearrangement of the stapelioid Ceropegieae (Apocynaceae–Asclepiadoideae). *Plant Systematics and Evolution* 234: 171–209.

Meve, U. and S. Liede. 2004. Subtribal division of Ceropegieae (Apocynaceae–Asclepiadoideae). *Taxon* 53: 61–72.

Moré, M., A.A. Cocucci, and R.A. Raguso. 2013. The importance of oligosulfides in the attraction of fly pollinators to the brood-site deceptive species *Jaborosa rotacea* (Solanaceae). *International Journal of Plant Sciences* 174: 863–876.

Müller, F. 1878. Über die Vortheile der Mimicry bei Schmetterlingen. *Zoologischer Anzeiger* 1: 54–55.

Nakonechnaya, O.V., V.S. Sidorenko, O.G. Koren, S.V. Nesterova, and Y.N. Zhuravlev. 2008. Specific features of pollination in the Manchurian birthwort, *Aristolochia manshuriensis*. *The Biological Bulletin* 35: 459–465.

Newman, E., B. Anderson, and S.D. Johnson. 2012. Flower colour adaptation in a mimetic orchid. *Proceedings of the Royal Society B* 279: 2309–2313.

Nickrent, D.L., A. Blarer, Y. Qiu, D.E. Soltis, P.S. Soltis, and M. Zanis. 2002. Molecular data place Hydnoraceae with Aristolochiaceae. *American Journal of Botany* 89: 1809–1817.

Nilssen, A.C., B.Å. Tømmerås, R. Schmid, and S.B. Evensen. 1996. Dimethyl trisulfide is a strong attractant for some calliphorids and a muscid but not for the reindeer oestrids *Hypoderma tarandi* and *Cephenemyia trompe*. *Entomologia Experimentalis et Applicata* 79: 211–218.

Ollerton, J. and R.A. Raguso. 2006. The sweet stench of decay. *New Phytologist* 172: 382–385.

Paczkowski, S. and S. Schütz. 2011. Post-mortem volatiles of vertebrate tissue. *Applied Microbiology and Biotechnology* 91: 917–935.

Pape, T. and H. Bänziger. 2000. Two new species of *Sarcophaga* (Diptera: Sarcophagidae) among pollinators of newly discovered *Sapria ram* (Rafflesiaceae). *The Raffles Bulletin of Zoology* 48: 201–208.

Patiño, S., T. Aalto, A.A. Edwards, and J. Grace. 2002. Is *Rafflesia* an endothermic flower? *New Phytologist* 154: 429–437.

Patiño, S., J. Grace, and H. Bänziger. 2000. Endothermy by flowers of *Rhizanthes lowii* (Rafflesiaceae). *Oecologia* 124: 149–165.

Phillips, R.D., T. Xu, M.F. Hutchinson, K.W. Dixon, and R. Peakall. 2013. Convergent specialization—The sharing of pollinators by sympatric genera of sexually deceptive orchids. *Journal of Ecology* 101: 826–835.

Pisciotta, S., A. Raspi, and M. Sajeva. 2011. First records of pollinators of two co-occurring Mediterranean Apocynaceae. *Plant Biosystems* 145: 141–149.

Plukenet, L. 1696. *Almagestum botanicum*. London: Privately published.

Proctor, M., P. Yeo, and A. Lack. 1996. *The Natural History of Pollination*. Oregon: Timber Press.

Sakai, S. 2002. A review of brood-site pollination mutualism: Plants providing breeding sites for their pollinators. *Journal of Plant Research* 115: 161–168.

Sakai, S. and T. Inoue. 1999. A new pollination system: Dung-beetle pollination discovered in *Orchidantha inouei* (Lowiaceae, Zingiberales) in Sarawak, Malaysia. *American Journal of Botany* 86: 56–61.

Schaefer, H.M. and G. Ruxton. 2008. Fatal attraction—Carnivorous plants roll out the red carpet to lure insects. *Biology Letters* 4: 153–155.

Schiestl, F.P., M. Ayasse, H.F. Paulus, C. Löfstedt, B.S. Hansson, F. Ibarra, and W. Francke. 1999. Orchid pollination by sexual swindle. *Nature* 399: 421–422.

Schiestl, F.P. and S.D. Johnson. 2013. Pollinator-mediated evolution of floral signals. *Trends Ecology and Evolution* 28: 307–315.

Schoenly, K. and W. Reid. 1987. Dynamics of heterotrophic succession in carrion arthropod assemblages—Discrete series or a continuum of change. *Oecologia* 73:192–202.

Scott, M.P. 1998. The ecology and behavior of burying beetles. *Annual Review of Entomology* 43: 595–618.

Seymour, R.S., M. Gibernau, and K. Ito. 2003a. Thermogenesis and respiration of inflorescences of the dead horse arum *Helicodiceros muscivorus*, a pseudo-thermoregulatory aroid associated with fly pollination. *Functional Ecology* 17: 886–894.

Seymour, R.S., E. Maass, and J.F. Bolin. 2009. Floral thermogenesis of three species of *Hydnora* (Hydnoraceae) in Africa. *Annals of Botany* 104: 823–832.

Seymour, R.S. and P. Schultze-Motel. 1999. Respiration, temperature regulation and energetics of thermogenic inflorescences of the dragon lily *Dracunculus vulgaris* (Araceae). *Proceedings of the Royal Society B* 266: 1975–1983.

Seymour, R.S., C.R. White, and M. Gibernau. 2003b. Heat reward for insect pollinators. *Nature* 426: 243–244.

Shirasu, M., K. Fujioka, S. Kakishima, S. Nagai, Y. Tomizawa, H. Tsukaya, J. Murata, Y. Manome, and K. Touhara. 2010. Chemical identity of a rotting animal-like odor emitted from the inflorescence of the Titan Arum (*Amorphophallus titanum*). *Bioscience, Biotechnology, and Biochemistry* 74: 2550–2554.

Shuttleworth, A. and S.D. Johnson. 2009. New records of insect pollinators for South African asclepiads (Apocynaceae: Asclepiadoideae). *South African Journal of Botany* 75: 689–698.

Shuttleworth, A. and S.D. Johnson. 2010. The missing stink: Sulphur compounds can mediate a shift between fly and wasp pollination systems. *Proceedings of the Royal Society B* 277: 2811–2819.

Simpson, S.J. and D. Raubenheimer. 2009. Macronutrient balance and lifespan. *Aging* 1: 875–880.

Skubatz, H., D.D. Kunkel, W.N. Howald, R. Trenkle, and B. Mookhergee. 1996. The *Sauromatum guttatum* appendix as an osmophore: Excretory pathways, composition of volatiles and attractiveness to insects. *New Phytologist* 134: 631–640.

Snow, W.E. 1957. Flies attracted to the giant star flower (*Stapelia gigantea* N.E. Br.). *Journal of Economic Entomology* 50: 693–694.

Sprengel, C.K. 1793. *Das entdeckte Geheimniss der Natur im Bau und in der Befruchtung der Blumen.* Berlin: Vieweg.

Stapelius, J.B. 1644. *Theophrasti eresii de historia plantarum.* Amsterdam: Privately published.

Statheropoulous, M., C. Spiliopoulou, and A. Apapiou. 2005. A study of volatile organic compounds evolved from the decaying human body. *Forensic Science International* 153: 147–155.

Stensmyr, M., I. Urru, M. Celander, and B.S. Hansson. 2002. Rotting smell of dead-horse arum florets. *Nature* 420: 625–626.

Stotz, G.C. and E. Gianoli. 2013. Pollination biology and floral longevity of *Aristolochia chilensis* in an arid ecosystem. *Plant Ecology & Diversity* 6: 181–186.

Stránský, K. and I. Valterová. 1999. Release of volatiles during the flowering period of *Hydrosme rivieri* (Araceae). *Phytochemistry* 52: 1387–1390.

Thien, L.B., H. Azuma, and S. Kawano. 2000. New perspectives on the pollination biology of basal angiosperms. *International Journal of Plant Sciences* 161 (6 Suppl.): S225–S235.

Thien, L.B., P. Bernhardt, M.S. Devall, Z.-D. Chen, Y.-B. Luo, J.-H. Fan, L.-C. Yuan, and J.H. Williams. 2009. Pollination biology of basal angiosperms (ANITA grade). *American Journal of Botany* 96: 166–182.

Thunberg, C.P. 1775. Beskrifning paa en ganska besynnerlig och obekant svamp, *Hydnora africana. Kongalia Vetenskaps Akademiens Handlingar* 36: 69–75.

Tomberlin J.K., T.L. Crippen, A.M. Tarone, B. Singh, K. Adams, Y.H. Rezenom, M.E. Benbow et al. 2012. Interkingdom responses of flies to bacteria mediated by fly physiology and bacterial quorum sensing. *Animal Behaviour* 84: 1449–1456.

Troje, N. 1993. Spectral categories in the learning behaviour of blowflies. *Zeitschrift für Naturforschung* 48c: 96–104.

Tuno, N. 1998. Spore dispersal of *Dictyophora* fungi (Phallaceae) by flies. *Ecological Research* 13: 7–15.

Uemura, S., K. Ohkawara, G. Kudo, N. Wada, and S. Higashi. 1993. Heat-production and cross-pollination of the Asian skunk cabbage *Symplocarpus renifolius* (Araceae). *American Journal of Botany* 80: 635–640.

Urru, I., M.C. Stensmyr, and B.S. Hansson. 2011. Pollination by brood-site deception. *Phytochemistry* 72: 1655–1666.

van der Niet, T., D.M. Hansen, and S.D. Johnson. 2011. Carrion mimicry in a South African orchid: Flowers attract a narrow subset of the fly assemblage on animal carcasses. *Annals of Botany* 107: 981–992.

van der Pijl, L. 1937. Biological and physiological observations on the inflorescence of *Amorphophallus*. *Recueil des Travaux Botaniques Neerlandais* 34: 157–167.

Vereecken, N.J. and J.N. McNeil. 2010. Cheaters and liars: Chemical mimicry at its finest. *Canadian Journal of Zoology* 88: 725–752.

Vogel, S. 1954. *Blütenbiologische Typen als Elemente der Sippengliederung, Dargestellt Anhand der Flora Südafrikas. Botanische Studien* 1: 1–338. Jena: Gustav Fischer Verlag. English translation by E. Pischtschan, 2012. Aachen: Shaker Verlag.

Vogel, S. 1996. Christian Konrad Sprengel's theory of the flower: the cradle of floral ecology. In: Lloyd, D.G. and S.C.H. Barrett (eds), *Floral Biology: Studies on Floral Evolution in Animal–Pollinated Plants.* New York: Chapman & Hall, pp. 44–63.

Waser, N.M. 2006. Specialization and generalization in plant–pollinator interactions: A historical perspective. In: Waser, N.M. and J. Ollerton (eds), *Plant–Pollinator Interactions: From Specialization to Generalization.* Chicago, London: University of Chicago Press, pp. 3–17

Weber, A. 2008. Pollination in the plants of the Golfo Dulce area. *Stapfia 88, zugleich Kataloge der oberösterreichischen Landesmuseen Neue Serie* 80: 509–538.

White, A. and B.L. Sloane. 1937. *The Stapelieae,* 2nd edition, Vols. 1–3. Pasadena: Abbey San Encino Press.

Woodward, C.L., P.E. Berry, H. Maas-van de Kamer, and K. Swing. 2007. *Tiputinia foetida,* a new mycoheterotrophic genus of Thismiaceae from Amazonian Ecuador, and a likely case of deceit pollination. *Taxon* 56: 157–162.

Wurdack, K.J. and C.C. Davis. 2009. Malpighiales phylogenetics: Gaining ground on one of the most recalcitrant clades in the angiosperm tree of life. *American Journal of Botany* 96: 1551–1570.

Young, H.J. 1986. Beetle pollination of *Dieffenbachia longispatha* (Araceae). *American Journal of Botany* 73: 931–944.

Zhang, L., S.C.H. Barrett, J.-Y. Gao, J. Chen, W.W. Cole, Y. Liu, Z.-L. Bai, and Q.-J. Li. 2005. Predicting mating patterns from pollination syndromes: The case of "sapromyiophily" in *Tacca chantrieri* (Taccaceae). *American Journal of Botany* 92: 517–524.

Zito, P., S. Guarino, E. Peri, M. Sajeva, and S. Colazza. 2013. Electrophysiological and behavioural responses of the housefly to "sweet" volatiles of the flowers of *Caralluma europaea* (Guss.) N.E. Br. *Arthropod-Plant Interactions* 7: 485–489.

17

Population Genetics and Molecular Evolution of Carrion-Associated Arthropods

Christine J. Picard, Jonathan J. Parrott, and John W. Whale

CONTENTS

17.1 Introduction

A population is defined as a group of organisms of the same species that inhabit a particular geographic area, but which are also capable of interbreeding with inhabitants of other geographic areas (Hartl and Clark 1997). Therefore, this discussion should begin with the overall goal of defining a population of carrion-frequenting insects. There are two classic models which would be appropriate to define populations of carrion insects: the island model (Wright 1951) and the stepping stone model (Kimura and Weiss 1964). The island model assumes equal migration in and out of defined populations (=islands), whereas the stepping stone model assumes a greater degree of exchange between neighboring populations in which isolation by distance is observed. The actual population structure is likely a mixture of these two models for carrion insects and is not well known or defined. This chapter will focus on describing the population structure of carrion insects using existing molecular genetic techniques.

Ecologically, an insect population may be individuals associated with a single or few items of carrion (=island), however, the scale is difficult to define. For example, and in the case of carrion insects, a single island could hold a large number of individuals and these individuals could make up a "population." These individuals will eventually complete their life cycle and emerge to become breeding adults in a (spatially and temporally) relatively small geographic area. Alternatively, a population could consist of a larger geographic area (e.g., ecoregion) in which many carrion islands are available for the colonizing insect community over space and time, and where adults emerge from different carrion resources to interbreed. Or, a population could be defined for carrion insects over an even larger geographic area (e.g., a continent), with gene flow among these geographic areas being necessarily generally limited. Each of these hierarchical levels of structuring could be viewed as a part of either type of ecological population model mentioned above, and thus could be investigated with population genetic and molecular evolution principles and methods.

For any given species, an individual or island-level population should be defined. For most carrion insects, the r-selection theory is most fitting (Norris 1965; MacArthur and Wilson 1967). That is, in an unpredictable environment (e.g., temperate regions), the ability to quickly and efficiently produce many offspring resulting in some surviving to the next generation is key. This strategy also implies that population numbers change very rapidly from generation to generation. Consequently, in some instances only a (genetically) small number of individuals would then be responsible for the next founder event (i.e., different carrion). Alternatively, K-selection favors an ability to successfully compete for limited resources, and some, typically larger species of carrion insects, possess these traits. For example, Sarcophagids give live birth to only a small number of offspring, but generally guarantee their survival by arriving early at carrion and depositing larvae that immediately begin feeding, a strategy which differs from that of the majority of egg-laying species that need at least one day for the eggs to hatch (Norris 1965).

From the above it should be obvious that for any given community of carrion insects, defining the scale of a population can be problematic. Is the insect community located in a city surrounded by rural farms, and does this particular city support a community of flies that are more "urban" in nature? Is there migration from neighboring communities? Is the entire community homogenous throughout the temporal season of insect colonization? The definition of a population thus generally will remain vague and arbitrary to some degree, albeit operational.

From an evolutionary standpoint, the unpredictable spatial and temporal availability of the necessary resource, in this case carrion, should select for individuals with the ability to quickly exploit a new resource. For example, did ectoparasitism (=myiasis) behavior evolve secondarily in carrion-breeding insects because earlier colonization of a live host confers a selective advantage? Several examples of flies that are both carrion breeders and agents of myiasis, though genetically identical at neutral markers, clearly exhibit a difference in behavior (Stevens 2003).

The goal of this chapter is to discuss available data on the population genetic structure of carrion insects in the context of ecological and evolutionary theory. Improvements in molecular tools now permit the simultaneous study of the ecology and genetics of any organism, and thus allowed researchers to begin to define populations retroactively based on ecological and/or molecular data. This approach, which had been previously termed landscape genetics (for a review, see Manel et al. 2003), resolves the population structure at finer spatial and temporal levels. Because carrion insects are dependent on an ephemeral resource, their population structure would additionally be defined by the resource characteristics at a given place. Just as in some plant–insect coevolutionary examples (Bernays and Graham 1988), carrion insects have adapted and evolved toward the appropriate detection response and utilization of this resource (Norris 1965).

17.2 Basic Population Genetic and Molecular Evolution Principles and Methods

Several methods exist to detect genetic variation, thus allowing population structure hypotheses to be tested. For a long time, genetic variation was detected by observing variation in proteins using electrophoresis (Hubby and Lewontin 1966). Enzymes were primarily used in which the separation matrix was soaked in a solution containing the enzyme substrate; therefore the reaction of the enzyme to the substrate could be easily detected. If the enzyme had an amino acid substitution, then its mobility through the matrix would change. These polymorphic enzymes were termed allozymes, and many natural populations of bacteria (Selander et al. 1986), plants (Hamrick and Godt 1990), or *Drosophila* (Diptera: Drosophilidae) (David and Capy 1988) and other insects have been studied. Unfortunately most of the studies done using allozymes was for identification of species rather than a population approach (Wallman and Adams, 2001), and this approach has been eclipsed by the advent of PCR-based methods.

Following allozymes, the use of restriction enzymes was employed to detect variation in fragment sizes of DNA (Roderick 1996). Because restriction enzymes cleave the DNA at specific DNA sequences, any change in sequence will result in a different fragment length. The presence or absence of a particular restriction fragment translates to differences in DNA sequences and can be used to identify groups of individuals based on similarities. Differences result in restriction fragment length polymorphisms (RFLP) that can be exploited. These are distributed throughout the genome, and thus were important in early population genetic surveys. Advantages to using RFLP markers are that no prior sequence knowledge is necessary, thus making it amendable to study a large array of species. Many examples below will highlight the use of RFLP in the population genetics and molecular identification of carrion insects.

The arrival of polymerase chain reaction (PCR) (Mullis and Faloona 1987) pushed the limit with regards to the amount of polymorphic DNA that can be amplified and detected, whether through length polymorphisms (such as microsatellites) or direct sequencing of single nucleotide polymorphisms (SNPs). Most advantageous to population genetics are loci that have a large number of allele variants each with relatively moderate frequencies (Hartl and Clark 1997). An example of this type of molecular marker includes microsatellites, which are easily attainable with PCR-based methods, easily separated and detected using capillary electrophoresis, and thus useful for studying the population structure.

Genetic variation can serve for determining the genetic relationships of individuals both within and among populations. The theory is that genetic polymorphisms are shared among different populations due to migration; therefore the amount of shared alleles is indicative of the rate of migration and gene flow between the populations (and, conversely, the relative isolation of populations). Following the same logic, ancestors will share alleles, and thus evolutionary histories of species, that is, ancestral relationships can be inferred from molecular data. Though the use of non-neutral genetic markers can be important for answering specific morphological and/or behavioral trait evolution, this chapter will focus on the use of mostly neutral loci for population genetic and molecular evolutionary studies.

17.3 Carrion Arthropod Population Genetics

17.3.1 Family Calliphoridae

The application of population genetics has facilitated the study of various families within the Diptera, however, the majority of research focusing on Drosophilidae (Hamblin and Veuille 1999; Falush et al. 2003). Nevertheless, many of the principles of *Drosophila* research can be applied to other fly families. Research involving SNPs associated with phenotypic traits (Lints and Gruwez 1972), selection experiments for specific traits (Malmendal et al. 2013; Wit et al. 2013), or genome resequencing (Turner et al. 2011) of differing populations all demonstrate useful methods for studying the population genetic structure of insects.

Population studies of insects associated with carrion, such as the Calliphoridae, Sarcophagidae, and Muscidae, will each be discussed further in this chapter. Population genetics can play an essential role not just in the examination of population or the metapopulation structure, gene flow and diversity, but can also be informative in the field of veterinary, urban, and pest entomology. Most recently, for example, an in-depth understanding of the population diversity of the family Calliphoridae has been examined due to applications in forensics.

Life history traits have been shown to depend on the population. For example, Gallagher et al. (2010) and Tarone et al. (2011) found that developmental times and body sizes varied among populations of North American *Lucilia sericata* (Meigen) originating from different geographic locations with genotype X environment interactions. This data demonstrate how important it is to determine the genetic effect of the population structure. For example, in the postmortem interval estimation of human remains based on the age of insects present on the body, the population structure that is unaccounted for can result in increased error in estimations if not considered or controlled for. Genotype environment interactions, like those suggested by these studies, can further be an indication of local adaptation in instances where the trait influences fitness (see Chapter 15).

An effective method for population structure analyses requires the use of polymorphic loci. One such method exploits variation in proteins via electrophoresis, thus termed allozyme variation. Wallman and Adams (1997, 2001) used electrophoresis to examine genetic differences among populations of Australian blow flies. For example, species, such as *Calliphora hilli hilli*, *Calliphora stygia*, and *Calliphora albifrontalis* on Kangaroo Island, Australia, were genetically isolated from the mainland populations due to genetic drift and natural selection. The distance between the mainland and island resulted in little to no genetic exchange between populations, creating a bottleneck situation on the island. Without the use of polymorphic markers, such an isolated population would not have been detected.

In recent years, large-scale genetic diversity and population structures of forensically important blow flies have been examined. Picard and Wells (2009, 2010) examined the genetic diversity of *Phormia regina* (Meigen) and *L. sericata* by way of amplified fragment length polymorphism (AFLP) profiles, which demonstrated the population structure on a small scale (locally) but no structure on a large scale (continentally). Samples of blow flies collected at the same time and location shared more alleles than what would be expected in a random sample. This suggests that individuals were related and breeding in close proximity to the bait at the time of collection. These results also have implications for the collection of adults for laboratory life history studies. As a result, the temporal and spatial scale of insect collection should be carefully planned for these sorts of experiments. Similarly, a worldwide study of *P. regina* samples demonstrated a stark contrast between North American and European specimens. Both Desmyter and Gosselin (2009) and Jordaens et al. (2013) found high intraspecific sequence differentiation (~ 4%) between populations of *P. regina* from Europe and North America. This important finding further clarifies the need for a more detailed population study of *P. regina*, especially if there is an expectation of similar life history traits.

A group of related species, in the tribe Luciliinae, has been studied thoroughly for intraspecific variation and the possibility of speciation and hybrids. Examining *Lucilia* populations with regard to gene flow among populations and the potential presence of speciation occurring is important. Speciation has long been debated in the genus *Lucilia*. *Lucilia sericata* and *Lucilia cuprina* are both facultative ectoparasites, and populations of these flies display different pathogenicity to livestock around the world. For example, in Australia (Foster et al. 1980) and South Africa (Zumpt 1965), *L. cuprina* is the main causal agent of myiasis in livestock, and *L. sericata* is rarely seen parasitizing livestock. In contrast, in the United Kingdom and North America, *L. cuprina* is not the primary myiasis causer, but rather *L. sericata* (Williams et al. 1985). Because of the substantial economic losses from damaged livestock due to myiasis, studying the population dynamics of these blow flies is important for pest management strategies. Stevens and Wall (1996, 1997b) used randomly amplified polymorphic DNA (RAPD) to determine the population structure of *L. sericata* and *L. cuprina*, both locally and globally. The RAPD profiles, although limited due to shortcomings of the technique itself, found no evidence of differentiation of populations across the world for *L. sericata*. In contrast, *L. cuprina* was observed to possess genetic population structuring, based on both RAPD data and mitochondrial sequencing. In particular, *L cuprina* from Hawaii, USA and Townsville, Australia, were assessed to possibly belong to a subspecies *L. cuprina cuprina* based on morphological characters by Norris (1990). Another subspecies, *L. cuprina dorsalis*, was also thought to differ from the majority type *L. cuprina*. Stevens et al. (2002) found that *L. cuprina* formed two subgroups, the majority type and a population from Oahu, Hawaii, USA (Figure 17.1). When nuclear 28S rRNA was examined, both subtypes formed a single *L. cuprina*

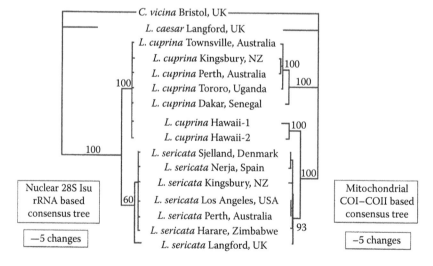

FIGURE 17.1 A demonstration of naturally occurring hybridization among and rapid evolution in nature of *L. sericata* and *L. cuprina* as detected by appropriate molecular markers. (From Stevens, J., R. Wall, and J.D. Wells. 2002. *Insect Molecular Biology* 11: 141–148.)

clade, however using mtDNA, the Hawaiian *L. cuprina* specimens were grouped with *L. sericata*. This was hypothesized to potentially have occurred when a male *L. cuprina* hybridized with a female *L. sericata*. These two species can interbreed in laboratory settings (Stevens and Wall 1995, 1997a). Because of this potential for interbreeding, and the apparent establishment of hybrid populations, the ability to control and manage pest control is completely dependent on identifying the susceptible population and being able to recognize the potential for other species or subtypes to increase in abundance and distribution.

Knowing the population structure of flies is important so that such information can be applied to the biological control of myiasis-causing flies, for instance by using the Sterile Insect Technique (SIT). This research has played an important role in the United States in the eradication of the New World screwworm fly *Cochliomyia hominivorax* (Coquerel) (Diptera: Calliphoridae). SIT relies on known genetic information and problems can arise if populations begin to isolate and diverge. Natural barriers, such as islands, can restrict gene flow and create distinct interbreeding populations when compared to connected populations (Krafsur and Whitten 1993). Lyra et al. (2009) employed PCR–RFLP methods of mitochondrial DNA fragments of *C. hominivorax* populations across 10 countries and found a complex population structure, with Caribbean island populations (i.e., restricted) having a significant population structure (ϕst = 0.5234) and low variability whereas mainland populations were relatively homogenous (ϕst = 0.0483). Simultaneously, microsatellites have been used for the examination of genetic variation within and among populations of *C. hominivorax* (Torres et al. 2004; Torres and Azeredo-Espin 2005). These authors developed markers for *C. hominivorax* and subsequently examined 10 sites on four Caribbean islands (Torres and Azeredo-Espin 2009), revealing stark population differences among island populations. This was hypothesized to be due to lack of gene flow and/or source-sink dynamics. Without knowing the population structure, it is possible that SIT attempts could fail in some of these locations because of potential differences in mate preferences and breeding.

Molecular and morphological data provided strong support for two distinct clades of the Old World screwworm fly *Chrysomya bezziana* (Villeneuve) (Diptera: Calliphoridae), a tragically serious pest of humans and livestock. One clade was made up of Sub Saharan Africa and the other the Gulf region of Asia (Hall et al. 2001). Additional support was found via sequencing of the cytochrome b mitochondrial DNA region which found a statistical correlation between haplotype and host species, thus indicating a genotype may lead to host preferences (Hall et al. 2009a). By studying a large number of geographically distinct samples for the population structure, then targeted control measures can be undertaken.

17.3.2 Other Dipteran Families

Another important family of Diptera that are associated with carrion are the Sarcophagidae. They are a speciose and diverse family of flies with a worldwide distribution (Pape 1996). Unfortunately, few population genetic studies have been conducted on this family, likely due to the difficulty in identification and its limited abundance at carrion. More is needed as this family comprises forensically important flies, which are primary colonizers of carcasses (Tabor et al. 2005; Oliveira and Vasconcelos 2010; Cherix et al. 2012) and cause myiasis (Sotiraki et al. 2010). The COI gene in the Sarcophagidae has been used for identification purposes and provides insight into intra and interspecific variation (Ratcliffe et al. 2003; Yadong et al. 2010). Two studies (Meiklejohn et al. 2011, 2012) found that intraspecific variation was largely within the barcoding threshold of ~1%, and values did not change significantly when including geographically isolated populations. This suggests the population genetic structure of Sarcophagidae is stable, though based only on a single, neutral molecular locus. Because Sarcophagidae are so morphologically distinct (Pape 1996) and possess much intraspecific variation, additional markers for identification would be useful to provide a more in-depth measure of gene flow and population structure of this group.

An economically important Sarcophagidae is the Wohlfarts' wound myiasis fly *Wohlfahrtia magnifica* (Schiner) (Diptera: Sarcophagidae). Studies of the population structure were done via the sequencing cytochrome b mitochondrial DNA (Hall et al. 2009b) and found that when sampling different populations around the Iberian Peninsula, central and northern barriers were determined to restrict migration and gene flow. This means that the lineages were at some point isolated, and recently introduced and this results in a higher prevalence of wohlfahrtiosis in that particular geographic area.

Another family of carrion-visiting flies are the Muscidae. Though Muscidae are more frequently found to be colonizing decomposing organic resources like dung and rotten plant material, they have been found on carrion. Despite their well-known interactions with human environments, few population studies have been performed on this family. Genetic variation was examined in the stable fly *Stomoxys calcitrans* (Linneaus) (Szalanski et al. 1996) documented the greatest variation within stable fly populations as well as extensive gene flow between populations, thus no significant Hardy–Weinberg departures from random mating. Marquez et al. (2007) also examined the stable fly, however, including a wider range of populations: 277 flies from 11 countries and 5 zoogeographical regions and found moderate (26%) regional differentiation, indicating a high degree of population structure on a macrogeographic scale (Palearctic, Nearctic and Neotropical). Black and Krafsur (1986) found no differences in allele frequencies between sexes and ages of *Musca domestica* L. (Diptera: Muscidae); however, frequencies varied depending on the season implying that environmental factors can have a major influence on allele frequencies (Krafsur 1985). This means that populations vary across very large geographic scales, but little on more local (continental) scales.

In summary, carrion insects, with their unique life history traits, make for an interesting and little studied group of insects for assessing genetic diversity, population structure, gene flow, phenotypic and genetic traits, and their correlations. In addition, because of their ease of collection, their diverse distribution, and the wealth of genetic and genomic tools available, their study will only continue to increase in this growing field of the basic science of carrion insects.

17.3.3 Families Silphidae and Staphylinidae

The study of beetle populations dates back at least to the mid-nineteenth century, initiated by interest of the great evolutionary biologist Charles Darwin. Since then, numerous new genera and species have been described; while many are restricted to a specific region or ecosystem, others are more widespread in nature. Beetle diversity, as exhibited by more than 350,000 known species (Beutel and Haas 2000), has maintained a high level of scientific interest, although only a very small proportion of these are associated with carrion.

The Silphidae, a family of burying beetles, are among the most common beetles at carrion (Byrd and Castner 2009). A classic species belonging to this group is the endangered American burying beetle, *Nicrophorus americanus* Olivier (Coleoptera: Silphidae), which once was widespread across the United States and three Canadian provinces, is now only found in six U.S. states (Bedick et al. 1999; Szalanski et al. 2000; Backlund et al. 2008). As a result, it has been the focus of much research especially in the field of conservation biology. Some of the initial analyses came in the form of DNA sequencing, such as of the internal transcribed spacer 1 (ITS1) region of rRNAs between *N. americanus* and other Silphidae species. Its closest relative, *N. orbicollis*, may not be a true sister species, and the current data indicate that sister species may have gone extinct. As this is a conservation effort, it appears that *N. orbicollis* is *N. americanus'* greatest congeneric competitor, and thus may be leading to its demise. The geographic distribution and subsequent isolation of this species has not yet resulted in a population structure that is indicative of long-term separation, however, more polymorphic markers may be needed to discover the differentiation of populations.

Nicrophorus americanus possesses low genetic variation based on RAPD analysis (Kozol et al. 1994) and satellite DNA (King and Cummings 1997) of isolated populations from Rhode Island and Arkansas and Oklahoma. Only ITS1 sequence data suggested differentiation of the Rhode Island (RI) population, which is likely due to its distance and separation from the other mainland populations (Szalanski et al. 2000) inhabiting the Midwestern states that are more likely to interbreed. There was no clear clustering observed between the Oklahoma, Arkansas, South Dakota, and Nebraska populations, nor a north–south pattern. This could be due to a small sample size (limited by the endangered status of the species), or result from the increased mixture of the Midwest populations. Low genetic variation, especially among populations of a declining species, can be evidence of small population sizes, bottleneck events, and inbreeding and is thus important for conservation practices.

Genetics plays a major role in population diversification, and this divergence can be enhanced by differing environmental factors and pressures upon different populations of the same species, such as increased individual beetle size as a result of residing at higher altitudes (Smith et al. 2000). Carrion beetles have also been observed to control brood size and perform asynchronous oviposition in competitive events to allow the maximum number of larvae to survive and avoid unnecessary overcrowding and inevitable mortality at the food source (Rauter and Rust 2012). This behavioral trait ultimately contributes to the success of the population and sustains greater genetic diversity and haplotypic variation. The majority of Nicrophinae species possess oral secretions that stunt bacterial growth on a carcass whereas most other Silphinae do not (larvae feed on maggots) (Hoback et al. 2004). This is clearly an evolutionary response to the behavior of carcass burial in several Nicrophinae species, and it is possible that the gene or genes controlling the production of these secretions differ across the subfamily, perhaps with particular genetic signals or sequences shared by species that prefer similarly sized carrion.

The diversification of species can be promoted by a number of factors such as geographic barriers, dramatic climatic change, or strong predatory events. The most effective mediators of differentiation are often geography and topography in the guise of islands and land masses separated by seas and oceans, mountain ranges, or deserts. Each island or ecoregion typically possesses a variable frequency and composition of beetle species. For example, there are differences observed among the Japanese islands (Ikeda et al. 2009; Ito et al. 2010). The Silphidae species *Silpha longicornis* and *Silpha imitator* are restricted to the central main island of Honshu in Japan, while *Silpha perforata* only occurs on the northern island of Hokkaido (Ikeda et al. 2009). These islands are separated by just 19 km of ocean at the narrowest point, but as a consequence of the flightless nature of these species, migratory events from one island to the other are unlikely. This inability to migrate large distances has also promoted the diversification and speciation of multiple species (Ikeda et al. 2012). Additionally, such migrations would probably mean moving away from a habitat to which they are well adapted, and travelling through unfamiliar regions where tolerance to these environments is likely much lower. Equally as effective as island groups in species diversification are isolations caused by rugged topography. Examples include *Sepedophilus castaneus* Horn (Coleoptera: Staphylinidae) of southern California (Caterino and Chatzimanolis 2008), *S. longicornis* (Portvein) (Coleoptera: Silphidae) of Honshu, Japan (Ikeda et al. 2009), and *N. encaustus* surrounded by *N. investigator* in the Himalayas (Sikes et al. 2008). In California, *S. castaneus* from San Gabriel, San Bernadino and San Jacinto, three adjacent mountain ranges, are closely related to each other, while populations of the Sierra Pelona and NW transverse Range (which flank the Central Transverse Range) are more closely related to each other than individuals obtained from the Central Transverse Range (Caterino and Chatzimanolis 2008), which may be due to variations in altitude or habitat preferences. Whereas increased genetic differentiation of *S. longicornis* on Honshu may be attributable to their wide geographical distribution on this large island, populations at the extreme peripheries of their range are unlikely to encounter each other, while the potential of intermediates of these extremes are more likely in the overlapping or shared region. These populations of *S. longicornis* possess many unique haplotypes, suggesting the absence of gene flow between them (Ikeda et al. 2009). These beetle species, as abundant and important as they are to carrion decomposition, are little studied, and thus warrant increased attention.

17.3.4 Other Coleopteran Families

There are coleopteran species outside the Silphidae and Staphylinidae groups that have been observed to feed on carrion. *Phanaeus* species (Scarabaeidae) primarily feed on animal excrement; however, several species have been observed to feed on carrion (Price and May 2009). An AFLP study of two species of this group (*Phanaeus vindex* [MacLeay] and *P. difformis* [LeConte]) possessed enough discriminatory power to differentiate between these closely related species featuring at least 18 unique alleles (Dickey 2006). However, according to Price and May (2009), these species have not been previously observed to feed on carrion. Changes in feeding behavior or food substrate preferences may be determined by quality and availability of main food substrates. Genetic studies aimed at understanding these behavioral decisions are becoming more economical to undertake due to the lower costs, advancements in the technologies available, and increased quantity and quality of sequence data they generate. To date, there are few studies of this nature on non-model organisms.

While genetic analyses among and between populations or species reveal relationships and often exhibit phylogenies and networks similar to those based upon morphological characters, each gene is subject to differing rates of evolution (Caterino et al. 2005; Sikes et al. 2006). As such, this rate is unlikely to be congruent between genes across the genome. Such variability has the potential to relocate species on a phylogenetic tree. To generate a comprehensive phylogeny, it would perhaps be more appropriate to analyze multiple genes or the entire genome; and, the latter becomes more of a distinct possibility as the generation and resequencing of genomes becomes more of an attainable goal.

17.4 Phylogenetic Studies of Carrion Arthropods

Studying molecular phylogenies of groups of species allows evolutionary hypotheses to be tested. An example, in which two different hypotheses are proposed to explain adaptive behaviors, can be tested using molecular methods. Either new mutations arise and are beneficial and thus selected for, or pre-existing genetic variation already in the population is selected for and its frequency increases over evolutionary time and selective pressures. A classic example is the peppered moth in the UK—the genotype that lead to the more beneficial phenotype was always present and maintained in the population, albeit at a very low level (Kettlewell 1955, 1973; Haldane 1956).

Adaptive traits and the underlying genotype are important because of the nature of the carrion resource and the competitive and selective nature of arthropods associated with carrion. Several strategies exist that lend an advantage to carrion insects. For example, ectoparasitism (arrive before the host dies, and you are almost guaranteed to obtain the resources necessary for immature growth), and predation/cannibalism (arrive later but feed on the other insects present) are two strategies that confer an advantage. Each of these behaviors has arisen several times independently in different lineages of carrion insects (Stevens 2003; Singh et al. 2010; Singh and Wells 2011).

There are several examples in the literature in which individuals of the same species (as defined morphologically or molecularly) vary in some major behavioral way. For example, there are several *Lucilia* species that display different behaviors depending on their geographic location, though to date no distinct underlying genetic differences have been uncovered. Both *L. sericata* and *L. cuprina* are major agricultural pests, but in North America they are almost exclusively carrion breeders (Stevens and Wall 1997a). Molecular systematic analyses do not show any differences between these individuals with different life history traits; however, a limitation may be that samples collected for the published studies were not collected from live sheep (Harvey et al. 2008). Some studies indicate some population differentiation (Stevens et al. 2002) and, interestingly, these two species appear to show introgression from hybridization and are capable of breeding with each other. Therefore, it is possible that the gene(s) leading to ectoparasitism were introgressed into *L. sericata* from *L. cuprina* (Stevens et al. 2002). Another example from this genus is *Lucilia silvarum* Meigen (Diptera: Calliphoridae), which in North America is an amphibian parasite, but is collected from carrion in Europe (Fremdt et al. 2012). In all three of these cases, due to limitations in data acquisition, the genetic basis of this behavior or adaptation to the environment (domestication of animals is relatively recent evolutionarily) remains unknown (Prinkkila and Hanski 1995; Bolek and Coggins 2002; Otranto and Stevens 2002; Stevens 2003; Adair and Kondratieff 2006).

Another example illustrating the usefulness of a molecular systematic approach to determine evolution of adaptive traits is demonstrated in a genus of blow flies, *Chrysomya*. Using a phylogenetic approach, morphological traits such as hairy maggots seem to have evolved independently at least two times (Singh et al. 2010). It appears that having tuberculate larvae is associated with (both intra and interspecific) predation on other larvae. This morphological change thus corresponds with a change in behavior, which in this case is an advantageous derived trait. In another example of the evolution of adaptive traits is the multiple appearance of ectoparasitism of mammals, as evident in two distinct phylogenetic lineages (*Chrysomya bezziana* and *Cochliomyia hominivorax*) (Singh and Wells 2011). The ability to parasitize a living mammal is advantageous and thus can be selected for in nature. This deviation suggests that genetic variation is present, likely at a very low frequency, and when selected for in case of environmental change, the frequency increases in the population (Figure 17.2).

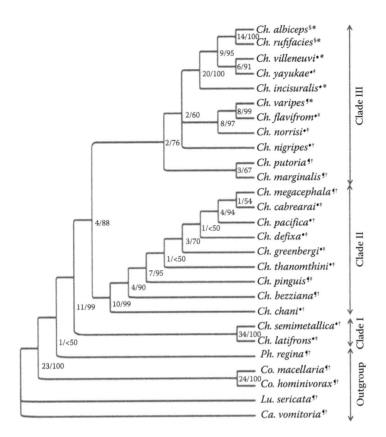

FIGURE 17.2 Multiple evolutionary events lead to behavioral changes important for adaptation. (*, species that produce hairy maggots; †, species that produce smooth maggots; ‡, species with unknown larval structure; §, monogenic spp.; ¶, amphogenic spp.; •, species with unknown sex determination system.) (From Singh, B., H. Kurahashi, and J.D. Wells. 2010. *Medical and Veterinary Entomology* 25: 126–134.)

While the molecular systematics of blow flies has been relatively well studied, that in carrion beetles, much more numerous and diverse, has not attracted same degree of effort. Though there has been research into the systematics and phylogenetic relationships of Coleoptera, early phylogenetic constructions were largely based upon Linnaeus' arrangement of beetle genera (Beutel et al. 2009), while more recent, and intricate morphological characteristics (wing venation and folding patterns and male genitalia in adults), and immature life stages, were used. Since the early 1990s, molecular identification of beetles has become more prevalent and influential due to the advancements in the technologies available.

One of the largest coleopteran lineages with around 60,000 currently described species (Beutel and Leschen 2005) are the Staphyliniformia, which consist of the Silphidae, Staphylinidae, and Histeridae beetle families. Coleopteran phylogenies have been well analyzed with attempts made to reconstruct accurate clades of related species from a complex arrangement of both morphological and molecular characters. Previous analyses of the Staphyliniformia by Lawrence and Newton (1982) and Hansen (1997) have been supported by the observation of this clade being monophyletic (Beutel and Leschen 2005). Constructing the coleopteran phylogenies based upon morphological characters is often an arduous, time-consuming process due to the vast number and diversity of features present or absent for both the immature and adult life stages throughout the order; thus utilization of molecular is becoming more common.

There are several coleopteran genera that are associated with decomposition and resulting applications in forensic entomology, with the most important belonging to Silphidae and Staphylinidae. While there are several cladistic studies of the Coleoptera, many focus on the order as a whole, covering multiple families of the infraorder Staphyliniformia which include mostly non-necrophageous species

(Archangelsky 1998; Caterino and Vogler 2002; Beutel et al. 2003; Beutel and Leschen 2005; Caterino et al. 2005). Other assemblies show the subtribe Staphylinina (which includes the forensically important Staphylinidae species, e.g., *Platydracus* spp.) to be monophyletic (Solodovnikov and Schomann 2009; Chatzimanolis et al. 2010), while the *Platydracus* spp. are polyphyletic and share a network branch with *Ontholestes murinus* and *Naddia* spp. (Chatzimanolis et al. 2010). One study by Palestrini et al. (1996) used larval characters, particularly those associated with the head (clypeus and components of and around the mouthparts; labrum, epipharynx, mandibles, and maxillae), as well as adult features within the Nicrophinae.

Recently, there has been a gradual shift in beetle phylogenetic classification toward molecular identification and techniques. Much of this movement has occurred during the last decade, not as a replacement for morphological taxonomy and systematics, but to facilitate identification of closely related species and to understand population structures. The need to accurately identify each individual specimen is paramount in forensic entomology, and the immature life stages are typically difficult to identify. Many beetle genera have been examined for species identification, the majority of them however belonging to herbivorous lineages and/or associated with pest control. Molecular identification of carrion beetles is still rare, but when examined both mitochondrial (Dobler and Muller 2000; Papadopoulou et al. 2009; Ito et al. 2010; Ikeda and Sota 2011; Zhuang et al. 2011; Ikeda et al. 2012; Tang et al. 2012) and nuclear genes (Kozol et al. 1994; King and Cummings 1997; Szalanski et al. 2000; Ikeda et al. 2012) have been investigated. The mtDNA COI, COII, and cytochrome b genes have frequently been used in species identification studies and in phylogeographic and population genetic species comparisons (Papadopoulou et al. 2009), due to their power of differentiating between species. Additionally the 12S, 16S, and 28S rRNA genes, Wingless (*Wg*), Topoisomerase I, Phosphenolpyruvate Carboxykinase (*PepCK*), and the *ITS1* gene have also been utilized (Szalanski et al. 2000; Chatzimanolis et al. 2010; Ikeda et al. 2012; Tang et al. 2012). However, difficulties are still encountered when discriminating between closely related species. Sufficient resolving power can in fact come from fairly short fragments: a <300 bp of the 16S gene is enough to differentiate between three Silphidae species of China (Tang et al. 2012).

The barcoding COI gene, which has been extensively used in insect identification studies, is able to differentiate between many families, species and in some cases individual specimens. Monophyly was observed for Staphylinidae and Silphidae species of China, and the Silphid subfamilies Nicrophinae and Silphinae were also monophyletic. *Creophilus maxillosus*, with moderate intraspecific variation (1.5%), formed two clades, in which individuals from a northeastern Chinese province (Heilongjing) were distinct in haplotypic differentiation (Cai et al. 2011). The Silphidae species *Thanatophilus sinuatus* from Chifeng, Inner Mongolia, has the potential to exhibit the population structure, as two clades were presented from this population, although very few specimens were examined (Cai et al. 2011). This study demonstrated the regional population differentiation of carrion beetles, and could be useful to infer postmortem movement. A more comprehensive study by Sikes et al. (2008) observed clear clustering of each species belonging to the morphologically similar *N. investigator* group when considering specimens from Sichuan province, China, Siberia, Russia, Hokkaido island, Japan, Honshu island, Japan, and Montana, USA. The hypothesis that *N. encaustus* from the Himalayas is derived from *N. investigator* is supported with its position surrounded by *N. investigator* populations from China and Siberia, while the hypothesis that *N. confusus* and *N. sepultor* are different species is not supported due to a very small genetic divergence between them in conjunction with a limited number of diagnostic characters to differentiate them (Sikes et al. 2008).

17.5 Conclusions and Implications

To date, systematics and population genetics of carrion arthropods have focused on nonforensic species, due to their economic value especially in cases of pest control (i.e., *Cochliomyia hominivorax*). However, when forensically important insects are studied, this is usually done with limitations on sample quantity and variation, geographic range, and genes. Additionally, the genes that have been analyzed are not always consistently covered, in that variable length sequence reads are used depending on the scale and objective(s) of the study. These shortcomings can usually be addressed with studies that focus on a family or group of closely related species that comprehensively identify and analyze the population structure and relationships. The geographic barriers that separate many populations from one another will continue to drive divergence and speciation. With the groundwork that has been done thus far, there is a promising outlook for this objective.

Because we are able to identify different individuals or species that possess a special phenotype, it is possible to delve deeper into the genetics of the organisms to find out the corresponding genetic basis. For example, are various *Lucilia* species more adapted to warmer temperatures (making them preadapted to thermal conditions they would experience in myiasis)? Have their odor reception genes changed to reflect the differences between odors on remains and odors on wounds? When pest control measures are employed, are these insects now resistant to some insecticides (and is the resistance gene linked to other adaptive genotypes)? Many of these questions remain to be answered, which can be done using a population genetic and molecular systematic approach.

REFERENCES

Adair, T.W. and B.C. Kondratieff. 2006. Three species of insects collected from an adult human corpse above 3300 m in elevation: A review of a case from Colorado. *Journal of Forensic Sciences* 51: 1164–1165.

Archangelsky, M. 1998. Phylogeny of Hydrophiloidea (Coleoptera: Staphyliniformia) using characters from adult and preimaginal stages. *Systematic Entomology* 23: 9–24.

Backlund, D.C., G.M. Marrone, C.K. Willimms, and K. Tilmon. 2008. Population estimate of the endangered American burying beetle, *Nicrophorus Americanus* Olivier (Coleoptera: Silphidae) in South Dakota. *Coleopterists Bulletin* 62: 9–15.

Bedick, J.C., B.C. Ratcliffe, W.W. Hoback, and L.G. Higley. 1999. Distribution, ecology, and population dynamics of the American burying beetle [*Nicrophorus Americanus* Olivier (Coleoptera: Silphidae)] in South-Central Nebraska, USA. *Journal of Insect Conservation* 3: 171–181.

Bernays, E. and M. Graham. 1988. On the evolution of host specificity in phytophagous arthropods. *Ecology* 69: 886–892.

Beutel, R.G., E. Anton, and M.A. Jäch. 2003. On the evolution of adult head structures and the phylogeny of hydraenidae (Coleoptera: Staphyliniformia). *Journal of Zoological Systematics and Evolutionary Research* 41: 256–275.

Beutel, R.G., F. Friedrich, and R.A. Leschen. 2009. Charles Darwin, beetles and phylogenetics. *Naturwissenschaften* 96: 1293–1312.

Beutel, R.G. and F. Haas. 2000. Phylogenetic relationships of the suborders of Coleoptera (Insecta). *Cladistics* 16: 103–141.

Beutel, R.G. and R.a.B. Leschen. 2005. Phylogenetic analysis of Staphyliniformia (Coleoptera) based on characters of larvae and adults. *Systematic Entomology* 30: 510–548.

Black, W.C. and E.S. Krafsur. 1986. Seasonal breeding structure in house fly *Musca Domestica* L., populations. *Heredity* 56: 289–298.

Bolek, M.G. and J.R. Coggins. 2002. Observations on myiasis by the calliphorid, *Bufolucilia silvarum*, in the eastern American toad (*Bufo americanus americanus*) from Southeastern Wisconsin. *Journal of Wildlife Diseases* 38: 598–603.

Byrd, J.H. and J.L. Castner. 2009. *Forensic Entomology: The Utility of Arthropods in Legal Investigations*. Boca Raton, FL: CRC Press.

Cai, J.F. et al. 2011. Identification of forensically significant beetles (Coleoptera: Staphylinidae) based on COI gene in China. *Romanian Journal of Legal Medicine* 19: 211–218.

Caterino, M.S. and S. Chatzimanolis. 2008. Conservation genetics of three flightless beetle species in southern California. *Conservation Genetics* 10: 203–216.

Caterino, M.S. T. Hunt, and A.P. Vogler. 2005. On the constitution and phylogeny of Staphyliniformia (Insecta: Coleoptera). *Molecular Phylogenetics and Evolution* 34: 655–672.

Caterino, M.S. and A.P. Vogler. 2002. The phylogeny of the Histeroidea (Coleoptera: Staphyliniformia). *Cladistics* 18: 394–415.

Chatzimanolis, S., I.M. Cohen, A. Schomann, and A. Solodovnikov. 2010. Molecular phylogeny of the mega-diverse rove beetle tribe Staphylinini (Insecta: Coleoptera: Staphylinidae). *Zoologica Scripta* 39: 436–449.

Cherix, D., C. Wyss, and T. Pape. 2012. Occurrences of flesh flies (Diptera: Sarcophagidae) on human cadavers in Switzerland, and their importance as forensic indicators. *Forensic Science International* 220: 158–163.

David, J.R. and P. Capy. 1988. Genetic variation of *Drosophila melanogaster* natural populations. *Trends in Genetics* 4: 106–111.

Dickey, A. 2006. Population genetics of *Phanaeus vindex* and *P. difformis* and congruence with morphology across a geographic species overlap. MS Thesis. University of Texas, Arlington.

Desmyter, S. and M. Gosselin. 2009. COI Sequence variability between Chrysomyinae of forensic interest. *Forensic Science International* 3: 89–95.

Dobler, S. and J.K. Muller. 2000. Resolving phylogeny at the family level by mitochondrial cytochrome oxidase sequences: Phylogeny of carrion beetles (Coleoptera: Silphidae). *Molecular Phylogenetics and Evolution* 15: 390–402.

Falush, D., M. Stephens, and J.K. Pritchard. 2003. Inference of population structure using multilocus genotype data: Linked loci and correlated allele frequencies. *Genetics* 164: 1567–1587.

Foster, G.G. et al. 1980. Cytogenetic studies of *Lucilia cuprina dorsalis* R.-D. (Diptera: Calliphoridae): Polytene chromosome maps of the autosomes and cytogenetic localization of visible genetic markers. *Chromosoma (Berlin)* 81: 151–168.

Fremdt, H., K. Szpila, J. Huijbreqts, A. Lindstrom, R. Zehner, and J. Amendt. 2012. *Lucilia silvarum* Meigen, 1926 (Diptera: Calliphoridae)—A new species of interest for forensic entomology in Europe. *Forensic Science International* 222: 335–339.

Gallagher, M.B., S. Sandhu, R. Kimsey. 2010. Variation in development time for geographically distinct populations of the common green bottle fly, *Lucilia sericata* (Meigen). *Journal of Forensic Sciences* 55: 438–442.

Haldane, J.B.S. 1956. The theory of selection for melanism in Lepidoptera. *Proceedings of the Royal Society B: Biological Sciences* 145: 303–306.

Hall, M.J., W. Edge, J.M. Testa, Z.J.O. Adams, and P.D. Ready. 2001. Old world screwworm, *Chrysomya bezziana*, occurs as two geographical races. *Medical and Veterinary Entomology* 15: 393–402.

Hall, M.J., A.H. Wardhana, G. Shahhosseini, Z.J.O. Adams, and P.D. Ready. 2009a. Genetic diversity of populations of old world screwworm fly, *Chrysomya bezziana*, causing traumatic myiasis of livestock in the Gulf Region and implications for control by sterile insect technique. *Medical and Veterinary Entomology* 23: 51–58.

Hall, M.J.R. et al. 2009b. Molecular genetic analysis of populations of Wohlfahrt's wound myiasis fly, *Wohlfahrtia magnifica*, in outbreak populations from Greece and Morocco. *Medical and Veterinary Entomology* 23: 72–79.

Hamblin, M.T. and M. Veuille. 1999. Population structure among African and derived populations of *Drosophila simulans*: Evidence for ancient subdivision and recent admixture. *Genetics* 153: 305–317.

Hamrick, J.L. and M.J.W. Godt. 1990. Allozyme diversity in plant species. In: A.L. Kahler (ed.), *Plant Population Genetics, Breeding, and Genetic Resources*, pp. 43–63. Sunderland, Massachusetts: Sinauer Associates Inc.

Hansen, M. 1997. Phylogeny and classification of the staphyliniform beetle families (Coleoptera). *Biologiske Skrifter, Det Kongelike Danske Videnskabernes Selskab* 48: 1–339.

Hanski, I. 1998. Metapopulation dynamics. *Nature* 396: 41–49.

Hartl, D.L. and A.G. Clark. 1997. *Principles of Population Genetics*. Sunderland, MA: Sinauer Associates, Inc.

Harvey, M.L., S. Gaudieri, M.H. Villet, and I.R. Dadour. 2008. A global study of forensically significant Calliphorids: Implications for identification. *Forensic Science International* 177: 66–76.

Hoback, W.W., A.A. Bishop, J. Kroemer, J. Scalzitti, and J.J. Shaffer. 2004. Differences among antimicrobal properties of carrion beetle secretions reflect phylogeny and ecology. *Journal of Chemical Ecology* 30: 719–729.

Hubby, J.L. and R.C. Lewontin. 1966. A molecular approach to the study of genic heterozygosity in natural populations. I. The number of alleles at different loci in *Drosophila pseudoobscura*. *Genetics* 54: 577–594.

Ikeda, H., K. Kubota, Y.B. Cho, H.B. Liang, and T. Sota. 2009. Different phylogeographic patterns in two Japanese *Silpha* species (Coleoptera: Silphidae) affected by climatic gradients and topography. *Biological Journal of the Linnean Society* 98: 452–467.

Ikeda, H., M. Nishikawa, and T. Sota. 2012. Loss of flight promotes beetle diversification. *Nature Communications* 3: 648.

Ikeda, H. and T. Sota. 2011. Macroscale evolutionary patterns of flight muscle dimorphism in the carrion beetle *Necrophila japonica*. *Ecology and Evolution* 1: 97–105.

Ito, N., T. Katoh, N. Kobayashi, and H. Katakura. 2010. Effects of straits as dispersal barriers for the flightless roving carrion beetle, *Silpha perforata* (Coleoptera: Silphidae: Silphinae). *Zoological Science* 27: 313–319.

Jordaens, K. et al. 2013. DNA barcoding and the differentiation between North American and West European *Phormia regina* (Diptera, Calliphoridae, Chrysomyinae). *ZooKeys* 365: 149–174.

Kettlewell, B. 1973. *The Evolution of Melanism. The Study of a Recurring Necessity; with Special Reference to Industrial Melanism in the Lepidoptera.* Oxford: Clarendon Press.

Kettlewell, H.B.D. 1955. Selection experiments on industrial melanism in the Lepidoptera. *Heredity* 9: 323–342.

Kimura, M. and G.H. Weiss. 1964. The stepping stone model of population structure and the decrease of genetic correlation with distance. *Genetics* 49: 561–576.

King, L.M. and M.P. Cummings. 1997. Satellite DNA repeat sequence variation is low in three species of burying beetles in the genus *Nicrophorus* (Coleoptera: Silphidae). *Molecular Biology and Evolution* 14: 1088–1095.

Kozol, A.J., J.F.A. Traniello, and S.M. Williams. 1994. Genetic variation in the endangered burying beetle *Nicrophorus americanus* (Coleoptera: Silphidae). *Annals of the Entomological Society of America* 87: 928–935.

Krafsur, E.S. 1985. Age composition and seasonal phenology of housefly (Diptera: Muscidae) populations. *Journal of Medical Entomology* 22: 515–523.

Krafsur, E.S. and C.J. Whitten. 1993. Breeding structure of screwworm fly populations (Diptera: Calliphoridae) in Colima, Mexico. *Journal of Medical Entomology* 30: 477–480.

Lawrence, J.F. and A.F. Newton. 1982. Evolution and classification of beetles. *Annual Review of Ecology and Systematics* 13: 261–290.

Lints, F.A. and G. Gruwez. 1972. What determines the duration of development in *Drosophila melanogaster. Mechanisms of Ageing and Development* 1: 285–297.

Lyra, M.L., L.B. Klaczko, and A.M. Azeredo Espin. 2009. Complex patterns of genetic variability in populations of the new world screwworm fly revealed by mitochondrial DNA markers. *Medical and Veterinary Entomology* 23: 32–42.

Macarthur, R.H. and E.O. Wilson. 1967. *The Theory of Island Biogeography.* Princeton, N.J.: Princeton University Press.

Malmendal, A. et al. 2013. Metabolomic analysis of the selection response of *Drosophila melanogaster* to environmental stress: Are there links to gene expression and phenotypic traits? *Naturwissenschaften* 100: 417–427.

Manel, S., M.K. Schwartz, G. Luikart, and P. Taberlet. 2003. Landscape genetics: Combining landscape ecology and population genetics. *Trends in Ecology and Evolution* 18: 189–197.

Marquez, J.G., M.A. Cummings, and E.S. Krafsur. 2007. Phylogeography of stable fly (Diptera: Muscidae) estimated by diveristy at ribosomal 16S and cytochrome oxidase I mitochondrial genes. *Journal of Medical Entomology* 44: 998–1008.

Meiklejohn, K.A., J.F. Wallman, S.L. Cameron, and M. Dowton. 2012. Comprehensive evaluation of DNA barcoding for the molecular species identification of forensically important Australian sarcophagidae (Diptera). *Invertebrate Systematics* 26: 515.

Meiklejohn, K.A., J.F. Wallman, and M. Dowton. 2011. DNA-Based Identification of forensically important Australian Sarcophagidae (Diptera). *International Journal of Legal Medicine* 125: 27–32.

Mullis, K.B. and F.A. Faloona. 1987. Specific synthesis of DNA *in vitro* via a polymerase-catalyzed chain reaction. *Methods in Enzymology* 155: 335–350.

Norris, K.B. 1965. The bionomics of blow flies. *Annual Reviews of Entomology* 10: 47–68.

Norris, K.R. 1990. Evidence for the multiple exotic origin of Australian populations of the sheep blowfly, *Lucilia cuprina* (Wiedemann) (Diptera: Calliphoridae). *Australian Journal of Zoology* 38: 635–648.

Oliveira, T.C. and S.D. Vasconcelos. 2010. Insects (Diptera) associated with cadavers at the Institute of Legal Medicine in Pernambuco, Brazil: Implications for forensic entomology. *Forensic Science International* 198: 97–102.

Otranto, D. and J.R. Stevens. 2002. Molecular approaches to the study of myiasis-causing larvae. *International Journal of Parasitology* 32: 1345–1360.

Palestrini, C., E. Barbero, M. Luzzatto, and M. Zucchelli. 1996. *Nicrophorus mexicanus* (Coleoptera; Silphidae: Nicrophorinae) larval morphology and phylogenetic considerations on the *N. investigator* group. Acta *Societatis Zoologicae Bohemicae* 60: 435–445.

Papadopoulou, A., A.G. Jones, P.M. Hammond, and A.P. Vogler. 2009. DNA taxonomy and phylogeography of beetles of the Falkland Islands (Islas Malvinas). *Molecular Phylogenetics and Evolution* 53: 935–947.

Pape, T. 1996. *Catalogue of the Sarcophagidae of the World (Insecta:Diptera)*. Gainsville, FL.: Associated Publishers.

Picard, C.J. and J.D. Wells. 2009. Survey of the genetic diversity of *Phormia regina* (Diptera: Calliphoridae) using amplified fragment length polymorphisms. *Journal of Medical Entomology* 46: 664–670.

Picard, C.J. and J.D. Wells. 2010. The population genetic structure of North American *Lucilia sericata* (Diptera; Calliphoridae), and the utility of genetic assignment methods for reconstruction of postmortem corpse relocation. *Forensic Science International* 195: 63–67.

Price, D.L. and M.L. May. 2009. Behavioral ecology of Phanaeus dung beetles (Coleoptera: Scarabaeidae): Review and new observations. *Acta. Zoologica Mexicana* 25: 211–238.

Prinkkila, M.-L. and I. Hanski. 1995. Complex competitive interactions in four species of *Lucilia* blowflies. *Ecological Entomology* 20: 261–272.

Ratcliffe, S.T., D.W. Webb, R.A. Weinzievr, and H.M. Robertson. 2003. PCR-RFLP identification of diptera (Calliphoridae, Muscidae and Sarcophagidae)—a generally applicable method. *Journal of Forensic Sciences* 48: 1–3.

Rauter, C.M. and R.L. Rust. 2012. Effect of population density on timing of oviposition and brood size reduction in the burying beetle *Nicrophorus pustulatus* Herschel (Coleoptera: Silphidae). *Psyche: A Journal of Entomology* 2012: 1–7.

Roderick, G.K. 1996. Geographic structure of insect populations: Gene flow, phylogeography, and their uses. *Annual Review of Entomology* 41: 325–352.

Selander, R.K. et al. 1986. Methods of multilocus enzyme electrophoresis for bacterial population genetics and systematics. *Applied and Environmental Microbiology* 51: 873–884.

Sikes, D.S., R.B. Madge, and S.T. Trumbo. 2006. Revision of *Nicrophorus* in part: New species and inferred phylogeny of the Nepalensis-Group based on evidence from morphology and mitochondrial DNA (Coleoptera: Silphidae: Nicrophorinae). *Invertebrate Systematics* 20: 305–365.

Sikes, D.S., S.M. Vamosi, S.T. Trumbo, M. Ricketts, and C. Venables. 2008. Molecular systematics and biogeography of *Nicrophorus* in part—the investigator species group (Coleoptera: Silphidae) using mixture model MCMC. *Molecular Phylogenetics and Evolution* 48: 646–666.

Singh, B., H. Kurahashi, and J.D. Wells. 2010. Molecular phylogeny of the blowfly genus *Chrysomya*. *Medical and Veterinary Entomology* 25: 126–134.

Singh, B. and J.D. Wells. 2011. Chrysomyinae (Diptera: Calliphoridae) is monophyletic: A molecular systematic analysis. *Systematic Entomology* 36: 415–420.

Smith, R.J. et al. 2000. Altitudinal variation in body size and population density of *Nicrophorus investigator* (Coleoptera: Silphidae). *Environmental Entomology* 29: 290–298.

Solodovnikov, A. and A. Schomann. 2009. Revised systematics and biogeography of "Quediina" of sub-Saharan Africa: New phylogenetic insights into the rove beetle tribe staphylinini (Coleoptera: Staphylinidae). *Systematic Entomology* 34: 443–466.

Sotiraki, S., R. Farkas, and M.J. Hall. 2010. Fleshflies in the flesh: Epidemiology, population genetics and control of outbreaks of traumatic myiasis in the mediterranean basin. *Veterinary Parasitology* 174: 12–18.

Stevens, J.R. 2003. The evolution of myiasis in blowflies (Calliphoridae). *International Journal of Parasitology* 33: 1105–1113.

Stevens, J. and R. Wall. 1995. The use of Random Amplified Polymorphic DNA (RAPD) analysis for studies of genetic variation in populations of the blowfly *Lucilia sericata* (Diptera: Calliphoridae) in Southern England. *Bulletin of Entomological Research* 85: 549–555.

Stevens, J. and R. Wall. 1996. Species, sub-species and hybrid populations of the blowflies *Lucilia cuprina* and *Lucilia sericata* (Diptera: Calliphoridae). *Proceedings of the Royal Society B* 263: 1335–1341.

Stevens, J. and R. Wall. 1997a. The evolution of ectoparasitism in the genus Lucilia (Diptera: Calliphoridae). *International Journal of Parasitology* 27: 51–59.

Stevens, J. and R. Wall. 1997b. Genetic variation in populations of the blowflies *Lucilia cuprina* and *Lucilia sericata* (Diptera:Calliphoridae). Random amplified polymorphic DNA analysis and mitochondrial DNA sequences. *Biochemical Systematics and Ecology* 25: 81–977.

Stevens, J., R. Wall, and J.D. Wells. 2002. Paraphyly in Hawaiian hybrid blowfly populations and the evolutionary history of anthropophilic species. *Insect Molecular Biology* 11: 141–148.

Szalanski, A.L., D.S. Sikes, R. Bischof, and M. Fritz. 2000. Population genetics and phylogenetics of the endangered American burying beetle, *Nicrophorus americanus* (Coleoptera: Silphidae). *Annals of the Entomological Society of America* 93: 589–594.

Szalanski, A.L., D.B. Taylor, and R.D. Perterson. 1996. Population genetics and gene variation of stable fly populations (Diptera: Muscidae) in Nebraska. *Journal of Medical Entomology* 33: 413–420.

Tabor, K.L., R.D. Fell, and C.C. Brewster. 2005. Insect fauna visiting carrion in Southwest Virginia. *Forensic Science International* 150: 73–80.

Tang, Z.C. et al. 2012. Identification of the forensically important beetles *Nicrophorus japonicus*, *Ptomascopus plagiatus* and *Silpha carinata* (Coleoptera: Silphidae) based on 16s rRNA gene in China. *Tropical Biomedicine* 29: 493–498.

Tarone, A.M., C.J. Picard, C. Spiegelman, and D.R. Foran. 2011. Population and temperature effects on *Lucilia sericata* (Diptera: Calliphoridae) body size and minimum development time. *Journal of Medical Entomology* 48: 1062–1068.

Torres, T.T. and M.L. Azeredo-Espin. 2005. Development of new polymorphic microsatellite markers for the new world screw-worm *Cochliomyia hominivorax* (Diptera: Calliphoridae). *Molecular Ecology Notes* 5: 815–817.

Torres, T.T. and M.L. Azeredo-Espin. 2009. Population genetics of new world screwworm from the Caribbean: Insights from microsatellite data. *Medical and Veterinary Entomology* 23: 23–31.

Torres, T.T., R.P.V. Brondani, E. Garcia, and M.L. Azeredo-Espin. 2004. Isolation and characterization of microsatellite markers in the new world screw-worm *Cochliomyia hominivorax* (Diptera: Calliphoridae). *Molecular Ecology Notes* 4: 182–184.

Turner, T.L., A.D. Stewart, A.T. Fields, W.R. Rice, and A.M. Tarone. 2011. Population-based resequencing of experimentally evolved populations reveals the genetic basis of body size variation in *Drosophila melanogaster*. *PLoS Genetics* 7: e1001336.

Wallman, J. and M. Adams. 2001. The forensic application of allozyme electrophoresis to the identification of blowfly larvae (Diptera: Calliphoridae) in Southern Australia. *Journal of Forensic Sciences* 46: 681–684.

Wallman, J.F. and M. Adams. 1997. Molecular systematics of Australian carrion-breeding blowflies of the genus Calliphora (Diptera: Calliphoridae). *Australian Journal of Zoology* 45: 337–356.

Williams, R.E., R.D. Hall, A.B. Broce, and P.J. Scholl. 1985. *Livestock Entomology*. New York: John Wiley & Sons.

Wit, J., T.N. Kristensen, P. Sarup, J. Frydenberg, and V. Loeschcke. 2013. Laboratory selection for increased longevity in *Drosophila melanogaster* reduces field performance. *Experimental Gerontology* 48: 1189–1195.

Wright, S. 1951. The genetical structure of populations. *Annals of Eugenics* 15: 323–354.

Yadong, G. et al. 2010. Identification of forensically important sarcophagid flies (Diptera: Sarcophagidae) based on COI gene in China. *Romanian Journal of Legal Medicine* 18: 217–224.

Zhuang, Q. et al. 2011. Molecular identification of forensically significant beetles (Coleoptera) in China based on COI gene. *Revista Colombiana De Entomologia* 37: 95–102.

Zumpt, F. 1965. *Myiasis in Man and Anaimals in the Old World*. London: Butterworths.

FIGURE 1.1 Salmon carcasses provide a significant input of nutrients and energy into streams of the Pacific Northwest as heterotrophically derived organic matter. (Photo by M.E. Benbow.)

FIGURE 1.2 Carrion patches become resources where complex interactions occur between species, as in this photo where a praying mantis is captured eating an adult fly on carrion. (Photo with permission from C.S. Ulrich and K. Black.)

FIGURE 1.3 Carrion is often quickly colonized by blow fly larvae that can take the form of huge masses that convert the carrion tissue into larval biomass, often with impressive efficiency. In this photo, an entire swine carcass was consumed by larvae that number well into the thousands: the larvae are beginning to disperse from the carcass location in order to find a place to pupate, and in this case they are even moving up a tree trunk. (Photo by M.E. Benbow.)

FIGURE 1.4 As part of the ecological process of carrion decomposition, many species of arthropods have evolved to efficiently use the resource using interesting adaptations such as dispersal away from larval conspecific competitors and predators in a dark place for pupation and the completion of their life cycle. (Photo by M.E. Benbow.)

(a)   (b)

FIGURE 1.5 Forensic entomologists can encounter different types of insects during an investigation. These insects can be (a) commonly encountered during investigations, as seen with the North American invasive *Chrysomya rufifacies* larva (Diptera: Calliphoridae), or (b) unexpected, such as these wood ants (*Camponotus* sp.) feeding on the open wound of a swine carcass. (Photos with permission by J.L. Pechal.)

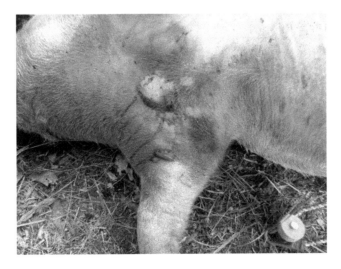

FIGURE 2.1 Swine decomposition showing the darkened colors of livor mortis beneath the skin. (Photo by S. Forbes.)

FIGURE 2.2 Swine decomposition showing autolytic degradation characterized by "marbling" in the torso as well as putrefactive degradation in the head. (Photo by S. Forbes.)

FIGURE 2.3 Swine decomposition showing partial skeletonization in the head and limbs and mummification of the remaining soft tissue. (Photo by S. Forbes.)

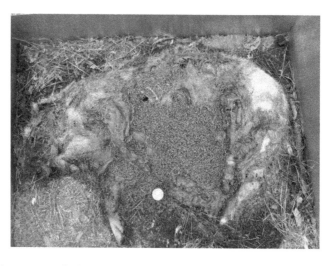

FIGURE 2.4 Larval masses contributing to rapid biomass loss after 6 days during swine decomposition. (Photo by S. Forbes.)

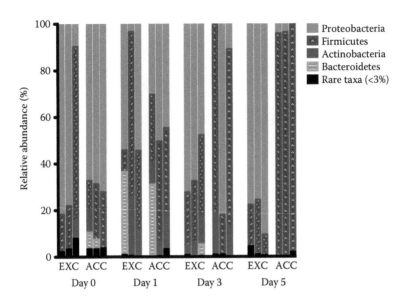

FIGURE 3.1 Successional changes in carrion bacterial communities over decomposition. Phylum-level taxonomic relative abundance of the microbial communities based on 454-pyrosequencing between carcasses where insect access was allowed (ACC) or excluded (EXC). Initial field placement (day 0) was followed by subsequent sample collections on days 1, 3, and 5. Rare taxa include any phyla with <3% of the total relative abundance. (Reprinted from Pechal, J.L. et al. 2013. *PLoS ONE* 8: e79035 with permission.)

FIGURE 4.1 Calliphoridae. Clockwise, from top right: adult *Calliphora vicina* Robineau-Desvoidy; adult *Cochliomyia macellaria* Fabricius; adult *Chrysomya rufifacies* (Macquart); adult *Lucilia coeruliviridis* Macquart; adult *Pollenia rudis* (Fabricius); and larva of Calliphoridae. (All photos by Steve Marshall.)

FIGURE 4.2   Sarcophagidae and Muscidae. Clockwise, from top right: adult *Sarcophaga rohdendorfi* Salem; adult and larvae of *Ravinia* spp.; larvae of *Musca domestica* L.; adult female *M. domestica* L.; and adult *Hydrotaea aenescens* (Wiedemann). (All photos by Steve Marshall.)

FIGURE 4.3   Fanniidae, Stratiomyidae, Sepsidae, Scathophagidae. Clockwise, from top right: larvae of *Fannia canicularis* L.; adult *F. canicularis* L.; adult *Hermetia illucens* L.; adult *Themira annulipes* (Meigen); and adult male and female *Scathophaga stercoraria* (L.). (Photo of *S. stercoraria* by Robert Armstrong; all other photos by Steve Marshall.)

FIGURE 4.4   Phoridae, Piophilidae, Heleomyzidae, Sphaeroceridae. Clockwise, from top right: larvae of Phoridae; adult *Megaselia* spp.; adult *Piophila casei* (L.); adult male and female *Neoleria* spp.; and adult *Lotophila atra* (Meigen). (Photos of phorid larvae by Richard Merritt; all other photos by Steve Marshall.)

FIGURE 4.5 Silphidae. Clockwise, from top right: larva of *Nicrophorus* spp. on pink salmon carcass; larva of *Nicrophorus* spp. on pink salmon carcass; adult *Nicrophorus tomentosus* Weber; and adult *N. investigator* Zetterstedt. (Photos by Robert Armstrong; except for the *Nicrophorus tomentosus* which was by Steve Marshall.)

FIGURE 4.6 Histeridae, Staphylinidae, Dermestidae, Scarabaeidae, Carabidae; Clockwise, from top left: adult of Histeridae (Photo by Richard W. Merritt.); adult *Creophilus maxillosus* (L.) feeding on Diptera larvae (Photo by Richard W. Merritt.); adult Staphylinidae feeding on Diptera eggs (Photo by Robert Armstrong.); adult *Dermestes lardarius* L. (Photo by Steve Marshall.); adult *Aphodius* spp. (Photo by Richard W. Merritt.); and adult *Pterostichus* spp. with mites (Photo by Robert Armstrong.)

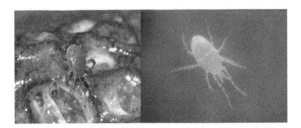

FIGURE 4.7 Acari. Left to right: deutonymph of Parasitidae on pink salmon carcass (Photo by Robert Armstrong.) and deutonymph of *Poecilochirus austrooasiaticus* Vitzthum on human corpse. (Photo by Alejandro Medino.)

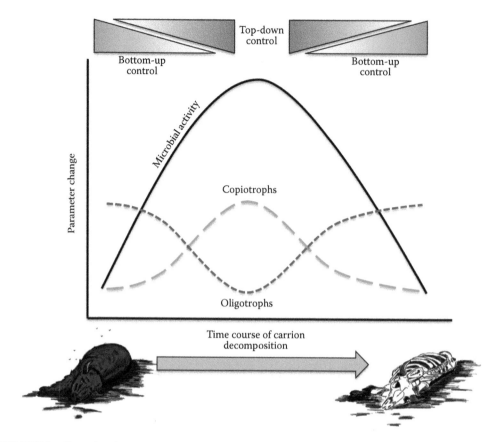

FIGURE 5.3 Change in soil community parameters during the time course of carrion decomposition. Microbial biomass and activity are likely to increase until active decay at which point it will begin to decrease. During this time, the typically oligotrophic soil community will give way to a more copitrophic one until carrion decomposition begins to slow. Finally, regulation of the microbial community will also change during carrion decomposition, with bottom-up regulation being more important during the early and later stages and top-down regulation being more important during active carrion decay.

FIGURE 6.1 Although few vertebrates are considered obligate scavengers, most species appear to utilize carrion resources facultatively. Results of experimental scavenging trials showing coyote, *Canis latrans* (top left) and black vulture, *Coragyps atratus* (top right) scavenging of a feral pig—*Sus scrofa*—carcass, scavenging of a cane toad—*Bubo marinus*—carcass in Hawaii by an invasive small Asian mongoose—*Herpestes javanicus* (bottom left), and scavenging of a rat carcass by a gray fox—*Urocyon cinereoargenteus* (bottom right).

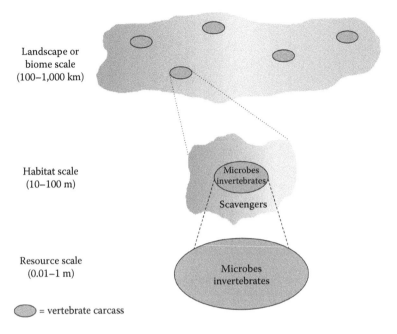

FIGURE 8.1 The scale at which an ecologist studies carrion decomposition is key to elucidating ecological interactions among the necrobiome or organisms using carrion such as microbes, invertebrates, and vertebrate scavengers. At the landscape or biome scale, there can be many resources (or carrion patches) available for consumption by members of the necrobiome. At the habitat scale, the focus is on a single resource within the landscape and its immediate (<100 m) surrounding habitat; there are fewer vertebrate scavengers present at this scale and thus there is a shift in focus towards the microbes and invertebrates members of the necrobiome. Finally, at the resource scale or a single vertebrate carcass, the emphasis is only on the interactions occurring between the microbes and invertebrates throughout decomposition.

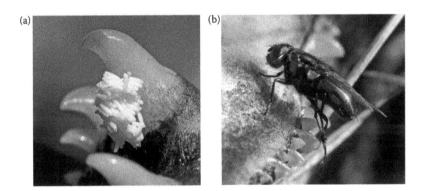

FIGURE 8.2 Blow flies (Calliphoridae) are often the first insects to colonize carrion: (a) Blow fly egg mass on the tooth of a salmon carcass and (b) An adult Callihporidae (*Calliphora* sp.). (Photo with permission from Bob Armstrong.)

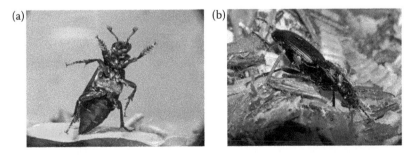

FIGURE 8.3 Coleoptera photographed with salmon carrion in southeast Alaska, carrying phoretic mites: (a) Carabidae and (b) Silphidae (*Nicrophorus* sp.). (Photo with permission from Bob Armstrong.)

FIGURE 8.4 There is a diverse community of vertebrate scavengers that can include vultures and hyenas, among many other species. (Photo with permission from Richard Merritt.)

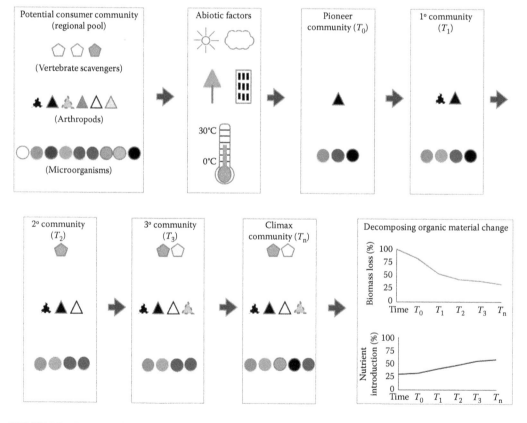

FIGURE 8.5 The succession of necrobiome members on decomposing organic material such as carrion is determined by abiotic (e.g., habitat type and temperature) and biotic interactions (e.g., competition, species sorting, and landscape patch dynamics). Specifically, the potential community of consumers including microorganisms, arthropods, and vertebrate scavengers will go through many community assemblages from the time a resource is made available (T_0) until climax community (T_n) is reached. The resulting rate of carrion resource biomass loss and nutrient introduce into the environment will depend on the consumer community assembly changes throughout time.

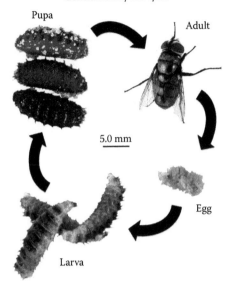

Generalized fly life cycle

Pupa

Adult

5.0 mm

Egg

Larva

FIGURE 10.2 Life cycle of a fly.

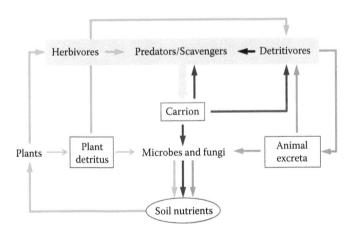

FIGURE 13.1 Simplified pathways of energy and nutrient flow in a carrion-centered detrital food web. The grey box represents all living animals and their potential contribution to the carrion pool in an ecosystem. Brown arrows depict the flow of energy and material from carrion resources through vertebrate scavengers and invertebrate detritivores and the channeling of carrion nutrients through microbes and fungi into the soil nutrient pool. Blue arrows show the transfer of energy from live animals to excreta into detritivores and microbes. The green arrows depict the flow of nutrients via plants, starting with the uptake of nutrients from soil and their flow through herbivores and their predators. Most plant biomass is not consumed by animals and gives the largest contribution to the detrital resource pool, which is also channeled through invertebrate detritivores and microbes. (Reproduced from Barton, P.S. et al. 2013a. *Oecologia* 171: 761–772. With permission.)

FIGURE 13.2 An example of how ecosystem context mediates localized responses of plant communities 1 year after addition of kangaroo carcasses. (a) Very little recolonization has occurred in grassland dominated by nutrient-sensitive species, whereas (b) recolonization has progressed significantly in grassland a few kilometers away but dominated by exotic nutrient-tolerant species after the same period of time. (Reproduced from Barton, P.S. et al. 2013b. *PLoS ONE* 8: e53961. With permission.)

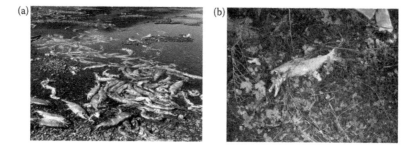

FIGURE 13.3 The annual spawning run of salmon (*Oncorhynchus* spp.) from the ocean into the upper reaches of inland rivers (a) and streams can deliver large quantities of marine-derived nutrients to riparian ecosystems (b). (Photo credits: M.E. Benbow.)

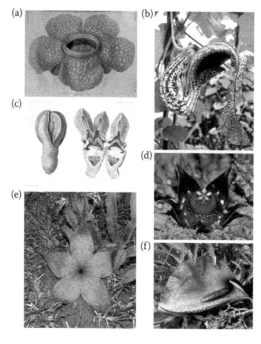

FIGURE 16.1 Examples of carrion-mimicking flowers/inflorescences. (a) *Rafflesia arnoldi* (Rafflesiaceae), reproduction of the original illustration by Franz Bauer used in the type description by Robert Brown (1822). (b) *Aristolochia cymbifera* (Aristolochiaceae; Photo: R. Roth). (c) *Hydnora africana* (Hydnoraceae), reproduction of an illustration by Ferdinand Bauer used by Robert Brown (1845) in his description of the structure of *H. africana*. (d) *Duvalia polita* (Apocynaceae; Photo: A. Shuttleworth). (e) *Stapelia gigantea* (Apocynaceae; Photo: A. Shuttleworth). (f) *Helicodiceros muscivorus* (Araceae; Photo: M. Stensmyr).

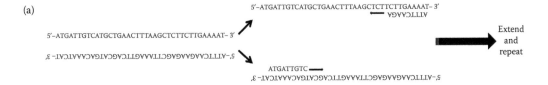

(a)

5′–ATGATTGTCATGCTGAACTTTAAGCTCTTCTTGAAAAT– 3′

5′–ATGATTGTCATGCTGAACTTTAAGCTCTTCTTGAAAAT– 3′

5′–ATTTCAAGAGAGGCTTAAAGTTCAGCGATGACAAATCAT–3′,

ATGATTGTC

5′–ATTTCAAGAGAGGCTTAAAGTTCAGCGATGACAAATCAT–3′

Extend
and
repeat

(b)

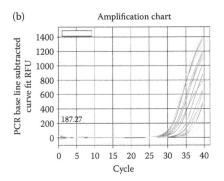

Amplification chart

PCR base line subtracted curve fit RFU

187.27

Cycle

FIGURE 18.1 PCR or polymerase chain reaction is a process of replicating double-stranded DNA. (a) Denaturing double-stranded DNA (1) by heating results in two separate but complementary strands (2). Binding of complementary primers to each of the two strands (3) and subsequent extension by a DNA polymerase enzyme generates two identical copies of the original molecule (4). The process is repeated (5) resulting in an exponential growth in target DNA concentration. (b) Graph showing an increase in DNA concentration per amplification cycle in a PCR reaction using a double-stranded DNA binding fluorescent dye. Colors represent individual PCR reactions with different starting concentrations of target DNA.

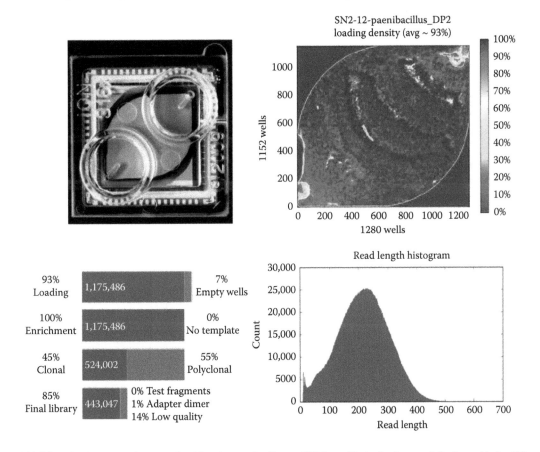

FIGURE 18.2 Representative example of data from an Ion Torrent PGM run. Clockwise from top left: disposable Ion 316 sequencing chip, chip loading density heat map, histogram of sequenced read lengths (200 bp chemistry), and run statistics.

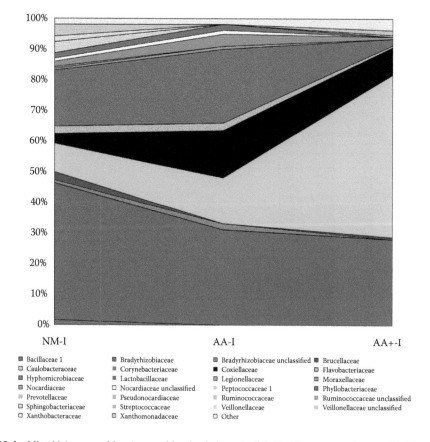

FIGURE 18.4 Microbial communities change with aging in lone star tick (*Amblyomma americanum*, L.). The number of bacterial families identified were on a NM-I, newly molted adult; AA-I, aged adult (60 days); AA+-I, aged adult (110 days); -I, respectively, raised indoors under climate-controlled conditions.

Legend for Figure 18.4:

- Bacillaceae 1
- Bradyrhizobiaceae
- Bradyrhizobiaceae unclassified
- Brucellaceae
- Caulobacteraceae
- Corynebacteriaceae
- Coxiellaceae
- Flavobacteriaceae
- Hyphomicrobiaceae
- Lactobacillaceae
- Legionellaceae
- Moraxellaceae
- Nocardiaceae
- Nocardiaceae unclassified
- Peptococcaceae 1
- Phyllobacteriaceae
- Prevotellaceae
- Pseudonocardiaceae
- Ruminococcaceae
- Ruminococcaceae unclassified
- Sphingobacteriaceae
- Streptococcaceae
- Veillonellaceae
- Veillonellaceae unclassified
- Xanthobacteraceae
- Xanthomonadaceae
- Other

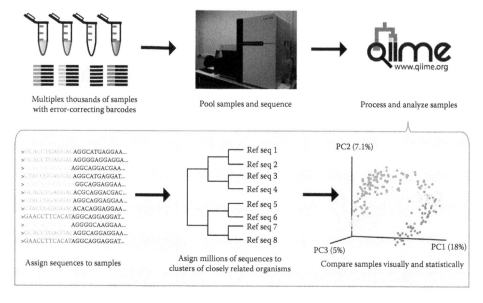

FIGURE 19.1 Overview of data generation, processing, and analysis workflow using QIIME (or other similar packages). After DNA is extracted, a target region (e.g., part of the 16S or 18S rRNA gene) is amplified using error-correcting, Golay barcoded primers (barcodes noted by sequences of different colors). Next, the multiplexed pool of amplicons is sequenced at a depth of millions of sequences. Finally, the open-access software package QIIME is used to demultiplex sequences, align sequences to the reference database (e.g., Greengenes or SILVA), and perform statistical analyses and generate visualizations (e.g., ANOSIM and PCoA plots). (From Hamady, M. et al. 2008. *Nature Methods* 5: 235–237. This figure is available at Figshare.com under the open-access Creative Commons-Attribution License (CC-BY).)

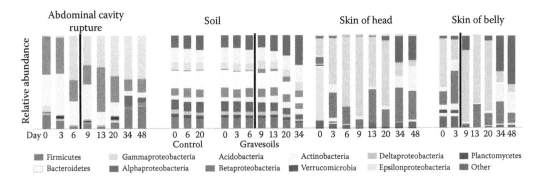

FIGURE 19.2 Relative abundance of phyla of bacteria over time for all body sites. Control soils were averaged across time points. For the abdominal site, day 0 includes cecum, fecal, and abdominal swab and liquid samples. For the soil site, control soils collected on days 0, 6, and 20 are shown on the left of the plot. (From Metcalf, J.L. et al. 2013. *eLife* 2: e01104 and is open access under the Creative Commons-Attribution License (CC-BY).)

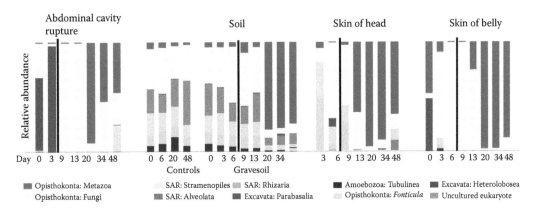

FIGURE 19.3 Relative abundance of microbial eukaryote taxa at the class level over time. Microbial eukaryotic community composition changes significantly and predictably over the course of decomposition. Eukaryotic community composition changes directionally and becomes dominated by the nematode *Oscheius tipulae*. (From Schoch, C.L. et al. 2012. *Proceedings of the National Academy of Sciences* 109: 6241–6246; Metcalf, J.L. et al. 2013. *eLife* 2: e01104 and is open access under the Creative Commons-Attribution License (CC-BY).)

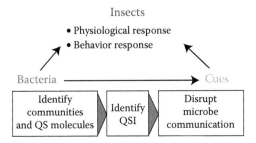

FIGURE 20.1 Conceptual diagram demonstrating the links between bacteria, their quorum sensing communication, and cues to insects that have been documented in the literature.

FIGURE 20.4 *E. coli* biofilm on a glass slide. Green (black as greyscale) fluorescent protein-tagged *E. coli* biofilm on a glass slide in a flow cell.

FIGURE 20.5 Consortium of two *E. coli* strains. Green (white as greyscale) *E. coli* cells were tagged with the green fluorescence protein and the red (dark grey as greyscale) *E. coli* cells were tagged with the red fluorescence protein. Cells were grown for 12 h in a microfluidic chamber and imaged via confocal microscopy and IMARIS software. (Process described more fully in Hong, S.H. et al. 2012. *Nature Communication* 3: 613.)

FIGURE 20.6 Carrion after several days of decomposition, when initial blow fly oviposition occurred within minutes of exposing the carcass. Note the mass of adult blow flies that have congregated between the hind legs, presumably at an area where the wall of the intestine was ruptured from larval feeding activity. What is cuing the adult flies to that area? (Photo by M.E. Benbow.)

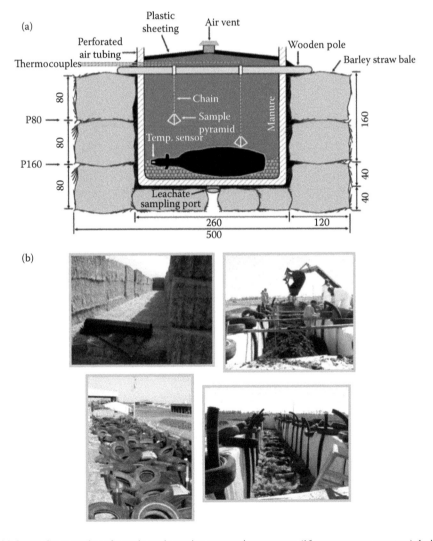

FIGURE 23.3 (a) Cross-section of a static cattle carrion composting structure (16 carcasses per structure) designed and equipped with biosecurity plastic sheeting, gas vents, leachate ports, and retrievable sampling cages. (Reprinted from Xu, W., T. Reuter, G.D. Inglis et al. 2009a. *Journal of Environmental Quality* 38: 437–450. With permission.) Units are in cm. (b) Pictures of construction of the biosecure composting structure.

18

Microbial Genetics and Systematics

Michael S. Allen and Michael G. LaMontagne

CONTENTS

Microbes are by far the most numerous organisms on earth, inhabiting every available niche. Bacteria alone number between 10^{29} and 10^{30} cells (Kallmeyer et al. 2012). Microbes facilitate nutrient turnover in all ecosystems, including playing an essential role in degradation of carrion. Which microbes are involved, whether they originate from the soil or the host, and what sorts of succession processes occur in microbial communities during the transition from death through decomposition are all active topics of research. Other chapters in this book address various aspects of these processes, including general ecological interaction of microbial communities (Chapter 3) and microbiome studies of carrion decomposition (Chapter 19). Here, we discuss the problems and considerations inherent in investigating microbial communities.

18.1 Introduction to Microbial Systematics

As their name implies, microbes are small in size. However, this distinction based on size is somewhat arbitrary, and the group includes representative of all three domains of life. In this chapter we will focus on the prokaryotes (bacteria and archaea), as they are the most numerically dominant and fastest growing organisms of the group, but the general guidelines and conclusions are also relevant to the microscopic forms of eukaryotes.

In addition to their small size, the overall physical appearance of the vast majority of prokaryotes allows them to be grouped into one of the three basic shapes: rod (bacillus), cocci (spherical), or spirillum (spiral-shaped). There remains little else morphologically to differentiate the astounding number of species. Early attempts at microbial taxonomy beyond this morphological limitation relied upon biochemical assays of isolates, and large numbers of tests were developed to classify a particular bacterium into a group based on how isolates performed on these tests. However, it turns out that microbes, especially bacteria, are very good at exchanging genes. In doing so, they can change their apparent phenotype and biochemical processes independent of direct lines of descent. While biochemical typing is still the predominant method for the identification of disease-causing organisms employed in clinical settings, its application to determining evolutionary relatedness is limited. For the latter, molecular techniques appear to be required.

18.1.1 Carl Woese, 16S, and the Third Domain

Any discussion on modern molecular-based microbial systematics must begin with the work of Carl Woese. In a landmark paper published in 1977, Woese and Fox first proposed the existence of a third domain of life: the archaebacteria (Woese and Fox 1977). The sole representatives of the group at that time were members of a poorly studied assemblage of single-celled organisms named the "methanogens" for their ability to anaerobically reduce CO_2 to methane. While fundamental biochemical differences between this group and other bacteria were known (e.g., the lack of a peptidoglycan-containing cell wall, unique methanogenic enzymes), the primary division into a wholly new domain of life was proposed based on the detected variations in the RNA component of the small ribosomal subunit (16S rRNA for prokaryotes, 18S for eukaryotes). Differences in the sequences among these microbes and all other previously described organisms indicated that they belonged to a completely new, and deeply branching, group of life. In most quarters, however, this paradigm-shifting three-pronged tree of life was met not with excitement but with scorn, derision, or disinterest (Morell 1997).

The general disregard with which these developments were initially greeted did not deter Woese and his students from continuing their line of research; though it is doubtful that he and his colleagues fully appreciated at the time the myriad of applications that would stem from their work. Of particular significance to the future studies in microbial phylogeny were the special traits of ribosomes. The ribosomes are suited to serve as a comparator in phylogenetic studies first and foremost due to their ubiquity within all self-replicating systems. Another key component of the ribosomes is ease of isolation. Depending on its growth phase, fully 90% of total RNA extracted from a cell can be ribosomal RNA (rRNA), which in turn is evenly distributed in archaea and bacteria among the 16S molecule of the small ribosomal subunit, and the 5S and 23S rRNA molecules of the large ribosomal subunit (Bremer and Dennis 1996). Thus, any preparation of total RNA is naturally enriched in 16S rRNA. By 1985, Lane et al. had used the large abundance of rRNA to their advantage by reverse transcribing and sequencing the molecules directly from total RNA extracts, thereby eliminating laborious isolation and cloning steps (Lane et al. 1985). Capitalizing on another key feature of 16S rRNA, its high evolutionarily conserved nature, the authors compared sequences to determine the relationships among bacterial isolates.

18.1.2 PCR

Notwithstanding the fundamental significance of the discovery of an entirely new domain of life, a second major development would be required to truly revolutionize microbial phylogenetics: the polymerase chain reaction (PCR). First developed in the early 1980s for sickle cell anemia testing by Kary Mullis and colleagues at Cetus Corporation (Saiki et al. 1985), the process was subsequently optimized and patented in 1987 and, in 1993, Mullis received the Nobel Prize in Chemistry for the development of PCR. Briefly, PCR involves the addition of two specific oligonucleotide primers that bind complementary strands flanking a section of DNA to be amplified, with the 3′ ends of the primer orientated toward the targeted region (Figure 18.1). Thermal denaturation of the DNA, followed by cooling in the presence of saturating concentrations of DNA primers allows each now-separated strand to bind its complementary primer. DNA-dependent DNA polymerase recognizes the free 3′ end of the primers, and in a mixture of nucleotide triphosphates and Mg^{2+}, generates a copy of each strand, resulting in a doubling of the abundance of a particular region of DNA. The use of a thermally stable polymerase such as *Taq* in the reaction facilitates repeating the process, which theoretically doubles the quantity of target DNA in each cycle.

(a)

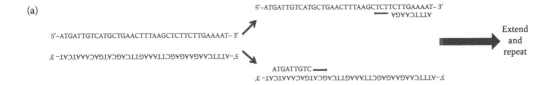

(b)

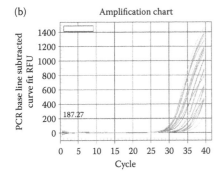

FIGURE 18.1 (See color insert.) PCR or polymerase chain reaction is a process of replicating double-stranded DNA. (a) Denaturing double-stranded DNA (1) by heating results in two separate but complementary strands (2). Binding of complementary primers to each of the two strands (3) and subsequent extension by a DNA polymerase enzyme generates two identical copies of the original molecule (4). The process is repeated (5) resulting in an exponential growth in target DNA concentration. (b) Graph showing an increase in DNA concentration per amplification cycle in a PCR reaction using a double-stranded DNA binding fluorescent dye. Colors represent individual PCR reactions with different starting concentrations of target DNA.

In short, a single specific copy of DNA can be amplified to over one million copies simply by repeating the process 20 times ($2^{20} = 1,048,576$).

18.2 Techniques and Applications

The two processes came together in a paper by Weisburg and colleagues in 1991 (Weisburg et al. 1991). In it, the authors described a group of phylogenetically specific primers along with what was later called "universal" primers and their application of these primers to the direct amplification of the 16S rRNA gene fragments from a variety of bacterial species in diverse samples without cultivation. These universal primers bind regions that, due to their contribution to the secondary structure of ribosomes, vary little. The ribosomal RNA secondary structure includes hairpins generated by complementary regions of single-stranded rRNA folding together. Regions involved in generation of these structures show little variation across sequences available in public databases. These conserved regions flank nine variable regions (V1–V9), the sequence of which allows for the classification of microbes. A dramatic expansion in applications ensued. The PCR-based 16S approach to microbial identification was rapidly incorporated into the study of difficult or impossible to culture pathogens (Anderson et al. 1991), identification of previously unknown groups of bacteria in unique environments (Liesack and Stackebrandt 1992), and the linkage of bacteria to environmental processes such as biodegradation (Herrick et al. 1993). We must recognize that the development of universal primers depends on the sequences available and so-called universal primers may not efficiently amplify fragments from unknown or poorly described taxa (Metcalf et al. 2013).

An alternative to sequencing, or at least an additional tool, was developed in parallel. It involved the hybridization of synthetic (usually fluorescently or radioactively labeled) oligonucleotide probes complementary to the 16S rRNA. Since cells contain thousands of ribosomes, targeting the rRNA improves the signal-to-noise ratio relative to detecting DNA. The application of fluorescently labeled 16S targeted probes has been used to both identify organisms by *in situ* hybridization, but more uniquely, to provide information on their spatial distribution relative to others in complex systems (Giovannoni et al. 1988; Amann et al. 1991). This probing technique was employed to identify a clade of bacteria first identified from cloned gene sequences from the Sargasso Sea and named SAR 11. The bacterium was later found in oceans all over the world and may represent the most ubiquitous and numerous living species on earth (Giovannoni et al. 1990). Interestingly, it would take an additional 12 years to learn to grow the bacterium in pure culture (Rappe et al. 2002), and a decade more to develop a defined medium for its cultivation (Paul et al. 2012), even with the complete genome in hand (Giovannoni et al. 2005).

The story of SAR 11, Candidatus *Pelagibacter ubique*, provides an interesting example of and insight into the "great plate count anomaly." The anomaly refers to early observations that the number of bacteria calculated by direct observation on a microscope was often 100X, the number that could be cultivated by standard plate count methods (Torsvik et al. 1996). The SAR11 case exemplifies the difficulty of isolating even those bacteria that appear to be numerically dominant from a specific environment, and the general limitations of traditional cultivation techniques. In the last decade high-throughput isolation strategies (Lagier et al. 2012) and *in situ* cultivation methods (Zengler et al. 2002; Bollmann et al. 2007) have made tremendous progress in increasing the coverage of culture collections, but given that the calculated number of microbial species may number well into the millions (Sogin et al. 2006), it is doubtful that such herculean efforts will ever be expended for the more obscure, less numerically dominant microbes among us. Modern microbiology must therefore continue to rely on molecular means to further its development. These are, however, fraught with difficulties that will be discussed later in this chapter.

A variation in the probing theme is the use of a microarray for hybridization of 16S rRNA genes or RNA isolated from the environment to an array of probes on a glass slide or a microchip. The technique can similarly provide a snapshot of the microbial community diversity and structure in a variety of natural systems at a higher resolution than "traditional" clone libraries (DeSantis et al. 2007). This phylogenetic microarray approach, as well as microarrays that detect functional genes, was recently reviewed (Dugat-Bony et al. 2012). Since efficacy of arrays depends on the sequences available for the probe design, and there are relatively few sequences available from carrion microbiomes, this chapter will focus on sequencing approaches.

18.3 Sequencing in (Massively) Parallel

18.3.1 Next-Generation Sequencing

Another important breakthrough for dissecting microbial communities was the advent of massively parallel sequencing technologies—the so-called next generation (next-gen) of sequencing techniques developed after Sanger sequencing. These technologies sequence by recording the incorporation of nucleotides or oligonucleotides into DNA. By spatially separating these polymerase reactions, next-generation technologies generate a sequence from millions of reactions at a time, while eliminating the need for fragment analysis following the sequencing reaction. The first next-gen platform was developed by Jonathan Rothberg at 454 Life Sciences Corp. Following the purchase of that company by Roche Diagnostics, the first commercial platform to the market was released in 2005 as the GS20 system. The basic system incorporated a flow cell containing 1.6 million wells individually interrogated by fiber optic bundles (Margulies et al. 2005). Within each well resided a nanoparticle bead containing clonal DNA. Sequencing occurred via a previously developed enzymatic cascade (Ronaghi et al. 1996). In this case, sequencing occurs by synthesis of the complementary strand of a single-stranded DNA template. In the presence of a suitable template, a primer binds and begins to extend using available nucleotides and a DNA-dependent DNA polymerase such as Taq. As a nucleotide is added to the growing polymer chain, a proton and pyrophosphate are released. An included sulfurylase enzyme catalyzes the conversion of the pyrophosphate to ATP, which is subsequently hydrolyzed in a luciferase-catalyzed reaction to generate a photon of light. By regulating the availability of nucleotides over the flow cell—adding each nucleotide sequentially and extensively washing between flows—a pattern of light across the flow cell can be detected that corresponds to the incorporation of a particular base at that site. In this way, the sequence can be determined via synthesis of the complementary strand in a process that has been termed "pyrosequencing" (Ronaghi et al. 1998).

A complication inherent in pyrosequencing is the inability to detect single photons in the complex matrix of the flow cell. To eliminate this problem, clonal amplification first occurs on the nanoparticle beads through a process called emulsion PCR. Briefly, DNA is diluted to a far-below saturation level of nanoparticle beads and added to a PCR reaction buffer containing polymerase, Mg^{2+}, and nucleotides. An emulsion is created in which a multitude of water droplets serve as micro-reactors, some of which will contain a single DNA fragment and replicate isoclonally, while others without DNA do not. Breaking the emulsion followed by template-specific purification results in a dramatically larger signal-to-noise ratio within the detection range of available photon counting devices (Margulies et al. 2005). After nearly a decade, the platform has continued to improve in the throughput and sequencing length, from roughly one hundred thousand, 100-base pair (bp) reads to over a million reads and up to 1000 bp in length. However, Roche announced in 2013 the phasing out of pyrosequencing, as the market appears to favor technologies that provide higher throughput.

Another popular high throughput approach applies sequencing-by-synthesis to detect incorporation of fluorophore-modified nucleotides. This method is employed in the next-gen platforms from Illumina. Oligonucleotide adapter sequences are annealed to size-fractionated DNA. The adapters are complementary to one of two oligonucleotides attached to the surface of the flow cell. Once a molecule anneals to the surface, the DNA undergoes PCR-based "Bridge Amplification," wherein the newly synthesized strand incorporates a complementary end for binding into the second oligonucleotide type present on the surface of the flow cell. Separation of the two strands (i.e., denaturing) yields two independent targets for amplification in close proximity. The process is repeated until discrete regions contain clusters with approximately 1000 copies of DNA. A clear advantage of the process is the random binding of the initial DNA to the flow cell surface, obviating the need for complex machined well microstructures and resulting in massively high potential throughput. The actual sequencing process is similar to that used in traditional Sanger sequencing reactions, in which any one of the four color-coded bases may be incorporated. As in the case of Sanger reaction, the presence of all four bases at the synthesis stage means that nucleotides compete for incorporation. This competition may enhance accuracy. Visualization of the incorporated base color indicates the sequence. Subsequent removal of the labeled terminator allows the process to be repeated. A downside to the technology is the relatively short sequencing length, which started in the mid 20 bp range. This has been extended to ≥70 bp on the HiSeq platforms by a second round of sequencing from the opposing end of the molecule (paired-end sequencing). Read lengths have improved in newer models with better chemistry (2×100 bp on the HiSeq 2500). However, the real advantages of the Illumina platforms continue to be their ability to generate enormous quantities of data (up to 600 Gigabases of data (Gb) per run on the HiSeq 2500, which provides increased depth of coverage in targeted sequencing applications and broader representation of complex metagenomic samples; data from www.illumina.com).

The third major player in the next-gen sequencing arena is Life Technologies with its SOLiD platform (Cummings et al. 2010). Recent advances with on flow cell isothermal amplification have eliminated the need for bead-based emulsion PCR and dramatically improved the sequencing density (i.e., the number of simultaneous sequencing reactions per unit of flowcell space) and throughput with a concomitant reduction in overall costs. This new and improved methodology called Wildfire is in many ways analogous to the approach used by Illumina for populating and amplifying DNA on flow cells (Ma et al. 2013). However, the similarities end there as the SOLiD platform sequences using a unique ligation process involving labeled oligonucleotides. The two-base pair encoding oligonucleotide design also results in a self-QC check insuring extreme accuracy while being immune from the complications in processing stretches of identical sequences. These "homopolymer stretches" can introduce errors in sequences generated with other next-generation sequencing technologies (Ross et al. 2013). The major drawback in the Wildfire system, however, is the relatively short read length (<50 bp), which complicates the assembly of genomes.

Short read lengths and high error rates, compared recently for various next-generation sequencing technologies (Loman et al. 2012; Bragg et al. 2013), create challenges for analysis of microbial communities by the widely used approach of sequencing PCR-amplified rRNA fragments. The read length relates to taxonomic resolution. While only ~100 bp of quality reads appear required for accurate species identification (Liu et al. 2007), identifying of higher taxonomic groups requires near full-length reads

(>900 bp; Yarza et al. 2014). Sequencing errors can lead investigators to overestimate diversity. For example, error rates average 2% for reads generated by pyrosequencing platforms (Mosher et al. 2013) but can reach 50% for homopolymer stretches (Luo et al. 2012). Since the threshold of 97% similarity is widely used to define operational taxonomic units (OTUs) in microbial community analysis, these errors could inflate richness, or α diversity, by leading to the identification of "pseudo-species." These errors, when combined with high sequencing depth, can also lead an inflation of β diversity, which compares the similarity between communities, as the greater number of reads increases the probability of a pseudo-species appearing in multiple samples (Smith and Peay 2014). The overall sequencing depth, the number of overlapping reads for each region sequenced, compensates for these limitations. For microbial community analysis through sequencing of PCR-amplified rRNA gene fragments, sequencing depth also allows investigators to describe the richness of microbial communities, as roughly 5000 quality sequencing reads are required to estimate the diversity of microbes in natural systems (Lundin et al. 2012).

18.3.2 Next–Next (3rd) Generation and Low-Cost Sequencing Platforms

Among the big three sequencing players, each has introduced a slimmed down, bench top sequencer designed for the smaller core or individual labs, with costs per run ranging from $500 to $2000. These include the Roche GS Jr., Illumina MiSeq, and Life Technologies' Ion Torrent. In most respects, the first two represent smaller versions of their higher throughput relatives. The MiSeq, however, has the added benefit of increased read length relative to Illumina's higher throughput platforms, which is advantageous for downstream assembly of DNA into larger DNA sequences. By contrast, the Ion Torrent Personal Genome Machine (developed by the same person as the Roche 454 platform but this time purchased by Life Technologies) deviates from other competitors since it does not use light during the reading process. Like the Roche 454 system, the Ion Torrent requires emulsion PCR and machined array flow chips. However, unlike the Illumina and Roche systems, the Ion Torrent technology does not rely on light but on the direct detection of protons released when bases are added to growing DNA strands. Sequential flow of nucleotides onto the flow cell and the direct measurement of pH (converted to voltage on the chip) results in a platform that requires only natural nucleotides without the need for enzymatic cascades or complex and expensive optics (Figure 18.2). As a result, the Ion Torrent represents the cheapest platform *per run*, although it still pales in comparison to the MiSeq in total cost *per base* sequenced because of the latter's significantly higher throughput.

Single-molecule technologies offer the advantage of long reads through a new generation of sequencing platforms, such as the PacBio RS II system from Pacific Biosciences. Using Single Molecule, Real Time (SMRT) cells, the system can sequence 150,000 single molecules at a time, with read lengths over 20,000 bp. The system utilizes a polymerase with a reduced process rate and directly monitors the incorporation of colored nucleotides during extension. While saddled with accuracy problems early on (some early reports of accuracy at ~85%), improvements in chemistry and techniques, such as circular sequencing, have dramatically improved accuracy. With circular sequencing, the DNA molecules are ligated with hairpin adapters, creating a single circular molecule, which can be sequenced multiple times, thereby increasing total accuracy. Error rates remain relatively high though, and even with circular sequencing, the longer reads generated with this single-molecule technology do not improve the accuracy of phylogenetic classification relative to pyrosequencing (Mosher et al. 2013).

18.4 Technical Considerations in Experimental Design

Defining phylogenetic groups allows comparisons of the structure and diversity within and among ecosystems. The species concept facilitates comparisons between studies; however, for microbial communities the question—what is a bacterial species?—requires discussion. Unlike eukaryotes that undergo sexual reproduction, the definition of species in eubacteria and archaea is less straight forward. The operational definition of a bacterial species has been given as "a category that circumscribes a (preferably) genomically coherent group of individual isolates/strains sharing a high degree of similarity in (many)

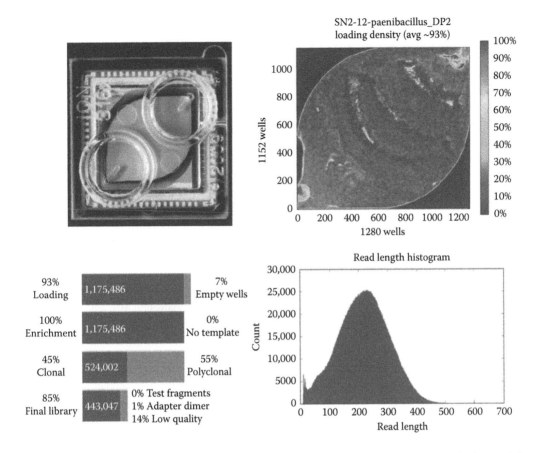

FIGURE 18.2 **(See color insert.)** Representative example of data from an Ion Torrent PGM run. Clockwise from top left: disposable Ion 316 sequencing chip, chip loading density heat map, histogram of sequenced read lengths (200 bp chemistry), and run statistics.

independent features, comparatively tested under highly standardized conditions" (from Stackebrandt et al. 2002 quoted in Gevers et al. 2005). Mechanistically this has been somewhat arbitrarily defined as 97% nucleotide sequence identity between 16S rRNA genes, which equates roughly to the previous standard of a 70% DNA–DNA hybridization ratio (Stackebrandt and Goebel 1994). The comparative ease of 16S sequencing, however, led to its rapid adoption over that of DNA–DNA hybridization in most instances. One facet of the 16S rRNA gene that is often overlooked is that it is not equally conserved across its entirety. Hot spots of variation, the so-called "variable regions," punctuate the gene providing different levels of conservation between samples. Although the authors have found no published systematic analysis of the percent variability among variable regions between two or more 16S genes that still fall within the 97% species classification at the full sequence level, Schloss notes in one article that "a difference of 10% between V6 fragments is comparable to a 3% difference over the full length of the gene ..." (Schloss 2009). Regardless of the similarity threshold set by the investigators, 16S rRNA gene sequences often lack the resolution to define ecologically significant phylogenetic groups. The current tendency to focus on only a subset of the gene including one or even a few specific variable regions merely exacerbates the problem (Huse et al. 2008). Isolates with identical 16S rRNA gene sequences may have widely divergent metabolisms (Jaspers and Overmann 2004). This suggests that sequences associated with a particular niche in one system may not be associated with a similar niche in another system. Conversely, artifacts in PCR-generated sequences can lead to false identification of bacterial species (discussed below). Variation in the 16S gene copy number among species also impacts quantitative community assessment, as does the presence of similar metabolic pathway genes among different species (see Chapter 19).

18.4.1 Sources of Error

To assess microbial community diversity by molecular means, the specific sources of experimental error must be addressed. Potential error sources include improper sampling and cross-contamination, differential or biased extraction of nucleic acids, errors or bias in PCR amplification or processing, database limitations, and many others. Here, a discussion of appropriate sample collection and archiving protocols is given, with highlights particular to carrion microbial communities, along with a discussion of major sources of error in phylogentic analyses of microbial communities.

18.4.2 DNA Extraction

The general work flow for analysis of microbial communities by DNA sequencing starts with DNA extraction. This statement belies the underlying complexity of the process. For example, are all bacteria lysed equally? The answer of course is "no." To understand the primary basis of this, it is first necessary to understand a bit of the biochemistry of bacteria. Most bacteria have historically been grouped into one of two camps: Gram positive or Gram negative. The name refers to a differential staining test developed by Christian Gram in 1884 (in Smith and Hussey 2005). Succinctly, most Gram-positive bacteria have thick outer cell walls made of multiple layers of cross-linked peptidoglycan. By contrast, most Gram-negative bacteria have dual cell walls consisting of a relatively thin inner peptidoglycan cell wall and an outer membrane studded with complex lipopolysaccharides. The fundamental difference as it relates to cell wall lysis is that Gram-negative cells are generally structurally weaker and more susceptible to detergents, whereas Gram-positive cells are more difficult to lyse. To further complicate matters, some Gram-positive cells are capable of forming endospores, which are dense, low-liquid structures containing DNA and wrapped in hard proteinaceous coats that can survive boiling, solvents, and extreme heat.

These physiological differences can lead to bias. In an early test of DNA extraction efficiency, Moré et al. compared repeated freeze–thaw cycles to bead mill homogenization of soil (Moré et al. 1994). As measured by viable cell counts, 2% of soil microbes survived bead-mill homogenization and 8% survived three rounds of freeze–thaw cycles. In soils spiked with endospores of *Bacillus subtilis*, 2% of endospores survived bead-mill homogenization and, in sharp contrast to the total community, 94% of the added spores survived repeated freeze–thaw cycles. This suggests both archiving and extraction protocols could create bias. For example, allowing samples to freeze and thaw during storage and transport could differentially lyse vegetative cells and aggressive extraction protocols appear necessary to recover DNA from endospores.

Most recent studies that applied PCR-based protocols to analyze microbial communities in natural systems directly extract DNA from environmental samples with commercial kits. These kits generally use mechanical agitation to lyse cells and an affinity column to concentrate and purify nucleic acids from other biomolecules and environmental contaminants, such as humic acids. As discussed above, bead-beating provides for the efficient lysis of endospores. This mechanical agitation and/or treatment with mutanolysin appear critical for efficient recovery of DNA from complex samples, such as those collected from the human microbiome (Yuan et al. 2012). Bead-beating protocols, and to a certain extent affinity columns and even pipetting, necessarily shear genomic DNA. Shearing does not apparently interfere with analyses that use PCR, direct sequencing, microarrays or small insert libraries; however, generation of large insert libraries, such as fosmid (cloning vectors based on the bacterial F-plasmid) libraries of environmental DNA (Stein et al. 1996), requires gentler extraction and purification methods. As discussed below these large insert libraries provide a valuable method for validating and interpreting the results.

The proper quantity of template DNA recovered from carrion microbiomes requires further examination. Carrion will contain animal degradation products, including eukaryote cells and associated DNA. This host-derived material could dilute microbial DNA in the PCR-template pool. PCR appears biased against proportionally detecting relatively rare templates in a multitemplate pool (Gonzalez et al. 2012). When evaluating template DNA preparation, investigators should assess the proportion of target to nontarget DNA with qPCR or similar approaches. At the very least, the bias introduced by host DNA should be assessed by creating artificial template mixtures that include representative host and microbe

DNA ratios. In addition, differential centrifugation and selective cell lysis can increase the proportion of microbial versus host DNA (Horz et al. 2010; Hunter et al. 2011).

Given the nature of carrion systems, the need for aggressive lysis protocols does not appear justified. Carrion communities necessarily select for rapidly growing microorganisms, as the available resources are rapidly depleted by decomposition and consumed by scavengers. These decomposer communities include many representatives of the Firmicutes phylum (Chapter 3) capable of forming endospores; however, there is no evidence that many of these microbes are in the endospore morphology. The rapid succession of microbial communities associated with carrion (Chapters 3 and 19) suggest that vegetative cells would dominate the system. Indeed, the few studies conducted of the carrion microbiome have used comparatively gentle lysis protocols (Metcalf et al. 2013; Pechal et al. 2014). Similar findings have been reported for intestinal biopsies (Carbonero et al. 2011).

18.4.3 PCR Bias and Error

Several sources of bias were recognized early in the application of PCR to microbial communities. Quality and quantity of DNA extracted from microbiome samples will influence the quality of PCR-based methods of community analysis. Humic acids and related compounds coextracted with nucleic acids from environmental samples can significantly interfere with PCR (Tebbe and Vahjen 1993). Many investigators have struggled to obtain robust amplification and addressed the issue by adding enhancers, such as bovine serum albumin (Kreader 1996), employing advanced purification methods (Tsai and Olson 1992; Jackson et al. 1997), or by simply attempting to dilute out inhibitors. Achieving robust amplification does not eliminate bias, as even if amplification occurs, diversity estimates will decrease with humic acid contamination (LaMontagne et al. 2002). Dilution does not provide a viable alternative either, as dilution of templates increases the bias inherent to multitemplate PCR (Chandler et al. 1997).

Variations in amplification efficiencies create bias in multitemplate PCR reactions (Suzuki and Giovannoni 1996; Polz and Cavanaugh 1998). These biases are well documented in clone libraries and recently examined with respect to community analyses that apply NextGen sequencing platforms (Pinto and Raskin 2012). Chimera, as the name suggests, are fusion molecules originating from two different source amplicons in a PCR reaction. First described by Liesack et al. with respect to microbial phylogenetics, the phenomenon results in an overestimation of community diversity and artifactual phylotypes (Liesack et al. 1991). In spite of a number of analytical software programs, Chimera detection remains challenging. The presence of chimerical sequences in public databases falsely suggests a high proportion of endemic bacterial species and confounds comparisons of bacterial communities between systems. Similarly, misincorporation of bases during amplification and sequencing errors can further inflate estimates of diversity.

Primer selection is perhaps the primary source of bias in microbial phylogenetic studies. The all-inclusive "universal primer" simply does not exist, as no primers will amplify all bacteria. Attempting to include the archaea or microbial eukaryotes further complicates studies (Baker et al. 2003). Recently, Klindworth et al. evaluated 175 primers and 512 primer pairs *in silico* for their ability to identify species in the SILVA dataset (Klindworth et al. 2013). The authors recommend 341F to 785R (*E. coli* 16S numbering) for the broadest possible coverage. Yet, the amplicon length of 464 bp may preclude certain available next-gen platforms. In the absence of perfect primers, decisions must be with respect to the likely target species present in a given sample. For example, the Verrucomicrobia, seemingly ubiquitous in soils, lack the 5′ region of the 16S rRNA gene targeted by many 16S primers (Bergmann et al. 2011). The limited number of studies performed on carrion leaves this an open question. However, the complexity of the host (e.g., the gut microbiome) added to the known complexity of the soil suggests that extreme care must be taken to fully characterize all taxa involved in carrion decomposition.

18.4.4 Sample Collection and Archiving

Appropriate methods for sample collection and archiving are also critical for microbial community analysis. At the microbial scale, carrion appears to contain a number of distinct environments that should support distinct microbiomes. Although this question is not well studied in carrion, we can assume that the

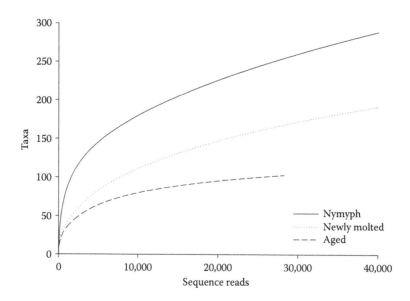

FIGURE 18.3 Rarefaction curve of new bacterial species (*y*-axis) versus number of sequencing reads (*x*-axis) at three different life stages in whole samples of surface-sterilized lone star tick *Amblyomma americanum*. (From Menchaca, A.C. et al. 2013. *PLoS ONE* 8: e67129.)

structure is somewhat analogous to the structure of microbiomes associated with live animals. In humans, microbiomes differ between different organs, such as the skin and gut (Gevers et al. 2012; Markowitz et al. 2012). A similar structure almost certainly exists in carrion, particularly shortly after death. Investigators should minimize sampling protocols that homogenize what are likely distinct communities, as the pooling of those communities will necessarily inflate diversity estimates and obfuscate analysis of the community structure. The size of the sample collected will also relate to diversity estimates. Species-area theory predicts that larger systems will contain more species (MacArthur and Wilson 1967). This concept is not well explored in microbiome research, but consistent with that prediction larger frogs have more diverse skin microbiomes (McKenzie et al. 2012). To reveal such differences investigators should control for the size of the sample collected. A species-area theory predicts, and rarefaction analysis confirms, that larger samples from the same system invariably yield higher richness estimates. Given the rapid succession of carrion microbiomes, researchers should also pay close attention to the timing of sample collections.

A rarefaction curve reveals how many samples are necessary to adequately assess a community. In Figure 18.3, the number of "species" (defined here as 97% 16S gene sequence identity) is compared to the sequencing effort (*x*-axis), in this case the different life stages of the lone star tick (Menchaca et al. 2013). As can be seen, the "Aged" adult tick bacterial diversity has begun to plateau, whereas the younger stages continue to identify new species, even after 30,000 sequence reads. This is surprising given the comparatively simple community composition expected from an organism that feeds solely on blood, and only rarely does that. Answers to questions such as "how many sequences are enough?" and "where is the point of diminishing returns?" therefore must be addressed on an individual case-by-case basis. The sequencing depth available with NextGen sequencing platforms provides extension to the rarefaction curves of microbial abundance (Figure 18.3) to include many of the rare as well as transient species. The extent to which these DNA signatures represent real species or technical artifacts remains an open question in the field. As discussed above, pseudo-species (Kunin et al. 2010) and chimeras, where fragments from two individual species are joined during PCR amplification (Haas et al. 2011) can inflate diversity estimates.

18.4.5 Error in Metagenomics

Amplicon-based studies are but one application of metagenomics. Another is shotgun metagenomics, in which all DNA extracted from an environment, is sequenced, and systematic information is extracted.

The first large-scale application of this approach was performed on the Sargasso Sea (Venter et al. 2004). This study was conducted using traditional Sanger sequencing and identified over 1800 microbial species from over a billion sequenced, nonredundant nucleotides. Two years later, Sogin and colleagues applied amplicon sequencing on ocean sediment and deep sea vent soils using a next-generation sequencing platform and identified microbial communities "…one to two orders of magnitude more complex than previously reported for any microbial environment" (Sogin et al. 2006). Later efforts would combine the higher throughput of next-generation platforms with the lower bias of direct sequencing of total DNA in order to capitalize on the best of both techniques. One study, in particular, investigated the concordance of shotgun metagenomics and rRNA amplicon-based metagenomics on next-gen platforms. Shakya et al. used synthetic communities to compare and contrast Illumina and 454-based shotgun metagenomics with amplicon-based approaches (Shakya et al. 2013). They reported that in both cases, shotgun metagenomics could outperform amplicon sequencing for quantifying the community, but the results were dependent upon both the platform used and analysis parameters.

18.4.6 Replication and Validation

Validation of DNA preparation requires assuring that the methods do not strongly bias the results. Simply, this can entail comparing several methods to determine the extent that diversity estimates depend on the methods. A systematic approach entails testing each step in the protocol. DNA extraction efficiency can be validated by calculating the microbial DNA in the sample from direct microscope counts and comparing the expected DNA from that sample to the recovered DNA (LaMontagne and Holden 2003). The purity of DNA is widely determined from the UV-spectra, where the spectra are compared to that of a DNA standard. Pure DNA has a distinctive peak at 260 nm and a trough at 230 nm. However, a number of common contaminants can obscure these readings. Alternatively, the quantity of DNA can be assessed by fluorescence using specific DNA-intercalating dyes, or by qPCR. Reagent blanks provide assurance that the DNA recovered comes from the sample, not the reagents. These controls should be instituted at the key steps of DNA extraction, purification, and amplification.

Reagent blanks, technical replicates, and model communities are keys for validating microbial community analyses. Technical replicates allow assessment of the reproducibility of community profiles. Most investigators run duplicate samples to set the threshold for detection limits. Model communities, including spiked samples, allow investigators to determine the proportion of recovery. Confirmation of the presence of a particular phylotype in a system from which the sequence was amplified can be accomplished by *in situ* hybridization and qPCR.

Validation of assumptions of the species associated with ribosomal sequences requires linking ribosomal genes to genomes. As discussed above, 16S rRNA sequences do not consistently relate to a particular bacterial species. For some phylogenetic groups such as methanogens and nitrifiers, there is a degree of correspondence between phylogeny based on ribosomal sequences and ecosystem function (Martiny et al. 2013), and recently the tools for predicting metabolic attributes of communities from ribosomal genes have improved (Okuda et al. 2012; Langille et al. 2013). However, until the microbe has been isolated, we often know little about it. As discussed above, shotgun metagenomics can complement surveys of ribosomal genes if the 16S and other genes can be linked. Single cell genomics, where numerically dominant microbes are separated from the environment to facilitate genomic analysis (Rodrigue et al. 2009; Woyke et al. 2010), provides a promising alternative approach to better describe species heretofore only known through ribosomal sequences and link-specific genes and pathways with particular phylogenetic groups.

18.5 Applications to the Microbial Ecology of Carrion

Given the nature of carrion, isolation of representative microbes from those systems appears feasible. The "great plate anomaly" hypothesis basically says that there are orders of magnitude difference between the number of microbes gathered from natural environments that are able to grow on agar media and the number countable by microscopic examination (Staley and Konopka 1985). This predicts little agreement between the sequences generated from microbiomes by metagenomic methods and those generated from isolates grown in culture. However, it should not be assumed that this prediction holds for carrion systems. Carrion provides a nutrient-rich resource that probably selects for rapidly growing microorganisms. Many of these opportunistic species may be relatively easy to culture. There appears to be a relationship between the percentage of the community that readily grows on standard laboratory media and nutrient availability in the systems (Simu et al. 2005). This suggests that it is possible to design a nutrient-rich media and isolate many of the microbial species associated with carrion. *In situ* cultivation methods, where microbes are cultivated in diffusion chambers, can dramatically increase the efficiency of isolation (Bollmann et al. 2007) and analysis of sequences generated with metagenomic surveys can improve and inform isolation strategies (Tyson and Banfield 2005).

The abundance of nutrients in carrion and the ephemeral, episodic nature of those systems allow us to predict general patterns of the ecology of the carrion microbiome. Species-energy theory, a corollary of species-area theory, predicts that diversity increases with available resources (Wright 1983). Thus, we find more species in the Amazon than the Arctic. In soil this concept appears to hold, as there is a general link between microbial biomass and diversity (LaMontagne et al. 2003). However, this theory holds mostly for mature systems, where a community has developed in a way that optimizes the influx of energy. The bonanza released in carrion decomposition appears to select for a relatively low diversity community (Tringe et al. 2005). In this regard, carrion microbiomes may be modeled as a disturbed system, where the community has not adapted to maximize resource-use efficiency.

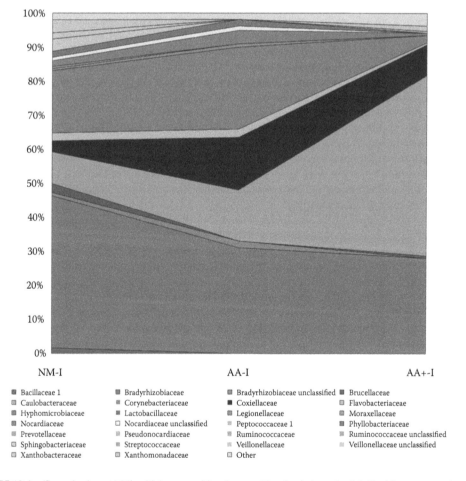

▣ Bacillaceae 1	▣ Bradyrhizobiaceae	▣ Bradyrhizobiaceae unclassified	▪ Brucellaceae
▫ Caulobacteraceae	▪ Corynebacteriaceae	▪ Coxiellaceae	▫ Flavobacteriaceae
▪ Hyphomicrobiaceae	▪ Lactobacillaceae	▣ Legionellaceae	▣ Moraxellaceae
▪ Nocardiaceae	▫ Nocardiaceae unclassified	▨ Peptococcaceae 1	▪ Phyllobacteriaceae
▨ Prevotellaceae	▨ Pseudonocardiaceae	▨ Ruminococcaceae	▨ Ruminococcaceae unclassified
▫ Sphingobacteriaceae	▨ Streptococcaceae	▨ Veillonellaceae	▨ Veillonellaceae unclassified
▫ Xanthobacteraceae	▨ Xanthomonadaceae	▫ Other	

FIGURE 18.4 (See color insert.) Microbial communities change with aging in lone star tick (*Amblyomma americanum*, L.). The number of bacterial families identified were on a NM-I, newly molted adult; AA-I, aged adult (60 days); AA+-I, aged adult (110 days); -I, respectively, raised indoors under climate-controlled conditions.

The low diversity of carrion microbiomes could also reflect dispersion limitations (Tsai and Olson 1992; Metcalf et al. 2013; Pechal et al. 2014). Dispersion appears an important factor in structuring microbial communities (Langenheder et al. 2012). It is reasonable to assume that certain species would adapt to thrive on carrion in systems where carrion is relatively abundant and that the ability of those species to reach fresh carcasses would be limiting, perhaps like those of salmon runs (see Chapter 22). There is little evidence to suggest, at least in terrestrial systems, that microbes have a method of moving from one carrion resource to another, although hitching a ride on carrion-adapted animals seems a reasonable strategy like those discussed in Chapter 3. In aquatic systems, detritus-associated microbes appear to use bioluminescence to attract grazers and subsequently fish (Zarubin et al. 2012). This suggests a strategy for microbes to associate with the gut, and eventually carrion. The high abundance and diversity of microbes in natural systems suggest dispersion in general does not limit establishment of microbial communities (Finlay 2002). As illustrated in Figure 18.4 for arachnid microbiomes and described recently for birds (Van Dongen et al. 2013), microbiomes associated with animals may lose diversity as the host ages. Indeed the diversity of microbial communities associated with carrion appears to decrease with time (see Chapter 19). This suggests that dispersion does not limit diversity in these systems, as we would predict that in a dispersal-limited system that diversity would increase with time, as colonization introduces transient species to the system. It appears that carrion selects for a community that thrives on the abundant nutrients. In this regard, carrion-associated microbial communities may resemble young microbial communities, which are dominated by transient species.

18.6 Data Analysis

The topic of data analysis is a treatise unto itself. For simple identification of single organisms, comparison of gene sequence identity by BLAST to the database at NCBI is generally sufficient. Accomplishing this feat on the scales of NextGen sequencing datasets is another matter entirely. Fortunately for the reader, a leading authority on bioinformatics and their applications to next-generation DNA sequencing datasets for the study of microbial communities has a chapter in this volume, and the reader is directed to Chapter 19 for further information.

18.7 Conclusions and Perspectives

In the last few decades, the application of molecular techniques revolutionized our understanding of microbial communities. These approaches have revealed diverse communities with predictable structures that fit testable theories. From these theories, there have been attempts to predict the structure of microbial communities associated with carrion and recommend appropriate methods to sample these communities. As noted, these methods are fraught with biases and artifacts. At this time no perfect method exists, and while method development continues at an impressive rate, it is not advised for the investigator to wait for techniques to improve. Instead we encourage researchers to proceed with testing hypotheses as to the structure of carrion microbial communities, with an acknowledgement of the known strengths and weaknesses of the methods applied, and make efforts to validate results while moving the field of carrion decomposition microbiology forward.

REFERENCES

Amann, R., N. Springer, W. Ludwig, H.-D. Gortz, and K.-H. Schleifer. 1991. Identification *in situ* and phylogeny of uncultured bacterial endosymbionts. *Nature* 351: 161–164.

Anderson, B.E., J.E. Dawson, D.C. Jones, and K.H. Wilson. 1991. *Ehrlichia chaffeensis*, a new species associated with human ehrlichiosis. *Journal of Clinical Microbiology* 29: 2838–2842.

Baker, G.C., J.J. Smith, and D.A. Cowan. 2003. Review and re-analysis of domain-specific 16S primers. *Journal of Microbiological Methods* 55: 541–555.

Bergmann, G.T. et al. 2011. The under-recognized dominance of *Verrucomicrobia* in soil bacterial communities. *Soil Biology and Biochemistry* 43: 1450–1455.

Bollmann, A., K. Lewis, and S.S. Epstein. 2007. Incubation of environmental samples in a diffusion chamber increases the diversity of recovered isolates. *Applied and Environmental Microbiology* 73: 6386–6390.

Bragg, L.M., G. Stone, M.K. Butler, P. Hugenholtz, and G.W. Tyson. 2013. Shining a light on dark sequencing: Characterising errors in Ion Torrent PGM Data. *PLoS Computational Biology* 9: e1003031.

Bremer, H. and P.P. Dennis. 1996. Modulation of chemical composition and other parameters of the cell by growth rate. In: F. Neidhardt, (ed.), *Escherichia coli and Salmonella typhimurium: Cellular and Molecular Biology.* Washington, DC: American Society for Microbiology, pp. 1553–1569.

Carbonero, F., G.M. Nava, A.C. Benefiel, E. Greenberg, and H.R. Gaskins. 2011. Microbial DNA extraction from intestinal biopsies is improved by avoiding mechanical cell disruption. *Journal of Microbiological Methods* 87: 125–127.

Chandler, D.P., J.K. Fredrickson, and F.J. Brockman. 1997. Effect of PCR template concentration on the composition and distribution of total community 16S rDNA clone libraries. *Molecular Ecology* 6: 475–682.

Cummings, C.A. et al. 2010. Accurate, rapid and high-throughput detection of strain-specific polymorphisms in *Bacillus anthracis* and *Yersinia pestis* by next-generation sequencing. *Investigative Genetics* 1: 5.

Desantis, T. et al. 2007. High-density universal 16S rRNA microarray analysis reveals broader diversity than typical clone library when sampling the environment. *Microbial Ecology* 53: 371–383.

Dugat-Bony, E. et al. 2012. Detecting unknown sequences with DNA microarrays: Explorative probe design strategies. *Environmental Microbiology* 14: 356–371.

Finlay, B.J. 2002. Global dispersal of free-living microbial eukaryote species. *Science* 296: 1061–1063.

Gevers, D. et al. 2005. Re-evaluating prokaryotic species. *Nature Reviews Microbiology* 3: 733–739.

Gevers, D. et al. 2012. The human microbiome project: A community resource for the healthy human microbiome. *PLoS Biology* 10: e1001377.

Giovannoni, S.J., T.B. Britschgi, C.L. Moyer, and K.G. Field. 1990. Genetic diversity in Sargasso Sea bacterioplankton. *Nature* 345: 60–63.

Giovannoni, S.J., E.F. Delong, G.J. Olsen, and N.R. Pace. 1988. Phylogenetic group-specific oligodeoxynucleotide probes for identification of single microbial cells. *Journal of Bacteriology* 170: 720–726.

Giovannoni, S.J. et al. 2005. Genome streamlining in a cosmopolitan oceanic bacterium. *Science* 309: 1242–1245.

Gonzalez, J.M., M.C. Portillo, P. Belda-Ferre, and A. Mira. 2012. Amplification by PCR artificially reduces the proportion of the rare biosphere in microbial communities. *PLoS ONE* 7: e29973.

Haas, B.J. et al. 2011. Chimeric 16S rRNA sequence formation and detection in Sanger and 454-pyrosequenced PCR amplicons. *Genome Research* 21: 494–504.

Herrick, J.B., E.L. Madsen, C.A. Batt, and W.C. Ghiorse. 1993. Polymerase chain reaction amplification of naphthalene-catabolic and 16S rRNA gene sequences from indigenous sediment bacteria. *Applied and Environmental Microbiology* 59: 687–694.

Horz, H.-P., S. Scheer, M.E. Vianna, and G. Conrads. 2010. New methods for selective isolation of bacterial DNA from human clinical specimens. *Anaerobe* 16: 47–53.

Hunter, S.J. et al. 2011. Selective removal of human DNA from metagenomic DNA samples extracted from dental plaque. *Journal of Basic Microbiology* 51: 442–446.

Huse, S.M. et al. 2008. Exploring microbial diversity and taxonomy using SSU rRNA hypervariable tag sequencing. *PLoS Genetics* 4: e1000255.

Jackson, C.R., J.P. Harper, D. Willoughby, E.E. Roden, and P.F. Churchill. 1997. A simple, efficient method for the separation of humic substances and DNA from environmental samples. *Applied and Environmental Microbiology* 63: 4993–4995.

Jaspers, E. and J. Overmann. 2004. Ecological significance of microdiversity: Identical 16S rRNA gene sequences can be found in bacteria with highly divergent genomes and ecophysiologies. *Applied and Environmental Microbiology* 70: 4831–4839.

Kallmeyer, J., R. Pockalny, R.R. Adhikari, D.C. Smith, and S. D'hondt. 2012. Global distribution of microbial abundance and biomass in subseafloor sediment. *Proceedings of the National Academy of Sciences* 109: 16213–16216.

Klindworth, A. et al. 2013. Evaluation of general 16S ribosomal RNA gene PCR primers for classical and next-generation sequencing-based diversity studies. *Nucleic Acids Research* 41: e1.

Kreader, C.A. 1996. Relief of amplification inhibition in PCR with bovine serum albumin or T4 Gene 32 protein. *Applied and Environmental Microbiology* 62: 1102–1106.

Kunin, V., A. Engelbrektson, H. Ochman, and P. Hugenholtz. 2010. Wrinkles in the rare biosphere: Pyrosequencing errors can lead to artificial inflation of diversity estimates. *Environmental Microbiology* 12: 118–123.

Lagier, J.-C. et al. 2012. Microbial culturomics: Paradigm shift in the human gut microbiome study. *Clinical Microbiology and Infection* 18: 1185–1193.

LaMontagne, M.G. and P.A. Holden. 2003. Comparison of free-living and particle-associated bacterial communities in a coastal lagoon. *Microbial Ecology* 46: 228–237.

LaMontagne, M.G., F.C. Michel, Jr., P.A. Holden, and C.A. Reddy. 2002. Evaluation of extraction and purification methods for obtaining PCR-amplifiable DNA from compost for microbial community analysis. *Journal of Microbiology Methods* 49: 255–264.

LaMontagne, M.G., J.P. Schimel, and P.A. Holden. 2003. Comparison of subsurface and surface soil bacterial communities in California grassland as assessed by terminal restriction fragment length polymorphisms of PCR-amplified 16S rRNA genes. *Microbial Ecology* 46: 216–227.

Lane, D.J. et al. 1985. Rapid determination of 16S ribosomal RNA sequences for phylogenetic analyses. *Proceedings of the National Academy of Sciences* 82: 6955–6959.

Langenheder, S., M. Berga, O. Ostman, and A.J. Szekely. 2012. Temporal variation of [Beta]-diversity and assembly mechanisms in a bacterial metacommunity. *ISME Journal* 6: 1107–1114.

Langille, M.G.I. et al. 2013. Predictive functional profiling of microbial communities using 16S rRNA marker gene sequences. *Nature Biotechnology* 31: 814–821.

Liesack, W. and E. Stackebrandt. 1992. Occurrence of novel groups of the domain bacteria as revealed by analysis of genetic material isolated from an Australian terrestrial environment. *Journal of Bacteriology* 174: 5072–5078.

Liesack, W., H. Weyland, and E. Stackebrandt. 1991. Potential risks of gene amplification by PCR as determined by 16S rRNA analysis of a mixed-culture of strict barophilic bacteria. *Microbial Ecology* 21: 191–198.

Liu, Z., C. Losupone, M. Hamady, F.D. Bushman, and R. Knight. 2007. Short pyrosequencing reads suffice for accurate microbial community analysis. *Nucleic Acids Research* 35: e120.

Loman, N.J. et al. 2012. High-throughput bacterial genome sequencing: An embarrassment of choice, a world of opportunity. *Nature Reviews Microbiology* 10: 599–606.

Lundin, D., I. Severin, J.B. Logue, O. Östman, A.F. Andersson, and E.S. Lindström. 2012. Which sequencing depth is sufficient to describe patterns in bacterial α- and β-diversity? *Environmental Microbiology Reports* 4: 367–372.

Luo, C., D. Tsementzi, N. Kyrpides, T. Read, and K.T. Konstantinidis. 2012. Direct comparisons of Illumina vs. Roche 454 sequencing technologies on the same microbial community DNA sample. *PLoS ONE* 7: e30087.

Ma, Z. et al. 2013. Isothermal amplification method for next-generation sequencing. *Proceedings of the National Academy of Sciences* 110: 14320–14323.

MacArthur, R.H. and E.O. Wilson. 1967. *The Theory of Island Biogeography*. New Jersey: Princeton University Press.

Margulies, M. et al. 2005. Genome sequencing in microfabricated high-density picolitre reactors. *Nature* 437: 376–380.

Markowitz, V.M. et al. 2012. Img/M-Hmp: A metagenome comparative analysis system for the human microbiome project. *PLoS ONE* 7: e40151.

Martiny, A.C., K. Treseder, and G. Pusch. 2013. Phylogenetic conservatism of functional traits in microorganisms. *ISME Journal* 7: 830–838.

McKenzie, V.J., R.M. Bowers, N. Fierer, R. Knight, and C.L. Lauber. 2012. Co-habiting amphibian species harbor unique skin bacterial communities in wild populations. *ISME Journal* 6: 588–596.

Menchaca, A.C. et al. 2013. Preliminary assessment of microbiome changes following blood-feeding and survivorship in the *Amblyomma americanum* nymph-to-adult transition using semiconductor sequencing. *PLoS ONE* 8: e67129.

Metcalf, J.L. et al. 2013. A microbial clock provides an accurate estimate of the postmortem interval in a mouse model system. *eLife* 2: e01104.

Moré, M.I., J.B. Herrick, M.C. Silva, W.C. Ghiorse, and E.L. Madsen. 1994. Quantitative cell lysis of indigenous microorganisms and rapid extraction of microbial DNA from sediment. *Applied and Environmental Microbiology* 60: 1572–1580.

Morell, V. 1997. Microbial biology: Microbiology's scarred revolutionary. *Science* 276: 699–702.

Mosher, J.J., E.L. Bernberg, O. Shevchenko, J. Kan, and L.A. Kaplan. 2013. Efficacy of a 3rd generation high-throughput sequencing platform for analyses of 16S rRNA genes from environmental samples. *Journal of Microbiological Methods* 95: 175–181.

Okuda, S., Y. Tsuchiya, C. Kiriyama, M. Itoh, and H. Morisaki. 2012. Virtual metagenome reconstruction from 16S rRNA gene sequences. *Nature Communications* 3: 1203.

Paul, C., S. Laura, B. Sara, and J.G. Stephen. 2012. Nutrient requirements for growth of the extreme oligotroph "*Candidatus pelagibacter ubique*" Htcc1062 on a defined medium. *ISME Journal* 7: 592–602.

Pechal, J. et al. 2014. The potential use of bacterial community succession in forensics as described by high throughput metagenomic sequencing. *International Journal of Legal Medicine* 128: 193–205.

Pinto, A.J. and L. Raskin. 2012. PCR biases distort bacterial and archaeal community structure in pyrosequencing datasets. *PLoS ONE* 7: e43093.

Polz, M.F. and C.M. Cavanaugh. 1998. Bias in template-to-product ratios in multitemplate PCR. *Applied and Environmental Microbiology* 64: 3724–3730.

Rappe, M.S., S.A. Connon, K.L. Vergin, and S.J. Giovannoni. 2002. Cultivation of the ubiquitous SAR11 marine bacterioplankton clade. *Nature* 418: 630–633.

Rodrigue, S. et al. 2009. Whole genome amplification and *de novo* assembly of single bacterial cells. *PLoS ONE* 4: e6864.

Ronaghi, M., S. Karamohamed, B. Pettersson, M. Uhlén, and P. Nyrén. 1996. Real-time DNA sequencing using detection of pyrophosphate release. *Analytical Biochemistry* 242: 84–89.

Ronaghi, M., M. Uhlén, and P. Nyrén. 1998. A sequencing method based on real-time pyrophosphate. *Science* 281: 363–365.

Ross, M.G., C. Russ, M. Costello, A. Hollinger, N.J. Lennon, R. Hegarty, C. Nusbaum, and D.B. Jaffe. 2013. Characterizing and measuring bias in sequence data. *Genome Biology* 14: R51.

Saiki, R. et al. 1985. Enzymatic amplification of beta-globin genomic sequences and restriction site analysis for diagnosis of sickle cell anemia. *Science* 230: 1350–1354.

Schloss, P.D. 2009. A high-throughput DNA sequence aligner for microbial ecology studies. *PLoS ONE* 4: e8230.

Shakya, M. et al. 2013. Comparative metagenomic and rRNA microbial diversity characterization using archaeal and bacterial synthetic communities. *Environmental Microbiology* 15: 1882–1899.

Simu, K., K. Holmfeldt, U.L. Zweifel, and A. Hagstrom. 2005. Culturability and coexistence of colony-forming and single-cell marine bacterioplankton. *Applied and Environmental Microbiology* 71: 4793–4800.

Smith, A.C. and M. Hussey. 2005. *Gram Stain Protocols*. MicrobeLibrary.org.

Smith, D.P. and K.G. Peay. 2014. Sequence depth, not PCR replication, improves ecological inference from next generation DNA sequencing. *PLoS ONE* 9: e90234.

Sogin, M.L. et al. 2006. Microbial diversity in the deep sea and the underexplored "Rare Biosphere." *Proceedings of the National Academy of Sciences* 103: 12115–12120.

Stackebrandt, E. and B.M. Goebel. 1994. Taxonomic note: A place for DNA–DNA reassociation and 16S rRNA sequence analysis in the present species definition in bacteriology. *International Journal of Systematic Bacteriology* 44: 846–849.

Stackebrandt, E. et al. 2002. Report of the ad hoc committee for the re-evaluation of the species definition in bacteriology. *International Journal of Systematic and Evolutionary Microbiology* 52: 1043–1047.

Staley, J.T. and A. Konopka. 1985. Measurement of *in situ* activities of nonphotosynthetic microorganisms in aquatic and terrestrial habitats. *Annual Review of Microbiology* 39: 321–346.

Stein, J.L., T.L. Marsh, K.Y. Wu, H. Shizuya, and E.F. Delong. 1996. Characterization of uncultivated prokaryotes: Isolation and analysis of a 40-kilobase-pair genome fragment from a planktonic marine archaeon. *Journal of Bacteriology* 178: 591–599.

Suzuki, M.T. and S.J. Giovannoni. 1996. Bias caused by template annealing in the amplification of mixtures of 16S rRNA genes by PCR. *Applied and Environmental Microbiology* 62: 625–630.

Tebbe, C.C. and W. Vahjen. 1993. Interference of humic acids and DNA extracted directly from soil in detection and transformation of recombinant DNA from bacteria and a yeast. *Applied and Environmental Microbiology* 59: 2657–2665.

Torsvik, V., R. Sørheim, and J. Goksøyr. 1996. Total bacterial diversity in soil and sediment communities—A review. *Journal of Industrial Microbiology & Biotechnology* 17: 170–178.

Tringe, S.G. et al. 2005. Comparative metagenomics of microbial communities. *Science* 308: 554–557.

Tsai, Y.L. and B.H. Olson. 1992. Rapid method for separation of bacterial DNA from humic substances in sediments for polymerase chain reaction. *Applied and Environmental Microbiology* 58: 2292–2295.

Tyson, G.W. and J.F. Banfield. 2005. Cultivating the uncultivated: A community genomics perspective. *Trends in Microbiology* 13: 411–415.

Van Dongen, W.F. et al. 2013. Age-related differences in the cloacal microbiota of a wild bird species. *BMC Ecology* 13: 11.

Venter, J.C. et al. 2004. Environmental genome shotgun sequencing of the Sargasso sea. *Science* 304: 66–74.

Weisburg, W.G., S.M. Barns, D.A. Pelletier, and D.J. Lane. 1991. 16S ribosomal DNA amplification for phylogenetic study. *Journal of Bacteriology* 173: 697–703.

Woese, C.R. and G.E. Fox. 1977. Phylogenetic structure of the prokaryotic domain: The primary kingdoms. *Proceedings of the National Academy of Sciences* 74: 5088–5090.

Woyke, T. et al. 2010. One bacterial cell, one complete genome. *PLoS ONE* 5: e10314.

Wright, D.H. 1983. Species-energy theory: An extension of species-area theory. *Oikos* 41: 496–506.

Yarza, P. et al. 2014. Uniting the classification of cultured and uncultured bacteria and archaea using 16S rRNA gene sequences. *Nature Reviews Microbiology* 12: 635–645.

Yuan, S., D.B. Cohen, J. Ravel, Z. Abdo, and L.J. Forney. 2012. Evaluation of methods for the extraction and purification of DNA from the human microbiome. *PLoS ONE* 7: e33865.

Zarubin, M., S. Belkin, M. Ionescu, and A. Genin. 2012. Bacterial bioluminescence as a lure for marine zooplankton and fish. *Proceedings of the National Academy of Sciences* 109: 853–857.

Zengler, K. et al. 2002. Cultivating the uncultured. *Proceedings of the National Academy of Sciences*. 99: 15681–15686.

19

Microbiome Studies of Carrion Decomposition

Jessica L. Metcalf, David O. Carter, and Rob Knight

CONTENTS

19.1 Introduction: Microbial Ecology—A Perspective from Three Domains of Life

Most of the biodiversity on earth is microbial. These microbes provide important functions for the planet, ranging from cycling of nutrients through the environment to metabolizing food within animal digestive tracts. Although we often associate the term "microbe" with bacteria, microbes are actually found in all three domains of life—bacteria, archaea, and eukaryotes (Woese et al. 1990; Pace 1997). Of the over 70 major eukaryotic groups, most of them are so small as to be invisible to the human eye (Keeling et al. 2005; Parfrey et al. 2006, 2011; Adl et al. 2012), and animals and plants only account for a small fraction of eukaryotic diversity. The ecology of any system, including decomposing carrion, thus requires consideration of the full tree of life because organisms from every domain may be contributing and interacting (described in Chapter 20 by Wood and colleagues).

Decomposing carrion is an ephemeral, nutrient-rich ecosystem (Carter et al. 2007), and major changes in microbial diversity occur in and on the carrion as well as in the surrounding soil. Scientists are just beginning to characterize these changes and the functional role of microbes in decomposition. For example, after the abdominal cavity bloats and ruptures, the environment shifts from anaerobic to aerobic, which is unfavorable for many gut bacteria with low oxygen tolerance, such those in the phylum Bacteroidetes (Metcalf et al. 2013). Additionally, as bacterial biomass increases in response to nutrient availability, bacterivorous microbial eukaryotic communities likely respond to new food resources supplied by an abundance of bacterial cells. Therefore, tracking the changes in bacterial, archaeal, and microbial eukaryotic communities can help researchers understand fundamental properties of the carrion ecosystem. For example, do microbial communities change in a consistent manner during decomposition? And are microbial decomposer communities universal, or specific to the host species or surrounding environment?

In this chapter, the goal will be to discuss recent advances and findings in microbiome research (the study of the genes of microbes in the context of complex microbial communities). It will begin by outlining recent advances in microbial community characterization using next-generation sequencing, focusing primarily on the approach of amplicon profiling. This will be followed by a description of why, in addition to sequencing, advances in bioinformatics tools and pipelines were necessary to push microbiome research forward. Within this context, it will be examined as to why and how microbiome research may be useful in providing further insight into the ecology of carrion decomposition. Finally, this chapter will conclude with a summary of recent findings from microbiome research directly related to carrion decomposition and a discussion of the potential of future studies.

19.2 Advances in Sequencing of Microbial Community DNA

The most efficient and cost-effective way to characterize microscopic communities is through culture-independent sequencing. It is estimated that less than 1% of bacterial diversity is culturable (Pace 1997). Because microbial diversity is high, with hundreds to thousands of species occupying a single habitat, quantifying diversity and identifying taxa have been historically challenging. Culture-independent techniques, such as sequencing of DNA directly from environmental samples, provide a snapshot of the full diversity by circumventing the need for culturing, which typically provides a view of only a few microbes that are relatively easy to grow in the laboratory. Early nonculture-based fingerprinting techniques such as Terminal Restriction Length Polymorphisms (TRFLPs), were useful for quantifying diversity (Liu et al. 1997), but did not allow for taxonomic characterization of microbial communities. Sequencing of the taxonomically informative 16S rRNA gene revolutionized the field of microbial ecology, allowing for the first accurate assessments of bacterial biodiversity and taxonomic composition in many systems, such as soils (Fierer and Jackson 2006), marine environments (Fuhrman and Davis 1997; Selje et al. 2004), and mammalian guts (Eckburg et al. 2005; Turnbaugh et al. 2007; Ley et al. 2008). Early surveys utilized clone library and Sanger sequencing approaches to isolate and sequence the 16S gene. Because these data sets were often limited to a few hundred sequences, it became clear that these studies were merely scratching the surface of the diversity present in some systems. Only in the last few years, with the revolution of high-throughput, massively parallel sequencing, it has been possible to characterize the microbial biodiversity on earth, allowing for the first broad and accurate assessments of the phylogenetic diversity of bacterial and archaeal communities across many different environments—from human skin to prairie dog fleas (Costello et al. 2009; Jones et al. 2010; Bates et al. 2011; Caporaso et al. 2011b; Huttenhower et al. 2012).

19.2.1 Advent of Next-Generation Sequencing

Over the last decade, the development of next-generation sequencing has revolutionized and democratized large-scale sequencing efforts. Using the new sequencing technologies (reviewed in Chapter 18), it is now possible to simultaneously generate thousands to millions of sequences per sample for hundreds of samples. These advances in technology have also dramatically decreased the cost of sequencing so that generating millions of sequences now costs less than generating hundreds of sequences would have a few years ago. Further capitalizing on the ability to generate millions of sequences, multiplexing samples of 16S rRNA amplicons has allowed hundreds to thousands of samples to be sequenced in a single sequencing run by using a unique, error-correcting barcode primer tag with each sample (Hamady et al. 2008; Hamady and Knight 2009). Once sequenced, bacterial DNA reads can be demultiplexed and associated with a sample computationally, resulting in phylogenetic and diversity data on hundreds of archaeal and bacterial communities (Figure 19.1).

An essential part of the microbiome sequencing revolution has been the development of computational pipelines and improvement and reduced cost of compute resources to process and analyze the resulting enormous data sets. Several pipelines are available for analyses of microbial community composition, including Quantitative Insights Into Microbial Ecology (QIIME) (Caporaso et al. 2010), Visualization and Analysis of Microbial Population Structures (VAMPS) (http://vamps.mbl.edu), and mothur (Schloss et al. 2009) among others (reviewed in Kuczynski et al. 2012), each of which enables quality filtering, demultiplexing samples, and down-stream analyses (Figure 19.1). QIIME, for example, incorporates phylogenetic community ecology statistics such as UniFrac (Lozupone and Knight 2005; Lozupone et al. 2011) and dimensionality reduction techniques such as Principle Coordinate Analyses (discussed in Gonzalez and Knight 2012), as well as other common ecological measures of community change

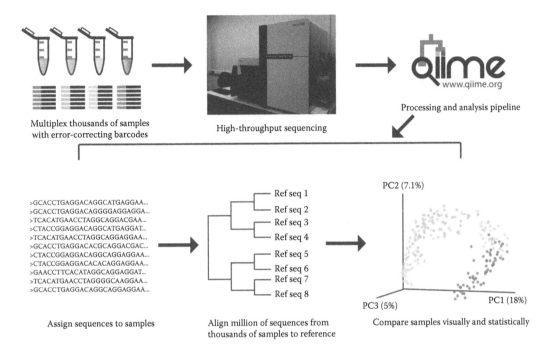

FIGURE 19.1 **(See color insert.)** Overview of data generation, processing, and analysis workflow using QIIME (or other similar packages). After DNA is extracted, a target region (e.g., part of the 16S or 18S rRNA gene) is amplified using error-correcting, Golay barcoded primers (barcodes noted by sequences of different colors). Next, the multiplexed pool of amplicons is sequenced at a depth of millions of sequences. Finally, the open-access software package QIIME is used to demultiplex sequences, align sequences to the reference database (e.g., Greengenes or SILVA), and perform statistical analyses and generate visualizations (e.g., ANOSIM and PCoA plots). (From Hamady, M. et al. 2008. *Nature Methods* 5: 235–237. This figure is available at Figshare.com under the open-access Creative Commons-Attribution License (CC-BY).)

and community difference. Popular software packages for additional statistical analyses include the open software of R's vegan package (Dixon 2003) and the commercial package PRIMER-E (Clarke and Gorley 2006). Many of the processing and analyses steps can be run on laptops or desktops, but additional computational resources such as computer clusters, supercomputers, and cloud computing resources such as Amazon's EC2 are useful for larger data sets and faster results. A discussion of processing and analysis software and the importance of open-source software can be found in Gonzalez and Knight (2012).

19.2.2 Amplicon Profiling the Tree of Life

Deep sequencing of amplicons generated from environmental DNA samples is becoming a widely used approach to profile communities not just for bacteria and archaea, but for many different branches of the tree of life. For example, Bates et al. (2013) surveyed the diversity of protists (unicellular eukaryotes) in a global set of soils representing many of the major biomes on earth. This approach utilized barcoded primers for 18S rRNA amplicons in a similar fashion as the more widely used, high-throughput 16S rRNA amplicon surveys for bacteria and archaea. Similarly, McGuire et al. (2013) utilized the fungal marker (Schoch et al. 2012) nuclear ribosomal internal transcribed spacer (ITS) to characterize fungal communities in New York City's parks and urban green roof soil communities. Amplicon-based approaches are not limited to ribosomal RNA, but can be used for any marker that provides taxonomic resolution for the group of interest and has an available reference database. For example, many scientists utilize the chloroplast markers trnH-psbA or trnl region to identify plants in environmental samples (Kress et al.

2005; Taberlet et al. 2007), and the mitochondrial marker Cytochrome Oxidase 1 for animals, including both invertebrates and vertebrates (Floyd et al. 2002; Hebert et al. 2003, 2004). A combination of these amplicon-based, community-profiling approaches could be applied to studies of carrion decomposition to better understand taxonomic and diversity changes across multiple trophic levels over time.

19.3 Insights from DNA Sequenced-Based Studies of Microbial Ecology

Together, these innovations have allowed researchers to characterize the microbial world across many habitats. This characterization in turn begins to reveal fundamental ecological properties of these communities, such as diversity and structuring across space and time. The structure or composition of bacterial communities is often consistent across major environments or sample types. Many of the first big-picture patterns obtained using bacterial microbiome data sets were from oceans and soils. The Venter et al. (2004) study of the marine microbial ecology of the Sargasso Sea and subsequent Global Ocean Sampling Initiative (Yooseph et al. 2007), along with continental-scale surveys of soils (Fierer and Jackson 2006; Lauber et al. 2009), has provided insights into the taxonomic and functional diversity of microbial communities on a large scale. For example, Fierer and Jackson (2006) surveyed soil microbial communities across North and South America, and discovered that community composition and diversity were not related to temperature and latitude, which commonly predict diversity for macroscopic taxa. Instead, pH drives microbial patterns at the ecosystem scale. Following up on these findings, King et al. (2010) used a nested sampling scheme, collecting soils at distances of 2–2000 m apart. As in Fierer and Jackson (2006), they found that pH was an important predictor of the relative abundance of major bacterial taxa, which has important implications for soil bacterial community changes associated with carrion decomposition; ruptured remains dramatically increase soil pH (King et al. 2010).

Time-series microbiome studies of different environments have allowed detection of patterns in temporal variability and stability, which can also be interpreted as patterns of succession and change. Caporaso et al. (2011a) found that human gut bacterial communities were quite variable over daily time-scales, but when compared to differences among body sites, remained stable (e.g., the variation in a person's gut is very small compared to the difference between gut and mouth bacterial communities). Primary succession has been detected in several environments, such as the infant gut (Koenig et al. 2011). Interestingly, in a study of the secondary succession of bacterial and fungal communities during forest restoration, Banning et al. (2011) uncovered patterns of succession in bacterial communities, but not fungal communities. Recently, Shade et al. (2013) performed a meta-analysis on published time-series bacterial microbiome data sets including air, aquatic, soils, plants, wastewater, and humans. The authors discovered that in many data sets, bacterial community change was often correlated with time, and temporal dynamics often predictable within a habitat. Importantly, the result was robust regardless of which diversity metric (phylogenetic diversity, richness, and evenness) was used, which suggests that studies using different diversity metrics are comparable. Given the results of these time-series studies, it is likely that decomposing carrion, which may represent a secondary succession ecosystem, may exhibit microbial succession over time. In fact, two recent studies by Pechal et al. (2014) and Metcalf et al. (2013), in which decomposing carrion were sampled over time, revealed consistent change in microbial communities during decomposition. These studies are discussed in depth later in the chapter.

Major sequencing initiatives, such as National Institutes of Health Human Microbiome Project (HMP—http://www.hmpdacc.org/), the European Commission's Metagenomics of the Human Intestinal Tract (MetaHIT—http://www.metahit.eu/), and the more recent Earth Microbiome Project (EMP—http://www.earthmicrobiome.org/), now enable scientists to understand far more about the microbial ecology of a wide range of ecosystems, from the human body to marine, freshwater, and terrestrial systems. For example, recent papers have demonstrated that healthy human body sites (e.g., oral, nasal, gut, and skin) have highly structured bacterial communities (Costello et al. 2009; Huttenhower et al. 2012; Yatsunenko et al. 2012), while others have shown that mammal gut communities, in general, are highly distinct (e.g., Ley et al. 2008; Ochman et al. 2010; Muegge et al. 2011) and distinguishable from other ecosystems (Lozupone and Knight 2007). The wealth of information now available for host-associated and soil microbes, in particular, provides a rich context for studies focusing on the microbes that are involved in carrion decomposition. Importantly, these large initiatives also provide standards for laboratory protocols, such as metadata formatting using MIMARKS (Yilmaz et al. 2011), DNA extraction methods, and primer selection (e.g., http://www.earthmicrobiome.org/emp-standard-protocols/). Although methods and sequencing technologies are always changing, efforts by research groups to use similar protocols can have a major pay off by allowing independent research studies to be easily comparable.

19.4 Building a Microbial Ecology Knowledgebase for Carrion Decomposition

One of the most exciting aspects of investigating the microbiome of carrion decomposition is that it will complement the knowledge generated from culture-based and biochemical studies. Although the understanding of carrion microbiology is not detailed, it has some breadth. Much of the understanding of carrion microbial communities comes from postmortem microbiology and food science. These fields have shown that the gut microbiota responds rapidly to the availability of a carcass. The metabolism of these microbes is responsible for putrefaction, bloating, and the release of purge fluids into the wider ecosystem (Gill-King 1997). Their ability to rapidly use a human corpse as a resource has consistently been a problem for pathologists, because putrefactive bacteria can migrate through tissues and confound postmortem examination (Kellerman et al. 1976). These microbes also are problematic for food science because their activity results in food spoilage (Ingram and Dainty 1971). Food spoilage is likely a result of fierce competition between microbes and larger carrion consumers that has resulted in bacteria-evolving competitive strategies to make the carrion source highly unattractive to other consumers (Janzen 1977). Food spoilage studies have been valuable because they show the types of culturable microbes that are present on and in a carcass during the early postmortem period (Gill and Newton 1977; Newton and Gill 1978; Lahellec and Colin 1979; Dainty and Mackey 1992). This work also provides some details of the biochemical studies conducted with an ecological perspective (Carter and Tibbett 2006).

Although carrion hosts a great concentration of internal and external microorganisms, primarily bacteria, relatively few carrion microbial ecology studies have been conducted to date. However, these studies have allowed recognition that carrion plays a crucial role in ecosystems by serving as a discrete nutrient patch that facilitates biodiversity and the dispersal of chemicals and biota throughout a habitat, including effects on ecosystem processes (Carter et al. 2007; Bump et al. 2009; Parmenter and MacMahon 2009; Hawlena et al. 2012; Barton et al. 2013). Three primary trends have been observed from this collection of studies: carrion availability triggers microbial activity, supports an increase in microbial abundance, and is associated with shifts in the microbial community structure as a carcass breaks down. These processes are supported by the high level of energy and nutrients present in carcasses, which are apparently the most nutritious resource available to decomposer organisms.

Many of the bacteria associated with carrion, including the genera *Acinetobacter*, *Clostridium*, *Proteus*, and *Streptococcus*, are present during life and contribute to carrion breakdown after death (Micozzi 1986; Howard et al. 2010). However, carrion microbial activity is not restricted to these endogenous microbial communities. Carter et al. (2010) have shown that soil microbial activity is triggered rapidly by the presence of a carcass. This activity is associated with an increase in microbial abundance (Carter et al. 2008), and these microbes release a significant amount of carbon dioxide (Putman 1978) and enzymes associated with the cycling of nitrogen and phosphorus (Carter et al. 2010). These biochemical observations are not surprising because a carcass contains many nutrients including nitrogen, phosphorus, sulfur, potassium, ammonium, nitrate, and phosphate (Vass et al. 1992; Towne 2003; Benninger et al. 2008; Carter et al. 2008; Bump et al. 2009; Parkinson et al. 2009; Parmenter and MacMahon 2009; Stokes et al. 2009; Damann 2010; Spicka et al. 2011; Anderson et al. 2013; Meyer et al. 2013).

More recent work using fatty acid methyl esters and phospholipid fatty acids as indices of the microbial community structure has shown that carrion microbial communities change during decomposition (Parkinson et al. 2009; Damann 2010). Little is known in detail about how these communities change, but it is logical that change should occur; changes in available nutrients typically stimulates a shift in the decomposer community structure (Swift et al. 1979). It appears that the intrinsic microbiota initiates carrion breakdown while the external microorganisms participate after carcasses and nutrients become available. Bacteria dominate the early stages of decomposition, and fungi tend to proliferate during later stages of breakdown (Tibbett and Carter 2003; Sagara et al. 2008). However, although microorganisms are the most abundant carrion decomposers, they typically do not consume the majority of a carcass. That service is provided by insects or other scavengers (Putman 1978). Microbes play a crucial role because their activity releases a diversity of compounds into the atmosphere that attract these decomposers. Thus, it seems that microorganisms facilitate carrion breakdown more than they directly carry out carrion breakdown. In turn, scavengers may bring their own communities of microbes, which they introduce to the carcass. Future studies on carrion decomposition ecology will illuminate some of these complex interactions among microscopic and macroscopic carrion consumers.

Much of the complexity in understanding carrion breakdown is in the relationships between decomposition, resource quality, physicochemical environment, and decomposer community (see Swift et al. 1979). The environment in which a carcass becomes available also has a significant impact on decomposition where temperature, moisture, pH, and oxygenation all modulate carrion microbial communities (Carter et al. 2007). An increase in microbial activity is associated with an increase in temperature (Carter et al. 2008) and moisture content, although optimal levels can be exceeded. Temperatures above 35°C are associated with enzyme denaturation and desiccation of carrion, which can inhibit microbial metabolism. Similarly, optimal moisture content can be exceeded where aerobic metabolism is depressed (Carter et al. 2010). Conversely, rupturing and insect tunneling allow oxygen into carrion and accelerate decomposition (Putman 1978).

The relationship between carrion breakdown and pH has been of great interest (Haslam and Tibbett 2009; Parkinson et al. 2009; Carter et al. 2010; Damann et al. 2012; Anderson et al. 2013; Meyer et al. 2013). As discussed earlier in this chapter, many microbes are greatly affected by the pH of their environment and the differences in diversity and richness of microbial diversity across ecosystems, for example, can be largely explained by pH (Fierer and Jackson 2006). Previous decomposition studies have shown that carrion tend to become acidic during decomposition (Clark et al. 2007) but associated soils tend to initially increase in alkalinity up to pH 8 or 9 (Carter et al. 2008, 2010); however this phenomenon appears related to soil type (Benninger et al. 2008; Van Belle et al. 2009). It will be very interesting to explore these relationships through analyzing the whole carrion microbiome.

19.4.1 Current State of Carrion Microbiome Research

Carrion decomposition research using TRFLPs (Parkinson et al. 2009) and other amplicon length polymorphism techniques (Moreno et al. 2011), quantitative PCR (Howard et al. 2010), sequencing of culture isolates (Howard et al. 2010), and Sanger-based DNA sequencing of amplicons cloned into *E. coli* (Dickson et al. 2011) have provided some general patterns and highlighted taxa that may be important in the decomposition process. For example, Parkinson et al. (2009) used both bacterial 16S rRNA gene and fungal ITS TRFLPs to investigate community change in soils associated with decomposing human cadavers. Their results suggested that both bacterial and fungal communities become progressively differentiated from starting communities during decomposition, and in some cases in a succession-like manner. Howard et al. (2010) observed that the abundance of lipolytic bacteria, which breakdown fats, increased in gravesoil during carcass decomposition while the abundance of proteolytic bacteria, which breakdown protein, decreased. These functional changes coincide with a putative decrease in nitrogen-fixing bacteria (Moreno et al. 2011), probably because the carcass itself acts as a source of nitrogen in the form of ammonium, although the nitrification of ammonium-N at gravesites in not well understood and may be more complex than originally assumed (Aitkenhead-Peterson et al. 2012). These studies, coupled with previously described culture-based and biochemical studies, suggest that microbial communities are responding with shifts in taxonomic composition and functional capacity to the ephemeral nutrient pulse made available by decomposing carrion. Therefore, microbial DNA sequence data at depths of thousands to millions of sequences per carrion-associated sample would likely shed additional light on the systems' ecology, as microbiome research has done for many other ecosystems.

Indeed, carrion microbiome studies utilizing next-generation sequencing techniques are being applied to better characterize these complex microbial communities by providing taxonomy via phylogenetic placement and estimating each taxon's relative abundance in the community. Recently, Pechal et al. (2014) demonstrated that skin and mouth bacterial communities followed a consistent trajectory of change during decomposition of three swine corpses over five days, confirming results of a succession-like pattern detected in earlier studies such as Parkinson et al. (2009). They discovered that overall taxa in the Phyla Firmicutes increased and Proteobacteria decreased. At the family level, they discovered that the taxa Enterobacteriaceae, Planococcaceae, Clostridiales *incertae sedis*, increased during decomposition.

In another recent microbiome study, Metcalf et al. (2013) utilized both 16S and 18S rRNA amplicon deep sequencing to characterize the postmortem microbial community changes in the abdominal cavity, associated soil, and on the skin in a mouse model system. They discovered dramatic, measurable, and repeatable changes in both bacterial and microbial eukaryotic communities at each site during decomposition. Because they used a mouse model system and destructive sampling, Metcalf et al. (2013) were able to test long-held hypotheses about the microbial ecology of the abdominal cavity. As it has long been assumed (Evans 1963), endogenous gut microbes such as members of the Phylum Firmicutes increase during bloating before rupture and then plummet in abundance postrupture at which point common environmental aerobic bacterial taxa from groups such as Alphaproteobacteria and Gammaproteobacteria increase in abundance. Similar to Pechal et al. (2014), they discovered that bacterial communities from each sample site changed significantly and consistently with time (Metcalf et al. 2013), demonstrating microbial succession on these ephemeral resources. However, in contrast to Pechal et al. (2014), taxa in the Phylum Firmicutes did not increase during decomposition, taxa in Proteobacteria increased (Figure 19.2). It is possible that this difference in abundant taxa associated with later stages of decomposition is due to the different environments of the respective studies. Early meat spoilage literature suggests that Clostridia taxa becomes abundant under warmer conditions (Ingram and Dainty 1971), which could explain why an increase in abundance of Firmicutes was detected in the study by Pechal et al. (2014), which was conducted outdoors during summer in Ohio. The study by Metcalf et al. (2013) was conducted in a laboratory setting with controlled ambient temperature (~70°F) and low humidity. Despite these different decomposition environments, there was an overlap in some of the taxa that both studies found as important indicators for the postmortem interval of the cadavers, including Xanthomonadaceae, Brucellaceae, and Lachnospiraceae. Therefore, it is possible that at least some components of the bacterial carrion decomposer community are universal across environments and host taxa.

Metcalf et al. (2013) also provided the first next-generation sequence-based characterization of microbial eukaryote community change during decomposition. Similar to bacterial communities, they found that microbial eukaryotes also changed significantly and consistently over decomposition time. The Rhabditidae nematode, *Oscheius tipulae*, dominated the community during the advanced decay stages of decomposition (Figure 19.3). The nematode bloom was likely a response to the increase in bacterial biomass that is associated with decomposition (Benninger et al. 2008; Carter et al. 2008, 2010; Parkinson et al. 2009; Damann et al. 2012), indicating that trophic interactions are likely important to take into account during decomposition. Interestingly, in a follow-up study in which mice were

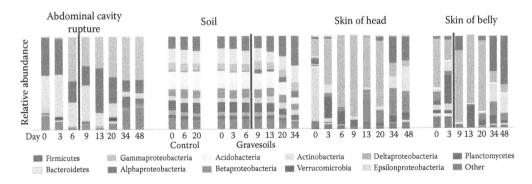

FIGURE 19.2 **(See color insert.)** Relative abundance of phyla of bacteria over time for all body sites. Control soils were averaged across time points. For the abdominal site, day 0 includes cecum, fecal, and abdominal swab and liquid samples. For the soil site, control soils collected on days 0, 6, and 20 are shown on the left of the plot. (From Metcalf, J.L. et al. 2013. *eLife* 2: e01104 and is open access under the Creative Commons-Attribution License (CC-BY).)

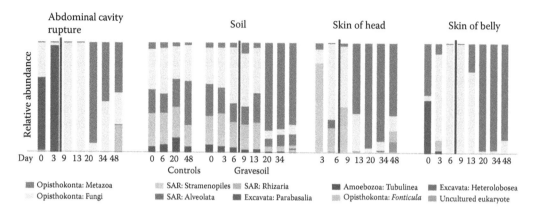

FIGURE 19.3 **(See color insert.)** Relative abundance of microbial eukaryote taxa at the class level over time. Microbial eukaryotic community composition changes significantly and predictably over the course of decomposition. Eukaryotic community composition changes directionally and becomes dominated by the nematode *Oscheius tipulae*. (From Schoch, C.L. et al. 2012. *Proceedings of the National Academy of Sciences* 109: 6241–6246; Metcalf, J.L. et al. 2013. *eLife* 2: e01104) and is open access under the Creative Commons-Attribution License (CC-BY).)

decomposed on three contrasting soils, nematodes dominate all late-stage decomposition soils (Metcalf et al. in prep). These results suggest that microbial eukaryotes may respond to carrion in a similar way across very different environments. One potential contributing factor to the nematode bloom may have been the fact that the laboratory-based mouse experiment may have prevented the entry of organisms preying on nematodes so it will be interesting to see if these patterns are detected in outdoor carrion experiments.

Overall, these microbiome studies suggest that both bacterial communities and microbial eukaryotic communities play key roles in decomposition, and that their successional patterns are highly reproducible. Surprisingly, archael taxa may not be major contributors, or at least were not detected at moderate or high abundance in Metcalf et al. (2013), and did not correlate with the successional stage. Integrating information across taxa, together with functional studies of the metabolic and regulatory networks involved in decomposition, will enable improved mechanistic models of the ecology of decomposition. This information will prove useful in applications such as the forensic sciences and estimating postmortem intervals.

19.5 Future Uses of Current Technologies for Research on Carrion Decomposition

A key unresolved question is: where do the microbes that become abundant during decomposition primarily come from—the host, the soil, the air, insects, scavengers, or elsewhere? This question can be addressed through a modified application of existing technology, where the goal is to sequence a few samples extremely deeply rather than getting broad coverage of a large number of samples. In this context, a crucial ecological property that is important to understand for bacteria is the ubiquity or endemism of the bacterial "seedbank." A long-standing hypothesis for bacteria is that "everything is everywhere, but the environment selects," meaning that all bacteria persist in a seedbank and the environment determines which taxa become abundant (Bass Becking 1934; O'Malley 2007). Supporting the idea of a persistent seedbank, recent work by Caporaso et al. (2012) showed that in the English Channel, which experiences major seasonal shifts in bacterial communities, >95% of all taxa surveyed over a six-year study could be detected in a single sample sequenced at a depth of ~10 million 16S reads. These results suggest that most bacteria may persist in environments at a very low abundance. In the case of carrion, decomposer bacterial communities may be universal across different environments (e.g., soil or substrate types) if a seedbank exists, although different taxa may become abundant under different conditions (e.g., temperature, humidity). As carrion microbiome studies emerge using different host taxa and across different environments (including aquatic), we will begin to understand how universal or specific these carrion decomposer communities are, and how to combine edaphic and climatic factors with host factors to improve predictive models.

Studies to date of microbial eukaryote communities involved in corpse decomposition have typically not focused on the fungi, yet the technology to do this exists, as do appropriate PCR primers. Understanding the role of fungi in corpse decomposition environments may improve both mechanistic understanding and predictive power.

The key challenge at this point is not so much in the sequencing technology, since powerful laboratory methods and software now exist for revealing and aiding understanding of microbial patterns and processes. Rather, we must create an extensive database of potential microbial sources, and of entire decomposition trajectories under different conditions, in order to understand what is specific to an individual event and what is generalizable and can provide a robust framework for understanding. Investment in such a decomposition database could have profound implications for forensic investigation, but potentially also for wildlife studies, monitoring of endangered species, and food safety.

REFERENCES

Adl, S.M. et al. 2012. The revised classification of eukaryotes. *Journal of Eukaryotic Microbiology* 59: 429–493.

Aitkenhead-Peterson, J.A., C.G. Owings, M.B. Alexander, N. Larison, and J.A. Bytheway. 2012. Mapping the lateral extent of human cadaver decomposition with soil chemistry. *Forensic Science International* 216: 127–134.

Anderson, B., J. Meyer, and D.O. Carter. 2013. Dynamics of ninhydrin-reactive nitrogen and pH in gravesoil during the extended postmortem interval. *Journal of Forensic Sciences* 58: 1348–1352.

Banning, N.C. et al. 2011. Soil microbial community successional patterns during forest ecosystem restoration. *Applied and Environmental Microbiology* 77: 6158–6164.

Barton, P.S., S.A. Cunningham, D.B. Lindenmayer, and A.D. Manning. 2013. The role of carrion in maintaining biodiversity and ecological processes in terrestrial ecosystems. *Oecologia* 171: 761–772.

Bass Becking, L.G.M. 1934. *Geobiologie of Inleiding Tot De Milieukunde*. The Netherlands: WPVan Stockum & Zoon.

Bates, S.T. et al. 2011. Examining the global distribution of dominant archaeal populations in soil. *ISME Journal* 5: 908–917.

Bates, S.T. et al. 2013. Global biogeography of highly diverse protistan communities in soil. *ISME Journal* 7: 652–659.

Benninger, L.A., D.O. Carter, and S.L. Forbes. 2008. The biochemical alteration of soil beneath a decomposing carcass. *Forensic Science International* 180: 70–75.

Bump, J.K. et al. 2009. Ungulate carcasses perforate ecological filters and create biogeochemical hotspots in forest herbaceous layers allowing trees a competitive advantage. *Ecosystems* 12: 996–1007.

Caporaso, J.G. et al. 2010. QIIME allows analysis of high-throughput community sequencing data. *Nature Methods* 7: 335–336.

Caporaso, J.G. et al. 2011a. Moving pictures of the human microbiome. *Genome Biology* 12: R50.

Caporaso, J.G. et al. 2011b. Global patterns of 16S rRNA diversity at a depth of millions of sequences per sample. *Proceedings of the National Academy of Sciences* 108: 4516–4522.

Caporaso, J.G., K. Paszkiewicz, D. Field, R. Knight, and J.A. Gilbert. 2012. The western English Channel contains a persistent microbial seed bank. *ISME Journal* 6: 1089–1093.

Carter, D.O. and M. Tibbett. 2006. Microbial decomposition of skeletal muscle tissue (*Ovis aries*) in a sandy loam soil at different temperatures. *Soil Biology & Biochemistry* 38: 1139–1145.

Carter, D.O., D. Yellowlees, and M. Tibbett. 2007. Cadaver decomposition in terrestrial ecosystems. *Naturwissenschaften* 94: 12–24.

Carter, D.O., D. Yellowlees, and M. Tibbett. 2008. Temperature affects microbial decomposition of cadavers (*Rattus rattus*) in contrasting soils. *Applied Soil Ecology* 40: 129–137.

Carter, D.O., D. Yellowlees, and M. Tibbett. 2010. Moisture can be the dominant environmental parameter governing cadaver decomposition in soil. *Forensic Science International* 200: 60–66.

Clarke, K.R. and Gorley, R.N. 2006. PRIMER v6. User manual/tutorial. Plymouth routine in mulitvariate eco-
    logical research. Plymouth Marine Laboratory.
Clark, M.A., M.B. Worrell, and J.E. Pless. 2007. Postmortem changes in soft tissues. In: W.D. Haglund and
    M.H. Sorg, (eds.), *Forensic Taphonomy: The Postmortem Fate of Human Remains*, pp. 151–164. Boca
    Raton, FL: CRC Press.
Costello, E.K. et al. 2009. Bacterial community variation in human body habitats across space and time.
    Science 326: 1694–1697.
Dainty, R.H. and B.M. Mackey. 1992. The relationship between the phenotypic properties of bacteria from
    chill-stored meat and spoilage processes. *Journal of Applied Bacteriology* 73: S103–S114.
Damann, F.E. 2010. Human decomposition ecology at the University of Tennessee Anthropology Research
    Facility [dissertation]. Knoxville, TN: Univ. of Tennessee.
Damann, F.E., A. Tanittaisong, and D.O. Carter. 2012. Potential carcass enrichment of the University of
    Tennessee anthropology research facility: A baseline survey of edaphic features. *Forensic Science
    International* 222: 4–10.
Dickson, G.C., R.T.M. Poulter, E.W. Maas, P.K. Probert, and J.A. Kieser. 2011. Marine bacterial succession
    as a potential indicator of postmortem submersion interval. *Forensic Science International* 209: 1–10.
Dixon, P. 2003. Vegan, a package of R functions for community ecology. *Journal of Vegetation Science* 14: 927–930.
Eckburg, P.B. et al. 2005. Diversity of the human intestinal microbial flora. *Science* 308: 1635–1638.
Evans, W.E.D. 1963. *The Chemistry of Death.* Springfield, IL: Charles C. Thomas.
Fierer, N. and R.B. Jackson. 2006. The diversity and biogeography of soil bacterial communities. *Proceedings
    of the National Academy of Sciences* 103: 626–631.
Floyd, R., E. Abebe, A. Papert, and M. Blaxter. 2002. Molecular barcodes for soil nematode identification.
    Molecular Ecology 11: 839–850.
Fuhrman, J.A. and A.A. Davis. 1997. Widespread archaea and novel bacteria from the deep sea as shown by
    16S rRNA gene sequences. *Marine Ecology Progress Series* 150: 275–285.
Gill, C.O. and K.G. Newton. 1977. Development of aerobic spoilage flora on meat stored at chill temperatures.
    Journal of Applied Bacteriology 43: 189–195.
Gill-King, H. 1997. Chemical and ultrastructural aspects of decomposition. In: W.D. Haglund, and M.H. Sorg,
    (eds.), *Forensic Taphonomy: The Postmortem Fate of Human Remains.* Boca Raton, FL: CRC Press,
    pp. 93–108.
Gonzalez, A. and R. Knight. 2012. Advancing analytical algorithms and pipelines for billions of microbial
    sequences. *Current Opinion in Biotechnology* 23: 64–71.
Hamady, M. and R. Knight. 2009. Microbial community profiling for human microbiome projects: Tools, tech-
    niques, and challenges. *Genome Research* 19: 1141–1152.
Hamady, M., J.J. Walker, J.K. Harris, N.J. Gold, and R. Knight. 2008. Error-correcting barcoded primers for
    pyrosequencing hundreds of samples in multiplex. *Nature Methods* 5: 235–237.
Haslam, T. and M. Tibbett. 2009. Soils of contrasting pH affect the decomposition of buried mammalian (*Ovis
    aries*) skeletal muscle tissue. *Journal of Forensic Sciences* 54: 900–904.
Hawlena, D., M.S. Strickland, M.A. Bradford, and O.J. Schmitz. 2012. Fear of predation slows plant-litter
    decomposition. *Science* 336: 1434–1438.
Hebert, P.D.N., S. Ratnasingham, and J.R. Dewaard. 2003. Barcoding animal life: Cytochrome C oxidase sub-
    unit 1 divergences among closely related species. *Proceedings of the Royal Society B-Biological Sciences*
    270: S96–S99.
Hebert, P.D.N., M.Y. Stoeckle, T.S. Zemlak, and C.M. Francis. 2004. Identification of birds through DNA
    barcodes. *PLoS Biology* 2: 1657–1663.
Howard, G.T., B. Duos, and E.J. Watson-Horzelski. 2010. Characterization of the soil microbial community
    associated with the decomposition of a swine carcass. *International Biodeterioration & Biodegradation*
    64: 300–304.
Huttenhower, C. et al. 2012. Structure, function and diversity of the healthy human microbiome. *Nature* 486:
    207–214.
Ingram, M. and R.H. Dainty. 1971. Changes caused by microbes in spoilage of meats. *Journal of Applied
    Bacteriology* 34: 21–39.
Janzen, D.H. 1977. Why fruits rot, seeds mold, and meat spoils. *American Naturalist* 111: 691–713.
Jones, R.T., R. Knight, and A.P. Martin. 2010. Bacterial communities of disease vectors sampled across time,
    space, and species. *ISME Journal* 4: 223–231.

Keeling, P.J. et al. 2005. The tree of eukaryotes. *Trends in Ecology & Evolution* 20: 670–676.

Kellerman, G.D., N.G. Waterman, and L.F. Scharfenberger. 1976. Demonstration *in vitro* of postmortem bacterial transmigration. *American Journal of Clinical Pathology* 66: 911–915.

King, A.J. et al. 2010. Biogeography and habitat modelling of high-alpine bacteria. *Nature Communications* 1 : Article 53.

Koenig, J.E. et al. 2011. Succession of microbial consortia in the developing infant gut microbiome. *Proceedings of the National Academy of Sciences* 108: 4578–4585.

Kress, W.J., K.J. Wurdack, E.A. Zimmer, L.A. Weigt, and D.H. Janzen. 2005. Use of DNA barcodes to identify flowering plants. *Proceedings of the National Academy of Sciences* 102: 8369–8374.

Kuczynski, J. et al. 2012. Experimental and analytical tools for studying the human microbiome. *Nature Reviews Genetics* 13: 47–58.

Lahellec, C. and P. Colin. 1979. Bacterial-flora of poultry–changes due to variations in ecological conditions during processing and storage. *Archiv für Lebensmittelhygiene* 30: 95–98.

Lauber, C.L., M. Hamady, R. Knight, and N. Fierer. 2009. Pyrosequencing-based assessment of soil pH as a predictor of soil bacterial community structure at the continental scale. *Applied and Environmental Microbiology* 75: 5111–5120.

Ley, R.E. et al. 2008. Evolution of mammals and their gut microbes. *Science* 320: 1647–1651.

Liu, W.T., T.L. Marsh, H. Cheng, and L.J. Forney. 1997. Characterization of microbial diversity by determining terminal restriction fragment length polymorphisms of genes encoding 16S rRNA. *Applied and Environmental Microbiology* 63: 4516–4522.

Lozupone, C. and R. Knight. 2005. Unifrac: A new phylogenetic method for comparing microbial communities. *Applied and Environmental Microbiology* 71: 8228–8235.

Lozupone, C.A. and R. Knight. 2007. Global patterns in bacterial diversity. *Proceedings of the National Academy of Sciences* 104: 11436–11440.

Lozupone, C., M.E. Lladser, D. Knights, J. Stombaugh, and R. Knight. 2011. Unifrac: An effective distance metric for microbial community comparison. *ISME Journal* 5: 169–172.

Mcguire, K.L. et al. 2013. Digging the New York City skyline: Soil fungal communities in green roofs and city parks. *PLoS One* 8: e58020.

Metcalf, J.L. et al. 2013. A microbial clock provides an accurate estimate of the postmortem interval. *eLife* 2: e01104.

Meyer, J., B. Anderson, and D.O. Carter. 2013. Seasonal variation of carcass decomposition and gravesoil chemistry in a cold (Dfa) climate. *Journal of Forensic Sciences* 58: 1175–1182.

Micozzi, M.S. 1986. Experimental study of postmortem change under field conditions: Effects of freezing, thawing, and mechanical injury. *Journal of Forensic Sciences* 31: 953–961.

Moreno, L.I. et al. 2011. The application of amplicon length heterogeneity PCR (LH-PCR) for monitoring the dynamics of soil microbial communities associated with cadaver decomposition. *Journal of Microbiological Methods* 84: 388–393.

Muegge, B.D. et al. 2011. Diet drives convergence in gut microbiome functions across mammalian phylogeny and within humans. *Science* 332: 970–974.

Newton, K.G. and C.O. Gill. 1978. Development of anaerobic spoilage flora of meat stored at chill temperatures. *Journal of Applied Bacteriology* 44: 91–95.

Ochman, H. et al. 2010 Evolutionary relationships of wild hominids recapitulated by gut microbial communities. *PLoS Biology* 8: e1000546.

O'Malley, M.A. 2007. The nineteenth century roots of "everything is everywhere." *Nature Reviews Microbiology* 5: 647–651.

Pace, N.R. 1997. A molecular view of microbial diversity and the biosphere. *Science* 276: 734–740.

Parfrey, L.W. et al. 2006. Evaluating support for the current classification of eukaryotic diversity. *PLoS Genetics* 2: 2062–2073.

Parfrey, L.W., D.J.G. Lahr, A.H. Knoll, and L.A. Katz. 2011. Estimating the timing of early eukaryotic diversification with multigene molecular clocks. *Proceedings of the National Academy of Sciences* 108: 13624–13629.

Parkinson, R.A. et al. 2009. Microbial community analysis of human decomposition in soil. In: K. Ritz, L.A. Dawson, and D. Miller, (eds.), *Criminal and Environmental Soil Forensics*, pp. 379–394. New York, NY, USA: Springer.

Parmenter, R.R. and J.A. Macmahon. 2009. Carrion decomposition and nutrient cycling in a semiarid shrub-steppe ecosystem. *Ecological Monographs* 79: 637–661.

Pechal, J.L., T. Crippen, M.E. Benbow, A. Tarone, and J.K. Tomberlin. 2014. The potential use of bacterial community succession in forensics as described by high throughput metagenomic sequencing. *International Journal of Legal Medicine* 128: 193–205.

Putman, R.J. 1978. Flow of energy and organic matter from a carcase during decomposition. 2. Decomposition of small mammal carrion in temperate systems. *Oikos* 31: 58–68.

Sagara, N., T. Yamanaka, and M. Tibbett. 2008. Soil fungi associated with graves and latrines: Toward a forensic mycology. In: M. Tibbett and D.O. Carter, (eds.), *Soil Analysis in Forensic Taphonomy: Chemical and Biological Effects of Buried Human Remains*, pp. 67–108. Boca Raton, FL, USA: CRC Press.

Schloss, P.D. et al. 2009. Introducing mothur: Open-source, platform-independent, community-supported software for describing and comparing microbial communities. *Applied and Environmental Microbiology* 75: 7537–7541.

Schoch, C.L. et al. 2012. Nuclear ribosomal internal transcribed spacer (ITS) region as a universal DNA barcode marker for fungi. *Proceedings of the National Academy of Sciences* 109: 6241–6246.

Selje, N., M. Simon, and T. Brinkhoff. 2004. A newly discovered roseobacter cluster in temperate and polar oceans. *Nature* 427: 445–448.

Shade, A., J.G. Caporaso, J. Handelsman, R. Knight, and N. Fierer. 2013. A meta-analysis of changes in bacterial and archaeal communities with time. *ISME Journal* 7: 1493–1506.

Spicka, A., R. Johnson, J. Bushing, L.G. Higley, and D.O. Carter. 2011. Carcass mass can influence rate of decomposition and release of ninyydrin-reactive nitrogen. *Forensic Science International* 209: 80–85.

Stokes, K.L., S.L. Forbes, and M. Tibbett. 2009. Freezing skeletal muscle tissue does not affect its decomposition in soil: Evidence from temporal changes in tissue mass, microbial activity and soil chemistry based on excised samples. *Forensic Science International* 183: 6–13.

Swift, M.J., O.W. Heal, and J.M. Anderson. 1979. *Decomposition in Terrestrial Ecosystems*. Oxford, UK: Blackwell Scientific.

Taberlet, P. et al. 2007. Power and limitations of the chloroplast *trnl* (UAA) intron for plant DNA barcoding. *Nucleic Acids Research* 35: e14 (epub: Apr 21, 2015).

Tibbett, M. and D.O. Carter. 2003. Mushrooms and taphonomy: The fungi that mark woodland graves. *Mycologist* 17: 20–24.

Towne, E.G. 2003. Prairie vegetation and soil nutrient responses to ungulate carcasses. *Oecologia* 122: 232–239.

Turnbaugh, P.J. et al. 2007. The human microbiome project. *Nature* 449: 804–810.

Van Belle, L.E., D.O. Carter, and S.L. Forbes. 2009. Measurement of ninhydrin reactive nitrogen influx into gravesoil during aboveground and belowground carcass (*Sus Domesticus*) decomposition. *Forensic Science International* 193: 37–41.

Vass, A.A., W.M. Bass, J.D. Wolt, J.E. Foss, and J.T. Ammons. 1992. Time since death determinations of human cadavers using soil solution. *Journal of Forensic Sciences* 37: 1236–1253.

Venter, J.C. et al. 2004. Environmental genome shotgun sequencing of the Sargasso Sea. *Science* 304: 66–74.

Woese, C.R., O. Kandler, and M.L. Wheelis. 1990. Towards a natural system of organisms—Proposal for the domains Archaea, Bacteria, and Eucarya. *Proceedings of the National Academy of Sciences* 87: 4576–4579.

Yatsunenko, T. et al. 2012. Human gut microbiome viewed across age and geography. *Nature* 486: 222–227.

Yilmaz, P. et al. 2011. Minimum information about a marker gene sequence (MIMARKS) and minimum information about any (X) sequence (Mixs) specifications. *Nature Biotechnology* 29: 415–420.

Yooseph, S. et al. 2007. The Sorcerer II global ocean sampling expedition: Expanding the universe of protein families. *PLoS Biology* 5: 432–466.

20

Interkingdom Ecological Interactions of Carrion Decomposition

Heather R. Jordan, Jeffery K. Tomberlin, Thomas K. Wood, and M. Eric Benbow

CONTENTS

20.1 Introduction

20.1.1 Ecological Interactions

Organisms interact in a myriad of ways. Some forms of interaction occur directly between organisms (e.g., predation or resource competition), whereas others are indirect such as modifications of an environment by one individual impacting another (e.g., a carcass changing the nutrient conditions of soil microbial communities). These interactions can be intraspecific or interspecific (see Chapter 11). Both can impact fitness and drive selection resulting in the evolution of some of the complex connections among species in food webs and ecosystems.

Such interactions can also occur between or among organisms that represent different kingdoms (interkingdom) or domains (interdomain) of life; however, these ecological relationships have been historically understudied compared to intra- and interspecific interactions. For this chapter, the direct and indirect relationships among organisms that represent broadly different taxonomic groups will be considered as interkingdom interactions, with the recognition that they occur both among kingdoms and domains of life.

All organisms rely on the ability to detect, process, and respond to stimuli within their environment. Multicellular organisms must differentiate evolutionarily important signals from background noise of a constantly changing environment. There is precedence that much of these stimuli are associated with unicellular communities, often described as the microbiome (Lowery et al. 2008; Bravo et al. 2011; Koch and Schmid-Hempel 2011; Ezenwa et al. 2012; Tomberlin et al. 2012). Many of these studies have focused on host-associated microbial communities or the importance of prokaryotic pathogens in mediating animal behavior. Few studies have investigated the mechanisms driving microbial-mediated animal behavior, and fewer still have experimentally manipulated microbial communities to test physiological and behavioral responses of eukaryotes (Li et al. 2009; Huttenhower et al. 2012). Next generation high-throughput metagenomic sequencing (NGS) technologies now allow scientists to advance the behavioral sciences by describing complex microbial communities, while genetically manipulating specific microbial species to pinpoint key pathways resulting in the production of compounds that serve as mechanisms regulating animal behavior.

In carrion systems, a carcass is no longer a living (collective) organism, but rather a resource patch that is composed of and attracts many species that represent all domains of life (i.e., Eukarya, Archaea, and Bacteria). At this patch, organismal interactions have been well studied and understood for some communities (e.g., insect succession) (see Chapters 4 and 8), while others have only recently been investigated as a response to the development of cost-effective molecular technologies (e.g., bacterial communities studied with NGS).

These ecological interactions in association with carrion have generally been studied only within the same kingdom. For instance, species succession on a carcass has a history of research that focused on the different arthropod species that colonize and sequentially replace each other throughout decomposition (Payne 1965; Schoenly and Reid 1987), with only cursory mention of the microbial communities that may be important for the initial colonization and changing resource quality that affects the fitness of these insects. Alternatively, there have been previous studies on bacteria and fungi associated with carcasses or the soil beneath them, but much of this research was limited to culturable species or the use of nongenomic techniques; approaches (Vass 2001; Carter and Tibbett 2003; Dickson et al. 2010) that could not provide comprehensive community information like that produced using NGS. Recent work has begun to evaluate multiple kingdoms during carrion decomposition (Ma et al. 2012; Tomberlin et al. 2012), uncovering new relationships that have potential ecological and evolutionary importance to carrion decomposition. The focus of this chapter will be on these relationships, the molecular and physiological mechanisms that control them, in addition to a review of similar interactions understood from other systems and their potential importance to carrion ecology, evolution, and their applications.

20.1.2 Carrion as a Model to Study Interkingdom Interactions

A key component for better understanding the influence of intra and interspecies actions on corresponding animal behaviors is to explore a model system that can be effectively used in replicated field experiments and representative of natural circumstances. The carrion system may be one such model that can be used to answer a fundamental question: *How do animals detect and respond to evolutionarily important signals in a complex environment?*

The carrion (decomposing heterotrophic biomass) system, that includes carrion-associated microbes and blow flies (Diptera: Calliphoridae) is an excellent model to test mechanisms of interkingdom interactions from the cellular to the ecosystem levels (Figure 20.1). As pointed out in many of the other chapters (Chapters 1, 7, 8, 10, 11) in this book, this system is exceptionally well suited to investigate physiological and behavioral components of interkingdom interactions because of the following:

  i. carrion decomposition is a ubiquitous, but ephemeral, ecosystem process;
  ii. blow flies and other necrophagous arthropods have been extensively studied;

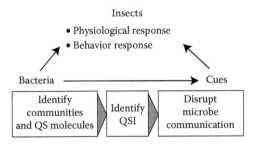

FIGURE 20.1 (See color insert.) Conceptual diagram demonstrating the links between bacteria, their quorum sensing communication, and cues to insects that have been documented in the literature.

iii. necrophagous arthropods and microbes interact on these resources;

iv. necrophagous arthropods and blow flies, in particular, are of human, animal, and environmental health importance;

v. the carrion model is a relatively uncomplicated system, usually involving only microbial and insect communities during early stages of decomposition;

vi. this system can be easily replicated both in the lab and field making it amenable to manipulations; and

vii. knowledge gaps exist in understanding the mechanistic role of how microbial communities interact and affect arthropod fitness.

Thus, the carrion system is ideal for studying interkingdom interactions during microbial decomposition and arthropod use of dead vertebrates, and so this chapter will review and discuss these important ecological interactions using the carrion model system.

Necrophagous insect succession was first identified over 150 years ago (Benecke 2001). Insect arrival to carrion follows a pattern of initial colonization by blow flies during early decomposition. Blow fly colonizers can be further categorized into early and late arrivers (Wells and Greenberg 1994; Anderson 2001). These species are followed by less common flies (e.g., Piophildae) during active decay and lastly by beetles (e.g., Coleoptera: Silphidae), which feed on remaining necrotic tissues or hairs and predate on fly larvae (Byrd and Castner 2001) (see Chapter 4). Questions remain as to the mechanisms governing priority effects and arrival patterns during this process (Brundage et al. 2014): *Why and how do blow flies detect and colonize carrion at temporally distinct intervals (priority effects)? What cue(s) indicate that the resource is suitable for maximizing larval survival and development (i.e., early vs. late colonization [oviposition] of the carrion resource)?*

Blow flies have been hypothesized, like many animals, to detect signals used for communication by microbial communities associated with carrion. As highly diverse and ubiquitous eukaryotes, a wealth of data exists that characterize insect physiological and behavioral responses to visual (Collett and Collett 2002), olfactory (Firestein 2001) and gustatory (Mullin et al. 1994) stimuli. And while microbial communities are ubiquitous in decomposition, no two naturally occurring microbial communities are thought to be the same (Curtis et al. 2002), suggesting that a cue(s) produced by the collective microbial community, rather than a specific microbial taxon or volatile organic compound (VOC) (see Chapter 9 on chemical ecology), is mediating insect response during carrion decomposition: in the natural environment mixed profiles of VOCs are the signal rather than only one or two compounds. Blow fly mechanisms for detecting carrion have previously been studied by characterizing behavioral responses to a single VOC or simple blends (e.g., 2–3 VOCs) collected from carrion (LeBlanc 2008; Frederickx et al. 2012). Yet, mechanisms governing blow fly behavioral responses to complex carrion-associated microbial communities, or epinecrotic microbial communities (Benbow et al. 2013), remain understudied.

20.2 Quorum Sensing and Ecological Eavesdropping

20.2.1 What Is Quorum Sensing?

One such interkingdom interaction that will be the focus of this chapter is the quorum sensing (QS) (Tuckerman et al. 2009) system of microbial species that also affects the behaviors of certain insect species. Microbial QS is defined as the regulation of gene expression of cells in response to surrounding cell density (Miller and Bassler 2001) (Figure 20.2). This regulation is driven by the production, detection, and release of autoinducer (AI) molecules that increase in response to higher cell densities and form the basis of intra and interspecific communication among bacterial cells (Miller and Bassler 2001). Prokaryotic organisms that utilize QS often act as an aggregate organism, performing activities and reproduction that no one individual cell can accomplish (Waters and Bassler 2005); thus, QS communication of microbes affords tremendous advantage for microbes living in natural, dynamic environments.

Microbial QS molecules are also known to influence animal behavior (Joint et al. 2007). For instance, in a plant–microbe–nematode tri-trophic interaction, a nematode that serves as a vector for bacteria to a plant facilitates the introduction of the bacteria to the plant, resulting in the emission of plant-induced VOCs, which act as a nematode aggregation cue (Horiuchi et al. 2005; Bais et al. 2008). Additionally, house flies, *Musca domestica* L. (Diptera: Muscidae), avoid unsuitable resources once semiochemicals produced by fungi are detected; these fungal communities will outcompete the fly larvae and reduce their fitness (Lam et al. 2010). These semiochemicals could be potentially QS related.

20.2.2 Ecological Eavesdropping via Quorum Sensing

Quorum sensing molecules can serve many roles outside of intra and interspecific communication between bacteria, including interactions with members from other kingdoms. Many QS compounds volatilize, generating complex VOC profiles. Resulting VOCs then serve as stimuli for other organisms. In terms of decomposition, these VOCs can serve to attract or repel arthropods that utilize carrion resources. Furthermore, these VOCs could indicate the presence of hosts in the case of parasitoids or prey; for example, predatory larvae of *Chrysomya rufifacies* (Macquart) (Diptera: Calliphoridae) are thought to cue to the VOC of other blow fly larvae on carrion (Flores 2013). Understanding the role of these compounds within the context of microbial and chemical ecology has already been covered in detail within this text (see Chapters 3 and 9, respectively). However, identifying (i.e., correlation vs. causation) these roles can be quite difficult. Concentration and makeup of the associated VOC plume can vary significantly (Verhulst et al. 2010b) depending on the temporal relationship between bacterial proliferation and metabolic function on the resource.

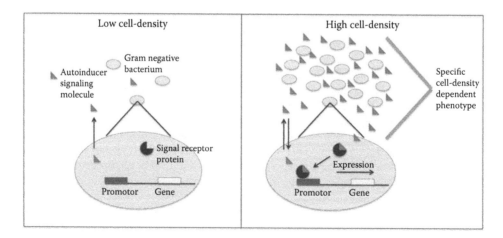

FIGURE 20.2 Simplified scheme of quorum sensing mediated gene expression in Gram-negative bacteria.

Many microbial-derived VOCs (e.g., QS related) are associated with the metabolic processes of the microbes and are secondarily used by arthropods as mechanisms to locate and utilize a resource or even avoid life-threatening hazards. *Staphylococcus epidermidis* is a common bacterium on human skin (Kloos and Musslewhite 1975; Otto 2009) that serves a key immunological role in protecting people from infections by other opportunistic pathogens (Duguid et al. 1992). However, VOCs produced by *S. epidermidis* attract the malaria vector *Anopheles gambiae* Giles (Diptera: Culicidae) (Meijerink et al. 2000; Verhulst et al. 2010a). Indole is one such compound produced by *S. epidermidis* that attract mosquitoes (Meijerink et al. 2000), though the primary role of this compound is to function as a QS molecule. While it is interesting that indole serves as an attractant for mosquitoes, the fact that this compound is also an important QS molecule is far more interesting, as it appears that decision making by the bacteria influences the behavioral response of the mosquito. Basically, the mosquito is demonstrating the principle of interkingdom communication where the mosquitoes "eavesdrop" on the communication of the bacteria. Thus, while the hypothesis that the microbiome influences the behavior of an animal is true (Ezenwa et al. 2012), Ma et al. (2012) and Tomberlin et al. (2012) suggest that what the microbes within the community are saying to one another is far more essential, since some bacteria can be present and not elicit a response until they "say" something that is of consequence to the arthropod. Zhang (2014) demonstrated this principle with the mosquito *Aedes aegypti aegypti* (Linnaeus) (Diptera: Culicidae) and *S. epidermidis*. She determined that approximately 75% of the mosquitoes in her experiments were attracted to blood-feeders treated with bacteria that could QS rather than the blood-feeder treated with the QS-inhibited bacteria. How VOC profiles vary between the wild type and mutant still remain unknown. But, such information could provide greater insight into how QS regulate arthropod behavior and what key VOC(s) are important in driving these behavioral responses.

The same interactions are known to occur with bacteria associated with a human after the individual ceases to live. *Proteus mirabilis* is a bacterium associated with decomposing human remains (Barnes et al. 2010). Recently, researchers were able to create a knockout library of *P. mirabilis* with mutants unable to exhibit swarming motility, which is a QS response (Ma et al. 2012). However, the swarming behavior of many of these mutants was rescued with known fly attracts demonstrating some relationship between fly attraction to a resource and bacterial behavior using a passive diffusion olfaction device (Figure 20.3). Furthermore, the researchers demonstrated that the level of attraction to a mutant of *P. mirabilis* by the blow fly, *Lucilia sericata* (Diptera: Calliphoridae), was significantly reduced when compared to their response to the wild type (swarming) (Ma et al. 2012; Tomberlin et al. 2012). While

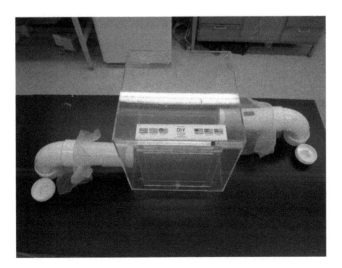

FIGURE 20.3 Olfaction device using passive diffusion to measure adult fly responses to two choices. Flies are placed in the center cube and allowed an allotted amount of time (4–8 h) to select and enter either the arm on the left or right of the device. A sticky trap inside the device will capture the flies thus preventing their escape. Choice of the fly is based on responses to volatiles emanating from treatments placed at the end of each arm. (Photo by J.K. Tomberlin.)

these studies demonstrated a potential link between QS and blow fly behavior, researchers are yet to demonstrate the evolutionary significance of such a relationship. Currently, it is not known if colonization of a carrion source by blow flies when the microbial community is at a given community structure has implications for resulting survivorship of their offspring. However, previous studies have demonstrated that bacteria in the exponential phase, rather than the stationary phase, are correlated with the production of VOCs that are volatilized forms of QS molecules (Kuzma et al. 1995; Bos et al. 2013).

In many instances, these QS molecules are by-products associated with the metabolism of essential amino acids (Liu 2014). Many other compounds known to attract invertebrates to carrion sources are also metabolic by-products of bacteria. For example, dimethyldisulfide, a known fly attractant (see Chapter 9 for a review of the associated chemical ecology) is a by-product of the degradation of methionine. Additionally, indole, which is a QS molecule for *Escherichia coli* (Lee et al. 2008), is a metabolite of tryptophan (Wang et al. 2001). And, phenylacetic acid, which is an antimicrobial compound produced by blow fly larvae (Sherman 2002), is a by-product of phenylalanine degradation (Erdmann and Khalil 1986; Nierop Groot and de Bont 1998). However, their link to QS activities is yet to be determined. But, in each case, the essential amino acid is critical for the production of the QS molecule that in turn serves as an attractant, or a repellent depending on concentration, of arthropods associated with decomposing vertebrate remains. Liu (2014) determined that the attraction of *L. sericata* to these compounds was physiological (e.g., male, gravid, or nongravid female) and dose dependent. Furthermore, she determined larval choice of which food to consume was dependent on the presence/absence of these essential amino acids. Most interestingly, she determined that if an antimicrobial were applied to the diets, the larvae avoided the diet deficient in methionine. However, without the antimicrobial, the larvae would consume this diet. She suspected that this behavior was due to the commensal bacterium, *P. mirabilis*, known to produce this amino acid, being suppressed in the larvae.

20.2.3 Ecological Effects of Eavesdropping

Quorum sensing compounds produced by microbes could also result in reduced competition with other organisms, such as invertebrates or vertebrates, attempting to consume a common resource. In the case of humans, the consumption of food products, such as fruit, milk, and unprocessed meats, by humans are simple examples demonstrating these indirect effects (Janzen 1977). However, it is not known if the compounds that repel human consumption of these materials are linked to QS.

While not decomposing vertebrate remains, these materials nonetheless constitute ephemeral resources that are only available for short periods of time to microbes and higher organisms. Microbes that colonize these food products are in direct competition with humans and other vertebrate animals. Their proliferation, alteration of the substrate, and associated production of QS molecules in part result in the food item spoiling and being recognized as such by vertebrate consumers. Furthermore, these compounds including the QS molecules volatilize into materials that inhibit the vertebrate consumption of the resource due to smell, taste, and possibly previous experiences resulting in illness. However, what might reduce competition with one set of animals allows for others to take advantage (Burkepile et al. 2006). Very similar ecological interactions occur for other vertebrates (DeVault et al. 2004) and invertebrates alike (Burkepile et al. 2006).

Quorum sensing compounds also serve as a mechanism enhancing dispersal ability of associated bacteria from an ephemeral resource declining in nutritional value. Fecal matter contains a large amount of *E. coli*, a common and ubiquitous bacterial species that produces indole. In return, indole produced by bacteria within a human host stimulates division and cues the bacteria to remain within the planktonic form; however, once the feces is deposited and the temperature decreases, indole stimulates biofilm formation while simultaneously attracting *M. domestica* which feed on the feces. Adult flies that emerge from the feces could potentially transport portions of biofilm with them resulting in their inoculating other resources, possibly of greater stability and quality, with the associated bacteria. Thus, the fitness value to the bacteria in using indole to induce biofilm formation on the feces may be that it increases the likelihood of dispersal to fresh resources.

Quorum sensing molecules produced by microbes associated with a host, either plant or arthropod, can attract predators, parasites, and parasitoids to a resource (Davis et al. 2013). An example from the plant world is Barley plants infected with the Barley yellow dwarf luteovirus (BYDV). Plants infected with the virus are more attractive to the associated vector, *Rhopalosiphum padi* L. (Homoptera: Aphidae), than those that are not infected (Jiménez-Martínez et al. 2004). This phenomenon has also been recorded between a beetle and fungus. The beetle, *Dendroctonus valens* LeConte (Coleoptera: Scoltidae), is more attracted to its host (i.e., pine tree) infected with the fungus *Leptographium procerum*. Researchers have shown that colonization of trees with *L. procerum* shifts VOCs produced by the tree with more 3-carene produced which also serves as the primary attractant for *D. valens* adults (Lu et al. 2011). However, additional research exploring this interaction is needed for a better understanding of the biological function of the fungus, as it is not known at this time if this compound is a true QS molecule.

Vertebrate scavengers depend on VOCs associated with vertebrate carrion (see Chapter 6 for review of vertebrate scavengers) (DeVault et al. 2003, 2004). The brown tree snake, *Boiga irrugularis*, utilizes VOC cues to locate mouse carrion (Shivik and Clark 1997). In fact, later research determined that volatiles associated with mouse decomposition are the result of microbe–resource interactions with *Enterobacter agglomerans* being the dominant bacterium present (Jojola-Elverum et al. 2001). This same bacterium is also responsible for partially attracting the Mexican fruit fly, *Anastrepha ludens* Loew (Diptera: Tephritidae), to resources (Robacker and Lauzon 2002). Putrescine, a known QS molecule (Patel et al. 2006), in association with carrion serves as an attractant for the red fox, *Vulpes fulpes* (Albone Eric et al. 1978). However, the link between microbial and vertebrate ecology as regulated by QS is yet to be investigated for many of these known relationships. This is an area of needed study.

Individuals that have evolved to consume a resource often manipulate the associated resource microbial communities (Barnes et al. 2010). This concept is something to be considered as it has potentially broad ramifications in behavioral ecology. Blow flies manipulate abiotic factors associated with a given environment (e.g., pH through oral excretions, temperature through maggot mass thermo-accumulation) resulting in select microbes proliferating (i.e., Gram-negative bacteria) while others are suppressed (i.e., Gram-positive bacteria) (Greenberg 1959a,b,c,d; Barnes et al. 2010). While doing so could reduce competition with other arthropods with a similar goal (Barnes et al. 2010) or be interpreted as "cultivating a microbial garden" to generate essential nutrients, these shifts result in alterations of the VOCs produced which can signal parasitoids and predators could be present (Stephen et al. 1993). These shifts are not restricted to bacteria but also have been identified with the presence of fungi and viruses as previously discussed. And, many of these compounds that serve as allomones or kairomones in the case of a mate, most likely are QS related (Davis et al. 2013).

Microbes can serve as an important component of immune response of their host or enhance an environment for survivorship of host offspring. Bacteria associated with house fly eggs suppress pathogenic microbes in the environment and enhance larval survivorship (Lam et al. 2009). Similar results have been found with microbe–plant interactions. Mycorrhization (*Glomus intraradices*) of *Medicago truncatula* Gaertn. induces indirect defenses against insect herbivores (Leitner et al. 2010). However, this influence depended on the plant's genetic makeup, which is hypothesized to explain the variable response of the plant to the fungus (Leitner et al. 2010).

In contrast, microbial symbionts assist their arthropod hosts with avoiding (Frago et al. 2012) or manipulating (Frago et al. 2012) defense mechanisms associated with the host resource. The relationship between the western corn rootworm, *Diabrotica virgifera virgifera* LeConte (Coleoptera: Chrysomelidae), and its symbiont, *Wolbachia*, is an example from the plant–insect community. Western corn rootworms feed on the roots of maize, and *Wolbachia* associated with the grub mediate the down-regulation of genes associated with maize defenses (Barr et al. 2010). Similarly, blow fly larvae that feed on necrotic tissue will introduce *P. mirabilis*, which is known to release mirabilicides that reduce bacterial diversity, specifically Gram-positive bacteria (Ceřovský et al. 2011), and suppresses some that could be pathogenic to the larvae (Greenberg 1968). This has also been demonstrated in nature with blow fly larvae increasing the pH associated with carrion allowing their offspring in conjunction with *P. mirabilis* to proliferate (Barnes et al. 2010).

Such close relationships between a host and its symbiont can be a driving force for speciation as both species depend on the QS molecules to make behavioral decisions. As in the case of the fungus *L. procerum*, its ability to induce such VOC shifts is population specific. In the instance of *L. procerum*, evidence exists that the fungus has evolved since its introduction to China and is now a better competitor for resources and attracting its vector, *D. valens* (Lu et al. 2011). Such genetic differences are interesting beyond implications for host adaptation. These data could also provide insight into other areas of phenotypic plasticity observed in the insect world such as genetic variation in the development rate of an insect from different regions of a country (Tarone and Foran 2006).

At the ultimate level, microbes can be a driver of natural selection and eventual speciation. A prime example is the *Drosophila paulistrorum* (Diptera: Drosophilidae) species complex that has a well described symbiotic relationship with *Wolbachia*. While the presence of *Wolbachia* in each semispecies provides fitness benefits, hybrids of the semispecies results in the *Wolbachia* becoming pathogenic resulting in embryonic lethality and male sterility (Miller et al. 2010). More recently, researchers determined the presence of the symbiotic bacteria *Lactobacillus* spp. including *L. plantarum* in some *Drosophila melanogaster* (Diptera: Drosophilidae) populations, and not others, reduced the likelihood of cross mating (Sharon et al. 2010). In this particular study, mechanisms governing mate selection due to the presence of the bacteria were related to a shift in VOC production or cuticular hydrocarbon sex pheromones (Sharon et al. 2010).

Quorum sensing is often employed among mixed microbial populations, or within the microbial–host marginal interface where there is often an exchange of biochemical signals among microbial populations or between the host and microbe. Resulting phenotypes most likely have profound effects on the microbial community structure, and associated host organisms. Therefore the fact that prokaryotes and eukaryotes have also evolved mechanisms to control the associated microbial community consortia by interfering with their cell-to-cell communication is not surprising. In this scenario, the competition operates through quenching of the QS system through a wide range of compounds. Indeed, prokaryotes and eukaryotes produce bioactive molecules to disrupt the QS process (termed quorum quenching—QQ) at different stages. Mechanisms governing QQ are discussed in the following section.

20.3 Quorum Quenching

20.3.1 Quorum Quenching Compound Activation and Screening

Microbes, and in particular, bacteria use QS (Tuckerman et al. 2009) to coordinate group behaviors in a cell-density-dependent matter. Many Gram-negative bacteria use small diffusible AI molecules, such as *N*-acyl homoserine lactone (AHL), that bind regulatory proteins, while most Gram-positive organisms use peptide signals that operate through membrane-bound receptor histidine kinases to induce gene expression leading to a variety of phenotypes such as biofilm formation, bioluminescence, motility, sporulation, root nodulation, and virulence factor production (Waters and Bassler 2005). Symbionts use QS to instigate mutually beneficial relationships with their host cells (Gonzalez and Marketon 2003; Studer et al. 2013). Inhibition of these QS processes and resulting phenotypes through QQ may occur by: (i) inhibition or reduction of the activity of the QS signal-producing gene; (ii) disruption of the structure of the signal molecule; (iii) modulation of the binding of the signal to the receptor site; or (iv) blocking the receptor site with antagonist signal analogs. The intra and interkingdom responses to QQ vary between pathogenic and beneficial bacteria.

Microbes may signal QQ compounds in response to certain antimicrobial compounds as a defensive strategy against threatening organisms. QQ compounds may be either AI-degrading enzymes that degrade AI molecules, or nonprotein-based compounds with structural similarity with AIs that act as competitor. They also may produce QQ agents as a survival strategy to gain benefits in a competitive environment and compete for space, nutrition, and ecological niches, though this may not be sufficient to explain all quorum quenching as some bacteria may use QQ molecules to inhibit QS-regulated harmful behaviors or may even be the result of incidental exposure.

Purification from crude extracts of candidate organisms and random high-throughput screening for QQ compounds from compound libraries have been used to identify QQ molecules among organisms and natural products. AHL and AI-2 bioreporters, genetically manufactured strains that express reporter genes in response to specific QS signals, are also valuable tools to identify a wide range of not only AHLs, but AHL analogues, mimics, or other inhibitors. Many reporter strains are engineered by fusing the QS-regulated promoter to the *lux* operon or *lacZ*. Although, these reporter strains have a functional regulator protein, they lack the AHL synthase enzyme. Promoter activity is induced by exogenous QS signals, resulting in a reporter that mimics the natural QS system with specific, easily identifiable phenotypes (Schipper et al. 2009). Quorum-sensing mutant strains have also been used to test mechanisms of QQ molecules on multiple QS signaling systems. For instance, *Vibrio harveyi* luminescence expression relies on the production and expression of two autoinducers, AI-1 and AI-2, in response to cell density (Mok et al. 2003). *V. harveyi* strains deficient in either system have been employed measuring luminescence to determine conditions which QQ molecules interact (Surette and Bassler 1998; Ren et al. 2001).

20.3.2 Bacterial Quorum Quenching Compounds

The ability to produce QQ compounds is maintained by bacterial species from many different prokaryotic phyla and include among others Proteobacteria: *Burkholderia* (Chan et al. 2011), *Pseudomonas* (Huang et al. 2003; Sio et al. 2006; Shepherd and Lindow 2009), *Agrobacterium* (Zhang et al. 2002a; Carlier et al. 2003), *Acinetobacter* (Kang et al. 2004), *Shewanella* (Morohoshi et al. 2008), *Comamonas* (Uroz et al. 2007), *Klebsiella* (Park et al. 2003), *Enterobacter* (Rajesh and Rai 2014), and *Variovorax* (Leadbetter and Greenberg 2000; Uroz et al. 2003); Actinobacteria: *Mycobacterium* (Afriat et al. 2006; Chow et al. 2009), *Microbacterium* (Wang et al. 2010), *Rhodococcus* (Park et al. 2006), *Arthrobacter* (Park et al. 2003), and *Streptomyces* (Park et al. 2005); Firmicutes: *Staphylococcus* (Chu et al. 2013), *Bacillus* (Dong et al. 2002; Lee et al. 2002; Ulrich 2004), and *Oceanobacillus* (Romero et al. 2011); Cyanobacteria: *Anabaena* (Romero et al. 2008); Bacteroidetes: *Tenacibaculum* (Romero et al. 2010). A list of known QQ compounds produced by these organisms as well as the QQ targets are listed in Table 20.1.

20.3.3 Quorum Quenching Compounds among Higher Organisms

Using some of the above-mentioned methods, several potential, naturally occurring QQ compounds have been discovered among prokaryotes and eukaryotes that include metabolites like halogenated furanones from the red algae, *Delisea pulchra* (Manefield et al. 1999) and other compounds from edible plants and fruits (Musthafa et al. 2010) that are similar to bacterial AHLs and inhibit QS-mediated activities in bacteria by competing with cognate AHL signals for their receptor site (LuxR) (Gram et al. 1996; Manefield et al. 2000; Ren et al. 2001, 2004). Several compounds isolated from the unicellular green alga *Chlamydomonas reinhardtii* were found to mimic bacterial signals and interfere with their QS system (Teplitski et al. 2000). Naturally occurring compounds with QQ activity have been reported from a large number of plants and include among many others the phenolic aldehyde vanillin from vanilla bean (Kappachery et al. 2010), furanones (Manefield et al. 1999; Jermy 2011), citrus flavonones (Truchado et al. 2012), the phenolic compound from turmeric, circumen (Rudrappa and Bais 2008), and the phenolic compound salicylic acid (Wang et al. 2007). Major limitations with natural QQ compounds are the small concentrations in which they are produced and the potential toxicity by some of the compounds. Therefore, chemically synthesizing these compounds has become an attractive alternative to circumvent these complications. Efforts to synthesize QQ compounds have included targeting the biosynthetic pathway leading to signal production, substitutions in signals, and changes in chain length, but despite this, only a few active synthetic analogs are known (Smith et al. 2003; Morohoshi et al. 2007; Roy et al. 2010).

Higher plants interfere with bacterial QS systems by producing low molecular weight compounds mimicking QS signals and acting as agonists or antagonists to bacterial QS systems. *Pisum sativum* (pea seedling) extracts have compounds that mimic AHL signals, and activity against violacein production, a QS phenotype in *Chromobacterium violaceum* (Teplitski et al. 2000). A study of the legume, *Medicago truncatula*, found 15–20 QS mimic compounds that either stimulated or inhibited QS activity from QS reporter bacterial strains, where production of mimicking compounds and activity was modulated during plant development (Gao et al. 2003).

TABLE 20.1

Known Quorum Quenching Compounds of Prokaryotes

Phylum	Genus	Quorum Quenching Enzyme	Target	Reference
Proteobacterium	*Burkholderia*	AHL oxidoreductase	N-(3-oxododecanoyl)-L-homoserine lactone (3-oxo-C_{12}-HSL)	Chan et al. (2011)
	Pseudomonas	AHL lactonase and AHL acylase	Long chain AHLs	Huang et al. (2003), Sio et al. (2006), Shepherd and Lindow (2009)
	Agrobacterium	AHL lactonase	AHLs	Zhang et al. (2002a), Carlier et al. (2003)
	Acinetobacter	AHL lactonase	AHLs	Kang et al. (2004)
	Shewanella	AHL acylase	Long chain AHLs	Morohoshi et al. (2008)
	Comamonas	AHL acylase	AHLs with acyl side chains	Uroz et al. (2007)
	Klebsiella	AHL lactonases	AHLs	Park et al. (2003)
	Enterobacter	AHL lactonase	AHLs	Rajesh and Rai (2014)
	Variovorax	Potential AHL lactonase	AHLs	Leadbetter and Greenberg (2000)
Actinobacteria	*Mycobacterium*	AHL lactonase	AHLs	Afriat et al. (2006), Chow et al. (2009)
	Microbacterium	AHL lactonase	AHLs	Wang et al. (2010)
	Rhodococcus	AHL lactonases and oxidoreductases	AHLs	Park et al. (2006)
	Arthrobacter	AHL lactonase	AHLs	Park et al. (2003)
	Streptomyces	AHL acylase	AHLs	Park et al. (2005)
Firmicutes	*Staphylococcus*	unknown	AHLs	Chu et al. (2013)
	Bacillus	AHL Lactonase	Short and long chain AHLs	Dong et al. (2002), Lee et al. (2002), Ulrich (2004)
	Oceanobacillus	AHL lactonase	AHLs	Romero et al. (2011)
Cyanobacteria	*Anabaena*	AHL acylase	Long chain AHLs	Romero et al. (2008)
Bacterioides	*Tenacibaculum*	AHL acylase and lactonase	AHLs	Romero et al. (2010)

QQ compounds have also been discovered within the Kingdom Animalia. Porcine kidney acylase I, an enzyme widely conserved in some animals such as mice, rats and zebrafish, was reported to have AHL-degradation activity, though its role as a QQ compound in these organisms has not been fully elucidated (Xu et al. 2003). In a mouse skin infection study, reactive oxygen and nitrogen intermediates including hypochlorous acid (HClO) and peroxynitrite (ONOO⁻), whose generation depends on NADPH oxidase, inactivate *S. aureus* AI peptides *in vitro* and *in vivo* (Rothfork et al. 2004). It was also shown that ground beef contains compounds that can inhibit AI-2-mediated bioluminescence in *V. harveyi* and negatively influenced biofilm formation in *E. coli* K12 (Soni et al. 2008).

Several of the QQ compounds that have been identified are enzymes. Most known QQ enzymes include lactonases, acylases, oxidoreductases, and mammalian paraoxonases (Leadbetter and Greenberg 2000; Zhang et al. 2002a,b; Yang et al. 2005; Camps et al. 2011). Lactonases hydrolyze the ester bond of AHL yielding *N*-acylhomoserine. Acyl-homoserine lactonase is a metallo-enzyme that requires metal ions most of the time, and target long- and short-chain AHLs (Dong et al. 2002; Liu et al. 2005; Thomas et al. 2005). AHL acylases hydrolyze amide bonds to yield homoserine lactone and the corresponding fatty acid chain, and exhibit substrate specificity based on the length of the acyl chain. AHL oxidoreductases usually catalyze a modification of AHLs (Chen et al. 2013).

20.3.4 Quorum Quenching Enzymes

20.3.4.1 AHL-Lactonases

AHL-lactonase production has been reported from many *Bacillus* spp. including among others, *B. cereus*, *B. anthracis*, *B. subtilis*, and *B. thuringiensis*, as well as other bacterial species (Dong et al. 2000, 2004; Ulrich 2004; Wang et al. 2004; Liu et al. 2008; Momb et al. 2008; Pan et al. 2008; Chan et al. 2010). The autoinducer inactivator AHL-lactonases (AiiA) produced by *Bacillus* species are closely related and share 90% sequence identity at their peptide level (Dong and Zhang 2005). AiiA nonspecifically targets AHL side chains to disrupt QS in Gram-negative bacteria. Because the Gram-positive *Bacillus* spp. rely on oligo-peptide-mediated QS, its own cell-to-cell communication is unaffected (Dong et al. 2000, 2002). AHL-lactonases produced by other bacterial species such as AttM and AhlK produced by *Agrobacterium tumefaciens* and *Klebsiella pneumoniae*, respectivel,y shows wide heterogeneity with only 30%–50% amino acid identity within its phylogenetic cluster (Dong and Zhang 2005). Other lactonases, AidH and BpiB, have recently been identified using an *A. tumefaciens* reporter strain (Schipper et al. 2009; Mei et al. 2010). Both AidH and BpiB were found to inhibit biofilm formation in *P. aeruginosa* (Schipper et al. 2009; Mei et al. 2010).

20.3.4.2 AHL-Acylases

AHL-acylase genes have been identified from a variety of bacterial species including *Streptomyces* (Park et al. 2005) and *Ralstonia* spp. (Lin et al. 2003). *Streptomyces* and *Ralstonia* spp. produce AhlM and AiiD, respectively that, in separate experiments, decreased the production of virulence factors (Lin et al. 2003; Park et al. 2005). AhlM produced by *Streptomyces* sp. is specific as it degrades AHLs with acyl chains having six or more carbons (Park et al. 2005). The AI inhibitor from the nitrogen-fixing cyanobacterium, Anabaena sp. PCC7120, AiiC, has broad acyl-length specificity whose role is thought to be in self-modulation of QS regulatory genes (Romero et al. 2008). The human pathogen, *P. aeruginosa* PAO1, uses two AHL-amalyases, PvdQ and QuiP, to modulate its own QS-dependent pathogenicity (Huang et al. 2006; Sio et al. 2006). Both PvdQ and QuiP degrade the *P. aeruginosa* QS molecule *N*-(3-oxododecanoyl)-l-homoserine lactone (3-oxo-C_{12}-HSL), and QuiP was shown to confer amylase activity against *E. coli* (Huang et al. 2006; Sio et al. 2006). A comparison of AhlM with AiiD and PvdQ AHL-acylases in *Ralstonia* sp. and *P. aeruginosa*, respectively, showed only 35% and 32% amino acid sequence identity, respectively (Lin et al. 2003; Park et al. 2005). AiiD and PvdQ are also quite diverse and share only 39% amino acid identity (Lin et al. 2003; Dong and Zhang 2005). *P. syringae* B728a produce two potent acylases, HacA and HacC, that degrade QS AHLs and influence colony morphology (Shepherd and Lindow 2009). Other acylases have been identified among marine bacterial community isolates and through genome data mining (Romero et al. 2011). Using a reporter strain, the enzyme acylase I from porcine kidneys was found to cleave acyl moieties of AHLs and reduced biofilm formation and membrane biofouling in a membrane bioreactor inoculated with bacterial communities from activated sludge (Xu et al. 2003; Dong and Zhang 2005). *Varivorax paradoxis* also uses an anti-AHL QQ tactic by acylase-mediated lactone ring opening.

20.3.4.3 Paraoxonases

Mammalian paraoxonases (PONs) are highly conserved calcium-dependent esterase/lactonases that hydrolyze a broad range of esters, including phosphotriesters, arylesters, and lactones, and have overlapping and also distinct, substrate specificities (Draganov et al. 2005). PON-lactonases differ from prokaryotic lactonases as mammalian PONs require calcium ion for their activity, and lack the HXHXDH motif (Charendoff et al. 2013). Mammalian PONs exhibit antioxidative properties and lead to protection from atherosclerosis in mouse models, though mechanisms for these activities are not known (Shih et al. 1998; Ng et al. 2001; Aviram and Rosenblat 2005; Horke et al. 2007). PONs consist of serum paraoxonases PON1, PON2, and PON3 (Draganov and La Du 2004), with the most-studied member of the family being PON1. Human PON1 and PON3 are synthesized in the liver and secreted into the blood, while

PON2 is expressed in many tissues and cell types, though its protein product is not found in the plasma (Ng et al. 2001; Draganov and La Du 2004). Furthermore, PONs are strongly expressed in lung epithelial cells in great part on the external cell membrane PONs, and particularly PON2 hydrolyzes *P. aeruginosa* 3-oxo-C_{12}-HSL in human and mouse serum, mouse lung and liver homogenates, and cultured human cell lysates (Teiber et al. 2008).

PONs are also able to hydrolyze AHL and interrupt quorum-sensing signals from bacteria located in proximity of the respiratory epithelium. An *in vitro* study found that airway epithelial cells could inactivate 3-oxo-C_{12}-HSL and that this capacity was present in cell membranes but not when PONs were secreted into the airway fluid (Chun et al. 2004; Stoltz et al. 2007). Exposure of cultured human lung epithelial cells to 3-oxo-C_{12}-HSL demonstrated that a lactonase was responsible for these outcomes (Ozer et al. 2005). Serum, an environment rich in PON1, also degraded 3-oxo-C_{12}-HSL and decreased *P. aeruginosa* biofilm growth, but this phenotype was negated from serum of PON1 knockout mice. Chemical complementation by the addition of purified PON1 to serum from PON1-KO mice restored the ability to degrade 3-oxo-C_{12}-HSL and also reduced biofilm formation (Draganov et al. 2005; Ozer et al. 2005). QQ enzymatic activity against 3-oxo-C_{12}-HSL has also been observed in serum from a number of mammals including cattle, goat, horse, mouse, and rabbit (Yang et al. 2005).

20.3.4.4 Oxidoreductases

Oxidoreductases catalyze the oxidation or reduction of the acyl side chain that leads to modification of the chemical structure of the signal, but does not lead to degradation (Uroz et al. 2005; Chowdhary et al. 2007). Two types of oxidoreductases, the P450/NADPH-P450 reductase and P450 monooxygenase, have been isolated from *B. megaterium* CYP102A1 that inactivate AHL signals (Chowdhary et al. 2007). Other studies involving cell extracts and whole cells from *R. erythropolis* strain W showed AHL degradation activity and reduced the pathogenicity of *Pectobacterium carotovorum* subsp. *carotovorum* in potato tubers. Investigations into the mechanism revealed an oxidoreductase that converts 3-oxo-HSLs to their corresponding 3-hydroxy derivatives, and an amidolytic activity that cleaves the amide bond linking the acyl chain to the HSL residue (Uroz et al. 2005).

20.3.4.5 QQ Compounds as Therapeutic Agents

Inhibition of QS by QQ compounds offers an attractive alternative to traditional antibiotics because this strategy is not bactericidal, and the general belief is that QQ compounds pose little or no selective pressure of bacteria, since these do not affect their growth, and the occurrence of bacterial resistance should therefore be reduced. QQ compounds also make pathogens potentially more susceptible to antibiotics and other drugs by disrupting biofilm formation and other virulence factors affecting pathogenicity. However, taking into account the multiplicity of receptors, it may be difficult to use a single molecule as a broad range QQ molecule. Also, some recent studies have demonstrated that cells evolve resistance to QQ compounds and mechanisms. Maeda et al. (2012) conducted an elegant study that was published in 2012 where selective pressure was created on *P. aeruginosa* cells using adenosine, a QS-dependent and physiologically relevant carbon source in minimal media (Maeda et al. 2012). Growth rate was measured where *P. aeruginosa* cells inhibited by the QQ compound C-30 (brominated furanone) were expected to grow more slowly, and cells resistant to QQ compounds were expected to grow more rapidly. Using the screening method, *P. aeruginosa* cells were isolated that were found to be resistant to C-30 due to mutations in *mexR* and *nalC* that encode repressors for the *MexAB-oprM* multidrug resistance operon (Maeda et al. 2012). These mutations lead to enhanced efflux of C-30. The mutants were also resistant to C-30 in a *Caenorhabditis elegans* model. To further demonstrate physiological relevance, the authors showed that *P. aeruginosa mexR* and *nalC* mutants isolated from chronic cystic fibrosis patients were also resistant to the QQ compound C-30 (Maeda et al. 2012). Following this, another study was conducted where the effects of C-30 and 5-fluorouracil on production of virulence factors alkaline protease, pyocyanin, and elastase of eight *P. aeruginosa* clinical strains from children were measured (Garcia-Contreras et al. 2013). Three strains were C-30 resistant, and two of these were also antibiotic resistant. Several of the strains were 5-fluorouracil resistant in some of the phenotypes (Garcia-Contreras et al. 2013).

20.4 Consequences of Interkingdom Interactions

Interactions between the host and microbe, either pathogenic or symbiotic, provoke a wide range of behavioral reactions especially in the presence of QS. Microbes and their associated hosts have coevolved, so it is not surprising that certain hosts appear to be able to sense QS signals, potentially allowing them to alter QS outcomes. Likewise, millions of years of microbe–host coevolution have presumably selected those microbial organisms that best exploit and manipulate their hosts. Host–microbe compounds are thought to have coevolved through molecular recognition and complementarity that selected biological molecules able to stabilize and interact functionally with each other with self-complementarity and across organisms (Root-Bernstein 2012).

Prokaryotic QS systems can regulate target genes by "eavesdropping" on exogenously provided AI signals produced by neighboring bacteria. For instance, *Salmonella enterica* and *E. coli* both encode a homologue of Lux R, SdiA, but lack signal-generating enzymes. The respective SdiAs instead respond to signals produced by other microbes (Michael et al. 2001; Ahmer 2004). The structure of *E. coli* SdiA was elucidated by Yao et al. (2006) who demonstrated that SdiA is capable of utilizing a variety of AHLs as switches for its folding (Yao et al. 2006). A recent study by Shimada et al. searched for effectors among synthesized AHL analogs that could be recognized *in vitro* by purified *E. coli* SdiA (Shimada et al. 2014). The researchers found three synthetic signal molecules that affected SdiA binding to its promoter. However, the researchers also found that SdiA interacts with multiple species of effector ligands and its promoter recognition specificity is modulated depending on the interacting effector AHL molecules (Shimada et al. 2014). In another example, QscR from *P. aeruginosa* auto-activated its own transcription in the presence of its own AHL signaling molecule (Ha et al. 2012). Results also showed that some non*P. aeruginosa* signals, produced by *P. fluorescens* and *B. vietnamiensis*, were able to preferentially activate QscR and the activated QscR could boost its own expression (Ha et al. 2012). These data support the idea that QS regulation can be modulated by differing species in a multispecies community.

Host response to AHL signals has also been demonstrated. A study of human lung epithelial cells showed changes in expression of 4347 genes in response to 3-oxo-C_{12}-HSL, several of which were involved in immune modulation (Bryan et al. 2010). This AHL is also known to elicit apoptosis and immunomodulatory responses in a broad array of mammalian cell lines (Smith et al. 2001, 2002b; Tateda et al. 2003; Shiner et al. 2006). In certain cell types the apoptotic effects of 3-oxo-C_{12}-HSL are mediated via a calcium-dependent signaling pathway, while some proinflammatory effects involve intracellular transcriptional regulators (Shiner et al. 2006). However, the mechanisms by which mammalian cells perceive and respond to 3-oxo-C_{12}-HSL are still not completely understood (Bryan et al. 2010).

The responses of eukaryotes to AHLs have been more extensively characterized in plants as opposed to mammalian systems. Plants respond to bacterial AHLS and regulate gene expression affecting important phenotypes. The different responses to QS signal molecules like AHLs in plant hosts, result in production of AHL decoy molecules, stimulation of root growth, uptake of signal molecules into shoots and even enhanced stomatal activity (Smith et al. 2002a; Bauer and Mathesius 2004; Fekete et al. 2010).

Alternatively, QS systems in epiphytic or endophytic bacteria are influenced by plant molecules and affect plant gene expression by AHLs. Plant-associated bacteria may enhance plant growth by producing metabolites and increasing available uptake. *Oryza sativa* rice plants contain molecules that activate different QS *N*-acyl HSL biosensors and are sensitive to *Bacillus* AiiA lactonase (Degrassi et al. 2007). Growth and defense of the flowering plant, *Arabidopsis thaliana*, can perceive AHLs as well as the length of the acyl chain and a particular functional group to modulate the plant's response. This signal recognition confers resistance toward biotrophic and hemibiotrophic pathogens by altered activation of AtmPk6, a mitogen activated protein kinase (MAPK) that directs response to an array of environmental stimuli (Schikora et al. 2011).

A subfamily of AHLs has been found in plant-associated bacteria that have lost the capacity to bind exogenous AHLs. These AHLs have instead evolved the ability to respond to low-molecular weight plant signaling compounds and bind to plant-produced compounds (Gonzalez et al. 2013; Venturi and Fuqua 2013). These LuxR compounds are very closely associated to relative prokaryotic QS LuxRs, and are widely distributed among plant-associated bacteria. Several of these have been studied including PsoR of *P. fluorescens* that responds to plant compounds of several plant species and controls *P. fluorescens* through transcriptional regulation in response to plant compounds (Subramoni et al. 2011). NesR of *Sinorhizobium* (Ferluga et al. 2007; Zhang et al. 2007; Ferluga and Venturi 2009; Subramoni et al. 2011; Chatnaparat et al. 2012) has been associated with survival under stress and utilization of various carbon sources (Patankar and Gonzalez 2009a,b). XccR from the plant pathogen, *Xanthomonas campestris* pv. *Campestris*, responds to an unidentified plant compound and regulates the neighboring pathogenicity-associated proline iminopeptidase (*pip*) gene (Zhang et al. 2007). OryR of the plant pathogen *X. oryzae* pv. *oryzae* is involved in virulence, and also responds to plant signals to activate expression of the neighboring *pip* and motility genes (Ferluga et al. 2007; Ferluga and Venturi 2009; Gonzalez et al. 2013) XagR of the soybean pathogen *X. axonopodis* pv. *glycines* is also involved in virulence by negative regulation of adhesion via *yapH* in response to plant compound(s) facilitating the spread of the pathogen in the plant (Chatnaparat et al. 2012). Like XccR and OryR, XagR also activates *pip* transcription that increases gradually after infection. Interestingly, homologs of OryR/XccR/XagR have only been found in major genera of plant-associated bacteria. This suggests a widespread unidirectional interkingdom signaling between plants and bacteria.

Microbial species composition in the GI tract has been hypothesized to drive homeostasis, and the nervous system drives the composition of microbes within the GI tract (Lyte 2010). Neural hormones, receptors, and neurotransmitter transporters are found in microorganisms with homology to those in humans including somatostatin (LeRoith et al. 1985), and gamma-aminobutyric acid (GABA) (Guthrie et al. 2000), and a sodium-dependent neurotransmitter transporter (Androutsellis-Theotokis et al. 2003). *S. typhimurium* has adapted its quorum-sensing regulator QseC to act as a receptor for the host hormone norepinephrine and thereby tie the regulation of virulence genes to the hormones present in tissue (Clarke et al. 2006). Also, norepinephrine, epinephrine, and dopamine increase the growth of commensal *E. coli* (Kinney et al. 2000), and norepinephrine stimulates enterotoxigenic *E. coli* growth and expression of virulence factors (Lyte et al. 1997a,b).

This widespread presence of neural hormones and receptors within microbes suggests that microorganismal systems are likely precursors to complex vertebrate systems. Therefore it can easily be imagined that changes in microbial diversity likely influence the function of components of host response as reflected in stress response, health, motor function, and sense perception. This has been shown in mice infection studies where infection with *Camyplobacter jejuni* leading to inflammation of the gastrointestinal tract leads to anxiety-like behavior and anorexia (Lyte et al. 1998; Goehler et al. 2005). Alternatively, commensal bacteria can have positive effects on brain function and behavior. Mice exposed to a diet supplemented with *Lactobacillus rhamnosis* for six weeks showed reduced stress-induced corticosterone and anxiety in standard lab tests. The researchers also found modulation in gene expression of neurotransmitter GABA receptors (Bravo et al. 2011).

Bacteria associated with nectar can change nectar chemistry to influence, either strengthening or weakening the plant-pollinator mutualism (Vannette et al. 2013; Arbaoui and Chua 2014). Also, it has been demonstrated that some bacteria and their VOCs influence recruitment of some invertebrates (Ponnusamy et al. 2008) and metamorphosis of some worms (Neumann 1979), but it is not yet understood how these higher organisms might influence the recruitment, persistence, signaling, and concentration of the associated microbial community.

A study of food web competition showed that some bacteria produce noxious metabolites (>90% found to be the nonesterified fatty acids Z-5 hexadecenoicacid, Z-9 octadecenoic acid, octadecanoic acid, hexadecanoic acid, and tetradecanoic) while feeding on carrion (Burkepile et al. 2006). The microbe-laden carrion was found to be four times more likely to be uncolonized by larger consumers than were fresh carrion. The authors concluded that the bacteria competed with larger scavengers by rendering the carion "chemically repugnant" (Burkepile et al. 2006).

Conversely, several commensal bacteria have been found on embryonic surfaces that provide protection against potential pathogens, suggesting recruitment of the bacteria by these animals. For example, embryos of the shrimp *Palaemon macrodactylus* are resistant to infection by the fungus *Lagenidium callinectes* due to inhibition caused by 2,3-indolinedione that is produced by an *Alteromonas* sp. on the embryo's surface (Gil-Turnes et al. 1989). The pea aphid, *Acyrthosiphon pisium*, also has resistance to colonization by the parasitoid wasp, *Aphidius ervi*, when infected with *Hamiltonella defensa*. This protection is brought about by a toxin-encoded bacteriophage that infects *H. defensa* and targets susceptible *A. ervi* (Oliver et al. 2005, 2009).

Nonetheless, the presence of replicating bacteria in the host, though often beneficial, can also come with a cost. For instance, for some insects, *Wolbachia* are mutualists and confer protection against pathogen infection, and pathogen-induced mortality (Moreira et al. 2009; Walker et al. 2011). However, for other insects, *Wolbachia* infection suppresses both host defense, and parasitoid counter-defense (Fytrou et al. 2006; Hedges et al. 2008). As mentioned previously, *Wolbachia* are also parasites that can influence their host reproductive outcomes by affecting male/female offspring ratios through various mechanisms (Werren et al. 2008). Genetic variability of the symbiont as well as the host background impacts all these associated phenotypes. Therefore, understanding these genotypic and phenotypic variables is essential to understanding microbial–host interactions.

20.5 Future Research Directions and Conclusions

20.5.1 Future Research Directions

An appreciation for bacterial QS is well established in the literature and this understanding has led to advances in synthetic biology. Biofilms have been engineered and imaged in the laboratory in microfluidic devices to manipulate QS regulatory networks in order to alter phenotypic and genotypic outcomes of biofilm formation and dispersal (Figure 20.4). This progress in synthetic biology has allowed researchers to create single species and polymicrobial biofilm constructs where specific biofilm patterning has been manipulated for the creation of QS circuits and switches for preferential displacement of an existing biofilm by the controlled formation of engineered cells (Figure 20.5). These applications will prove valuable to combat effects of QS-mediated microbial biofilm formation such as biofouling, biocorrosion, and to combat diseases.

FIGURE 20.4 **(See color insert.)** *E. coli* biofilm on a glass slide. Green (black as greyscale) fluorescent protein-tagged *E. coli* biofilm on a glass slide in a flow cell.

FIGURE 20.5   **(See color insert.)** Consortium of two *E. coli* strains. Green (white as greyscale) *E. coli* cells were tagged with the green fluorescence protein and the red (dark grey as greyscale) *E. coli* cells were tagged with the red fluorescence protein. Cells were grown for 12 h in a microfluidic chamber and imaged via confocal microscopy and IMARIS software. (Process described more fully in Hong, S.H. et al. 2012. *Nature Communication* 3: 613.)

The link between the ecological context of these compounds and their significance to higher-order ecosystem functions, however, is yet to be established. Within the context of vertebrate carrion systems, bacteria play a large role in regulating the decomposition process. While this role might not be the actual recycling of nutrients back into the larger ecosystem, they do release many of the compounds that regulate their behavior and function while simultaneously serving as a mechanism to attract, or repel, other vertebrate consumers of said resource. Additionally research is needed to provide greater detailed analysis of VOCs associated with vertebrate carrion and determining the role, if one exists, of each compound in regards to bacterial QS. Furthermore, understanding the context of these QS molecules within the context of a greater scale—not just on the vertebrate carrion itself but within the larger ecosystem whether it be a field, forest, or lake.

Researchers are just beginning to tie together microbial ecology with higher organism responses. While it is appreciated that microbiomes influence animal behavior, in many instances, how they do so is not known. VOC profiles of microbes that are exhibiting QS-linked behaviors are needed to explain their relationship with attraction, or repellence, of consumers. Doing so could offer some insight on a larger scale of animal foraging behavior and the resulting benefits/detriments of such responses to microbe as well as the animal.

20.6 Conclusions

With rapid advances in molecular and sensing (e.g., remote and nano-scale sensing) technologies comes the potential to better explore more discrete forms of interactions among organisms that are separated taxonomically. Early research in parasitology and more recent endeavors in disciplines such as disease ecology have framed interkingdom interactions within the context of host–pathogen dynamics, focusing mostly on vertebrate animals and viral, bacterial, or fungal pathogens. However, the recent advances in high-throughput metagenomic sequencing are uncovering a plethora of other such interactions with biological, ecological, and evolutionary importance; many of these were outlined in this chapter. However, one mainstream example of some of these advances that can represent the future of more ecologically oriented science would be the many human health-related advances of understanding the human microbiome (Turnbaugh et al. 2007).

FIGURE 20.6 **(See color insert.)** Carrion after several days of decomposition, when initial blow fly oviposition occurred within minutes of exposing the carcass. Note the mass of adult blow flies that have congregated between the hind legs, presumably at an area where the wall of the intestine was ruptured from larval feeding activity. What is cuing the adult flies to that area? (Photo by M.E. Benbow.)

The Human Microbiome Project (HMP) provided the first catalog of bacterial genomes associated with the healthy human organism (The Human Microbiome Consortium 2010) and later provided a framework for future work with the human microbiome (The Human Microbiome Consortium 2012). The studies that have either been a part of the original HMP or have continued and expanded similar lines of research are reporting significant health importance of the human bacterial community, often documented through direct communication of prokaryotic cells with the cells of the human organism (e.g., intestine or skin) (Costello et al. 2009; Dave et al. 2012; Schommer and Gallo 2013). Indeed, the collective HMP research generally considers the human organism an ecosystem, with projections that the microbial communities follow general community ecology principles, including intra and interspecies communication (Fierer et al. 2012; Morgan et al. 2013). Thus, documented interkingdom communication between human and bacterial cells documented in research of the human microbiome provides a provocative precedent that similar interactions occur more ubiquitously than previously considered. While not comprehensively tested to date within the bioscience literature, the more generally role of ecological eavesdropping on microbial QS communication systems offers a potentially transformative avenue of inquiry for diverse disciplines to identify previously unknown connections in ecologically complex systems, like that of carrion decomposition.

To conclude this chapter, there is evidence that interkingdom communication and associated eavesdropping does occur and can have significant effects on animal fitness, mostly documented in laboratory studies (Ma et al. 2012; Tomberlin et al. 2012). With an in-depth understanding of if and how these interactions occur in nature, scientists may begin to answer questions such as, "Why are adult blow flies attracted to certain areas of the carrion body even though larval masses are dense and will outcompete any new early instars?" The answer to this example might be that the adults are cueing on new microbial VOC profiles that would typically represent early decomposition: that is, enteric bacteria in the intestines become exposed later in decomposition, in part because of the larval feeding activity of the first generation (Figure 20.6).

A rich history of literature in such disciplines as parasitology, arthropod-borne diseases, and human health suggest that such interkingdom interactions are widespread in nature and that prokaryotes and eukaryotes share a long evolutionary history that include many linkages that scientists are yet to fully understand. The age of "omics" is allowing many disciplines to bridge theoretical and contextual areas of science that have been traditionally separated by limitations of technology and dialogue. To that end, the carrion decomposition system can provide a unique opportunity to test new and undiscovered interactions between microbes, invertebrates, and vertebrates with potential applications of such new discoveries in such areas as human and environmental health and forensics. The next century should bring a wealth of new frontiers and dimensions to carrion decomposition ecology, evolution, and their applications.

REFERENCES

Afriat, L., C. Roodveldt, G. Manco, and D.S. Tawfik. 2006. The latent promiscuity of newly identified microbial lactonases is linked to a recently diverged phosphotriesterase. *Biochemistry* 45: 13677–13686.

Ahmer, B.M. 2004. Cell-to-cell signalling in *Escherichia coli* and *Salmonella enterica*. *Molecular Microbiology* 52: 933–945.

Albone, E.S., P.E. Gosden, G.C. Ware, D.W. Macdonald, and N.G. Hough. 1978. Bacterial action and chemical signalling in the red fox (*Vulpes vulpes*) and other mammals. In: *Flavor Chemistry of Animal Foods*, Bullard, R.W. (ed.), pp. 78–91. American Chemical Society.

Anderson, G.S. 2001. Insect succession on carrion and its relationship to determining time of death. In: J.H. Byrd, and J.L. Castner, (eds.), *Forensic Entomology: The Utility of Arthopods in Legal Investigations*, pp. 143–175. Boca Raton, FL: CRC Press LLC.

Androutsellis-Theotokis, A., N.R. Goldberg, K. Ueda, T. Beppu, M.L. Beckman, S. Das, J.A. Javitch, and G. Rudnick. 2003. Characterization of a functional bacterial homologue of sodium-dependent neurotransmitter transporters. *The Journal of Biological Chemistry* 278: 12703–12709.

Arbaoui, A.A. and T.H. Chua. 2014. Bacteria as a source of oviposition attractant for *Aedes aegypti* mosquitoes. *Tropical Biomedicine* 31: 134–142.

Aviram, M. and M. Rosenblat. 2005. Paraoxonases and cardiovascular diseases: Pharmacological and nutritional influences. *Current Opinion in Lipidology* 16: 393–399.

Bais, H., C. Broeckling, and J. Vivanco. 2008. Root exudates modulate plant—Microbe interactions in the rhizosphere. In: P. Karlovsky, (ed.), *Secondary Metabolites in Soil Ecology*, pp. 241–252. Berlin, Heidelberg: Springer.

Barnes, K.M., D.E. Gennard, and R.A. Dixon. 2010. An assessment of the antibacterial activity in larval excretion/secretion of four species of insects recorded in association with corpses, using *Lucilia sericata* Meigen as the marker species. *Bulletin of Entomological Research* 100: 635–640.

Barr, K.L., L.B. Hearne, S. Briesacher, T.L. Clark, and G.E. Davis. 2010. Microbial symbionts in insects influence down-regulation of defense genes in maize. *PLoS One* 5: e11339.

Bauer, W.D. and U. Mathesius. 2004. Plant responses to bacterial quorum sensing signals. *Current Opinion in Plant Biology* 7: 429–433.

Benbow, M.E., A. Lewis, J.K. Tomberlin, and J.L. Pechal. 2013. Seasonal necrophagous insect community assembly during vertebrate carrion decomposition. *Journal of Medical Entomology* 50: 440–450.

Benecke, M. 2001. A brief history of forensic entomology. *Forensic Science International* 120: 2–14.

Bos, L.D.J., P.J. Sterk, and M.J. Schultz. 2013. Volatile metabolites of pathogens: A systematic review. *PLoS Pathogen* 9: e11003311.

Bravo, J.A., P. Forsythe, M.V. Chew, E. Escaravage, H.M. Savignac, T.G. Dinan, J. Bienenstock, and J.F. Cryan. 2011. Ingestion of *Lactobacillus* strain regulates emotional behavior and central GABA receptor expression in a mouse via the vagus nerve. *Proceedings of the National Academy of Sciences* 108: 16050–16055.

Brundage, A.L., M.E. Benbow, and J.K. Tomberlin. 2014. Priority effects on the life-history traits of two carrion blow fly (Diptera, Calliphoridae) species. *Ecological Entomology* 39: 447–539.

Bryan, A., C. Watters, L. Koenig, E. Youn, A. Olmos, G. Li, S.C. Williams, and K.P. Rumbaugh. 2010. Human transcriptome analysis reveals a potential role for active transport in the metabolism of *Pseudomonas aeruginosa* autoinducers. *Microbes and Infection/Institut Pasteur* 12: 1042–1050.

Burkepile, D.E., J.D. Parker, C.B. Woodson, H.J. Mills, J. Kubanek, P.A. Sobecky, and M.E. Hay. 2006. Chemically mediated competition between microbes and animals: Microbes as consumers in food webs. *Ecology* 87: 2821–2831.

Byrd, J.H. and J.L. Castner. 2001. *Forensic Entomology: The Utility of Arthropods in Legal Investigations*. Boca Raton, FL: CRC Press.

Camps, J., I. Pujol, F. Ballester, J. Joven, and J.M. Simo. 2011. Paraoxonases as potential antibiofilm agents: Their relationship with quorum-sensing signals in Gram-negative bacteria. *Antimicrobial Agents and Chemotherapy* 55: 1325–1331.

Carlier, A., S. Uroz, B. Smadja, R. Fray, X. Latour, Y. Dessaux, and D. Faure. 2003. The Ti plasmid of *Agrobacterium tumefaciens* harbors an AttM-paralogous gene, AiiB, also encoding *N*-acyl homoserine lactonase activity. *Applied Environmental Microbiology* 69: 4989–4993.

Carter, D.O. and M. Tibbett. 2003. Taphonomic mycota: Fungi with forensic potential. *Journal of Forensic Sciences* 48: 168–171.

Ceřovský, V., J. Slaninova, V. Fucik, L. Monincova, L. Bednarova, P. Malon, and J. Stokrova. 2011. Lucifensin, a novel insect defensin of medicinal maggots: Synthesis and structural study. *ChemBioChem* 12: 1352–1361.

Chan, K.G., S. Atkinson, K. Mathee, C.K. Sam, S.R. Chhabra, M. Camara, C.L. Koh, and P. Williams. 2011. Characterization of *N*-acylhomoserine lactone-degrading bacteria associated with the *Zingiber officinale* (Ginger) rhizosphere: Co-existence of quorum quenching and quorum sensing in *Acinetobacter* and *Burkholderia*. *BMC Microbiology* 11: 51.

Chan, K.G., C.S. Wong, W.F. Yin, C.K. Sam, and C.L. Koh. 2010. Rapid degradation of *N*-3-oxo-acylhomoserine lactones by a *Bacillus cereus* isolate from Malaysian rainforest soil. *Antonie van Leeuwenhoek* 98: 299–305.

Charendoff, M.N., H.P. Shah, and J.M. Briggs. 2013. New insights into the binding and catalytic mechanisms of *Bacillus thuringiensis* lactonase: Insights into *B. thuringiensis* AiiA mechanism. *PLoS One* 8: e75395.

Chatnaparat, T., S. Prathuangwong, M. Ionescu, and S.E. Lindow. 2012. XagR, a LuxrR homolog, contributes to the virulence of *Xanthomonas axonopodis* pv. *glycines* to soybean. *Molecular Plant-Microbe Interactions* 25: 1104–1117.

Chen, F., Y. Gao, X. Chen, Z. Yu, and X. Li. 2013. Quorum quenching enzymes and their application in degrading signal molecules to block quorum sensing-dependent infection. *International Journal of Molecular Sciences* 14: 17477–17500.

Chow, J.Y., L. Wu, and W.S. Yew. 2009. Directed evolution of a quorum-quenching lactonase from *Mycobacterium avium* subsp. *Paratuberculosis* K-10 in the amidohydrolase superfamily. *Biochemistry* 48: 4344–4353.

Chowdhary, P.K., N. Keshavan, H.Q. Nguyen, J.A. Peterson, J.E. Gonzalez, and D.C. Haines. 2007. *Bacillus megaterium* CYP102A1 oxidation of acyl homoserine lactones and acyl homoserines. *Biochemistry* 46: 14429–14437.

Chu, Y.Y., M. Nega, M. Wolfle, L. Plener, S. Grond, K. Jung, and F. Gotz. 2013. A new class of quorum quenching molecules from *Staphylococcus* species affects communication and growth of Gram-negative bacteria. *PLoS Pathogen* 9: e1003654.

Chun, C.K., E.A. Ozer, M.J. Welsh, J. Zabner, and E.P. Greenberg. 2004. Inactivation of a *Pseudomonas aeruginosa* quorum-sensing signal by human airway epithelia. *Proceedings of the National Academy of Sciences* 101: 3587–3590.

Clarke, M.B., D.T. Hughes, C. Zhu, E.C. Boedeker, and V. Sperandio. 2006. The QsecC sensor kinase: A bacterial adrenergic receptor. *Proceedings of the National Academy of Sciences* 103: 10420–10425.

Collett, T.S. and M. Collett. 2002. Memory use in insect visual navigation. *National Reviews Neuroscience* 3: 542–552.

Costello, E.K., C.L. Lauber, M. Hamady, N. Fierer, J.I. Gordon, and R. Knight. 2009. Bacterial community variation in human body habitats across space and time. *Science* 326: 1694–1697.

Curtis, T.P., W.T. Sloan, and J.W. Scannell. 2002. Estimating prokaryotic diversity and its limits. *Proceedings of the National Academy of Sciences of the United States of America* 99: 10494–10499.

Dave, M., P.D. Higgins, S. Middha, and K.P. Rioux. 2012. The human gut microbiome: Current knowledge, challenges, and future directions. *Translational Research* 160: 246–257.

Davis, T.S., T.L. Crippen, R.W. Hofstetter, and J.K. Tomberlin. 2013. Microbial volatile emissions as insect semiochemicals. *Journal of Chemical Ecology* 39: 840–859.

Degrassi, G., G. Devescovi, R. Solis, L. Steindler, and V. Venturi. 2007. *Oryza sativa* rice plants contain molecules that activate different quorum-sensing *N*-acyl homoserine lactone biosensors and are sensitive to the specific AiiA lactonase. *FEMS Microbiology Letters* 269: 213–220.

DeVault, T.L., J.I.L. Brisbin, and J.O.E. Rhodes. 2004. Factors influencing the acquisition of rodent carrion by vertebrate scavengers and decomposers. *Canadian Journal of Zoology* 82: 502–509.

DeVault, T.L., J.O.E. Rhodes, and J.A. Shivik. 2003. Scavenging by vertebrates: Behavioral, ecological, and evolutionary perspectives on an important energy transfer pathway in terrestrial ecosystems. *Oikos* 102: 225–234.

Dickson, G.C., R.T.M. Poulter, E.W. Maas, P.K. Probert, and J.A. Kieser. 2010. Marine bacterial succession as a potential indicator of postmortem submersion interval. *Forensic Science International* 209: 1–10.

Dong, Y.H., A.R. Gusti, Q. Zhang, J.L. Xu, and L.H. Zhang. 2002. Identification of quorum-quenching *N*-acyl homoserine lactonases from *Bacillus* species. *Applied Environmental Microbiology* 68: 1754–1759.

Dong, Y.H., J.L. Xu, X.Z. Li, and L.H. Zhang. 2000. AiiA, an enzyme that inactivates the acylhomoserine lactone quorum-sensing signal and attenuates the virulence of *Erwinia carotovora*. *Proceedings of the National Academy of Sciences* 97: 3526–3531.

Dong, Y.H. and L.H. Zhang. 2005. Quorum sensing and quorum-quenching enzymes. *Journal of Microbiology* 43: 101–109.

Dong, Y.H., X.F. Zhang, J.L. Xu, and L.H. Zhang. 2004. Insecticidal *Bacillus thuringiensis* silences *Erwinia carotovora* virulence by a new form of microbial antagonism, signal interference. *Applied Environmental Microbiology* 70: 954–960.

Draganov, D.I. and B.N. La Du. 2004. Pharmacogenetics of paraoxonases: A brief review. *Naunyn-Schmiedeberg's Archives of Pharmacology* 369: 78–88.

Draganov, D.I., J.F. Teiber, A. Speelman, Y. Osawa, R. Sunahara, and B.N. La Du. 2005. Human paraoxonases (Pon1, Pon2, and Pon3) are lactonases with overlapping and distinct substrate specificities. *Journal of Lipid Research* 46: 1239–1247.

Duguid, I.G., E. Evans, M.R.W. Brown, and P. Gilbert. 1992. Growth-rate-independent killing by ciprofloxacin of biofilm-derived *Staphylococcus epidermidis* evidence for cell-cycle dependency. *Journal of Antimicrobial Chemotherapy* 30: 791–802.

Erdmann, G.R. and S.K.W. Khalil. 1986. Isolation and identification of two antibacterial agents produced by a strain of *Proteus mirabilis* osolated from larvae of the screwworm (*Cochliomyia hominivorax*) (Diptera: Calliphoridae). *Journal of Medical Entomology* 23: 208–211.

Ezenwa, V.O., N.M. Gerardo, D.W. Inouye, M. Medina, and J.B. Xavier. 2012. Animal behavior and the microbiome. *Science* 338: 198–199.

Fekete, A., C. Kuttler, M. Rothballer, B.A. Hense, D. Fischer, K. Buddrus-Schiemann, M. Lucio, J. Muller, P. Schmitt-Kopplin, and A. Hartmann. 2010. Dynamic regulation of *N*-acyl-homoserine lactone production and degradation in *Pseudomonas putida* IsoF. *FEMS Microbiology Ecology* 72: 22–34.

Ferluga, S., J. Bigirimana, M. Hofte, and V. Venturi. 2007. A LuxR homologue of *Xanthomonas oryzae* pv. *oryzae* is required for optimal rice virulence. *Molecular Plant Pathology* 8: 529–538.

Ferluga, S. and V. Venturi. 2009. OryR is a LuxR-family protein involved in interkingdom signaling between pathogenic *Xanthomonas oryzae* Pv. oryzae and rice. *Journal of Bacteriology* 191: 890–897.

Fierer, N., S. Ferrenberg, G.E. Flores, A. Gonzalez, J. Kueneman, T. Legg, R.C. Lynch et al. 2012. From animalcules to an ecosystem: Application of ecological concepts to the human microbiome. *Annual Review of Ecology, Evolution, and Systematics* 43: 137–155.

Firestein, S. 2001. How the olfactory system makes sense of scents. *Nature-London* 413: 211–218.

Flores, M. 2013. Life-history traits of *Chrysomya rufifacies* (Macquart) (Diptera: Calliphoridae) and its associated non-consumptive effects on *Cochliomyia macellaria* (Fabricius) (Diptera: Calliphorirdae) behavior and development. PhD, Texas A&M University.

Frago, E., M. Dicke, and H.C.J. Godfray. 2012. Insect symbionts as hidden players in insect plant interactions. *Trends in Ecology & Evolution (Personal edition)* 27: 705–711.

Frederickx, C., J. Dekeirsschieter, F.J. Verheggen, and E. Haubruge. 2012. Responses of *Lucilia sericata* Meigen (Diptera: Calliphoridae) to cadaveric volatile organic compounds. *Journal of Forensic Sciences* 57: 386–390.

Fytrou, A., P.G. Schofield, A.R. Kraaijeveld, and S.F. Hubbard. 2006. *Wolbachia* infection suppresses both host defence and parasitoid counter-defence. *Proceedings of the Royal Society B: Biological Sciences* 273: 791–796.

Gao, M., M. Teplitski, J.B. Robinson, and W.D. Bauer. 2003. Production of substances by *Medicago truncatula* that affect bacterial quorum sensing. *Molecular Plant-Microbe Interactions* 16: 827–834.

Garcia-Contreras, R., M. Martinez-Vazquez, N. Velazquez Guadarrama, A.G. Villegas Paneda, T. Hashimoto, T. Maeda, H. Quezada, and T.K. Wood. 2013. Resistance to the quorum-quenching compounds brominated furanone C-30 and 5-fluorouracil in *Pseudomonas aeruginosa* clinical isolates. *Pathogens and Disease* 68: 8–11.

Gil-Turnes, M.S., M.E. Hay, and W. Fenical. 1989. Symbiotic marine bacteria chemically defend crustacean embryos from a pathogenic fungus. *Science* 246: 116–118.

Goehler, L.E., R.P. Gaykema, N. Opitz, R. Reddaway, N. Badr, and M. Lyte. 2005. Activation in vagal afferents and central autonomic pathways: Early responses to intestinal infection with *Campylobacter jejuni*. *Brain, Behavior, and Immunity* 19: 334–344.

Gonzalez, J.E. and M.M. Marketon. 2003. Quorum sensing in nitrogen-fixing rhizobia. *Microbiology and Molecular Biology Reviews* 67: 574–592.

Gonzalez, J.F., M.P. Myers, and V. Venturi. 2013. The inter-kingdom solo OryR regulator of *Xanthomonas oryzae* is important for motility. *Molecular Plant Pathology* 14: 211–221.

Gram, L., R. de Nys, R. Maximilien, M. Givskov, P. Steinberg, and S. Kjelleberg. 1996. Inhibitory effects of secondary metabolites from the red alga *Delisea pulchra* on swarming motility of *Proteus mirabilis*. *Applied Environmental Microbiology* 62: 4284–4287.

Greenberg, B. 1959a. Persistence of bacteria in the developmental stages of the housefly: I. Survival of enteric pathogens in the normal and aseptically reared host. *American Journal of Tropical Medicine and Hygiene* 8: 405–411.

Greenberg, B. 1959b. Persistence of bacteria in the developmental stages of the housefly: II. Quantitative study of the host-contaminant relationship in flies breeding under natural conditions. *American Journal of Tropical Medicine and Hygiene* 8: 412–416.

Greenberg, B. 1959c. Persistence of bacteria in the developmental stages of the housefly: III. Quantitative distribution in prepupae and pupae. *American Journal of Tropical Medicine and Hygiene* 8: 613–617.

Greenberg, B. 1959d. Persistence of bacteria in the developmental stages of the housefly: IV. Infectivity of the newly emerged adult. *American Journal of Tropical Medicine and Hygiene* 8: 618–622.

Greenberg, B. 1968. Model for destruction of bacteria in the, idgut of blow fly maggots. *Journal of Medical Entomology* 5: 31–38.

Guthrie, G.D., C.S. Nicholson-Guthrie, and H.L. Leary, Jr. 2000. A bacterial high-affinity GABA binding protein: Isolation and characterization. *Biochemical and Biophysical Research Communications* 268: 65–68.

Ha, C., S.J. Park, S.J. Im, S.J. Park, and J.H. Lee. 2012. Interspecies signaling through QscR, a quorum receptor of *Pseudomonas aeruginosa*. *Molecules and Cells* 33: 53–59.

Hedges, L.M., J.C. Brownlie, S.L. O'neill, and K.N. Johnson. 2008. *Wolbachia* and virus protection in insects. *Science* 322: 702.

Hong, S.H., M. Hegde, J. Kim, X. Wang, A. Jayaraman, and T.K. Wood. 2012. Synthetic quorum-sensing circuit to control consortial biofilm formation and dispersal in a microfluidic device. *Nature Communication* 3: 613.

Horiuchi, J., B. Prithiviraj, H.P. Bais, B.A. Kimball, and J.M. Vivanco. 2005. Soil nematodes mediate positive interactions between legume plants and rhizobium bacteria. *Planta* 222: 848–857.

Horke, S., I. Witte, P. Wilgenbus, M. Kruger, D. Strand, and U. Forstermann. 2007. Paraoxonase-2 reduces oxidative stress in vascular cells and decreases endoplasmic reticulum stress-induced caspase activation. *Circulation* 115: 2055–2064.

Huang, J.J., J.I. Han, L.H. Zhang, and J.R. Leadbetter. 2003. Utilization of acyl-homoserine lactone quorum signals for growth by a soil pseudomonad and *Pseudomonas aeruginosa* PAO1. *Applied Environmental Microbiology* 69: 5941–5949.

Huang, J.J., A. Petersen, M. Whiteley, and J.R. Leadbetter. 2006. Identification of QuiP, the product of gene *Pa1032*, as the second acyl-homoserine lactone acylase of *Pseudomonas aeruginosa* PAO1. *Applied Environmental Microbiology* 72: 1190–1197.

Human Microbiome Project Consortium. 2012. Structure, function and diversity of the healthy human microbiome. *Nature* 486: 207–214.

Huttenhower, C., D. Gevers, R. Knight, S. Abubucker, J. Badger, A. Chinwalla et al. 2012. Structure, function and diversity of the healthy human microbiome. *Nature* 486: 207–214.

Janzen, D.H. 1977. Why fruits rot, seeds mold, and meat spoils. *American Naturalist* 111: 691–713.

Jermy, A. 2011. Antimicrobials: Disruption of quorum sensing meets resistance. *Nature Reviews Microbiology* 9: 767.

Jiménez-Martínez, E.S., N.A. Bosque-Pérez, P.H. Berger, R.S. Zemetra, H. Ding, and S.D. Eigenbrode. 2004. Volatile cues influence the response of *Rhopalosiphum padi* (Homoptera: Aphididae) to barley yellow dwarf virus–infected transgenic and untransformed wheat. *Environmental Entomology* 33: 1207–1216.

Joint, I., J.A. Downie, and P. Williams. 2007. Bacterial conversations: Talking, listening and eavesdropping. An introduction. *Philosophical Transactions of the Royal Society B: Biological Sciences* 362: 1115–1117.

Jojola-Elverum, S.M., J.A. Shivik, and L. Clark. 2001. Importance of bacterial decomposition, and carrion substrate to foraging, brown treesnakes. *Journal of Chemical Ecology* 27: 1315–1331.

Kang, B.R., J.H. Lee, S.J. Ko, Y.H. Lee, J.S. Cha, B.H. Cho, and Y.C. Kim. 2004. Degradation of acyl-homoserine lactone molecules by *Acinetobacter* sp. strain C1010. *Canadian Journal of Microbiology* 50: 935–941.

Kappachery, S., D. Paul, J. Yoon, and J.H. Kweon. 2010. Vanillin, a potential agent to prevent biofouling of reverse osmosis membrane. *Biofouling* 26: 667–672.

Kinney, K.S., C.E. Austin, D.S. Morton, and G. Sonnenfeld. 2000. Norepinephrine as a growth stimulating factor in bacteria—mechanistic studies. *Life Sciences* 67: 3075–3085.

Kloos, W.E. and M.S. Musslewhite. 1975. Distribution and persistence of *Staphylococcus* and *Micrococcus* species and other aerobic bacteria on human skin. *Applied Microbiology* 30: 381–395.

Koch, H. and P. Schmid-Hempel. 2011. Socially transmitted gut microbiota protect bumble bees against an intestinal parasite. *Proceedings of the National Academy of Sciences* 108: 19288–19292.

Kuzma, J., M. Nemecek-Marshall, W.H. Pollack, and R. Fall. 1995. Bacteria produce the volatile hydrocarbon isoprene. *Current Microbiology* 30: 97–103.

Lam, K., K. Thu, M. Tsang, M. Moore, and G. Gries. 2009. Bacteria on housefly eggs, *Musca domestica*, suppress fungal growth in chicken manure through nutrient depletion or antifungal metabolites. *Naturwissenschaften* 9: 1127–1132.

Lam, K., M. Tsang, A. Labrie, R. Gries, and G. Gries. 2010. Semiochemical-mediated oviposition avoidance by female house flies, *Musca domestica*, on animal feces colonized with harmful fungi. *Journal of Chemical Ecology* 36: 141–147.

Leadbetter, J.R. and E.P. Greenberg. 2000. Metabolism of acyl-homoserine lactone quorum-sensing signals by *Variovorax paradoxus*. *Journal of Bacteriology* 182: 6921–6926.

LeBlanc, H.N. 2008. Olfactory stimuli associated with the different stages of vertebrate decomposition and their role in the attraction of the blowfly *Calliphora vomitoria* (Diptera: Calliphoridae) to carcasses. The University of Derby, UK.

Lee, S.J., S.Y. Park, J.J. Lee, D.Y. Yum, B.T. Koo, and J.K. Lee. 2002. Genes encoding the *N*-acyl homoserine lactone-degrading enzyme are widespread in many subspecies of *Bacillus thuringiensis*. *Applied Environmental Microbiology* 68: 3919–3924.

Lee, J., X.-S. Zhang, M. Hegde, W.E. Bentley, A. Jayaraman, and T.K. Wood. 2008. Indole cell signaling occurs primarily at low temperatures in *Escherichia coli*. *International Society of Microbial Ecology Journal* 2: 1007–1023.

Leitner, M., R. Kaiser, B. Hause, W. Boland, and A. Mithöfer. 2010. Does mycorrhization influence herbivore-induced volatile emission in *Medicago truncatula*? *Mycorrhiza* 20: 89–101.

Leroith, D., W. Pickens, A.I. Vinik, and J. Shiloach. 1985. *Bacillus subtilis* contains multiple forms of somatostatin-like material. *Biochemical and Biophysical Research Communications* 127: 713–719.

Li, W., S.E. Dowd, B. Scurlock, V. Acosta-Martinez, and M. Lyte. 2009. Memory and learning behavior in mice is temporally associated with diet-induced alterations in gut bacteria. *Physiology & Behavior* 96: 557–567.

Lin, Y.H., J.L. Xu, J. Hu, L.H. Wang, S.L. Ong, J.R. Leadbetter, and L.H. Zhang. 2003. Acyl-homoserine lactone acylase from *Ralstonia* strain XJ12B represents a novel and potent class of quorum-quenching enzymes. *Molecular Microbiology* 47: 849–860.

Liu, W. 2014. Chemical and nutritional ecology of *Lucilia sericata* (Meigen) (Diptera: Calliphoridae) as related to volatile organic compounds and associated essential amino acids. PhD, Texas A&M University.

Liu, D., B.W. Lepore, G.A. Petsko, P.W. Thomas, E.M. Stone, and D. Ringe. 2005. Three-dimensional structure of the quorum-quenching *N*-acyl homoserine lactone hydrolase from *Bacillus thuringiensis*. *Proceedings of the National Academy of Sciences* 102: 11882–11887.

Liu, D., J. Momb, P.W. Thomas, A. Moulin, G.A. Petsko, W. Fast, and D. Ringe. 2008. Mechanism of the quorum-quenching lactonase (AiiA) from *Bacillus thuringiensis*. *Biochemistry* 47: 7706–7714.

Lowery, C.A., T.J. Dickerson, and K.D. Janda. 2008. Interspecies and interkingdom communication mediated by bacterial quorum sensing. *Chemical Society Reviews* 37: 1337–1346.

Lu, M., M.J. Wingfield, N. Gillette, and J.-H. Sun. 2011. Do novel genotypes drive the success of an invasive bark beetle–fungus complex? Implications for potential reinvasion. *Ecology* 92: 2013–2019.

Lyte, M. 2010. The microbial organ in the gut as a driver of homeostasis and disease. *Medical Hypotheses* 74: 634–638.

Lyte, M., B. Arulanandam, K. Nguyen, C. Frank, A. Erickson, and D. Francis. 1997a. Norepinephrine induced growth and expression of virulence associated factors in enterotoxigenic and enterohemorrhagic strains of *Escherichia coli*. *Advances in Experimental Medicine and Biology* 412: 331–339.

Lyte, M., A.K. Erickson, B.P. Arulanandam, C.D. Frank, M.A. Crawford, and D.H. Francis. 1997b. Norepinephrine-induced expression of the K99 pilus adhesin of enterotoxigenic *Escherichia coli*. *Biochemical and Biophysical Research Communications* 232: 682–686.

Lyte, M., J.J. Varcoe, and M.T. Bailey. 1998. Anxiogenic effect of subclinical bacterial infection in mice in the absence of overt immune activation. *Physiology & Behavior* 65: 63–68.

Ma, Q., W. Liu, A.T. Fields, M.L. Pimsler, A.M. Tarone, T.L. Crippen, J.K. Tomberlin, and T.K. Wood. 2012. *Proteus mirabilis* interkingdom swarming signals attract blow flies. *International Society of Microbial Ecology Journal* 6: 1356–1366.

Maeda, T., R. Garcia-Contreras, M. Pu, L. Sheng, L.R. Garcia, M. Tomas, and T.K. Wood. 2012. Quorum quenching quandary: Resistance to antivirulence compounds. *International Society of Microbial Ecology Journal* 6: 493–501.

Manefield, M., R. de Nys, N. Kumar, R. Read, M. Givskov, P. Steinberg, and S. Kjelleberg. 1999. Evidence that halogenated furanones from *Delisea pulchra* inhibit acylated homoserine lactone (AHL)-mediated gene expression by displacing the AHL signal from its receptor protein. *Microbiology* 145 (Part 2): 283-291.

Manefield, M., L. Harris, S.A. Rice, R. De Nys, and S. Kjelleberg. 2000. Inhibition of luminescence and virulence in the black tiger prawn (*Penaeus monodon*) pathogen *Vibrio harveyi* by intercellular signal antagonists. *Applied Environmental Microbiology* 66: 2079–2084.

Mei, G.Y., X.X. Yan, A. Turak, Z.Q. Luo, and L.Q. Zhang. 2010. AidH, an alpha/beta-hydrolase fold family member from an *Ochrobactrum* sp. strain, is a novel *N*-acylhomoserine lactonase. *Applied Environmental Microbiology* 76: 4933–4942.

Meijerink, J., A.H. Braks, A.A. Brack, W. Adam, T. Dekker, M.A. Posthumus, T.A. Va Beek, and J.J. Van Loon. 2000. Identification of olfactory stimulants for *Anopheles gambiae* from human sweat samples. *Journal of Chemical Ecology* 26: 1367–1382.

Michael, B., J.N. Smith, S. Swift, F. Heffron, and B.M. Ahmer. 2001. SdiA of *Salmonella enterica* is a LuxR homolog that detects mixed microbial communities. *Journal of Bacteriology* 183: 5733–5742.

Miller, M.B. and B.L. Bassler. 2001. Quorum sensing in bacteria. *Annual Reviews in Microbiology* 55: 165–199.

Miller, W.J., L. Ehrman, and D. Schneider. 2010. Infectious speciation revisited: Impact of symbiont-depletion on female fitness and mating behavior of *Drosophila paulistorum*. *PLoS Pathogens* 6: e1001214.

Mok, K.C., N.S. Wingreen, and B.L. Bassler. 2003. *Vibrio harveyi* quorum sensing: A coincidence detector for two autoinducers controls gene expression. *EMBO Journal* 22: 870–881.

Momb, J., C. Want, D. Liu, P.W. Thomas, G.A. Petsko, H. Guo, D. Ringe, and W. Fast. 2008. Mechanism of the quorum-quenching lactonase (AiiA) from *Bacillus thuringiensis*. 2. Substrate modeling and active site mutations. *Biochemistry* 47: 7715–7725.

Moreira, L.A., I. Iturbe-Ormaetxe, J.A. Jeffery, G. Lu, A.T. Pyke, L.M. Hedges, B.C. Rocha et al. 2009. A *Wolbachia* symbiont in *Aedes aegypti* limits infection with dengue, chikungunya, and *Plasmodium*. *Cell* 139: 1268–1278.

Morgan, X.C., N. Segata, and C. Huttenhower. 2013. Biodiversity and functional genomics in the human microbiome. *Trends in Genetics* 29: 51–58.

Morohoshi, T., S. Nakazawa, A. Ebata, N. Kato, and T. Ikeda. 2008. Identification and characterization of *N*-acylhomoserine lactone-acylase from the fish intestinal *Shewanella* sp. strain MIB015. *Bioscience, Biotechnology, and Biochemistry* 72: 1887–1893.

Morohoshi, T., T. Shiono, K. Takidouchi, M. Kato, N. Kato, J. Kato, and T. Ikeda. 2007. Inhibition of quorum sensing in *Serratia marcescens* AS-1 by synthetic analogs of *N*-acylhomoserine lactone. *Applied Environmental Microbiology* 73: 6339–6344.

Mullin, C., S. Chyb, H. Eichenseer, B. Hollister, and J. Frazier. 1994. Neuroreceptor mechanisms in insect gustation: A pharmacological approach. *Journal of Insect Physiology* 40: 913–931.

Musthafa, K.S., A.V. Ravi, A. Annapoorani, I.S. Packiavathy, and S.K. Pandian. 2010. Evaluation of anti-quorum-sensing activity of edible plants and fruits through inhibition of the *N*-acyl-homoserine lactone system in *Chromobacterium violaceum* and *Pseudomonas aeruginosa*. *Chemotherapy* 56: 333–339.

Neumann, R. 1979. Bacterial induction of settlement and metamorphosis in the planula larvae of *Cassiopea andromeda* (Cnidaria, Scyphozoa, Rhizostomeae). *Marine Ecology Progress Series* 1: 21–28.

Ng, C.J., D.J. Wadleigh, A. Gangopadhyay, S. Hama, V.R. Grizalva, M. Navab, A.M. Fogelman, and S.T. Reddy. 2001. Paraoxonase-2 is a ubiquitously expressed protein with antioxidant properties and is capable of preventing cell-mediated oxidative modification of low density lipoprotein. *Journal of Biological Chemistry* 276: 44444–44449.

Nierop Groot, M.N. and J.a.M. De Bont. 1998. Conversion of phenylalanine to benzaldehyde initiated by an aminotransferase in *Lactobacillus plantarum*. *Applied and Environmental Microbiology* 64: 3009–3013.

Oliver, K.M., P.H. Degnan, M.S. Hunter, and N.A. Moran. 2009. Bacteriophages encode factors required for protection in a symbiotic mutualism. *Science* 325: 992–994.

Oliver, K.M., N.A. Moran, and M.S. Hunter. 2005. Variation in resistance to parasitism in aphids is due to symbionts not host genotype. *Proceedings of the National Academy of Sciences* 102: 12795–12800.

Otto, M. 2009. *Staphylococcus epidermidis*—the "accidental" pathogen. *Nature Reviews Microbiology* 7: 555–567.

Ozer, E.A., A. Pezzulo, D.M. Shih, C. Chun, C. Furlong, A.J. Lusis, E.P. Greenberg, and J. Zabner. 2005. Human and murine paraoxonase 1 are host modulators of *Pseudomonas aeruginosa* quorum-sensing. *FEMS Microbiology Letters* 253: 29–37.

Pan, J., R. Huang, F. Yao, Z. Huang, C.A. Powell, S. Qiu, and X. Guan. 2008. Expression and characterization of *AiiA* gene from *Bacillus subtilis* Bs-1. *Microbiological Research* 163: 711–716.

Park, S.Y., B.J. Hwang, M.H. Shin, J.A. Kim, and J.K. Lee. 2006. *N*-acylhomoserine lactonase producing *Rhodococcus* spp. with different AHL-degrading activities. *FEMS Microbiology Letters* 261: 102–108.

Park, S.Y., H.O. Kang, H.S. Jang, J.K. Lee, B.T. Koo, and D.Y. Yum. 2005. Identification of extracellular *N*-acylhomoserine lactone acylase from a *Streptomyces* sp. and its application to quorum quenching. *Applied Environmental Microbiology* 71: 2632–2641.

Park, S.Y., S.J. Lee, T.K. Oh, J.W. Oh, B.T. Koo, D.Y. Yum, and J.K. Lee. 2003. AhlD, an *N*-acylhomoserine lactonase in *Arthrobacter* sp., and predicted homologues in other bacteria. *Microbiology* 149: 1541–1550.

Patankar, A.V. and J.E. Gonzalez. 2009a. An orphan LuxR homolog of *Sinorhizobium meliloti* affects stress adaptation and competition for nodulation. *Applied Environmental Microbiology* 75: 946–955.

Patankar, A.V. and J.E. Gonzalez. 2009b. Orphan LuxR regulators of quorum sensing. *FEMS Microbiology Reviews* 33: 739–756.

Patel, C.N., B.W. Wortham, J.L. Lines, J.D. Fetherston, R.D. Perry, and M.A. Oliveira. 2006. Polyamines are essential for the formation of plague biofilm. *Journal of Bacteriology* 188: 2355–2363.

Payne, J.A. 1965. A summer carrion study of the baby pig *Sus scrofa* Linnaeus. *Ecology* 46: 592–602.

Ponnusamy, L., N. Xu, S. Nojima, D.M. Wesson, C. Schal, and C.S. Apperson. 2008. Identification of bacteria and bacteria-associated chemical cues that mediate oviposition site preferences by *Aedes aegypti*. *Proceedings of the National Academy of Sciences* 105: 9262–9267.

Rajesh, P.S. and V.R. Rai. 2014. Molecular identification of *AiiA* homologous gene from endophytic *Enterobacter* species and *in silico* analysis of putative tertiary structure of AHL-lactonase. *Biochemical and Biophysical Research Communications* 443: 290–295.

Ren, D., L.A. Bedzyk, R.W. Ye, S.M. Thomas, and T.K. Wood. 2004. Differential gene expression shows natural brominated furanones interfere with the autoinducer-2 bacterial signaling system of *Escherichia coli*. *Biotechnology and Bioengineering* 88: 630–642.

Ren, D., J.J. Sims, and T.K. Wood. 2001. Inhibition of biofilm formation and swarming of *Escherichia coli* by (5z)-4-bromo-5-(bromomethylene)-3-butyl-2(5h)-furanone. *Environmental Microbiology* 3: 731–736.

Robacker, D. and C. Lauzon. 2002. Purine metabolizing capability of *Enterobacter agglomerans* affects volatiles production and attractiveness to Mexican fruit fly. *Journal of Chemical Ecology* 28: 1549–1563.

Romero, M., R. Avendano-Herrera, B. Magarinos, M. Camara, and A. Otero. 2010. Acylhomoserine lactone production and degradation by the fish pathogen *Tenacibaculum maritium*, a member of the Cytophaga-Flavobacterium-Bacterioides (CFB) group. *FEMS Microbiology Letters* 304: 131–139.

Romero, M., S.P. Diggle, S. Heeb, M. Camara, and A. Otero. 2008. Quorum quenching activity in *Anabaena* sp. pcc 7120: Identification of AiiC, a novel AHL-acylase. *FEMS Microbiology Letters* 280: 73–80.

Romero, M., A.B. Martin-Cuadrado, A. Roca-Rivada, A.M. Cabello, and A. Otero. 2011. Quorum quenching in cultivable bacteria from dense marine coastal microbial communities. *FEMS Microbiology Ecology* 75: 205–217.

Root-Bernstein, R. 2012. A modular hierarchy-based theory of the chemical origins of life based on molecular complementarity. *Accounts of Chemical Research* 45: 2169–2177.

Rothfork, J.M., G.S. Timmins, M.N. Harris, X. Chen, A.J. Lusis, M. Otto, A.L. Cheung, and H.D. Gresham. 2004. Inactivation of a bacterial virulence pheromone by phagocyte-derived oxidants: New role for the NADPH oxidase in host defense. *Proceedings of the National Academy of Sciences* 101: 13867–13872.

Roy, V., J.A. Smith, J. Wang, J.E. Stewart, W.E. Bentley, and H.O. Sintim. 2010. Synthetic analogs tailor native AI-2 signaling across bacterial Species. *Journal of the American Chemical Society* 132: 11141–11150.

Rudrappa, T. and H.P. Bais. 2008. Curcumin, a known phenolic from *Curcuma longa*, attenuates the virulence of *Pseudomonas aeruginosa* PAO1 in whole plant and animal pathogenicity models. *Journal of Agricultural and Food Chemistry* 56: 1955–1962.

Schikora, A., S.T. Schenk, E. Stein, A. Molitor, A. Zuccaro, and K.H. Kogel. 2011. *N*-acyl-homoserine lactone confers resistance toward biotrophic and hemibiotrophic pathogens via altered activation of AtMPK6. *Plant Physiology* 157: 1407–1418.

Schipper, C., C. Hornung, P. Bijtenhoorn, M. Quitschau, S. Grond, and W.R. Streit. 2009. Metagenome-derived clones encoding two novel lactonase family proteins involved in biofilm inhibition in *Pseudomonas aeruginosa*. *Applied Environmental Microbiology* 75: 224–233.

Schoenly, K.G. and W. Reid. 1987. Dynamics of heterotrophic succession in carrion arthropod assemblages: Discrete seres or a continuum of change? *Oecologia* 73: 192–202.

Schommer, N.N. and R.L. Gallo. 2013. Structure and function of the human skin microbiome. *Trends in Microbiology* 21: 660–668.

Sharon, G., D. Segal, J.M. Ringo, A. Hefetz, I. Ziiber-Rosenbert, and E. Rosenberg. 2010. Commensal bacteria play a role in mating preference of *Drosophila melanogaster*. *Proceedings of the National Academy of Sciences of the United States of America* 46: 20051–20056.

Shepherd, R.W. and S.E. Lindow. 2009. Two dissimilar *N*-acyl-homoserine lactone acylases of *Pseudomonas syringae* influence colony and biofilm morphology. *Applied Environmental Microbiology* 75: 45–53.

Sherman, R.A. 2002. Maggot therapy for foot and leg wounds. *Lower Extremity Wounds* 1: 135–142.

Shih, D.M., L. Gu, Y.R. Xia, M. Navab, W.F. Li, S. Hama, L.W. Castellani, L.G. Costa, A.M. Fogelman, and A.J. Lusis. 1998. Mice lacking serum paraoxonase are susceptible to organophosphate toxicity and atherosclerosis. *Nature* 394: 284–287.

Shimada, T., K. Shimada, M. Matsui, Y. Kitai, J. Igarashi, H. Suga, and A. Ishihama. 2014. Roles of cell division control factor SdiA: Recognition of quorum sensing signals and modulation of transcription regulation targets. *Genes to Cells* 19: 405–418.

Shiner, E.K., D. Terentyev, A. Bryan, S. Sennoune, R. Martinez-Zaguilan, G. Li, S. Gyorke, S.C. Williams, and K.P. Rumbaugh. 2006. *Pseudomonas aeruginosa* autoinducer modulates host cell responses through calcium signalling. *Cellular Microbiology* 8: 1601–1610.

Shivik, J.A. and L. Clark. 1997. Carrion seeking in brown tree snakes: Importance of olfactory and visual cues. *Journal of Experimental Zoology* 279: 549–553.

Sio, C.F., L.G. Otten, R.H. Cool, S.P. Diggle, P.G. Braun, M. Daykin, M. Camara, P. Williams, and W.J. Quax. 2006. Quorum quenching by an *N*-acyl-homoserine lactone acylase from *Pseudomonas aeruginosa* PAO1. *Infection and Immunity* 74: 1673–1682.

Smith, K.M., Y. Bu, and H. Suga. 2003. Induction and inhibition of *Pseudomonas aeruginosa* quorum sensing by synthetic autoinducer analogs. *Chemistry & Biology* 10: 81–89.

Smith, R.S., E.R. Fedyk, T.A. Springer, N. Mukaida, B.H. Iglewski, and R.P. Phipps. 2001. Il-8 production in human lung fibroblasts and epithelial cells activated by the *Pseudomonas* autoinducer *N*-3-oxododecanoyl homoserine lactone is transcriptionally regulated by NF-Kappa B and activator protein-2. *Journal of Immunology* 167: 366–374.

Smith, R.S., S.G. Harris, R. Phipps, and B. Iglewski. 2002a. The *Pseudomonas aeruginosa* quorum-sensing molecule *N*-(3-oxododecanoyl) homoserine lactone contributes to virulence and induces inflammation *in vivo*. *Journal of Bacteriology* 184: 1132–1139.

Smith, R.S., R. Kelly, B.H. Iglewski, and R.P. Phipps. 2002b. The *Pseudomonas* autoinducer *N*-(3-oxododecanoyl) homoserine lactone induces cyclooxygenase-2 and prostaglandin E2 production in human lung fibroblasts: Implications for inflammation. *Journal of Immunology* 169: 2636–2642.

Soni, K.A., P. Jesudhasan, M. Cepeda, K. Widmer, G.K. Jayaprakasha, B.S. Patil, M.E. Hume, and S.D. Pillai. 2008. Identification of ground beef-derived fatty acid inhibitors of autoinducer-2-based cell signaling. *Journal of Food Protection* 71: 134–138.

Stephen, F.M., C.W. Berisford, D.L. Dahlsten, P. Feen, and J.C. Moser. 1993. Invertebrate and microbial associates. In: T.D. Schowalter, and G.M. Filip, (eds.), *Beetle-Pathogen Interactions in Conifer Forests*, pp. 129–153. San Diego: Academic Press.

Stoltz, D.A., E.A. Ozer, J.M. Yu, S.T. Reddy, A.J. Lusis, N. Bourquard, M.R. Parsek, J. Zabner, and D.M. Shih. 2007. Paraoxonase-2 deficiency enhances *Pseudomonas aeruginosa* quorum sensing in murine tracheal epithelia. *American Journal of Physiology. Lung Cellular and Molecular Physiology* 292: L852–L860.

Studer, S.V., J.A. Schwartzman, J.S. Ho, G.D. Geske, H.E. Blackwell, and E.G. Ruby. 2013. Non-native acyl-ated homoserine lactones reveal that LuxIR quorum sensing promotes symbiont stability. *Environmental Microbiology* 8: 2623–2634.

Subramoni, S., J.F. Gonzalez, A. Johnson, M. Pechy-Tarr, I. Paulsen, J.E. Loper, C. Keel, and V. Venturi. 2011. Bacterial subfamily of LuxR regulators that respond to plant compounds. *Applied Environmental Microbiology* 77: 4579–4588.

Surette, M.G. and B.L. Bassler. 1998. Quorum sensing in *Escherichia coli* and *Salmonella typhimurium*. *Proceedings of the National Academy of Sciences* 95: 7046–7050.

Tarone, A.M. and D.R. Foran. 2006. Components of developmental plasticity in a Michigan population of *Lucilia sericata* (Diptera: Calliphoridae). *Journal of Medical Entomology* 43: 1023–1033.

Tateda, K., Y. Ishii, M. Horikawa, T. Matsumoto, S. Miyairi, J.C. Pechere, T.J. Stanford, M. Ishiguro, and K. Yamaguchi. 2003. The *Pseudomonas aeruginosa* autoinducer *N*-3-oxododecanoyl homoserine lactone accelerates apoptosis in macrophages and neutrophils. *Infection and Immunity* 71: 5785–5793.

Teiber, J.F., S. Horke, D.C. Haines, P.K. Chowdhary, J. Xiao, G.L. Kramer, R.W. Haley, and D.I. Draganov. 2008. Dominant role of paraoxonases in inactivation of the *Pseudomonas aeruginosa* quorum-sensing signal *N*-(3-oxododecanoyl)-L-homoserine lactone. *Infection and Immunity* 76: 2512–2519.

Teplitski, M., J.B. Robinson, and W.D. Bauer. 2000. Plants secrete substances that mimic bacterial *N*-acyl homoserine lactone signal activities and affect population density-dependent behaviors in associated bacteria. *Molecular Plant-Microbe Interactions* 13: 637–648.

The Human Microbiome Consortium. 2010. A catalog of reference genomes from the human microbiome. *Science* 328: 994–999.

The Human Microbiome Consortium. 2012. A framework for human microbiome research. *Nature* 486: 215–221.

Thomas, P.W., E.M. Stone, A.L. Costello, D.L. Tierney, and W. Fast. 2005. The quorum-quenching lactonase from *Bacillus thuringiensis* is a metalloprotein. *Biochemistry* 44: 7559–7569.

Tomberlin, J.K., T.L. Crippen, A.M. Tarone, B. Singh, K. Adams, Y.H. Rezenom, M.E. Benbow et al. 2012. Interkingdom responses of flies to bacteria mediated by fly physiology and bacterial quorum sensing. *Animal Behaviour* 84: 1449–1456.

Truchado, P., J.A. Gimenez-Bastida, M. Larrosa, I. Castro-Ibanez, F.A. Tomas-Barberan, M.T. Garcia-Conesa and A. Allende. 2012. Inhibition of quorum sensing (Qs) in *Yersinia enterocolitica* by an orange extract rich in glycosylated flavanones. *Journal of Agricultural and Food Chemistry* 60: 8885–8894.

Tuckerman, J.R., G. Gonzalez, E.H. Sousa, X. Wan, J.A. Saito, M. Alam, M.A. and Gilles-Gonzalez. 2009. An oxygen-sensing diguanylate cyclase and phosphodiesterase couple for c-di-GMP control. *Biochemistry* 48: 9764–9774.

Turnbaugh, P.J., R.E. Ley, M. Hamady, C.M.R. Knight, and J.I. Gordon. 2007. The human microbiome project. *Nature* 449: 804–810.

Ulrich, R.L. 2004. Quorum quenching: Enzymatic disruption of *N*-acylhomoserine lactone-mediated bacterial communication in *Burkholderia thailandensis*. *Applied Environmental Microbiology* 70: 6173–6180.

Uroz, S., S.R. Chhabra, M. Camara, P. Williams, P. Oger, and Y. Dessaux. 2005. *N*-acylhomoserine lactone quorum-sensing molecules are modified and degraded by *Rhodococcus erythropolis* W2 by both amido-lytic and novel oxidoreductase activities. *Microbiology* 151: 3313-3322.

Uroz, S., C. D'Angelo-Picard, A. Carlier, M. Elasri, C. Sicot, A. Petit, P. Oger, D. Faure, and Y. Dessaux. 2003. Novel bacteria degrading *N*-acylhomoserine lactones and their use as quenchers of quorum-sensing-regulated functions of plant-pathogenic bacteria. *Microbiology* 149: 1981–1989.

Uroz, S., P. Oger, S.R. Chhabra, M. Camara, P. Williams, and Y. Dessaux. 2007. *N*-acyl homoserine lactones are degraded via an amidolytic activity in *Comamonas* sp. strain D1. *Archives of Microbiology* 187: 249–256.

Vannette, R.L., M.P. Gauthier, and T. Fukami. 2013. Nectar bacteria, but not yeast, weaken a plant-pollinator mutualism. *Proceedings of the Royal Society B: Biological Sciences* 280: 20122601.

Vass, A.A. 2001. Beyond the grave: Understanding human decomposition. *Microbiology Today* 28: 190–192.

Venturi, V. and C. Fuqua. 2013. Chemical signaling between plants and plant-pathogenic bacteria. *Annual Review of Phytopathology* 51: 17–37.

Verhulst, N.O., R. Andriessen, U. Groenhagen, G. Bukovinszkiné Kiss, S. Schulz, W. Takken, J.J.A. van Loon, G. Schraa, and R.C. Smallegange. 2010a. Differential attraction of malaria mosquitoes to volatile blends produced by human skin bacteria. *PLoS One* 5: e15829.

Verhulst, N.O., W. Takken, M. Dicke, G. Schraa, and R.C. Smallegange. 2010b. Chemical ecology of interactions between human skin microbiota and mosquitoes. *FEMS Microbiology Ecology* 9999: 1–9.

Walker, T., P.H. Johnson, L.A. Moreira, I. Iturbe-Ormaetxe, F.D. Frentiu, C.J. McMeniman, Y.S. Leong et al. 2011. The wMel *Wolbachia* strain blocks dengue and invades caged *Aedes aegypti* populations. *Nature* 476: 450–453.

Wang, D., X. Ding, and P.N. Rather. 2001. Indole can act as an extracellular signal in *Escherichia coli*. *Journal of Bacteriology* 183: 4210–4216.

Wang, W.Z., T. Morohoshi, M. Ikenoya, N. Someya, and T. Ikeda. 2010. AiiM, a novel class of *N*-acylhomoserine lactonase from the leaf-associated bacterium *Microbacterium testaceum*. *Applied Environmental Microbiology* 76: 2524–2530.

Wang, L.H., L.X. Weng, Y.H. Dong, and L.H. Zhang. 2004. Specificity and enzyme kinetics of the quorum-quenching *N*-acyl homoserine lactone lactonase (AHL-lactonase). *The Journal of Biological Chemistry* 279: 13645–13651.

Wang, Y.D., J.C. Wu, and Y.J. Yuan. 2007. Salicylic acid-induced taxol production and isopentenyl pyrophosphate biosynthesis in suspension cultures of *Taxus chinensis* var. *mairei*. *Cell Biology International* 31: 1179–1183.

Waters, C.M. and B.L. Bassler. 2005. Quorum sensing: Cell-to-cell communication in bacteria. *Annual Review of Cell and Developmental Biology* 21: 319–346.

Wells, J.D. and B. Greenberg. 1994. Resource use by an introduced and native carrion flies. *Oecologia* 99: 181–187.

Werren, J.H., L. Baldo, and M.E. Clark. 2008. *Wolbachia*: Master manipulators of invertebrate biology. *Nature Reviews Microbiology* 6: 741–751.

Xu, F., T. Byun, H.J. Deussen, and K.R. Duke. 2003. Degradation of *N*-acylhomoserine lactones, the bacterial quorum-sensing molecules, by acylase. *Journal of Biotechnology* 101: 89–96.

Yang, F., L.H. Wang, J. Wang, Y.H. Dong, J.Y. Hu, and L.H. Zhang. 2005. Quorum quenching enzyme activity is widely conserved in the sera of mammalian species. *FEBS Letters* 579: 3713–3717.

Yao, Y., M.A. Martinez-Yamout, T.J. Dickerson, A.P. Brogan, P.E. Wright, and H.J. Dyson. 2006. Structure of the *Escherichia coli* quorum sensing protein SdiA: Activation of the folding switch by acyl homoserine lactones. *Journal of Molecular Biology* 355: 262–273.

Zhang, X. 2014. Effect of quorum sensing by *Staphylococcus epidermidis* on the attraction response by female adult yellow fever mosquito, *Aedes aegypti aegypti* (Linnaeus) (Diptera: Culicidae) to a blood-feeding source. MS thesis, Texas A&M University.

Zhang, L., Y. Jia, L. Wang, and R. Fang. 2007. A proline iminopeptidase gene upregulated in planta by a Luxr homologue is essential for pathogenicity of *Xanthomonas campestris* Pv. campestris. *Molecular Microbiology* 65: 121–136.

Zhang, R.G., K.M. Pappas, J.L. Brace, P.C. Miller, T. Oulmassov, J.M. Molyneaux, J.C. Anderson, J.K. Bashkin, S.C. Winans, and A. Joachimiak. 2002b. Structure of a bacterial quorum-sensing transcription factor complexed with pheromone and DNA. *Nature* 417: 971–974.

Zhang, H.B., L.H. Wang, and L.H. Zhang. 2002a. Genetic control of quorum-sensing signal turnover in *Agrobacterium tumefaciens*. *Proceedings of the National Academy of Sciences* 99: 4638–4643.

21

Ecology of African Carrion

Sarah C. Jones, Eli D. Strauss, and Kay E. Holekamp

CONTENTS

21.1 Introduction

Many areas of Africa are inhabited by a rich abundance of organisms that produce or consume carrion. In this chapter, carrion is defined as any dead animal or parts of dead animals, and carrion consumers are divided into scavengers and decomposers. Scavengers feed on carrion by ingesting it and often physically removing pieces of it, whereas decomposers, such as microbes, are largely saprophytic or saprozoic and utilize carrion as a food source exclusively *in situ* (Putnam 1983). The focus of this chapter will largely be on scavengers, as they profoundly influence carrion degradation in most African ecosystems.

The carrion ecology of Africa is unique for several reasons, the most important of which is the continent's extraordinary diversity of mammals and birds. Herbivorous mammals in Africa range in size from elephants down to small rodents (see Appendix for scientific names). Of particular relevance to carrion ecology is the abundance and diversity of medium- and large-bodied herbivores; these include gazelles, impala, buffalo, zebra, wildebeest, and many others (Owen-Smith 2013). Africa is also unique in its abundance of mega-herbivores, which are herbivores weighing more than 1000 kg as adults; these include elephants, hippos, rhinoceros, and giraffes (Owen-Smith 1988). In areas where these herbivores are abundant, they can provide large amounts of carrion for scavengers and decomposers. Africa is also unique in that its guild of top predators remains largely intact; this guild includes lions, leopards, cheetahs, wild dogs, hyenas, jackals, and many smaller carnivores. The majority of these carnivores are known to scavenge opportunistically, and some, such as striped and brown hyenas, acquire most of their food by scavenging. Hunting and scavenging by African carnivores have major effects on carrion availability and persistence.

The majority of the African continent lies between the Tropic of Cancer and the Tropic of Capricorn (Figure 21.1). Therefore, most of the continent is tropical, and experiences little seasonal variation in temperature. However, many areas experience distinct wet and dry seasons as the intertropical convergence zone moves back and forth across the equator. Thus, most of Africa is characterized by a warm climate with low and/or variable rainfall. However, there is still considerable variation in climate throughout the continent, and both ecosystem structure and carrion ecology vary accordingly among African biomes.

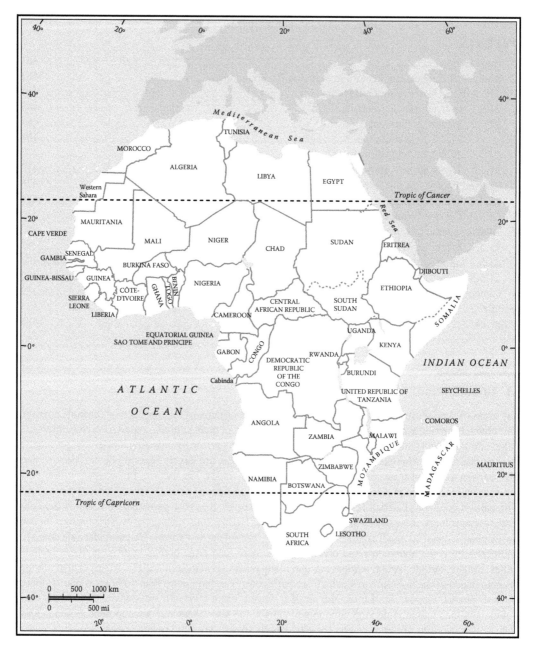

FIGURE 21.1 The majority of the African continent lies in the tropics.

Like the animals that inhabit them, African habitats are extremely diverse. Although the central and some western parts of Africa are covered by dense rain forest, vast deserts cover much of northern Africa as well as the southwestern edge of the continent. Much of Africa is covered by semiarid land, including vast grasslands and open woodlands that are collectively referred to as savanna. Savanna habitats surround the tropical forest region in the center of the continent like a giant horseshoe and cover roughly two-thirds of the land area in Africa (Adams 1996). Because most data germane to carrion ecology were collected in savanna regions of Eastern and Southern Africa, the focus of this chapter will be on the carrion ecology of African savannas. The primary characteristics of savannas are seasonal precipitation, a fairly continuous cover of grasses and shrubs that are tolerant of both intense sunlight and seasonality in rainfall, and light tree cover that does not form a closed canopy (Adams 1996). African savannas support a particularly high biomass of herbivores, and a higher diversity of ungulates (46 species) than found in any other biome on earth. Most savanna regions in Africa support resident herbivore populations, but some also support vast herds of migratory herbivores. These African savannas offer a unique carrion ecology characterized by occasional periods of high carcass availability from their ungulate guilds. In fact, anthropologists have conducted some of the most detailed work on savanna carrion ecology while investigating the role of scavenging in the evolution of our hominid ancestors, who lived in African savannas (e.g., O'Connell et al. 1988; Blumenschine 1989; Tappen 1995).

The study of carrion ecology encompasses processes and organisms affecting carrion availability and persistence in the environment, and how associated nutrients and energy are transformed within the ecosystem. Here the factors influencing the abundance, nature, and persistence of carrion in Africa will be addressed by considering the major groups of animals that consume carrion, how important carrion is for their survival, and what roles these groups of animals play in carrion degradation. These issues will first be addressed in regard to African ecosystems, in general, followed by a more detailed examination of the Serengeti-Mara savanna ecosystem of East Africa. Finally, because increasing human activity affects carrion and its consumers in many ways, there will be an evaluation of anthropogenic influences on the carrion ecology of Africa.

21.2 Sources of Carrion and Factors Influencing Its Abundance

Direct measurements of the abundance of carrion available in African ecosystems are rare. In general, carrion is a patchy resource that is available infrequently and unpredictably. Few studies give estimates of carrion availability, and none of these studies specifically focus on the amount of carrion present in the environment (O'Connell et al. 1988; Skinner et al. 1995; Butler and du Toit 2002; Wiesel 2006; Owen-Smith and Mills 2008). Furthermore, they often rely on sporadic human searches or patrols to record carrion opportunistically as it is found (Butler and du Toit 2002; Owen-Smith and Mills 2008). Thus, they likely underestimate the amount of carrion that is encountered by scavengers, which spend much time searching for food and are well adapted to searching large areas. Nevertheless, these studies demonstrate the great variability in carrion availability among study sites, showing carrion abundances ranging from ~30.27 kg/km²/year in mixed savanna woodland in northern Tanzania (estimated from O'Connell et al. 1988) to ~386 kg/km²/year in the Sengwa Wildlife Research Area in Zimbabwe (estimated from Butler and du Toit 2002). The highest concentration of carrion recorded is ~8105 kg in just 0.1 km²/year in coastal Namibia, where nearly 800 carcasses of juvenile Cape fur seal can be found seasonally on a 1.5 km stretch of beach (estimated from Skinner et al. 1995; Wiesel 2006), providing scavengers such as jackals and brown hyenas with immense quantities of food in a very small area. By contrast, Hadza hunter-gatherers in Tanzania experience a relatively low availability of carrion. Although 20% of the carcasses acquired by the Hadza are scavenged, the carcasses are rarely intact and are available unpredictably (O'Connell et al. 1988). Similarly, Cooper (1990) found only 18 herbivores carcasses in 100 km² of Chobe National Park, Botswana, over the course of a 2-year study. The great variation in carrion availability among studies is probably influenced by several key factors addressed below, including variation in rates of carrion degradation; these have been reported to vary considerably, even between areas relatively close to one another (Wells 1989).

In Africa, ungulate species make up the majority of wild animal biomass. Consequently, most carrion occurs in the form of ungulate carcasses (Braack 1986; O'Connell et al. 1988; Butler and du Toit 2002; Owen-Smith and Mills 2008), but ungulate abundance varies widely across the continent. Ungulate population density is largely determined by annual rainfall in African ecosystems (Coe et al. 1976), so carrion is rare in Africa's desert regions. At the other end of the spectrum are lush grasslands such as the Serengeti, which support enormous populations of herbivores, including over a million migratory ungulates. However, even in ungulate-rich areas of Africa, large numbers of carcasses are rarely available throughout the year.

The source of mortality among herbivores has a profound effect on how much of their body mass enters the ecosystem as carrion. Herbivores that are limited by predation (top-down regulation) are likely to be killed and rapidly consumed by predators, which leave only small amounts of carrion for scavengers and decomposers. By contrast, herbivores that are limited by nutrient or water availability (bottom-up regulation) leave their entire body mass as carrion when they perish (Houston 1974b; Blumenschine 1989). Predation pressure is intense in areas characterized by high ratios of resident predators to resident prey, such as many national parks in Eastern and Southern Africa. As a result, very few ungulates die from causes unrelated to predation, and therefore these areas have low levels of carrion abundance. For example, carcasses uneaten by carnivores are rarely found in Kruger National Park, except during severe drought years or outbreaks of anthrax (Owen-Smith and Mason 2005).

Ungulate body size influences both cause of death and the rate of decomposition. Sinclair et al. (2003) found that there is a negative sigmoid relationship between predation rates and body size among non-migratory adult ungulates, with predation rates dropping off dramatically for ungulates over 150 kg. This pattern suggests that most small ungulates are quickly consumed by predators, whereas a greater proportion of large ungulate biomass remains available to scavengers. However, the role of small carcass abundance in carrion ecology in Africa remains poorly understood. Because of their small size, these carcasses are hard to find and difficult to observe, so they have largely gone unstudied (Wells 1989). Some invertebrate species, including certain species of flesh flies (family Sarcophagidae), may specialize on small carcasses, which suggests that there may be significant numbers of small animal carcasses available in some African habitats (Braack 1987). These invertebrate scavengers likely require specific adaptations to locate and utilize small carrion (Braack 1987). Small carcasses tend to decompose faster than large ones, suggesting that invertebrate scavengers and decomposers may be the primary consumers of small isolated patches of carrion. This chapter focuses instead on the carcasses of medium- and large-bodied herbivores, specifically because they represent the majority of carrion biomass available to consumers in Africa.

Carcasses of mega-herbivores may persist for long periods of time because of their immense body size, often putrefying before being completely consumed by scavengers (Pereira et al. 2014). Elephant and giraffe carcasses persist over 13 times longer than do wildebeest or zebra carcasses, which are generally consumed in less than a day in the Serengeti (Blumenschine 1989). Because mega-herbivore carcasses persist longer in the environment than do smaller carcasses and provide more biomass for scavengers, scavenging opportunities may generally be greater in areas of Southern Africa, where mega-herbivores make up a relatively larger proportion of the herbivore population than in Eastern Africa (Houston 1979; East 1984; Cooper et al. 1999), than in Eastern Africa. On the other hand, although mega-herbivores make up between 40 and 70% of the large ungulate biomass in many African national parks (Owen-Smith 1988), mega-herbivores are long lived and experience low mortality rates, and these factors may reduce their contribution to overall carrion levels. In any case, the contribution made by mega-herbivores to carrion abundance in African savannas is usually patchy in both space and time.

Catastrophes such as drought and disease epizootics can produce huge spikes in carrion abundance in African ecosystems. For instance, millions of kilograms of carrion may occur per square kilometer during droughts that cause mass starvation among elephants (e.g., Coe 1978). Similarly, several thousand migratory wildebeest may drown over the course of a few days while trying to cross a large river, resulting in a pileup of many thousands of kilograms of carrion (e.g., de Pastino 2007). Because of their ephemeral nature and spatial and temporal variability, these massive die-offs can be conceptualized as resource pulses. Previous work has demonstrated that, even though resource pulses may be brief and occur infrequently, they can nevertheless have long-lasting impacts on ecosystem structure (Yang 2008). Carrion from such catastrophes may persist in the environment for a relatively long time due to scavenger satiation, allowing for increased deposition of key nutrients such as nitrogen. Resource pulses can also have long-lasting effects on local consumer populations, especially opportunistic scavengers who may benefit from increased access to largely intact carcasses (Capaldo and Peters 1995). For instance, early hominids may have taken advantage of mass drownings of migratory ungulates in the Serengeti, which result in many whole carcasses spread out over a large area in a short period of time. Such a distribution pattern would result in reduced inter- and intraspecific competition among scavengers and thus offer huge nutrient reserves, which might be gained with little energy expenditure (Capaldo and Peters 1995). (The ecology of mass drownings will be discussed further in Section 21.4.2, which addresses carrion availability in the Serengeti ecosystem.) In conclusion, although natural disasters, such as epizootics and droughts, and the mass mortality events they cause are generally rare, they may still have significant effects on the environment and carrion consumers in the area. Overall, their significance for African carrion ecology is poorly understood, so more research is needed in this area.

Several environmental variables affect the abundance and persistence of carrion in African habitats. Climatic variables such as temperature and rainfall directly affect carrion decomposition, and they also affect carrion availability indirectly by influencing ecosystem structure. In many studies, ambient temperature is the most significant factor in determining how quickly carrion decomposes due to its effects on the metabolic activity of invertebrate scavengers and decomposers (DeVault et al. 2003). Arthropods and microbes consume carrion at faster rates in warm, tropical areas near the equator than in cooler, temperate regions at higher latitudes (DeVault et al. 2003). Rainfall also has important effects on carrion abundance and decomposition, as moisture can enhance invertebrate abundance and microbial activity, leading to faster decomposition rates. Carrion dries out more quickly in arid regions, making consumption more difficult for both decomposers and scavengers, so rainfall affects the length of time during which carrion is usable by scavengers and decomposers. Lower rainfall also reduces plant cover, creating exposed areas where carcasses degrade more quickly and are more easily found by scavengers (DeVault et al. 2003). Thus, rainfall, temperature, and vegetation cover may interact to affect rates at which carrion is consumed. Because 60% of Africa has an arid or semiarid climate, primary productivity is heavily limited by rainfall, and consequently even slight variation in rainfall across years, seasons, or even weeks, can affect herbivore abundance, distribution, and mortality (Duncan et al. 2012). In general, more herbivores die during dry than wet seasons (Sinclair 1979a; Pereira et al. 2014). In addition to dying from water deprivation per se, herbivores aggregate in small areas where surface water remains available during periods of low rainfall, making them more easily found and killed by predators. Because much of Africa experiences distinct seasonality in rainfall, carrion availability and persistence tends to vary seasonally (Tappen 1995).

Finally, in addition to climatic variables, local predator/prey ratios, herbivore body size, and mortality sources among herbivores, the abundance and persistence of carrion in African ecosystems is importantly affected by the size and composition of scavenging guilds. In addition to microbes, there are three important guilds of carrion consumers: invertebrates, mammals, and birds. There is considerable competition for carrion both within and among these guilds, and the intensity of this competition also affects the abundance and persistence of carrion.

21.3 The African Scavenging Guilds

21.3.1 Invertebrate Scavengers

Other than microbes, arthropods are typically the most species-rich and abundant organisms found on carrion (Barton et al. 2013). Almost 300 species of carrion-associated arthropods have been recorded in Africa. These 300 species represent 40 taxonomic families, with the vast majority belonging to class Insecta (Braack 1987; Villet 2011). This diversity is roughly equivalent to that seen in other parts of the world such as the Holarctic and Neotropics (Villet 2011). Invertebrate carrion communities in Africa are distinctive at the species level, but at the family level they are strikingly similar to other such communities worldwide (Villet 2011, but see Barton et al. 2013). Such similarity in family-level diversity is likely due to the ability of arthropods to travel with humans as well as disperse far on their own (Villet 2011).

Beetles (Order Coleoptera) and flies (Order Diptera) are the predominant arthropods associated with carcasses in Africa. For example, Braack (1987) found that a 65-kg impala carcass in South Africa can support over 115,000 fly larvae and 20,600 histerid beetles. The situation is likely similar throughout Africa, as there is minimal geographic variation in taxonomic composition at the family level within these two groups (Hanski 1987; Ellison 1990; Villet 2011). Other groups of arthropods that are associated with African carrion include ants, wasps, mites, and moths. Some of these arthropods appear on carrion only to prey on or parasitize species consuming the carrion itself. Both Villet (2011) and Braack (1987) provide detailed reviews of the important carrion-related arthropod communities in Africa.

At most carcasses, flies (Order Diptera) are numerically dominant, the most abundant families being Calliphoridae (blow flies) and Sarcophagidae (flesh flies). The larvae of many of these species feed on carrion, though adults often utilize other food sources. The distribution of blow fly species varies across the African continent, due largely to variation in physiological tolerance to climatic variables like humidity and temperature (Richards et al. 2009). Species-specific sensitivity to temperature in blow flies is such that they compete indirectly with one another through larvae-generated heat at carcasses. Species with a higher tolerance for heat are often able to utilize more of a carcass because their large numbers generate heat, thus driving less heat-tolerant species to the exterior of the carcass (Richards et al. 2009). Like blow flies, flesh flies utilize carcasses for breeding, but tend to prefer smaller carrion (Braack 1987; Villet 2011). In addition, there are at least 14 other fly families that are consistently associated with carcasses in Africa, 12 of which are also found on carrion in South America and the Holarctic (Villet 2011).

Beetles are less numerous on carcasses but more species-rich than flies. Roughly 24 families of beetles have been found on African carrion, but only 10 families and 90 species occur predictably. This diversity is less impressive than that in South America, where 221 beetles belonging to up to 15 families are found on carrion (Villet 2011). The beetle families most commonly and abundantly represented on African carrion are Histeridae (24 species), Trogidae (6), Scarabaeidae (46), Dermestidae (1), and Cleridae (1) (Braack 1987). Some beetle species are necrophages but others, including many in the Histeridae family, are predators that are drawn to carcasses by the presence of other arthropods, especially blow fly species such as *Chrysomya regalis* (Braack 1987).

Dung beetles (Family Scarabaeidae) are one of the beetle families commonly associated with carrion in Africa. Dung beetles are numerous and speciose in Africa, with as many as 3000 species, mostly concentrated in the tropics (Hanski 1987). However, only 46 of these species are consistently associated with dead vertebrates (Midgley et al. 2012). Of these, it seems the majority of species are attracted to the rumen contents of artiodactyl carcasses rather than to the soft tissues of the ungulates themselves (Tshikae et al. 2008). Some authors posit that so few dung beetles in Africa are carrion specialists because of intense competition with large scavenging carnivores and vultures (Putnam 1983; Martin-Piera and Lobo 1996; Tshikae et al. 2008). On the other hand, Hanski (1987) argues that this is due to an abundance of dung rather than a shortage of carrion. The fact that there is a dung beetle carrion specialist (*Metacatharsius opacus*) in the Kalahari desert, an area with abundant scavengers and low densities of carrion, suggests that competition with vertebrate scavengers does not preclude the evolution of dung beetles that specialize on carrion (Putnam 1983; Tshikae et al. 2008).

Many of the invertebrate families that attend carcasses preferentially arrive at a specific phase of the carcass degradation sequence. The timing of arrival and length of stay at a carcass often depend on which resource the arthropod species consumes, be it skin, keratin, fat deposits, fluids, soft tissues, or other carrion-attendant arthropods. In particular, necrophagous families such as Calliphoridae, Sarcophagidae, Silphidae, Dermestidae, and Trogidae tend to have predictable timing of appearance at carrion sites (Villet 2011). Calliphorid species are usually the first to arrive at a carcass, with sarcophagids arriving a day or two later. The larvae of these flies have an important accelerating effect on carcass decomposition, as they secrete digestive enzymes, which dissolve muscle and other soft tissue (Braack 1987). The length of larval activity is the most intense period of insect activity at a carcass (Braack 1987; Richards et al. 2009), and the fly larvae facilitate the arrival of other carrion-associated arthropods. For instance, they expose portions of the carcass such as rumen contents, which attract dung beetles. Parasitoid and predatory species, including many beetles, also visit carcasses to prey upon the larvae. In fact, blow flies are also subject to predation from members of their own family. For example, *Chrysomya albiceps* larvae in their second or third larval stages often predate heavily upon *C. regalis* larvae (Braack 1987).

By the time most fly larvae have departed, most soft tissues have been consumed and the carrion dries out. Arthropod activity then decreases, as only species able to take advantage of dry carcasses remain, including dermatophages (i.e., dermestid and trogid beetles). Adult dermestid beetles consume mainly (nearly dry) skin and ligamentous tissues, and trogid beetles consume hair and some of the keratin in horns, hooves, and skin (Braack 1987). Finally, only the keratophages such as tineid moths and some trogid beetles remain (Braack 1987; Villet 2011). Larvae of two tineid moth species (*Trichophaga swinhoeie* and *Ceratophaga vastella*) feed on keratin and are often the last animals to make use of a carcass. *C. vastella*, which bores holes in ungulate horns to oviposit in them, arrives at the carcass several weeks to months after the ungulate has died (Coe 1978; Braack 1987). It should be noted that vertebrates may arrive at any point to feed on fly larvae or the carcass itself; this disrupts the sequence of events, which then becomes less predictable.

As in the rest of the world, arthropods can play a large part in the decomposition of carrion in Africa. Carcasses of small–medium-bodied antelope in Kruger National Park, South Africa, can be quickly colonized by arthropods, which are able to remove all soft tissue from the remains within days (Braack 1987). Similarly, fly larvae can consume 2–10 kg mammal carcasses within 3 days in riverine forests in Tanzania (Houston 1985), and invertebrates and microbes can remove all soft tissue from an elephant carcass in 21 days (Coe 1978). Arthropods also facilitate microbial colonization of a carcass by transporting microbes between carrion sites and by aerating intact carcasses (Putnam 1983). Byproducts of microbial decomposition can make undiscovered carrion unsuitable for consumption by late-arriving vertebrate scavengers (Janzen 1977), thus allowing invertebrates and microbes to monopolize older carcasses that have not mummified.

Arthropods degrade carrion more quickly in warm, wet environments than in cool, dry habitats. The length of fly larvae activity, which is responsible for removing the majority of soft tissue biomass from carrion attended only by invertebrates, is dependent on temperature and humidity (Braack 1987; Richards et al. 2009). The speed of carrion decomposition in warm weather is largely due to the rapid rate of development of blow fly larvae (Braack 1987). In South Africa, blow flies can remove all soft tissue from a medium–large-sized mammalian carcass in 4–5 days in warm weather, but take 14 days to do the same in cool weather (Braack 1987). Rainfall is also an important determinant of carrion decomposition in many areas of Africa due to its strong seasonality and its effects on invertebrates. For instance, White and Diedrich (2012) found elephant carrion lasted much longer in dry than wet seasons, a difference that may be attributed to a lack of insect activity during dry periods. In particular, skin may remain on a carcass for many months during periods of low rainfall due to mummification and decreased activity by dermestid beetles (Coe 1978).

The general process of carrion decomposition by invertebrates in Africa is similar to that seen in many parts of the world, including North America and Europe (Villet 2011). However, the local activity of vertebrate scavengers can have a large effect on the importance of invertebrates in carcass degradation; in many African ecosystems, avian and mammalian scavengers quickly find and devour carrion, which does not allow time for the slower-working insects to consume much of the carrion. For instance, vertebrate scavengers in Kruger National Park, South Africa, only fail to discover 10%–20% of carcasses of medium- and large-bodied antelope, and they discover fewer carcasses in wooded areas than on open grassland (Richardson 1980; Braack 1984). Theoretically, vertebrate scavengers should monopolize an even higher proportion of carcasses in more open ecosystems, such as savannas. Although invertebrates clearly play a significant role in carrion decomposition in Africa as they do worldwide, vertebrate scavengers play uniquely important roles in African carrion ecology.

21.3.2 Vertebrate Scavengers

There is a distinct set of vertebrate scavengers on each major continent, and the African vertebrate scavenger guild is one of the most diverse (Barton et al. 2013). In most African ecosystems studied to date, vertebrate scavengers are responsible for consumption of the vast majority of carrion (Putnam 1983). As relatively large scavengers with high energetic demands, vertebrates play a key role in carrion persistence in the environment. They can open carcasses, thus providing access to smaller vertebrates, invertebrates, and aerobic microbes (Putnam 1983). They can disrupt the decomposition process by disturbing and eating blow fly larvae and other invertebrate scavengers. They remove large portions of a carcass and scatter them across the landscape. Finally, they are capable of devouring carrion very quickly, thus drastically decreasing its persistence in the environment. For example, a group of spotted hyenas can reduce a 200 kg antelope from a living creature to nothing more than a pile of rumen contents and a bloody patch on the ground in as little as 13 min (Holekamp, unpublished).

Few vertebrate species rely heavily on carrion as a food source, but scavenging can yield myriad benefits to vertebrate predators. These include avoiding time and energy lost finding vulnerable prey, and avoiding injury during hunting. Large African predators such as spotted hyenas are known to go up to several weeks without food due to the difficulty of finding vulnerable prey (Kruuk 1972; Mills 1990; Hofer and East 1993a,b). Hunting demands a great deal of time and energy. For instance, black-shouldered kites in South Africa expend 61% of their energy budget engaged in hunting behavior (Tarboten 1978). Furthermore, the costly behavior of hunting does not always pay off. Hunting success varies, with spotted hyenas experiencing success rates of 30%–35% (Holekamp et al. 1997), lions rates of 27%–30% (Funston et al. 2001; Van der Meer et al. 2011), and wild dogs rates of 44%–91% (Van der Meer et al. 2011). Success rates are similar among avian predators, with 15%–50% for Lanner Falcons (South Africa; Kemp 1993), 21.7% for Taita Falcons (Zimbabwe; Hartley et al. 1993), 8.8%–18.9% for black-shouldered kites (Tarboten 1978), and 15.5%–21.2% for Peregrine falcons in South Africa (Jenkins 2000). Even if a predator successfully kills a prey animal, it risks loss of its kill to kleptoparastism, or the aggressive usurpation of a fresh carcass by another predator (Höner et al. 2002). Gregarious animals risk members of their own social group taking most of their kill, and many predators, both solitary and gregarious, often lose kills to species of larger body size. For instance, wild dogs lose about 50% of their kills to larger predators (Kruuk 1972; Van der Meer et al. 2011). It is, therefore, not surprising that many vertebrates are known to take advantage of scavenging opportunities when they arise. Nevertheless, there is much variation in scavenging behavior among both mammalian and avian species.

21.3.2.1 Mammalian Scavengers

Table 21.1 estimates the proportion of the diet derived from scavenging by each of several African mammals. Many African mammals are known to scavenge occasionally, including porcupines, warthogs, honey badgers, hippos, baboons, wildcats, caracals, servals, mongooses, bat-eared foxes, civets, and genets (Blumenschine 1989; Ellison 1990; Dudley 1998; Ray and Sunquist 2001; Klare et al. 2010; Braczkowski et al. 2012). Although information about their dietary preferences is scarce, these African mammals all appear to scavenge a very minor portion of their diet, and only consume small portions of carcasses. For instance, Butler and du Toit (2002) found that small carnivorous mammals, including honey badgers, white-tailed mongooses, and genets, consume less than 1% of carcass biomass in rural Zimbabwe. Large carnivores consume far more carrion in African savannas than do any other mammals. Competition among mammalian scavengers is often fierce, and larger species can usually drive smaller species off carcasses unless they are greatly out-numbered. Larger species such as lions (110–180 kg) and spotted hyenas (45–80 kg) are thus able to usurp carcasses from smaller species such as cheetahs (35–55 kg), wild dogs (17–20 kg), and jackals (6–15 kg) (Schaller and Lowther 1969; Kingdon 1977). Leopards (35–55 kg) can generally avoid kleptoparasitism from lions and spotted hyenas because they store the carcasses of their prey in trees.

Virtually all mammalian carnivores are known to take advantage of scavenging opportunities when they arise (Kruuk 1972; Houston 1979), so a surprising diversity of mammalian carnivores may be found at a single carcass. The mammalian scavengers most commonly seen at carcasses include jackals, spotted hyenas, and lions (Hunter et al. 2007). Brown hyenas and striped hyenas are known to take a large proportion of their diet from carrion (Table 21.1); however, brown hyenas have a restricted range and striped hyenas occur at low densities, so they are seen less often at carrion sites than are more gregarious species or those with larger ranges. Some other African predators, such as leopards and cheetahs, rarely scavenge (Hunter et al. 2007; but see Butler and du Toit 2002), and even the mammals recognized as prominent scavengers sometimes derive only a small proportion of their diet from scavenging (e.g., Table 21.1). The proportion of diet scavenged often varies in both space and time, suggesting that carnivores can switch facultatively between scavenging and hunting.

TABLE 21.1

Percent of Diet Obtained from Scavenging by Several African Mammal and Bird Species

Species	% Diet Obtained from Scavenging	Study Area	Source
Lions	5.6–53	Etosha National Park (NP), Namibia; Serengeti NP, Tanzania (plains)	Stander (1991), Schaller (1972)
Wild dogs	2.9–5	Serengeti NP, Tanzania	Estimated from Creel and Creel (2002), Schaller (1969, 1972)
Cheetahs	<1	Serengeti NP, Tanzania	Caro (1994), Kruuk (1972), Bertram (1979)
Leopards	5–10	Serengeti NP, Tanzania	Bertram (1979)
Brown hyenas	94	Kalahari Desert, Namib Desert	Mills (1990), Owens and Owens (1978), Goss (1986)
Striped hyenas	51.6–61	Laikipia, Kenya	Kruuk (1976), Wagner (2006)
Spotted hyenas	5–99	Studies across range	See Table 21.2
Black-backed jackals	3[a]–18%	Serengeti NP, Tanzania	Wyman (1967)
Side-striped jackals	<20[a]	Highveld commercial farmland, Zimbabwe	Atkinson et al. (2002)
Golden jackals	3[a]–19	Ngorongoro Crater, Tanzania; Serengeti NP, Tanzania	Wyman (1967), Moehlman (1978)
Honey badgers	0.5	Kgalagadi Transfrontier Park, South Africa and Botswana	Begg et al. (2003)
Humans (Hazda hunter-gatherers)	14	Tanzania	O'Connell et al. (1988)
White-backed vultures	100	Serengeti-Mara	Houston (1974b)
Ruppell's griffon vultures	100	Serengeti-Mara	Houston (1974b)

Source: From DeVault, T.L., O.E. Rhodes Jr, and J.A. Shivik. 2003. *Oikos* 102(2): 225–234.

[a] Estimates from fecal analysis, which is an inaccurate way of distinguishing scavenged carrion from hunted prey; this method generally underestimates the proportion of an organism's diet obtained from scavenging.

One such facultative scavenger is the jackal, a fox-sized generalist that often opportunistically exploits whichever food source is most prevalent in its environment (Kaunda and Skinner 2003). Three jackal species are found in Africa: black-backed, side-striped, and golden jackals. Their ranges vary, but multiple jackal species occur sympatrically in areas of Eastern Africa (Sillero-Zubiri et al. 2004). When the three species are present together, black-backed jackals are most commonly seen at kills made by larger African carnivores (Wyman 1967). Black-backed jackals are more likely to try feeding from kills made by lions and hyenas than are the other species (Estes 1991; Sillero-Zubiri et al. 2004). The prominence of black-backed jackals at kills may in part be due to an aggressive disposition, as they are known to aggressively displace side-striped jackals at carcasses despite their smaller body size.

Although most studies suggest that only a minor portion of their diet comes from carrion, all three jackal species are known to take advantage of scavenging opportunities. For instance, black-backed jackals scavenge the remains of beached marine mammals in Namibia, where they are also known to follow brown hyenas to find carcasses (Wiesel 2006). Golden jackals subsist largely on scavenged carrion and garbage in some areas (Poché et al. 1987). Despite this, it is clear that jackals are capable of hunting the vast majority of their food because they thrive in areas where scavenging opportunities are rare (Klare et al. 2010). Even in areas where scavenging opportunities are available, jackals may not be a major presence at carcasses. For example, jackals are present at only 14% of cheetah kills in the Serengeti (Hunter et al. 2007), and although side-striped jackals are present at many carcasses in the Sengwa Wildlife Research Area of Zimbabwe, they rarely feed on them (Butler and du Toit 2002). This is likely due to their small body size compared to that of the other scavengers present. Overall, jackals in Africa clearly do not rely on carrion as a primary dietary staple.

African lions are important scavengers in savanna ecosystems (Houston 1979) and are known to scavenge throughout their range (see Table 21.1; Trinkel and Kastberger 2005). In the Sengwa Wildlife Research Area, lions consume 31% of carrion biomass, a bigger proportion than that consumed by any other wild mammal (Butler and du Toit 2002). Of all the carcasses on which Serengeti lions were found feeding by Grant et al. (2005), 24.5% were scavenged; however, the same study estimated that only a small proportion of the biomass of the lions' diet was scavenged. When scavenging, lions have the benefit of being substantially larger than, and thus physically dominant to, all other carnivores, allowing them to obtain a large portion of their scavenged food through kleptoparasitism. They are known to drive spotted hyenas, cheetahs, and wild dogs away from kills (Kruuk 1972; Cooper et al. 1999; Creel and Creel 2002). In fact, when both lions and spotted hyenas are found feeding from the same carcass, more often than not (53%–93% of the time) the lions are scavenging from a spotted hyena kill (Kruuk 1972; Schaller 1972). Despite these advantages with respect to scavenging, lions in most populations obtain the vast majority of their food from prey they hunt themselves (Schaller 1972; Stander 1991), and they are often less abundant than other carnivores in African ecosystems. Therefore, lions may not even be present at the majority of carcasses in an area. For example, lions are present at only 3.3% of cheetah kills in the Serengeti, far less often than vultures, jackals, or spotted hyenas (Hunter et al. 2007). Even so, lions can be formidable competitors for sympatric scavengers due to their large size and sociality.

Of all the mammalian carnivores in Africa, brown, striped, and spotted hyenas (hereafter referred to simply as hyenas) derive the largest portion of their diet from scavenging. This is not surprising, as they show many unique adaptations specifically for scavenging. They have a unique body posture and a "rocking-horse" gait (Eloff 1964; Tilson and Henschel 1986; Hofer and East 1993a; Frank 1996) that allows them to cover long distances in search of carrion and prey. They also have strong forequarters and necks, which allow them to drag or carry large portions of carcasses away from kill-sites when competing with conspecifics or other scavengers. Spotted hyenas are able to detect sounds of predators killing prey or feeding on carcasses over distances of up to 10 km (Mills 1990). Spotted hyenas are also particularly resistant to disease, allowing them to take advantage of periodic epidemics that cause massive die-offs of prey species. For example, they have frequently been found feeding on carcasses of ungulates that have died of anthrax, yet they show no ill effects of the disease (Pienaar 1969; Lindeque 1981; Gasaway et al. 1991).

All three hyena species exhibit a suite of craniodental features that enable them exert tremendous bite forces useful in durophagy, or the consumption of hard foods like bones. Their powerful jaws are able to crack open bones as large as elephant leg bones (Haynes 1988), and these unique adaptations allow them to make extremely efficient use of carrion. Furthermore, their tooth enamel is unusually resistant to cracking and chipping (Stefen 1997; Rensberger 1999; Rensberger and Wang 2005; Rensberger and Stefen 2006). Their bone-cracking ability allows hyenas to access fat-rich portions of carcasses that are inaccessible to other predators, namely the brain and bone marrow. Bone marrow is an excellent source of fat (yellow marrow), water (red marrow), protein, and vitamins and minerals, including calcium, phosphorus, iron, zinc, vitamin A, niacin, and thiamin (Van Valkenburgh 1996; Receveur et al. 1998). Yellow marrow is particularly useful, as fat-rich foods are otherwise rare for carnivores; fat usually comprises less than 5% of the body mass of wild ungulates (Ledger 1968). The water content in red bone marrow can be particularly important in arid and semiarid areas, where spotted and brown hyenas may obtain most of their water from their prey (Green et al. 2004). Overall, access to bone marrow can be surprisingly profitable; just 100 g of ungulate bone marrow can contain over 700 kcal from fat and 23 g of protein (in comparison, 100 g of sirloin steak contains about 201 kcal of fat and 35 g of protein) (Receveur et al. 1998; United States Department of Agriculture (USDA), 2013).

Adaptations for durophagy also allow hyenas access to nutrients in carcasses that are too desiccated to benefit any other carrion consumers, including microbes. For instance, spotted hyenas are known to eat old, dried out carcasses and are often the last vertebrate scavenger to feed on carrion (Holekamp et al. 1997; Cooper et al. 1999; White and Diedrich 2012). However, though desiccated carcasses contain significant amounts of protein, carbohydrates, and minerals (Wambuguh 2007), their nutrient quality and water content are considerably lower than those of fresh carcasses. Desiccated carcasses may also be harder to find than fresh carrion, as other animals are not generating sounds as they compete over them, and they do not have a strong odor (Cooper et al. 1999). Dried carcasses do not contribute significantly to the diet of spotted hyenas (Wambuguh 2007), but they may nevertheless be important during times of prey scarcity, or for low-ranking, immature, or injured animals that cannot compete effectively for fresh kills (Cooper et al. 1999).

Of all three duraphagous hyena species, spotted hyenas are the most widespread and best studied. They are also unique among the hyenas in that they are both efficient scavengers and effective hunters. Although widely believed to subsist mainly on carrion, as do striped and brown hyenas, the largest portion of the diet of spotted hyenas in almost all studied populations is acquired from predation (Table 21.2); on average, two-thirds of the hyenas' diet consists of prey they have killed themselves and one-third is obtained by scavenging. Nevertheless, spotted hyenas show great flexibility in their diet, and some populations engage in far more scavenging than others; hyenas inhabiting Kenya's Masai Mara National Reserve obtain only 5% of their diet from scavenging (Cooper et al. 1999), while Serengeti hyenas gain over 50% of their diet from scavenging (Kruuk 1972), and the semiurban spotted hyenas in Harar, Ethiopia, scavenge virtually all of their food (Baynes-Rock 2013). Like lions, spotted hyenas often scavenge via kleptoparasitism. For example, spotted hyenas are present at 15.5% of cheetah kills and are responsible for the largest loss of kills by cheetah (Hunter et al. 2007). Spotted hyenas can even drive lions away from kills if they significantly outnumber them and if no adult male lions are present (Cooper 1991). Overall, spotted hyenas are capable predators as well as scavengers. Due to their dietary flexibility, their widespread distribution on the African continent and their sheer numbers, spotted hyenas offer a useful model system in which to study the ecology of carrion usage.

Clearly, scavenging is not the primary foraging strategy utilized by most carnivorous mammals in Africa. Even generalist species with substantial dietary flexibility, such as jackals and spotted hyenas, usually acquire the majority of their food by hunting. This is likely due to the scarcity and poor nutrient quality of carrion compared to live prey. For example, spotted hyenas in Etosha National Park, Namibia, encounter only 0.6 items of carrion per hour foraging, compared to 2.8 groups of live prey (Gasaway et al. 1991). Reliance on carrion requires searching large areas that are often characterized by limited visibility. Even if a mammal can find carrion, it is rarely as nutritionally rich as live prey. Of 35 carrion items found by the hyenas in Etosha National Park, 34 had little or no flesh, and 25 of the 35 items were desiccated and rarely fed upon by hyenas (Gasaway et al. 1991). Carrion is also likely to contain less water, which is particularly disadvantageous to carnivores inhabiting the arid and semiarid regions of Africa. Overall, an animal that primarily scavenges has to deal with lower quality food as well as more travel between food patches (Wambuguh 2008).

Carnivore foraging patterns may shift as a result of varying hunting costs. Hunting costs can change seasonally, decreasing when prey animals are weak from malnourishment or drought, or when there are

TABLE 21.2

Percent of Spotted Hyena Diet Obtained from Scavenging across Different Populations

Study Site	Country	% Spotted Hyena Diet Obtained from Scavenging	Source
Chobe NP	Botswana	25	Cooper (1990)
Serengeti NP	Tanzania	57	Kruuk (1972)
Aberdares NP	Kenya	35	Sillero-Zubiri and Gotelli (1992)
Etosha NP	Namibia	25	Gasaway et al. (1991)
Kruger NP	South Africa	<50	Henschel and Skinner (1990)
Namib Desert	Namibia	50	Tilson et al. (1980)
Kalahari Desert	South Africa	39	Mills (1990)
Masai Mara National Reserve	Kenya	5	Cooper et al. (1999)
Ngorongoro Crater 1960s	Tanzania	18	Kruuk (1972)
Ngorongoro Crater 1990s	Tanzania	31	Höner et al. (2002)
Harar	Ethiopia	>99[a]	Baynes-Rock, Pers. Comm.

[a] It is unclear how much of this 99% consists of carrion versus other human refuse/trash.

infants present in the prey population (Pereira et al. 2014). At times of year when vulnerable prey are scarce, some carnivores may depend more heavily on carcasses left behind or stolen from other predators (Henschel and Skinner 1990). This may be less true for ambush predators, such as lions and leopards, which are more likely to take down healthy prey, than for pursuit hunters (Periera et al. 2014). Note, however, that scavenging is only one of several alternatives to hunting strong prey. Animals such as jackals and brown and striped hyenas may increase the proportion of fruit in their diet when vulnerable prey are rare, and other carnivores may focus upon smaller prey when their preferred prey species are in good condition or otherwise difficult to catch (Mills 1990; Periera et al. 2014).

The amount of scavenging in which a particular carnivore population engages often depends on the density of other predators and scavengers in that area. Spotted hyenas and lions have largely overlapping dietary preferences and commonly scavenge from one another (Kruuk 1972). Spotted hyenas are more likely to scavenge from lions in areas with lower lion densities. The reasons for this are twofold. First, at lower densities, lions are more likely to leave larger portions of a carcass behind. Secondly, spotted hyenas need to outnumber lions in order to usurp their food at kills, and when male lions are present, they may be unable to accomplish this in any case (Cooper 1991). Therefore, spotted hyenas are better able to scavenge when lions are present in lower densities and when there are fewer males in the lion population (Cooper 1991; Trinkel and Kastberger 2005; Watts and Holekamp 2008). Finally, larger predators such as spotted hyenas and lions may scavenge more in areas inhabited by high densities of smaller predators such as cheetahs and wild dogs, from which they can easily steal carcasses.

In conclusion, a variety of carnivores shift facultatively between hunting and scavenging. More research is needed to better estimate what proportion of carnivore diets are derived from carrion by different species and populations. More research is also needed to determine exactly what factors contribute to shifts in foraging strategy seen in generalists such as spotted hyenas and jackals.

21.3.2.2 Avian Scavengers

Like carnivorous mammals, most carnivorous birds in Africa rely on hunting as their primary food source, but also scavenge opportunistically. Omnivorous birds, such as many corvids, survive on a highly varied diet acquired by hunting and scavenging, but also by feeding on vegetable matter. Although the number of avian species that consume carrion is large, there are far fewer species that depend on scavenged remains as a crucial part of their diet. Avian scavengers must cope with the same challenges as mammalian scavengers; however, their ability to fly enables them to search over large areas relatively cheaply, which makes scavenging a more feasible strategy. Mammals such as spotted hyenas and lions are more likely to find antelope carcasses in open areas covered with short grass than in bushy or forested habitat (Grant et al. 2005; Hunter et al. 2007), which suggests that locating carrion by sight is very helpful. The vistas surveyed from the air by avian scavengers can be enormous, which suggests that avian species should have a significant advantage over mammals when searching for carrion. Thus, it should not be surprising that more avian than mammalian species rely exclusively on carrion in African ecosystems.

Of the avian scavengers that rely on carrion, vultures are the most dominant at carcasses. They also exhibit the most specialized adaptations for scavenging and they consume the most carrion (Houston 1974b, 1979). Although all extant vultures exhibit a common suite of morphological and behavioral characteristics, vultures have evolved three separate times (Seibold and Helbig 1995; Wink 1995). The convergent evolution of a suite of traits that includes carrion dependence, soaring flight, a featherless head, a highly acidic stomach, and large body size suggests that these traits are specialized adaptations for feeding on carrion. Soaring flight and large body size allow vultures to spend long periods of time traveling in search of carrion over enormous distances, whereas featherless heads and highly concentrated stomach acid appear to be adaptations for maintaining health and vigor while also feeding on carrion that may be infected with potential disease agents (Houston 1974a,b; Houston and Cooper 1975).

Of 23 extant species of vultures, 16 comprise the polyphyletic group of Old World vultures. Nine out of 16 Old World vultures are found in Africa, and many of them share overlapping ranges. Accordingly, it is not uncommon to find multiple species of vultures feeding on a single carcass. Early work by Kruuk (1967) and Pennycuick (1972) on sympatric vultures in the Serengeti-Mara ecosystem suggests that most

African vultures can be divided into three categories based on beak morphology and foraging behavior. The lappet-faced and the white-headed vultures have strong, hooked beaks capable of tearing off tougher parts of a carcass; these birds thus play an important role in allowing other avian scavengers to access the softer insides of the carcass. These two vulture species occur in small numbers at carcasses, and pairs are often territorial. Unlike most other African vultures, these large vultures are sedentary, perhaps due to differences in wing loading (or the ratio between body weight:wing area) (Pennycuick 1972).

The Egyptian and hooded vultures are the smallest of the African vultures. They have slender beaks that are better for pecking up small scraps of tissue than for tearing off sizeable pieces of meat. Although they roost solitarily, they can gather in large numbers at carcasses or garbage dumps. They boast a more versatile diet than many other Old World vultures, feeding on dung, carcass scraps, and sometimes insects (Zimmerman et al. 1996). These smaller vultures are often forced to utilize lower-quality foraging areas because they are rarely the first scavengers to arrive at carcasses and they are easily dominated by larger vultures (Kendall 2012).

The vultures in the genus *Gyps* are large-bodied vultures that forage and roost communally and appear to be the most carrion-dependent of the Old World vultures (Houston 1974b). They have sharp beaks, distinctive long, bare necks, and stick their heads deep into carcasses to feed on muscle and viscera. *Gyps* vultures forage in a web-like pattern in which each vulture searches for carrion individually but remains within sight of neighboring foragers. When one vulture locates a carcass and descends to begin feeding, the others follow, resulting in congregations of large numbers of *Gyps* vultures that can competitively exclude other species from feeding (Houston 1974a,b). The cliff-nesting Eurasian griffon is found mostly in Europe and Asia with a small subset of its range in Northern Africa. The Cape griffon is also a cliff nesting bird and is found only in Southern Africa. The tree-nesting white-backed and cliff-nesting Ruppell's griffons are both found in Central and Eastern Africa. Houston (1974a,b) suggests that *Gyps* vultures evolved to exploit carcasses of migratory ungulates. Kendall et al. (2014), however, found that vultures only cluster around migratory herds during the dry season when ungulate mortality is highest and depend heavily on scavenging from resident ungulates during the wet season.

The lammergeier and the palm-nut vulture do not fit into any of the categories described above, but have both evolved uniquely specialized feeding behavior. The palm-nut vulture, although a consumer of carrion, feeds primarily on fruits of the oil and Raphia palms. It is also known to hunt fish, small reptiles, and crustaceans. The lammergeier is a large, solitary, cliff dwelling vulture that specializes in consuming bone marrow. Small bones are consumed whole, whereas larger bones are smashed by dropping them onto rocks. The lammergeier occurs at low densities in mountainous regions of Northern, Eastern and Southern Africa, as well as Eurasia, and the Middle East.

Although vultures are the most specialized and diverse group of avian scavengers in Africa, a few other birds also rely heavily on carrion. The marabou stork feeds primarily on carrion, although its nestlings are fed small vertebrates (Kahl 1966). The marabou's long sharp bill is well adapted for hunting small vertebrates, but not for ripping flesh from a carcass. Accordingly, marabou storks generally consume pieces of carrion dropped or ignored by vultures and other scavengers (Kahl 1966). Marabou storks are particularly common around human settlements.

Although primarily hunters, bateleur and tawny eagles frequently consume carrion (Kendall 2012). Instead of soaring at high altitudes like vultures, these smaller scavengers fly closer to the ground in search of carrion and potential prey. As a result, they are often the first scavengers to arrive at a carcass (Kendall 2012). However, despite often arriving first, due to their small body size, these raptors are poor competitors at carcasses and often abandon them to other species once they arrive. Therefore, neither of these two eagles is likely to consume significant quantities of African carrion.

21.3.3 The Role of Vertebrate Scavengers in African Carrion Ecology

Although carrion represents a relatively small component of the diets of most vertebrate species in Africa, vertebrates play a hugely important role in carcass availability and degradation. In most African ecosystems, either large carnivores or vultures consume the majority of all carrion. Vultures appear to play a particularly large role in carrion degradation in areas where herbivores are controlled by bottom-up processes. Here they have an advantage over mammalian scavengers with respect to locating carrion and they are often among the initial consumers of carrion. In fact, spotted hyenas and lions often follow vultures to find carcasses (but see Hunter et al. 2007). In the experimental study by Butler and du Toit (2002), vultures were the most successful wild scavenger at finding carcasses in Zimbabwe, and they consumed more carrion biomass (37%) than any mammalian scavenger. In Kenya, the absence of vultures at a carcass nearly triples the time for decomposition, the number of scavenging mammals present, and the time these mammals spend at the carcass (Ogada et al. 2012). Houston (1979) placed 64 experimental carcasses in the Serengeti, all of which were fed upon at some point by vultures, compared to only 16% that were fed upon at some point by mammalian scavengers. Vultures can clearly play an important role in carcass degradation, decreasing persistence time and the probability that other scavenger guilds can make use of a carcass.

Although vultures consume the vast majority of carrion in some ecosystems, mammalian scavengers can often outcompete vultures. Because vultures must forage during the day, nocturnal mammals such as brown and striped hyenas, which rely heavily upon olfaction to detect carcasses, can monopolize carrion at night. Vultures are also less likely to exploit carrion produced by predation. Vultures are smaller than most mammalian scavengers, and thus can easily be driven off kills by the original hunters and by other mammalian scavengers. As such, areas where the majority of ungulate mortality is caused by predation have less carrion available for avian scavengers. For instance, in Ngorongoro crater, where mammalian predators cause most ungulate mortality, vultures were present at 21 carcasses but were only able to feed on one (Blumenschine 1989). Thus, whether most of a carcass is consumed by vultures or mammalian predators varies among areas, depending on whether ungulate mortality is more frequently caused by top-down or bottom-up processes.

In general, vertebrate scavengers can consume carrion much more rapidly than can invertebrates, particularly when they feed in groups. For instance, a group of *Gyps* vultures can consume 50 and 100 kg carcasses in 8 and 30 min, respectively (Houston 1974b), and groups of large carnivores can reduce a 2000 kg hippo to a skull and spine in only 4 days (http://msuhyenas.blogspot.ca/2008/11/day-3-hyena-heaven_19.html) (Figure 21.2). Even more impressive, vertebrate scavengers can reduce a 3000 kg elephant to nothing more than skin and bones in only 3 days (Richardson 1980). However, even among vertebrate scavengers, species vary in the speed at which they feed on carrion and their ability to access and consume the various parts of a carcass. Although avian scavengers are capable of consuming soft tissues quickly, they often cannot open carcasses of large, tough skinned mammals such as buffalo or elephant. Blumenschine (1989) found that three-fourths of the fleshy parts of very large carcasses cannot be exposed by vultures feeding in the absence of mammalian scavengers, suggesting that mammalian carnivores serve to greatly hasten destruction of large carcasses while also providing access for smaller scavengers like birds, insects, and microbes (Braack 1987).

Within the mammalian guild, lions and spotted hyenas play the most important roles in carrion degradation. As the largest and most gregarious carnivores, lions and spotted hyenas can overpower other vertebrate scavengers present at carcasses. For instance, White and Diedrich (2012) found that lions and spotted hyenas were largely responsible for consuming an elephant carcass over the course of 2 weeks during the dry season in Zambia. Hyenas are especially efficient consumers of carrion because they can ingest portions of carcasses that are inaccessible to other species. On average, only 7% of an adult wildebeest carcass remains after it is abandoned by spotted hyenas (Blumenschine 1989). By contrast, lions leave 55.5% and vultures leave 52.5% because they are unable to consume the bones and head contents. Furthermore, although a wildebeest carcass initially consumed by lions persists for an average of 34.5 h, a carcass initially consumed by spotted hyenas only persists for 8.2 h (Blumenschine 1989). Overall, high densities of hyenas in an ecosystem lead to low carcass abundance and relatively brief carcass persistence. Although other vertebrates like lions and vultures are likely to leave significant portions of carrion in the environment, they often leave only unpalatable parts, which are of little use to most scavengers.

The impact of vertebrate scavengers on African carrion is reduced in areas characterized by low predator to prey ratios and after catastrophic events, such as mass drownings of migratory ungulates and herbivore die-offs caused by severe droughts. For example, during the 1971 Tsavo drought, massive numbers of herbivores died suddenly, causing carrion to be abundant due to scavenger satiation (Wells 1989). The impact of vertebrate scavengers is also reduced in habitats characterized by dense vegetation,

FIGURE 21.2 Photographic documentation of the fate of an adult hippo carcass, located in the Masai Mara National Reserve in Kenya, over the course of 4 days. (a) Day 1: lions feed on the carcass a few hours after killing it. (b) Day 2: vultures feeding on the remains. Flesh on the head, lower legs/feet, and spinal cord is the only soft tissue left. (c, d) Day 3: Nearly all soft tissue has been consumed, leaving only bones, tough hide, and small amounts of flesh clinging to the skeleton. (e) Day 4: Only head, spinal column, pelvis, hide, and a few ribs remain. (Photos courtesy of Kate Shaw Yoshida.)

where carcasses tend to be smaller and more difficult to find; in these areas, most of the available carrion should theoretically be consumed by invertebrate and microbial scavengers.

It is important to note that vertebrate predators play a dual role in carrion availability, both by competing for naturally occurring carrion with other scavengers and decomposers and by adding carrion to the environment through predation. The latter is especially true for mammalian predators, as they often hunt prey much larger than themselves. They can therefore add large carcasses to the environment and may not completely consume them before microbes, invertebrates, or other vertebrates move in. However, predators generally waste as little as possible of the carcasses yielded by their kills and they have evolved strategies to prevent loss to competitors or scavengers. Therefore, predators only provide scavenging opportunities to others if they can be overpowered or if prey/carcass abundance is high enough for them to abandon less palatable portions of their kill.

21.4 Carrion Ecology in the Serengeti Ecosystem

The Serengeti-Mara ecosystem in Eastern Africa, hereafter referred to as the Serengeti, is one of the most intensively studied African ecosystems and offers the opportunity to develop a more detailed understanding of carrion dynamics in African savanna. The Serengeti remains one of the most pristine ecosystems in Africa and is unusual in that it supports one of the few remaining large ungulate migrations in the world, a phenomenon that was widespread in recent evolutionary history but has since become rare (Harris et al. 2009). Because large migrations are a huge source of carrion for scavengers, understanding carrion ecology in the Serengeti can potentially shed light on the role of carrion in both contemporary and ancient ecosystems.

21.4.1 The Migration

The seasonal migration of herds of roughly 1.2 million wildebeest (Mduma et al. 1999; Holdo et al. 2011) and many thousands of zebra is the defining feature of the Serengeti ecosystem. This is a savanna area of 25,000 km^2 at a mean elevation of 1525 m, bounded by natural topographic features that limit animal movements. Rainfall patterns vary throughout the ecosystem due to the interaction of global and local weather patterns, and this heterogeneity in rainfall gives rise to substantial spatiotemporal variation in availability of water and forage for herbivores (Pennycuick 1975; Sinclair 1979a; Mduma et al. 1999). For example, although the southern Serengeti has only one long wet season each year, the northernmost portion of the ecosystem has two relatively brief wet seasons. Wildebeest and zebra are effective at finding high-quality forage patches across this vast landscape, and this gives them a demographic edge over sedentary herbivores (Holdo et al. 2011).

The migration occurs each year in a clockwise triangular pattern (Figure 21.3). The herds spend the wet season (November–May) on the southern plains of Serengeti National Park, Tanzania, where soil phosphorus and nitrogen concentrations are high (Ben-Shahar and Coe 1992). With the dry season comes a shortage of surface drinking water and reduced grass growth that cannot sustain large numbers of ungulates, so the herds move to the northwest to find better forage. The northwestern grasslands are usually exhausted by late June or early July, when the herds move northeast into the Masai Mara National Reserve in Kenya. They return to the southern plains in October each year, and the cycle starts anew.

The seasonal ungulate migration has profound effects on the abundance and persistence of carrion. Because large predators defend fixed territories and often require denning areas to reproduce, they cannot follow migratory herds as they move through the Serengeti ecosystem. Furthermore, migratory behavior allows herds of wildebeest and zebra to reach larger sizes, as they can simply move once they have depleted an area of resources. As a result, the ratio of large predators to large-bodied prey in the Serengeti is relatively small compared to that found in areas without large migrations. For example, the predator/prey ratio in the Serengeti is between 1/4 and 1/10 of that in Ngorongoro crater, an ecosystem with the exact same habitat and species representation as the Serengeti, but lacking a migration (Blumenschine 1989). Because of this comparatively low predator/prey ratio, many Serengeti ungulates die of malnutrition, disease, drowning, and other causes unrelated to predation. Unsurprisingly, the large numbers of migratory ungulates that die from these other causes represent a huge source of carrion for scavengers.

21.4.2 Carrion Availability

An animal that dies of causes other than predation provides an order of magnitude more carrion than does a carcass abandoned by a satiated predator. Mortality in migratory populations tends to be driven by bottom-up processes (Hopcraft et al. 2013), with predation rates of only around 25% (Mduma et al. 1999). By contrast, sedentary wildebeest populations appear to be regulated by top-down processes, experiencing ~90% predation rates (Sinclair et al. 2003). Table 21.3 lists the species of ungulates found in the Serengeti and approximates their contribution to carrion abundance by estimating the proportion of each species'

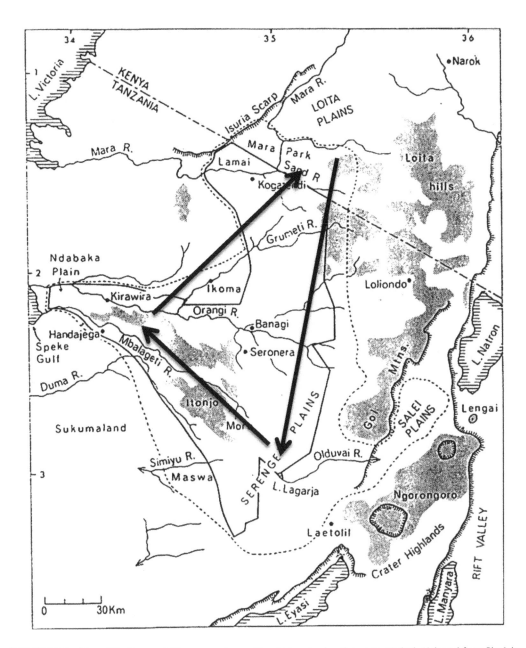

FIGURE 21.3 Map of the Serengeti-Mara ecosystem, with wildebeest migration route marked. (Adapted from Sinclair, A.R.E. 1979a. *Serengeti: Dynamics of an Ecosystem.* Chicago: University of Chicago Press, pp. 1–30. Arrows added by the current authors.)

mortality due to predation as well as the amount of time a carcass of each species persists in the environment; carcasses that persist longer contribute more to carrion abundance and are therefore more important resources for scavengers. Houston (1979) estimates that 70% of ungulate mortality (or 26,000,000 kg of carrion/year) in the Serengeti is due to causes other than predation; this theoretically amounts to one carcass a day every 33 km^2 in the southern plains when the migratory herds are present. Because most ungulate mortality is caused by bottom-up processes and because carcasses produced by predation leave little soft tissue available for scavengers, this section focuses on whole carcasses not resulting from predation events.

TABLE 21.3

Adult Live Weights for Ungulates Inhabiting the Serengeti Ecosystem, Percent of Their Mortality due to Predation, Carcass Persistence, and Proportion of Their Carcasses Colonized by Invertebrates

Species	Adult Live Weight (kg)[a]	% of Mortality due to Predation	Adult Carcass Persistence (h) Mean (sd)[b]	% of Carcasses That Are Colonized by Invertebrates[b]
Thomson's Gazelle	20[a]	—	4.4 (4.1)	0
Grant's Gazelle	50[a]	—	4.4 (4.1)	0
Impala	50[a]	100[a]	6.5 (2.9)	3.8
Topi	120[a]	100[a]	20.8 (26.3)	1.9
Wildebeest	170[a]	86.8[a] (resident) 25[c] (migratory)	20.8 (26.3)	1.9
Zebra	250[a]	73.4[a]	20.8 (26.3)	1.9
Buffalo	450[a]	23.3[a]	100.1 (61.3)	42.9
Giraffe	800[a]	5[a]	272 (63)	100
Rhino	1200[a]	0[a]		—
Hippo	2000[a]	0[a]		—
Elephant	3000[a]	0[a]	272 (63)	100

Note: The samples sizes from Blumenschine (1986) are small for some species.
[a] From Sinclair, A.R.E., S. Mduma, and J.S. Brashares. 2003. *Nature* 425: 288–290.
[b] From Blumenschine, R.J. 1986. *Early Hominid Scavenging Opportunities: Implications of Carcass Availability in the Serengeti and Ngorongoro Ecosystems* (Vol. 283). British Archaeological Reports.
[c] From Mduma, S. A., Sinclair, A., and R. Hilborn. 1999. *J. Anim. Ecol.* 68(6), 1101–1122.

Like carrion in most ecosystems around the world, carrion in the Serengeti is patchily distributed in space and time. Heterogeneous rainfall dictates the distribution of ungulates and thus also carrion. As migratory ungulates move about the Serengeti, carnivores currently living near the herds experience a dramatic increase in food availability. For example, Cooper et al. (1999) found that carnivores in the Talek region of the Masai Mara National Reserve in Kenya (Figure 21.3) experienced a 3.5-fold increase in prey density when the migratory herds were present. Thus, carrion is likely to be most abundant in areas where the migratory herds spend the most time. In some years with unusually high levels of rainfall in the south, migratory herds do not ever leave the southern plains, resulting in low carrion availability for scavengers in more northern regions of the ecosystem. Carrion also varies temporally with rainfall. As the dry season progresses, suitable forage becomes scarce, and more herbivores die of malnutrition and disease. Mduma et al. (1999) found that 64% of observed adult migratory wildebeest deaths occurred during the 4-month dry season in the Serengeti. In general, carrion is more abundant at the end of the dry season than at the beginning (Houston 1985; Pereira et al. 2014). Interestingly, carrion also occurs most often in the early morning (Houston 1974b; Mduma et al. 1999; Kendall et al. 2012) and is consumed by vultures and mammalian scavengers throughout the day. Heterogeneity in carrion availability is also apparent across years, with long-term shifts in rainfall and thus food supply causing up to 50% change in carcass numbers (Mduma et al. 1999; Wilson and Wolkovich 2011).

Resource pulses caused by mass drownings of migratory wildebeest as they cross rivers and lakes in search of greener pasture illustrate the extreme heterogeneity in carrion availability in the Serengeti ecosystem. Drownings produce a variable amount of carrion—hundreds to thousands of wildebeest can die in a single drowning event or smaller numbers can drown in several drowning events when migratory herds are on the move. Small drowning events occur practically every year in the Serengeti ecosystem, and Sinclair (1979b) reports that wildebeest drowning in rivers in the Serengeti woodlands are quite common, especially during times when water levels are high. Sinclair (1979b) reported drowning events occurring 2–3 times a year and involving between 20 and 500 wildebeest. Mass drownings of at least 1000 and up to 5000 wildebeest were reported in the small soda lakes of Masek and Ndutu in the Serengeti plains in 1984 and 1988, years when lake levels were high. Vertebrate scavengers make use of carcasses after all these drowning events, but may utilize them differently depending on the nature of the drowning. Capaldo and Peters (1995) inferred from studies of bone deposition that large drownings result in more densely congregated and complete wildebeest skeletons. This is likely due to scavenger satiation and reduced competition over remains. In contrast, smaller drownings result in high competition between scavengers, which utilize carcasses more completely and may move choice pieces of carcasses away from the original drowning site, so they may eat without interference from other predators. Overall, drowning events represent a variable but potentially rich source of carrion for Serengeti scavengers.

Of the roughly 26,000,000 kg (Houston 1979, 1974b) of carrion estimated to occur in the Serengeti ecosystem each year, the majority is consumed by vertebrate scavengers. Multiple studies have investigated the time-course of scavenger consumption of various types of carcasses in both naturally occurring and experimental situations. The most thorough of these studies, by Blumenschine (1989), documented the fates of 232 natural carcasses in the Serengeti and Ngorongoro crater and found that 96% were completely consumed by vertebrate scavengers. It appears that invertebrate scavengers in the Serengeti ecosystem generally play a significant role only in the consumption of extremely large carcasses that cannot quickly be consumed by vertebrates or in the large pulses of carrion that result from mass drowning events or epidemics, as discussed above.

Vultures are by far the greatest consumers of carrion in the Serengeti. The near-monopolization of carrion by vultures is mostly due to their large population size and their ability to locate and consume carrion quickly (Houston 1974a,b). In Blumenschine's (1989) experimental study, 65% of 232 carcasses were fed on by vultures, and reports of mass wildebeest drownings indicate that scavenger birds are the first and primary consumers of the resulting carrion (Capaldo and Peters 1995). The most important consumers of carrion among the five species of vulture regularly seen in the Serengeti are the *Gyps* vultures (Ruppell's griffon and the white-backed vulture), which comprise roughly 90% of all vulture sightings (Houston 1979; Kendall 2012). Because of their gregarious foraging behavior, *Gyps* vultures accumulate at a carcass soon after it is discovered by the first individual and proceed to consume large amounts of soft tissue very quickly. These birds are so adept at finding carrion that other scavengers pay close attention to them. For instance, Serengeti lions use vultures to find roughly 11% of the carcasses they scavenge (Schaller 1972). The reliability of vulture presence as an indicator of carrion presence has similarly led a number of researchers to use descending birds to locate carrion in the course of their studies (Blumenschine 1989; Kendall 2012).

Of the mammalian carnivores present in the Serengeti, hyenas are the best adapted to scavenging and they consume the most carrion (Kruuk 1972; Blumenschine 1989; Mills 1990). Brown hyenas do not occur in the Serengeti, and although carrion comprises roughly 60% of the diet of striped hyenas (Wagner 2006), striped hyenas occur in the Serengeti ecosystem at low density (Kruuk 1976). Therefore, the small numbers of striped hyenas make them relatively insignificant players in the carrion ecology of the Serengeti. Instead, spotted hyenas, which occur at a high density in this ecosystem, seem to be the major mammalian consumers of carrion.

The extent to which spotted hyenas ingest carrion varies considerably among populations (Table 21.2) and depends in part on the intensity of local intra- and interspecific competition for food. For spotted hyenas inhabiting the southern part of the Serengeti, estimates of the percentage of the diet derived from kills made by the hyenas themselves range from 43 to 69% (Kruuk 1972; Höner et al. 2002), but estimates hover around 95% for hyenas inhabiting the northernmost portion of this ecosystem (Cooper et al. 1999). Perhaps due to the relatively low predator/prey ratios throughout the ecosystem, spotted hyenas often fail to consume carcasses in their entirety. Kruuk (1970) reported that Serengeti hyenas leave bones and other less palatable parts of carcasses behind, unlike hyenas in Ngorongoro Crater, which compete among themselves for food much more intensively than do hyenas in the Serengeti, and which consume carcasses completely. In the Masai Mara, spotted hyenas often leave large portions of ungulate carcasses behind during those months of the year when migratory herds are present and feeding competition is relaxed due to the superabundance of food. However, Mara hyenas consume every bit of each carcass during the months when the migratory ungulates are absent. Despite the variability in their scavenging behavior, spotted hyenas often have the largest and most obvious effect on carrion in the environment. For instance, spotted hyenas cause the most damage, relative to other scavengers, to wildebeest skeletons after drowning events. This is especially true for drownings involving fewer wildebeest, where the hyenas are more likely to utilize entire carcasses (Capaldo and Peters 1995).

Although many other vertebrates scavenge opportunistically, they appear to be of little importance in the carrion ecology of the Serengeti. Jackals, tawny eagles, and bateleur eagles, for example, are regularly seen at carrion sites, but their effects on carrion persistence are negligible, as they consume trivial amounts of carrion (Blumenschine 1989). Although Serengeti lions scavenge 5%–53% of their diet (Table 21.1), 84% of scavenged items are obtained by kleptoparasitism from smaller predators, so the lions end up consuming very little carrion that would not otherwise have been consumed immediately by the original predator (Schaller 1972).

To summarize, the Serengeti is home to a multitude of scavengers, supported primarily by the whole carcasses of migratory and, to a lesser extent, resident ungulates. The majority (55.25%) of the soft tissue of this carrion is consumed by vultures (Houston 1979), in particular, the gregarious *Gyps* vultures. Large mammalian carnivores are the next most important consumers of carrion, usually arriving after vultures and sometimes alerted to carcass whereabouts by descending birds. Spotted hyenas are the primary mammalian scavengers, aided by their ability to consume all parts of a carcass as well as the voracious pace at which groups of them can consume even very large carcasses. Desiccated carcasses are consumed only by hyenas and some invertebrates (e.g., dermestid beetles, tineid moths) and may play a role in maintaining hyenas as the most numerous large carnivores in the Serengeti. Mega-herbivores like buffalo, rhino, elephants, and giraffes serve a unique function in that they suffer little to no predation and their carcasses persist for days, providing carrion for scavengers that would otherwise be outcompeted by vultures, hyenas, and lions. Because the carcasses of mega-herbivores are too large for vertebrate scavengers to consume in a short period of time, these carcasses often host a large number of invertebrate carrion consumers and may be of particular importance in their ecology. Resource pulses in the form of mass drownings and epidemics, like the carcasses of mega-herbivores, provide large quantities of carrion that provide important nutritional resources for vertebrate scavengers. Finally, as in many other ecosystems, the ecology of carrion from smaller-bodied animals in the Serengeti remains largely a mystery.

21.5 Anthropogenic Disturbances

21.5.1 Effects of Human Activity on Carrion Availability

Human population growth is drastically changing the carrion ecology of many African ecosystems via effects on major carrion consumers and producers. Anthropogenic disturbances, such as global climate change, habitat destruction or fragmentation, and hunting, are causing a widespread decline in numbers of African carnivores and herbivores (Ogutu et al. 2005; UNEP 2008; Ogutu et al. 2009). Wild ungulates are losing foraging areas as agriculture spreads, and poaching has caused drastic declines in several species of herbivores, particularly elephants (Ottichilo et al. 2001; Wittemyer et al. 2013). Increased drought and desertification due to climate change are also major threats, as many herbivore populations in Africa are limited by rainfall. Global warming may result in longer dry periods, causing more herbivores to die of malnutrition during these months (Pereira et al. 2014). Finally, as predators decline due to anthropogenic causes, they are less able to control prey populations. This results in prey populations being controlled by bottom-up rather than top-down processes and puts them at risk of sudden changes in population size (i.e., mass mortality from disease epidemics) (Wilson and Wolkovitch 2011). Wilson and Wolkovitch (2011) suggest that long-term maintenance of carrion and scavenging populations may be dependent on healthy predator guilds, which in turn keep prey populations stable. Overall, wild ungulate carrion may increase in the short term due to increased mortality in the face of anthropogenic disturbance, but should decrease dramatically in the long term as herbivore population sizes continue to shrink.

Vertebrate scavengers are also declining in Africa, and the major carrion consumers there are facing a unique set of challenges. As pastoralist communities infringe on natural areas, many predators are turning to livestock for food. Villagers commonly retaliate against predators that kill livestock by lacing carcasses with poison, such as the insecticide Furadan. Carcass poisoning can cause mass casualties among scavenging animals, and has contributed to recent population declines of vultures, lions, and brown, striped, and spotted hyenas (Arumugam et al. 2008; Wiesel et al. 2008; Virani et al. 2011). Scavengers are also commonly killed in predator control programs and hunted for bush meat and medicinal purposes (Sillero-Zubriri et al. 2004; Thiollay 2006; Höner et al. 2008; Wiesel et al. 2008; Arumugam et al. 2008). As important species of vertebrate scavengers decline, the process of carrion degradation is likely to change such that less threatened guilds (such as invertebrates and microbes) may come to play more important roles in decomposition, as they do on continents with less intact large carnivore guilds.

Human activity can also negatively influence scavengers by moderating the amount of carrion available to them. Aside from their negative effects on numbers of wild herbivores, humans affect the carrion available to scavengers in several ways. Hunting and culling by people can restrict the number of prey animals that die of old age, malnourishment, and disease (Wilson and Wolkovitch 2011; Pereira et al. 2014), eliminating the most bountiful source of carrion in Africa. Interference in seasonal herbivore migrations in areas such as the Serengeti, for example, by the installation of fences and roads, can severely reduce carcass availability in areas normally frequented by herds of migratory ungulates (Holdo et al. 2011).

Although it is clear that human activity is likely to affect the availability of carrion in African systems, few studies have examined this in detail. Butler and du Toit (2002) found that carrion is becoming less available to wild scavengers due to competition from domestic dogs. Domestic dog populations are increasing with human populations, growing at a rate of 6.5% per annum in Africa (Butler and du Toit 2002). Butler and du Toit (2002) found that domestic dogs consume 60% of small-medium carcass biomass in the boundary area between Sengwa Wildlife Research Area and Gokwe Communal Land in Zimbabwe (an area representative of many communal/reserve boundary areas in Africa). The authors speculate that the high densities of dogs, their ability to dominate vultures (the second most successful scavenger in this area), and their tolerance of human proximity allow dogs to monopolize much of the carrion that becomes available.

Despite the negative effects of anthropogenic disturbance on carrion abundance in Africa, human activity may increase carrion availability in other ways. As mentioned previously, humans' negative effect on wild herbivore populations can temporarily increase carrion availability. Though hunting, poaching, and culling can decrease the number of animals dying from old age, these activities can also produce carrion as a byproduct (Wilson and Wolkovitch 2011). This is especially true if only part of the animal is collected, such as when poachers collect elephant tusks or rhinoceros horn. Furthermore, numbers of animals killed on roads are increasing rapidly in many parts of Africa. In South Africa, studies have found that 0.48–5.44 mammals and 0.12–1.14 birds are killed per 100 km of road (Siegfried 1965; Dean and Milton 2003; Bullock et al. 2011). These studies probably underestimate the number of animals killed, as roadside carcasses of small animals may be consumed quickly, and injured animals may leave the road before dying. Road kill may provide a highly visible source of carrion for some scavengers, but unfortunately consuming carrion on roads is a dangerous activity, as the scavengers themselves may also be hit by cars.

Humans can also increase carrion availability by introducing livestock. In 1999, livestock made up over 90% of the ungulate biomass in Africa (Du Toit and Cumming 1999), and this percentage has undoubtedly only increased since then. Livestock thus represent an important potential source of carrion. Humans can provide carrion for scavengers by discarding livestock remains and other uneaten carcasses (e.g., dead animals resulting from research, poaching, disease, etc.). Butler and du Toit (2002) found that domestic species contribute 95.5% of the carcasses and 84.6% of the biomass in the Gokwe Communal Land bordering the Sengwa Wildlife Research Area in Zimbabwe, whereas inside the reserve, domestic species comprise 20% of the carcasses but less than 1% of the biomass.

In conclusion, anthropogenic disturbance can be expected to have negative effects on scavenger populations and, in the long term, decrease the amount of carrion available in the environment. Though scavengers may benefit *temporarily* as wild herbivores experience increased mortality in the face of human disturbance, it is clear that increases in carrion caused by humans are short-lived or conditional on how humans discard waste and care for livestock. Therefore, they do not represent a stable source of food for most scavengers, and increasing human populations are likely to be associated with decreasing production and consumption of carrion by other vertebrates.

21.5.2 Spotted Hyenas and Human Disturbance

Though their range is shrinking, spotted hyenas remain the most abundant large carnivore in Africa, and are listed as a species of "least concern" by the IUCN (Höner et al. 2008). This is despite a host of anthropogenic influences on their mortality and behavior (Hofer et al. 1996; Höner et al. 2008; Holekamp and Dloniak 2010). Scavenging livestock remains and garbage from human settlements may help hyenas survive in the face of human population growth. For example, although spotted hyenas living in prey-rich regions like the Masai Mara hunt the majority of their own food, they are still known to scavenge from local refuse pits. This is especially true for individuals of low social status and during periods of low abundance of natural prey (Kolowski and Holekamp 2007). Scavenging from humans is particularly important to spotted hyenas in the horn of Africa, a region that includes Ethiopia, which has virtually no remaining natural prey. Spotted hyenas in parts of Ethiopia depend exclusively on anthropogenic food sources, consisting mostly of livestock remains (Abay et al. 2011; Yirga et al. 2012). Hyenas in the horn of Africa also experience times of "feast" during famine, drought, war, and epidemics, tragedies that have occurred throughout the history of the area and can result in massive losses of both human life and livestock (Gade 2006). In this region, it is clear that their ability to derive large portions of their diet from scavenging has allowed spotted hyenas to thrive despite great human disturbance. However, the horn of Africa has some unique characteristics that greatly benefit spotted hyenas. For instance, in Ethiopia many people follow strict religious restrictions on which parts of livestock they can eat, even in times of famine. Therefore, livestock remains may be more abundant here than in other areas (Yirga et al. 2012). Spotted hyenas are also unusually well tolerated in parts of Ethiopia, despite occasionally preying on livestock and even killing people in surrounding areas. In Harar, Ethiopia multiple groups of spotted hyenas travel and feed regularly within the city walls and are tolerated for their ability to clean the streets of biological waste and for their religious significance, as they are thought to drive away evil spirits (Baynes-Rock, 2013). Spotted hyenas in other parts of Africa are far less likely to be able to take advantage of human presence in this way.

In certain environments spotted hyenas are capable of taking advantage of human population growth, but are other mammalian carnivores able to do so as well? This does not seem to be the case in the horn of Africa, as most other large carnivores have been completely extirpated from the region. Hyena species may be better able to take advantage of human-generated carrion due to their adaptations for scavenging. Because of their extremely efficient use of carrion, they can better utilize livestock carcasses from which humans have already taken the preferred parts. They can also eat carcasses even in advanced stages of putrification, which other carnivores shun. Furthermore, spotted hyenas may be better able to take advantage of herbivore deaths from epidemics due to an immune system uniquely able to deal with disease-infected meat (Pienaar 1969; Lindeque 1981; Gasaway et al. 1991). Similarly, striped hyenas are thought to take large proportions of their diet by scavenging from humans in Israel and northern Kenya, and this may also be due in part to their unique adaptations for consuming carrion (Leakey et al. 1999).

Finally, spotted hyenas demonstrate astounding behavioral flexibility that may buffer them from some sources of conflict with humans. For instance, their ability to shift from a diurnal to a nocturnal feeding strategy allows them to consume remains left near human dwellings without having to come into direct contact with the humans themselves (Gade 2006). Their ability to switch easily between consuming live prey and carrion is also helpful, as seen when they quickly shift from scavenging livestock remains to preying on donkeys in Ethiopia during periods of religious fasting (Yirga et al. 2012). Other scavengers with flexible diets such as jackals and brown and striped hyenas may also be able to take advantage of human resources. Because jackals are generalists with flexible diets, they are reported to do well in some human-dominated landscapes (Poché et al. 1987). However, jackals are small and can therefore be driven off refuse easily by domestic dogs, making them poorly suited for coexistence with other human populations. More studies are needed to determine the extent to which scavenging in particular (as opposed to general behavioral flexibility) contributes to the survival of mammalian carnivores such as jackals and hyenas in various regions of Africa.

21.5.3 Avian Scavengers and Human Disturbance

Avian scavengers are another group that may sometimes benefit from growing human populations. This is seen in the few studies focusing on urban environments. Tropical cities in developing countries often have open food markets, large garbage disposal areas, and may also have poor sanitation (Mundy et al. 1992). These characteristics make them particularly suitable habitat for some avian scavengers. Certain species of birds are more likely than others to scavenge from human settlements. In particular, corvids and vultures predominate in tropical cities (Borrow and Demey 2001). In some areas, they are seen as useful contributors to city sanitation and thus may be tolerated despite occasionally stealing bits of food and the negative superstitions associated with scavenging birds. Campbell (2009) found that hooded vultures and pied crows were more common in urban than rural areas in Ghana, and that their population sizes were positively correlated with human numbers. Scavenging birds such as marabou storks, pied crows, hooded vultures, and black kites are also seen in towns and cities across Uganda (Pomeroy 1973). A study near Rwenzori National Park, Uganda, found that people augment the diets of avian scavengers by discarding unwanted parts of livestock carcasses and scraps remaining after gutting/cleaning fish. Maribou storks eat the majority of waste at fish factories, where over half the mass of each daily catch remains after filleting (Pomeroy 1973). Finally, many birds, such as crows, are frequently found feeding on road kill (Dean et al. 2006; Bullock et al. 2011).

Some authors suggest that greater mobility in the air, tolerance of people, and large body size are particularly adaptive traits for avian scavengers in urban environments (Gbogbo and Awotwe-Pratt 2008). Like mammalian carnivores, bird species with broader and more flexible diets will likely be most able to take advantage of urban environments as these continue to expand in Africa. Campbell (2009) found that pied crows were able to scavenge in more areas of cities than were vultures due to their broader diets. It is also clear that vultures in Africa are not doing well in the face of growing human populations (Virani et al. 2011). Therefore, the ability to scavenge opportunistically may be helpful, but specialized reliance on scavenging may be very costly in the face of anthropogenic disturbance.

Vultures are declining around the world, and population crashes have been recorded recently in West and East Africa. In West Africa all vulture species except the hooded vulture have decreased by an average of 95% in the last several decades (Ogada et al. 2012). In East Africa, vulture populations have declined by 70% over three years in north central Kenya (Ogada and Keesing 2010), and in the Masai Mara vulture populations have declined by an average of 62% over the last 30 years (Virani et al. 2011; Kendall et al. 2012). A number of factors may be causing these declines, including hunting, carcass-poisoning, killing for medicinal purposes, and habitat loss or degradation (Thiollay 2006; Virani et al. 2011; Ogada et al. 2012). Carcass poisoning has been shown to be a particularly important factor in the decline of vultures in several areas of the world, including the Masai Mara in Kenya. Green et al. (2004) showed that poisoning less than 1% of carcasses could explain annual mortality rates of 22%–50% in *Gyps* vultures. Furthermore, the benefits vultures might accrue from human refuse have decreased due to changing livestock husbandry practices (Thiollay 2006; Ogada et al. 2012) and growing feral dog populations (Butler and du Toit 2002). Overall, it appears that dietary and behavioral flexibility, rather than scavenging per se, may buffer avian and mammalian scavengers from anthropogenic disturbance.

21.6 Conclusions and Future Directions

Although much is known about herbivore population dynamics and the major guilds of African scavengers, there is a general dearth of detailed, quantitative studies on carrion in African ecosystems. Blumenschine's (1989) study in the Serengeti National Park remains the most thorough study on carrion abundance and persistence in Africa. Quantitative studies comparing carrion abundance across ecosystems varying in such factors as predator/prey ratio, climate, and herbivore/mega-herbivore abundance are needed to determine the relative importance of these variables in carrion ecology.

More detailed studies on the relative importance of scavenged and hunted items in the diets of several avian and mammalian scavengers would also be useful. It is clear from work on well-studied scavengers, such as the spotted hyena, that scavenging activity can vary dramatically among populations in different parts of Africa. Though spotted hyenas represent an extreme example, other scavengers may also show significant variation in diet among populations. Developing a clearer picture of the importance of scavenged items in the diet will allow us to more fully explore the factors that promote scavenging behavior.

APPENDIX

Common Name	Latin Name
Class Mammalia	
Baboon, olive	*Papio anubis*
Bat-eared fox	*Otocyon megalotis*
Buffalo	*Syncerus caffer*
Caracal	*Felis caracal*
Cheetah	*Acinonyx jubatus*
Civet	e.g., *Civetticti scivetta*
Dik-dik	*Madoqua* spp.
Domestic dog	*Canis lupus familiaris*
Elephant, African	*Loxodanta africana*
Gazelles	
Thomson's	*Eudorcas thomsonii*
Grant's	*Nanger granti*
Genet, Rusty-spotted	*Genetta maculata*
Giraffe	*Giraffa camelopardalis*
Hippopotamus	*Hippopotamus amphibus*
Honey badger	*Mellivora capensis*
Hyenas	
Spotted	*Crocuta crocuta*
Striped	*Hyaena hyaena*
Brown	*Parahyaena brunnea*
Impala	*Aepyceroc melampus*
Jackals	
Side-striped	*Canis adustus*
Black-backed	*Canis mesomelas*
Golden	*Canis aureas*
Leopard	*Panthera pardus*
Lion	*Panthera leo*
Mongoose, White-tailed	*Ichneumia albicauda*
Porcupine	*Hystrix* spp.
Rhinoceros	
Black	*Diceros bicornis*
White	*Ceratotherium simum*
Seal, Cape fur	*Arctocephalus pusillus pusillus*
Serval	*Felis serval*
Warthog, common	*Phacochoerus africanus*
Wild dog	*Lycaon pictus*
Wildcat, African	*Felis sylvestris*
Wildebeest	*Connochaetes* spp.
Zebra	*Equus* spp.
Class Aves	
Eagles	
Bateleur	*Terathopius ecaudatus*
Tawny	*Aquila rapax*
Falcon	
Lanner	*Falco biarmicus*
Peregrine	*Falco peregrinus*
Taita	*Falco fasciinucha*
Kite, black-shouldered	*Elanus axillaris*
Marabou stork	*Leptoptilos crumeniferus*

(Continued)

Common Name	Latin Name
Vultures	
Egyptian	*Neophron percnopterus*
Hooded	*Necrosyrtes monachus*
Lammergeier	*Gypaetus barbatus*
Lappet-faced	*Torgos tracheliotus*
Palm-nut	*Gypohierax angolensis*
White-headed	*Trigonoceps occipitalis*
Vultures (Griffon)	
Cape	*Gyps scoprotheres*
Eurasian	*Gyps fulvus*
Ruppell's	*Gyps ruppellii*
White-backed	*Gyps africanus*

Despite the current paucity of systematic studies, it is nevertheless evident that African savannas have a rich and unique carrion ecology, with carrion being exploited as a food source by three substantial guilds of scavengers. African savannas are inhabited by several vertebrate species that are specifically adapted for scavenging, such as vultures and hyenas, as well as diverse and abundant herbivores whose carcasses comprise the vast majority of available carrion. Studies on the carrion ecology of other African ecosystems, such as tropical forests and deserts, are currently scarce, but such studies would yield useful comparative data.

An interesting area of future research might examine the species that are faring best in the face of human threats, including spotted hyenas and corvid species. It may be that behavioral and dietary flexibility, paired with at least some degree of human tolerance, can buffer scavenging species from anthropogenic disturbance. As populations of vertebrate scavengers and large ungulates change in Africa, the major drivers of carrion ecology will begin to look different than they do today. As wild populations of birds and mammals continue to shrink, it is likely that carrion will generally come to occur in the form of livestock, which tend to be of smaller body-size than many wild herbivores, and the primary consumers of carrion will be domestic dogs and avian scavengers able to thrive in urban environments. In the future, sanitation and livestock husbandry practices in Africa will undoubtedly change. Changes in livestock husbandry practices, in particular, such as systematic vaccinations and easier transport and marketing of livestock, may greatly reduce the abandonment of weak and sickly animals that occurs so commonly today (Thiollay 2006). Such changes will undoubtedly reduce amounts of human-generated carrion available to scavengers. However the largest changes in carrion ecology are likely to come from global warming, with its associated desertification of many savanna ecosystems. Whereas carrion abundance currently varies seasonally in most savannas, desert conditions will be inhospitable to both herbivores and scavengers throughout the year. Thus the abundance of carrion, its distribution in time and space, and its persistence will all likely change considerably in the future in many parts of Africa.

REFERENCES

Abay, G., H. Bauer, K. Gebrihiwot, and J. Deckers. 2011. Peri-urban spotted hyena (*Crocuta crocuta*) in northern Ethiopia: Diet, economic impact, and abundance. *European Journal of Wildlife Research* 57: 759–765.

Adams, M.E. 1996. Savanna environments. In: Adams, W., A. Goudie, and A. Orme (eds), *The Physical Geography of Africa*. Oxford: Oxford University Press, pp. 197–210.

Arumugam, R., A. Wagner, and G. Mills. 2008. *Hyaena hyaena*. IUCN 2013 Red List of Threatened Species. Version 2013.1. www.iucnredlist.org (accessed August 1, 2013).

Atkinson, R.P.D., W. Macdonald, and R. Kamizola. 2002. Dietary opportunism in side-striped jackals *Canis adustus*. *Journal of Zoology, London* 257: 129–139.

Barton, P.S., S.A. Cunningham, D.B. Lindenmayer, and A.D. Manning. 2013. The role of carrion in maintaining biodiversity and ecological processes in terrestrial ecosystems. *Oecologia* 171: 761–772.

Baynes-Rock, M. 2013. Life and death in the multispecies commons. *Social Science Information* 52(2): 210–227.

Begg, C.M., K.S. Begg, J.T. Du Toit, and M.G.L. Mills. 2003. Sexual and seasonal variation in the diet and foraging behavior of a sexually dimorphic carnivore, the honey badger (*Mellivora capensis*). *Journal of Zoology, London* 260: 301–316.

Ben-Shahar, R. and M.J. Coe.1992. The relationships between soil factors, grass nutrients and the foraging behaviour of wildebeest and zebra. *Oecologia* 90(3): 422–428.

Bertram, B.C.R. 1979. Serengeti predators and their social systems. In: Sinclair, A.R.E. and M. Norton Griffiths (eds), *Serengeti: Dynamics of an Ecosystem*. Chicago: University of Chicago Press, pp. 263–286.

Blumenschine, R.J. 1986. Early hominid scavenging opportunities: Implications of carcass availability in the Serengeti and Ngorongoro ecosystems (Vol. 283). British Archaeological Reports.

Blumenschine, R.J. 1989. A landscape taphonomic model of the scale of prehistoric scavenging opportunities. *Journal of Human Evolution* 18: 345–371.

Borrow, N. and R. Demey. 2001. *The Birds of Western Africa*. London: Christopher Helm.

Braack, L.E.O. 1984. An ecological investigation of the insects associated with exposed carcasses in the northern Kruger National Park. PhD Dissertation, University of Natal, South Africa.

Braack, L.E.O. 1986. Arthropods associated with carcasses in the northern Kruger National Park. *South African Journal of Wildlife Research* 16(3): 91–98.

Braack, L.E.O. 1987. Community dynamics of carrion-attendant arthropods in tropical African woodland. *Oecologia* 72: 402–409.

Braczkowski, A., L. Watson, D. Coulson, J. Lucas, B. Peiser, and M. Rossi. 2012. The diet of caracal, *Caracal caracal*, in two areas of the southern cape, South Africa as determined by scat analysis. *South African Journal of Wildlife Research* 42(2): 111–116.

Bullock, K.L., G. Malan, and M.D. Pretorius. 2011. Mammal and bird road mortalities on the Upington to Twee Rivieren Main Road in Southern Kalahari, South Africa. *African Zoology* 46(1): 60–71.

Butler, J.R.A. and J.T. du Toit. 2002. Diet of free-ranging domestic dogs (*Canis familiaris*) in rural Zimbabwe: Implications for wild scavengers on the periphery of wildlife reserves. *Animal Conservation* 5: 29–37.

Campbell, M. 2009. Factors for the presence of avian scavengers in Accra and Kumasi, Ghana. *Area* 41(9): 341–349.

Capaldo, S.D. and C.R. Peters. 1995. Skeletal inventories from wildebeest drownings at Lakes Masek and Ndutu in the Serengeti ecosystem of Tanzania. *Journal of Archaeological Science* 22: 385–408.

Caro, T.M. 1994. *Cheetahs of the Serengeti Plains: Group Living in an Asocial Species*. Chicago: The University of Chicago Press.

Coe, M. 1978. The decomposition of elephant carcasses in the Tsavo (East) National Park, Kenya. *Journal of Arid Environments* 1: 71–86.

Coe, M.J., D.H. Cumming, and J. Phillipson. 1976. Biomass and production of large African herbivores in relation to rainfall and primary production. *Oecologia* 22:341–354.

Cooper, S.M. 1991. Optimal hunting group size: The need for lions to defend their kills against loss to spotted hyaenas. *African Journal of Ecology* 29: 130–136.

Cooper, S.M., K.E. Holekamp, and L. Smale. 1999. A seasonal feast: Long-term analysis of feeding behavior in the spotted hyaena (*Crocuta crocuta*). *African Journal of Ecology* 37: 149–160.

Creel, S. and N.M. Creel. 2002. *The African Wild Dog: Behavior, Ecology, and Conservation*. Englewood Cliffs, NJ: Princeton University Press.

Dean, W.R.J., S.J. Milton, and M.D. Anderson. 2006. Use of road kills and roadside vegetation by pied and cape crows in semi-arid South Africa. *Ostrich* 77(1–2): 102–104.

Dean, W.R.J. and S.J. Milton. 2003. The importance of roads and road verges or raptors and crows in the Succulent and Nama-Karoo, South Africa. *Ostrich* 74(3–4): 181–186.

dePastino, B. 2007. 10,000 Wildebeest drown in migration "pileup." National Geographic News. http://news. nationalgeographic.com/news/2007/10/071001-wildebeest.html (accessed September 1, 2013).

DeVault, T.L., O.E. Rhodes Jr, and J.A. Shivik. 2003. Scavenging by vertebrates: Behavioral, ecological, and evolutionary perspectives on an important energy transfer pathway in terrestrial ecosystems. *Oikos* 102(2): 225–234.

Dudley, J.P. 1998. Reports of carnivory by the common hippo *Hippopotamus amphibius*. *South African Journal of Wildlife Research* 28(2):58–59.

Duncan, C., A.L.M. Chauvenet, L.M. McRae, and N. Pettorelli. 2012. Predicting the future impact of droughts on ungulate populations in arid and semi-arid environments. *PLoS ONE* 7(12): e51490.

Du Toit, J.T. and D.H.M. Cumming. 1999. Functional significance of ungulate diversity in African savannas and the ecological implications of the spread of pastoralism. *Biodiversity and Conservation* 8: 1643–1661.

East, M.L. 1984. Rainfall, soil nutrient status and biomass of large African savanna mammals. *African Journal of Ecology* 22: 245–270.

Ellison, G.T.H. 1990. The effect of scavenger mutilation on insect succession at impala carcasses in southern Africa. *Journal of Zoology, London* 220: 679–688.

Eloff, F.C. 1964. On the predatory habits of lions and hyaenas. *Koedoe* 7: 105–112.

Estes, R.D. 1991. *The Behavior Guide to African Mammals*. Berkeley, CA: University of California Press.

Frank, L.G. 1996. Female masculinization in the spotted hyena: Endocrinology, behavioral ecology and evolution. In: Gittleman, J.L. (ed.), *Carnivore Behavior, Ecology and Evolution*. Ithaca: Cornell University Press, pp. 78–131.

Funston, P.J., M.G.L. Mills, and H.C. Biggs. 2001. Factors affecting the hunting success of male and female lions in Kruger National Park. *Journal of Zoology* 253: 419–431.

Gade, D.W. 2006. Hyenas and human in the horn of Africa. *Geographical Reviews* 96: 609–632.

Gasaway, W.C., K.T. Mossestad, and P.E. Stander. 1991. Food acquisition by spotted hyaenas in Etosha National Park, Namibia: Predation versus scavenging. *African Journal of Ecology* 29: 64–75.

Gbogbo, F. and V.P. Awotwe-Pratt. 2008. Waste management and hooded vultures on the Legon Campus of the University of Ghana in Accra, Ghana. *West Africa Vulture News* 58: 16–22.

Goss, R.A. 1986. The influence of food source on the behavioural ecology of brown hyaenas *Hyaena brunnea* in the Namib Desert. MSc Thesis, University of Pretoria, Pretoria.

Grant, J., C. Hopcraft, A.R.E. Sinclair, and C. Packer. 2005. Planning for success: Serengeti lions seek prey accessibility rather than abundance. *Journal of Animal Ecology* 74: 559–566.

Green, R.E., I. Newton, S. Shultz, A.A. Cunningham, M. Gilbert, D.J. Pain, and V. Prakash. 2004. Diclofenac poisoning as a cause of vulture population declines across the Indian subcontinent. *Journal of Applied Ecology* 41(5): 793–800.

Hanski, I. 1987. Nutritional ecology of dung- and carrion-feeding insects. In: Slansky Jr, F. and J.G. Rodriguez (eds), *Nutritional Ecology of Insects, Mites, and Spiders*. New York: John Wiley and Sons, pp. 837–884.

Harris, G., S. Thirgood, J.G.C. Hopcraft, J. Cromsigt, and J. Berger. 2009. Global decline in aggregated migrations of large terrestrial mammals. *Endangered Species Research* 7(1): 55–76.

Hartley, R.R., G. Bodington, Dunkley, and A. Groenewald. 1993. Notes on the breeding biology, hunting behavior, and ecology of the taita falcon in Zimbabwe. *Journal of Raptor Research* 27:133–142.

Haynes, G. 1988. Longitudinal studies of African elephant death and bone deposits. *Journal of Archaeological Science* 15(2): 131–157.

Henschel, J.R. and J.D. Skinner. 1990. The diet of spotted hyaenas *Crocuta crocuta* in Kruger National Park. *African Journal of Ecology* 28: 69–82.

Hofer, H., K.L.I. Campbell, M.L. East, and S.A. Huish. 1996. The impact of game meat hunting on target and non-target species in the Serengeti. In: Taylor, V.J. and N. Dunstone (eds), *The Exploitation of Mammal Populations*. London: Chapman & Hall, pp. 117–146.

Hofer, H. and M.L. East. 1993a. The commuting system of Serengeti spotted hyaenas: How a predator copes with migratory prey. II. Intrusion pressure and commuters' space use. *Animal Behaviour* 46: 559–574.

Hofer, H. and M.L. East. 1993b. The commuting system of Serengeti spotted hyaenas: How a predator copes with migratory prey: III. Attendance and maternal care. *Animal Behaviour* 46: 575–589.

Holdo, R.M., J.M. Fryxell, A.R. Sinclair, A. Dobson, and R.D. Holt. 2011. Predicted impact of barriers to migration on the Serengeti wildebeest population. *PLoS ONE* 6(1): e16370.

Holekamp, K.E. and S.M. Dloniak. 2010. Intraspecific variation in the behavioral ecology of a tropical carnivore, the spotted hyena. *Advances in the Study of Behavior* 42: 189–228.

Holekamp, K.E., L. Smale, R. Berg, and S.M. Cooper. 1997. Hunting rates and hunting success in the spotted hyaena. *Journal of Zoology, London* 242: 1–15.

Höner, O., K.E. Holekamp, and G. Mills. 2008. *Crocuta crocuta*. IUCN 2013 Red List of Threatened Species. Version 2013.1. www.iucnredlist.org (accessed August 1, 2013).

Höner, O., P. Wachter, M.L. East, and H. Hofer. 2002. The response of spotted hyenas to long-term changes in prey populations: Functional response and interspecific kleptoparasitism. *Journal of Animal Ecology* 71: 236–246.

Hopcraft, J.G.C., A. Sinclair, R.M. Holdo, E. Mwangomo, S. Mduma, S. Thirgood et al. 2013. Why are wildebeest the most abundant herbivore in the Serengeti ecosystem? In: Sinclair, A.R.E., S.A.R. Mduma, and C. Packer (eds), *Serengeti IV: Sustaining Biodiversity in a Coupled Human–Natural System*. Chicago: University of Chicago Press. In press.

Houston, D.C. 1974a. Food searching behaviour in griffon vultures. *East African Wildlife Journal* 12: 63–77.

Houston, D.C. 1974b. The role of griffon vultures *Gyps africanus* and *Gyps ruppellii* as scavengers. *Journal of Zoology* 172(1): 35–46.

Houston, D.C. 1979. The adaptations of scavengers. In: Sinclair, A.R.E. and M. Norton-Griffiths, (eds), *Serengeti: Dynamics of an Ecosystem*. Chicago: University of Chicago Press, pp. 263–286.

Houston, D.C. 1985. Evolutionary ecology of a frotropical and neotropical vultures in forests. *Ornithological Monographs* 36: 856–864.

Houston, D.C. and J.E. Cooper. 1975. The digestive tract of the whiteback griffon vulture and its role in disease transmission among wild ungulates. *Journal of Wildlife Diseases* 11(3): 306–313.

Hunter, J.S., S.M. Durant, and T.M. Caro. 2007. Patterns of scavenger arrival at cheetah kills in Serengeti National Park Tanzania. *African Journal of Ecology* 45: 275–281.

Janzen, D.H. 1977. Why fruits rot, seeds mold, and meat spoils. *American Naturalist* 111 (980): 691–713.

Jenkins, A.R. 2000. Hunting mode and success of African peregrines *Falco peregrines minor*: Does nesting habitat quality affect foraging efficiency? *Ibis* 142: 235–248.

Kahl, M.P. 1966. A contribution to the ecology and reproductive biology of the Marabou stork (*Leptoptilos crumeniferus*) in East Africa. *Journal of Zoology, London* 148(3): 289–311.

Kaunda, S.K.K. and J.D. Skinner. 2003. Black-backed jackal diet at Mokolodi Nature Reserve, Botswana. *African Journal of Ecology* 41: 39–46.

Kemp, A.C. 1993. Breeding biology of lanner falcons near Pretoria, South Africa. *Ostrich* 64: 26–31.

Kendall, C.J. 2012. Alternative strategies in avian scavengers: How subordinate species foil the despotic distribution. *Behavioral Ecology and Sociobiology* 67: 1–11.

Kendall, C.J., Virani, M.Z., and K.L. Bildstein. 2012. Assessing mortality of African vultures using wing tags and GSM-GPS transmitters. *Journal of Raptor Research* 46(1): 135–140.

Kendall, C.J., Virani, M.Z., Hopcraft, J.G.C., Bildstein, K.L., and D.I. Rubenstein. 2014. African vultures don't follow migratory herds: Scavenger habitat use is not mediated by prey abundance. *PLoS ONE* 9(1): e83470.

Kingdon, J. 1977. *East African Mammals Volume IIIA: Carnivores*. Chicago: University of Chicago Press.

Klare, U., J.F. Kameler, U. Stenkewitz, and D.W. Macdonald. 2010. Diet, prey selection, and predation impact of black-backed jacks in South Africa. *Journal of Wildlife Management* 74(5): 1030–1042.

Kolowski, J.M. and K.E. Holekamp. 2007. Effects of an open refuse pit on space use patterns of spotted hyenas. *African Journal of Ecology* 46: 341–349.

Kruuk, H. 1967. Competition for food between vultures in East Africa. *Ardea* 55: 171–193.

Kruuk, H. 1970. Interactions between populations of spotted hyaenas (*Crocuta crocuta* Erxleben) and their prey species. In: Watson, A. (ed.), *Animal Populations in Relation to their Food Resources*. Oxford: Blackwell Scientific Publications, pp. 359–374.

Kruuk, H. 1972. *The Spotted Hyena: A Study of Predation and Social Behavior*. Chicago: University of Chicago Press.

Kruuk, H. 1976. Feeding and social behaviour of the striped hyaena (Hyaena vulgaris Desmarest). *East African Wildlife Journal* 14: 91–111.

Leakey, L.N., Milledge, A.H., Leakey, S.M., Edung, J., Haynes, P., and D.K. Kiptoo. 1999. Diet of striped hyaena in northern Kenya. *African Journal of Ecology* 37: 314–326.

Ledger, H.P. 1968. Body composition as a basis for comparative study of some East African mammals. *Symposia of the Zoological Society of London* 21: 289–310.

Lindeque, M. 1981. Reproduction in the spotted hyaena, *Crocuta crocuta* (Erxleben). PhD Dissertation, University of Pretoria, South Africa.

Martin-Piera, F. and J.M. Lobo. 1996. A comparative discussion of trophic preferences in dung beetle communities. *Miscel-lània Zoológica* 19(1): 13–31.

Mduma, S.A., A. Sinclair, and R. Hilborn. 1999. Food regulates the Serengeti wildebeest: A 40-year record. *Journal of Animal Ecology* 68(6), 1101–1122.

Midgley, J.M., I.J. Collett, and M.H. Villet. 2012. The distribution, habitat, diet, and forensic significance of the Scarab (*Frankenbergerius forcipatus*) (Harold, 1881) (Coleoptera: Scarabaeidae). *African Invertebrates* 53(2): 745–749.

Mills, M.G.L. 1990. *Kalahari Hyaenas: The Comparative Behavioural Ecology of Two Species*. London: Unwin Hyman.

Moehlman, P.D. 1978. Jackals in the Serengeti. *Wildlife News* 13(3): 2–6.

Mundy, P., D. Butchart, J. Ledger, and S. Piper. 1992. *The Vultures of Africa*. Johannesburg: Acorn Press Russel Friedman Books.

O'Connell, J.F., K. Hawkes, and N.B. Jones. 1988. Hazda scavenging: Implications for Plio/Pleistocene hominid subsistence. *Current Anthropology* 29(2): 356–363.

Ogada, D.L. and F. Keesing. 2010. Decline of raptors over a three-year period in Laikipia, central Kenya. *Journal of Raptor Research* 44: 43–49.

Ogada, D.L., F. Keesing, and M.Z. Virani. 2012. Dropping dead: Causes and consequences of vulture population declines worldwide. *Annals of the New York Academy of Sciences* 1249: 57–71.

Ogutu, J.O., N. Bhola, and R. Reid. 2005. The effects of pastoralism and protection on the density and distribution of carnivores and their prey in the Mara ecosystem of Kenya. *Journal of Zoology, London* 265: 281–293.

Ogutu, J.O., H.P. Piepho, H.T. Dublin, N. Bhola, and R. Reid. 2009. Dynamics of Mara–Serengeti ungulates in relation to land use changes. *Journal of Zoology, London* 278: 1–14.

Ottichilo, W.K., J. de Leeuw, and H.H.T. Prins. 2001. Population trends of resident wildebeest [*Connochaetes taurinus hecki* (Neumann)] and factors influencing them in the Masai Mara ecosystem, Kenya. *Biological Conservation* 97: 271–282.

Owens, M.J. and D. Owens. 1978. Feeding ecology and its influence on social organization in brown hyaenas (*Hyaena brunnea*, Thunberg) of the central Kalahari Desert. *East African Wildlife Journal* 16: 113–135.

Owen-Smith, N. 1988. *Megaherbivores: The Influence of Very Large Body Size on Ecology*. Cambridge: Cambridge University Press.

Owen-Smith, N. 2013.Contrasts in the large herbivore faunas of the southerner continents in the late Pleistocene and the ecological implications for human origins. *Journal of Biogeography* 40(7): 1215–1224.

Owen-Smith, N. and D.R. Mason. 2005. Comparative changes in adult versus juvenile survival affecting population trends of African ungulates. *Journal of Animal Ecology* 74: 762–773.

Owen-Smith, N. and M.G.L. Mills. 2008. Predator–prey size relationships in an African large-mammal food web. *Journal of Animal Ecology* 77: 173–183.

Pennycuick, C.J. 1972. Soaring behavior and performance of some East African birds, observed from a motorglider. *Ibis* 114(2): 178–218.

Pennycuick, C.J. 1975. Mechanics of flight. *Journal of Avian Biology* 5: 1–75.

Pereira, L.M., N. Owen-Smith, and M. Moleón. 2014. Facultative predation and scavenging by mammalian carnivores: Seasonal, regional and intra-guild comparison. Mammal Review 44(1): 44–55.

Pienaar, U.D.V. 1969. Predator–prey relationships amongst the larger mammals of Kruger National Park. *Koedoe* 12: 108–176.

Poché, R.M., S.J. Evans, P. Sultana, M.E. Haque, R. Sterner, and M.A. Siddique. 1987. Notes on the golden jackal (*Canis aureus*) in Bangladesh. *Mammalia* 51:259–270.

Pomeroy, D.E. 1973. Birds as scavengers of refuse in Uganda. *Ibis* 117: 69–81.

Putnam, R.J. 1983. *Carrion and Dung: Decomposition of Animal Wastes*. Studies in Biology No. 156. London: Edward Arnold.

Ray, J.C. and M.E. Sunquist. 2001. Trophic relations in a community of African rainforest carnivores. *Oecologia* 127: 395–408.

Receveur, O., Kassi, N., Chan, H.M., Berti, P.R., and H.V. Kuhnlein. 1998. *Yukon First Nation's Assessment of Dietary Benefit/Risk*. Quebec, Canada: Centre for Indigenous Peoples' Nutrition and Environment.

Rensberger, J.M. 1999. Enamel microstructural specialization in the canine of the spotted hyena (*Crocuta crocuta*). *Scanning Microscopy* 13: 343–361.

Rensberger, J.M. and C. Stefen. 2006. Functional differentiation of the microstructure in the upper carnassial enamel of the spotted hyena. *Palaeontographica Abteilung A* 278: 1–6.

Rensberger, J.M. and X. Wang. 2005. Microstructural reinforcement in the canine enamel of the hyaenid *Crocuta crocuta*, the felid *Puma concolor* and the Late Miocene canid *Borophagus secundus*. *Journal of Mammalian Evolution* 12: 379–403.

Richards, C.S., B.W. Price, and M.H. Villet. 2009. Thermal ecophysiology of seven carrion-feeding blowflies in southern Africa. *Entomologia Experimentalis et Applicata* 131: 11–19.

Richardson, P.R.K. 1980. The natural removal of ungulate carcasses, and the adaptive features of the scavengers involved. MSc Thesis, University of Pretoria.

Schaller, G.B. 1972. *The Serengeti Lion*. Chicago: University of Chicago Press.

Schaller, G.B. and G.R. Lowther. 1969. The relevance of carnivore behavior to the study of early hominids. *Southwestern Journal of Anthropology* 25(4): 307–341.

Seibold, I. and A.J. Helbig. 1995. Evolutionary history of New and Old World vultures inferred from nucleotide sequences of the mitochondrial cytochrome *b* gene. *Philosophical Transactions of the Royal Society B: Biological Sciences* 350(1332): 163–178.

Shaw, K. 2008. Notes from Kenya: MSU Hyena Research. Michigan State University. http://msuhyenas. blogspot.com/search/label/hippo (accessed September 8, 2013).

Siegfried, W.R. 1965. A survey of wildlife mortality on roads in the Cape Province. Cape Department of Nature Conservation Investigational Report No. 6, Cape Town, South Africa.

Sillero-Zubiri, C. and M.D. Gottelli. 1992. Feeding ecology of spotted hyaena (Mammalia: *Crocuta crocuta*) in a mountain forest habitat. *Journal of African Zoology* 106: 169–176.

Sillero-Zubiri, C., M. Hoffmann, and D.D.W. Macdonald (eds), 2004. *Canids: Foxes, Wolves, Jackals and Dogs: Status Survey and Conservation Action Plan*, Vol. 62. Gland, Switzerland: International Union for the Conservation of Nature.

Sinclair, A.R.E. 1979a. Dynamics of the Serengeti ecosystem. In: Sinclair, A.R.E. and M. Norton-Griffiths (eds), *Serengeti: Dynamics of an Ecosystem*. Chicago, IL: University of Chicago Press, pp. 1–30.

Sinclair, A.R.E. 1979b. The eruption of the ruminants. In: Sinclair, A.R.E. and M. Norton-Griffiths (eds), *Serengeti: Dynamics of an Ecosystem*. Chicago, IL: University of Chicago Press, pp. 82–103.

Sinclair, A.R.E., S. Mduma, and J.S. Brashares. 2003. Patterns of predation in a diverse predator–prey system. *Nature* 425: 288–290.

Skinner, J.D., R.J. van Aarde, and R.A. Goss. 1995. Space and resource use by brown hyenas *Hyaena brunnea* in the Namib Desert. *Journal of Zoology, London* 237: 123–131.

Stander, P.E. 1991. Foraging dynamics of lions in a semi-arid environment. *Canadian Journal of Zoology* 70: 8–21.

Stefen, C. 1997. Differences in Hunter-Schreger bands of carnivores. In: Koenigswald, W.V. and P.M. Sander (eds), *Tooth Enamel Microstructure*. Rotterdam: Balkema Press, pp. 123–136.

Tappen, M. 1995. Savanna ecology and natural bone deposition: Implications for early hominid site formation, hunting, and scavenging. *Current Anthropology* 36: 223–260.

Tarboten, W.R. 1978. Hunting and the energy budget of the black-shouldered kite. *Condor* 80(1): 88–91.

Thiollay, J.M. 2006. The decline of raptors in West Africa: Long-term assessment and role of protected areas. *Ibis* 148: 240–254.

Tilson, R.L. and J.R. Henschel, 1986. Spatial arrangement of spotted hyaena groups in a desert environment, Namibia. *African Journal of Ecology* 24: 173–180.

Tilson, R.L., F. von Blottnitz, and J.R. Henschel. 1980. Prey selection by spotted hyaena (*Crocuta crocuta*) in the Namib Desert. *Madoqua* 12: 41–49.

Trinkel, M. and G. Kastberger. 2005. Competitive interactions between spotted hyenas and lions in Etosha National Park, Namibia. *African Journal of Ecology* 43: 220–224.

Tshikae, B.P., A.L.V. Davis, and C.H. Scholtz. 2008. Trophic associations of a dung beetle assemblage (Scarabaeidae: Scarabaeinae) in a woodland savanna of Botswana. *Environmental Entomology* 37(2): 431–441.

United Nations Environment Programme (UNEP). 2008. *Africa: Atlas of Our Changing Environment*. http:// www.unep.org/dewa/africa/africaAtlas.

United States Department of Agriculture (USDA), Agricultural Research Service. 2013. National nutrient database for standard reference release 26. http://ndb.nal.usda.gov/ndb/foods/show/4122? qlookup=13929& max=25&man=&lfacet=&new=1 (accessed September 6, 2013).

Van der Meer, E., M. Moyo, G.S.A. Rasmussen, and H. Fritz. 2011. An empirical and experimental test of risk and cost of kleptoparasitism for African wild dogs (*Lycaon pictus*) inside and outside a protected area. *Behavioral Ecology* 22(5): 985–992.

Van Valkenburgh, B. 1996. Feeding behavior in free-ranging large African carnivores. *Journal of Mammalogy* 77: 240–254.

Villet, M.H. 2011. African carrion ecosystems and their insect communities in relation to forensic entomology. *Pest Technology* 5(1): 1–15.

Virani, M.Z., C. Kendall, P. Njoroge, and S. Thomsett. 2011. Major declines in the abundance of vultures and other scavenging raptors in and around the Masai Mara ecosystem, Kenya. *Biological Conservation* 144: 746–752.

Wagner, A.P. 2006. Behavioral ecology of the striped hyena (*Hyaena hyaena*). PhD Dissertation, Montana State University, USA.

Wambuguh, O. 2008. Dry wildebeest carcasses in the African savannah: The utilization of a unique resource. *African Journal of Ecology* 46(4): 515–522.

Watts, H.E. and K.E. Holekamp. 2008. Interspecific competition influences reproduction in spotted hyenas. *Journal of Zoology, London* 276: 302–410.

Wells, M.P. 1989. The use of carcass data in the study of management of African elephants: A modeling approach. *African Journal of Ecology* 27: 95–110.

White, P.A. and C.G. Diedrich. 2012. Taphonomy story of a modern African elephant carcass on a lakeshore in Zambia (Africa). *Quaternary International* 276–277: 287–296.

Wiesel, I. 2006. Predatory and foraging behaviour of brown hyenas (*Parahyaena brunnea* (Thunberg, 1820)) at Cape Fur Seal (*Arctocephalus pusillus pusillus* Schreber, 1776) Colonies. PhD Dissertation, University of Hamburg, Germany.

Wiesel, I., G. Maude, D. Scott, and G. Mills. 2008. *Hyaena brunnea*. IUCN 2013 Red List of Threatened Species. Version 2013.1. www.iucnredlist.org (accessed August 1, 2013).

Wilson, E.E. and E.M. Wolkovitch. 2011. Scavenging: How carnivores and carrion structure communities. *Trends in Ecology & Evolution* 26(3): 129–135.

Wink, M. 1995. Phylogeny of old and new world vultures (Aves: Accipitridae and Cathartidae) inferred from nucleotide sequences of the mitochondrial cytochrome b gene. *Zeitschrift für Naturforschung C. Journal of Bioscience* 50(11): 868–882.

Wittemyer, G., D. Daballen, and I. Douglas-Hamilton. 2013. Comparative demography of an at-risk African elephant population. *PLoS ONE* 8(1): 1–10.

Wyman, J. 1967. The jackals of the Serengeti. *Animals* 10:79–83.

Yirga, G., H.H. De Iongh, H. Leirs, K. Gebrihiwot, J. Deckers, and H. Bauer. 2012. Adaptability of large carnivores to changing anthropogenic food sources: Diet change of spotted hyena (*Crocuta crocuta*) during Christian fasting period in northern Ethiopia. *Journal of Animal Ecology* 81(5): 1052–1055.

Zimmerman, D.A., D.A. Turner, and D.J. Pearson. 1996. *Birds of Kenya and Northern Tanzania*. Princeton: Princeton University Press, p. 740.

Section IV

Applications of Carrion Decomposition

22

Carrion Communities as Indicators in Fisheries, Wildlife Management, and Conservation

M.D. Hocking and S.M. O'Regan

CONTENTS

22.1 Introduction: Why Carrion-Dependent Species as Indicators in Conservation?

In traditional food-web ecology, animal carcasses are relegated to the detrital subweb, where insects, fungi, and microbes are represented as the only decomposers (Swift et al. 1979). Further, these decomposers are typically assigned to the second trophic position, as equivalent to primary consumers. However, scavenging of animal carcasses by omnivores and carnivores is a widespread and important behavior in both aquatic and terrestrial ecosystems and may be involved in up to 45% of food-web links (Wilson and Wolkovich 2011). Though it has yet to be regularly recognized as a distinct component of food webs (DeVault et al. 2003), carrion scavenging represents a unique and significant form of energy transfer between trophic levels. It allows scavengers access to high-quality food with no additional prey death and without expenditure of energy to capture prey (Wilson and Wolkovich 2011). Additionally, because scavengers feed on multiple prey species, they increase the level of branching in food webs, which suggests they may provide an important stabilizing force in ecosystems (var der Zanden and Vadeboncoeur 2002).

In this chapter, we present evidence from ecosystems across the globe that scavengers often occur at high trophic positions, can be sensitive to anthropogenic change, and thus are good candidate indicators in conservation. In Figure 22.1, we present a conceptual diagram, which presents the hypothesis of how the traditional food web or ecological pyramid supports a similar pyramid of detritivores and scavengers across a range of trophic levels. Because larger-bodied animals, predators, and species that occupy high trophic positions are often the most sensitive to human stressors throughout the globe, the carrion-dependent species that are involved in their decomposition can be useful as indicators of environmental change.

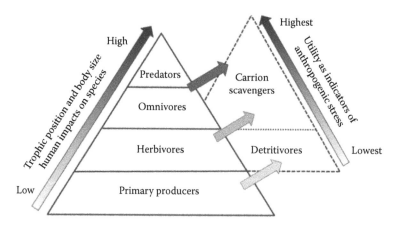

FIGURE 22.1 Conceptual diagram of the standard ecological pyramid of trophic position and biomass that supports a similar pyramid consisting of detritivores and scavengers. While many obligate scavenger species do not discriminate across carcasses of different types, some can occur at very high trophic positions because they specialize using the carcasses of species that occur high in the standard food web. Large-bodied species and predators are often the most impacted by human activities and thus scavengers responsible for their decomposition are good candidate indicators in conservation.

Many scavengers are generalists—omnivores and carnivores that can switch between food or prey types and consume carrion opportunistically (DeVault et al. 2003). In terrestrial systems, arthropods are often the most diverse and abundant organisms found at carrion, and among these, ants (Hymenoptera: Formicidae) are the most abundant generalist scavengers (Braack 1987). However, because of the unpredictable and ephemeral availability of carrion and its patchy distribution, arriving early at a carcass is key if species are to exploit the resource before it is exhausted (Barton et al. 2013). Indeed, unlike for predators, whose key constraint is the ability to capture prey, for scavengers it is detecting and moving to carrion (Ruxton and Houston 2003). As a result, many scavengers have evolved specialized life histories to more effectively compete for carrion as a food and reproductive resource.

Specialized traits associated with early colonization and spatial monopolization of carrion include large body size, flight capability, and keen sensory abilities. Large body size allows scavengers to consume more carrion when it is discovered and to carry greater reserves. Flying is faster than terrestrial movement, allows more ground to be searched rapidly, and increases the ability to encounter carrion across greater distances (Ruxton and Houston 2004). Barton et al. (2013) observed that beetle (Insecta: Coleoptera) species at carrion were larger and had higher wing loading than at control sites. By contrast, generalist ants showed no significant difference in body size between sampling sites and were not flight-capable, making them less able to be the first to exploit carrion resources. All beetle species belonging to the necrophagous Nicrophorinae subfamily of the Silphidae, are capable of flight, whereas in the other subfamily, Silphinae, loss of flight capability in some species may have accompanied the adoption of a carnivorous life history and the reduced need to search widely for food (Ikeda et al. 2006). *Onthophagus* (Coleoptera: Scarabaeidae) necrophagous dung beetles from Africa and Southeast Asia that appear to specialize in using freshly dead millipede (Diplopoda) carcasses for food are attracted to the repellent defensive secretions of the millipedes (Krell 2004).

Among terrestrial vertebrates, obligate scavengers are very rare (Houston 2001). Only two separate bird lineages (Cathartidae and Accipitridae), comprising 23 extant species of vulture, have evolved an obligate scavenging life history. All vulture species possess the energetically advantageous large body size and wing area and soar on air rather than flying by flapping (Ruxton and Houston 2004). These traits are not compatible with the ability to kill prey because predatory birds must be able to land accurately on a target, show agility at low flight speeds, and rapidly escape failed attacks on prey (Ruxton and Houston 2004). It is likely that there has been no similar strong selection pressure for mammals to become obligate scavengers because it is not necessary for mammalian predators to lose the ability to kill prey and also exploit carrion (Ruxton and Houston 2004).

Among the suite of obligate carrion scavengers, many species discriminate very little in carrion type, size, time of year or even habitat, while others have much more specialized niches and are thus more sensitive to changes in specific carrion resources. Flies (e.g., Diptera: Calliphoridae) are dominant members of many carrion communities, particularly in temperate regions, and although competition can structure niche differentiation among fly species on carrion (Kneidel 1984), when provided the opportunity many flies do not discriminate among carrion of different sizes or types (Kuusela and Hanski 1982). In contrast, *Nicrophorus* (Coleoptera: Silphidae) burying beetles often show preferences for carcasses of different sizes and specialized biparental or co-operative breeding to compete with the competitively dominant flies (Scott 1998; Hocking et al. 2006). Either way, like top predators, these obligate carrion scavengers often occupy high trophic positions due to animals constituting a large proportion of their diets (e.g., Nyssen et al. 2005; Bennett and Hobson 2009). Therefore, because of their trophic position and dependence on carrion, specialized scavenges can be sensitive to anthropogenic disturbances that affect the availability of carrion resources.

The objectives of this chapter are threefold: (1) to discuss the important role of scavengers in ecosystems and how they may represent ideal indicators of ecosystem health; (2) to illustrate how information about carrion-dependent species and communities can be applied in fisheries, wildlife management, and conservation; and (3) to explore how anthropogenic change such as global warming and overfishing may impact carrion communities and their ability to sustain the ecological services they perform. The following sections show examples of these phenomena across a range of ecosystems (also see Table 22.1).

22.2 Carrion- and Dung-Dependent Communities as Indicators of Habitat Fragmentation and Loss

Forest loss is accelerating at a rate that threatens biodiversity across the globe. It is increasingly imperative that researchers and managers be able to detect early changes in the structure and function of forest communities in response to disturbance (Andresen 2003), predict impacts of anthropogenic land-use on native biota (Gardner et al. 2009), and identify regions to prioritize for conservation (Wilson et al. 2006). Collecting data on the distribution, richness, or levels of endemism of species underpins each of these tasks; however, it is impossible to comprehensively survey all species, particularly in tropical regions. An alternative approach to gathering information on the ecological consequences of human activities is to identify and monitor ecological disturbance indicators (Gardner 2010)—specific taxa that provide relevant measures of the ecological integrity of an ecosystem with less expense and effort.

A candidate ecological disturbance indicator group must satisfy three basic criteria (Gardner 2010): (1) the field, laboratory, and taxonomic expertise must exist to make them *viable* for study; (2) they must provide *reliable* responses to gradients of anthropogenic disturbance at spatial scales relevant to management and be measurable using a cost-effective standardized protocol; and (3) there must be sufficient ecological knowledge about the relationship between observed patterns of abundance and occupancy and changes in environmental variables must exist so as to make the data *interpretable*.

Nicrophorine carrion beetles (Coleoptera: Silphidae) and Scarabaeine dung beetles (Coleoptera: Scarabaeidae) are strong candidates for use as ecological disturbance indicators in tropical forest ecosystems (e.g., Halffter and Favila 1993; Spector 2006; Nichols and Gardner 2011). They are *viable* indicators because they may be representatively sampled within a few days using standardized protocols, primarily with inexpensive baited pitfall traps (Spector 2006). Further, they are taxonomically well defined and have a broad geographic distribution, with more than 5900 dung beetle species described across all continents (other than Antarctica) (ScarabNet 2013).

Carrion and dung beetles are *reliable* indicators of environmental disturbance for three reasons. First, they are sensitive to natural environmental gradients, displaying different patterns of occurrence and beetle assemblage structure at both regional and local spatial scales, which are likely in response to underlying climatic and edaphic preferences (Nichols and Gardner 2011). For example, in Borneo, dung beetle assemblages showed clear differences in species composition in riverine and interior rainforest, but also fine-scale subdivision within riverine forest among river-edge, river-bank, and riverine nonedge/bank microclimates (Davis et al. 2001). Second, they show different species-response patterns

TABLE 22.1

Links between Anthropogenic Disturbance, Responses of Carrion Dependent Communities, and Their Use as Potential Indicators in Conservation from Ecosystems Throughout the Globe

Disturbance Type	Habitat	Carrion Community Affected	Responses or Effects	Indicator of
Habitat fragmentation, habitat loss	Tropical forests, temperate forests	Carrion and dung beetles	Altered carrion species composition, shift to more generalist species	Biodiversity loss (e.g., native mammals and birds), altered nutrient cycling, and rates of decomposition
Loss of apex predators	Grasslands, forests, marine ecosystems	Vertebrate and invertebrate scavengers	Mesopredator release, altered carrion input, and distribution, changes in carrion communities, changes in carrion stoichiometry	Altered trophic pathways, altered ecosystem functioning
Climate change	Terrestrial and marine ecosystems	Temperate zone scavengers, abyssal carrion communities, polar carrion communities	Altered carrion species composition, abundance, and distribution	Increasing temperatures, changing marine conditions, altered trophic pathways, altered ecosystem functioning, altered marine to terrestrial resource subsidies
Overfishing, fisheries bycatch	Marine ecosystems, islands, beaches, coastal temperate watersheds	Abyssal carrion communities, terrestrial vertebrate, and invertebrate scavengers	Altered carrion species composition and distribution, declines in scavenger abundance, changes in reproductive timing and behavior	Declines in fish biomass (particularly large fish), declines in Pacific salmon abundance, altered marine to terrestrial resource subsidies, localized increases in bycatch mortality
Competition with non-native species	Forests, grasslands	Carrion beetles, vertebrate, and invertebrate scavengers	Altered carrion species composition, introduced and more generalist scavengers outcompete native species	Altered trophic pathways, altered ecosystem functioning

to land-use change and disturbance. A meta-analysis of 26 tropical dung beetle studies revealed that total species richness declines relative to intact forest in every type of disturbed habitat investigated (selectively logged forest, secondary forest and agroforests, to tree plantations, annual crops, pasture, and clear-cuts) (Nichols et al. 2007). In another study by Díaz et al. (2010), 87% of forest species in Los Tuxtlas, Mexico, were unable to leave forest fragments, indicating that there is a strong edge effect between forest and pasture. However, responses to disturbance differ between beetle assemblages (Díaz et al. 2010; Caballero and León-Cortés 2012) and there is ongoing debate by researchers on whether forests regenerated on degraded land can effectively offset species loss from primary forests or support dung beetle communities with similar community attributes (Nichols et al. 2007; Gardner et al. 2008). It is likely that other rarely measured variables, such as fragment age, changes in vegetation structure, and carrion or dung availability, also play a role in determining beetle response to fragmentation (Nichols et al. 2007).

Third, carrion and dung beetles provide a valuable gauge of ecological integrity because they are specialized scavengers and are sensitive to resource availability. Carrion beetles feed and reproduce primarily on the carcasses of small birds and mammals (Gibbs and Stanton 2001), and dung beetles feed on the microorganism-rich liquid component of mammalian dung (and less frequently that of other vertebrates, as well as rotting fruit, fungus and carrion) and use the more fibrous material to brood their larvae (Nichols et al. 2008). Land-use conversion associated with agriculture and exploitation negatively impacts mammals and vertebrates (Corlett 2007; Peres and Palacios 2007) upon which carrion and dung beetles depend. *Circellium bacchus*, a large, flightless, dung beetle that prefers elephant dung for feeding and buffalo dung for nesting has disappeared almost entirely from its southern African range as a combined consequence of habitat loss and replacement of native herbivores with livestock (Scholtz et al. 2006). In conjunction with direct effects of habitat loss and fragmentation, generalist scavengers that compete with dung and carrion beetles for resources may be more abundant in fragmented landscapes. In Central Amazonia, forest fragments have larger populations of rodents than in continuous forest. As seed predators, rodents compete with beetles for the dung of frugivores to consume the seeds it contains (Andresen 2003). However, these processes also occur in temperate forests. For example, in Syracuse, New York, USA, carrion beetle, *Nicrophorus* sp. (Coleoptera: Silphidae), species richness and abundance in fragmented forests was two-thirds and one-third, respectively, that of large forest tracts (Gibbs and Stanton 2001; Figure 22.2). The authors attributed these declines to habitat loss, drier microclimates in fragmented forests, soil changes—especially compaction and loss of organic matter, which are common to fragmented forests and may hinder the ability of beetles to bury carcasses or to bury them to a sufficient depth as to keep them from other scavengers—and competition with exotic earthworms. Earthworms are generalist scavengers, which are 10 times more abundant in the soil of fragmented forests than large forest tracts in the northeastern United States.

Knowledge of carrion and dung beetle ecology and the two drivers of change in beetle assemblages—changes in vegetation and associated microclimates, and changes in food and breeding resources—increases ecologists' ability to interpret changes in community structure and translate them into an understanding of human impacts on ecological processes and ecological integrity (Nichols and Gardner 2011). Carrion and dung beetles drive numerous ecological processes through their consumption and relocation of these detrital resources (as reviewed in Nichols et al. 2008). For instance, by burying dung below the ground, dung beetles may enhance soil fertility and plant growth by increasing nitrogen availability and bioturbating soil and are often key pollinators of decay-scented flowers (Andresen 2003). They also participate in secondary seed dispersal by accidentally burying seeds present in dung, which

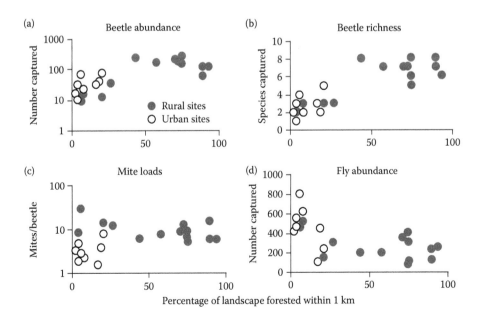

FIGURE 22.2 Trends in communities of carrion beetles and their associates along a forest fragmentation gradient, Onondaga County, New York. (From Gibbs, J.P. and E.J. Stanton. 2001. *Ecological Applications* 11: 79–85.)

reduces seed predation and mortality due to pathogens and provides seeds with favorable microclimates for germination (Andresen 2003). Species-level life-history traits such as beetle body size and nesting strategy strongly determine the role species play in these processes. For instance, evidence suggests large-bodied dung beetles consume disproportionately more dung (Nichols et al. 2008) and bury more seeds (Andresen 2003) than smaller species. Importantly, the mean size of beetles increases with increasing forest area (Andresen 2003). Loss of beetle biodiversity with habitat loss and fragmentation and the loss of large mammals may thus precipitate loss of functional capacity (Slade et al. 2007; Nichols et al. 2009).

In summary, the dependency of carrion and dung beetles on intact vertebrate communities and specific microhabitats, in addition to the key services they perform, means that these species are excellent candidate indicators in forest management and biodiversity conservation (Spector 2006; Nichols and Gardner 2011; Table 22.1).

22.3 Declining Populations of Large Terrestrial Vertebrate Predators and the Implications for Carrion-Dependent Species

Across the globe, there have been major declines in the populations of large predators such as wolves, *Canis lupus*, jaguars, *Panthera onca*, lions, *Panthera leo*, and wild dogs, *Lycaon pictus*, due to habitat loss and direct predator culls to alleviate concerns about human and livestock safety (Prugh et al. 2009). In many regions, large terrestrial mammalian carnivores have now declined by 95%–99% (Berger et al. 2001). As a result, smaller predators whose abundance is controlled through direct lethal encounters with apex predators, as well as through adjustments in their behavior and distribution to avoid direct encounters (Ritchie and Johnson 2009), have expanded in distribution and density in what is termed mesopredator release (Brashares et al. 2010). Prugh et al. (2009) calculated that although all North American apex predator ranges have contracted, 60% of mesopredator ranges have expanded. However, mesopredators promoted to the top of food chains are not ecologically identical to the extirpated apex predators, but have different, cascading effects on ecosystems (Prugh et al. 2009). For example, the failure to consider the importance of predators in suppressing mesopredators has triggered the collapse of marine ecosystems (e.g., Myers et al. 2007). In this section, we discuss how communities of carrion-dependent species can be useful indicators of these shifts in top predator species.

The abundance of large vertebrate predators in terrestrial ecosystems strongly influences the spatial and temporal availability of carrion resources for scavengers. In Yellowstone National Park, Wyoming, USA, gray wolves, *Canis lupus*, were reintroduced 10 years ago. They now subsidize 13 species of scavengers during the late fall, winter, and early spring by provisioning elk, *Cervus elaphus*, carcasses that in the absence of wolves had been abundant only during times of deep snow (Wilmers et al. 2003a,b). In Chile's Aysén District in central Patagonia, Elbroch and Wittmer (2012) have documented 12 vertebrate scavengers at puma kills, including the threatened Andean condor, *Vultur gryphus*, a carrion-dependent species.

Large, solitary felids play a particularly unique keystone role in scavenger communities (Elbroch and Wittmer 2012). Although wolves hunting in larger packs may efficiently consume carcasses before scavengers arrive, and will defend their kills from competitors (Smith et al. 2006), solitary felids generally retreat to minimize conflicts with competitors (Ruth and Murphy 2010) and are thus quite susceptible to food stealing by dominant scavengers (Elbroch and Wittmer 2012; Krofel et al. 2012). Food stealing, or kleptoparasitism, is a ubiquitous behavior across taxa and ecosystems (Iyengar 2008) and can have a large impact on both apex predators and dominant scavengers. In the Holarctic region, brown bears, *Ursus arctos*, are the largest terrestrial scavengers and important kleptoparasites of Eurasian lynx, *Lynx lynx* (Krofel et al. 2012). Eurasian lynx are specialized predators of small ungulates and have a feeding time that lasts many days. Brown bears find about one-third of lynx deer kills and as lynx are only able to compensate for 59% of their losses by increasing kill rate, bear kleptoparasitism may have pronounced effects on lynx populations already threatened by anthropogenic factors. In addition, because bears consume carcasses so rapidly, scavenger community structure in areas with high bear densities can be impacted by decreases in the availability of carrion (Krofel et al. 2012).

As the populations of apex predators decline throughout the globe, the subsequent release of meso-predators has consequences for other species that rely on carrion. The American burying beetle, *Nicrophorus americanus* (Coleoptera: Silphidae), an endangered species in the United States, has likely declined in large part due to the population release of raccoons and other mesopredators and the success of these mesopredators in monopolizing carrion in highly altered landscapes (DeVault et al. 2011). Human development has aggravated the effects of mesopredator release in fragmented habitats by adding to the resources available to mesopredators, such as trash, crops, and livestock (Prugh et al. 2009). For example, coyotes, *Canis latrans*, are major predators of several vertebrates listed as threatened and endangered by the U.S. Fish and Wildlife Service, and in the absence of gray wolves, they place intense pressure on their typical prey. In the American West, coyote predation may have been a key driver of the dramatic decline in leporids (rabbits and hares) (Ripple et al. 2013). Livestock carrion available for coyotes to scavenge has been increasing in this region and has likely helped maintain coyote abundance.

Apex predators not only subsidize resources for diverse scavengers and keep mesopredator abundance in check, but community stability theory predicts such keystone species (Power et al. 1996) and high levels of species diversity (Paine 1969) may also be key to maintaining resilient communities in the face of climate change. As global surface temperatures increase, northern latitude and high elevation areas will experience shorter, milder winters and earlier, more rapid snow melt. In Yellowstone, carrion from elk mortality without wolves was plentiful during severe winters, but smaller snow packs and earlier melt means that elk are less stressed, have easier access to food, and have decreased energy expenditure (Wilmers and Getz 2005). Scavengers that rely on late winter carrion such as grizzly bears, *Ursus arctos*, will therefore experience the greatest decline in carrion resources (Wilmers and Post 2006; Figure 22.3). Reintroduced wolves are now the primary source of winter carrion and buffer the effects of climate

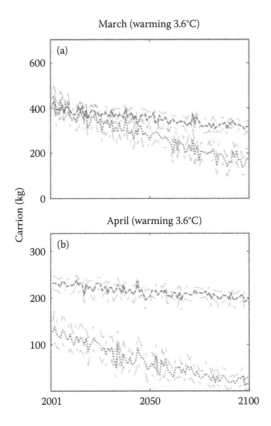

FIGURE 22.3 Simulated average March and April carrion abundances from 2000 to 2100 in scenarios with (dashed line) and without (dotted line) wolves assuming moderate levels of global warming. Light grey lines represent standard errors. (Modified from Wilmers, C.C. and E. Post. 2006. *Global Change Biology* 12: 403–409.)

change on the timing and abundance of carrion available to carrion-dependent species (Wilmers and Post 2006). For bears, scavenging of prey from felid predators appears to be especially important in spring, when little other food is available and their energy stores are depleted after hibernation. This may become even more important with climate change (Krofel et al. 2012), as bear denning periods are predicted to shorten (Pigeon 2011).

In summary, the composition of communities that depend on large vertebrate carrion can be a useful indicator of the composition of the apex predator communities that generate this carrion. Moving forward, large predator management will need to consider not only the cascading effects of apex predator removal or reintroduction on prey species, but also incorporate the strength and structure of interactions among apex predators, mesopredators, prey species, and scavengers (Prugh et al. 2009; Table 22.1).

22.4 Abyssal Carrion Communities and Relationships to Fisheries and Climate Change

Abyssal ecosystems cover a massive 54% of the Earth's surface (Smith et al. 2008). They are characterized by an absence of *in situ* primary production (except at rare hydrothermal vents and cold seeps). As a result, abyssal communities are considered to be resource-limited because benthic production depends on the input of organic matter produced in the euphotic zone thousands of meters above (Smith et al. 2008). This organic matter primarily consists of phytodetritus, but also of larger epipelagic nekton carcasses.

Top predators at abyssal depths are primarily fish, and similar to terrestrial predators, they may exploit two trophic pathways: (1) phytodetritus is consumed by deposit feeders, which are then consumed by primary carnivores, and so on up the food chain; and (2) scavenging on the allochthonous supply of epipelagic nekton carcasses (Drazen et al. 2008). Abyssal predators often bypass the classic phytodetrital food web and are strongly reliant on epipelagic nekton for their nutrition (Drazen et al. 2008; Yeh and Drazen 2009; Boyle et al. 2012; Figure 22.4). In oligotrophic regions, these scavenging communities appear to be dominated by ophidiid fish, and in eutrophic regions, by macourid fish (Yeh and Drazen 2009; Fleury and Drazen 2013).

There are fundamental differences between how carcasses are cycled through abyssal and terrestrial ecosystems (as reviewed in Beasley et al. 2012). First, terrestrial and marine food webs have different spatial dynamics. Although carrion on land remains relatively unmoved throughout the decomposition and scavenging process, water allows for carrion to move, spreading the activities associated with scavenging across a larger, three-dimensional scale. Additionally, carcasses sink rapidly to depths inaccessible to pelagic scavengers, and this added vertical dimension imposes a time limit on the length of time they may feed on the carcass. This has facilitated the evolution of a diverse assemblage of benthic scavengers. Second, in terrestrial systems, warm weather reduces scavenging rates by terrestrial vertebrates and maximizes the activity of insects and bacteria. By contrast, temperature fluctuations in abyssal systems are minimal. Therefore, carrion communities are not regulated by climate to the degree that they are on land and carcasses can provide food to vertebrates for years rather than days. There is perhaps no better example of this than a whale-fall ecosystem.

Whale carcasses represent a huge, but initially localized, food fall to the abyssal sea floor. The organic carbon contained in a 40-ton whale ($\sim 2 \times 10^6$ g C) is equivalent to that which usually sinks from the euphotic zone to a hectare of abyssal sea floor over 100–200 years (Smith and Baco 2003). Time-series studies of natural and experimentally planted whale carcasses have demonstrated that a single carcass can support an amazing diversity of carrion-dependent species (nearly 200 species) over three successional stages (as reviewed in Smith and Baco 2003): (1) the *mobile-scavenger stage*—during which soft tissue attached to the carcass is removed by species of successively smaller body size (e.g., sleeper sharks, hagfishes and macrourids, then amphipods and crabs, and ending with calanoid copepods). For megafauna at 1200–1800 m off California, this stage lasts 4–5 months to 1.5–2.0 years, depending on carcass size; (2) the *enrichment-opportunist stage*—where dense assemblages of heterotrophic macrobenthos (dominated by polychaete worms *Vigtorniella* sp. and *Ophryotrocha* sp., and cumacean *Cumella* sp.) colonize the organic-rich sediments and bones; and (3) the *sulfophilic stage*—after removal of the soft tissue by scavengers, anaerobic microbes decompose the large lipid reservoirs within the bones,

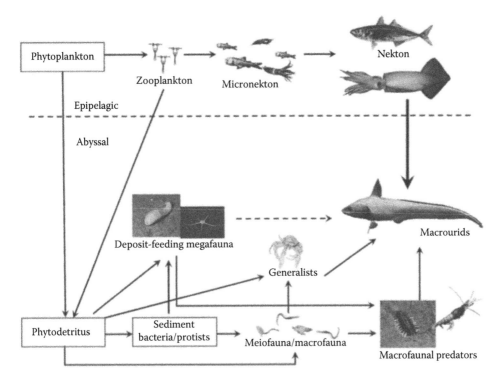

FIGURE 22.4 Simplified trophic pathways from primary production to macrourid fishes. The classical phytodetrital pathway and its bypass by the allochthonous input of carrion are illustrated. For pathways leading to the macrourids, more important paths are indicated by thick arrows and minor ones are indicated by dotted lines. (From Drazen, J.C. et al. 2008. *Limnology and Oceanography* 53: 2644–2654.)

which may constitute roughly 5%–8% of the total whale body mass. This process generates an efflux of sulfide from the bones, which attracts species that can exploit sulfide-based chemoautotrophic production or that can tolerate elevated sulfide concentrations. A number of these species may be whale-fall specialists as they have rarely, or never, been seen in other habitats.

Because of the strong link between abyssal populations and epipelagic nekton, seasonal and interannual fluxes in nekton result in altered sediment community activity and macrofaunal abundance, biomass, and changes in the average size and trophic status of taxa (Ruhl et al. 2008). At Station M, a long-studied abyssal site in the eastern North Pacific, Drazen et al. (2012) found that an increase in abundance and size of grenadiers over a 15-year period correlated with yearly spawning stock biomass of Pacific hake (a proxy for carrion flux), the most abundant nekton species in the region and the target of the largest commercial fishery off the west coast. In the northeast Pacific, changes in the abundance of mobile epibenthic megafauna, which inhabit the sediment–water interface and contribute significantly to benthic respiration and biogeochemistry (Bergmann et al. 2011) are also correlated to climate fluctuation dominated by El Niño/La Niña (Ruhl and Smith 2004). This growing body of literature demonstrates that fishing and climate change may have important impacts on the abyssal ecosystem.

Fishing and climate change have been reported to be associated with reduced populations and altered the community composition of epipelagic nekton (Brodeur et al. 2003; Sibert et al. 2006). Commercial fishing of predators produces carrion as a byproduct that is left in the marine ecosystem. Levels of bycatch can be extremely high, resulting in millions of tons of carrion added to oceans annually (Hall 1996). Moreover, fisheries that harvest only part of animals, such as shark-fin fisheries, leave the remaining biomass for scavengers. In the short term, this can increase carrion flux to abyssal communities (Catchpole et al. 2006; Kaiser and Hiddink 2007). However, as the average trophic level of fisheries continues to decline and the highest quality biomass is removed (Pauly et al. 1998), it decreases the quality of naturally occurring carrion in systems (Wilson and Wolkovich 2011). Further, the nonrandom spatial distribution of bycatch discards may affect scavenger communities and nutrient cycling (Hall 1996). Twentieth-century whaling has altered the rates and geographic distribution of whale falls (Smith and Baco 2003) and may have caused extinctions of specialist fauna dependent on whale carrion (Smith et al. 2006). Changes in the size distribution and location of whale falls due to whaling can also alter abyssal community composition. For example, a minimum degree of calcification is required for a whale skeleton to sustain chemoautotrophic communities for years. Therefore, although large whale skeletons may support sulphophilic communities for at least 50 years, juvenile whales have smaller-sized bones and bone-lipid reservoirs and do not sustain such complex trophic structure (Smith and Baco 2003; Lundsten et al. 2010).

Climate change will induce a host of biogeochemical changes in oceans that will alter the availability of food to abyssal ecosystems. Increased sea-surface temperatures and thermal stratification, coupled with reductions in nutrient upwelling (Gregg et al. 2003; Fischlin et al. 2007) are expected to reduce primary production in the euphotic zone and the resulting quality and quantity of organic matter that sinks to the seafloor (Smith et al. 2008). Evidence of decreasing food supplies to megafaunal organisms has been found at the Arctic deep-sea observatory HAUSGARTEN. At this observatory, Bergmann et al. (2011) observed a decrease in megafaunal densities correlated with an increase in bottom-water temperatures and a decline in the total organic content and microbial biomass of the sediments. Epipelagic nekton biomass may decline in abundance with reduced primary production, but also shift in distribution because their mobility allows them to seek resources elsewhere (Perry et al. 2005; Polovina et al. 2011). As a result, scavenger community composition may change (Drazen et al. 2012) and the provision of ecosystem services by the abyss, such as nutrient recycling, carbon burial, and dissolution of calcium carbonate, may be impacted (Snelgrove et al. 2004).

In summary, the composition and distribution of scavenging communities in the deep sea can indicate much about the state of our fisheries, the impacts of climate change, and, ultimately, the health of the world's oceans (Table 22.1).

22.5 Carrion Communities as Indicators of Resource Subsidies across Ecosystem Boundaries

Many animals possess complex life cycles that include life-stage transitions across ecosystem boundaries. The death of these animals can represent an important transport of resources between ecosystems, with potentially strong effects on consumers (Hoekman et al. 2011). For example, in northeast Iceland, adult midges (Diptera: Chironomidae) synchronously emerge from lakes in a massive pulse after having completed their aquatic larval and pupal development (Hoekman et al. 2011). Moving on land to mate, many die and become food for terrestrial arthropods. Dreyer et al. (2012) found arthropod abundance along lakeshores was positively correlated with an increasing gradient of midge emergence (Figure 22.5). Lower-order omnivorous predators, such as spiders, Opiliones, and beetles (Coleoptera: Staphylinidae), showed the strongest positive responses to midges, perhaps because they were able to take advantage of both living and dead midges. Experimentally simulating the input of dead midges on a lakeshore, Hoekman et al. (2011) also found consistent increases in terrestrial arthropod abundance. Here, strong numerical responses primarily by detrivorous Diptera and soil microarthropods to midge deposition persisted for more than 1 year.

Carrion subsidies can allow species to become more abundant than they would be if supported only by *in situ* production and stabilize populations by allowing them to persist even when *in situ* production fluctuates (Polis et al. 2004; Bicknell et al. 2013). For small coastal islands, marine biomass in the form of carcasses of animals that live in the sea but also use land, such as sea turtles and seabirds, and marine carrion that has drifted to shore, can make up a large component of the terrestrial energy budget (Polis et al. 2004). These resources support large populations of opportunistic intertidal and supralittoral detrivores and predators, who may also play a significant role as scavengers in exchanging nutrients across ecosystem boundaries (Laidre 2013). Polis et al. (2004) found supralittoral ground and flying arthropods can be 10 and 100 times more abundant, respectively, than their arthropod counterparts of

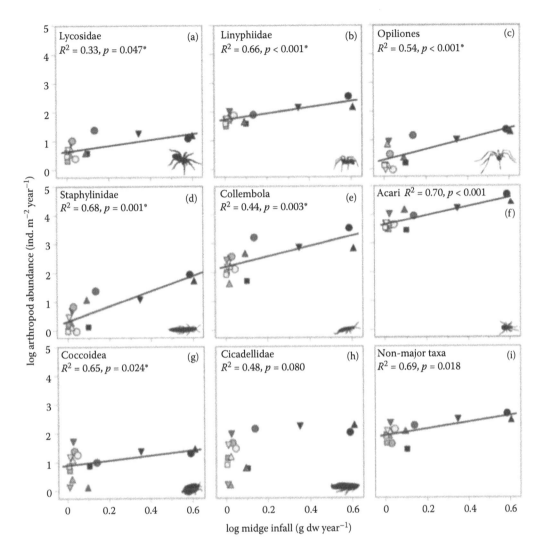

FIGURE 22.5 Relationship of terrestrial arthropod (average no m^{-2} year^{-1}) taxa collected around four northern Icelandic lakes and midge "infall" (g dw year^{-1}) on land. (From Dreyer, J., D. Hoekman, and C. Gratton. 2012. *Oikos* 121: 252–258.)

inland regions. In Baja California, Mexico, fish, invertebrates, seabirds, and marine mammals such as California sea lions, *Zalophus californianus*, common dolphin, *Delphinus delphis*, and the bottlenose dolphin, *Tursiops truncatus* regularly drift ashore. As a result, coastal coyote populations subsidized by these resources reach densities 2.4–13.7 times greater than those in adjacent inland areas that do not receive marine input (Rose and Polis 1998). Subsidies may in turn have cascading effects on (1) food webs, for example, as increased consumer densities depress terrestrial resources (Rose and Polis 1998); (2) competitive dynamics, for example, by increasing the abundance of generalists that also use terrestrial resources and reducing the availability of terrestrial resources for those that exclusively consume them; and (3) community diversity, particularly on islands by impacting probabilities of consumer persistence and extinction (Polis et al. 2004).

Large carrion resource pulses not only have strong effects on consumer biomass and community assemblage, but can deliver large inputs of limiting nutrients such as nitrogen and phosphorus to the soil once aboveground scavengers are satiated (Dreyer et al. 2012). In North American deciduous forests, periodical cicadas (Hemiptera: Cicadidae) are the most abundant herbivores in both number and biomass. After 17 years belowground, adults emerge, mate, and die in huge pulses that litter the forest floor in carcasses. Field experiments by Yang (2004) in which cicada carcasses were applied to forest plots demonstrated that the release of nitrogen into the ecosystem from the cicada decomposition process can be rapidly used in plant growth and reproduction. In another example from grasslands, Hawlena et al. (2012) show that the carcasses of a dominant herbivore, grasshoppers, alter belowground ecosystem functioning by increasing soil nutrient concentrations, plant growth, and plant litter decomposition. However, the protein content of these grasshopper carcasses decreases when top predators, spiders, are abundant, which then retards belowground ecosystem functioning compared to when spiders are absent (Hawlena et al. 2012).

These examples show how resource subsidies across habitat boundaries affect consumers including scavengers in recipient habitats and how the functioning of these recipient communities can change as a result. Carrion communities supported by resource subsidies can thus be an indicator of local dynamics and community structure (Rose and Polis 1998; Polis et al. 2004). However, these carrion communities should also be useful as indicators of changes in the *donor* habitat. Although this has not often been tested, a broad example is described in the previous section on abyssal carrion communities, as these are essentially benthic indicators of change in the marine pelagic food web (Drazen et al. 2008; Smith et al. 2008). In fact, environmental impacts in one habitat may be propagated to another, where the abundance of recipient consumers could be used as an indicator of disturbance in the donor food web. One example for this process comes from streams where chronic stream pollution reduces insect emergence and subsequently the abundance of top predators in riparian food webs (Paetzold et al. 2011). Another classic example more relevant to scavengers as indicators involves populations of anadromous Pacific salmon, *Oncorhynchus* spp., whose carcasses subsidize numerous terrestrial species. Changes in the functioning of marine food webs through climate change and/or overharvesting of salmon is predicted to translate into shifts in the functioning of terrestrial food webs (e.g., Darimont et al. 2010; Hocking and Reynolds 2011; Table 22.1). This remarkable resource subsidy is discussed further in the next section.

22.6 The Case Study of Pacific Salmon Carcasses: An Essential Resource for Terrestrial Scavengers

Millions of Pacific salmon weighing 1 to over 20 kg each return from the ocean to spawn in hundreds of thousands of lakes, streams, and tributaries throughout the Pacific Rim from Japan and Russia through Alaska to California, USA. Salmon are born in freshwater, migrate as juveniles to the ocean where they grow for 2–5 years and migrate thousands of kilometers before they return to their natal streams to spawn and die. This incredible migration deposits millions of kilograms of marine-derived biomass into freshwater and terrestrial food webs each year and alters terrestrial ecosystem functioning (Janetski et al. 2009; Hocking and Reynolds 2011; Schindler et al. 2013).

Live and dead salmon provide a predictable pulsed resource for hundreds of freshwater and terrestrial species of vertebrates and invertebrates (Willson and Halupka 1995; Cederholm et al. 1999; Gende et al. 2002; Hocking et al. 2009). Most of these species that utilize salmon as food scavenge on salmon carcasses or eggs or can switch between predating on live salmon or scavenging the carcasses. This means that although some dominant predators like bears (*Ursus* spp.) can exert strong selective pressures on salmon (Carlson et al. 2007), most of the food web links are scavenging links or include predation on fish that have already spawned and thus exert little selective influence on salmon but often strong effects on salmon consumers. For example, the reproductive cycles and distribution of many of these species are tied to the seasonal pulse of salmon. In female mink, *Mustela vison*, the timing of lactation is shifted to coincide with the seasonal availability of salmon, which provides a selective advantage for offspring survival (Ben-David 1997). The energy requirements for hibernation in coastal bears are often hugely dependent on salmon, with some studies indicating that 90% or more of dietary protein in bears comes from salmon (Hilderbrand et al. 1999). Many species of coastal birds also depend on salmon, including gulls, corvids, ducks, raptors, and even songbirds (e.g., Christie and Reimchen 2008). The distribution, abundance, and reproductive success of bald eagles, *Haliaeetus leucocephalus*, is affected by the availability of salmon (Hansen 1987; Cederholm et al. 1999). In streams, salmon carcasses and dislodged dead eggs supplement the diets of a range of aquatic invertebrates and stream fish such as sculpins, *Cottus* spp., Cutthroat trout, *O. clarki*, steelhead, *O. mykiss*, and juvenile coho salmon, *O. kisutch* (Bilby et al. 1998; Janetski et al. 2009).

Bears and wolves are the most dominant large predators of salmon in terrestrial environments. In times of resource plenty and when the stream habitat facilitates access to the fish (e.g., shallow water), these carnivores will feed selectively on energy-rich parts of the salmon like the head, brain or roe, and leave the remnants for a suite of smaller terrestrial scavengers (Reimchen 2000; Gende et al. 2001). Bears, in particular, have been described as a "keystone species" because of their role in distributing salmon carcasses to the forests beside streams. In fact, bears can transfer more than 50% of the salmon run to the forest with some fish ending up over 100 m from the stream banks (Reimchen 2000; Hocking and Reimchen 2006). These carcasses deposited on the forest floor support a diverse community of small mammals, birds, and amphibians, but are dominated by carrion invertebrates. For example, a total of 60 species of invertebrates were collected from salmon carcasses from two streams in the Great Bear Rainforest of British Columbia, Canada, which include obligate carrion specialists and more opportunistic scavengers or predators on salmon carcasses (Hocking et al. 2009).

There are three main groups of carrion specialists that dominate salmon carcass decomposition in terrestrial habitats: (1) the Diptera consisting of several dominant families, the Calliphoridae and Dryomyzidae, and many additional ones (e.g., Heleomyzidae, Sphaeroceridae, Phoridae, Muscidae); (2) the Coleoptera, dominated by the Silphidae, Staphylinidae, and Leiodidae; and (3) parasitic Hymenoptera (Ichneumonidae, Braconidae, Figitidae) (Hocking and Reimchen 2006; Hocking et al. 2009; Lisi and Schindler 2011). All three of these salmon specialist groups include species that could be used as indicators in salmon conservation. This is because these species: (1) exist at a high trophic position because of their diet of salmon; (2) are correlated in abundance to both the density of salmon and activity of bears that transfer carcasses to the riparian zone; and (3) support specialized life histories and reproductive timing which coincide with the rich salmon resource. For example, the flies are the most numerically dominant consumers of salmon and have been shown to time their summer emergence to the timing of sockeye salmon spawning in Alaska, which also secondarily affects the bloom timing of some plant species (Lisi and Schindler 2011). Specialist fly larvae that have consumed salmon carcasses can also escape the size spectrum of the forest soil community (Figure 22.6; Hocking et al. 2013). Fly larvae abundance is transiently far higher than their body size would predict within the size–abundance constraints of the forest soil food web because their energy is derived from a subsidized resource—salmon carcasses. Among the Coleoptera, the burying beetles, in particular, the species *Nicrophorus investigator* (Coleoptera: Silphidae), exhibits specialized communal breeding of many females per carcass to produce brood complexes of up to 750 larvae (Hocking et al. 2006, 2007). This adaptation likely enables *N. investigator* to compete effectively with flies on salmon carcasses that are an order of magnitude larger than the carcasses typically buried by species in the Genus *Nicrophorus*. Finally, the populations of several species of parasitic wasps (Ichneumonidae: *Atractodes* sp., and Braconidae: *Alysia alticola*), which parasitize the fly larvae that consume salmon carcasses, may be the most specialized of all. The abundance of these wasp species is tightly correlated to the abundance of salmon carcasses in the riparian zone during spawning, while no wasps have been caught on nonsalmon streams, nor have any been caught on any streams in early summer before salmon spawn (M.D. Hocking and J.D. Reynolds unpublished data). Further, stable isotope analysis of these species indicates that they occur at very high trophic positions equivalent to or higher than sea lions and orca whales (Hocking et al. 2009).

Salmon populations have declined in many regions of the Pacific Rim, primarily as a result of overfishing, freshwater habitat degradation, and changing marine conditions. Some areas like Washington, Oregon, and California have experienced declines in salmon biomass of over 90%. While key salmon populations have been continuously monitored for abundance in many streams that support fisheries, there generally has not been the same continuous effort in monitoring the animal species that scavenge or prey on salmon. For this reason, there is very little time-series data that show how the populations of salmon consumers may track declines in salmon. In contrast, there is a growing amount of data that can substitute space for time. For example, the population sizes of scavenging birds and some songbirds increase in density both within and across watersheds in response to spatial variation in salmon abundance (Christie and Reimchen 2006, 2008; Field and Reynolds 2011, 2013). Thus, as fisheries managers

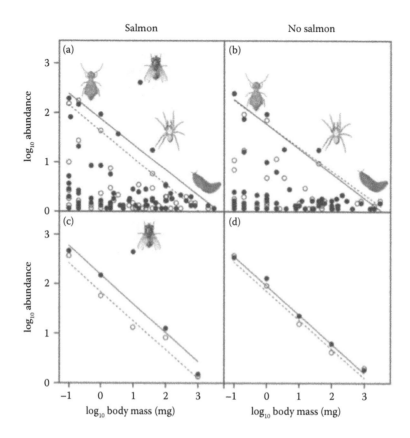

FIGURE 22.6 Salmon subsidize an escape from a forest soil size spectrum. Species (a, b) and individual (c, d) body size (M)—abundance (N) relationships in forest soil communities beside streams with spawning Pacific salmon (a, c) and above waterfall barriers without salmon (b, d). When salmon carcasses are available, we observe systemic increases in the intercept, and specialist species—fly larvae that have consumed salmon carcasses—escaping the size spectrum of this forest soil community. Surveys occurred in two time periods: in early summer before salmon have arrived (open circles and dotted lines), and in the autumn when salmon spawn (solid lines and filled circles). Pictures relate to the nearest data points and highlight several of the major species/groups sampled along the gradient in body mass. (From Hocking, M.D. et al. 2013. *Proceedings of the Royal Society of London—Biological Sciences* 280: 1–8.)

begin incorporating ecosystem values and indicators into salmon management, it is becoming increasingly relevant to include salmon consumers in monitoring efforts. These species affect crucial functions in coastal ecosystems such as predator–prey dynamics (Darimont et al. 2008), nutrient cycling (Hocking and Reimchen 2006), primary production (Hocking and Reynolds 2011), and plant pollination (Lisi and Schindler 2011), and therefore they are good candidate indicators in salmon fisheries management and conservation (e.g., Levi et al. 2012).

22.7 Stable Isotopic Evidence for High Trophic Positions of Carrion-Dependent Species

Most food-web models underestimate the importance of scavenging as a mechanism of energy transfer and relegate recycling of animal carcasses to the detrital sub web (DeVault et al. 2011). However, depending on carcass type and availability, scavengers may in fact represent a wide range of trophic levels (Bennett and Hobson 2009), and even occupy trophic positions equivalent or higher to those of predators (e.g., Nyssen et al. 2005).

Identifying the relative trophic level of species in both terrestrial and aquatic food webs has been advanced by the use of measurements of naturally occurring stable isotopes of nitrogen ($^{15}N/^{14}N$) (Hobson et al. 2002). Because enrichment in ^{15}N with increasing trophic level is a relatively constant, step-wise effect, it can be used to model relative trophic position. Nitrogen isotope analysis has revealed that scavengers may often have a high relative trophic position. For example, in a community of boreal forest arthropods, carrion beetles had higher $\delta^{15}N$ values compared to the predaceous ladybugs (Coleoptera: Coccinellidae), ground beetles (Coleoptera: Carabidae), and tiger beetles (Coleoptera: Cicindelidae) (Bennett and Hobson 2009). Ikeda et al. (2006) similarly interpreted higher $\delta^{15}N$ values in flight-capable Silphinae beetles than flightless species to indicate that flight-capable species consume more vertebrate carcasses.

In the marine environment, Nyssen et al. (2005) classified 11 species of Antarctic amphipods into different trophic categories using stable-isotope and fatty-acid signatures. The four scavenger species among these had high trophic positions, as determined by high $\delta^{15}N$ values (Figure 22.7). Isotopic survey of the North Water Polynya food web identified that scavengers (e.g., glaucous gulls, *Larus hyperboreus*) had average to high $\delta^{15}N$ values, along with half of the predators assessed (Hobson et al. 2002). However, the authors noted that some species that have been observed to scavenge in other regions (e.g., Northern fulmars, *Fulmarus glacialis*, ivory gulls, *Pagophila eburnean*) had unexpectedly low trophic positions, which is perhaps indicative that the nutritional importance of scavenging may vary within a species, as well as geographically. Such analyses have also been used to confirm the occurrence of specialized scavenging life histories in new deep-sea species. For example, stable isotope analyses on tissue and gut contents of abyssal provannid-like snails led to the discovery of a new genus, *Rubyspira* (Caenogastropoda: Abyssochrysoidea), whose main source of nutrition is whale bone (Johnson et al. 2010).

Stable isotopes have also been commonly used to identify predators and scavengers of Pacific salmon. Because adult salmon occupy between the third and fourth trophic level, terrestrial species that feed on salmon occur at a trophic position between four and five, at least for the period of their life history when salmon are a central diet item (Welch and Parsons 1993; Darimont and Reimchen 2002; Hocking et al. 2009). For example, flies, burying beetles, and parasitic wasps collected beside pristine salmon streams in British Columbia, Canada, have stable isotope signatures that indicate both a salmon diet and high

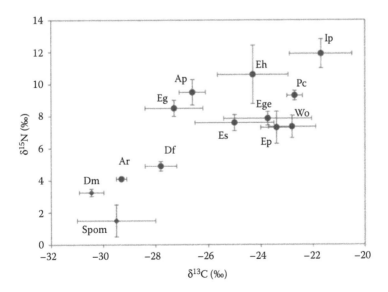

FIGURE 22.7 Carbon and nitrogen isotopic signatures of 11 species of Antarctic amphipods. The four scavenger species are Ap: *Abyssorchomene plebs*, Eg: *Eurythenes gryllus*, Pc: *Pseudorchomene coatsi*, and Wo: *Waldeckia obesa*. SPOM: suspended particulate organic matter; Dm: brown macroalgae *Desmarestia mensiezii*. (From Nyssen, F. et al. 2005. *Marine Ecology Progress Series* 300: 135–145.)

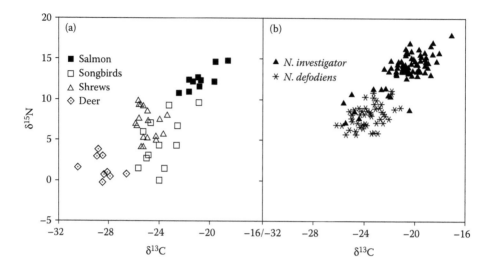

FIGURE 22.8   δ¹⁵N and δ¹³C stable isotope signatures in (a) a range of carrion sources including salmon (*O. keta* and *O. gorbusha*), the shrew *Sorex monticolus*, songbirds (*Troglodytes troglodytes, Catharus ustulatus, Catharus guttatus*), and black-tailed deer (*Odocoileus hemionus hemionus*) and in (b) adult burying beetles *Nicrophorus investigator* and *Nicrophorus defodiens* captured in summer from the Clatse River, British Columbia. (From Hocking, M.D. et al. 2007. *Canadian Journal of Zoology* 85: 437–442.)

trophic position (Hocking and Reimchen 2006; Hocking et al. 2006, 2007, 2009; Figure 22.8). Many individuals of these species derive all of their larval nutrition from salmon.

22.8 Conclusions and Management Implications

In this chapter, there was a discussion of how carrion communities, in particular, scavengers that feed off of animal remains, can be good candidate indicators in resource management and conservation. Although traditional food-web models typically assign species that feed on dead plants and animals into a broad category called decomposers, we present the hypothesis that detritivores and scavengers form their own biomass pyramid that is supported by the traditional one (Figure 22.1). Although many scavengers can switch between scavenging and predatory lifestyles or feed opportunistically across a range of carrion types and sizes, we show evidence from a range of biomes including tropical rainforests, grasslands, coastal temperate areas, and the abyssal plain that many scavengers can have quite specialized roles in food webs and occur at high trophic positions. These two attributes make carrion-dependent species and communities sensitive to environmental change and biodiversity loss (Table 22.1).

Biodiversity monitoring traditionally involves the collection of direct observational data or indices of species presence such as feces, footprints, or nesting sites. However, this is cost-, time-, and effort-intensive work that require the involvement of many experts. Recently, Calvignac-Spencer et al. (2013) identified that blow and flesh flies, which are generalist scavengers that feed on carrion, open wounds of living animals and/or fecal matter, frequently contain mammalian DNA from a broad variety of taxa. This carrion fly-derived DNA can be retrieved and assigned to species, enabling it to be used as an indicator of biodiversity. Using this method in Côte d'Ivoire, the authors were even able to detect rare and endangered species such as Jentink's duiker, *Cephalophus jentinki*, whose entire population is estimated to be <3500 individuals. As anthropogenic impacts such as habitat loss, overfishing, and climate change continue to rapidly alter ecosystem dynamics, such emerging applications of carrion ecology may provide novel tools for more efficient monitoring and response strategies.

REFERENCES

Andresen, E. 2003. Effect of forest fragmentation on dung beetle communities and functional consequences for plant regeneration. *Ecography* 26: 87–97.

Barton, P.S., S.A. Cunningham, B.C.T. Macdonald, S. McIntyre, D.B. Lindenmayer, and A.D. Manning. 2013. Species traits predict assemblage dynamics at ephemeral resource patches created by carrion. *PLoS ONE* 8: e53961.

Beasley, J.C., Z.H. Olson, and T.L. Devault. 2012. Carrion cycling in food webs: Comparisons among terrestrial and marine ecosystems. *Oikos* 121: 1021–1026.

Ben-David, M. 1997. Timing of reproduction in wild mink: The influence of spawning Pacific salmon. *Canadian Journal of Zoology* 75: 376–382.

Bennett, P. and K.A. Hobson. 2009. Trophic structure of a boreal forest arthropod community revealed by stable isotope ($\delta^{13}C$, $\delta^{15}N$) analyses. *Entomological Science* 12: 17–24.

Berger, J., P.B. Stacey, L. Bellis, and M.P. Johnson. 2001. A mammalian predator–prey imbalance: Grizzly bear and wolf extinction affect avian neotropical migrants. *Ecological Applications* 11: 947–960.

Bergmann, M., T. Soltwedel, and M. Klages. 2011. The interannual variability of megafaunal assemblages in the Arctic deep sea: Preliminary results from the HAUSGARTEN observatory (79°N). *Deep Sea Research Part I: Oceanographic Research Papers* 58: 711–723.

Bicknell, A.W.J., D. Oro, K.C.J. Camphuysen, and S.C. Votier. 2013. Potential consequences of discard reform for seabird communities. *Journal of Applied Ecology* 50: 649–658.

Bilby, R.E., B.R. Fransen, P.A. Bisson, and J.K. Walter. 1998. Response of juvenile coho salmon (*Oncorhynchus kisutch*) and steelhead (*Oncorhynchus mykiss*) to the addition of salmon carcasses to two streams in southwestern Washington, U.S.A. *Canadian Journal of Fisheries and Aquatic Sciences* 55: 1909–1918.

Boyle, M.D., D.A. Ebert, and G.M. Cailliet. 2012. Stable-isotope analysis of a deep-sea benthic-fish assemblage: Evidence of an enriched benthic food web. *Journal of Fish Biology* 80: 1485–1507.

Braack, L. 1987. Community dynamics of carrion-attendant arthropods in tropical African woodland. *Oecologia* 72: 402–409.

Brashares, J.S., L.R. Prugh, C.J. Stoner, and C.W. Epps. 2010. Ecological and conservation implications of mesopredator release. In: Terborgh, J. and J.A. Estes (eds), *Trophic Cascades: Predators, Prey, and the Changing Dynamics of Nature*. USA: Island Press, pp. 221–240.

Brodeur, R.D., W.G. Pearcy, and S. Ralston. 2003. Abundance and distribution patterns of nekton and micronekton in the Northern California Current Transition Zone. *Journal of Oceanography* 59: 515–535.

Caballero, U. and J.L. León-Cortés. 2012. High diversity beetle assemblages attracted to carrion and dung in threatened tropical oak forests in Southern Mexico. *Journal of Insect Conservation* 16: 537–547.

Calvignac-Spencer, S., K. Merkel, N. Kutzner, H. Kühl, C. Boesch, P.M. Kappeler, S. Metzger, G. Schubert, and F.H. Leendertz. 2013. Carrion fly-derived DNA as a tool for comprehensive and cost-effective assessment of mammalian biodiversity. *Molecular Ecology* 22: 915–924.

Carlson, S.M., R. Hilborn, A.P. Hendry, and T.P. Quinn. 2007. Predation by bears drives senescence in natural populations of salmon. *PLoS ONE* 2: e1286.

Catchpole, T.L., C. Frid, and T.S. Gray. 2006. Importance of discards from the English *Nephrops norvegicus* fishery in the North Sea to marine scavengers. *Marine Ecology Progress Series* 313: 215–226.

Cederholm, C.J., M.D. Kundze, T. Murota, and A. Sibatani. 1999. Pacific salmon carcasses: Essential contributions of nutrients and energy for aquatic and terrestrial ecosystems. *Fisheries* 24: 6–15.

Christie, K.S. and T.E. Reimchen. 2006. Post-reproductive Pacific salmon as a major nutrient source for large aggregations of gulls. *Canadian Field-Naturalist* 119: 202–207.

Christie, K.S. and T.E. Reimchen. 2008. Presence of salmon increases passerine density on Pacific Northwest streams. *Auk* 125: 51–59.

Corlett, R.T. 2007. The impact of Hunting on the mammalian fauna of tropical Asian Forests. *Biotropica* 39: 292–303.

Darimont, C.T. and T.E. Reimchen. 2002. Intra-hair stable isotope analysis implies seasonal shift to salmon in gray wolf diet. *Canadian Journal of Zoology* 80: 1638–1642.

Darimont, C.T., H.M. Bryan, S.M. Carlson, M.D. Hocking, M. MacDuffee, P.C. Paquet, M.H.H. Price, T.E. Reimchen, J.D. Reynolds, and C.C. Wilmers. 2010. Salmon for terrestrial protected areas. *Conservation Letters* 3: 379–389.

Darimont, C.T., P.C. Paquet, and T.E. Reimchen. 2008. Spawning salmon disrupt tight trophic coupling between wolves and ungulate prey in coastal British Columbia. *BMC Ecology* 8: 14.

Davis, A.J., J.D. Holloway, H. Huijbregts, J. Krikken, A.H. Kirk-Spriggs, and S.L. Sutton. 2001. Dung beetles as indicators of change in the forests of northern Borneo. *Journal of Applied Ecology* 38: 593–616.

DeVault, T.L., Z.H. Olson, J.C. Beasley, and J. Rhodes, Olin E. 2011. Mesopredators dominate competition for carrion in an agricultural landscape. *Basic and Applied Ecology* 12: 268–274.

DeVault, T.L., J. Rhodes, E. Olin, and J.A. Shivik. 2003. Scavenging by vertebrates: Behavioral, ecological, and evolutionary perspectives on an important energy transfer pathway in terrestrial ecosystems. *Oikos* 102: 225–234.

Díaz, A., E. Galante, and M.E. Favila. 2010. The effect of the landscape matrix on the distribution of dung and carrion beetles in a fragmented tropical rain forest. *Journal of Insect Science* 10: 1–16.

Drazen, J.C., D.M. Bailey, H.A. Ruhl, J. Smith, and L. Kenneth. 2012. The role of carrion supply in the abundance of deep-water fish off California. *PLoS ONE* 7: e49332.

Drazen, J.C., B.N. Popp, C.A. Choy, T. Clemente, L. De Forest, and K.L.J. Smith. 2008. Bypassing the abyssal benthic food web: Macrourid diet in the eastern North Pacific inferred from stomach content and stable isotopes analyses. *Limnology and Oceanography* 53: 2644–2654.

Dreyer, J., D. Hoekman, and C. Gratton. 2012. Lake-derived midges increase abundance of shoreline terrestrial arthropods via multiple trophic pathways. *Oikos* 121: 252–258.

Elbroch, L.M. and H.U. Wittmer. 2013. Table scraps: Inter-trophic food provisioning by pumas. *Biology Letters* 8: 776–779.

Field, R.D. and J.D. Reynolds. 2011. Sea to sky: Impacts of residual salmon-derived nutrients on estuarine breeding bird communities. *Proceedings of the Royal Society of London B* 278: 3081–3088.

Field, R.D. and J.D. Reynolds. 2013. Ecological links between salmon, large carnivore predation, and scavenging birds. *Journal of Avian Biology* 44: 9–16.

Fischlin, A., G.F. Midgley, J. Price, R. Leemans, B. Gopal, C. Turley, M. Rounsevell, P. Dube, J. Tarazona, and A. Velichko. 2007. Ecosystems, their properties, goods, and services. In: Parry, M.L., O. Canziani, J. Palutikof, P. van der Linden, and C. Hanson (eds), *Climate Change 2007: Impacts, Adaptation and Vulnerability*. Contribution of Working Group II to the Fourth Assessment Report of the Intergovernmental Panel on Climate Change. Cambridge, UK: Cambridge University Press, pp. 211–272.

Fleury, A.G. and J.C. Drazen. 2013. Abyssal scavenging communities attracted to sargassum and fish in the Sargasso Sea. *Deep Sea Research Part I: Oceanographic Research Papers* 72: 141–147.

Gardner, T. 2010. *Monitoring Forest Biodiversity: Improving Conservation Through Ecologically Responsible Management*. London: Earthscan Publications Ltd.

Gardner, T.A., J. Barlow, R. Chazdon, R.M. Ewers, C.A. Harvey, C.A. Peres, and N.S. Sodhi. 2009. Prospects for tropical forest biodiversity in a human-modified world. *Ecology Letters* 12: 561–582.

Gardner, T.A., M.I.M. Hernández, J. Barlow, and C.A. Peres. 2008. Understanding the biodiversity consequences of habitat change: The value of secondary and plantation forests for neotropical dung beetles. *Journal of Applied Ecology* 45: 883–893.

Gende, S.M., R.T. Edwards, M.F. Willson, and M.S. Wipfli. 2002. Pacific salmon in aquatic and terrestrial ecosystems. *BioScience* 52: 917–928.

Gende, S.M., T.P. Quinn, and M.F. Willson. 2001. Consumption choice by bears feeding on salmon. *Oecologia* 127: 372–382.

Gibbs, J.P. and E.J. Stanton. 2001. Habitat fragmentation and arthropod community change: Carrion beetles, phoretic mites, and flies. *Ecological Applications* 11: 79–85.

Gregg, W.W., M.E. Conkright, P. Ginoux, J.E. O'Reilly, and N.W. Casey. 2003. Ocean primary production and climate: Global decadal changes. *Geophysical Research Letters* 30.

Halffter, G. and M.E. Favila. 1993. The Scarabaeinae (Insecta: Coleoptera) an animal group for analyzing, inventorying and monitoring biodiversity in tropical rainforest and modified landscapes. *Biology International* 27: 15–21.

Hall, M.A. 1996. On bycatches. *Reviews in Fish Biology and Fisheries* 6: 319–352.

Hansen, A.J. 1987. Regulation of bald eagle reproductive rates in southeast Alaska. *Ecology* 68: 387–392.

Hawlena, D., M.S. Strickland, M.A. Bradford, and O.J. Schmitz. 2012. Fear of predation slows plant-litter decomposition. *Science* 336: 1434–1438.

Hilderbrand, G.V., C.C. Schwartz, C.T. Robbins, M.E. Jacoby, T.A. Hanley, S.M. Arthur, and C. Servheen. 1999. The importance of meat, particularly salmon, to body size, population productivity, and conservation of North American brown bears. *Canadian Journal of Zoology* 77: 132–138.

Hobson, K.A., A. Fisk, N. Karnovsky, M. Holst, J.-M. Gagnon, and M. Fortier. 2002. A stable isotope (δ^{13}C, δ^{15}N) model for the North Water food web: Implications for evaluating trophodynamics and the flow of energy and contaminants. *Deep Sea Research Part II: Topical Studies in Oceanography* 49: 5131–5150.

Hocking, M.D., C.T. Darimont, K.S. Christie, and T.E. Reimchen. 2007. Niche variation in burying beetles associated with marine and terrestrial carrion. *Canadian Journal of Zoology* 85: 437–442.

Hocking, M.D., N.K. Dulvy, J.D. Reynolds, R.A. Ring, and T.E. Reimchen. 2013. Salmon subsidize an escape from a size spectrum. *Proceedings of the Royal Society of London—Biological Sciences* 280: 1–8.

Hocking, M.D. and T.E. Reimchen. 2006. Consumption and distribution of salmon (*Oncorhynchus* spp.) nutrients and energy by terrestrial flies. *Canadian Journal of Fisheries and Aquatic Sciences* 63: 2076–2086.

Hocking, M.D. and J.D. Reynolds. 2011. Impacts of salmon on riparian plant diversity. *Science* 331: 1609–1612.

Hocking, M.D., R.A. Ring, and T.E. Reimchen. 2006. Burying beetle *Nicrophorus investigator* reproduction on Pacific salmon carcasses. *Ecological Entomology* 31: 5–12.

Hocking, M.D., R.A. Ring, and T.E. Reimchen. 2009. The ecology of terrestrial invertebrates on Pacific salmon carcasses. *Ecological Research* 24: 1091–1100.

Hoekman, D., J. Dreyer, R.D. Jackson, P.A. Townsend, and C. Gratton. 2011. Lake to land subsidies: Experimental addition of aquatic insects increases terrestrial arthropod densities. *Ecology* 92:2063–2072.

Houston, D.C. 2001. *Vultures and Condors*. Colin Baxter, Granton-on-Spey.

Ikeda, H., K. Kubota, T. Kagaya, and T. Abe. 2006. Flight capabilities and feeding habits of silphine beetles: Are flightless species really "carrion beetles"? *Ecological Research* 22: 237–241.

Iyengar, E.V. 2008. Kleptoparasitic interactions throughout the animal kingdom and a re-evaluation, based on participant mobility, of the conditions promoting the evolution of kleptoparasitism. *Biological Journal of the Linnean Society* 93: 745–762.

Janetski, D.J., D.T. Chaloner, S.D. Tiegs, and G.A. Lamberti. 2009. Pacific salmon effects on stream ecosystems: A quantitative synthesis. *Oecologia* 159: 583–595.

Johnson, S.B., A. Warén, R.W. Lee, Y. Kano, A. Kaim, A. Davis, E.E. Strong, and R.C. Vrijenhoek. 2010. *Rubyspira*, new genus and two new species of bone-eating deep-sea snails with ancient habits. *The Biological Bulletin* 219: 166–177.

Kaiser, M.J. and J.G. Hiddink. 2007. Food subsidies from fisheries to continental shelf benthic scavengers. *Marine Ecology Progress Series* 350: 267–276.

Kneidel, K.A. 1984. Competition and disturbance in communities of carrion-breeding Diptera. *Journal of Animal Ecology* 53: 849–865.

Krell, F.-T. 2004. East African dung beetles (Scarabaeidae) attracted by defensive secretions of millipedes. *Journal of East African Natural History* 93: 69–73.

Krofel, M., I. Kos, and K. Jerina. 2012. The noble cats and the big bad scavengers: Effects of dominant scavengers on solitary predators. *Behavioral Ecology and Sociobiology* 66: 1297–1304.

Kuusela, S. and I. Hanski. 1982. The structure of carrion fly communities: The size and type of carrion. *Holarctic Ecology* 5: 337–348.

Laidre, M.E. 2013. Foraging across ecosystems: Diet diversity and social foraging spanning aquatic and terrestrial ecosystems by an invertebrate. *Marine Ecology* 34: 80–89.

Levi, T., C.T. Darimont, M.M. MacDuffee, M. Mangel, P.C. Paquet, and C.C. Wilmers. 2012 Using grizzly bears to assess economic and ecosystem tradeoffs in Pacific salmon management. *PLoS Biology* 10: e1001303.

Lisi, P.J. and D.E. Schindler. 2011. Spatial variation in timing of marine subsidies influences riparian phenology through a plant-pollinator mutualism. *Ecosphere* 2: 1.1.

Lundsten, L., K.L. Schlining, K. Frasier, S.B. Johnson, L.A. Kuhnz, J.B.J. Harvey, G. Clague, and R.C. Vrijenhoek. 2010. Time-series analysis of six whale-fall communities in Monterey Canyon, California, USA. *Deep Sea Research Part I: Oceanographic Research Papers* 57: 1573–1584.

Myers, R.A., J.K. Baum, T.D. Shepherd, S.P. Powers, and C.H. Peterson. 2007. Cascading effects of the loss of apex predatory sharks from a coastal ocean. *Science* 315: 1846–1850.

Nichols, E. and T.A. Gardner. 2011. Dung beetles as a candidate study taxon in applied biodiversity conservation research. In: Simmons, L.A. and J. Ridsill-Smith (eds), *Dung Beetle Ecology and Evolution*. Oxford, UK: Wiley-Blackwell Publishing Ltd.

Nichols, E., T.A. Gardner, C.A. Peres, S. Spector, and T.S.R. Network. 2009. Co-declining mammals and dung beetles: An impending ecological cascade. *Oikos* 118: 481–487.

Nichols, E., T. Larsen, S. Spector, A.L. Davis, F. Escobar, M. Favila, and K. Vulinec. 2007. Global dung beetle response to tropical forest modification and fragmentation: A quantitative literature review and meta-analysis. *Biological Conservation* 137: 1–19.

Nichols, E., S. Spector, J. Louzada, T. Larsen, S. Amezquita, and M.E. Favila. 2008. Ecological functions and ecosystem services provided by Scarabaeinae dung beetles. *Biological Conservation* 141: 1461–1474.

Nyssen, F., T. Brey, P. Dauby, and M. Graeve. 2005. Trophic position of Antarctic amphipods—Enhanced analysis by a 2-dimensional biomarker assay. *Marine Ecology Progress Series* 300: 135–145.

Paetzold, A., M. Smith, P.H. Warren, and L. Maltby. 2011. Environmental impact propagated by cross-system subsidy: Chronic stream pollution controls riparian spider populations. *Ecology* 92: 1711–1716.

Paine, R.T. 1969. A note on trophic complexity and community stability. *The American Naturalist* 103: 91–93.

Pauly, D., V. Christensen, J. Dalsgaard, R. Froese, and J. Torres, F. 1998. Fishing down marine food webs. *Science* 279: 860–863.

Peres, C.A. and E. Palacios. 2007. Basin-wide effects of game harvest on vertebrate population densities in Amazonian forests: Implications for animal-mediated seed dispersal. *Biotropica* 39: 304–315.

Perry, A.L., P.J. Low, J.R. Ellis, and J.D. Reynolds. 2005. Climate change and distribution shifts in marine fishes. *Science* 308: 1912–1915.

Pigeon, K. 2011. Denning behaviour and climate change: Linking environmental variables to denning of grizzly bears in the Rocky Mountains and boreal forest of Alberta, Canada. In: Stenhouse, G. and K. Graham (eds), *Foothills Research Institute Grizzly Bear Program*. 2010 Annual Report. Alberta: Hinton, pp. 11–25.

Polis, G.A., F. Sánchez-Piñero, P.T. Stapp, W.B. Anderson, and M.D. Rose. 2004. Trophic flows from water to land: Marine input affects food webs of islands and coastal ecosystems worldwide. In: Polis, G.A., M.E. Power, and G.R. Huxel (eds), *Food Webs at the Landscape Level*. Chicago, IL: The University of Chicago Press, pp. 200–216.

Polovina, J.J., J.P. Dunne, P.A. Woodworth, and E.A. Howell. 2011. Projected expansion of the subtropical biome and contraction of the temperate and equatorial upwelling biomes in the North Pacific under global warming. *ICES Journal of Marine Science* 68: 986–995.

Power, M.E., D. Tilman, J.A. Estes, B.A. Menge, W.J. Bond, L.S. Mills, G. Daily, J.C. Castilla, J. Lubchenco, and R.T. Paine. 1996. Challenges in the quest for keystones. *Bioscience* 46: 609–620.

Prugh, L.R., C.J. Stoner, C.W. Epps, W.T. Bean, and W.J. Ripple. 2009. The rise of the mesopredator. *Bioscience* 59: 779–791.

Reimchen, T.E. 2000. Some ecological and evolutionary aspects of bear-salmon interactions in coastal British Columbia. *Canadian Journal of Zoology* 78: 448–457.

Ripple, W.J., A.J. Wirsing, C.C. Wilmers, and M. Letnic. 2013. Widespread mesopredator effects after wolf extirpation. *Biologial Conservation* 160: 70–79.

Ritchie, E.G. and C.N. Johnson. 2009. Predator interactions, mesopredator release and biodiversity conservation. *Ecology Letters* 12: 982–998.

Rose, M.D. and G.A. Polis. 1998. The distribution and abundance of coyotes: The effects of allochthonous food subsidies from the sea. *Ecology* 79: 998–1007.

Ruhl, H.A., J.A. Ellena, J. Smith, and L. Kenneth. 2008. Connections between climate, food limitation, and carbon cycling in abyssal sediment communities. *Proceedings of the National Academy of Sciences of the United States of America* 105: 17006–17011.

Ruhl, H.A. and K.L. Smith. 2004. Shifts in deep-sea community structure linked to climate and food supply. *Science* 305: 513–515.

Ruth, T. and K. Murphy. 2010. Competition with other carnivores for prey. In: Hornocker, M. and S. Negri (eds), *Cougar Ecology and Conservation*. Chicago, IL: University of Chicago Press, pp. 163–174.

Ruxton, G.D. and D.C. Houston. 2003. Could *Tyrannosaurus rex* have been a scavenger rather than a predator? *Proceedings of the Royal Society of London—Biological Sciences* 270: 731–733.

Ruxton, G.D. and D.C. Houston. 2004. Obligate vertebrate scavengers must be large soaring fliers. *Journal of Theoretical Biology* 228: 431–436.

Scarabnet. 2013. ScarabNet Global Taxon Database.

Schindler, D.E., J.B. Armstrong, K.T. Bentley, K. Jankowski, P.J. Lisi, and L.X. Payne. 2013. Rising the crimson tide: Mobile terrestrial consumers track phonological variation in spawning of an anadromous fish. *Biology Letters* 9: 20130048.

Scholtz, C.H., K.S. Cole, R. Tukker, and U. Kryger. 2006. Biology and ecology of *Circellium bacchus* (Fabricius 1781) (Coleoptera Scarabaeidae), a South African dung beetle of conservation concern. *Tropical Zoology* 19: 185–207.

Scott, M.P. 1998. The ecology and behavior of burying beetles. *Annual Review of Entomology* 43: 595–618.

Sibert, J., J. Hampton, P. Kleiber, and M. Maunder. 2006. Biomass, size, and trophic status of top predators in the Pacific Ocean. *Science* 314: 1773–1776.

Slade, E.M., D.J. Mann, J.F. Villanueva, and O.T. Lewis. 2007. Experimental evidence for the effects of dung beetle functional group richness and composition on ecosystem function in a tropical forest. *The Journal of Animal Ecology* 76: 1094–1104.

Smith, C.R. 2006. Bigger is better: The role of whales as detritus in marine ecosystems. In: Estes, J.A., D.P. DeMaster, R.L. Brownell Jr., D.F. Doak, and T.M. Williams (eds), *Whales, Whaling and Ocean Ecosystems*. Berkeley, CA: University of California Press, pp. 286–301.

Smith, C.R. and A.R. Baco. 2003. Ecology of whale falls at the deep-sea floor. *Oceanography and Marine Biology: An Annual Review* 41: 311–354.

Smith, C.R., F.C. De Leo, A.F. Bernardino, A.K. Sweetman, and P.M. Arbuzu. 2008. Abyssal food limitation, ecosystem structure and climate change. *Trends in Ecology and Evolution* 23: 518–528.

Smith, D.W., D.R. Stahler, and D.S. Guernsey. 2006. *Yellowstone Wolf Project: Annual Report, 2005*. National Park Service, Yellowstone Center for Resources, USA.

Snelgrove, P., M. Austen, S. Hawkins, T. Iliffe, R. Kneib, L. Levin, J. Weslawski, R. Whitlatch, and J. Garey. 2004. Vulnerability of marine sedimentary ecosystem services to human activities. Pages 161–183 *in* D. Wall, editor. *Sustaining Biodiversity and Ecosystem Services in Soils and Sediments*. Washington, DC: Island Press.

Spector, S. 2006. Scarabaeine dung beetles (Coleoptera: Scarabaeidae: Scarabaeinae): An invertebrate focal taxon for biodiversity research and conservation. *The Coleopterists Bulletin* 5: 71–83.

Swift, M.J., O.W. Heal, and J.M. Anderson. 1979. *Decomposition in Terrestrial Ecosystems*. Berkeley, CA: University of California Press.

Vander Zanden, M.J. and Y. Vadeboncoeur. 2002. Fishes as integrators of benthic and pelagic food webs in lakes. *Ecology* 83: 2152–2161.

Welch, D.W. and T.R. Parsons. 1993. $\delta^{13}C$-^{15}N values as indicators of trophic position and competitive overlap for Pacific salmon (*Oncorhynchus* spp.). *Fisheries Oceanography* 2: 11–23.

Willson, M.F. and K.C. Halupka. 1995. Anadromous fish as keystone species in vertebrate communities. *Conservation Biology* 9: 489–497.

Wilmers, C.C., R.L. Crabtree, D.W. Smith, K.M. Murphy, and W.M. Getz. 2003a. Trophic facilitation by introduced top predators: Grey wolf subsidies to scavengers in Yellowstone National Park. *Journal of Animal Ecology* 72: 909–916.

Wilmers, C.C. and W.M. Getz. 2005. Gray wolves as climate change buffers in Yellowstone. *PLoS Biology* 3: e92.

Wilmers, C.C. and E. Post. 2006. Predicting the influence of wolf-provided carrion on scavenger community dynamics under climate change scenarios. *Global Change Biology* 12: 403–409.

Wilmers, C.C., D.R. Stahler, R.L. Crabtree, D.W. Smith, and W.M. Getz. 2003b. Resource dispersion and consumer dominance: Scavenging at wolf- and hunter-killed carcasses in Greater Yellowstone, USA. *Ecology Letters* 6: 996–1003.

Wilson, K.A., M.F. McBride, M. Bode, and H.P. Possingham. 2006. Prioritizing global conservation efforts. *Nature* 440: 337–340.

Wilson, E.E. and E.M. Wolkovich. 2011. Scavenging: How carnivores and carrion structure communities. *Trends in Ecology and Evolution* 26: 129–135.

Yang, L.H. 2004. Periodical cicadas as resource pulses in North American forests. *Science* 306: 1565–1567.

Yeh, J. and J.C. Drazen. 2009. Depth zonation and bathymetric trends of deep-sea megafaunal scavengers of the Hawaiian Islands. *Deep Sea Research Part I: Oceanographic Research Papers* 56: 251–266.

23

Composting as a Method for Carrion Disposal in Livestock Production

Shanwei Xu, Tim Reuter, Kim Stanford, Francis J. Larney, and Tim A. McAllister

CONTENTS

23.1 Introduction

Livestock and resulting products are an important food source for the majority of the 7.5 billion people on earth. However, the disposal of associated wastes generated during the course of livestock production represents a potential economic and environmental liability. In general, carcasses and slaughter wastes are the primary constituents of carrion waste from livestock farming and subsequent meat processing (Arvanitoyannis and Ladas 2008; CAST 2009). The carrion arising from livestock production can consist of the entire carcass or portions of the carcass that have been deemed unsuitable for human consumption.

These portions may include blood, hair, tail, horns, bones, feathers, or specified risk materials (SRMs, i.e., spinal cord, skull, brain, vertebral column, eyes, tonsils, trigeminal and dorsal root, distal ileum) that have been removed from the human food chain (Arvanitoyannis and Ladas 2008). According to the Food and Agriculture Organization of the United Nations (2013), there were approximately 1.5×10^9 cattle and buffaloes, 1.9×10^9 sheep and goats, 9.6×10^8 pigs, and 2.2×10^{10} poultry produced globally in 2011 with an estimated gross production value of US \$1.3 trillion (Figure 23.1a). Given these large numbers, a significant volume of carrion is generated on a continual basis, either from death by natural causes, disease, accident, or from slaughter.

Proper disposal of carrion wastes from livestock production is important to prevent environmental pollution and protect the health of humans and animals. Improper carrion disposal can result in foul

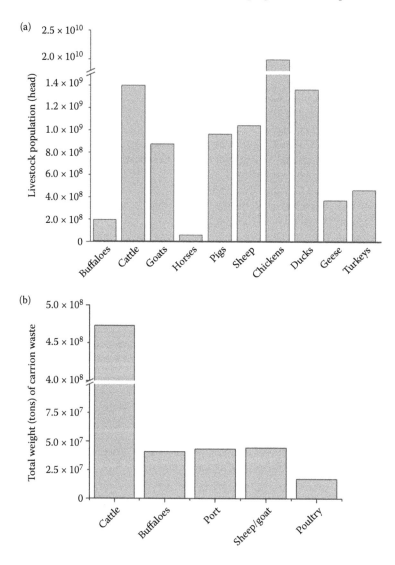

FIGURE 23.1 (a) Global livestock population in 2011. Data were cited from Food and Agriculture Organization of the United Nations (2013); (b) An estimated weight (tons) of carrion waste generated globally in 2011 for each livestock category when the animals were slaughtered and processed. Data were calculated on a basis of livestock population in 2011 shown in (a) and an average waste weight per slaughtered animal. (From the British Columbia Ministry of Agriculture BCMA. 2009. Slaughterhouse waste and specified risk material (SRM) management study. British Columbia Ministry of Agriculture. Final report for Vancouver Island. Prepared by the Ground Up Resource Consultants Inc., Courtenay, BC. http://www.agf.gov.bc.ca/resmgmt/SRM_Program/index.htm. Accessed May 8, 2013.)

odor, pathogen transmission, and nutrient-enriched leachate entering soil, surface, or groundwater. Thus, government agencies have established rules that require removal and disposal of animal carcasses using proper and timely methods. In Canada, animal carcasses need to be disposed of within 48 h of death, while this time is reduced to 24 h in Minnesota, Texas, and Indiana, USA (Kalbasi et al. 2005).

A variety of approaches are used for the disposal of carrion waste products from abattoirs, on-farm, and road kill, each with its own advantages and disadvantages (Table 23.1). In North America, the majority of carrion generated from slaughter plants that is banned from the food chain is rendered, dehydrated, and disposed of in landfills (Xu et al. 2010; Figure 23.2), whereas in Europe, incineration is the primary method of disposal (Paisley and Hostrup-Pedersen 2005). If conducted properly, incineration is effective at preventing the spread of pathogens, but it derives little to no value from carrion wastes and can adversely impact air quality. Livestock production typically occurs over wide geographical areas, and transportation distances frequently make rendering or incineration impractical as a disposal method for livestock species other than poultry. Furthermore, alternatives to rendering or incineration such as burial or natural decomposition can increase the risk of pathogen transmission and environmental contamination. While other methods, such as alkaline hydrolysis, anaerobic digestion, and gasification, are being assessed as methods of carrion disposal (Fedorowicz et al. 2007; Kalbasi-Ashtari et al. 2008), their employment requires considerable capital investment and energy expenditure.

In North America, composting is gaining popularity as a means of disposing of livestock carrion as it is relatively simple, environmentally sound, and economical. Furthermore, composting negates the need

TABLE 23.1

Advantages and Disadvantages of Methods for the Disposal of Carrion Wastes

Disposal Methods	Advantages	Disadvantages	Current Application[a]
Natural disposal	Easy and inexpensive	Not legal in all areas Air, water, and soil contamination No disease control Increase predation of animals	Permitted in some provinces in Canada, but only in sparsely populated areas
Burial	Permanent containment for disease outbreaks Good choice for large land bases	Need sites ready for winter Possible groundwater contamination Tracking of sites required Costly for multiple burial pits Limitation of degradation	Banned in Europe, but permitted in Canada and United States with specified rules
Rendering	Good for on-farm disease control No residues left Simple—just call for pick up Hide and tallow recycled	Long distance required to transport Minimum weights for pickup required Not available everywhere Excessive time may elapse before pick up	Main disposal method for slaughter wastes in Canada and United States
Incineration	Superior disease control High volume of waste reduction Energy generation	Expensive, equipment, and fuel required Time required Ash requires disposal Gas emissions	Main application for specified risk material disposal in Europe
Anaerobic digestion	Renewable energy production Pathogen reduction Organic fertilizer production	Expensive infrastructure Technical Economies of scale	Not permitted within Europe without prior rendering process
Alkaline hydrolysis	Fewer gas emission Pathogen reduction Beneficial by-product	Relative expensive High level of effluent with a high pH Limited capacity for large volume	Tested in Europe Small lab-scale units in North America

[a] Adapted from Gwyther C.L. et al. 2011. *Waste Management* 31: 767–778 and NABC. 2004. *Carcass Disposal: A Comprehensive Review*. National Agriculture Biosecurity Centre Consortium. Kansas State University. Manhattan, KS.

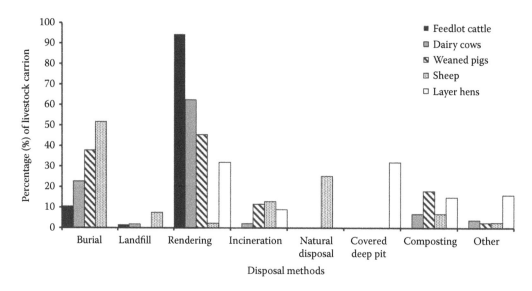

FIGURE 23.2 Percentage (%) of livestock carrion disposed of by various methods in United States. Note: values may not total 100% as more than one disposal method may be used within an operation. (Adapted from NABC. 2004. *Carcass Disposal: A Comprehensive Review.* National Agriculture Biosecurity Centre Consortium. Kansas State University. Manhattan, KS.)

for transport as it can generally be conducted at the site where the carrion is generated. Moreover, composting can also serve as a posttreatment of other carrion disposal methods such as anaerobic digestate, which helps to ensure more complete tissue degradation and disease control (Franke-Whittle and Insam 2013). This chapter focuses on using composting as a method of disposing of livestock carrion.

23.2 Defining the Source of Livestock Carrion

Sources of intact livestock carrion can be classified as either routine or catastrophic. Routine on-farm carrion represents a small proportion of the livestock population, but numbers can fluctuate widely if disease outbreaks occur (CAST 2009). For example, cervid farms in Canada have approximately 145,000 animals with overall mortality rates ranging from 4.8 to 29.8% for adult deer and fawns, respectively (Haigh et al. 2005; Canadian Cervid Alliance 2009). In the cattle industry, the mortality rates of newborn calves are normally 2%–3%, but can exceed 20% during disease outbreaks or unfavorable climatic conditions (Uetake 2013).

Natural disasters, epidemic disease, or road kills can also generate large quantities of carrion. For instance, in 1999, Hurricane Floyd in the United States led to the death of more than 2,860,000 chickens, 28,000 swine, and 600 cattle in North Carolina (Ellis 2001). At the height of bovine spongiform encephalopathy (BSE) outbreak in the United Kingdom (1988–1998), approximately 174,000 cases of BSE were diagnosed (WOAH 2013) and more than 1.6 million BSE-suspect cattle killed (Croston 1997). Moreover, road kills can also generate a significant amount of carrion, even though they are not viewed as being livestock. In Canada, approximately 45,000 animal–vehicle collisions occur yearly (Huijser et al. 2009), generating a significant amount of carrion. For example, in between 1996 and 2003, 5585 wild deers were killed in vehicular collisions in Nova Scotia, Canada, generating at least 500 tons of carrion (Transport Canada 2012). In New York, USA, road kills of deer, moose, raccoons, skunks, coyotes, and foxes is thought to generate over 2000 tons of carrion annually with only a fraction of this disposed of by the State Department of Transportation (Bonhotal et al. 2007).

A much larger volume of carrion waste is generated from the slaughter of livestock for human consumption. Depending on the livestock species, between 1.7×10^7 and 4.7×10^8 tons of carrion were produced globally in 2011 as a result of livestock slaughter and processing (Figure 23.1b). As a result of BSE, regulations have been imposed in Canada and USA to prevent the introduction of SRM into the food chain (Dewell et al. 2008; CFIA 2013). Although SRM typically account for less than 10% of the waste in abattoirs (BCMA 2009), it is not always separated from other slaughter wastes at source, increasing the proportion of slaughterhouse wastes that must be treated as SRM. In Canada, it is estimated that 250,000 tons of SRM are generated annually from the slaughter of cattle and sheep (Gilroyed et al. 2010).

Feathers are also generated in large quantities during poultry production. According to the Food and Agriculture Organization of the United Nations (2013), the consumption of chicken meat exceeded 89 million tons in 2011. Chicken feathers represent 5%–7% of body weight and thus approximately 5 million tons of waste feathers are generated annually (Mărculescu et al. 2012). As European legislation restricts the use of feathers as animal feed, the disposal of feather wastes is becoming an increasing challenge (Forgács et al. 2013).

23.3 Application of Composting in Livestock Carrion Disposal

23.3.1 Concept of Composting

Composting has been used for centuries to recycle agricultural and horticultural wastes, such as animal manure and plant residues, and is being increasingly used as method for the disposal of organic urban wastes (Rynk 1992; Kutzner 2000). Composting is a natural, primarily aerobic biological process of decomposition that leads to the stabilization of organic matter (Keener et al. 1993). It can be divided into two major stages: active and curing, with the active stage being a period of accelerated decomposition of organic matter and the formation of an environment unsuitable for the survival of most pathogens (Xu et al. 2009a). The active stage can persist for days or weeks depending on the characteristics and quantity of the substrate and the availability of oxygen (Haug 1993). Upon depletion of degradable organic matter, microbial activity and compost temperatures decline and the curing phase begins, culminating in the formation of mature and stable compost.

The active composting stage can be further divided into mesophilic and thermophilic stages. Mesophilic generally refers to periods where the temperature of compost is below 40°C, and thermophilic to temperatures above 40°C (Miller 1996). In general, at the start of the mesophilic stage, diverse populations of mesophilic and/or thermo-tolerant bacteria and fungi are responsible for the degradation of soluble and easily degradable carbon sources (Kutzner 2000). Subsequently, when temperatures rise above 40°C, microbial diversity decreases and thermophilic bacteria and fungi dominate the composting process (Fogarty and Tuovinen 1991). These thermophilic microbes produce a broad range of enzymes that degrade the vast array of substrates present in compost (Ryckeboer et al. 2003). Upon completion of the thermophilic phase, temperatures return to the mesophilic range, and mesophilic microbes once again predominate as the compost enters the curing phase. Ryckeboer et al. (2003) described more than 70 genera of bacteria and 150 genera of fungi during the thermophilic and mesophilic phases of composting of a range of wastes including garden residues, animal manure, municipal solid waste, wood-by products, and food and kitchen wastes. Recently, molecular sequencing techniques have been used to characterize bacterial and fungal populations during composting (de Gannes et al. 2013; Neher et al 2013; Tkachuk et al. 2014). Other microbial communities including archaea (Yamamoto et al. 2012) and invertebrates such as nematodes (Steel et al. 2013) are also being investigated for their roles in the composting process.

23.3.2 Current Uses of Composting for Carrion Disposal

Livestock carrion wastes are typically composed of about 70% water with 30% dry matter which consists of 52% protein, 41% fat, and 6% ash (NABC 2004). Composting of carrion using windrows, static piles, and bins or vessels has been investigated for poultry, swine, sheep, cattle, and road-killed deer (Table 23.2). Stanford et al. (2000) reported that sheep carrion including keratinized wool were completely degraded

TABLE 23.2

Summary of Carrion Waste Degradation during Composting

Category	Carrion Tissue	Degradation	Initial Amount	Co-Composting Materials	Compost System	Time	Compost Temperature	Type of Aeration	Compost Moisture (Wet Basis)	Compost Porosity	C/N Ratio	pH	Reference
Cattle	Carcass	Void of soft tissue; large bones present	5488 kg (wet basis)	Cattle manure	Biosecure structure	147 days	Average 45°C	Passive aeration	48%–70%	—	13–26	8.3–9.5	Xu et al. (2009a)
	Fat	Up to 92% (dry basis)	~3.4 kg (dry basis)	Wastewater sludge and wood chips	30-L insulated vessel	50 days	~30–63°C; thermophilic composting for 21 days	Forced aeration & mixing	44%–55%	Air-filled porosity 40%–75%	—	7.0 at the start	Ruggieri et al. (2008)
	Brain	>90% (dry matter)	5 g (wet basis)	Cattle manure	Biosecure structure	7 days	Average 49°C	Passive aeration	48%–70%	—	13–26	8.3–9.5	Xu et al. (2009a)
	Brain	73% (dry matter)	50 g (wet basis)	Cattle manure and wood shavings	110-L composter	28 days	Maximum 74°C; >50°C for 4 days	Passive aeration & turning	57%–67%	Bulk density 480–510 kg m^{-3}	15–22	7.7–9.4	Xu et al. (2013a)
	Hoof	>80% (dry matter)	5 g (wet basis)	Cattle manure	Biosecure structure	56 days	Average 56°C	Passive aeration	48%–70%	—	13–26	8.3–9.5	Xu et al. (2009a)
	Rib bone	13% (dry matter)	10 g (wet basis)	Cattle manure	Biosecure structure	147 days	Average 45°C	Passive aeration	48%–70%	—	13–26	8.3–9.5	Xu et al. (2009a)
	Femur and Tibia	45% (wet basis)	~2.4 kg (wet basis)	Cattle manure and straw	Windrow	11 weeks	Average 46°C	Passive aeration & turning	55% at end compost	—	16–20	—	Stanford et al. (2009)
Swine	Carcass	0.04% (dry basis) residual bone in final compost	2845 kg (wet basis)	Swine manure and straw	Compost piles	338 days	~25–75°C; >55°C from days 45–115 prior to turning	Passive aeration & turning	40%–65%	—	15–69	—	Fonstad et al. (2003)
	Carcass	66%–86% (wet basis)	241–259 kg (wet basis)	Corn stalks, oat straw and silage	Biosecure structure	16 weeks	Peaked at 50–60°C; dropped to 20–40°C	Passive aeration	23.3%–75.1% at end compost	Bulk density 163–918 kg m^{-3}	—	—	Ahn et al. (2007), Glanville et al. (2007)

(Continued)

TABLE 23.2 (Continued)

Summary of Carrion Waste Degradation during Composting

Category	Carrion Tissue	Degradation	Initial Amount	Co-Composting Materials	Compost System	Time	Compost Temperature	Type of Aeration	Compost Moisture (Wet Basis)	Compost Porosity	C/N Ratio	pH	Reference
Chicken	Carcass	48.6% (wet weight reduction of compost mixtures)	Carcasses/compost materials = 1/6.35 (wet basis)	Caged layer manure and hay	Bin (1.2 × 1.2 × 1.2 m)	114 days	Peaked at 70°C; remained >55°C for 44 days	Passive aeration and mixing	23%–60%	—	20–25 at the start	—	Sivakuma et al. (2008)
	Feather keratin	24% (calculated from CO_2 production)	160 g (wet basis)	Three-month old compost	8.4-L composter	30 days	Reached 55°C within 4 days; remained at 50–55°C	Forced aeration and mixing	Maintained at 50%	—	10.3 at the end	—	Barone and Arikan (2007)
Turkey	Carcass	Flesh not visible	—	Turkey litter and sunflower hulls	Static compost piles	40–65 days	≥55°C for 3–10 days	Passive aeration and turning	27.6%–51.3%	—	13.2–17.2	7.8–9.3	Rahman (2012)
Lamb	Carcass	Virtually no bones remained	100 kg	Sheep manure and barley straw	Bin (2.4 × 2.4 × 2.4 m)	118 days	20–70°C; remained >60°C for up to 41 days	Passive aeration	34.7% at end compost	—	12.7 at the end	7.1 at the end	Stanford et al. (2000)
Sheep	Carcass	Only skulls and pelvic girdles intact	585 kg	Cured sheep manure compost and barley straw	Bin (2.4 × 2.4 × 2.4 m)	163 days	20–55°C in primary bin; >60°C in secondary bin	Passive aeration	49.4% at end compost	—	12.2 at the end	7.1 at the end	Stanford et al. (2000)
Horse	Carcass	Soft tissues gone; very few large bones remained	500 kg	Horse manure and wood shavings	Bin (3 × 6 m W × L)	9 months	40–71°C; remained >55°C for several days or weeks	Passive aeration and turning	43%–52% at end compost	—	7.7 at the end	42–46 at the start; 19–24 at the end	Mukhtar et al. (2003)
Deer	Carcass	Effective carcass degradation; useable end product after 12 months	4 road killed deer per compost pile	Wood chips	Static compost piles	12 months	>40°C within 2 weeks and remained for 9–75 days	Passive aeration	—	—	—	34.5–71.2	Schwarz et al. (2010)

during composting, with temperatures exceeding 60°C for a period of 41 days. Using a ratio of five parts manure to one part cattle carrion in a windrow composting trial, Stanford et al. (2009) demonstrated that <1% of bone remained in the cured compost. Xu et al. (2009a) reported that more than 90% of bovine brain tissues decomposed after 7 days and 80% of bovine hooves decomposed after 56 days of composting. Moreover, in the same composting system, Xu et al. (2009b) measured a >93% reduction in bovine mitochondrial DNA, suggesting that soft tissues were almost completely decomposed after 147 days of composting. In these conditions, decomposition of the carcasses occurred without attracting carrion flies, provided that the dry matter content of compost was such that no effluent was formed. These studies suggest that under optimal conditions, composting can be an effective method of disposing of livestock carrion.

23.3.3 Current Procedures and Systems for Carrion Composting

An appropriate composting system and employment of proper composting procedures are essential for the complete conversion of carrion wastes to high-quality compost. Berge et al. (2009) recommended that carrion composting systems should meet four requirements: (1) protect ground and surface water; (2) minimize risk of pathogen transmission; (3) prevent access of scavengers; and (4) have no adverse impact on air quality. However, introduction of carrion into the composting matrix can create regions of nutrient nonuniformity making it difficult to achieve optimal composting conditions. Currently, composting windrows, piles, bins, and vessels have been assessed for their ability to dispose of livestock carrion.

23.3.3.1 Windrow Composting

Windrow composting is employed with the use of defined width and variable length as needed for the disposal of a given volume of carrion waste. This system is suitable for composting a large number of whole carcasses (>227 kg; Berge et al. 2009), but is also suitable for the disposal of slaughterhouse wastes (Hao et al. 2009) and poultry feathers (Bohacz and Korniłowicz-Kowalska 2009). Compost windrows containing animal carrion are usually constructed using cocomposting materials available on site such as animal manure, straw, wood-by products, or poultry litter. Initial construction includes an absorptive layer of at least 60 cm of carbonaceous material such as cereal straw, wood shavings, chips, or sawdust. Subsequently, carcasses are placed within layers of the cocomposting matrix with a final covering of the matrix placed over the surface of the entire windrow or pile (Gwyther et al. 2011). Previous research has shown that large carcasses such as cattle or deer can be effectively composted if they are placed in single or double layers within windrows (Stanford et al. 2009; Schwarz et al. 2010), while small carcasses such as poultry can be placed in multiple layers (AARD 2011).

In most instances, windrows are open and are not surrounded by either a roof or walls. However, the uncovered materials can attract scavengers, such as skunks, racoons, rats, and coyotes, resulting in potential transmission of pathogens to the surrounding environment. In regions of high precipitation, infiltration of excessive moisture into the compost pile can negatively impact carrion decomposition during composting. Under these circumstances, covering of the windrow or piles with a water-proof barrier or roof may be necessary. Glanville et al. (2006) reported that at least 45–60 cm of cocomposting material was required to cover the top of cattle carcasses and minimize the risk of pathogen transmission. More detailed recommendations for thickness of the final cocomposting layer and dimensions of the windrow for composting various sizes of animal carcasses have been established (NABC 2004; Table 23.3). However, for enhanced biosecurity, Xu et al. (2009a) designed a field-scale composting structure using straw bales, with carcasses contained in plastic sheeting and aeration supplied by plastic piping within a fenced area (Figure 23.3). Glanville et al. (2007) employed a similar composting system to dispose of swine carcasses.

Most windrow composting of carrion employs natural ventilation processes (Berge et al. 2009). To reduce the likelihood of pathogen transmission, windrows are not turned during biosecure composting, but if infectious disease control is not a priority, carrion compost can be turned immediately after the decline from thermophilic peak temperature. For example, Stanford et al. (2009) turned and mixed

TABLE 23.3

Dimensions and Recommended Thicknesses of Matrix Layers for Composting Windrows
to Dispose of Various Categories of Livestock Carcasses

Carcass Size	Bottom Width (m)	Top Width (m)	Height (m)	Thickness of Co-Composting Material Layer (cm)
Small (<23k g)	3.6	1.5	1.8	30
Medium (23–11 4kg)	3.9	0.3	1.8	45
Large and very large (114–227 kg and >227 kg)	4.5	0.3	2.1	60

Source: Adapted from NABC. 2004. *Carcass Disposal: A Comprehensive Review.* National Agriculture Biosecurity Centre Consortium. Kansas State University. Manhattan, KS.

Note: Composting windrow length need to be adjusted to accommodate the number of carcasses to be composted.

windrows containing cattle carcasses using a tractor equipped with a loader or a grinding bucket mounted on a backhoe after composting for 90 and 200 days, respectively. Turning often results in complete degradation of carrion after 300 days of composting (Hao et al. 2001; Xu et al. 2007). Although the initial cost for windrow composting is less than bin composting (NABC 2004), more intense management may be required owing to the open nature of the windrows.

23.3.3.2 Bin and In-Vessel Composting

Compost bins are typically used for small- or medium-sized carcasses such as swine, sheep, and poultry (Mescher et al. 1997; Stanford et al. 2000). Composting bins consist of a floor and walls generally constructed from wood, although the floor may also consist of concrete or gravel. Carrion and cocomposting materials are confined within a bin which may or may not be covered by a roof (NABC 2004). However, bins should be sized to accommodate the equipment used to handle the compost. Simple and economical bins can also be constructed using large hay or straw bales to adequately confine carrion compost (Kalbasi et al. 2005). Similar to windrow composting, a bottom layer of carbonaceous material and the layering of carrion are recommended within bin systems. Upon completion of the initial heating cycle, composted carrion can be transferred to a secondary bin, with soft tissues typically being completely decomposed during the primary phase (Keener et al. 2000). According to NABC (2004), every kilogram of carrion waste needs approximately 300 cm^3 of combined bin capacity for optimal composting. Keener et al. (2000) developed models for calculating the optimal volume of compost bins for composting poultry carrion.

In-vessel composting systems have been used for the disposal of various carrion wastes, such as small livestock carcasses (<100 kg), SRM, poultry feathers, and animal fats (Tiquia et al. 2005; Ruggieri et al. 2008; Xu et al. 2010). However, these systems are generally considered to be unsuitable for the composting of large carcasses unless they are sectioned or ground prior to composting (NABC 2004). During in-vessel composting, compost materials are contained within an insulated structure that is passively aerated by vents or actively aerated through the use of fans. However, due to the relatively short thermophilic composting period of in-vessel systems, composting can be continued in windrows or bins until decomposition is complete (Rynk 2003). Compost vessels allow a high degree of control and consistency in the composting process, but are restricted to relatively small quantities of material (Mason and Milke 2005). Therefore, these systems are also employed in laboratories to investigate the degradation of carrion tissues or inactivation of pathogens associated with the carrion wastes during composting (Ruggieri et al. 2008; Xu et al. 2013a). In practice, in-vessel systems generally need more capital and operating costs and have a limited capacity for disposal of large volumes of carrion. Forced aeration systems may also increase the risk of pathogen transmission. However, they have advantages over compost windrows or bins in that they largely exclude the impact of the environment on composting and are also an effective barrier to scavengers.

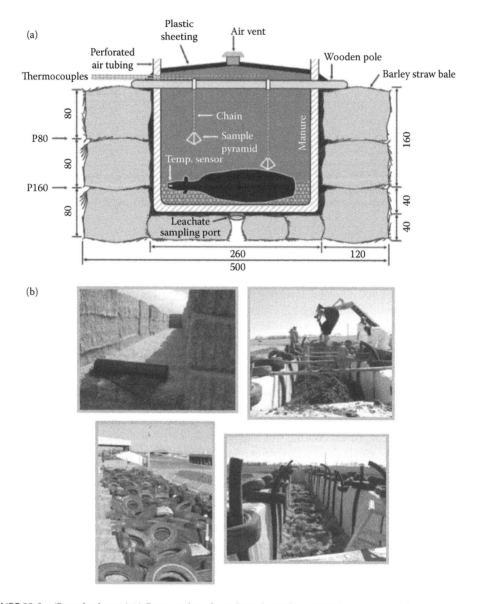

FIGURE 23.3 **(See color insert.)** (a) Cross-section of a static cattle carrion composting structure (16 carcasses per structure) designed and equipped with biosecurity plastic sheeting, gas vents, leachate ports, and retrievable sampling cages. (Reprinted from Xu, W., T. Reuter, G.D. Inglis et al. 2009a. *Journal of Environmental Quality* 38: 437–450. With permission.) Units are in cm. (b) Pictures of construction of the biosecure composting structure.

23.4 Biodegradation of Livestock Carrion during Composting

23.4.1 Differential Degradation of Carrion Tissues

Once carrion wastes enter compost, a variety of physical, chemical, and biological processes lead to their decomposition. Stentiford (1993) demonstrated that the rate of decomposition of organic matter varies according to its chemical composition. Usually, sugars, starch, lipids, amino acids, and nucleic acids are rapidly degraded, while hemicellulose, cellulose, chitin, lignin, and lignocellulose being degraded at much slower rates. Likewise, the rate of decomposition of carrion in compost depends on the nature of the tissues of which it is composed (Table 23.2). For example, Xu et al. (2009a) observed that the rate of decomposition for bovine tissues ranked as brain > hoof > bone over 230 days of composting. Previous research showed that when conducted properly, muscle and intestinal tissues decompose during the active composting phase, while hooves and fur are further degraded in the curing phase. Large bones (e.g., femur) and teeth are highly recalcitrant and may remain in finished compost and require removal by screening or other means if their application to land is to be avoided.

Proteins in carrion can also differ substantially in their rates of decomposition. Structural proteins such as collagen, elastin, and keratin provide tensile strength and rubber-like resilience to animal tissues and vary in their abundance in tissues (Suzuki et al. 2006). Keratins are the primary structural proteins in feathers, hooves, horns, and fur. Keratin proteins are stabilized by cross-linked disulfide bonds and beta-sheets which enhance their resistance to enzymatic degradation (Fuchs 1995). Consequently, feathers, hooves, horns, and fur, along with bones are the components of carrion that are most difficult to completely decompose during composting. Some eubacteria (e.g., *Bacillus licheniformis*), actinobacteria (e.g., *Streptomyces fradiae*), and fungi such as *Trichophyton* spp., *Microsporum* spp., *Arthroderma* spp., *Nannizia* spp., *Chrysosporium* spp., and *Myceliophtora* spp. are capable of degrading native keratin (Korniłłowicz-Kowalska and Bohacz 2010). Therefore, enrichment of compost for these keratinolytic microorganisms has been one approach to increasing the degradation of this recalcitrant substrate (see Section 23.4.4).

23.4.2 Relationship between Duration of Temperature Exposure and Carrion Degradation

Compost temperature reflects microbial activity, thus composts that exhibit high temperatures for sustained periods are generally more effective at decomposing carrion. Xu et al. (2009a) showed that the majority of SRM (>90%) was degraded within the first 7 days of thermophilic composting. Although Berge et al. (2009) demonstrated that decomposition of carrion still occurred at lower temperatures, the degradation process was prolonged. Stanford et al. (2009) found that the time required for degradation of bovine long bones (i.e., tibia and femur) was positively correlated with compost temperature. In practice, thermophilic composting (>55°C) is recommended for disposal of carrion. Excessively high temperatures (i.e., >70°C) can inactivate enzymes involved in the decomposition of carrion (Kutzner 2000), but as these temperatures are seldom achieved, this is unlikely to impede the composting process.

The time required for carrion decomposition also depends on the weight of the carcasses, with adult cattle requiring 9–12 months (Fleming and MacAlpine 2006; Stanford et al. 2007) and swine 4–10 months to decompose (Fonstad et al. 2003; Ahn et al. 2007). Smaller carcasses such as poultry required 2–3 months for decomposition (Sivakumar et al. 2008). Age of animal carcasses also impacts the time required for composting, as bones contain less connective tissue and become more ossified with age. Consequently, neonatal bones decompose more rapidly than those from adult animals (Stanford et al. 2000).

Season of the year, preparation procedures, and management can also affect the time required for decomposition of carrion in compost. NABC (2004) demonstrated that under optimal conditions, composting time is generally shorter at higher ambient temperatures. However, if high ambient temperatures reduce the moisture content of compost to suboptimal levels, the duration required for decomposition of carrion can increase (Sivakumar et al. 2008). Periodic mixing of the compost can also accelerate the decomposition of carrion as this process enhances aeration, redistributes moisture, and breaks up aggregates, exposing substrates to microbial attack. This enhanced microbial activity is reflected by the increase in compost temperature that often occurs after a mixing event.

23.4.3 Other Critical Factors Required for Carrion Degradation

Keener et al. (1993) reported that more than 20 factors can affect the decomposition of organic matter in compost and a number of these are summarized in Table 23.2.

23.4.3.1 Moisture

Appropriate moisture content is essential to achieve the thermophilic temperatures that are optimal for carrion decomposition during composting. Compost moistures ranging from 40 to 65% are desirable for optimal composting (Rynk 1992), but effective carrion composting has also been reported at 60%–80% moisture (Ahn et al. 2008). During composting, moisture contents are always higher in the vicinity of carrion, while outer matrix materials are often much dryer. Moreover, the longer period required for carrion degradation in windrow composting and its open nature increases moisture loss when compared to in-vessel composting (Xu et al. 2014). However, Ahn et al. (2008) recommended not applying water directly to carrion compost as it could generate leachate and transport pathogens into the broader environment. These authors recommended that moisture should be adjusted to optimal levels through the appropriate mixture of carrion and matrix materials prior to composting. The placement of bulking agents under the carrion can also be used to absorb any leachate produced.

However, moisture content alone may not reflect the bioavailability of water for microbial metabolism. One approach to identifying the proper moisture range for composting is to measure water activity (a_w), which measures the amount of freely available water for microbial populations. Ultimate levels of water activities ≥ 0.92 for bacteria and ≥ 0.70 for fungi ensure that there is adequate water available to promote and optimize microbial activity during composting (Reuter et al. 2010).

23.4.3.2 Oxygen

Typically, oxygen concentrations as a result of natural diffusion into compost piles are optimal at or near the surface of compost piles, but may be deficient in the core, where carrion is located (Berge et al. 2009). For optimal composting, oxygen concentrations should be above 5% with levels greater than 10% being more desirable (Kutzner 2000). Considering the volume of carrion wastes, maintenance of adequate aeration during composting is critical for complete degradation of tissues and avoidance of the offensive odors associated with anaerobic degradation or the production of methane, a potent greenhouse gas. Although forced aeration using fans or compressed air can provide sufficient oxygen to facilitate the complete degradation of organic matter in compost, natural convection has been shown to be more economical and suitable for carrion composting, especially when infectious pathogens are associated with the waste (Xu et al. 2010, 2009). Keener et al. (2000) also recommended the transfer of compost from a primary to a secondary compost pile or bin to ensure adequate aeration and mixing of compost. Furthermore, the nature of the composting matrix mixed with carrion can affect the availability of oxygen for microbes involved in carrion decomposition. Ahn et al. (2007) observed that cocomposting of swine carrion with silage significantly impeded aeration and slowed its degradation when compared to cornstalks or oat straw.

23.4.3.3 Porosity

Porosity, bulk density, particle density, and free air space are used to describe the physical nature of the composting matrix. Since the availability of substrates for microbial growth is dependent on particle surface area, the larger surface area of smaller particles usually accelerates the composting process. However, smaller particles can also pack more densely, reducing porosity and possibly promoting the development of anaerobic conditions that can impede composting (Agnew and Leonard 2003).

For optimal composting, free air space should account for 20%–30% of the total volume of compost (Kutzner 2000) with particle sizes ranging from 3 to 50 mm (Rynk 1992). Most carrion wastes exceed this optimal particle size unless they are chopped or ground prior to composting. However, these processes can create aerosols that increase the risk of pathogen transmission. Consequently, such practices are discouraged if the carrion is known to have arisen as a result of mortalities from infectious disease. Bulking agents such as wood by-products (e.g., chips, shavings, sawdust, and postpeelings) and crop residues can be used to increase the compost porosity and make excellent matrixes for the composting of carrion. Kalbasi et al. (2005) recommended that mixing carbonaceous materials with carrion at a ratio of 1:1 (v v⁻¹) can create a suitable level of porosity for windrow or pile composting.

23.4.3.4 Carbon to Nitrogen (C/N) Ratio

Carbon and nitrogen are the primary nutrients required by the microbes involved in carrion decomposition. Carr et al. (1998) suggested a range of C/N ratios of 20:1 to 35:1 for efficient carrion composting. However, most livestock carrion has excessive nitrogen. Thus, bulking agents such as wood shavings, sawdust, and straw can be added as carbon sources (Table 23.2). Bohacz and Korniłłowicz-Kowalska (2009) suggested addition of rye straw or pine bark to obtain an optimum C/N ratio level of 25:1 for composting chicken feathers. Xu et al. (2010) mixed wood shavings or barley straw with manure to adjust initial compost C/N ratio for effective composting of SRM.

23.4.3.5 pH

Compost pH can affect the growth of microorganisms and the inactivation of pathogens within carrion compost (Ryckeboer et al. 2003; Xu et al. 2013a). Optimal composting of carrion occurs over a pH range of 6.5–8.0 (Langston et al. 2002). However, if adequate buffering capacity is present within the matrix, composting may occur over a broader pH range.

Frequently, compost pH changes over the composting period. Xu et al. (2009a) observed that pH increased from 8.3 to 9.4 during the composting of cattle carrion. The pH is influenced by the degradation products arising from substrates as well as other factors such as the degree of aeration. Anaerobic conditions can produce organic acids which lower compost pH (Beck-Friis et al. 2000). In general, pH frequently decreases shortly after the initiation of composting as the oxygen demand of aerobic microorganisms involved in decomposition often exceeds oxygen availability, leading to localized anaerobic areas where organic acids are produced. However, as decomposition proceeds, pH may increase as ammonia is released into the matrix as a result of deamination of amino acids arising from carrion protein and the metabolism of organic acids (Xu et al. 2009a). During the latter stages of carrion composting, volatilization of ammonia and nitrification may lead to a decline in compost pH (Xu et al. 2010).

23.4.4 Microbial Communities Associated with Carrion Degradation during Composting

The decomposition of carrion during composting relies on a diverse microbial community that is naturally present within the matrix materials or the carrion. Xu et al. (2010) observed that the most rapid period of degradation of SRM during composting corresponded with the thermophilic period where the most rapid increase in microbial biomass was also observed. However, microbial community succession occurs during the composting process as determined by the selective pressures of temperature, moisture, nutrient supply, and substrate availability. Microbial diversity and a defined succession of populations is a prerequisite to adequate degradation of organic matter during composting (Ryckeboer et al. 2003; Tkachuk et al. 2014).

The nature of microbial communities associated with the degradation of carrion during composting has not been widely investigated. Most previous research has focused on quantitative and qualitative changes in microbial communities, with few studies directly measuring microbial activity. For instance, Xu et al. (2010, 2013b) used phospholipid fatty acid and denaturing gradient gel electrophoresis (DGGE) to reveal that a wide variety of bacteria and fungi were responsible for the degradation of SRM in compost. Given the complexity of its composition, carrion requires a myriad of functional microbial communities for its complete decomposition during composting. Accordingly, Xu et al. (2013b) suggested that the bacterial and fungal communities associated with the degradation of SRM differ from those involved in the degradation of other matrix materials. Cayuela et al. (2009) also observed that the composition of microbial communities differed between compost matrices containing either bone and meat meal or horn and hoof meal. Moreover, during composting of poultry wastes, Bohacz and

Korniłłowicz-Kowalska (2007) reported that including recalcitrant feathers in the matrix stimulated the growth of more specialized keratinolytic fungi that facilitated the degradation of keratin. These keratinolytic fungi could also facilitate the breakdown of other carrion tissues, as they have the ability to hydrolyze various types of animal proteins including casein, bovine serum albumin, and elastin (Friedrich and Kern 2003). In agreement with this hypothesis, Xu et al. (2013a, 2014) further observed that inclusion of feathers in manure compost enhanced the degradation of both SRM and scrapie prion proteins after 14 days of composting. However, knowledge of the ecological role of these functional taxa within compost microbial communities remains rudimentary.

A wide variety of bacteria and fungi have been isolated from different phases of composting (Ryckeboer et al. 2003). However, characterization of these microbes associated with carrion decomposition in compost has rarely been undertaken. Using DGGE with sequencing analysis, Xu et al. (2013b) reported that bacteria related to the genera Saccharomonospora, Thermobifida, Thermoactinomycetaceae, Thiohalospira, Pseudomonas, Actinomadura, and Enterobacter, and fungi related to the genera Dothideomycetes, Cladosporium, Chaetomium, and Trichaptum were involved in SRM degradation during composting. Sanabria-León et al. (2007) utilized a commercial Biolog system to characterize several bacterial genera including *Pasteurella, Aquaspirillum, Alcaligenes, Bacillus, Vibrio,* and *Providencia* involved in the composting of slaughterhouse wastes. Furthermore, Cayuela et al. (2009) used microarrays to confirm the presence of more than 60 bacterial species during composting of meat and hoof wastes. Using high-throughput sequencing, we recently found that the bacterial phyla Actinobacteria, Firmicutes, Proteobacteria, Bacteroidetes, and Synergistetes were associated with the degradation of beef carrion in a field scale composting system (Tkachuk et al. 2014). Similar bacterial phyla were also documented during the natural decomposition of swine carrion under ambient conditions (Pechal et al. 2014), but the predominant genera differed from those in compost. The specific substrates that these microbes degraded during the composting of carrion have yet to be identified.

Effective degradation of carrion suggests that compost could be a potential reservoir of microbes (e.g., *Actinomadura keratinilytica* sp. nov.) with the ability to degrade recalcitrant proteins (Puhl et al. 2009). Lin et al. (1999) isolated a strain of *B. licheniformis* from composted canola meal that could effectively degrade feather wastes. Subsequently, Ichida et al. (2001) observed that degradation of feathers was more complete in compost inoculated with *B. licheniformis* and enriched with feathers. Thus, further characterization of these highly active proteolytic bacteria and fungi in carrion compost could enhance our understanding of the specific role of these microbes in carrion biodegradation.

23.5 Environmental and Health Risk Associated with Composting Livestock Carrion

23.5.1 Potential Pathogens

Carrion generated during livestock farming and from meat processing plants and road kills could serve as a reservoir or as substrate for the enrichment of pathogenic bacteria, viruses, prions, or parasites. Pathogens in carrion can be mobilized through surface water and leach into soil or groundwater during decomposition, potentially spreading infective agents to animals or humans (Figure 23.4). Another potential route of pathogen acquisition may be via solid or liquid waste generated in rendering plants and slaughterhouses that are unknowingly processing infected carrion (Figure 23.4; Franke-Whittle and Insam 2013). A wide variety of bacteria such as *Escherichia coli* O157:H7, *Salmonella, Listeria,* and *Campylobacter,* and viruses including hepatitis A and avian influenza can be associated with livestock and infect both animals and humans (McAllister and Topp 2012; Chen et al. 2013).

Furthermore, emergency disposal of livestock carrion wastes related to the catastrophic disease requires more precautions and hygienic measures to prevent pathogen transmission. In 1984, the outbreak of avian influenza in Virginia, USA, cost poultry farms $40 million and required the disposal of 5700 tons of poultry carrion with 88% of this material disposed of on-site in burial trenches. However, these sites were associated with the contamination of groundwater up until the late 1990s (Flory and Peer 2010). Fortunately, this contamination was not associated with the transmission of an infectious disease.

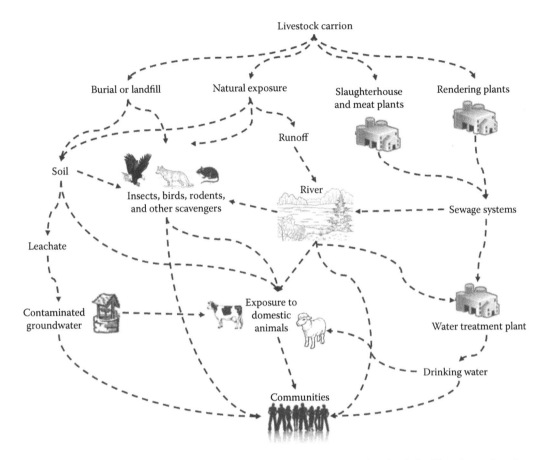

FIGURE 23.4 Avenues of flow of pathogens arising from the improper disposal of carrion derived from livestock production and road kills. Routine disposals include (1) natural exposure; (2) burial or landfill; and (3) rendering. Note that proper disposal of carrion wastes plays a critical role in preventing the transmission of infectious pathogens to both animals and humans. (Adapted from McAllister, T.A. and E. Topp. 2012. *Animal Frontiers* 2(2): 17–27.)

23.5.2 Pathogen Inactivation during Composting

Composting is a particularly effective method for the disposal of carrion as temperatures commonly exceed 55°C, inactivating most pathogens (Table 23.4) including *Listeria* (Erickson et al. 2009a), Shiga-toxigenic *Escherichia coli* (Xu et al. 2009a), *Salmonella* (Erickson et al. 2009b), Giardia and *Cryptosporidium* (Van Herk et al. 2004), and viruses associated with avian influenza, Newcastle disease (Guan et al. 2009), and foot-and-mouth disease (Guan et al. 2010).

23.5.2.1 Bacterial Pathogens

Many bacterial pathogens can survive outside of a host (Berge et al. 2009), and the degree and duration of exposure to elevated temperature are important determinants of their degree of inactivation during composting. Guidelines from CCME (Canadian Council of Ministers of the Environment 2005) and USEPA (United States Environmental Protection Agency 1995) suggest that compost temperatures should exceed 55°C for at least 15 consecutive days in windrows that are turned three times, and 3 consecutive days in static or in-vessel composters. These recommendations differ among countries, with Austria requiring a minimum temperature of 65°C for at least 6 consecutive days or two 3-day periods >65°C (Franke-Whittle and Insam 2013).

TABLE 23.4

Inactivation of Selected Pathogens Associated with Carrion Wastes during Composting

Pathogen	Pathogen Survivability	Method Used	Substrate Analyzed	Cocomposting Material	Compost System	Compost Temperature	Compost Moisture (Wet Basis)	pH	Reference
Bacteria *Salmonella* spp.	Complete inactivation after average of 4 days	Culturing	Dairy manure	Manure, straw, cottonseed	15-L composter	Maximum 57°C; >50°C for 3 days	50%–60%	7.5–9.2	Erickson et al. (2009b)
E. coli O157:H7	Absent after 3 days at the bottom location	Culturing	Dairy manure	Manure, vegetable waste	Compost pile	Maximum 60°C; >55°C for 8 days	45%–14%	7–9	Shepherd Jr. et al. (2010)
Listeria moncytogenes	Complete inactivation after 5 days	Culturing	Dairy manure	Manure, straw, cottonseed	15-L composter	Maximum 56°C; increased to 40°C within 1 day	50%–60%	7.1–8.4	Erickson et al. (2009a)
Clostridum perfringens	No decreased after 24 months	qPCR	Biosolid	Biosolid, wood debris	Windrow	—	—	—	Karpowicz et al. (2010)
Campylobacter jejuni	No decrease with the time	Culturing and qPCR	Bovine manure	Bovine manure	Windrow	>55°C for 62 days	Average 42.5%	7.9–8.4	Inglis et al. (2010)
M. avium subsp. *paratuberculosis* (Johne's disease)	Still detectable after 250 days	qPCR	Infectious lymph tissues	Cattle carcasses, manure, straw	Biosecure structure	10–64°C; remained >50°C for 35–60 days	48%–80%	7.5–9.0	Tkachuk et al. (2013)
Virus Avian influenza (H6N2)	Virus inactivated by day 7; viral RNA degraded by day 21	Culturing and qPCR	Chicken manure, litter, tissue	Chicken manure, corn silage	Compost barrel and bin	40–50°C by day 3; 50–60°C by day 7	65% at the start	—	Guan et al. (2009)
Bovine viral diarrhea virus	Virus and viral RNA inactivated at >40°C	Culturing and qPCR	Bovine splenic tissue	Manure, straw, cattle carcasses	Compost bin	>50°C at least 14 days	65% at the start	—	Guan et al. (2012)
Foot-and-mouth disease virus	Virus inactivated by day 10; viral RNA degraded by day 21	qPCR and ELISA	Swine carcass tissues	Chicken manure, wood shavings	Compost bin	>50°C by day 10; 70°C by day 19	65% at the start	—	Guan et al. (2010)
Newcastle disease virus	Virus inactivated by day 7; viral RNA degraded by day 21	Culturing and qPCR	Chicken manure, litter, tissues	Chicken manure, corn silage	Compost barrel and bin	40–50°C by day 3; 50–60°C by day 7	65% at the start	—	Guan et al. (2009)

However, relying solely on time–temperature criteria for bacterial pathogen inactivation does not take into consideration other factors that may impact compost safety. Factors, such as moisture, C/N ratio, pH, available nutrients, ammonia volatilization, and competition among indigenous microflora can also affect the survival of bacterial pathogens. Desiccation of compost can induce water stress in bacterial pathogens, whereas water saturation can create anoxic conditions that slow the heating of compost and prolong pathogen survival. Moreover, the pH of manure-based compost is high (>8) owing to its ammonia content and is thus unfavorable for many bacterial pathogens. Ammonia volatilization during thermophilic composting may also contribute to pathogen inactivation (Himathongkham and Riemann 1999; Singh et al. 2012). Singh et al. (2011) observed that compost with optimal C/N ratio was more effective at inactivating *E. coli* O157:H7 than compost with a suboptimal C/N ratio. Lafond et al. (2002) found that fecal *Streptococci* and total Coliforms were more rapidly inactivated in compost with C/N ratio of 33:1 than with a C/N ratio of 68:1.

Some bacteria such as *Bacillus* and *Clostridium* can form spores that remain viable even under extreme thermophilic conditions(Berge et al. 2009). As temperatures decline, these spores can germinate, re-establishing vegetative bacteria within compost. Competition for nutrients between pathogenic and nonpathogenic bacteria could further contribute to a decline in the survival of bacterial pathogens during composting. Sidhu et al. (2001) found that the growth of *Salmonella* in mature compost declined as organic carbon and nitrogen were depleted, whereas the densities of other bacteria remained relatively stable. However, the inhibitory effect of indigenous microorganisms may be greater when sufficient nutrients are available. Optimized nutrient conditions can increase the activity of indigenous microorganisms, enhancing their antagonism towards pathogens (Sidhu et al. 2001). In addition, some highly active bacteriophages inherently present in compost may also contribute to the inactivation of bacterial pathogens (Heringa et al. 2010).

23.5.2.2 *Viral Pathogens*

Most viruses are heat-sensitive and elevated temperature is the major factor responsible for their inactivation during composting. The lack of an envelope in some viruses renders them even more thermally intolerant as is the case with the virus responsible for foot-and-mouth disease (Xu et al. 2009a). Inactivation of viruses is accelerated at temperatures $\geq 50°C$ (Bertrand et al. 2012), with Guan et al. (2009) demonstrating that the avian influenza virus H6N2 was inactivated after 7 days of composting at temperatures over 50°C. Further studies by Elving et al. (2012) found H7N1 strain was rendered nonviable after 24 h of composting.

Other factors can also impact inactivation of viral pathogens in compost. Guan et al. (2010) reported that the time required for inactivation of foot-and-mouth disease virus was positively correlated with the initial concentration of the virus in infected tissues. Guan et al. (2009) also observed that degradation of avian influenza virus was influenced by the nature of the matrix during composting with inactivation being more rapid in chicken litter (i.e., a mixture of chicken manure, oat straw, and wood shavings) than chicken manure or tissues, probably due to a greater microbial activity induced by an optimal C/N ratio in chicken litter. Moreover, compost pH, microbial antagonism, toxicity from catabolites arising from the degradation of organic matter such as ammonia, sulfides, or organic acids or even proteolytic enzymes produced by microbes could influence the infectivity of viruses.

23.5.2.3 *Prions*

Transmissible spongiform encephalopathies (TSE) are a class of fatal neurodegenerative diseases and include scrapie in sheep and goats, chronic wasting disease (CWD) in deer and elk, BSE in cattle, and Creutzfeldt-Jackob disease in humans. TSE arise as a result of a conformational change in the normal prion protein (PrP[C]) to an infectious version designated as PrP[TSE].

Differing from bacterial and viral pathogens, PrP[TSE] have been shown to be extremely resistant to inactivation by a wide range of methods commonly used to kill microorganisms and denature nucleic acids and proteins (Taylor 2000). Several rigorous practices are currently used for inactivation of PrP[TSE] including a two-stage incineration at 850 and 1000°C for 16 h, 1-h of hydrolysis in 2 N sodium hydroxide, autoclaving in saturated steam at 134°C for 60 min, thermal hydrolysis at 180°C under 12 atmospheric pressure for at least 40 min, and alkaline hydrolysis at 150°C and 4 atmospheric pressure with 15% sodium hydroxide or 19% potassium hydroxide (w w^{-1}) added for no less than 180 min (Brown et al. 2004; BCMA 2009). Consequently, inactivation of prions in carrion by composting was thought to be infeasible (Franke-Whittle and Insam 2013).

However, several microbial proteinases can degrade recalcitrant PrP^{TSE} in minutes or hours (Suzuki et al. 2006) and these bacterial species have been shown to be present in compost (Ryckeboer et al. 2003). The microbial consortia in compost may biodegrade PrP^{TSE}, owing to the wide range of proteolytic enzymes produced by these complex microbial communities. Moreover, the period of time that PrP^{TSE} could be exposed to enzymatic activity during composting is much longer than the duration required for purified proteases to degrade PrP^{TSE}.

Huang et al. (2007) showed that scrapie prions in carrion were degraded to below detectable levels by western blot after 108–148 days of composting. More recently, our research group observed a reduction of 3 log$_{10}$ in CWD and 1 log$_{10}$ in BSE prions in carrion after 28 days of laboratory-scale composting and at least a 4.8 log$_{10}$ reduction of scrapie 263 K prions after 230 days of field-scale composting using a hamster bioassay (Xu et al. 2014). These results support that composting can result in the effective degradation of PrP^{TSE} in carrion.

23.5.3 Challenges to Safe Disposal of Livestock Carrion by Composting

Despite the extensive literature available on the inactivation of pathogens during carrion composting, few studies have investigated the survival of pathogens in infected tissues. Some recalcitrant pathogens, such as *Mycobacterium avium* subsp. *Paratuberculosis*, remain viable after composting, especially if they are embedded within infected tissues (Tkachuk et al. 2013; Table 23.4). In some laboratory-scale composting studies, optimized compost conditions are well controlled with a homogenous matrix, leading to the complete inactivation of most pathogens (Table 23.4). However, carrion compost being a heterogeneous mixture of animal tissues and other matrix components makes it difficult to avoid spatial variability in microbial activity. Pathogens can survive and carrion can persist in regions of the pile where microbial activity is compromised. If moisture levels become excessive, leachate or percolate from compost can enter the soil or groundwater along with nutrients that can support the growth of pathogens (Kim et al. 2009). Other factors, such as the nature of other components of the composting matrix, weather conditions, and seasonal changes, can also affect the efficiency of carrion decomposition and pathogen inactivation (Shepherd Jr. et al. 2010; Glanville et al. 2013). Therefore, these variables increase the challenges to safe disposal of carrion by composting.

23.6 Current Regulations and Guidelines for Carrion Composting

Globally, composting for the disposal of livestock carrion has not gained popularity in some continents such as Asia, Africa, or South America, probably due to the reuse of carrion as human food or animal feed. However, this option is being reinvestigated by intensive farming communities in Australia, New Zealand, and North America (Wilkinson 2007). Currently, composting has been approved in several states in the United States as a method for routine or emergency disposal of livestock carrion, while others have not established regulations or prohibit the practice (Berge et al. 2009). In Minnesota, USA, composting of poultry, swine, sheep, and goats is allowed without a permit, whereas disposal of other livestock species requires a permit (Morse 2009). In Canada, government agencies are being encouraged to build the guidelines for accepting composting as a method for disposal of carrion wastes. These guidelines also vary among provinces, but most have specified the minimum distance that carrion composting sites can be established in relation to watercourses, wells, and residences as well as the parameters required for optimal composting (Cleary et al. 2010). The Canadian Food Inspection Agency (CFIA 2011) has allowed the limited use of composting for the disposal of SRM through issuing of temporary permits, and although Europe has investigated this method of carrion disposal, they have yet to approve it (Gwyther et al. 2011).

Generally, the final product of composting is sufficiently stable and free of viable pathogens and weed seeds that it is suitable for land application. However, the application of finished carrion compost to land may be restricted. In the United States, although some states allow carrion composting, they do not have specific regulations with regard to the land application of the finished product (Morse 2009). Application of finished product onto land owned by the composting site may be allowed as long as it does not pose a public health hazard (Bass et al. 2012). In Canada, the government does not recommend the direct spread of carrion compost onto pasture land or land that is used for crop production, but does allow its use for land reclamation or tree farming, provided that the land is not grazed for at least 5 years (OMAF 2010). Moreover, removal of carrion compost from the farm of origin is allowable with a permit issued by CFIA, but the direct sale of carrion compost is prohibited (CFIA 2011).

23.7 Conclusions

Carrion composting is a natural biological process that effectively reduces the volume of carrion and inactivates most carrion-associated pathogens. This method of disposal has the advantages of being low cost, applicable on farm, achievable using a wide range of composting matrices and resulting in the production of a valuable soil amendment.

Adequate design and operation of carrion-composting systems are prerequisites to ensure that carrion is completely degraded and pathogens inactivated. Tissue degradation and pathogen inactivation may not be achieved under suboptimal composting conditions. Although some recalcitrant pathogens such as infectious prions or bacterial spores may not be completely destroyed by composting, potential for the spread of these pathogens at infectious doses after composting is likely negligible owing to dilution and biological inactivation. Composting has considerable merit as a method for the disposal of livestock carrion as opposed to natural decomposition.

REFERENCES

AARD. 2011. Poultry mortality composting. Agdex 450/29-1. Alberta Agriculture and Rural Development, Edmonton, AB. http://www1.agric.gov.ab.ca/$department/deptdocs.nsf/all/agdex6117 (accessed May 23, 2013).

Agnew, J.M. and J.J. Leonard. 2003. The physical properties of compost. *Compost Science & Utilization* 11(3): 238–264.

Ahn, H.K., T.D. Glanville, B.P. Crawford, J.A. Koziel, and N. Akdeniz. 2007. Evaluation of the biodegradability of animal carcasses in passively aerated bio-secure composting system. Paper presented at the meeting of ASABE (Paper No. 074037), Minneapolis, MN.

Ahn, H.K., T.L. Richard, and T.D. Glanville. 2008. Optimum moisture levels for biodegradation of mortality composting envelope materials. *Waste Management* 28: 1411–1416.

Arvanitoyannis, I.S. and D. Ladas. 2008. Meat waste treatment methods and potential uses. *International Journal of Food Science & Technology* 43: 543–559.

Barone, J.R. and O. Arikan. 2007. Composting and biodegradation of thermally processed feather keratin polymer. *Polymer Degradation and Stability* 92: 859–867.

Bass, T., D. Colburn, J. Davis et al. 2012. Livestock mortality composting for large and small operations in the semi-arid west. SKU EB0205. Montana State University, Bozeman, MT. http://store.msuextension.org/Products/Livestock-Mortality-Composting-for-Large-and-Small-Operations-in-the-Semi-Arid-West__EB0205.aspx (accessed August 14, 2014).

BCMA. 2009. Slaughterhouse waste and specified risk material (SRM) management study. British Columbia Ministry of Agriculture. Final report for Vancouver Island. Prepared by the Ground Up Resource Consultants Inc., Courtenay, BC. http://www.agf.gov.bc.ca/resmgmt/SRM_Program/index.htm (accessed May 8, 2013).

Beck-Friis, B., M. Pell, U. Sonesson, H. Jönsson, and H. Kirchmann. 2000. Formation and emission of N_2O and CH_4 from compost heaps of organic household waste. *Environmental Monitoring and Assessment* 62: 317–331.

Berge, A.B., T.D. Glanville, P.D. Millner, and D.J. Klingborg. 2009. Methods and microbial risks associated with composting of animal carcasses in the United States. *Journal of the American Veterinary Medical Association* 234: 47–56.

Bertrand, I., J.F. Schijven, G. Sánchez et al. 2012. The impact of temperature on the inactivation of enteric viruses in food and water: A review. *Journal of Applied Microbiology* 112: 1059–1074.

Bohacz, J. and T. Korniłłowicz-Kowalska. 2007. Choice of maturity indexes for feather-plant material composts on a base of selected microbiological and chemical parameters—Preliminary research. *Polish Journal of Environmental Studies* 16(2A): 719–725.

Bohacz, J. and T. Korniłłowicz-Kowalska. 2009. Changes in enzymatic activity in composts containing chicken feathers. *Bioresource Technology* 100: 604–612.

Bonhotal, J., E.Z. Harrison, and M. Schwarz. 2007. Composting road kill. Cornell Cooperative Extension. Cornell Waste Management Institute. Ithaca, NY: Cornell University. http://cwmi.css.cornell.edu/tirc. htm (accessed August 3, 2013).

Brown, P., E.H. Rau, P. Lemieux, B.K. Johnson, A.E. Bacote, and D.C. Gajdusek. 2004. Infectivity studies of both ash and air emissions from simulated incineration of scrapie-contaminated tissues. *Environmental Science & Technology* 38: 6155–6160.

Canadian Cervid Alliance. 2009. Canadian overview: The cervid industry in Canada. http://www.cervid.ca/en/ canadiancervidalliancecanadianoverview.php (accessed March 11, 2012).

Carr, L., H.L. Broide, H.M. John, G.W. Malcone, D.H. Palmer, and N. Zimmermann. 1998. Composting catastrophic event poultry mortalities. Fact Sheet 723. University of Maryland & Maryland Cooperative Extension. College Park, MD. http://www.enst.umd.edu/compost/resources (accessed June 7, 2013).

CAST. 2009. Ruminant carcasses disposal options for routine and catastrophic mortality. The science source for food, agricultural, and environmental issues paper. No. 41. Council for Agricultural Science and Technology, Ames, IA. http://www.cast-science.org/publications/?ruminant_carcass_disposal_options_ for_routine_and_catastrophic_mortality&show=product&productID=2944 (accessed August 14, 2014).

Cayuela, M.L., C. Mondini, H. Insam, T. Sinicco, and I. Franke-Whittle. 2009. Plant and animal wastes composting: Effects of the N source on process performance. *Bioresource Technology* 100: 3097–3106.

CCME. 2005. Guidelines for compost quality. http://www.ccme.ca/ourwork/waste.html?category_id=132 (accessed May 2, 2012).

CFIA. 2011. Specified risk materials—Requirements for fertilizers and supplements. http://www.inspection. gc.ca/plants/fertilizers/registration-requirements/srm/eng/1320613799112/1320615608072 (accessed July 20, 2013).

CFIA. 2013. Meat hygiene manual of procedures. Chapter 17: Ante and post-mortem procedures, dispositions, monitoring and controls—Red meat species, ostriches, rheas and emus—Annex D [online]. Canadian Food Inspection Agency. http://www.inspection.gc.ca/english/fssa/meavia/man/ch17/annexde. shtml#a1(accessed July 3, 2013).

Chen, Y., W. Liang, S. Yang et al. 2013. Human infectious with the emerging avian influenza A H7N9 virus from wet market poultry: Clinical analysis and characterization of viral genome. *Lancet* 381(9881): 1916–1925.

Cleary, B.A., R.J. Gordon, R.C. Jamieson, and C.B. Lake. 2010. Waste management of typical livestock mortalities in Canada: An overview of regulations and guidelines. *Canadian Biosystems Engineering* 52: 6.11–6.18.

Croston, D. 1997. Update on progress with BSE (mad-cow disease) in the UK. *Livestock Production Science* 52(3): 235–237.

de Gannes, V., G. Eudoxie, and W.J. Hickey. 2013. Prokaryotic successions and diversity in composts as revealed by 454-pyrosequencing. *Bioresource Technology* 133: 573–580.

Dewell, R.D., T.W. Hoffman, D.R. Woerner et al. 2008. Estimated compliance for removal of specified risk materials from 18 U.S. beef packing plants. *Journal of Food Protection* 71(3): 573–577.

Ellis, D.B. 2001. Carcass disposal issues in recent disasters, accepted methods, and suggested plan to mitigate future events. Master Dissertation, Southwest Texas State University.

Elving, J., E. Emmoth, A. Albihn, B. Vinnerås, and J. Ottoson. 2012. Composting for avian influenza virus elimination. *Applied and Environmental Microbiology* 78(9): 3280–3285.

Erickson, M.C., J. Liao, L. Ma, X.P. Jiang, and M.P. Doyle. 2009a. Pathogen inactivation in cow manure compost. *Compost Science & Utilization* 17(4): 229–236.

Erickson, M.C., J. Liao, L. Ma, X.P. Jiang, and M.P. Doyle. 2009b. Inactivation of *Salmonella* spp. in cow manure composts formulated to different initial C:N ratios. *Bioresource Technology* 100: 5898–5903.

Fedorowicz, E.M., S.F. Miller, and B.G. Miller. 2007. Biomass gasification as a means of carcass and specified risk materials disposal and energy production in the beef rendering and meatpacking industries. *Energy Fuels* 21: 3225–3232.

Fleming, R. and M. MacAlphine. 2006. On-farm composting of beef mortalities. Paper presented at the meeting of CSBE/SCGAB (Paper No. 06110), Edmonton, AB.

Flory, G.A. and R.W. Peer. 2010. Verification of poultry carcass composting research through application during actual avian influenza outbreaks. *ILAR Journal* 51(2):149–157.

Fogarty, A.M. and O.H. Tuovinen. 1991. Microbial degradation of pesticides in yard waste composting. *Microbiological Reviews* 55(2): 225–233.

Fonstad, T.A., D.E. Meier, L.J. Ingram, and J. Leonard. 2003. Evaluation and demonstration of composting as an option for dead animal management in Saskatchewan. *Canadian Biosystems Engineering* 45: 6.19–6.25.

Food and Agriculture Organization of the United Nations. 2013. http://faostat.fao.org (accessed April 23, 2013).

Forgács G., M. Lundin, M.J. Taherzadeh, and I.S. Horváth. 2013. Pretreatment of chicken feather waste for improved biogas production. *Applied Biochemistry and Biotechnology* 169: 2016–2028.

Franke-Whittle, I.H. and H. Insam. 2013. Treatment alternatives of slaughterhouse wastes, and their effect on the inactivation of different pathogens: A review. *Critical Reviews in Microbiology* 39(2): 139–151.

Friedrich, J. and S. Kern. 2003. Hydrolysis of native proteins by keratinolytic protease of *Doratomyces microsporus*. *Journal of Molecular Catalysis B: Enzymatic* 21: 35–37.

Fuchs, E. 1995. Keratins and the skin. *Annual Review of Cell and Development Biology* 11: 123–153.

Gilroyed, B.H., T. Reuter, A. Chu, X. Hao, W. Xu, and T. McAllister. 2010. Anaerobic digestion of specified risk materials with cattle manure for biogas production. *Bioresource Technology* 101(15): 5780–5785.

Glanville, T.D., Ahn, H.K., Koziel, J.A., Akdeniz, N. and Crawford, B.P. 2007. Performance evaluation of a passively-aerated plastic-wrapped composting system designed for emergency disposal of swine mortalities. Paper presented at the meeting of ASABE (Paper No. 074038), Minneapolis, MN.

Glanville, T.D., H.K. Ahn, T.L. Richard, J.D. Harmon, D.L. Reynolds, and S. Akinc. 2013. Effect of envelope material on biosecurity during emergency bovine mortality composting. *Bioresource Technology* 130: 545–551.

Glanville, T.D., T.J. Richard, J.D. Harmon, D.L. Reynolds, H.K. Ahn, and A. Akinc. 2006. Environmental impacts and biosecurity and composting for emergency disposal of livestock mortalities. Agricultural and Biosystems Engineering Project Reports. Ames, IA: Iowa State University. http://works.bepress.com/jay_harmon/56 (accessed May 23, 2013).

Guan, J., M. Chan, B.W. Brooks, J.L. Spencer, and J. Algire. 2012. Comparing *Escherichia coli* O157:H7 phage and bovine viral diarrhea virus as models for destruction of classical swine fever virus in compost. *Compost Science & Utilization* 20: 8–23.

Guan, J., M. Chan, C. Grenier, D.C. Wilkie, B.W. Brooks, and J.L. Spencer. 2009. Survival of avian influenza and Newcastle disease viruses in compost and at ambient temperatures based on virus isolation and real-time reverse transcriptase PCR. *Avian Diseases* 53(1): 26–33.

Guan, J., M. Chan, C. Grenier et al. 2010. Degradation of foot-and-mouth disease virus during composting of infected pig carcasses. *Canadian Journal of Veterinary Research* 74(1): 40–44.

Gwyther C.L., A.P. Williams, P.N. Golyshin, G. Edwards-Jones, and D.L. Jones. 2011. The environmental and biosecurity characteristics of livestock carcass disposal method: A review. *Waste Management* 31: 767–778.

Haigh, J., J. Berezowski, and M.R. Woodbury. 2005. A cross-sectional study of the causes of morbidity and mortality in farmed white-tailed deer. *The Canadian Veterinary Journal* 46: 507–512.

Hao, X, C. Chang, F.J. Larney, and G.R. Travis. 2001. Greenhouse gas emissions during cattle feedlot manure composting. *Journal of Environmental Quality* 30: 376–386.

Hao, X., K. Stanford, T.A. McAllister, F.J. Larney, and S. Xu. 2009. Greenhouse gas emissions and final compost properties from co-composting bovine specified risk material and mortalities with manure. *Nutrient Cycling in Agroecosystems* 83: 289–299.

Haug, R.T. 1993. *The Practical Handbook of Compost Engineering*. Ann Arbor: Lewis Publications.

Heringa, S.D., J. Kim, X. Jiang, M.P. Doyle, and M.C. Erickson. 2010. Use of a mixture of bacteriophages for biological control of *Salmonella enterica* strains in compost. *Applied and Environmental Microbiology* 76(15): 5327–5332.

Himathongkham, S. and H. Riemann. 1999. Destruction of *Salmonella typhimurium*, *Escherichia coli* O157:H7 and *Listeria monocytogenes* in chicken manure by drying and/or gassing with ammonia. *FEMS Microbiology Letters* 171: 179–182.

Huang, H., J.L. Spencer, A. Soutyrine, J. Guan, J. Rendulich, and A. Balachandran. 2007. Evidence for degradation of abnormal prion protein in tissues from sheep with scrapie during composting. *Canadian Journal of Veterinary Research* 71: 34–40.

Huijser, M.P., J.W. Duffield, A.P. Clevenger, R.J. Ament, and P.T. McGowen. 2009. Cost–benefit analyses of mitigation measures aimed at reducing collisions with large ungulates in the United States and Canada: A decision support tool. *Ecology and Society* 14(2): 15.

Ichida, J.M., L. Krizova, C.A. LeFevre, H.M. Keener, D.L. Elwell, and E.H. Burtt Jr. 2001. Bacterial inoculum enhances keratin degradation and biofilm formation in poultry compost. *Journal of Microbiological Methods* 47: 199–208.

Inglis, G.D., T.A. McAllister, F.J. Larney, and E. Top. 2010. Prolonged survival of Campylobacter species in bovine manure compost. *Applied and Environmental Microbiology* 76(4): 1110–1119.

Kalbasi, A., S. Mukhtar, S.E. Hawkins, and B.W. Auvermann. 2005. Carcass composting for management of farm mortalities: A review. *Compost Science & Utilization* 13(3): 180–193.

Kalbasi-Ashtari, A., M.M. Schutz, and B.W. Auvermann. 2008. Carcass rendering systems for farm mortalities: A review. *Journal of Environmental Engineering and Science* 7: 199–211.

Karpowicz, E., A. Novinscak, F. Bärlocher, and M. Filion. 2010. qPCR quantification and genetic characterization of *Clostridium perfringens* populations in biosolids composted for 2 years. *Journal of Applied Microbiology* 108: 571–581.

Keener, H.M., D.L. Edwell, and M.J. Monnin. 2000. Procedures and equations for sizing of structures and windrows for composting animal mortalities. *Applied Engineering in Agriculture* 16(6): 681–692.

Keener, H.M., C. Marugg, R.C. Hansen, and H.A.J. Hoitink. 1993. Optimizing the efficiency of the composting process. In: Hoitink, H.A.J. and H.M. Keener (eds), *Science and Engineering of Composting*. Worthington: Renaissance Publications, pp. 55–94.

Kim, H., M.W. Shepherd Jr, and X. Jiang. 2009. Evaluating the effect of environmental factors on pathogen regrowth in compost extract. *Microbial Ecology* 58: 498–508.

Korniłłowicz-Kowalska, T. and J. Bohacz. 2010. Dynamics of growth and succession of bacterial and fungal communities during composting of feather waste. *Bioresource Technology* 101: 1268–1276.

Kutzner, H.J. 2000. Microbiology of composting. In: Klein, J. and J. Winter (eds), *Biotechnology: A Multi-Volume Comprehensive Treatise*. Vol. 11c, 2nd edition. Weinheim: Wiley-VCH, pp. 35–91.

Lafond, S., T. Paré, H. Dinel, M. Schnitzer, J.R. Chambers, and A. Jaouich. 2002. Composting duck excreta enriched wood shavings: C and N transformation and bacterial pathogen reductions. *Journal of Environmental Science and Health, Part B* 37: 173–186.

Langston, J., D. Carman, K. Van Devender, and J.C. Boles Jr. 2002. Disposal of swine carcasses in Arkansas. MP397-5M-9-97N. Cooperative Extension Service. University of Arkansas. Little Rock, AR.

Lin, X., G.D. Inglis, L.J. Yanke, and K.-J. Cheng. 1999. Selection and characterization of feather-degrading bacteria from canola meal compost. *Journal of Industrial Microbiology and Biotechnology* 23: 149–153.

Mărculescu, C., C. Stan, and A. Badea. 2012. Energetic potential assessment of poultry waste processing industry. *Environmental Engineering and Management Journal* 9: 1567–1572.

Mason, I.G. and M.W. Milke. 2005. Physical modelling of the composting environment: A review. Part 1: Reactor systems. *Waste Management* 25: 481–500.

McAllister, T.A. and E. Topp. 2012. Role of livestock in microbiological contamination of water: Commonly the blame, but not always the source. *Animal Frontiers* 2(2): 17–27.

Mescher, T., K. Wolfe, S. Foster, and R. Stowell. 1997. Swine composting facility design. Fact Sheet AEX-713-97. The Ohio State University, Columbus, OH. http://ohioline.osu.edu/aex-fact/0713.html (accessed May 24, 2013).

Miller, F.C. 1996. Composting of municipal solid waste and its components. In: Palmisano, A.C. and M.A. Barlaz (eds), *Microbiology of Solid Waste*. Boca Raton: CRC Press, pp. 115–154.

Morse, D.E. 2009. Composting animal mortalities. Minnesota Department of Agriculture, St. Paul, MN. http://www.mda.state.mn.us/animals/livestock/composting-mortalities.aspx (accessed July 21, 2013).

Mukhtar, S., B.W. Auvermann, K. Heflin, and C.N. Boriack. 2003. A low maintenance approach to large carcass composting. Paper presented at the meeting of ASAE (Paper No. 032263), Las Vegas, NV.

NABC. 2004. *Carcass Disposal: A Comprehensive Review*. National Agriculture Biosecurity Centre Consortium. Manhattan, KS: Kansas State University.

Neher, D.A., T.R. Weicht, S.T. Bates, J.W. Leff, and N. Fierer. 2013. Changes in bacterial and fungal communities across compost recipes, preparation methods, and composting times. *PLoS ONE* 8(11): e79512.

OMAF. 2010. Composting of cattle on-farm. Factsheet ISSN 1198-712X. Written by B. Hawkins. Ontario Ministry of Agriculture and Food, Guelph, ON. http://www.omafra.gov.on.ca/english/engineer/facts/10-063.htm (accessed July 20, 2013).

Pechal, J.L., T.L. Crippen, M.E. Benbow, A.M. Tarone, S. Dowd, and J.K. Tomberlin. 2014. The potential use of bacterial community succession in forensics as described by high throughput metagenomic sequencing. *International Journal of Legal Medicine* 128: 193–205.

Paisley, L.G. and J. Hostrup-Pedersen. 2005. A quantitative assessment of the BSE risk associated with fly ash and slag from the incineration of meat and bone meal in a gas-fired power plant in Denmark. *Preventive Veterinary Medicine* 68: 263–275.

Puhl, A.A., L.B. Selinger, T.A. McAllister and G.D. Inglis. 2009. *Actinomadura keratinilytica* sp. nov., a keratin-degrading actinobacterium isolated from bovine manure compost. *International Journal of Systematic and Evolutionary Microbiology* 59: 828–834.

Rahman, S. 2012. Suitability of sunflower-hulls-based turkey litter for on-farm turkey carcass composting. *Canadian Biosystems Engineering* 54: 6.1–6.8.

Reuter, T., T.W. Alexander, W. Xu, K. Stanford, and T.A. McAllister. 2010. Biodegradation of genetically modified seeds and plant tissues during composting. *Journal of the Science of Food and Agriculture* 90: 650–657.

Ruggieri, L., A. Artola, T. Gea, and A. Sanchez. 2008. Biodegradation of animal fats in a co-composting process with wastewater sludge. *International Biodeterioration & Biodegradation* 62: 297–303.

Ryckeboer, J., J. Mergaert, K. Vaes et al. 2003. A survey of bacteria and fungi occurring during composting and self-heating processes. *Annals of Microbiology* 53: 349–410.

Rynk, R. 1992. *On-farm Composting Handbook*. Ithaca: Northeast Regional Agricultural Engineering Service.

Rynk, R. 2003. Large animal mortality composting goes mainstream. *BioCycle* 44(6): 44–49.

Sanabria-León, R., L.A. Cruz-Arroyo, A.A. Rodríguez, and M. Alameda. 2007. Chemical and biological characterization of slaughterhouse wastes compost. Waste Management 27: 1800–1807.

Schwarz, M., J. Bonhotal, E. Harrison, W. Brinton, and P. Storms. 2010. Effectiveness of composting road-killed deer in New York State. *Compost Science & Utilization* 18(4): 232–241.

Shepherd Jr, M.W., R. Singh, J. Kim, and X. Jiang. 2010. Effect of heat-shock treatment on the survival of *Escherichia coli* O157:H7 and *Salmonella enterica* Typhimurium in dairy manure co-composted with vegetable wastes under field conditions. *Bioresource Technology* 101: 5407–5413.

Sidhu, J., R.A. Gibbs, G.E. Ho, and I. Unkovich. 2001. The role of indigenous microorganisms in suppression of Salmonella regrowth in composted biosolids. *Water Research* 35(4): 913–920.

Singh, R., J. Kim, and X. Jiang. 2012. Heat inactivation of *Salmonella* spp. in fresh poultry compost by simulating early phase of composting process. *Journal of Applied Microbiology* 112: 927–935.

Singh, R., J. Kim, M.W. Shepherd Jr, F. Luo, and X. Jiang. 2011. Determining thermal inactivation of *Escherichia coli* O157:H7 in fresh compost by simulating early phases of the composting process. *Applied and Environmental Microbiology* 77(12): 4126–4135.

Sivakumar, K., V. Ramesh Saravana Kumar, P.N. Richard Jagatheesan, K. Viswanathan, and D. Chandrasekaran. 2008. Seasonal variations in composting process of dead poultry birds. *Bioresource Technology* 99: 3708–3713.

Stanford, K., X. Hao, S. Xu, T.A. McAllister, F. Larney, and J.J. Leonard. 2009. Effects of age of cattle, turning technology and compost environment on disappearance of bone from cattle mortality compost. *Bioresource Technology* 100: 4417–4422.

Stanford, K., F.J. Larney, A.F. Olson, L.J. Yanke, and R.H. McKenzie. 2000. Composting as a means of disposal of sheep mortalities. *Compost Science & Utilization* 8: 135–146.

Stanford, K., V. Nelson, B. Sexton, T.A. McAllister, X. Hao, and F.J. Larney. 2007. Open-air windrows for winter disposal of frozen cattle mortalities: Effects of ambient temperature and mortality layering. *Compost Science & Utilization* 15: 257–266.

Steel, H., F. Verdoodt and A. Čerevková et al. 2013. Survival and colonization of nematodes in a composting process. *Invertebrate Biology* 132(2): 108–119.

Stentiford, E.I. 1993. Diversity of composting systems. In: Hoitink, H.A.J. and H.M. Keener (eds), *Science and Engineering of Composting: Design, Environmental, Microbiological and Utilization Aspects*. Worthington: Renaissance Publications, pp. 95–110.

Suzuki, Y., Y. Tsujimoto, H. Matsui, and K. Watanabe. 2006. Review: Decomposition of extremely hard-to-degrade animal proteins by thermophilic bacteria. *Journal of Bioscience and Bioengineering* 102: 73–81.

Taylor, D.M. 2000. Inactivation of transmissible degenerative encephalopathy agent: A review. *The Veterinary Journal* 159: 10–17.

Tiquia, S.M., J.M. Ichida, H.M. Keener, D.L. Elwell, E.H. Burtt, and F.C. Michel. 2005. Bacterial community profiles on feathers during composting as determined by terminal restriction fragment length polymorphism analysis of 16S rDNA genes. *Applied Microbiology and Biotechnology* 67: 412–419.

Tkachuk, V.L., T.A. McAllister, and K. E. Buckly et al. 2013. Assessing the inactivation of *Mycobacterium avium* subsp. *paratuberculosis* during composting of livestock carcasses. *Applied and Environmental Microbiology* 79(10): 3215–3224.

Tkachuk, V.L., D.O. Krause, N.C. Knox et al. 2014. Targeted 16S rRNA high-throughput sequencing to characterize microbial communities during composting of livestock mortalities. *Journal of Applied Microbiology* 116(5): 1181–1194.

Transport Canada. 2012. Update of data source on collisions involving motor vehicles and large animals in Canada—Statistical review. https://www.tc.gc.ca/eng/motorvehiclesafety/tp-tp14798-1289.htm#e5.1 (accessed December 29, 2013).

Uetake, K. 2013. Newborn calf welfare: A review focusing on mortality rates. *Animal Science Journal* 84: 101–105.

USEPA. 1995. Chapter 7 of EPA's decision-maker's guide to solid waste management. http://www.epa.gov/epawaste/conserve/composting/pubs/ (accessed August 14, 2014).

Van Herk, F.H., C.L. Cockwill, N. Guselle, F.J. Larney, M.E. Olson, and T.A. McAllister. 2004. Elimination of Giardia cysts and Cryptosporidium oocysts in beef feedlot manure compost. Compost Science & Utilization 12: 235–241.

Wilkinson, K.G. 2007. The biosecurity of on-farm mortality composting. *Journal of Applied Microbiology* 102: 609–618.

WOAH. 2013. Number of cases of bovine spongiform encephalopathy (BSE) reported in the United Kingdom. World Organisation for Animal Health. http://www.oie.int/?id = 504 (accessed May 7, 2013).

Xu, S., X. Hao, K. Stanford, T. McAllister, and F.J. Larney. 2007. Greenhouse gas emissions during co-composting of cattle mortalities with manure. *Nutrient Cycling in Agroecosystem* 78: 177–187.

Xu, S., T.A. McAllister, J.J. Leonard, O.G. Clark, and M. Belosevic. 2010. Assessment of microbial communities in decomposition of specified risk material using a passively aerated laboratory-scale composter. *Compost Science & Utilization* 18(4): 255–265.

Xu, S., T. Reuter and B.H. Gilroyed et al. 2013a. Biodegradation of specified risk material and fate of scrapie prions in compost. *Journal of Environmental Science and Health, Part A* 48(1): 26–36.

Xu, S., T. Reuter and B.H. Gilroyed et al. 2013b. Microbial communities and greenhouse gas emissions associated with the biodegradation of specified risk material in compost. *Waste Management* 33(6): 1372–1380.

Xu, S., T. Reuter, B.H. Gilroyed et al. 2014. Biodegradation of prions in compost. *Environmental Science & Technology* 48: 6909–6918.

Xu, W., T. Reuter, G.D. Inglis et al. 2009a. A biosecure composting system for disposal of cattle carcasses and manure following infectious disease outbreak. *Journal of Environmental Quality* 38: 437–450.

Xu, W., T. Reuter, Y. Xu et al. 2009b. Use of quantitative and conventional PCR to assess biodegradation of bovine and plant DNA during cattle mortality composting. *Environmental Science & Technology* 43: 6248–6255.

Yamamoto, N., R. Oishi, Y. Suyama, C. Tada, and Y. Nakai. 2012. Ammonia-oxidizing bacteria than ammonia-oxidizing archaea were widely distributed in animal manure composts from field-scale facilities. *Microbes and Environments* 27(4): 519–524.

24

Human Decomposition and Forensics

Gail S. Anderson

CONTENTS

24.1 Introduction

An understanding of carrion insect ecology has very practical value in the world of forensic science and human homicide. It has long been recognized that certain species of insects utilize carrion for a part or all of their life cycle and that many of these species colonize in a predictable pattern and sequence. Understanding this sequence allows a forensic entomologist to predict many details about the circumstances surrounding the death, including tenure of insects, location of wounds, and whether a body has been disturbed or not. Although carrion ecology involves much more than just insects, it is the fauna of the cadaver that has been most greatly studied and applied and will be discussed here.

Forensic entomology in its broadest sense includes all aspects of the study of insects as it pertains to legal matters; however, forensic or more correctly, medico-legal entomology, as it will be discussed here, is the study of the insects which colonize a body over time, with a view to elucidating information about the death in order to assist law enforcement agencies in solving crimes. As carrion insects have very specific feeding requirements and occupy very specific ecological habitats and geographical niches, an analysis of the entomofauna of a cadaver can result in a great deal of valuable information about the death. Many carrion insects colonize a cadaver in a very predictable sequence, and once present, they develop in a predictable manner. This allows a forensic entomologist to estimate the tenure of the insects on the cadaver and hence indicate a minimum period of insect colonization, thereby inferring the minimum elapsed time since death. This is extremely valuable in a homicide investigation as it can support or refute a suspect's alibi, eliminate suspects who do not fit that time frame, and allow police officers to focus the investigation on the time line preceding death.

The proliferation of television shows featuring forensic entomology has given the public the impression that this is a very new science. However, it is possibly one of the oldest forensic disciplines, first documented in the tenth century in China (Greenberg and Kunich 2002). Forensic entomology in its more modern understanding began in France in the mid-nineteenth century and was used in South America and Europe in the early part of the twentieth century, but was not commonly used in North America until the 1970s. Since then, interest from both academics and law enforcement agencies has grown exponentially around the globe, resulting in a tremendous increase in the number of researchers involved in the field and the number of publications and investigations in which forensic entomology is being used.

24.2 The Use of Insect Development Rates to Estimate Postcolonization Intervals

Insect associations with carrion have been divided into two main phases, the precolonization interval and the postcolonization interval (Tomberlin et al. 2011, p. 403). The precolonization interval begins at the moment of death and includes "exposure" where insects have not yet detected the cadaver, "detection," where insects first locate the cadaver, and "acceptance," when insects assess the quality of the resource for potential oviposition sites. The postcolonization interval begins at the moment of colonization and includes the "consumption phase" and "dispersal phase" (Tomberlin et al. 2011, p. 403) during which the insects colonize and feed on the cadaver until all nutrients are removed. It is this latter phase, also referred to as the Period of Insect Activity (PIA) (Amendt et al. 2007, p. 95) with which forensic entomologists are often most concerned.

Certain species of insects are attracted to carrion very shortly after death, as long as conditions are conducive for insect activity. These early colonizers are primarily Diptera in the families Calliphoridae (blow flies) and, in some areas, Sarcophagidae (flesh flies). During the precolonization phase, the adult flies detect and locate carrion, then assess it as a suitable oviposition medium. The postcolonization stage begins when eggs, or in the case of Sarcophagidae, larvae are laid at an appropriate site on the carcass. The eggs hatch into first instar larvae, which feed primarily on liquid protein such as blood or open wounds, or on the mucosal layer. As most first instar larvae cannot pierce adult human skin, the adult female chooses oviposition sites located near wounds or exposed orifices, offering the delicate early instar a suitable food source. The first instar molts into a larger second instar usually within 24–48 h, then molts again to a third instar larva. This stage feeds voraciously and frequently forms large aggregates. After a period of feeding, the third instar larvae enter a wandering, nonfeeding stage, and in most cases leave the food source in active search of an appropriate area in which to pupariate. Once pupariation has occurred, the insect is sessile, so it cannot avoid predation or a change in abiotic conditions such as flooding. Therefore, the choice of pupariation site is very important. Larvae frequently burrow several centimeters into the soil to avoid predators. When leaving a human cadaver, they may also pupate in the hair, in or under clothing, carpeting, or furniture.

The distance the postfeeding third instar larvae may crawl from the remains varies greatly depending on many factors such as species, substrate, predation, competition, endogenous rhythms, and abiotic parameters (Lewis and Benbow 2011). It is very important that forensic entomologists understand the postfeeding behavior of carrion flies and the factors that impact this behavior so that they can correctly interpret a crime scene to ensure that the stages no longer on the cadaver are still located and collected. If it is not possible for the forensic entomologist to attend the crime scene, it is vital that the police officers or crime scene technicians at the scene are instructed in the areas they must search for later stages of the life cycle.

Diptera undergo a complete metamorphosis in which larval tissues are completely broken down and rebuilt, with the only tissues not assimilated being the chitinized cephalopharyngeal skeleton or mouthparts, and the exterior of the posterior spiracular slits. When the fly has completed its metamorphosis, it pushes open the top of the puparium using a hemalymph-filled sac, the ptilinum, as a muscle, breaking the operculum into two separate caps. These caps are frequently lost in the soil, being small and dark, but they are important because the third-instar cephalopharyngeal skeleton is attached to the inside of the lower cap. Puparia alone are very difficult to identify, since most bear few diagnostic features, but the mouthparts are often diagnostic and so should be recovered if possible. When freshly emerged, the insect is unable to fly, so it crawls up into foliage and, using the hemalymph as a muscle, it expands its cuticle and wings before it dries. Such flies are referred to as teneral and are very useful if collected as, due to their inability to fly, they are inextricably related to the carrion, so their age can be used to estimate a PIA. Once dried and capable of flight, adult blow flies are of little value in estimating the PIA, since they may have just emerged but equally may have just flown in to the cadaver minutes before.

Insects are cold-blooded, so their entire life cycle is temperature-dependent. As temperature increases, their developmental rate increases and the time spent in each developmental stage is shortened. During optimal temperatures, this relationship is linear, making it predictable. In order to estimate the minimum PIA, the forensic entomologist must know the temperatures to which the insects were exposed, the identity of the species of insects colonizing the cadaver and how far through the life cycle each species has progressed.

Temperature data are usually acquired from the nearest and most appropriate government weather station. Sometimes the closest weather station is not the most appropriate because altitude, shading, or other conditions at the crime scene may be much different than the weather station. Therefore, care must be taken in selecting the weather station for data analysis and interpretation.

In most cases, the species of insect can be identified using adult or larval keys. Adults are usually easier to identify, so the forensic entomologist will raise larval specimens collected from the remains to adulthood to facilitate identification. In some situations, when only parts of an insect or only preserved early instars are available, such as may occur in a cold case, DNA may be used to identify a specimen.

The oldest stage of insects associated with the body may be easily identified if proper collection techniques are used. It is very important that samples of both live and correctly preserved insects are collected. Preserving larval specimens halts development at the time of preservation and maintains them in this stage indefinitely, so they can be later examined, measured, and catalogued by the forensic entomologist and any future forensic entomologists who may wish to examine them. This allows the forensic entomologist to know exactly what stage the insects were at the scene, at the time of collection, without having to take into account temperature changes during transport. Living specimens are collected, so they can be raised to adulthood to facilitate identification.

Once the species and oldest stage of insect have been identified, together with the temperatures to which the insects have been exposed, the forensic entomologist estimates how long it would take these species to reach these stages at these temperatures, based on literature reports of developmental rates. From these four pieces of data: species, life stage, temperature, and published developmental data, the forensic entomologist can estimate the minimum age of the oldest insects, hence the PIA. From this, the minimum elapsed time since death can be inferred.

Forensic entomologists can only indicate the tenure of the insects on a body and not the actual elapsed time since death. Death may have preceded insect colonization by only a few minutes. However, if the remains were protected from insects for a period of time, or the time of day or season was not conducive to colonization, death may precede colonization by hours, days, or months. If the precolonization conditions are known, the entomologist may be able to offer some guidance as to the time elapsed prior to colonization, however, in many situations, these conditions are not known. As well, most entomologists utilize the minimum or mode of developmental times in their calculations. This allows entomologists to be sure of their estimates, as some specimens in a cohort will take longer to develop than others, but this also means that the answer will probably be an underestimate of the actual elapsed time since death. This is desirable as it is always better to err on the side of caution as a person's freedom or even life may rest on the analysis. It is therefore important for the forensic entomologist to explain that the estimate of the tenure of the insects on the cadaver may underestimate the time since death.

24.2.1 Case History Example

A body was found below cliffs in western Canada in early summer (Anderson and Cervenka 2001). Feeding larvae were collected from the body, and a few pupae were collected from the surrounding area. The nearest Environment Canada weather station indicated that the weather had been cool for the preceding 3 weeks, with a mean of 15.2°C and highs reaching the low 20s on occasion. Three species of immature blow flies were collected from the remains: *Phormia regina* (Meigen), *Protophormia terraenovae* (R.-D.) and *Calliphora vomitoria* (L.) (Diptera: Calliphoridae), but the oldest insects were live pupae of *P. regina*. Very few pupae were collected, suggesting that pupation had only just begun.

Phormia regina takes a minimum of 18.4 days to reach the beginning of the pupariation stage at a constant temperature of 16.1°C (Anderson 2000). This is slightly warmer than the crime scene and if applied directly, also does not take into account natural diurnal temperature fluctuations. So these data can be converted to thermal units or degree days (or degree hours if the data are hourly). This is based on the hypothesis that insect development is linear between upper and lower development thresholds, above and below which no development occurs. If development is converted to a thermal unit, it can be applied to any temperature over the linear portion of the graph. In other words, if *P. regina* takes a minimum of 18.4 days to reach the pupariation stage at 16.1°C, then this can also be expressed in thermal units in that it requires a minimum of 296 (18.4 × 16.1) thermal units or degree days (DD) to reach the beginning of the pupariation stage. The number of accumulated degree days (ADD) available is determined by a reverse summation process, adding the number of thermal units available per day, starting from the day of collection, back until the required ADD are reached. In this case, based on local temperature data, 296 ADD would have been accumulated 20 days earlier, indicating that oviposition must have occurred on or earlier than 20 days prior to discovery. It was later determined that the victim had last been seen alive in the evening, 21 days earlier (Anderson and Cervenka 2001).

24.3 The Use and Understanding of Successional Ecology to Estimate Postcolonization Interval

A body provides a very rich but ephemeral resource, which supports a dynamic sequence of carrion fauna. The soft tissue of the body decomposes through a predictable sequence of stages from the moment of death until there are no nutrients remaining. The body undergoes a relatively rapid sequence of biological, physical, and chemical changes and each of these stages of decomposition offers differing nutrients, thereby attracting and supporting different carrion feeders at different stages of the decomposition cycle. The early colonizers, such as blow flies, are only attracted to fresh remains and are unable to consume later, dryer tissues, whereas the later arrivals which thrive on dry tissue would be unable to survive on the wetter fresh tissue. As each group of insects colonizes, it gradually changes the cadaver, making it less appropriate for their own species and more appropriate for the subsequent insects. The carcass supports necrophagous insects, which feed directly on the cadaver, as well as omnivores, parasites, and predators (see Chapter 4 in this volume).

The sequence of insects that colonizes a body is predictable, within specific geographic regions, seasons, and habitats; therefore, an analysis of the insect fauna of a cadaver, in comparison with field data that are locally generated and seasonally appropriate, can allow a forensic entomologist to estimate the minimum PIA on the remains, and hence infer a minimum elapsed time since death.

24.3.1 Case History Example

In the far north of Canada, a teenage girl was last seen alive in March and, based on other evidence, the police investigators strongly suspected that the girl's teenage boyfriend had killed her (Anderson and Cervenka 2001). In late May of the following year, the young man's parents declared their intention to move. After hearing that they would be moving, the young man locked himself in his bedroom and the family heard ominous hammering noises emanating from the room. After a period of time, the young man ran away. Concerned, the father broke into his son's room. The room was almost entirely occupied by a homemade waterbed frame. The mattress portion had been removed and emptied, but the frame and liner were intact, and the young man had apparently been sleeping beside the bed. The father removed part of the frame, revealing the mummified, fully clothed remains of the missing victim.

The remains were very well preserved and had been colonized by several later colonizing insect species. The insect evidence included the cast larval skins of dermestid beetles, Dermestidae frass and peritrophic membrane, as well as pupae and empty puparia of flies belonging to Fanniidae and Sphaeroceridae (Diptera) (Anderson and Cervenka 2001). Peritrophic membrane lines the midgut of insects and, in most species, is continuously produced and lost with excreta, but in some situations, usually indoors, or inside vehicles, dermestid frass and periotrophic membrane can build up, as it did in this case. If protected well enough, these materials can survive for years, and when no live Dermestidae are present, they are indicative that the body has passed through the stages attractive to dermestids. The species and number of later colonizing insects on the remains indicated that the remains had been under the bed during an insect season. However, when the body was discovered in May, it was still too early in the year for insect colonization, with day time temperatures in single digits and night time temperatures dropping below 0°C. Therefore, the presence of the insects indicated that the victim had been dead and under the bed during the previous summer.

In this case, more interesting than the species that were present, were the species that were not. No evidence of the earlier presence of blow fly activity was found. The remains were not in a state of decomposition which would have been attractive to blow flies when recovered, but if the victim had died the previous summer, then blow flies would have colonized. It was evident that this had not occurred for several reasons. Firstly, the remains were intact and very well preserved with very little loss of tissue, and had mummified. If blow fly larvae had fed on the remains, the soft tissue would have been removed and the stench of putrefaction would have been very noticeable, especially inside an inhabited house. Secondly, if blow fly larvae had been present, they would have emerged as adult flies, leaving behind empty puparia. In an outdoor situation, only a small percentage of puparia may be recovered and many may be missed, despite extensive searches, as the insects have evolved so that their puparia blend into soil very well. However, in an indoor environment, they are easily seen and the area was small and exhaustively searched, so no blow fly puparia were present. Finally, had blow flies colonized the remains, very large numbers of adult flies would have left the body and emerged into the house, which would have been noticed by the occupants. It could be suggested that blow flies were unable to reach the remains inside the basement bedroom, however, as other insects colonized, this is extremely unlikely. It is quite possible that there could have been a delay in colonization, possibly of a few days but had the victim died in the previous summer, then the remains should have shown evidence of blow fly colonization.

The only explanation for the absence of early colonizers and the presence of later colonizers is that the victim died weeks or more before the insect season began, allowing time for mummification to occur in the warm, dry, centrally heated residence, which encouraged the colonization of later colonizers once the insect season began. This is consistent with death occurring in March of the previous year when she was last seen alive. The suspect later came forward and confessed to killing her at that time.

24.4 Factors That Impact Insect Colonization and Development

24.4.1 Initial Attraction to the Carcass and Oviposition

If conditions are appropriate, blow flies will be attracted to a corpse very rapidly after death. There are many factors that are involved in insect attraction to remains and subsequent oviposition; not all blow fly species are immediately attracted to a fresh cadaver, and their arrival times and preferences appear to varying geographically (Anderson 2009a,b).

A killer may hide a body in an effort to prevent discovery. Female carrion flies are extremely adept at locating such a rich but ephemeral resource and can usually access the carcass, although oviposition may be delayed. Wrapping an animal carcass has been shown to prevent fly access for two and a half days in Hawai'i, USA (Goff 1992) and between one and 13 days in Malaysia, depending on fly species (Ahmad et al. 2011). As well, cadavers inside a residence are often not colonized for several days

(Reibe and Madea 2010; Anderson 2011). Colonization is also delayed when a cadaver is inside a car trunk, with the actual delay depending on whether the cars had firewalls between the passenger area and the trunk. However, once colonized, decomposition and larval development may be accelerated due to increased temperatures inside the vehicle (Miller 2009). Colonization was also delayed on carcasses in passenger compartments of vehicles (Voss et al. 2008).

Insecticides are sometimes used in suicide but with the increase in television shows featuring forensic entomology, killers may try to kill or repel insects from their victim by spraying them with an insecticide. Malathion sprayed on rabbit carcasses delayed oviposition by 1–3 days (Mahat et al. 2009); malathion also delayed and excluded some taxa from a suspected suicide (Gunatilake and Goff 1989). It is important for the forensic entomologist to be aware of the possible presence of such repellents, as not only may they delay colonization, but they also may impact development. Malathion in rabbits not only delayed insect colonization, but also slowed Calliphoridae development by 12–26 h (Yan-Wei et al. 2010).

The presence of other insect species, such as red imported fire ants, *Solenopsis invicta*, Buren (Hymenoptera: Formicidae) and wasps (Hymenoptera), have frequently been reported to delay fly colonization and change colonization and succession patterns (Anderson 2009a,b; Lindgren et al. 2011). Timing of death can also impact colonization, as blow flies are diurnal and studies globally have shown that nocturnal oviposition does not usually occur under natural conditions. Finally, insects are limited to activity in warm weather, so season can greatly inhibit insect activity. Also, low temperatures occurring during warmer seasons may delay colonization as, in general, it is considered that most blow flies will not oviposit below 12°C (Erzinclioglu 1996). Therefore, there are many parameters that may delay or inhibit colonization and an understanding of biotic and abiotic parameters of the crime scene is imperative to understand the entire picture.

24.4.2 Macro- and Microclimates

When using insect development to calculate the minimum PIA, knowledge of the temperatures to which the insects have been exposed is fundamental in calculating the age of the oldest larvae on the cadaver. However, this can never be precisely known but only inferred from local weather station data. It is good practice to place a temperature datalogger at the crime scene for a period of time after the cadaver has been removed (Amendt et al. 2007). The data recorded at the crime scene can then be statistically compared with data for the same time frame from the weather station to determine whether there is a correlation, and the subsequently generated equation can be used to predict the temperatures at the crime scene, based on those from the weather station. For example, in Manitoba, human remains were found in a densely wooded area. The nearest weather station indicated a mean temperature of $15.7 \pm 3.8°C$ from the time the victim was last seen alive to the time of discovery. However, as the remains were found in a heavily shaded area, it was suspected that the scene might have been cooler than the weather station. A SmartReader© datalogger was placed at the scene and recorded scene temperatures for 6 weeks after the victim had been removed. These data were then statistically compared with temperature data for the same time period recorded by the weather station. A regression analysis showed a very significant correlation between the two datasets ($R^2 = 0.84$, significance 1.65^{-18}), indicating that the data generated by the datalogger could be used to predict the scene temperatures with a high level of confidence. The equation of the regression model indicated that the crime scene was slightly cooler than the weather station site, with a mean temperature of 14.7°C. Although this weather station was the closest to the scene, concerns from defense counsel prompted the forensic entomologist to analyze data from all weather stations within 100 km of the crime scene, with almost identical results (Anderson and Cervenka 2001). In Australia, retrospective correlation methods improved the accuracy of predicting crime scene temperatures in experimental cases by 92% (Archer 2004). Further work also showed that accuracy was not impeded by season, length of time the logger was at the scene, nor distance between station and scene as long as it was less than 15 km (Johnson et al. 2011).

Factors such as amount of shade, habitat, exposure, and clothing must also be taken into account. As well, the body itself and its dynamic carrion ecosystem may impact the micro-climate and this is much more difficult to predict. When fresh, the remains and its fauna will be close to ambient temperature, but as bacterial and larval insect activity increase, temperature and humidity will be altered locally. The presence of large aggregations of fly larvae can increase local temperatures by as much as 30°C above ambient (Anderson and VanLaerhoven 1996). The actual temperature increase is dependent on the size of the mass as well as the stage and species of insects which comprise the mass (Rivers et al. 2010). Further, each mass will only raise the local temperature, and different masses in different areas of the body will often be of different temperatures.

Another consideration is the effect of refrigeration on larval development as it is common practice for a body to be refrigerated prior to autopsy. It is always preferable to collect insects at the scene when possible. However, it is often the case that a forensic entomologist is not contacted until after the remains have been removed to the morgue. In such cases, the forensic entomologist will be obliged to collect the insects after a period of refrigeration. When only a small number of larvae are present on the remains, refrigeration should stop development or slow it to a negligible rate (Johl and Anderson 1996). However, when larger numbers of larvae are present, they may maintain a temperature above the refrigerator temperature for several hours (Haskell et al. 1997; Huntington et al. 2007). The impact that refrigeration will have on development has also been shown to vary greatly depending on what stage is refrigerated (Myskowiak and Doums 2002). This means that it cannot be assumed that refrigerated larvae do not develop during at least part of the refrigeration time and this must be included in the analysis.

24.4.3 Impact of Larval Aggregations

Many insect species form aggregations at certain stages of their life cycle (Parrish and Edelstein-Keshet 1999; Prokopy and Roitberg 2001), and there has been much discussion on the benefits and costs associated with such aggregations. Calliphoridae larval aggregates, or masses, however are different from most others in that they generate a great deal of heat by elevated and concentrated individual larval metabolism as a dense collective. The increased heat may facilitate more rapid development up to a point, as well as protect the larvae in cool periods (Campobasso et al. 2001). Movement in and out of the mass may also allow the larvae some level of control over homeostasis. However, temperatures inside the mass often reach or exceed the upper temperature development threshold and result in proteotoxic stress (Rivers et al. 2011). In experiments with a calliphorid, *P. terraenovae* and a sarcophagid, *Sarcophaga bullata* Parker (Diptera: Sarcophagidae), both species were found to develop more rapidly during the immature feeding stages as size, and consequently, temperature of experimental maggot masses increased, but once the larvae had left the food source, duration of puparial stages actually increased and puparial weight and ability to eclose decreased (Rivers et al. 2010). These effects were correlated with the expression of heat shock proteins in the larval brains, suggesting that larvae in the larger masses are experiencing proteotoxic stress (Rivers et al. 2010). *Calliphora vomitoria* also exhibited increased developmental rates with overcrowding but the larvae were undersized and produced smaller adults (Ireland and Turner 2006).

There has been much speculation as to why such larval aggregates form, as there appear to be both benefits and costs to such behavior. Possible benefits include: the increased breakdown of cadaver tissue due to large amounts of proteolytic enzyme being secreted by many insects, a reduced risk of parasitism or predation (Parrish and Edelstein-Keshet 1999; Rohlfs and Hoffmeister 2004), temperature regulation, and protection against cold weather and increased metabolic rate (Rivers et al. 2011). Conversely, possible costs include increased attraction of predators and parasitoids due to chemo-attraction, heat shock or stress, overcrowding and competition (Rivers et al. 2011).

Taking larval mass into consideration when analyzing larval development is very difficult as the actual temperatures to which the insects are exposed are unknown. Internal temperatures of the mass should be measured when maggots are collected, but there is a large temperature range between the outside and inside of a large mass. It is traditionally thought that larvae move somewhat randomly within the mass from hotter to cooler areas to thermoregulate, breathe, and feed. However, recent studies using newly developed methods have allowed individual larvae to be tracked throughout a mass and have shown that movement is not random, with larvae actively seeking out the hottest part of the mass, although different species have different preferred optimal temperatures (Johnson 2013). Another consideration is the fact that mass temperature does not remain consistently high but fluctuates by as much as 35°C over a 24 h period, exhibiting a much greater range than ambient temperature (Anderson and

Van Laerhoven 1996). As well, temperature increases only begin with second instars and reach their highest temperatures during active decay when large numbers of third instars are feeding. This means that only some stages of larvae are impacted by the temperature increase. Finally, in some situations, the oldest larvae may leave the carcass before masses form, and hence be unaffected by increased heat (Dillon 1997).

24.4.4 Upper and Lower Developmental Thresholds

Developmental temperature thresholds are the temperatures above and below which insects can no longer develop. Developmental thresholds should be taken into consideration when calculating the developmental rate of insects on the body. However, both upper and lower developmental thresholds vary between different species and have not been identified for all blow fly species. This should be relatively easy to correct with further research; however, thresholds have also been shown to vary within the same species at different stages (Davies and Ratcliffe 1994; Warren and Anderson 2013) and also between the same species collected from different geographical regions (Ames and Turner 2003). This makes evolutionary sense as many cosmopolitan species may have been exposed to very different temperature ranges. It would be beneficial for members of a species exposed to consistently cooler temperatures to adapt to lower temperatures than conspecifics living in warmer conditions. However, this means extrapolating developmental data generated from one geographic area to another is problematic. It is best to use developmental data generated from populations close to the crime scene.

As well, some researchers have reported that blow flies do develop, albeit extremely slowly, at temperatures considerably below the calculated lower developmental threshold (Davies and Ratcliffe 1994; Myskowiak and Doums 2002; Ames and Turner 2003). Recent studies have also shown a sex-specific difference in developmental rates within species, with males developing more rapidly than females during the third instar (Picard et al. 2013) and emerging as adults one day earlier than females (Beuter and Mendes 2013).

24.4.5 Impact of Geographical Region

Decomposition and insect succession are greatly impacted by geography. Although general trends occur, the families and species of insects that colonize, and their tenure on the remains, can differ dramatically in different parts of the world. See Anderson (2009a,b) for a review and Chapter 4 in this volume. For example, Piophilidae or skipper flies are later colonizers, but their arrival and development on remains has been shown to vary greatly, depending on region. In Europe, Piophilidae larvae are reported on remains 3–6 months after death (Smith 1986), but adults have been collected 33–36 days after death in Hawai'i (Goff and Flynn 1991) and larvae only 26 days after death in human cases in British Columbia (BC) (Anderson 1995) and 29 days after death from pig carcasses in open grassland (Anderson and VanLaerhoven 1996) and a forest (Dillon and Anderson 1995).

Geography impacts blow fly colonization times but also has been shown to impact development. Developmental data have been recorded for many species of blow flies and it is generally considered that such data are valid for all members of that species. However, several studies have shown that there can be variation in development rates between geographically different populations of the same species (Saunders and Hayward 1998; Ames and Turner 2003; Boatright and Tomberlin 2010; Gallagher et al. 2010). These differences may be due to genetic differences in geographically distinct populations (Tarone et al. 2011) or may relate to experimental conditions such as food moisture, destructive sampling, and disturbing postfeeding larvae (Tarone and Foran 2006). Differences in cuticular hydrocarbons between geographically different populations of *P. regina* have also been recorded (Byrne et al. 1995). Such geographical variations highlight the need for developing local data for forensically important fly species as well as standardization of rearing methods to eliminate concerns with applying data from a geographically distinct population.

A few years ago, only a few carrion studies from isolated parts of the world existed and most were examined from a purely ecological point of view, with no thought that such data may be of value in a forensic context. However, in the last few years, there has been an exponential increase in the number of carrion studies worldwide, providing valuable local data for many regions and hence allowing local forensic entomologists to use geographically appropriate data in analyzing cases.

24.4.6 Impact of Season

Season impacts meteorological conditions, flora, and fauna, so can be expected to have a major impact on insect colonization and succession; therefore, data generated in one season may not necessarily be applicable to other seasons in the same geographic region. Season impacts which species will colonize as well as relative abundance of each species, with some species being considered season specific.

Blow fly colonization is greatly impacted by season, with some species being much less abundant or even excluded in some seasons. Later colonizers are also impacted by season and this may relate to species abundance or differences in decomposition, nutrient content, and moisture of the cadaver. In some cases, the same families may still colonize the remains in the same sequence in different seasons, but the individual species within the families will vary.

Some species are considered to be seasonally distinct and thus could potentially be used to estimate a season of death. For instance, in Ohio, USA, the relative abundance of species varied greatly and indicator species could be identified for each stage of decomposition in summer, fall, and winter (Benbow et al. 2013). If the local insect carrion ecology is known, such data can even be applied to ancient remains. In Belgium, puparia of *P. terraenovae* found with Pleistocene era mammals were used to indicate the season of death (Germonpre and LeClercq 1994) and in New Brunswick, Canada, Calliphoridae, Muscidae, and Helomyzidae puparia associated with two human burials dated between 2000 and 2500 years prior were used to indicate a time lapse between death and interment (Teskey and Turnbull 1979).

24.4.7 Impact of Habitat

Many insects show distinct habitat preferences. Whether the cadaver decomposes in sun or shade will impact the temperature and humidity of the site, and consequently, the speed of decomposition of the remains as well as the insect fauna. Decomposition is accelerated in carcasses exposed to direct sunlight due to increased temperatures. However, shade has been shown to promote vertebrate scavenging in some cases, which can create greater numbers of oviposition sites and therefore, increased blow fly activity (Dillon and Anderson 1995). In Alberta, Canada, species abundance was much greater in sunlit carcasses but similar species dominated in both sun and shade (Hobischak et al. 2006), whereas in Saskatchewan, Canada, greater diversity was found on sunlit versus shaded carcasses (Sharanowski et al. 2008).

Some carrion insects are highly synanthropic, utilizing human garbage as a carrion resource and are more frequently found in urban areas, whereas other species are more commonly found in rural areas on natural carrion, and yet others are ubiquitous, being found in any habitat (Anderson 1995). In a study in England in which a continuum of habitats was sampled from a densely urban area, through a suburban area into a rural area, three distinct habitats were identified with very different blow fly communities (Hwang and Turner 2005).

Some species are more likely to colonize cadavers found inside residences or vehicles, whereas other species will not enter enclosed areas. Case histories and studies have shown that the blow flies *Lucilia sericata* (Meigen), *P. regina*, and *Calliphora vicina* (R.-D.) are commonly recovered from indoor cadavers (Anderson 1995), and in a study in Alberta, Canada, comparing colonization of pig carcasses inside and outside a residence, of the several blow flies which colonized, only *P. terraenovae* and *Lucilia illustris* (Meigen) were found exclusively outdoors (Anderson 2005).

In a retrospective analysis of 50 human death cases in Malaysia, a greater diversity of fly species was observed on cadavers found indoors versus those found outside, with Calliphoridae species being found very commonly in both situations, but Sarcophagidae, Phoridae, and Muscidae being found most commonly on cadavers found indoors (Kumara et al. 2012). In Germany, a pig (*Sus scrofa* L.) carcass inside an occupied house was almost exclusively colonized by *C. vicina* but a paired outside carcass was colonized by *C. vicina*, *C. vomitoria*, *L. sericata*, *L. caesar* (L.), and *L. illustris* (Diptera: Calliphoridae)

(Reibe and Madea 2010). In Hawai'i, USA, an analysis of 35 human death cases also revealed greater Diptera diversity on cadavers inside residences as well as a greater diversity of Coleoptera on cadavers found outdoors (Goff 1991). Some species, including *Stomoxys calcitrans* (L.) (Muscidae), were considered to be so characteristic of indoor deaths as to be considered indicator species (Goff 1991). This highlights the importance of developing local databases as such species may only be indicators of an indoor decomposition in that region. *Stomoxys calcitrans*, for instance, is commonly referred to as the stable fly and is found around domestic animals. Although its common name suggests it readily enters buildings such as stables, it is also breeds in animal dung found in pastures, so is not restricted to indoor scenarios outside Hawai'i, USA.

Burial impacts the species of insects which colonize, excluding some and encouraging others. In British Columbia (BC), Canada, pig carcasses buried in shallow graves were colonized by low numbers of Calliphoridae, with Muscidae dominating in the early decompositional stages, in contrast with aboveground carcasses. Consequently, no larval masses formed, and carcass temperatures were very similar to soil temperatures (VanLaerhoven and Anderson 1999). However, in Tennessee, USA, experimentally buried human cadavers were colonized by Calliphoridae and a rise in temperature was noted, although this varied with depth (Rodriguez and Bass 1985). Burial may also delay colonization and in some cases, the same taxa are attracted to both buried and unburied carcasses but their time of arrival may vary (VanLaerhoven and Anderson 1999). In BC, immature *Fannia cannicularis* (L) (Diptera: Fanniidae) were collected 2 weeks after death from buried pig carcasses in the subboreal Spruce zone and 6 weeks after death in the Coastal Western Hemlock Zone, much earlier than on exposed carcasses (VanLaerhoven and Anderson 1999).

24.4.8 Effects of Different Food Substrates

When insects feed on a body, they feed on a variety of tissue types with a range of nutritional values. Primary sites of colonization and feeding are facial orifices, often leading into the brain, or wound sites, which might lead to muscle tissue or abdominal organs. Therefore, larvae feeding on a cadaver may frequently feed on a range of different tissue types. Developmental standards are usually studied under controlled, laboratory conditions and the most common food substrate used is beef liver. Recently, there have been a number of studies which have shown that larval development, mortality, and size can be impacted by substrate. Significantly, some studies comparing development on isolated tissues have shown that development is often retarded on liver in comparison with other tissues (Kaneshrajah and Turner 2004; Day and Williams 2006; El-Moaty and Kheirallah 2013). It has been suggested that this relates to the lower lipid content of liver compared with more fatty tissue, suggesting that greater energy is expended in order to metabolize higher protein food such as liver (Day and Williams 2006; El-Moaty and Kheirallah 2013). Conversely, other studies have shown liver to be a good developmental substrate with 0.2 g of pig liver sufficient for one larva to complete development, in comparison with 0.5 g of muscle tissue and 1 g of brain tissue, suggesting that brain is much lower in nutrient content (Ireland and Turner 2006). All the above studies single out tissue types for comparison, but in reality, larvae on a cadaver may feed on a range of tissues. In the only study to compare calliphorid larval development on various beef organs with an entire animal (*Rattus* sp.), *P. terraenovae* development on liver was shown to be most similar to that on the whole carcass (Warren and Anderson 2009).

Age of tissue has also been shown to impact larval development with decomposed liver retarding development rates of young *C. vicina* larvae, although larger third-instar larvae developed normally. It was speculated that older larvae are naturally adapted to feed on decomposed tissue as, if laid on a fresh cadaver, the tissue would be decomposing by the time they reached that age (Richards et al. 2013). Hence, in most cases, this is unlikely to impact PIA estimations as the oldest stage of larvae is analyzed. However, it could have an impact in situations in which the insects are not able to access the remains immediately and so colonize an already decomposing cadaver, such as when death occurs inside a residence or vehicle (Richards et al. 2013). The results of this study contradict earlier work with this species that suggested *C. vicina* was preferentially attracted to older tissue (Erzinclioglu 1996). As baseline development data are fundamental to estimating the PIA on a cadaver, it is important that further research on larval nutritional requirements is conducted.

24.4.9 Considerations for Succession Studies

Although many succession studies have been carried out around the world, the experimental design, the extent of the experiments, the number of replicates, and the choice of carcass vary greatly between studies, affecting comparison across regions and applicability into real-world cases. Three main areas of concern have been identified: lack of appropriate replication, with 78% showing pseudoreplication or design error, lack of independence of experiments, and studies which did not cover characteristic ranges of normal variability (Michaud et al. 2012) (and Chapter 7).

The carcass chosen and the number of replicates depend on funding, space, and availability. Carcasses from turtles to elephants have been studied and this may affect comparison between studies and application to human cases. Pig carcasses have become the traditional model for human decomposition studies in the absence of human cadavers and are considered closest to humans in size, skin type and gut fauna and are usually readily available (Catts and Goff 1992). Some early studies using human cadavers were conducted at the Anthropological Research Facility at the University of Tennessee in Knoxville, the first in North America to study human decomposition (Rodriguez and Bass 1983); however, a restriction with using human cadavers is that large numbers of replicates are not possible, and even if such numbers could be obtained, there will still be a lack of uniformity in former life styles, diet, mechanism of death, nutritional state, and muscle mass. There has been a recent increase in the number of "body farms" in North America, in which human cadavers are decomposed, which may remove this limitation.

Size of carcass may also impact succession as some species do not colonize very small carcasses, due to lack of sufficient resources. Some succession studies have used carcasses as small as mice and lizards, whereas some have considered carcasses as large as elephants (Barnes 2013). When pig carcasses are used, they are usually chosen to represent an adult human cadaver torso and head (the major site of decomposition) so usually range from 22 kg or more (Catts and Goff 1992).

The condition of the carcasses used in studies also varies, such as whether fresh, or frozen/thawed carcasses are used, or whether the carcasses are clothed or nude.

Although it has been shown that there were differences in decomposition and disarticulation between fresh and frozen/thawed rats (Micozzi 1986), a more recent study using pig muscle tissue showed that freeze/thawing had no effect on decomposition or on soil chemistry (Stokes et al. 2009). Fortunately, freeze/thawing has been shown to have no impact on fly attraction, colonization, and larval development or beetle colonization on whole pig carcasses (Bugajski et al. 2011). Clothing can also impact colonization and decomposition, providing protection for larval masses, maintaining moisture in the tissue, and in some cases, prolonging decomposition stages (Dillon 1997; Kelly et al. 2009; Voss et al. 2011).

Sampling methods also vary between studies and could have an effect on decomposition and succession. De Jong and Hoback (2006), in Colorado, USA, compared traditional repeated qualitative subsampling methods, such as are used in most successional studies, with single subsampling on a specified day followed by destructive sampling and found no differences between the repeated subsampling and the single subsampling, with both methods only missing rare or very cryptic species in comparison with destructive sampling. Similarly, Michaud and Moreau in New Brunswick Canada (2013) studied the effects of sampling intensity and found that sampling ~5% of the fauna, as conducted in most succession experiments, did not adversely impact succession.

An analysis of 23 datasets of carrion–insect succession from a variety of geographical regions showed that although 80% of taxa only spent a single time period on the carcass, some taxa recurred on the carcass at a later time, which could potentially confuse temporal analyses and again must be understood for a given region (Schoenly 1992).

Tomberlin et al. (2012) reviewed ~75 succession publications and found tremendous variability across studies. They identified 13 criteria that they suspected influence succession and recommend that these criteria should be standardized and recorded in all studies to allow comparisons. These criteria include method of euthanasia, method and length of storage prior to placement, time between storage and placement, time of day carcasses exposed, number of replicates, season, location, dates, carcass species, time of death, time of first arthropod contact, and time of actual colonization (Tomberlin et al. 2012). Overall, the increase in succession research around the world is highly commendable and further work and standardization in methodology will improve the value of this technique.

24.5 Other Practical Applications of Entomology in Human Death Investigations

24.5.1 Wound Identification

Fly larvae feed primarily by piercing the carcass with mouth-hooks and secreting proteolytic enzymes onto the food source, liquefying the nutrients before ingesting. First instars have very small and delicate mouthparts making it difficult to pierce adult human tissue. Consequently, eggs are usually laid in open wounds or orifices where tissues are easier to pierce, and early instars frequently utilize softer tissue such as brain (Kaneshrajah and Turner 2004). As the early larval environment is so important, female blow flies spend several minutes on a carcass testing and tasting the substrate in order to locate the best possible oviposition site. Wounds, with easy access to moist muscle tissue, organs, and liquid protein in the form of blood are, therefore, extremely attractive as oviposition sites. A consequence of this natural activity is that wounds, when present, are usually the first areas to be colonized. In the absence of wounds, it is usually the natural orifices which are first colonized. Once decomposition progresses, it may be difficult to identify wound sites visually but intense larval activity at a site distant from the orifices, may suggest a wound site. It is not up to the forensic entomologist to identify the area as a wound site, but rather the forensic pathologist, who can testify to an actual wound, but it is up to the entomologist to point out to investigators that the area of larval activity is not a usual oviposition site in the absence of a wound. This can be extremely valuable in determining whether manner of death is suicide, homicide, or accident.

In a case in Maryland, a young woman was found with large masses of larvae on her chest, but also on the palms of her hands. Unfortunately, no insect evidence was collected and a forensic entomologist was not involved. The death was ruled to be a suicidal drug overdose and the victim was buried. Later concern by one of the investigators led to photographs of the victim being re-examined and the atypical colonization pattern was noted (Lord 1990). Colonization primarily in the chest region suggests a wound, but colonization of the palms of the hands which are covered by particular tough skin, suggested not only trauma to this area, but also the mechanism of such trauma in that damage to the palms of the hands frequently occur when a victim attempts to defend themselves from a knife attack. Fortunately, the remains had been buried not cremated and a more thorough autopsy of the exhumed remains revealed multiple stab wounds on the skeleton, and the death was ruled a homicide.

24.5.2 Body Relocation

It is very common for a murderer to relocate a body sometime after the killing in an effort to hide the death. This may occur immediately after death, but in many cases may occur hours or days later. If the body remains at one site long enough for insect colonization, relocating the remains will also relocate the colonizing insects. Although many carrion insects are ubiquitous, some species are very specific to certain areas such as rural or urban areas, or dense shade or open sunlit areas, as discussed earlier. When the remains are found at the second site, they may be colonized by the local fauna, but they may also exhibit evidence of older insects that are not endemic to the crime scene. Understanding the normal habitat preferences and ranges of local species can allow a forensic entomologist to relate to investigators that the older insects indicate that the remains had been in a very different scenario from the death scene for a certain period and were then relocated at a certain time and recolonized by local species prevalent at the crime scene. This indicates that the body has been moved which, on its own, suggests foul play and may also indicate a likely original crime scene. For instance, in a linear trapping series from central London, England, into the rural countryside, three regions were identified as having distinct fly populations: urban, typified by *L. sericata*, *L. illustris*, and *C. vicina*, rural grassland, typified by *L. caesar*, and rural woodland, typified by *C. vomitoria* as well as species of Muscidae, Heliomyzidae, and Dryomyzidae (Hwang and Turner 2005). Similarly, in Poland, a study of carrion fauna in rural open and rural forested regions indicated several species that were considered specific to only one habitat (Matuszewski et al. 2013). Such specific populations may help in indicating the location from which remains have been moved.

However, it is imperative to have local databases available before such conclusions can be made as such habitat preferences may vary geographically. For instance, in the above studies, *L. sericata*, found in England to typify urban habitats (Hwang and Turner 2005) was considered to be exclusive to open rural habitats in Poland (Matuszewski et al. 2013). As well, *C. vomitoria* is usually considered to be a rural species but was the dominant species in an urban study in Austria and was collected with *Chrysomya albiceps* (Wiedemann), normally considered to be a tropical or subtropical species, not usually found in northern Europe (Grassberger and Frank 2004).

More recently, intraspecific genetic (Tarone et al. 2011) and biochemical (Byrne et al. 1995) differences have been reported in blow flies which could, in the future, perhaps identify where a body was killed even in ubiquitous fly species such as *P. regina*.

24.5.3 Body Disturbance

Murderers frequently return to the scene of the crime and disturb the body, perhaps to fantasize about the crime, or to ensure that the remains are still undiscovered. In rare situations, the disturbance may have affected the insects, allowing an entomologist to indicate that the remains have been disturbed and the possible timing of the disturbance. Such might be the case where a body has been buried, disturbed, then reinterred. In BC, the skeletonized remains of a man were found in a shallow grave (Anderson 1999). The excavation took several days and at first the insect evidence was confusing. The remains were cleanly skeletonized and showed no evidence of vertebrate scavenging, indicating that tissue removal was most certainly due to blow fly activity, but there was a paucity of empty puparia found in the surrounding area which was carefully searched and sifted. Blow fly puparia can often survive for decades or more or may be broken down quite quickly depending on soil conditions. Local experience indicated that puparia could survive for several years although, in these soil types, were usually degraded in less than 25 years, but time was not a factor in this case as the victim had been missing for less than a year. However, as the excavation continued, a very large number of puparia were found associated with only the lower half of the remains. This suggested that the upper body had decomposed at the same site as the lower half, but had perhaps become exposed, so the remains had been moved slightly and reburied by someone. The insect evidence suggested this disturbance had occurred a few weeks after the original buried.

24.5.4 Entomotoxicology

Insects bioaccumulate toxins ingested from the cadaver. Consequently, they can be used as alternate toxicological specimens when traditional specimens are too degraded. Toxins have been recovered not just from the primary insects feeding directly on the tissue, but also from other insects which fed on these primary insects (Nuorteva and Nuorteva 1982). Toxins have also been recovered from Calliphoridae and Coleoptera puparia (Bourel et al. 2001). As fly puparia can survive for decades or even millennia in protected circumstances, it might be possible to perform a toxicological analysis decades or even centuries after death.

A wide variety of drugs and other toxins have been recovered from insects, primarily Calliphoridae and Sarcophagidae, in human death cases as well as in experimental settings. See Goff and Lord (2009) for a recent review. At first, it was hoped that toxins could be identified and described both qualitatively and quantitatively from insects but, although numerous studies have shown the value of insects in qualitatively identifying a wide range of toxins and their metabolites in the cadaver host, the results for quantifying the toxins have been very mixed and it is, therefore, recommended that insects are used to simply identify the presence of a toxin and not to attempt quantification (Goff and Lord 2009). Caution must be exercised when using such insects to infer PIA as insects which have fed on drug-laden tissue have been shown to exhibit delayed or accelerated development, depending on drug, stage, and species of insect (Goff and Lord 2009). Gunshot residue can also be recovered from larvae. Gunshot residue contains barium, lead, and antimony and all three were recovered from insects fed on meat laced with gunshot residue although only barium appeared to bioaccumulate (Roeterdink et al. 2004). Such an analysis would be extremely valuable in identifying a bullet wound after decomposition has obscured it, or in distinguishing between a stab wound and a bullet wound (LaGoo et al. 2010). Insects can, therefore, be valuable toxicological specimens when no other viable specimen is available, in some cases, decades or more after death.

24.5.5 Abuse and Neglect in Living Victims

Myiasis is the colonization of a living animal with Diptera larvae and many of the flies that normally feed on a dead body will also feed on necrotic tissue on living people, such as an untended wound or untreated bed sores. Some species will only colonize living victims, but will continue to develop if the victim dies. In forensic cases of cutaneous myiasis, the forensic entomologist is still estimating the PIA, but to infer the length of time of neglect or minimum time since injury, rather than death. This is a growing area and many cases of abuse and neglect have been reported, particularly in the elderly or very young or in people unwilling or incapable of looking after themselves.

Diaper rash and the presence of dirty diapers are extremely attractive to gravid female flies. In Hawai'i, USA, a toddler was found dumped in a rural area and taken to hospital where she was found to have second instars of *Chrysomya megacephala* (F.) in her diaper, feeding on areas of diaper rash as well as on lesions at her wrists indicative of binding. This information was crucial in estimating the length of time the child had been abandoned and in her mother's subsequent conviction. The forensic entomologist had a great deal of experience with the local carrion fauna and was aware that this species is not usually attracted to living tissue, but instead to feces and carrion, so cautioned that if the child had died before she was found and the forensic entomologist had not been informed of the exact collection site, the dirty diaper, then the time of death may have been greatly overestimated (Goff et al. 1991). Similarly, in BC, an unconscious man was found with severe head injuries. The injured areas together with the mouth were heavily colonized by third-instar *C. vomitoria* and *L. sericata*. The victim succumbed to his injuries in hospital, but had he died on the streets and not been found until after death, the colonization patterns of wounds and orifices would have given no clue that colonization had preceded death by several days (Anderson and Cervenka 2001).

Neglect is often difficult to prove, but forensic entomology can provide quantifiable evidence of the length of time a person has been neglected. Several undernourished and neglected children were taken to hospital for assessment and found to be colonized in the anal and genital areas with 4–5-day-old blow fly larvae. This was the only quantitative data on the minimum length of time of neglect (Lord 1990). The thought of a living person colonized by larvae is extremely repugnant to most people, hence, such evidence is extremely probative in court.

24.5.6 Using Entomology to Link a Suspect to Scene or Victim

An understanding of the ecology, feeding behavior, and life history of a variety of insects and their relatives can be used to link a suspect with a victim or a crime scene. The earliest published case in North America was a short description of the entomological aspects of the Mercure trial in New Brunswick in which a man was accused of the fatal beating of a home owner during a break and enter. Canadian 25¢ notes, known as "shinplasters," were found in his possession and microscopic examination revealed plumose hairs. These were taken to an entomologist who identified them as bumble bee hairs (*Bombus* sp., Hymenoptera: Apidae). Investigators discovered that the victim kept his bank notes in a drawer from which a dead bee was recovered, resulting in the suspect's subsequent conviction.

In southern California, after processing a crime scene in which the naked body of a young woman was found, investigators noticed small, very distinctive lesions on their own bodies. These were caused by bites of the immature stages of the blood-feeding mite, *Eutrombicula belkini* Gould (Acari: Tromiculidae), also known as chiggers (Webb et al. 1983). When a suspect was identified and examined, it was noted that he also displayed similar lesions. Public health authorities conducted extensive field trapping studies in the area of the crime scene and surrounding areas. It was concluded that chiggers of this species were rare throughout the area, but were found in large numbers in a very small ecozone, which included the crime scene. In this case, the crime scene was bordered by an area disturbed by agriculture and an undisturbed area. Chigger ecology indicates that they prefer very specific areas such as this and are often found in such clumps. Large numbers of unfed chiggers were found in this area alone, as well as their common vertebrate hosts, which also prefer such disturbed areas. No chiggers were collected in the areas in which the suspect claimed to have been. The progression of the bite lesions was also observed on investigators and researchers and a very specific pathology was noted which allowed aging of the lesions. This allowed the investigator to conclude that the suspect must have been in the area of the crime scene in order to be bitten and that his lesions had developed to a level which indicated he had been bitten in the time frame of the homicide. This evidence was crucial in his trial and subsequent conviction (Webb et al. 1983).

An understanding of the uni-voltine life cycle of a weevil (Coleoptera: Superfamily: Curculionoidea) was pivotal in the conviction of a masked rapist. A suspect was identified and a search of his home uncovered a ski mask, but he claimed he had not worn the mask since the previous winter. However, plant material, including two cockleburs, were found adhering to the mask and an examination of the cockleburs by a forensic entomologist revealed the presence of live immature cocklebur weevils which have a one year life cycle in which the larvae grow and pupariate in the cocklebur, and adults emerge in late summer to enter diapause until the following spring. The larvae are only present in the cocklebur during the summer, the time of the rape. The presence of live weevil larvae, which were at the same level of development as those in the area in which the rape had occurred, indicated that the mask had to have been worn that same summer. The suspect was convicted (Greenberg 1985).

Many species of insect blood feed and several studies have shown that host DNA, in these cases, the suspect or the victim, can be individualized. Lice (Phthiraptera) and adult fleas (Siphonaptera) are blood-feeding insects which also excrete partially digested host blood. Host DNA has been successfully isolated and individualized from human crab louse (*Pthirus pubis* L. Phthiraptera: Pthiridae) excreta (Replogle et al. 1994) and blood meals (Lord et al. 1998) as well as from human body lice, *Pediculus humanis* L. and could be detected for 12–24 h after feeding, dependent on amplicon (Davey et al. 2007). If a suspect is supporting a louse or flea infestation, then they may shed engorged insects at a homicide or rape scene, as well as the insect's excreta, which could be analyzed to individualize the host, in this case, the offender.

Human host DNA has also been recovered and identified from bed bugs (*Cimex lectularius* L. (Hemiptera: Cimicidae) (Szalanski et al. 2006a,b). This is of particular interest as, unlike fleas, lice, and mosquitoes, bed bugs are associated with a particular site rather than a specific host. Therefore, it is possible that human DNA isolated from bed bugs inhabiting a crime scene could be used to identify the perpetrator in any sort of crime, including serious crimes such as homicide and rape, but also crimes in which entomology is rarely used, such as robbery and break and enter, as they would indicate that a person had been present in that room. Experiments have shown that human DNA can be detected up to 7 days after feeding using a mtDNA marker and up to 60 days using an STR (short tandem repeat) marker (Szalanski et al. 2006a,b).

Human DNA can also be recovered from larvae feeding on a body. Larvae store a large quantity of predigested food in the crop of the foregut and many studies have shown that human DNA can be recovered and individualized. Of course, usually the body is still present, so tissue from the bones or teeth can be utilized, but in rare situations DNA from the larvae may be vital. For instance, if the perpetrator removes the remains to dispose of them after a period of time, feeding larvae may fall off the remains or postfeeding larvae may remain at the original site. The human DNA in the crops may be the only evidence linking a death to the scene or linking remains dumped in a rural area to a residence. Also, in many cases, other carrion, or even a second victim, may be found close by. It may be vital to prove which insects came from which food source. In murder/rape cases, eggs may be laid in the genitalia and larvae may feed on semen, potentially destroying valuable evidence. However, it is possible to extract DNA from an offender's semen found in a larva crop (Clery 2001).

The first actual case to identify human DNA from the gut contents of larvae occurred in Mexico in which a burned body was discovered with larvae colonizing the remaining tissue. The larval gut contents were analyzed using STR typing and compared with the putative father's profiles resulting in a 99.7% probability of paternity, thus identifying the victim (Chavez-Briones et al. 2013). Human DNA has also been recovered and individualized from the empty puparia of *L. sericata* in two forensic cases (Marchetti et al. 2013) which means that it may be possible to identify a decedent long after death.

Human host DNA can even be recovered from fecal and regurgitation spots from adult flies feeding on a body (Durdle et al. 2013). Although this may have value in identifying a victim whose remains have been disposed of, it may also confuse investigators as such spots can be confused with blood spatter patterns (Benecke and Barksdale 2003). Much more unusually, an understanding of the postfeeding behavior of necrophagous caterpillars of the casemaking clothes moth, *Tinea pellionella* L. (Lepidoptera: Tineidae), was valuable in an indoor death in Texas. Postfeeding caterpillars leave the food source to seek an appropriate pupariation site and then construct a species distinctive larval shelter and attach it to a ceiling or wall. In this case, human hairs from the decedent were found incorporated in the shelter and mtDNA isolated from the hair was used to identify the decedent (Bucheli et al. 2010).

Carrion ecology, therefore, has a very real role in the criminal justice system and research in this area continues to expand its applications in forensic settings.

REFERENCES

Ahmad, A. et al. 2011. Cadaver wrapping and arrival performance of adult flies in an oil palm plantation in northern Peninsular Malaysia. *Journal of Medical Entomology* 48: 1236–1246.

Amendt, J. et al. 2007. Best practice in forensic entomology—Standards and guidelines. *International Journal of Legal Medicine* 121: 90–104.

Ames, C. and B. Turner. 2003. Low temperature episodes in development of blowflies: Implications for post-mortem interval estimation. *Medical and Veterinary Entomology* 17: 178–186.

Anderson, G.S. 1995. The use of insects in death investigations: An analysis of forensic entomology cases in British Columbia over a five year period. *Canadian Society of Forensic Science Journal* 28: 277–292.

Anderson, G.S. 1999. Forensic entomology in death investigations. In: S. Fairgreave (ed.), *Forensic Osteological Analysis: A Book of Case Studies*. Springfield, IL: Charles C. Thomas, pp. 303–326.

Anderson, G.S. 2000. Minimum and maximum developmental rates of some forensically significant calliphoridae (Diptera). *Journal of Forensic Sciences* 45: 824–832.

Anderson, G.S. 2009a. Factors that influence insect succession on carrion. In: Byrd, J. and E. Castner (eds), *Forensic Entomology: The Utility of Arthropods in Legal Investigations*. Boca Raton, FL: CRC Press, pp. 201–250.

Anderson, G.S. 2009b. Forensic entomology. In: James, S.H. and J. Nordby (eds), *Forensic Science, an Introduction to Scientific and Investigative Techniques*. Boca Raton, FL: CRC Press, pp. 137–165.

Anderson, G.S. 2011. Comparison of decomposition rates and faunal colonization of carrion in indoor and outdoor environments. *Journal of Forensic Sciences* 56: 136–142.

Anderson, G.S. and S.L. VanLaerhoven. 1996. Initial studies on insect succession on carrion in Southwestern British Columbia. *Journal of Forensic Sciences* 41: 617–625.

Anderson, G.S. and V.J. Cervenka. 2001. Insects associated with the body: Their use and analyses. In: Haglund, W. and M. Sorg (eds), *Advances in Forensic Taphonomy. Methods, Theory and Archeological Perspectives*. Boca Raton, FL: CRC Press, pp. 174–200.

Archer, M.S. 2004. The effect of time after body discovery on the accuracy of retrospective weather station ambient temperature corrections in forensic entomology. *Journal of Forensic Sciences* 49: 553–559.

Barnes, K.M. 2013. Forensic entomology. In: Cooper, J.E. and M.E. Cooper (eds), *Wildlife Forensic Investigation; Principles and Practice*. Boca Raton, FL: CRC Press, Taylor Francis Group, pp. 149–160.

Benbow, M.E., A.J. Lewis, J.K. Tomberlin, and J.L. Pechal. 2013. Seasonal necrophagous insect community assembly during vertebrate carrion decomposition. *Journal of Medical Entomology* 50: 440–450.

Benecke, M. and L. Barksdale. 2003. Distinction of bloodstain patterns from fly artifacts. *Forensic Science International* 137: 152–159.

Beuter, L. and J. Mendes. 2013. Development of *Chrysomya albiceps* (Wiedemann) (Diptera: Calliphoridae) in different pig tissues. *Neotropical Entomology* 42: 426–430.

Boatright, S.A. and J.K. Tomberlin. 2010. Effects of temperature and tissue type on the development of *Cochliomyia macellaria* (Diptera: Calliphoridae). *Journal of Medical Entomology* 47: 917–923.

Bourel, B. et al. 2001. Morphine extraction in necrophagous insects remains for determining ante-mortem opiate intoxication. *Forensic Science International* 120: 127–131.

Bucheli, S.R., J.A. Bytheway, and D.A. Gangitano. 2010. Necrophagous caterpillars provide human MtDNA evidence. *Journal of Forensic Sciences* 55: 1130–1132.

Bugajski, K.N., C.C. Seddon, and R.E. Williams. 2011. A comparison of blow fly (Diptera: Calliphoridae) and beetle (Coleoptera) activity on refrigerated only versus frozen-thawed pig carcasses in Indiana. *Journal of Medical Entomology* 48: 1231–1235.

Byrne, A.L., M.A. Camann, T.L. Cyr, E.P. Catts, and K.E. Espelie. 1995. Forensic implications of biochemical differences among geographic populations of the Black blow fly, *Phormia regina* (Meigen). *Journal of Forensic Sciences* 40: 372–377.

Campobasso, C.P., G. Di Vella, and F. Introna. 2001. Factors affecting decomposition and diptera colonization. *Forensic Science International* 120: 18–27.

Catts, E.P. and M.L. Goff. 1992. Forensic entomology in criminal investigations. *Annual Review of Entomology* 37: 253–272.

Chavez-Briones, M.D.L. et al. 2013. Identification of human remains by DNA analysis of the gastrointestinal contents of fly larvae. *Journal of Forensic Sciences* 58: 248–250.

Clery, J. 2001. Stability of prostate specific antigen (Psa), and subsequent Y-Str typing, of *Lucilia (Phaencia) sericata* (Meigen) (Diptera: Calliphoridae) maggots reared from a simulated postmortem sexual assault. *Forensic Science International* 120: 72–76.

Davey, J.S., C.S. Casey, I.F. Burgess, and J. Cable. 2007. DNA detection rates of host MtDNA in bloodmeals of human body lice (*Pediculus humanus* L., 1758). *Medical and Veterinary Entomology* 21: 293–296.

Davies, L. and G.G. Ratcliffe. 1994. Development rates of some pre-adult stages in blowflies with reference to low temperatures. *Medical and Veterinary Entomology* 8: 245–254.

Day, D.M. and J.F. Williams. 2006. Influence of substrate tissue type on larval growth in *Calliphora augur* and *Lucilia cuprina* (Diptera: Calliphoridae). *Journal of Forensic Sciences* 51: 657–663.

De Jong, G.D. and W.W. Hoback. 2006. Effect of investigator disturbance in experimental forensic entomology: Succession and community composition. *Medical and Veterinary Entomology* 20(2): 248–258.

Dillon, L.C. 1997. Insect succession on carrion in three biogeoclimatic zones in British Columbia. MSc, Department of Biological Sciences, Simon Fraser University, Burnaby, BC, 76 pp.

Dillon, L.C. and Anderson, G.S. 1995. Forensic entomology: The use of insects in death investigations to determine elapsed time since death. Canadian Police Research Centre, TR-05-95. Ottawa, Ontario.

Durdle, A., R.a.H. Oorschot, and R.J. Mitchell. 2013. The morphology of fecal and regurgitation artifacts deposited by the blow fly *Lucilia cuprina* fed a diet of human blood. *Journal of Forensic Sciences* 58: 897–903.

El-Moaty, Z.A. and A.E.M. Kheirallah. 2013. Developmental variation of the blow fly *Lucilia sericata* (Meigen, 1826) (Diptera: Calliphoridae) by different substrate tissue types. *Journal of Asia-Pacific Entomology* 16: 297–300.

Erzinclioglu, Z. 1996. *Blowflies.* Slough: Richmond Publishing Co. Ltd.

Gallagher, M.R., S. Sandhu, and R. Kimsey. 2010. Variation in developmental time for geographically distinct populations of the common green bottle fly, *Lucilia sericata* (Meigen). *Journal of Forensic Sciences* 55: 438–442.

Germonpre, M. and M. LeClercq. 1994. Pupae of *Protophormia terraenovae* associated with pleistocene mammals in the Flemish valley (Belgium). *Bulletin de l'Institut Royal des Sciences Naturelles de Belgique Sciences de la Terre* 64: 265–268.

Goff, M.L. 1991. Comparison of insect species associated with decomposing remains recovered inside dwellings and outdoors on the Island of Oahu, Hawaii. *Journal of Forensic Sciences* 36: 748–753.

Goff, M.L. 1992. Problems in estimation of postmortem interval resulting from wrapping of the corpse: A case study from Hawaii. *Journal of Agricultural Entomology* 9: 237–243.

Goff, M.L. and M.M. Flynn. 1991. Determination of postmortem interval by arthropod succession: A case study from the Hawaiian Islands. *Journal of Forensic Sciences* 36: 607–614.

Goff, M.L., S. Charbonneau, and W. Sullivan. 1991. Presence of fecal matter in diapers as potential source of error in estimations of postmortem intervals using arthropod development patterns. *Journal of Forensic Sciences* 36: 1603–1606.

Goff, M.L. and W.D. Lord. 2009. Entomotoxicology. Insects as toxicological indicators and the impact of drugs and toxins on insect development. In: Byrd, J.H. and J.L. Castner (eds), *Forensic Entomology: The Utility of Arthropods in Legal Investigations.* Boca Raton, FL: CRC Press, pp. 427–436.

Grassberger, M. and C. Frank. 2004. Initial study of arthropod succession on pig carrion in a central European urban habitat. *Journal of Medical Entomology* 41: 511–523.

Greenberg, B. 1985. Forensic entomology: Case studies. *Bull. Entomol. Soc. Am.* 31: 25–28.

Greenberg, B. and J.C. Kunich. 2002. *Entomology and the Law: Flies as Forensic Indicators.* Cambridge, UK: Cambridge University Press.

Gunatilake, K. and M.L. Goff. 1989. Detection of organophosphate poisoning in a putrifying body by analyzing arthropod larvae. *Journal of Forensic Sciences* 34: 714–716.

Haskell, N.H., R.D. Hall, V.J. Cervenka, and M.A. Clark. 1997. On the body: Insects' life stage presence and their postmortem artifacts. In: Haglund, W.D. and M.H. Sorg (eds), *Forensic Taphonomy. The Postmortem Fate of Human Remains.* Boca Raton: CRC, pp. 415–448.

Hobischak, N.R., S.L. VanLaerhoven, and G.S. Anderson. 2006. Successional patterns of diversity in insect fauna on carrion in sun and shade in the boreal forest region of Canada, near Edmonton, Alberta. *Canadian Entomologist* 138: 376–383.

Huntington, T.E., L.G. Higley, and F.P. Basendale. 2007. Maggot development during morgue storage and its effect on estimating the post-mortem interval. *Journal of Forensic Sciences* 52: 453–458.

Hwang, C. and B.D. Turner. 2005. Spatial and temporal variability of necrophagous Diptera from urban to rural areas. *Medical and Veterinary Entomology* 19: 379–391.

Ireland, S. and B. Turner. 2006. The effects of larval crowding and food type on the size and development of the blowfly, *Calliphora vomitoria. Forensic Science International* 159: 175–181.

Johl, H.K. and G.S. Anderson. 1996. Effects of refrigeration on development of the blow fly, *Calliphora vicina* (Diptera: Calliphoridae) and their relationship to time of death. *Journal of Entomological Society of British Columbia* 93: 93–98.

Johnson, A.P. 2013. Novel approaches to the investigation of thermogenesis in the maggot masses of blow flies (Diptera: Calliphoridae). PhD School of Biological Sciences, University of Wollongong, NSW, Australia, 160 pp.

Johnson, A.P., J.F. Wallman, and M.S. Archer. 2011. Experimental and casework validation of ambient temperature corrections in forensic entomology. *Journal of Forensic Sciences* 57: 215–221.

Kaneshrajah, G. and B.D. Turner. 2004. *Calliphora vicina* larvae grow at different rates on different body tissues. *International Journal of Legal Medicine* 118: 242–244.

Kelly, J.A., T.C. Van Der Linde, and G.S. Anderson. 2009. The influence of clothing and wrapping on carcass decomposition and arthropod succession during the warmer seasons in central South Africa. *Journal of Forensic Sciences* 54: 1105–1112.

Kumara, T.K. et al. 2012. Occurrence of oriental flies associated with indoor and outdoor human remains in the tropical climate of North Malaysia. *Journal of Vector Ecology* 37: 62–68.

LaGoo, L., L.S. Schaeffer, D.W. Szymanski, and R.W. Smith. 2010. Detection of gunshot residue in blowfly larvae and decomposing porcine tissue using inductively coupled plasma mass spectrometry (ICP-MS). *Journal of Forensic Sciences* 55: 624–632.

Lewis, A.J. and M.E. Benbow. 2011. When entomological evidence crawls away: *Phormia regina* en masse larval dispersal. *Journal of Medical Entomology* 48: 1112–1119.

Lindgren, N.K., S.R. Bucheli, A.D. Archambeault, and J.A. Bytheway. 2011. Exclusion of forensically important flies due to burying behavior by the red imported fire ant (*Solenopsis invicta*) in southeast Texas. *Forensic Science International* 204: e1–e3.

Lord, W.D. 1990. Case histories of the use of insects in investigations. In: Catts, E.P. and N.H. Haskell (eds), *Entomology and Death: A Procedural Guide.* Clemson, SC: Joyce's Print Shop, pp. 9–37.

Lord, W.D. et al. 1998. Isolation, amplification, and sequencing of human mitochondrial DNA obtained from human crab louse, *Pthirus pubis* (L.), blood meals. *Journal of Forensic Sciences* 43: 1097–1100.

Mahat, N.A., Z. Zafarina, and P.T. Jayaprakash. 2009. Influence of rain and malathion on the oviposition and development of blowflies (Diptera: Calliphoridae) infesting rabbit carcasses in Kelantan, Malaysia. *Forensic Science International* 192: 19–28.

Marchetti, D., E. Arena, I. Boschi, and S. Vanin. 2013. Human DNA extraction from empty puparia. *Forensic Science International* 229: e26–e29.

Matuszewski, S., M. Szafałowicz, and M. Jarmusz. 2013. Insects colonising carcasses in open and forest habitats of central Europe: Search for indicators of corpse relocation. *Forensic Science International* 231: 234–239.

Michaud, J.-P., K.G. Schoenly, and G. Moreau. 2012. Sampling flies or sampling flaws? Experimental design and inference strength in forensic entomology. *Journal of Medical Entomology* 49: 1–10.

Micozzi, M.S. 1986. Experimental study of postmortem changes under field conditions: Effects of freezing, thawing and mechanical injury. *Journal of Forensic Sciences* 31: 953–961.

Miller, S.L. 2009. Effects of confinement in a vehicle trunk and arson on the progression and survivability of forensic entomological evidence MA, School of Criminology, Simon Fraser University, Burnaby (BC). 99 pp.

Myskowiak, J.-B. and C. Doums. 2002. Effects of refrigeration on the biometry and development of
Protophormia terraenovae (Robineau-Desvoidy) (Diptera: Calliphoridae) and its consequences in estimating post-mortem interval in forensic investigations. *Forensic Science International* 125: 254–261.

Nuorteva, P. and S.L. Nuorteva. 1982. The fate of mercury in sarcosapropahgous flies and in insects eating
them. *Ambio* 11: 34–37.

Parrish, J.K. and L. Edelstein-Keshet. 1999. Complexity, pattern, and evolutionary trade-offs in animal aggregation. *Science* 284: 99–101.

Picard, C.J. et al. 2013. Increasing precision in development-based postmortem interval estimates: What's sex
got to do with it? *Journal of Medical Entomology* 50: 425–431.

Prokopy, R.J. and B.D. Roitberg. 2001. Joining and avoidance behavior in nonsocial insects. *Annual Review of
Entomology* 46: 631–665.

Reibe, S. and B. Madea. 2010. How promptly do blowflies colonise fresh carcasses? A study comparing indoor
with outdoor locations. *Forensic Science International* 195: 52–57.

Replogle, J., W.D. Lord, B. Budowle, T.L. Meinking, and D. Taplin. 1994. Identification of host DNA by
amplified fragment length polymorphism analysis: Preliminary analysis of human crab louse (Anoplura:
Pediculidae) excreta. *Journal of Medical Entomology* 31: 686–690.

Richards, C.S., C.C. Rowlinson, L. Cuttiford, R. Grimsley, and M.J. Hall. 2013. Decomposed liver has a significantly adverse affect on the development rate of the blowfly *Calliphora vicina*. *International Journal
of Legal Medicine* 127: 259–262.

Rivers, D.B., C. Thompson, and R. Brogan. 2011. Physiological trade-offs of forming maggot masses by
necrophagous flies on vertebrate carrion. *Bulletin of Entomological Research* 101: 599–611.

Rivers, D.B., T. Ciarlo, M. Spelman, and R. Brogan. 2010. Changes in development and heat shock protein expression in two species of flies (*Sarcophaga bullata* [Diptera: Sarcophagidae] and *Protophormia
terraenovae* [Diptera: Calliphoridae]) reared in different sized maggot masses. *Journal of Medical
Entomology* 47: 677–689.

Rodriguez, W.C. and W.M. Bass. 1983. Insect activity and its relationship to decay rates of human cadavers in
East Tennessee. *Journal of Forensic Sciences* 28: 423–432.

Rodriguez, W.C. and W.M. Bass. 1985. Decomposition of buried bodies and methods that may aid in their location. *Journal of Forensic Sciences* 30: 836–852.

Roeterdink, E.M., I.R. Dadour, and R.J. Watling. 2004. Extraction of gunshot residues from the larvae of
the forensically important blowfly *Calliphora dubia* (Macquart) (Diptera: Calliphoridae). *International
Journal of Legal Medicine* 118: 63–70.

Rohlfs, M. and T.S. Hoffmeister. 2004. Spatial aggregation across ephemeral resource patches in insect communities: An adaptive response to natural enemies? *Oecologia* 140: 654–661.

Saunders, D.S. and S.a.L. Hayward. 1998. Geographical and diapause-related cold tolerance in the blow fly,
Calliphora vicina. *Journal of Insect Physiology* 44: 541–551.

Schoenly, K. 1992. A statistical analysis of successional patterns in carrion-arthropod assemblages: Implications
for forensic entomology and determination of the postmortem interval. *Journal of Forensic Sciences* 37:
1489–1513.

Sharanowski, B.J., E.G. Walker, and G.S. Anderson. 2008. Insect succession and decomposition patterns on
shaded and sunlit carrion in Saskatchewan in three different seasons. *Forensic Science International* 179:
219–240.

Smith, K.G.V. 1986. *A Manual of Forensic Entomology*. London: Trustees of The British Museum (Nat. Hist.)
and Cornell University Press.

Stokes, K.L., S.L. Forbes, and M. Tibbett. 2009. Freezing skeletal muscle tissue does not affect its decomposition in soil: Evidence from temporal changes in tissue mass, microbial activity and soil chemistry based
on excised samples. *Forensic Science International* 183: 6–13.

Szalanski, A.L. et al. 2006a. Time course analysis of bed bug, *Cimex lectularius* L. (Hemiptera: Cimicidae)
blood meals with the use of polymerase chain reaction. *Journal of Agricultural and Urban Entomology*
23: 237–241.

Szalanski, A.L. et al. 2006b. Isolation and characterization of human DNA from bed bug, *Cimex lectularius*
L. (Hemiptera: Cimicidae) blood meals. *Journal of Agricultural and Urban Entomology* 23: 189–194.

Tarone, A.M., C.J. Picard, C. Spiegelman, and D.R. Foran. 2011. Population and temperature effects on *Lucilia sericata* (Diptera: Calliphoridae) body size and minimum development time. *Journal of Medical Entomology* 48: 1062–1068.

Tarone, A.M. and D.R. Foran. 2006. Components of developmental plasticity in a Michigan population of *Lucilia sericata* (Diptera: Calliphoridae). *Journal of Medical Entomology* 43: 1023–1033.

Teskey, H.H. and C. Turnbull. 1979. Diptera puparia from pre-historic graves. *Canadian Entomologist* 111: 527–528.

Tomberlin, J.K., J. Byrd, J.R. Wallace, and M.E. Benbow. 2012. Assessment of decomposition studies indicates need for standardized and repeatable methods in forensic entomology. *Journal of Forensic Research* 3.

Tomberlin, J.K., R. Mohr, M.E. Benbow, A.M. Tarone, and S.L. VanLaerhoven. 2011. A roadmap for bridging basic and applied research in forensic entomology. *Annual Review of Entomology* 56: 401–421.

VanLaerhoven, S.L. and G.S. Anderson. 1999. Insect succession on buried carrion in two biogeoclimatic zones of British Columbia. *Journal of Forensic Sciences* 44: 32–43.

Voss, S.C., D.F. Cook, and I.R. Dadour. 2011. Decomposition and insect succession of clothed and unclothed carcasses in Western Australia. *Forensic Science International* 211: 67–75.

Voss, S.C., S.L. Forbes, and I.R. Dadour. 2008. Decomposition and insect succession on cadavers inside a vehicle environment. *Forensic Science, Medicine and Pathology* 4: 22–32.

Warren, J.-A. and G.S. Anderson. 2009. A comparison of development times for *Protophormia terraenovae* (R-D) reared on different food substrates. *Canadian Society of Forensic Science Journal* 42: 161–171.

Warren, J.A. and G.S. Anderson. 2013. Effect of fluctuating temperatures on the development of a forensically important blow fly, *Protophormia terraenovae* (Diptera: Calliphoridae). *Environmental Entomology* 42: 167–172.

Webb, J.P.J., R.B. Loomis, M.B. Madon, S.G. Bennett, and G.E. Greene. 1983. The chigger species *Eutrombicula belkini* Gould (Acari: Trombiculidae) as a forensic tool in a homicide investigation in Ventura County, California. *Bulletin of the Society for Vector Ecology* 8: 141–146.

Yan-Wei, S., L. Xiao-Shan, W. Hai-Yang, and Z. Run-Jie. 2010. Effects of malathion on the insect succession and the development of *Chrysomya megacephala* (Diptera: Calliphoridae) in the field and implications for estimating postmortem interval. *American Journal of Forensic Medicine & Pathology* 31: 46–51.

25

Frontiers in Carrion Ecology and Evolution

M. Eric Benbow, Jeffery K. Tomberlin, and Aaron M. Tarone

CONTENTS

Death is as much a part of life as the interactions living animals have with each other and their ecosystems. Documenting the significance of their remains is vital at its most basic level for understanding ecosystem health, sustainability, and conservation. But, as shown in this book, such information is critical in many applied areas of science as well. Unfortunately, decomposition ecology has been historically limited primarily to decomposing plant organic matter with relatively less research attention on organic matter of animal remains. Consequently, animal tissue decomposition and its associated nutrient recycling have been understudied. This paucity in research on carcass decomposition is in large part due to technological limitations for studying bacterial and other microbial components of the process, but also because such research is esthetically unappealing and sometimes dangerous.

Major technological innovations and advancements within the past decade have facilitated the capacity of scientists to explore dynamic and complex segments of the carrion community and its relevance to ecosystems. These include developments in computation power and analyses, as well as techniques for exploring the system on a smaller scale including important microbial mechanisms and interactions with invertebrates and vertebrates that drive carrion decomposition. Other technological advances now allow researchers to better study larger-scale ecological associations of carrion, such as monitoring the seasonal movements of scavengers or the effects of landscape configuration (e.g., forest fragmentation) on the diversity of necrophagous animals that use carcasses as a food resource. These advances in the capacity of scientists to better pursue various aspects of carrion ecology and evolution have provided a surge in recent research efforts that have also improved the applications of provocative basic science in decomposition.

This book provides a much needed synthesis of these more recent research endeavors and brings together a team of scientists that have often worked in isolation, typically within more specific disciplines ranging from microbiology to ecology to engineering associated with rendering carcasses as part of industrial food production. Basic biological questions related to the ecology and evolution of carrion decomposition are clearly relevant to forensics, as well as for human and environmental health applications. By bringing together such a large group of scientists derived from numerous distinct disciplines, this book provides an introduction to the myriad and significant aspects of carrion decomposition while delivering one of the largest collections of relevant literature on this topic to the broader scientific community.

The importance of carrion decomposition in the natural environment is appreciated. However, mechanistic understandings and quantitative assessments of the individual behavior, population biology, community structure, species interactions, and other biological dynamics important to carrion-associated nutrient, and energy cycling are lacking. The primary goals of this book were to: (1) fill a large literature gap of a relatively understudied subdiscipline of ecology and (2) provide a synthesis and direction for future directions important for studies of carrion decomposition that improve the general understanding of decomposition in ecosystems. To this end, this book provides a foundation of knowledge and synthesis that can be used to develop novel hypotheses and studies relevant to general ecological and evolutionary theory, while also having direct translational relevance to wide reaching applications.

25.1 Concluding Thoughts

The carrion system represents a discrete model that can be easily replicated and manipulated to effectively test and understand multiscale mechanisms and multivariate processes that dictate carrion decomposition and define broader aspects of ecological and evolutionary theory. These studies can range from laboratory-based experiments that test bacterial signaling effects on carrion fly fitness, small-scale manipulations of colonization timing on necrophagous community assembly, meso-scale field experiments testing the role of mass emerging insect (e.g., cicadas of mayflies) density on soil nutrient concentrations and shifts in predation to scavenging behavior of rodents, to ecosystem level surveys of ungulate carcasses and associated scavenging on forest biodiversity. The potential for new and innovative scientific inquiry using the carrion system model is at an exciting stage of intellectual development, taking advantage (for once) that there is one inescapable connection within the web of life—death.

To our knowledge, this is one of the first books on carrion decomposition to bring together multiple basic scientific disciplines in a way that can be used in so many applications, from human mass disasters to habitat/ecosystem conservation. The goal of this undertaking was to identify new areas of inquiry that can drive the scientific understanding of carrion decomposition forward and enable its enhanced application in a variety of settings and circumstances. With such a synthesis of multiple disciplines associated with carrion ecology and evolution now available, the frontiers of this emerging discipline are beginning to be realized and expanded. We look forward to the coming years with excitement, anticipating continued advancements in carrion ecology. We hope that the effort put into this book by all the authors will be useful for our colleagues that are driving this exciting field forward into previously unexplored or newly discovered areas of inquiry.

Index

A

Abyssal carrion communities, 502–504
Acari, 78; *see also* Arthropod community
    domestic mites, 79
    phoretic mites, 78–79
Acceptance phase, 168–170; *see also* Precolonization
        interval (pre-CI)
Accumulated degree days (ADD), 170, 544; *see also*
        Carrion-arthropod succession analysis
    models, 139
Accumulated degree hours (ADH), 155
Acid, 190; *see also* Volatiles
Acyl-homoserine lactonase, 442; *see also* Quorum
        quenching (QQ)
N-Acyl homoserine lactone (AHL), 440; *see also* Quorum
        quenching (QQ); Quorum quenching enzymes
    -acylases, 443
    host response to AHL signals, 445
    -lactonases, 443
    in plant-associated bacteria, 445–446
ADD, *see* Accumulated degree days (ADD)
Adenosine triphosphate (ATP), 14
ADH, *see* Accumulated degree hours (ADH)
Adipocere, 46; *see also* Microbial interaction; Vertebrate
        decomposition
    formation, 25–26
AFLP, *see* Amplified fragment length polymorphism (AFLP)
Africa, 461–462
    climate, 462
    habitats of, 463
African carrion ecology, 461, 483, 485; *see also* Carrion
        ecology
    avian scavengers, 472–473
    avian scavengers and human disturbance, 482–483
    carrion names, 484–485
    factors affecting carrion abundance, 463–465
    human activity on carrion availability, 480–481
    hyenas, 470
    hyenas and human disturbance, 482
    invertebrate scavengers, 465–467
    mammalian scavengers, 468–472
    scavenging diet of hyena, 471
    vertebrate scavengers, 467–468, 473–475
    vultures, 473
Aggregation model of coexistence, 159; *see also* Carrion
        community ecology
Agyrtidae, 76; *see also* Coleoptera
AHL, *see N*-Acyl homoserine lactone (AHL)
AI, *see* Autoinducer (AI)
AiiA, *see* Autoinducer inactivator AHL-lactonases (AiiA)
Aldehydes, 191; *see also* Volatiles
Alkaline hydrolysis, 519
Allee effect, 310; *see also* Ecological genetics

Allozyme, 388
    variation, 389
Amorphophallus titanum, 377; *see also* Carrion flower
Amplified fragment length polymorphism (AFLP), 389
Anaerobes, 194
Anaerobic digestion, 519
Anthomyiidae, 73; *see also* Diptera
Anthropogenic activities, 108
Anthropogenic disturbance and carrion communities, 498
Anthropogenic toxicants, 120–121; *see also* Vertebrate
        scavengers
Apex predators, 119, 501–502; *see also* Vertebrate
        scavengers
Apidae, 78; *see also* Hymenoptera
Apneumone production, 166; *see also* Carrion community
        ecology
Aquatic arthropods, 258; *see also* Aquatic carrion
        decomposition
Aquatic carrion decomposition, 247
    acid stress, 251
    anthropogenic impacts, 263
    applications and implications, 260
    arthropods in freshwaters, 258
    bacteria on implanted pig carcasses, 257
    carrion decomposition on swine carcasses, 259
    carrion detritus quality, 254–255
    carrion-island ecosystems, 254
    changes in macroinvertebrate FFGs, 262
    chemical factors, 250–251
    collector filterers, 258
    community organization and dynamics, 255–259
    crustaceans, 258
    dissolved oxygen, 249
    diversity in food-web complexity, 260
    ecological impacts, 261
    energy transfer mechanisms, 252
    forensic applications, 261–263
    gas exchange, 249
    hydrology, 249–250
    invertebrate scavengers, 259
    lotic systems, 248
    macroorganism scavengers, 259
    nutrient limitations, 250
    parameters affecting, 248
    perspectives, 263–264
    PMSI estimation, 261
    postspawning, 253
    properties, 248
    pulse perturbations and nutrient cycling, 252–254
    salinity, 251
    salmon subsidies, 250, 253
    seafloor geology, 259
    stages, 256
    temperature, 248–249

Milton Keynes UK
Ingram Content Group UK Ltd.
UKHW052025141024
449569UK00016B/705